DENDRIMERS AND HYPERBRANCHED POLYMERS

SYNTHESIS TO APPLICATIONS

DENDRIMERS AND HYPERBRANCHED POLYMERS

SYNTHESIS TO APPLICATIONS

NIRANJAN KARAK
SUKUMAR MAITI

MD Publications Pvt Ltd
New Delhi

Published by
MD Publications Pvt Ltd
"MD House", 11, Darya Ganj, New Delhi-110002
Tel. : +91-11-41562846, 41563325
E-mail : contact@mdppl.com
Website : www.mdppl.com

ISBN : 978-81-7533-104-4

Price : Rs. 1995.00

Printed and Published by **MD Publications Pvt Ltd** at Times Press, New Delhi

CONTENTS

Chapter *Pages*

Preface xv

1. BASIC CONCEPT OF DENDRITIC POLYMERS 1—58

1.1 Introduction 1

1.1.1 Definition 2

1.1.2 Structural Aspects 4

1.1.2.1 Theoretical Consideration 8

1.1.2.2 Effect of Core and Branching Multiplicity 8

1.1.3 Historical Background 9

1.1.4 Significance 10

1.1.5 Comparison with Linear and Star Branched Polymers 11

1.1.6 Nomenclature and Structural Representation 13

1.1.7 Classification 16

1.2 Synthetic Methodology 16

1.2.1 Divergent Approach 17

1.2.2 Convergent Approach 18

1.2.3 Accelerated Approach 20

1.2.3.1 Hypercore and Branched Monomer 20

1.2.3.2 Double Exponential and Mixed Growth 21

1.2.4 Single Step Approach 22

1.3 Characterization Techniques 23

1.3.1 Structural Elucidation Techniques 23

1.3.1.1 Techniques for Determination of Element Composition and Functional Groups by IR, UV-Visible, and Elemental Analysis 23

1.3.1.2 Techniques for Determination of Structure by NMR Spectroscopy 24

1.3.2 Techniques for Determination of Homo-geneity, Purity and Molecular Weight & Its Distribution by Chromatographic Techniques and Mass spectrometry 25

1.3.3 Techniques for Determination of Morphology by SAXS, AFM, EPR, STM, etc. Spectroscopy 27
1.3.4 Techniques for Determination of Thermal Characteristics by TGA and DSC Analysis 29
1.4 Properties 29
1.4.1 Physical Properties 31
1.4.1.1 Solubility 32
1.4.1.2 Viscosity 32
1.4.1.3 Density 33
1.4.1.4 Crystallinity 34
1.4.1.5 Optical Properties 35
1.4.2 Mechanical and Rheological Properties 36
1.4.3 Chemical Properties 37
1.4.4 Thermal Properties 37
1.4.5 Electrical Properties 38
1.4.6 Amphiphilic Behavior 38
1.4.7 Host-Guest Interaction Chemistry 39
1.5 Applications 40
1.5.1 Biological and Medicinal 41
1.5.2 Catalysis 42
1.5.3 Energy and Electron Transfer 43
1.5.4 Encapsulation 44
1.5.5 Unimolecular Micelles 44
1.5.6 Liquid Crystal Display (LCD) 45
1.5.7 Surface Coatings 45
1.5.8 Multipurpose Polymeric Additives and Others 45
1.6 Summary 46
References 47

2. DENDRITIC POLYHYDROCARBONS 59—118

2.1 Introduction 59
2.2 Synthesis, Characterization and Properties of Different Types of Hydrocarbon Dendrimers 59
2.2.1 Polyphenylenes 59

2.2.2 Stilbeniods 70
2.2.3 Poly(phenyl acetylene)s 77
2.2.4 Dendronized Polyhydrocarbons 91
2.2.5 Phenylcarbazole Ethynylene Dendrons 92
2.2.6 Iptycenes 96
2.3 Synthesis, Characterization and Properties of Different Types of Hyperbranched Polyhydro-carbons 96
2.3.1 Polyphenylenes 96
2.3.2 Polyethynylenes 99
2.3.3 Poly(phenylene vinylene)s 99
2.3.4 Polystyrenes 100
2.3.5 Others 103
2.4 Applications of Dendritic Polyhydro-carbons 104
2.4.1 Sensors 104
2.4.2 Nanocomposites 105
2.4.3 Electronics 106
2.4.4 Polymeric Additives 112
2.4.5 Catalysis 113
2.5 Summary 113
References 114

3. DENDRITIC POLYAMIDES **119—162**

3.1 Introduction 119
3.2 Synthesis, Characterization and Properties of Different types of Polyamide Dendrimers 119
3.2.1 Polypeptides 120
3.2.1.1 Poly(amino acid)s 120
3.2.1.2 Poly(nucleic acid)s 126
3.2.1.3 Other Polypeptides 127
3.2.2 Glycodendrimers 130
3.2.3 Redoxactive 136
3.2.4 Photoactive 138
3.2.5 Chiral 140

3.2.6 Dendronized 141
3.2.7 Poly(ether amide)s 143
3.3 Synthesis, Characterization and Properties of Hyperbranched Polyamides 146
3.4 Applications of Dendritic Polyamides 151
3.4.1 Biomedical 152
3.4.2 Encapsulation 154
3.4.3 Self-assembly 156
3.4.4 Others 157
3.5 Summary 157
References 158

4. DENDRITIC POLY(AMIDO AMINE)S 163—235

4.1 Introduction 163
4.2 Synthesis, Characterization and Properties of Poly(amido amine) Dendrimers 164
4.2.1 Synthesis 164
4.2.2 Characterization 167
4.2.3 Properties 178
4.3 Modification of Poly(amido amine) dendrimers 184
4.3.1 Photoactive 185
4.3.2 Redoxactive 189
4.3.3 Liquid Crystalline 192
4.3.4 Glycodendrimers 194
4.3.5 Amphiphilic 196
4.3.6 Copolymers and Polymer Blends 198
4.3.7 Conducting 201
4.3.8 Bioactive 203
4.3.9 Fullerodendrimers 206
4.3.10 Others 207
4.4 Synthesis, Characterization and Properties of Hyperbranched Poly(amido amine)s 208

4.5 Applications of Dendritic Poly(amido amine)s 210
    4.5.1 Catalysis 210
    4.5.2 Biological and Medicinal 212
    4.5.3 Electronics 221
    4.5.4 Nanocomposites/Nanoparticles 222
    4.5.5 Sensor Materials 224
    4.5.6 Others 225
4.6 Summary 225
    References 226

5. DENDRITIC POLY(ALKYLENEIMINE)S 237—282

5.1 Introduction 237
5.2 Synthesis, Characterization and Properties of Different types of Poly(alkyleneimine) Dendrimers 238
    5.2.1 Poly(propylene imine) 238
        5.2.1.1 Synthesis 238
        5.2.1.2 Characterization 239
        5.2.1.3 Properties 240
        5.2.1.4 Modifications 243
            5.2.1.4.1 Photoactive 243
            5.2.1.4.2 Redoxactive 248
            5.2.1.4.3 Amphiphilic 251
            5.2.1.4.4 Liquid Crystalline 254
            5.2.1.4.5 Self-assembly 256
            5.2.1.4.6 Glycodendrimers 259
            5.2.1.4.7 Host-Guest Chemistry 260
            5.2.1.4.8 Others 263
    5.2.2 Poly(ethylene imine) 268
5.3 Applications of Dendritic Poly(alkyleneimine)s 269
    5.3.1 Electronics and Nanoparticles 269
    5.3.2 Host-guest and Controlled Release Systems 270
    5.3.3 Biomedical 274
    5.3.4 Unimolecular Micelles 275
    5.3.5 Catalysis 276

5.4 Summary 277
References 277

6. DENDRITIC POLYETHERS **283—380**

6.1 Introduction 283
6.2 Synthesis, Characterization and Properties of Different types of Polyether Dendrimers 283
6.2.1 Aliphatic 283
6.2.2 Aromatic 288
6.2.3 Poly(ether-urethane)s 295
6.2.4 Liquid Crystalline (LC) 297
6.2.5 Redoxactive 300
6.2.6 Amphiphilic 305
6.2.7 Photoactive 308
6.2.8 Self-assembly 318
6.2.9 Crosslinked 320
6.2.10 Chiral 322
6.2.11 Fullerene Containing Polyethers 330
6.2.12 Copolymers 333
6.2.13 Others 335
6.3 Modification of Polyether dendrimers 341
6.3.1 Surface 342
6.3.2 Core 342
6.4 Synthesis, Characterization and Properties of Different types of Hyperbranched Polyethers 346
6.4.1 Aliphatic 347
6.4.2 Aromatic 352
6.4.3 Liquid Crystalline 355
6.4.4 Others 355
6.5 Applications of Dendritic Polyethers 359
6.5.1 Catalysis 359
6.5.2 Macromolecular Initiators 365
6.5.3 Electronics /Nanocomposites Materials 366
6.5.4 Energy Transfer Materials 367

6.5.5 Biomedical 369

6.5.6 Micelles 369

6.5.7 Encapsulation 370

6.6 Summary 371

References 372

7. DENDRITIC POLYESTERS **381—425**

7.1 Introduction 381

7.2 Synthesis, Characterization and Properties of Different Types of Polyester Dendrimers 381

7.2.1 Aliphatic 381

7.2.2 Aromatic 388

7.2.3 Copolymers 397

7.3 Modification of Polyester Dendrimers 404

7.4 Synthesis, Characterization and Properties of Different Types of Hyperbranched Polyesters 404

7.4.1 Aliphatic 405

7.4.2 Aromatic 411

7.4.3 Copolymers 417

7.5 Applications of Dendritic Polyesters 418

7.5.1 Coatings 418

7.5.2 Binders and Crosslinkers 420

7.5.3 Catalysis 420

7.5.4 Encapsulating Materials 421

7.5.5 Electronics 421

7.6 Summary 422

References 422

8. SILICON BASED DENDRITIC POLYMERS **427—495**

8.1 Introduction 427

8.2 Synthesis, Characterization and Properties of Silicon Based Different types of Dendrimers 428

8.2.1 Carbosilanes 428

8.2.2 Carbosiloxanes 450
8.2.3 Siloxanes 455
8.2.4 Carbosilazanes 457
8.2.5 Siltranes 459
8.2.6 Silsesquioxanes 460
8.2.7 Silanes 463
8.2.8 Others 466
8.3 Synthesis, Characterization and Properties of Silicon Based Different types of Hyperbranched Polymers 468
8.3.1 Polycarbosilanes 468
8.3.2 Polycarbosiloxanes 474
8.3.3 Polysiloxanes 476
8.3.4 Polysilanes 477
8.3.5 Polyalkoxysilanes 478
8.4 Applications of Silicon Based Dendritic Polymers 480
8.4.1 Catalysis 480
8.4.2 Xerogels 483
8.4.3 Medicinal 483
8.4.4 Micelles 484
8.4.5 Others 484
8.5 Summary 485
References 486

9. PHOSPHOROUS CONTAINING DENDRITIC POLYMERS 497—549

9.1 Introduction 497
9.2 Synthesis, Characterization and Properties of Phosphorous Containing Dendrimers 497
9.2.1 Synthesis 498
9.2.2 Characterization 507
9.2.3 Properties 508
9.3 Modification of Phosphorous Containing Dendrimers 509
9.3.1 Modification through Surface Groups 510

9.3.2 Modification through Internal Layers 528
9.3.3 Modification through Core 533
9.4 Synthesis, Characterization and Properties of Phosphorous Containing Hyperbranched Polymers 538
9.5 Applications of Phosphorous Containing Dendritic Polymers 541
9.6 Summary 544
References 545

10. OTHERS DENDRITIC POLYMERS 551—610

10.1 Introduction 551
10.2 Synthesis, Characterization and Properties of Other Dendrimers 551
10.2.1 Polycarbonates 551
10.2.2 Quaternary Ammonium Ions 552
10.2.3 Carbohydrates 557
10.2.4 Organometallics 561
10.2.5 Polyureas and Polyurethanes 574
10.2.6 Dendrimers of Mixed Chemical Linkages 575
10.2.6.1 Poly(ester amide)s 576
10.2.6.2 Poly(ester amine)s 578
10.2.6.3 Poly(ether amine)s 578
10.2.6.4 Poly(ether imide)s 579
10.2.6.5 Poly(alkylnyl ester-amide)s 581
10.2.6.6 Others 582
10.2.7 Polyamines 583
10.2.8 Polyhydrazones 586
10.2.9 Others 587
10.3 Synthesis, Characterization and Properties of Other Hyperbranched Polymers 593
10.3.1 Polycarbonates 593
10.3.2 Carbohydrates 595
10.3.3 Polyureas and Polyurethanes 596

10.3.4 Polyimides 599
10.3.5 Poly(ether imide)s 601
10.3.6 Poly(arylene ether phosphineoxide)s 602
10.3.7 Poly(ether sulfone) 603
10.3.8 Polyamines 604
10.4 Summary 606
References 606

PREFACE

"Dendrimer and hyperbranched Polymers – Synthesis to Applications" is the first book in India on this topic. The purpose of the book is to encourage the research in this field and to aware about the uses of dendritic polymers to the scientific community. The researchers in the fields of chemical sciences and engineering, materials science and engineering, medical science and technology, physical sciences, environmental science and engineering, biological sciences, biochemical engineering, biotechnology, etc. can use this book as a reference material in their respective professions.

Dendritic polymers are one of the advanced frontier macromolecules of Polymer Science and Technology. They are novel and versatile because of their unique architectural features of highly branched structures with large numbers of active surface functionalities. Because of their unique structural features like no entanglement, high surface functionalities, etc. they possess a large variety of unusual properties, like low viscosity, high reactivity, interior cavity, etc. The chemical modification or functionalization of these dendritic polymers resulted a wide range of variation in physical and chemical properties. These unique characteristics of these macromolecules render them in a wide number of valuable applications, including medicinal to polymeric additives. They play important roles in various applications such as medicinal and diagnosis, adhesive and coatings, gene therapy and chemical sensors, drug delivery systems, light harvesting materials, catalysts, electronic materials, separating agents, and many more. The dendritic polymers of unit ratio of branch sites to repeat unit have emerged as an important new concept in macromolecular design. Finally, it is one of the most exciting and young field of researches and need to be maturated, so there is lot of scopes to start or more involvement in this field of works. Their importance in advanced frontier Polymer Science and Technology could be easily understood from the number of publications coming out in each year in this field.

The book is aiming to blend the basic ideas along with the advance understanding about this important class of macromolecules, so that it becomes a valuable document for not only to researchers, faculty members and advanced undergraduate and graduate students in this field but industrial professionals too.

This book is dedicated to the memory of my elder sister, late *Anjali* without whom I may not be in this position (N.Karak). The authors are grateful to their students especially Sibdas, Jaytishmoy, Nandini, Suvangshu and Harekrishna, family members specially to Susmita and Soham (wife and son respectively of N. Karak), colleagues and friends those who directly or indirectly help in various ways to complete this book. We also acknowledge with thanks to all the researchers and publishers who actually performed and published the works, and without which this book never would come out. Lastly but not the least, we feel honour to offer our sincere thanks to the Vice-chancellor, Tezpur University, Director, Indian Institute of Technology, Kharagpur and other organizations and institutes for using their libraries and other related facilities for the preparation of this book.

NIRANJAN KARAK
SUKUMAR MAITI

ABBREVIATIONS

η	intrinsic viscosity
18-C-6	18-crown-6-ether
1B4M	2-(*p*-isothiocyanatobenzyl)-6-methyl-diethylene triaminepenta acetic acid
2,3-dpp	2,3-*bis*(2-pyridyl)pyrazine
2,3-metpp	2-[2-(1-methylpyridiniumyl)]-3-(2-pyridyl) pyrazine
2E PyMeTf	N-methyl-2-ethnylpyridinium trifluoromethanesulfonate
3D	three dimensional
5DSA	5 doxylstearic acid
6FDA	2,2-*bis*(3,4-dicarboxy phenyl) hexafluoropropane dianhydride
9-BBN	9-borabicyclo[3.3.1]nonane
Ac	acetyl
acac	acetylacetonate
acacH	pentane-2,4-dione
AFM	Atomic Force Microscopy
AIBN	2,2'-azobisisobutyronitrile
Anhyd.	anhydrous
APCI	atmospheric pressure chemical ionization
aq.	aqueous
Ar	aryl
ATRP	atom transfer radial polymerization
B3OH	1,3,5-trihydroxy benzene
BAM	Brewster angle microscope

BINOL	1,1'-bi-2-naphthol
bis-MPA	2,2-*bis*(hydroxymethyl)propionic acid
bis-PVA	2,2-*bis*(4-hydroxy phenyl) pivalonic acid
Bn	benzyl
Boc	*tert*-butoxycarbonyl
BOP-Cl	*bis*(2-oxo-3-oxazolidinyl)phosphinic chloride
BPA	bisphenol-A
BPO	benzoyl peroxide
BPP34C10	bis-*p*-phenylene-34-crown-10
bpy	2,2'-bipyridyl
BR	Bengal Rose
Bu	butyl
Bz	benzoyl
C_{60}	fullerene
CAS	Chemical Abstracts Service
cat.	catalyst
CB	cucurbituril
Cbz	benzyloxycarbonyl
CD	cyclodextrin
CDI	1,1'-carbonyl diimidazole
Cl_2PyNO	2,6-dichloropyridine-N-oxide
CMPD	critical molecular design parameter
CMV	human cytomegalovirus
cod	cyclooctadiene
conc.	concentrated
cot	cyclooctatetraene
Cp	cyclopentadienyl
CPMAS	cross polarization magic angle spinning

CSA	camphorsulfonic acid
CTB	cholera toxin B subunit
CV	cyclic voltammetry
Cy	cycloalkyl
d	day(s)
Da	dalton(s)
DAAQ	diamino anthraquinone
DAB	1,4-diamino butane
DABCO	1,4-diazabicyclo[2.2.2]octane
dansyl	5-dimethylamino-1-naphthalene sulphoamido
DB	degree of branching
dba	dibenzylidene acetone
DB24C8	dibenzo-[24]-crown-8
DBN	1,5-diazabicyclo[4.3.0]non-5-ene
DBOP	diphenyl-(2,3-dihydro-2-thioxo-3-benzoxazolyl)phosphate
DBU	1,8-diazabicyclo[5.4.0]undec-7-ene
DCC	N,N-dicyclohexylcarbodiimide
DDB	1,4-*bis*(dimethylamino)-2,3-dimethoxy butane
DDQ	2,3-dichloro-5,6-dicyano-1,4-benzoquinone
DDTB	1,4-*bis*[4,4''-bis(1,1-dimethylethyl)-2,2':6',2'' tetrapyridine-4´-yl]benzene
DEAD	diethyl azodicarboxylate
dendr	dendrimer without core unit and terminal functional groups
DET	diethyl tartrate
DIBALH	diisobutylaluminum hydride
dil.	dilute
DIPEA	N,N-diisopropylethylamine

DIPT	diisopropyl tartrate
DLCP	dendritic liquid crystalline polymers
DLS	dynamic light scattering
DMAc	N,N-dimethy acetamide
DMAP	4-(dimethylamino)pyridine
DMB	3,4-dimethoxybenzyl
DME	1,2-dimethoxyethane
DMF	dimethylformamide
DMH	2,6-dimethyl-4-heptnol
DMPC	dimyristoylphosphatidylcholine
DMPU	1,3-dimethyl-3,4,5,6-tetrahydro-2(1H)-pyrimidinone
DMSO	dimethyl sulfoxide
DMT	5'-dimethoxy trityl
dNbpy	dinonyl bipyridine
DNC	dendrimer nanocomposties
DNS	5-(dimethylamino)-1-naphthalenesulfonic acid
DOTA	1,4,7,10-tetra-azacyclododecane–N,N′,N′′,N′′′-tetra acetic acid
DPPA	diphenylphosphory azide
dppe	*bis*(diphenylphosphino)ethane
dppm	*bis*(diphenylphosphino)methane
DPTS	4-(dimethyl amino)pyridinium 4-toluene sulfonate
DRI	differential refractometer index
DSC	differential scanning calorimetry
DSDA	3,3′,4,4′-biphenyl sulfonates tetracarboxylic dianhydride
DTAB	dodecyltetramonium bromide

DTPA	diethylenetriamine penta acid
DTPA	*trans*-cyclohexane-1,2-diamine-N,N′,N′′,N′′-pentaacetic acid
DuPhOS	1,2-*bis*(diethylphospholano)benzene
DVB	divinylbenzene
EDA	ethylene diamine
EDC	ethyl-N,N-dimethyl amino-propylcarbodiimide
EDTA	ethylenediaminetetraacetic acid
ee	enantiomer excess
EEDQ	2-ethoxy-4-ethoxy carbonyl-1,2-dihydroquinone
EET	excitation energy transfer
EGFR	epidermal growth factor receptor
EL	electro-luminescence
ELLA	enzyme-linked lectin-assays
EPR	Electro Paramagnetic Resonance
EQCM	electrochemical quarts crystal microbalance
equiv	equivalent(s)
ESEM	environmental scanning electron microscopy
ESI	Electron Spray Ionization
Et	ethyl
Et-DuPhO	1,2-*bis*(diethylphospholano)benzene
EXAFS	extended X-ray absorption fine structure
FL	fluorescence
FLE	fluorescence excitation
Fmoc	9-fluorenylmethoxycarbonyl
FRET	Fluorescence resonance excitation energy transfer
fs	femto second
FTIR	Fuorier Transform Infra-red Spectroscopy

g	gram
GalNAc	α-D-N-acid-glucosamine
GFC	Gel Filtration Chromatography
GO_x	glucose oxidase
GPC	Gel Permeation Chromatography
h	hour(s)
H_2TCPP	polyglutamic free base meso-tetra-4-carboxy-phenyl porphyrin
HB	hyperbranched
HEAP	1-[N,N-*bis*(2-hydroxyethyl)amino]-2-propanol
HMDA	hexamethylene diamine
HMPA	hexamethylphosphoric triamide
HOBT	hydroxybenzotriazole
HOPG	highly oriented pyrolytic graphite
HOTf	trifluoromethan sulfonic acid
HPLC	high performance liquid chromatography
HSV	herpes simplex viruses
HTBO	4-(hydroxymethyl)-2,6,7-trioxalbicyclo-[2,2,2]octane
IC_{50}	concentration causing 50% inhibition
^iPr	isopropyl
L	liter(s)
LALLS	Low Angle Laser Light Scattering
LC	liquid crystalline
LCAA-CPG	long chain alkylamine-controlled pore glass
LDA	lithium diisopropylamide
LIOB	laser-induced optical breakdown
LLS	laser light scattering
LSIMS	liquid secondary ion mass spectrometry

LTMP	lithium 2,2,6,6-tetramethylpiperidide
M	molecular weight
MALDI-TOF	Matrix Assisted Laser Desorption / Ionization –Time of Flight
MAP	multiple antigen peptide
MB	methylene blue
MBROP	multibranched ring opening polymerization
m-CPBA	*m*-chloroperoxybenzoic acid
MD	molecular dynamics
Me	methyl
min	minute(s)
MLCT	metal to ligand charge transfer
mmol	millimole
MO	methyl orange
mol	mole(s)
MRI	magnetic resonance imaging
Ms	methanesulfonyl (mesyl)
MV^{2+}	1,1'-dimethyl-4,4'-bipyridinium dication
nbd	norbornadiene
NBS	*N*-bromosuccinimide
n-Bu	normal butyl
NCN-Br	4-amino substituted 2,6-*bis*(dimethylamino)-methyl)-1-bromobenzene
NCS	*N*-chlorosuccinimide
NHS	N-hydroxysuciimide
NIS	N-iodo succinimide
NMP	N-methyl pyrolidone
NMR	Nuclear Magnetic Resonance Spectroscopy
Nu	nucleophile

OLED	organic light emitting diode
OPM	optical polarized microscopy
OPV	oligo phenylene-vinylene
OS	organosilicon
PAMAM	poly(amidoamine)
PCC	pyridinium chlorochromate
PDA	10,12-pentacosadiynoic acid
PDC	pyridinium dichromate
PDI	polydispersity index
PdTCPP	Pd-meso-tetra-4-carboxy-phenyl prophyrin
PEG	poly(ethylene glycol)
PEI	poly(ethyleneimine)
PEO	polyethylene oxide
Ph	phenyl
Phen	phenanthroline
Ph-TCBD	phenyl-tetracynobutadienyl
phth	phthalate
PL	photoluminescence
PMB	4-methoxyphenyl
PMDA	pyromellitic dianhydride
PMI	perylene dicarboxymonoimide
PMMA	poly(methylmethacrylate)
PNB	4-nitrobenzyl
POSS	polyhedral oligomeric silsesquioxanes
PPA	poly(phosphoric acid)
PPI	poly(propylene imine)
PPTS	pyridinium *p*-toluenesulfonate
PPV	poly(*p*-phenylene vinylene)
Pr	propyl

PS	polystyrene
ps	pico second
P-SCN-Bz-DOTA	2-(4-isothiocyanatobenzyl)-1,4,7,10-tetraazacyclododecane-N,N,N,N'-tetraacetate
PSUF	polymer supported ultrafitration
p-TsOH/PTSA	*para* toluenesulfonic acid
PU	Polyurethane
Pv	pivaloyl
PVC	poly(vinyl chloride)
Py	pyridine
QCMB	quartz crystal microbalance
Rg	radius of gyration
RPIA	radial partition immunoassayed
RT	room temperature
s	second(s)
SAM	self assemble monolayer
SANS	Small Angle Neutron Scattering
SAXS	Small Angle X-ray Scattering
SBD-TSA	starburst dendrimer template surfactant aggregates
s-Bu	secondary butyl
SCMF	self-consistent mean field model
SCVP	self-condensation vinyl polymerization
SDS	sodium dodecyl sulfate
SEC	size exclusion chromatography
SEM	Scanning Electron Microscopy
SFM	Scanning Force Microscopy
SPC	self-polycondensation
SPCR	Suzuki polycondensation reaction

STM	Scanning Tunneling Microscopy
TADDOL	α, α, α' α'-tetraaryl-1,3-dioxalane-4,5-dimethanol
TAPA	*tris*(4-aninophenyl)amine
TBAB	tetrabutylammoniumbromide
TBAC	tetra-n-butylammonium chloride
TBAF	tetrabutyl ammonium fluoride
TBDMS	*tert*-butyldimethylsilyl
TBDMS-Cl	tetra-butyl dimethyl silyl–chloride
TBDMS-OBB	tetra-butyl dimethyl silyl–oxy benzyl bromide
TBDPS	*tert*-butyldiphenylsilyl
TBBPA	tetrabromobisphenol-A
TBTU	O-(Benzotriazole-1-yl)-N,N,N',N'-tetramethyl uranium tetrafluoroborate
t-Bu	*tert*-butyl
TCNE	tetracyanoethylene
TDI	tetraethynyl terrylene tetracarboxydiimide
TEM	Transmission Electron Microscopy
TEMPO	2,2,6,6-tetramethyl piperidinyloxy
TEO	tri(ethyleneoxy methyl ether)
Teoc	trimethylsilyl ethyloxy carboxyl
Tf	trifluoromethanesulfonyl (triflyl)
TFA	trifluoroacetic acid
TFAA	trifluoroacetic anhydride
TFFH	N,N,N',N'-tetramethyl-2-fluoroformamidinium hexafluorophosphate
T_g	glass transition temperature
TGA	thermogravemetric analysis
THF	tetrahydrofuran

THP	tetrahydropyran-2-yl
tht	tetrahydrothiophene
TIPDS	1,1,3,3,-tetraisopropyldisiloxane-1,3-diyl
TIPS	triisopropylsilyl
TLCV	thin layer cyclic voltammetry
T_m	melting point
TM-AFM	tapping mode atomic force microscopy
TMEDA	1,1,2,2-tetramethyl ethyene diamine
TMEDA	*N,N,N',N'*-tetramethyl-1,2-ethylenediamine
TMP	trimethylol-propane
TMS	tetramethylsilane
TMSOTf	trimethylsilyl trifluoromethanesulfonate
TOPM	thermal optical polarized microscopy
TPP	triphenyl phosphite / pyridine
tpy	2,2:6,2-terpyridine
Tr	triphenylmethyl (trityl)
TRFAM	time resolved fluorescence anisotropy measurement
TRIS	tris(hydroymethyl) methylamine
Ts	tosyl, 4-toluenesulfonyl
TsCl	*p*-toluene sulfony chloride
TTF	tetrathiafulvalene
UV-vis.	Ultaviolet-Visible Spectrophotometer
viologen	4,4'-bipyridinium
VLE	vapor-liquid equilibrium
VOC	volatile organic compound
XANES	X-ray absorption near edge structure
XPS	X-ray photoelectron spectroscopy
XRD	X-ray diffraction

CHAPTER 1

BASIC CONCEPT OF DENDRITIC POLYMERS

1.1 INTRODUCTION

Most of the natural things possess three dimensional structural symmetry. Nature also manipulates and controls three dimensional architectural symmetric molecules. For example, enzymes, which are very specific as well as effective in their functions are due to their controlled structural symmetry. Scientists also, therefore, attempt to mimic nature in this regard to develop highly regular molecular three-dimensional (3D) architecture. These three dimensional synthetic molecules, especially the macromolecules, may act as building blocks for molecular devices and other nanotechnologies [1], where the specificity and effectiveness to their functions are almost similar to enzymes. One of such 3D macromolecules of regular architectural symmetry has sparked significant interest is termed as 'dendritic polymers'[2]. This symmetry of the building blocks dramatically affects the ultimate properties of these polymers. The explosion activity in this area can be easily understood from the fact that number of publications relating to the dendritic polymers has increasing steadily in each coming year since the starting of the actual journey in 1985 (Fig.1.1).

Thus dendritic polymers are presently one of the most intensely studied new materials because of the most interesting and unique properties. They can be described beautifully as "a jungle of branches traversed by winding trails, which lead to sweet fruits and bright blossoms". These fruits and blossoms are photochemically, electrochemically, or synthetically addressable units, catalytically active sites, etc. Therefore, they have a wide range of potential applications in different fields. The details of these applications for different dendritic polymers are discussed in the subsequent chapters.

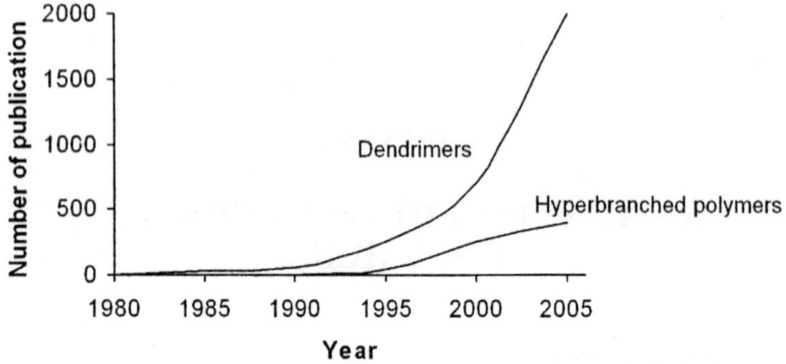

FIG.1.1 Number of publications appearing in each year (obtained from CAS)

However, it is difficult to place, dendritic polymers, especially the dendrons and dendrimers, definitely in any branch of chemistry. They are sometimes described in terms of supramolecular chemistry or and polymer chemistry, although in some cases they are neither supramolecules nor polymers in reality. So, perhaps they may be 'supermolecules' just like 'superman' in fantasy. In this book, however, they are treated as macromolecules.

In accordance with the great attention that this type of macromolecules has recently been drawn, a few books [2-5], chapters of books [6-9], a large number of review articles that summarized whole field [10-26] and important parts [27-36] have been published.

1.1.1 Definition

The dendritic polymer includes both dendrimer/dendron and hyperbranched polymer. The term 'dendrimer' is a Greek word [37], 'dendros' meaning tree like and 'meros' meaning units or parts. Thus dendrimers are tree like in growth, structure and appearance. These 3D macromolecules have a highly regular-branched architecture in which all repeating units constitute a branch point and all chain ends originate from a central core. A generally accepted definition of dendrimer is a monodisperse macromolecule with perfectly branched regular structure, free from major defects and having at least one branched junction at each repeat unit [6] (Fig. 1.2). Thus the dendrimers are a new class of 3D regular structural man-made macromolecules produced by unusual synthetic routes, which incorporate repetitive branching sequences to create a unique and novel architecture. The exceptional features of this dendritic architecture include a high degree of structural symmetry and a density gradient from core to periphery. Dendritic polymers are nanometer-sized (10^{-9} m) spherical type molecules linked in a branched manner to produce dense spherical structure with no entanglement and coiling that can be described as globular macromolecules.

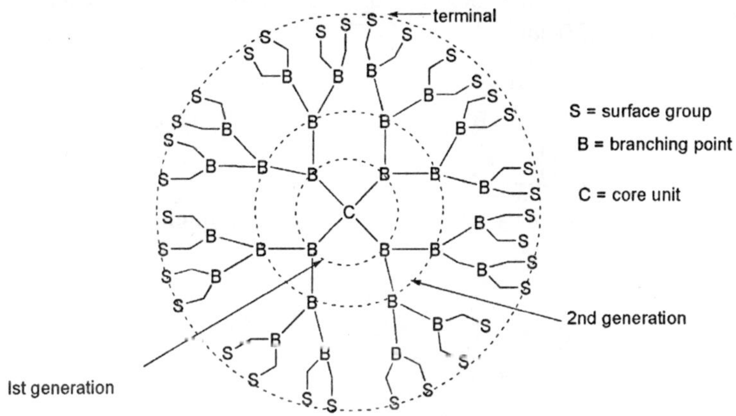

FIG.1.2 The ideal general structure of a dendrimer

Ideally, the structures of the dendrimers are perfectly monodisperse with regular and highly branched 3D structures, which are produced in iterative sequence of different reaction steps, each additional iteration leading to a higher generation dendrimer. Their unique molecular architecture, as well as the large number of terminal functionality that become available with these dendritic structures can be utilized as the tools to display desired functions such as well defined shape, internal voids or a variable surface functionalization. Whereas the dendrons are the wedges of dendrimers and have ideal open branched assemblies with a focal point. The dendrimers are also called as 'starburst dendrimers', 'cascade molecules', 'arborols', 'molecular scaffolds' and 'dendritic polymers' [11].

The hyperbranched polymers [33-36, 38-40] are closely related to highly branched macromolecules with significantly less regular structural architecture and contain a large number of structural defects and missing branches (Fig.1.3). They do not have any distinguishable interior layer or periphery in their structures. However, they have almost equal potential for different applications as they are almost similar in properties like dendrimers. The most important advantage of these HB polymers over more structurally regular dendrimers is their easier preparation by single step processes, which allow their large scale production, are less expensive and hence likely to have more commercial value. Thus they open up a new avenue in highly branched macromolecular fields.

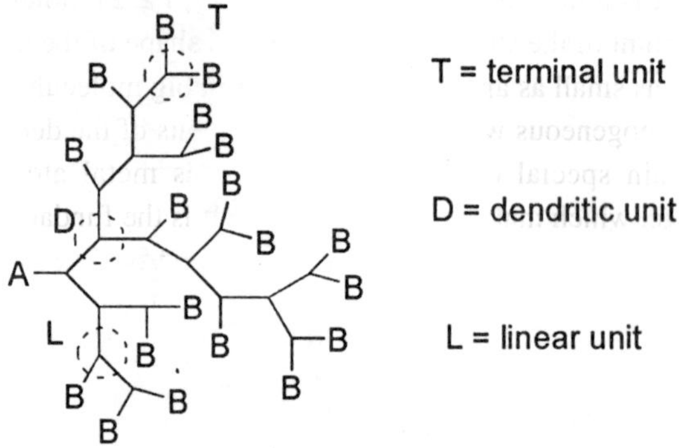

FIG.1.3 The general structure of a HB polymer

Now a days a large number of HB polymers have been synthesized from not only AB_m ($m \geq 2$) type monomers but also AB* and latent AB_m ($m \geq 2$) monomers [33]. A few research groups including us also used A_2 and B_3 type starting compounds, which formed AB_2 or A_2B monomer [41-42a-i] *in-situ* during the polymerization process. They are generally synthesized by self-polycondensation reaction (SPCR), self-condensation vinyl polymerization (SCVP) and multibranched ring opening polymerization (MBROP) techniques [33].

1.1.2 Structural Aspects

The dendrimers are unique in structural respect as they possess three distinguishing architectural features in their chemical structures. These are (i) a central multifunctional core (C), (ii) interior layers (generation, G) composed of repeating units or propagating units (B) radially attached to the central core, and (iii) an exterior layer or surface of terminal functionalities (S) attached to the outer most generation

(Fig.1.2). By definition and construction wise these three components of dendrimer are independent and reflect a unique molecular genealogy. The dendrimer begins with the central core, the molecular details are sequentially transcribed and stored to produce interior layers or generations and ultimately exterior features, which are the characteristics of it.

(i) Central Core Unit (C)

The core unit is a multifunctional (functionality, $f \geq 2$) moiety present at the focal point of the structure and affects the shape of the dendrimer. This may be as small as an atom or as large as a big molecule (Fig.1.4). It may be homogeneous with the other components of the dendrimer or it may contain special design features such as metal atoms, chromophores, etc. which makes it heterogeneous. It is the fundamental unit to which the wedges of dendrimer known as 'dendrons' (an ideal open branched assemblies) are tethered; so one may thought this, as a

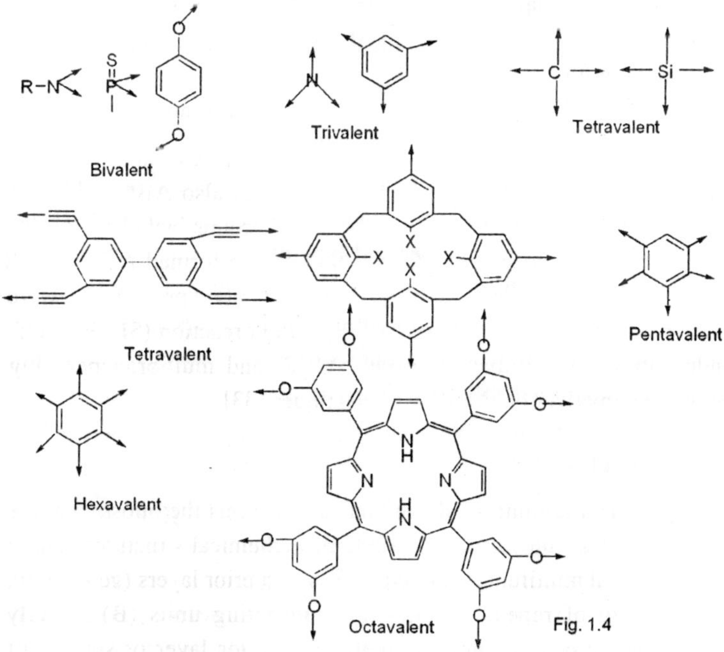

Fig. 1.4

FIG.1.4 The possible structures of different core units

primitive abiotic gene [43]. It's size, shape, multiplicity and specialized functional group will clearly influence the dendrimer throughout its construction. The core unit design may include chelating groups, molecular recognition moieties or hydrophilic/ hydrophobic domains for desired host-guest properties. Thus the selection and synthesis of central core has of a great importance.

(ii) Interior Layers or Generations (G)

The interior layers are produced by the repetitive attachment of the repeating or propagating units or building blocks (B) to the core unit or previous layer of the dendrimer. These layers or tires are called generation of the dendrimer. The repeating or propagating unit or building block is an AB_x type ($x \geq 2$) monomer, which coupled with central core to yield exponential growth in both molecular weight and end group functionality of the dendrimer. The symmetric properties of the building blocks dramatically affect the overall properties of the dendrimers. If the branch lengths are all equal then the terminal groups are placed on the surface of an ideal sphere and hence the shape of the dendrimer will be spherical.

(iii) Terminal Groups or Surface Functionalities (S)

The terminal or surface functional groups are of central importance for many applications of the dendrimers. The effectiveness of the dendrimer depends on the number and type of such functional groups as they are exposed in the outer layer. The number of such terminal groups depends on the multiplicity of core unit and the repeating units of the dendrimer. These functional groups can also be transformed to other desirable functional groups as per the necessity for their particular applications by the appropriate and simple chemical methods (Scheme 1.1).

Thus in a geometrical context, according to group theory [44] dendrimer may be viewed as triads of symmetrical trees rooted to a common central core. Each tree consists of a root i.e. central core, with connectivity to vertices i.e. branch junctures through edges i.e. branch segments, which expand symmetrically according to a geometric progression to produce generations containing predictable numbers of

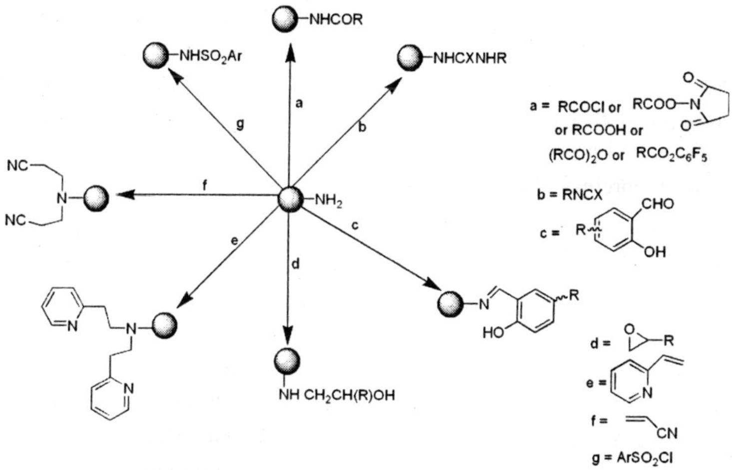

SCHEME 1.1 Chemical transformations of terminal amine groups of dendritic polymer into different other functionalities

vertices and edges i.e. terminal groups. Again, from molecular perspective, dendrimers are simple assemblies of dendrons attached to a common central core or a variety of initiator cores in the same molecule. Starting with the central core, these branched assemblies are developed by growing dendrons in a stepwise radial manner according to multiplicity rules. It is evident from many studies [45-48] that upon end group modification of the dendrimer, phase segregation between the dendritic core and the end groups can take place. In the chemical structure of dendritic polymers, the presence of secondary interactions such as π-π interaction, electrostatic interaction, hydrophobic interaction or hydrogen bonding interactions makes it possible to assemble the end groups at the periphery of the dendrimer. Back folding is thereby precluded, yielding in an homogeneous density distribution over the dendritic macromolecule and decrease in flexibility.

However, the HB polymers do not possess such distinct structural features. These dendritic polymers are distinguished from the dendrimers in that the former have a degree of branching of less than 100% in their structures. A typical structure of a HB polymer obtained from AB_m ($m \geq 2$) type monomer is shown in Fig.1.3. This structure has three different possible structural mode viz. dendritic (D), linear (L) and terminal (T) linkages. The properties of HB polymers are strongly depending on the degree of branching and the structural characteristics. They also have globular relative compact shape structures without any chain entanglement like dendrimers.

1.1.2.1 Theoretical Consideration

One of the first theoretical reports by de Genes *et al.* [49] considered the position of the end groups of the dendrimers. The authors used a self-consistent field model in which the building blocks of each generation are assumed to be fully elongated and in which the end groups of the dendrimer are grouped in concentric circles around the central core. The mathematical prediction by de Gennes *et al.* showed that there is a limit beyond which branching can no longer occur in ideal [49] fashion and the concept of starburst dense packing arrived [50-51]. It also showed that the core of the dendritic macromolecule has the lower density then periphery. However, Lescanec and Muthukur predicts a monotonic decrease in density on going from the center of the dendrimer to its periphery [52]. This indicates that the ends of the branches are not positioned at the surface but are severely back folded. Mansfield and Klushin [53] quantitatively showed the similar results by Monte Corlo simulation. Muret and Grest [54] further showed an increase of back folding with generation number and a strong effect of solvent polarity on the mean radius of gyration of the dendrimers by molecular dynamic study. Again, by help of a self-consistent mean field model (SCMF), Brris *et al.* [55] predicted that the density is highest in the core and the end groups are distributed throughout the volume of the dendritic polymers. Thus almost all the theoretical studies discussed above, except Gennes and Hervet study [49] predicted back folding branches in the dendrimeric structure.

1.1.2.2 Effect of Core and Branching Multiplicity

The number of building blocks or branch cells, number of surface or terminal functional groups, molar mass and degree of polymerization of the dendrimer are directly related to the functionality or multiplicity of the central core (f) and building blocks (b) as a function of generation number (n). The values can be predicted mathematically for an ideal system according to the following equations [56].

Number of terminal groups, $x = f.b^n$

Number of branch cells, $y = f\,(b^n-1) / (b-1)$

Number of repeating unit or degree of polymerization,

$$z = f\,(b^{n+1}-1) / (b-1)$$

Theoretical molar mass,

$$M = M_c + f\,[\{M_{ru}(b^{n+1}-1) / (b-1)\} + M_t\,b^{n+1}] \text{ or}$$
$$= M_c + f\,[\{M_{bc}(b^{n+1}-1) / (b-1)\} + M_t\,b^n]$$

where M_c, M_{ru} or M_{bc} and M_t are the molar masses of the central core, repeating units or branch cells and terminal units respectively. The size, shape, and central core multiplicity dramatically influence the ultimate critical molecular design parameters (CMPDs) of the dendrimer.

1.1.3 Historical Background

The concept of the three-dimensional network structural growths was initially introduced by Flory [57] and Zimm and Stockmayer [58]. In 1943, Flory introduced the term network cell, which they defined as the most fundamental unit in a molecular network structure [59]. Graessley [60] took the notion one step further by describing ensembles of these network cells as micro-networks. However, there were no well-characterized synthetic examples of such architecture, as these species were only conceptual. Although it is proposed that such symmetrical, ordered architecture for some biological molecules such as amylopectin but later on Erlander and French [61], as well as by Burchard *et al*, [62] observed the structure of amylopectin and glycogen are not regular rather they are random and highly branched architecture. In 1978, Vogtle *et al.* [63] reported a single isolated example of a protection / deprotection scheme involving Michael addition reaction of an amine to acrylonitrile, followed by reduction of the nitrile group to an amino group. In this reaction, scheme as notion of assembling branches upon branches involved, though no subsequent reports in this work is found. Later on Mulhaupt [64] and de Brabander [65] reported the synthesis of polypropyleneimine dendrimers based on the original work of Vogtle [63]. In 1979 Tomalia's group found that a variety of amines could be added to methyl acrylate and the product subsequently amidate with α,ω-diaminoalkanes to the 'cascade products' but it was not until the mid eighty's when the same group [16] and Newkome's group [66] reported independently their successful syntheses and details characterization of starburst dendrimers and arborol (Latin word 'arbor' meaning a tree) structures. In 1990, Frechet [67] introduced the convergent approach for the synthesis of aromatic polyether dendrimers. Tomalia's group from Dow laboratories in Midland, Michigan (USA) showed by a number of subsequent publications [36, 68-72] that these starburst system offered a wide range of synthetic option for controlling the five CMDPs, viz. size, shape, interior topology, flexibility and surface chemistry at the molecular level. The monodispersed macromolecular products were reported by Denkewalter *et al.* [73] in 1982, by synthesizing a series of dendrimers possessing unsymmetrical branch lengths. The authors patented [74-75] the synthesis of lysine dendrimer however, other than size exclusion chromatographic results [76], no detail characterization was published.

Even though Kricheldorf reported [77] the copolymerization of AB and AB_2 type of monomers to form highly branched polyester, in 1982. However, HB polymers have lately attracted considerable attention after Kim and Webster reported [78] the homopolymerization of AB_m ($m \geq 2$) type monomers to form HB polyphenylene.

1.1.4 Significance

The dendritic polymers are unique in that they possess distinguishable architectural features in their structures. For example, dendrimers having three distinguishing structural units namely (a) a central core, (b) interior layers (generation, G) and (c) an exterior layer or surface or terminal functionalities, as already described earlier. These surface groups can also be changed from one functionality to other by appropriate and simple method. Thus the nature and the properties can be controlled easily as per the necessity. Because of their unique structural features like no entanglement, high surface functionalities, etc. they possess a large variety of unusual properties, like low viscosity, high reactivity, high solubility, interior cavity, etc. The dendrimers are monodisperse, which is the most important and unique feature for them. The most significant feature of dendritic polymer is the concept of densely packing structure [50-51]. The basis of this concept is that once a specific generation number is reached, the space available for attachment of the next layer is less than that required by all the monomer units that will fill that layer. Therefore the surface becomes "densly packed'' and regular growth is stopped with incomplete functionalization of the terminal layer. Thus there is significant void space within the molecule. These voids consist of channels and cavities, as the exterior parts are densely packed but the interior parts are loosely packed for this type of molecules. This unique geometry has a number of practical interesting consequences and potentially useful properties. For example, the densely packed dendrimer would be capable of entrapping guest molecules that could be slowly released through the congested surface layer, affording (i) a novel drug delivery system [79-81], (ii) controlled release, pharmaceuticals and pesticides [82], fertilizers [83], and mosquito repellants [84]. This will also offer a special property such as adhesive characteristic in a very effective way.

These macromolecules do not obey the well-known Mark-Houwink equation, $[\eta] = kM^a$, especially at higher molecular weight and in fact diverge markedly from it. It has become evident from a number of different studies in their enhanced solubility when compared to functionally analogous to their linear polymer. One of the most interesting topological aspects of dendritic polymers are the exponential increase in number of end groups and molecular mass as a function of generation with nominal increase of size.

Another significance of the highly branched macromolecules is that they bring the active functional groups in close proximity. They are constrained by the architecture of the molecule to be at high local concentration. The dendritic polymer is shown to have more functionality in closer proximity to the surface where a chemical or a physical process takes place. In fact, Chen *et al.*[85] shown that biocide moieties work more effectively when attached to a dendritic molecule compared to free state in solution.

1.1.5 Comparison with Linear and Star Branched Polymers

These three-dimensional, non-entangled and uncoiled highly branched architectural dendritic polymers differ from linear polymers, which possess random coil structures with high degree of conformational freedom. The linear polymers are generally prepared by chain or step growth polymerization process, whereas the dendritic polymers are synthesized by convergent, divergent or single step process. The dendritic polymers are monodisperse or having very low polydispersity index, whereas linear polymers having high polydispersity index. The dendritic polymers also have lower viscosity, high solubility and higher number of surface functionalities compared to the linear analogs. On the other hand, pre-organized supramolecular receptor molecules might contain internal cavity like dendritic polymers, but they lack the presence of a distinct microenvironment suitable for complexation of multiple molecules. The occupied volume of a single dendritic molecule increased cubically with generation in solution, whereas its mass increases exponentially.

There are differences in behavior of dendritic polymer from their linear analogs. Hawker *et al.* [86] reported a study where the dendritic polyethers are compared with their linear isomers, having exactly the same number of repeat units and end groups (Fig. 1.5). This study shows that the hydrodynamic volume of fifth generation polyether dendrimer is ~30 % lesser than that of its linear analog. This is due to the fact that the dendritic polymer has a more compact, and back folded globular structure. In addition, the fifth generation dendrimer is

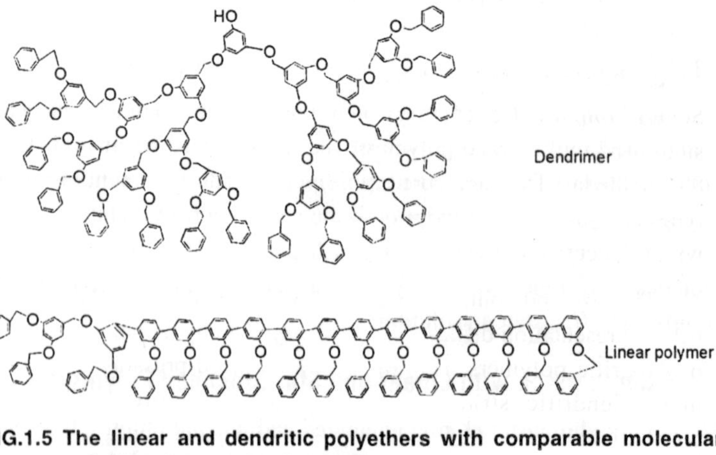

FIG.1.5 The linear and dendritic polyethers with comparable molecular weight and same structural units

completely amorphous and soluble in a variety of solvents, whereas the linear polymer is highly crystalline and shows poor solubility in organic solvents.

Although star-branched polymers have some similarities to a certain extent with respect to central core, radial branching or even telechelic functionality [87] with dendritic polymers but there are significant differences present between these two categories of polymers. The most important is the branch segment densities, which are just opposite. In the former case, the branch segment densities are greater in the central core and decrease in a concentric radial fashion (Fig.1.6) in contrast to trend of the same in case of dendritic polymers (Fig.1.2). The star-branched polymers are asymmetrically branched, whereas dendritic polymer especially the dendrimers are symmetrically branched.

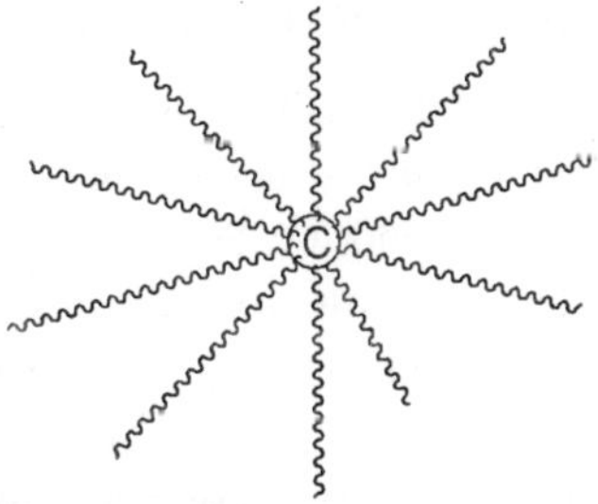

FIG.1.6 The general structure of a star polymer

1.1.6 Nomenclature and Structural Representation

Several important documents on systematic nomenclature for linear single and multi-strand polymers have been published since 1986 [88-89]. In the last 15 years there has been a rapid development of different types of non-linear macromolecules for which a little attention has paid with respect to their nomenclature.

The size and complexity of structures of dendritic polymers make their representation difficult both on paper and by name. The depiction of dendritic polymers may be somewhat more pressing problem, as many dendritic structures simply can't be portrayed in more conventional ways. Although, bent and elongated bonds can often be used to drag the structures of the dendritic polymer. When this practice becomes unacceptable, two possibilities are left, i) either 'cartoon' representation or ii) the structure drawn in part can be used (Fig. 1.7). In this figure the entire structure of the third generation dendrimer implied by drawing just one branch from core to surface.

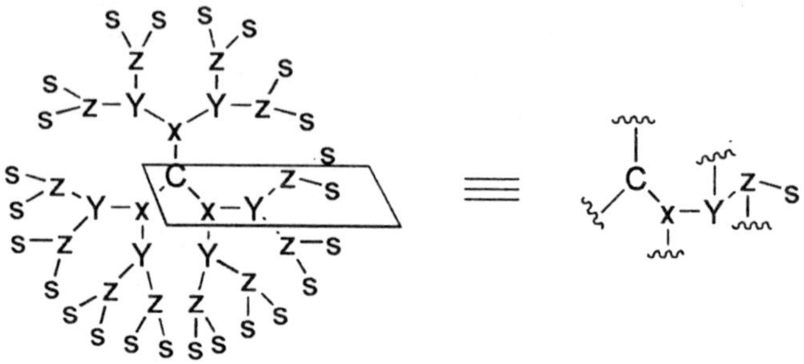

(A)

FIG.1.7 Representation of part of a dendritic structure

The cartoon representations are being used by many researchers for the illustration of macromolecular structures on the printed page. In this representation each of the dendritic branch units is represented by truncated wedge shapes with different shadings. Reactive functionalities are shown as dark shadings, while passive functionalities by light shadings. These illustrations allow very large structures to be drawn concisely and help to clarify subsequent synthetic schemes and diagrams.

Newkome *et al.* [90-91] and also Baker *et al.* [92] have proposed a systematic structure-based nomenclature for dendritic polymers. Although, Newkome *et al.* [90-91] proposed a systematic nomenclature for dendrimers but that has not been generally well accepted. Instead of that the short hand versions nomenclatures are well accepted and used by various researchers for the sake of brevity. Although a number of short hand versions have been used in practice but they only fall into two main categories. In the first one, the total number of monomers or building blocks are counted [93] and the second one, where the number and nature of surface functionalities are defined, followed by the generation number and the functionality at the focal point (center core of the dendron) or core of the dendrimer [94]. For example, a dendritic fragment having sixteen cyano groups at the chain ends, four layers of building blocks, i.e. 4th generation dendron and a hydroxyl group at the central core (Fig.1.8), may be represented as $(CN)_{16}$–[G-4]-OH. Another representation in case of poly(propylene imine) dendrimer has been shown as DAB-dendr-$(NH_2)_x$, where DAB stands for the core molecule 1,4-diamino butane, 'dendr' stands for the poly(propylene imine) dendrimer with 'x' number of terminal primary amine groups and the number 'x' varies with number of generation, e.g. 4, 8, 16, 32, 64----for 1st, 2nd, 3rd, 4th, 5th ---generation, respectively.

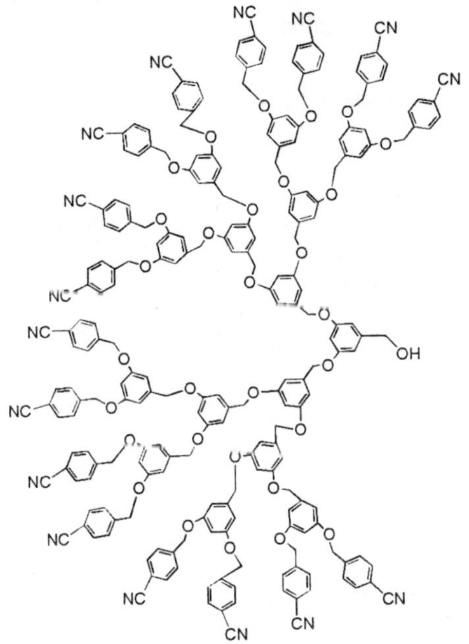

FIG.1.8 The structure of a 4th generation dendrimer with 16-CN groups

Wilks also discussed the nomenclature and structure representation for HB polymers [95] and dendrimers [96]. However, Chemical Abstracts Service (CAS) has not yet addressed the problem of indexing dendritic polymers structurally. IUPAC has formed a working committee to address dendritic polymer nomenclature. The structural based representation for a polymer is generally supposed to be superior to source based representation because only the former conveys structural information not the latter. However, there are reports [97] on source-based representations for some dendritic polymers also.

1.1.7 Classification

It is indeed a real problem to classify the dendritic polymers in a clear-cut way. A considerable confusion and ambiguity still exist regarding classification of this novel class of macromolecules. There is no single well-accepted classification for the dendritic polymers. An attempt has been made in this section to classify dendritic polymers into different categories (Table 1.1) based on their mode of synthesis, presence of main element, main functional linkage, physical properties, applications, etc. However, this is also not a very useful classification, but it is the first attempt.

TABLE 1.1 Classification of dendritic polymers

Basis of classification	Categories
Mode of Synthesis	Convergent, Divergent, Accelerated and Single step
Main Functionality	Hydrocarbon, Amide, Amido-amine, Imine, Ether, Ester, Amine ether, Carbosilane, Carbosiloxane, Carbosilazane, Carbosileses-quinoxane, Urethane, Phosphazine, Metalloid, etc.
Physical Properties and Application	Photoactive, Redox active, Liquid crystalline, Amphiphilic Light harvesting, Catalytic, Bioactive, Chiral, etc.

In the subsequent chapters, all of above categories of dendritic polymers are discussed in the appropriate places.

1.2 SYNTHETIC METHODOLOGY

The most important factors, one has to keep in mind for the synthesis of this novel class of materials, is to control over molecular shape, size and disposition of the building blocks. By the proper choice of the synthetic technique, it is possible to control size, shape, topology, flexibility and surface chemistry referred as critical molecular design parameters (CMDPs) at the molecular level. The control of these CMDPs allows the development of features and functions inherent in all living systems namely, their ability to reproduce, to pass on genetic information, i.e. a prototype blue print, to their progeny, to perform various specialized functions and finally to grow to pre-specified limited dimensions usually with some characteristic morphology.

In functional coiled and entangled macromolecules, it is difficult to obtain high yield and efficient reactions. Because of their coiled and entangled structure, the propagating end faced difficulty to access by the reactive monomer unit. But unlike this situation, in case of dendritic polymers, the radial orientation of the terminal functionalities expose in a favorable fashion which are easily accessible to the incoming reactive monomers / building blocks. For this reason, the feasible reaction can proceed equally well for the first few generations. However, at higher generation the surface become densely packed and the reaction kinetics and stoichiometry would expect to be changed dramatically. The synthetic methodologies employed for the synthesis of the dendritic polymers are mainly three types, though now-a-days other accelerated techniques are used frequently, which are further developed and resulted method of the above three techniques. These techniques are discussed in the subsequent sub-sections.

1.2.1 Divergent Approach

Historically the initial attempt to synthesize dendrimers was the divergent approach. The name comes from the mode formation of the dendrimers. In this approach the dendrimer starts from the central core and then diverged stepwise manner to the periphery through the repetitive activation of the surface functional groups and addition of branching monomer units [98]. Thus to synthesize a dendrimer by this approach two basic units are required: (i) the central core unit and (ii) the propagating units / building blocks. The core unit contains more than two same types of functional groups, which could be used for the branch growth. The propagating unit always has two different functional groups, which are non-reacting to each other and any of these groups does not react with central core functionality. So this group, which is not reactive to the functionalities of core unit is either a new functional group or a protected form of the core functional group, but in any case it has to be converted easily to the reactive group. A schematic representation of the divergent growth approach is shown in scheme 1.2. In the first step, 'R' of core reacts with 'R' of progating unit to form a symmetric growth compound with two or more times the number of function 'P'. Then each function 'P' is converted to 'R', in order to construct the first generation (G_1). The resulting first generation acts then as a next core and the reaction of the first generation are repeated to yield the next generation. In the same way, the reactions are repeated to get higher generation (G_n, n > 2).

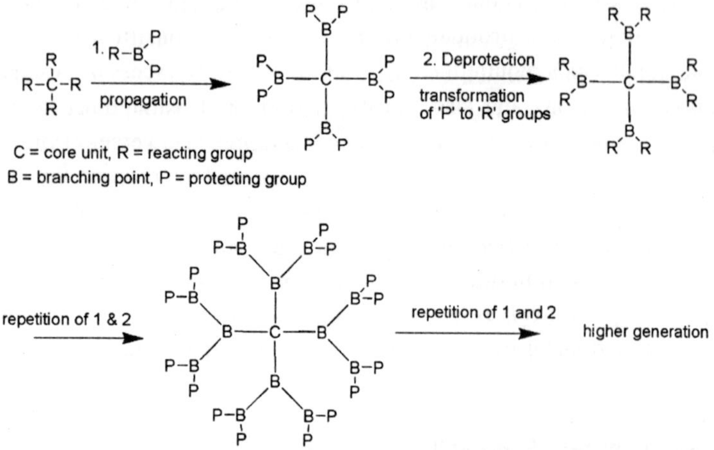

SCHEME 1.2 Schematic representation of a divergent growth approach

Here, it should be noted that all the reactions involved in each growth step should be quantitative and without any side reaction, which causes structural defects and results polydisperse dendrimer. It is also important to note that the purity of each generation should be very high for the subsequent growth of the dendrimer.

This approach is successful for the production of large quantities of dendrimers as in each step the molar mass of the product increased exponentially. So very large dendrimers with very high molecular weight are also possible to synthesize by this approach. But incomplete growth steps and side reactions especially at higher generations, lead to the imperfect structure of the dendrimers. In fact, virtually it is very difficult to isolate pure product from their side products by this approach. The synthetic macromolecular chemists, therefore, must rely on extremely efficient reactions in order to ensure low polydispersity of the synthesized dendrimers.

1.2.2 Convergent Approach

Even though the divergent growth approach has proved extremely successful for the synthesis of dendrimers, but suffers from a serious drawback of relatively uncontrolled reaction steps. Even at lower generations, the number of terminal groups become very large and thus their selective functionalization is extremely difficult. Also, the large number of simultaneous reactions required fully to functionalize these terminal groups introduce difficulties, because purification of the macromolecules require their separation from large excess of monomer and / or reagents and also leads to problems in maintaining 'perfect growth'. Therefore, to overcome the weaknesses of divergent synthetic approach, an alternative strategy called 'convergent approach' was introduced. In this approach [98] the dendrimer starts from the periphery and proceeds to the central core, so it took opposite course for the construction of the dendrimer/dendron. Similar to the divergent growth it also involves two steps: (i) the attachment of the outer most layers functional groups to the next inner generation and (ii) attachment of the final generation to the central core to get the desired dendrimer. A schematic representation of convergent growth approach for synthesis of dendrons and dendrimers are shown in scheme 1.3.

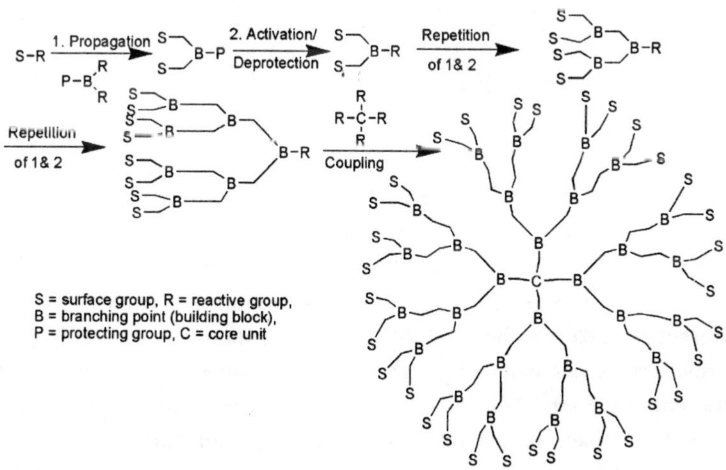

SCHEME 1.3 Schematic representation of a convergent growth approach

The advantages of convergent growth over the divergent approach are (i) allowing more precise control over the dendritic architecture, (ii) providing direct access to structurally uniform macromolecules as well as compounds with differential periphery functionalities, and the most importantly, the purification of the perfect dendrimer is much simpler. Also growth reaction do not have to be so stringently efficient like divergent approach. Moreover, unlike divergent approach, the convergent approach is characterized by the continuous presence of a single functional group at the focal point, which can be exploited to generate a variety of novel macromolecular architectures. Another consequence of the single focal point is that generation growth requires only a single activation reaction and usually only two coupling reactions. Whereas divergent approach requires an increasing number of both activation and coupling reaction for generation growth. Thus the chances of creation of defect during generation growth are much less in convergent approach than the divergent one.

But there are some shortcomings also in the convergent method. The number of steps required to build up a large structure is not reduced compared to the divergent approach, yet a more starting material is required. It also suffers from low yields in the synthesis of large structures and dendrons of higher generations encounter serious steric problems in the reactions of their focal points with the core unit.

1.2.3 Accelerated Approach

Although the divergent and convergent methods are largely used to synthesize various dendrimers successfully, but both are suffering from common multistep procedure problem. So they are time consuming, laborious processes and low over all yield that make the materials more expensive. Recently, a number of accelerated approaches are being used for the synthesis of dendrimers/dendrons, which required less number of steps for the synthesis of higher generation of dendrimers. These are double exponential growth technique, hypercore technique etc. The common feature of all these accelerated techniques is that the number of steps for the synthesis of desired dendrimer is drastically reduced, which greatly speed up the synthetic process and thereby decrease the work load and ultimately the cost of the materials.

1.2.3.1 Hypercore and Branched Monomer

In the early 1990, Frechet and his co-workers [99-100] demonstrated the utilization of 'hypercores' and 'branched monomers' for accelerated synthesis of dendrimers. These approaches involve the pre-assembly of oligomeric species, which can then be linked together to offer the desired dendrimers in fewer steps with higher overall yields. A schematic representation of this approach is shown in scheme 1.4.

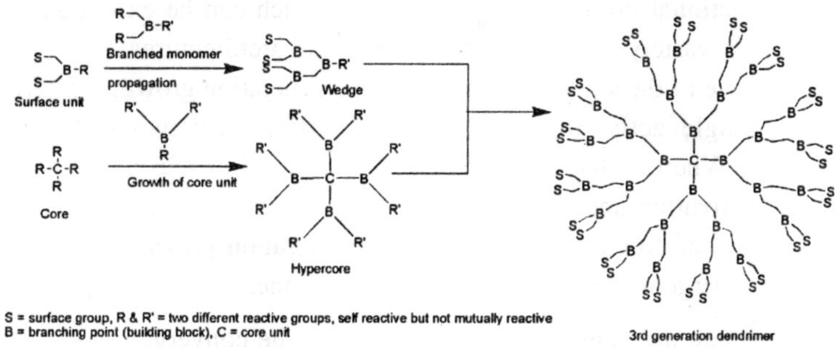

SCHEME 1.4 Schematic representation of a hypercore growth approach

However, this approach is more like a convergent approach with less number of steps requirement for the generation growth. Although an interesting comparison of convergent, divergent and hypercores approaches for the synthesis of poly(phenyl acetylene) dendrimer was shown by Moore and his co-workers [101]. But the solubility problems in the divergent approach here made the convergent approach favorable (the details are discussed in chapter 2).

1.2.3.2 Double Exponential and Mixed Growth

The most recent fundamental breakthrough in the practice of dendrimer synthesis has come forward with the concept and implications of 'double exponential and mixed growth' approach [101] (Scheme 1.5). In this approach an AB_2 type monomer (P_2-B-P`) with orthogonal protecting groups 'A' with 'B' is used. This approach allows the preparation of monomers for both convergent and divergent growth from a single starting material by selective reactions at the surface and at the focal point separately. These two deprotected compounds are reacted together to give an orthogonal protected trimer, which may be used to repeat the growth process again (Scheme 1.5). The first use of a versatile building block in this way performed by Shinkai [102-103] without realizing the full potential of the idea, but Moore and his coworkers [101] first coined this term 'double exponential'. This type of approach has also been used by Frechet [104], Sharpless [105], Roy [106] and others for the synthesis of their dendrimers.

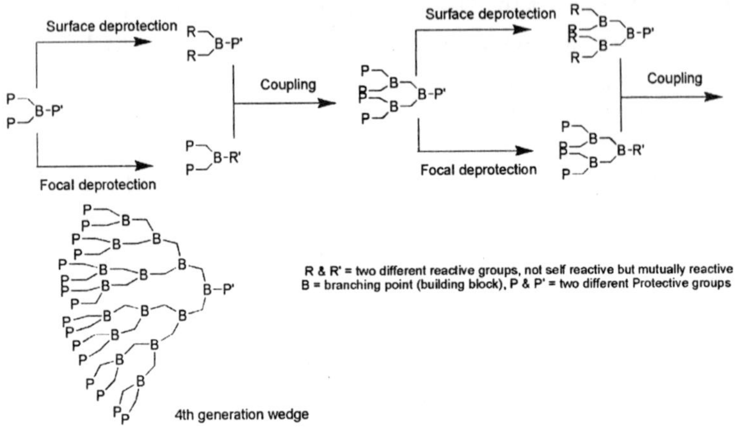

SCHEME 1.5 Schematic representation of a double exponential growth approach

The strength of double exponential growth is more to build large dendrimers in relatively few steps. In fact, double exponential growth is so fast that it can be repeated two or three times before further growth becomes almost impossible. This methodology provides a means whereby a dendritic fragment can be extended in either the convergent or the divergent direction as required. Thus the positive aspects of both approaches can be accessed without the necessity to blow their shortcomings.

1.2.4 Single Step Approach

Instead of synthesizing the dendritic polymer in a stepwise iterative technique by divergent or convergent approach, some attempts have been made to synthesize the similar type dendritic polymers by a less laborious and time consuming single step approach [107]. However, as there is a possibility to form more defects on the structure of the dendritic polymers, because of involvement of large number of reactions on the same molecules in a single step, so one will only get HB polymers not dendrimer or dendron by this technique. Thus the synthesis of HB polymers are much simpler compared to dendrimers, as the former synthesis does not require the use of protection/ deprotection or multistep procedure. This is due to the fact that HB polymers are allowed to contain some linear chains in their structures. Hence, so much accurate control of polymerization reactions is not required unlike the case of the synthesis for dendrimers.

Although all these above approaches are being used presently by the researchers working in this field but currently only the divergent approach is attractive for the production of kilogram quantities of dendrimers and only a few dendrimers such as poly(amido amine), poly(propylene imine), etc. are commercially available [65, 108]. Whereas the single step process is well accepted for synthesis of HB polymers and a hyperbranched polyester has already been commerciallized by this technique.

1.3 CHARACTERIZATION TECHNIQUES

The characterization of dendritic polymers can not be easily carried out like small molecules or even high molecular weight linear polymers, because of their unique HB structural characteristics along with a large number of functional groups at the periphery. Also the possibility of structural defects at any generation of the dendritic polymers causes further complexity in their final structures. The dendritic polymers are being characterized by conventional techniques like IR, UV-Visible, NMR, GPC, viscosity measurement etc. along with some special techniques like MALDI-TOF / FAB/ ESI- mass spectrometry, AFM, STM, OPM, CV, etc. depending upon the requirement. Mainly the following dendritic features have been studied by the different spectroscopic and analytic techniques. These are structural characteristics, elemental composition, molecular weight and its distribution, homogeneity or purity, dimension, thermal characteristics, optical behavior, etc. of the dendritic polymers.

1.3.1 Structural Elucidation Techniques

The structure of the dendritic polymers can be elucidated by different techniques, as described below.

1.3.1.1 Techniques for Determination of Element Composition and Functional Groups by IR, UV-Visible, and Elemental Analysis

Infra-red (IR), Ultaviolet-Visible (UV-vis.) Spectroscopies and Elemental Analysis

The IR and UV-visible spectroscopies are largely used to determine the functional groups present in the structures of the dendritic polymers like linear polymers. The UV-vis. spectroscopy is not only used to determine the presence of chromophoric groups or conjugation in the dendritic structures but also largely used to characterize the photophysical properties of the photoactive dendrimers. Elemental analysis such as CHN analyzer, hetroatomanalyzer, etc. are used to determine the percentage of elements present in the structure, but elemental composition of the dendritic polymer does not vary too much for different generations. Also it is difficult to determine the exact percent of elements present in the structure as there are some structural defects always present in the structures, especially at higher generation dendrimers/dendrons and HB polymers.

1.3.1.2 Techniques for Determination of Structure by NMR Spectroscopy

Nuclear Magnetic Resonance (NMR) Spectroscopy

NMR can play an important role to help, to understand the structures and properties of dendritic polymers. Mainly one-dimensional NMR (1D-NMR) techniques have been used for the characterization of these novel materials. Indeed, ^{1}HNMR and ^{13}CNMR spectra of dendrimers can be surprisingly simple and contain a great deal of information about defects and impurities in their structures. But NMR has not been preferred technique for studying dendrimers, despite the wealth of information available from measurable NMR parameters. This is largely due to the result of the limited ability to resolve the resonances from the many unique, but very similar groups in these molecules. The most thorough NMR studies have been performed on PAMAM type dendrimers. Lambert and his co-workers applied ^{29}Si–^{29}Si INADEQUATE 2D-NMR to confirm the structure of the first generation dendritic polysilane [109]. It demonstrated the clear connectivity for each of the three types of Si–Si bonds in the molecules. However, this experiment can only be used for silane dendrimers, and INADEQUATE experiment is a very insensitive method. Because of their high cost and the difficulties in preparation of isotropically labeled synthetic polymers including dendrimers, the NMR technique used must be very sensitive and give efficient suppress of undesired signal components. Of course now there is a report [110] on the use of ^{1}H, ^{13}C, ^{29}Si triple resonance 3D-NMR combined with pulse field gradient techniques [111], which can be used to fully characterize carbosilane dendrimers without resorting to isotopic labeling. They also demonstrated [112] the application of the double-resonance 3D HMQC–TOCSY NMR experiment [113] for the complete characterization poly(propylene imine) dendrimers.

NMR spectroscopy is a very useful tool for determination of DB in HB polymers. The degree of branching is determined by the following methods. From Fig.1.3, it has been found that the structure of HB polymer contains dendritic (D) and terminal (T) linkages like dendrimer along with linear (L) linkages like linear polymer. In 1991, Frechet and his co-workers [114] described the degree of branching (DB) as a factor to explain the structure of HB polymers as

$$DB = (D + T) / (D + T + L)$$

However, Frey and his coworkers [115] given a modified equation for DB based on the growth direction as

$$DB = (2D) / (2D + L) = (D + T - N) / (D + T + L - N)$$

where 'N' is the number of molecules. Basically, both the equations are same as for high molecular weight polymers, 'N' is negligible. Ideally DB is equal to 1 for dendrimers or dendrons, zero for linear polymers and it is close to 0.5 for HB polymers. As it becomes close to unity, the architecture of the HB polymer is eventually the same as that of the corresponding dendron/dendrimer. The NMR spectroscopy like ^{1}HNMR, ^{13}CNMR, ^{15}NNMR, ^{19}FNMR, ^{29}SiNMR and ^{31}PNMR, etc. are largely used to determine DB for the HB polymers [33]. However, when the HB polymers are composed by degradable linkages like –COOR or –O-C(=O)-O-, DB can also be calculated from the quantitative analyses of the products after degradation [116-117].

1.3.2 Techniques for Determination of Homo-geneity, Purity and Molecular Weight & Its Distribution by Chromatographic Techniques and Mass spectrometry

Size Exclusion Chromatography (SEC)

Both Gel Filtration Chromatography (GFC) and Gel Permeation Chromatography (GPC) are being used as SEC. In both the cases the molecules are separated according to their size in solution. In GFC, aqueous solutions of the water soluble polymers are used, whereas in GPC organic solvents are used. To determine molecular mass, polydispersity and hydrodynamic diameter SEC, Low Angle Laser Light Scattering (LALLS), and intrinsic viscosity measurement have been widely used in case of dendritic polymers. The hydrodynamic diameter of the dendritic polymer has been determined by the application of the Hester and Mitchell equation [118]. This equation relates the molar mass, the intrinsic viscosity and molecular hydrodynamic diameter as

$$d = (240/\pi N_A)^{1/3} (M [\eta])^{1/3}$$

where d = hydrodynamic diameter in cm, N_A = Avogadro number, M = molar mass, and $[\eta]$ = intrinsic viscosity in dL/g.

Small Angle Neutron Scattering (SANS) and viscometry measurement suggested [119] a linear relationship between the radius of gyration of the PPI dendrimer and its generation number. The SEC along with intrinsic viscosity measurements is also being used to determine the conformation behavior of PAMAM dendrimers [68, 120]. These studies suggested that the dendrimers have a hollow core and a densely packed outer layer.

The SEC has been used for the calculation of radius of gyration, hydrodynamic radii [118], polymer average molecular weight and its distribution. However, it has a drawback, as here only approximate molecular weights of the dendritic polymers have been determined by comparing with polystyrene standard but the relationship between hydrodynamic volume and molecular weight is not the same for all the polymers. So to get the exact average molecular weight and it's distribution, mass spectrometry has been used.

The chromatographic techniques such as high performance liquid chromatography (HPLC), column chromatography, etc. are also sometimes used to check the purity of the dendritic polymers.

Mass Spectrometry

It actually measures the hydrodynamic volume of the polymer chain in the solution. This gives a relative measure of molecular weight, because obviously, higher the molecular weight of a polymer, the greater its hydrodynamic volume. Mass spectrometry is a valuable technique for the characterization of dendritic polymers. The mass spectrometry of dendritic polymers has benefited from the soft ionization techniques developed for the study of large biomolecules. Many results have been obtained by the use of Matrix Assisted Laser Desorption / Ionization-Time of Flight (MALDI-TOF) [121-122] and Electron Spray Ionization (ESI) [123-124] mass spectrometric methods. MALDI-TOF mass spectrometry measures the mass more accurately, because it does not compare the polymer being measured to anything. It gives an absolute measurement of mass. Here the molecular weight of the polymer has been determined by transferring them into charge species (cation or anion) under vacuum after mixing with strong UV-absorber (e.g. trans cinnamic acid or 2,5-dihydroxy benzoic acid) with the help of UV-laser. ESI-mass spectrometry has also been used to characterize dendritic polymers. All the generations with amine or nitrile end groups of PPI dendrimers have been analyzed by ESI-mass spectrometry to quantitatively determine the degree of various side reactions [125]. Often mass spectrometry is the best evidence for determination of dendritic structures, even when NMR spectroscopy is ambiguous.

1.3.3 Techniques for Determination of Morphology by SAXS, AFM, EPR, STM, etc. Spectroscopy

Small Angle X-ray Scattering (SAXS)

A small angle X-ray scattering [126-127] is used to characterize the single particle scattering factors produced by different types of dendrimers in dilute solutions. The higher generation (5th to 10th) PAMAM dendrimers exhibited higher order scattering features, which indicate a nearly monodisperse, spherical particles with essentially uniform internal segment densities [127]. Whereas for dilute HB polymer solution exhibited scattering more indicative of the inherent irregularity of internal segment densities and overall sizes to be expected within these systems. The study indicated from the measurement of radius of gyration as a function of generation (molecular weight) that lower generation dendrimers are less dense than their higher generations relatives, but after the first few generation (M_w

> 50,000 Da), the average density of the dendrimer appears to be roughly independent of generation. But HB polymers have approximately twice the radius of gyration compared to the dendrimers of comparable molecular weight. However, because of their relatively feature-lessness of the particle scattering curves and inherent polydispersity, it is difficult to conclude about their internal HB structure. SAXS and dynamic light scattering (DLS) studies suggested that the dendritic structure possesses a flexible core and a dense shell that will have internal cavities available for guest molecules [126].

Atomic Force Microscopy (AFM)

AFM is a very useful tool for characterization of polymer surface and thin films. It provides a topographic map of the polymer surface with high spatial resolution, in a non-invasive way and without additional sample treatments. For example, it is used to analyze self-assembly of alkyl substituted polyphenylene dendrimers on graphite surface. It is the first scanning probe microscope, which overcomes the limitation of STM in imaging thin film samples on electrically conductive materials. In addition to the favorable image conditions and highly resolution, AFM offers the possibility to characterize mechanical properties of the sample on the nanometer scale.

Electro Paramagnetic Resonance (EPR)

EPR has been used to characterize dendrimer for their host–guest interaction. It provides the binding and structural properties of host molecules and surfaces in the case of various spin labels and spin probes as guest molecules. Turro and his co-workers used this technique to characterize PAMAM dendrimers [128-129].

Scanning Tunneling Microscopy (STM)

STM is a suitable tool to investigate thin film or individual molecule deposited on an electrically conducting surface down to the atomic scale. It provides information on the conformation of the adsorbed molecules, which can't be accessed by other methods. For example, as dendritic polymers are mostly amorphous in nature, they can't be accessed by X-ray crystallography, but can be accessed by STM [130-131]. Moreover, surface assembly processes can give insight into the nature and strength of intermolecular and surface molecule interactions. Hermann *et al.* [130] investigated the conformation of chiral dendrimers on a Pt (100) surface by STM technique. The STM image of the chiral polyether dendrimer at low temperature (-196°C) showed better resolution that at ambient conditions. A comparison of doubly branched dendrimers shows a difference of only in chirality compared to the chiral dendrimers, indicates that STM is a valuable tool in providing information on the cause of the unexpected case of distereoselectivity. Merz *et al.* [131] also succeed to image single, physicosorbed porphyrin dendrimers on noble metal (Pt / Ir : 9/11) surfaces with STM.

Other techniques are also used for some special purposes such as electron microscopy has been of great use in the visualization of dendrimers [16, 132-133] and their aggregates [134-136]. The electrochemical behavior of the redox active dendrimers are being studied by cyclic voltammetry (CV) in homogeneous solution, as well as that confined to electrode surface, where the dendrimers serve as electrode modifiers. The liquid crystalline behavior of the dendritic polymers is studied by optical polarized microscopy (OPM) and X-ray diffraction (XRD) spectroscopy.

1.3.4 Techniques for Determination of Thermal Characteristics by TGA and DSC Analysis

The thermal characterizations of the dendritic polymers are generally carried out by thermogravemetric analysis (TGA) and differential scanning calorimetry (DSC) techniques. The thermal stability of the dendritic polymers [137-138] along with the nature of the degradation in different environmental conditions are being studied by TGA technique. Whereas DSC is used to get their glass transition temperature (T_g), and melting temperature (T_m) along with enthalpy changes during the different transition or degradation stages [139].

1.4 PROPERTIES

One of the obvious objectives of the development of architectural symmetric monodisperse (dispersity index close to unity) dendritic polymers is to attain the new challenging bulk properties of these macromolecules with regard to their wide range of applications. Although, structure-property relationship is very useful in this regard, but still there is lacuna to get the exact picture of it for these novel class of materials. This may partly due to synthetic difficulties, availability and partly due to incomplete studies. A brief overview of the properties of dendritic polymers is presented here, although individual properties of the respective dendritic polymers are discussed in the subsequent chapters.

Because of complexity, particularly at higher generations, a variety of defects may arise in the structure of the dendritic polymers. For example, in the divergent synthetic approach several hundred reactions are involved on the same molecular fragment for the higher generation dendrimers, so statistical defects in the final product are a reality. These defects may arise from two types of synthetic reaction complexity viz. intra-dendritic events and inter-dendritic bridging. The former causes branching defects, whereas the latter brings more polydispersity in the system. These defects have to be minimized or eliminated to get ideality in the structure, which will offer tailor made properties. This can be done by the optimization of the process of the synthetic strategy.

The branch defects may be due to (i) incomplete reactions resulting from the surface groups deactivation or inappropriate reaction conditions, (ii) branch-juncture fragmentation, (iii) abnormal branch-juncture development, or (iv) sterically induced stoichiometry. The branching defects and intra-molecular cyclization lead to reduce surface stoichiometry and affect ultimate symmetric properties of the polymers.

In addition to intermolecular bridging or looping, the polydispersity may also occur due to (i) incomplete removal of reagents (which may function as initiator cores), and (ii) dendrimer fragmentation (to produce molecular entities, which may subsequently become initiator species).

The significance of the side reactions has been calculated by using an iterative computing process. In case of a fifth generation amine functionalized PPI dendritic polymer, the computer simulation indicated a polydispersity of 1.002 and a dendritic purity 23 % only, without any defect. This can be related to an average selectivity of 99.4 % per reaction step, since 248 reactions are required (Scheme 1.6) to obtain 5th generation dendrimer with 64 terminal groups (0.994^{248} = 0.23). If the iterative multi-step reaction sequence is replaced by a one step procedure, branched macromolecules are obtained with a high degree of branching, and a large molecular weight distribution, which are coined as 'HB polymer'.

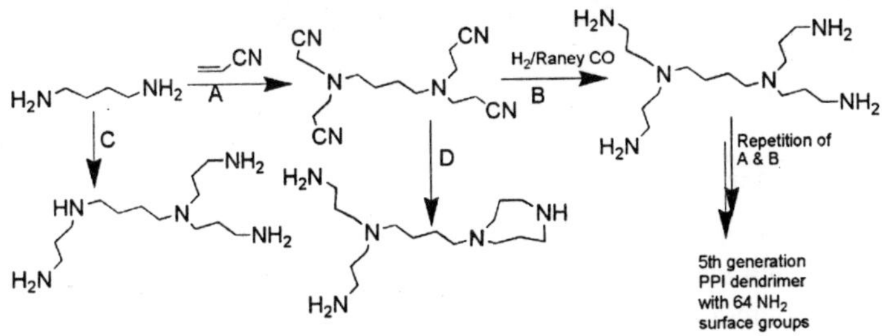

SCHEME 1.6 Synthesis of a 5th generation poly(propylene imine) dendrimer with 64 terminal groups

1.4.1 Physical Properties

The significant feature of dendritic polymers that has sparked of considerable interest is the belief that they would adopt a globular shape depending on the core molecule, branching multiplicity and length of the branch segment, in both solution and solid state. This behavior is similar to a number of natural systems, much different from traditional one-dimensional linear polymers that adopt a random coiled structure. This globular structure will definitely affect the physical and chemical properties of these unusual macromolecules. This globular structure is also free from chain entanglements, which is in direct contrast to linear polymers where physical properties are dominated by chain entanglements.

The high molecular weight protein molecules also possess globular structure, which exhibit low viscosity and higher solubility like dendritic polymers. Even conventional linear polymers also form globular like structure in a relatively bad solvent, where solvent-polymer interaction is less and hence the surface area of the polymers is low. But in both the cases, these globular structures are not same as in case of dendritic polymers. This is due to the fact that both protein and linear polymers in bad solvent do not have the unique architectural features like dendritic polymers.

The surface functionality of these globular dendritic polymers can be specifically designed and transformed from one functionality to another. Because the surrounding building blocks can be controlled, a unique microenvironment can be created around the functionality leading to a structure that may mimic the active site of an enzyme. The new improved properties have compelled much research as exciting new applications are being anticipated in areas ranging from drug delivery and diagnostic tools to rheology control and micro fabrication.

1.4.1.1 Solubility

The solubility characteristics of the dendritic polymers are more useful for their end applications as they exhibit enhanced solubility compared to their functionally analogs and equivalent molecular weight linear polymers. This may be due to the globular shape and the presence of large number of end functionalities in the former compared to the latter. The enhancement of solubility for dendritic polyphenylenes compared to linear polyparaphenylene is found to be 10^5 times [137]. As the solubility largely depends on the nature of the end groups and thus highly polar carboxylate groups containing dendrimer is even water soluble. All the polyphenylene dendrimers have good solubility in common organic solvents like THF, toluene, CH_2Cl_2, $CHCl_3$, etc. The solubility of a polyether dendrimer in THF is approximately 50 times higher than that of its linear analog polymer [140]. A comparative study on the solubility of the isomeric polyesters based on 3,5-dihydroxy benzoic acid that has benzyl ether end groups and meta linkages with the analogs linear polyester having same number of benzyl ether end groups and comparable molecular weight showed [141] that the dendrimer has extremely high solubility (1.15 g/ml in THF) compared to the linear polymer (0.025 g/ml in THF). This study further confirms that the highly branched dendritic structures dramatically affect the solubility.

1.4.1.2 Viscosity

The solution intrinsic viscosity, [η] of the linear or star branched polymers varies according to the Mark-Houwink equation, $[\eta] = kM^a$, where 'k' and 'a' are constants for a particular polymer-solvent system at a particular temperature and M is the viscosity average molecular weight of the polymer. However, the dendritic polymers do not obey this equation, and in fact they diverge markedly from it [142]. It has been found that the plot of log [η] vs. log [M] resulted a bell shape curve with a maximum at certain intermediate generation (molecular weight) and not the expected fast rising curve (Fig.1.9). This can be explained by the standard theories if it is assumed that dendrimers have spherical shape. Thus during generation growth their volume increased cubically ($V = 4/3\ \pi r^3$) while their mass effectively doubles and therefore increases exponentially ($M \sim 2^{n-1}$, n = generation number). As the intrinsic viscosity effectively reflects the ratio of volume/mass, a maximum is obviously expected. The maximum in the intrinsic viscosity can also be corrected with a shape change from an extended to a more compact globular structure at high molecular weights. A shape transition also experimentally observed in the 3rd to 4th generation of polyether dendrimer [143] by a series of solvatochromic measurements. The study of the melt viscosity behavior of the dendrimer has also led to similar conclusions [144]. Thus the dendritic polymers have low melt and solution viscosities compared to their linear analogs (Fig.1.9).

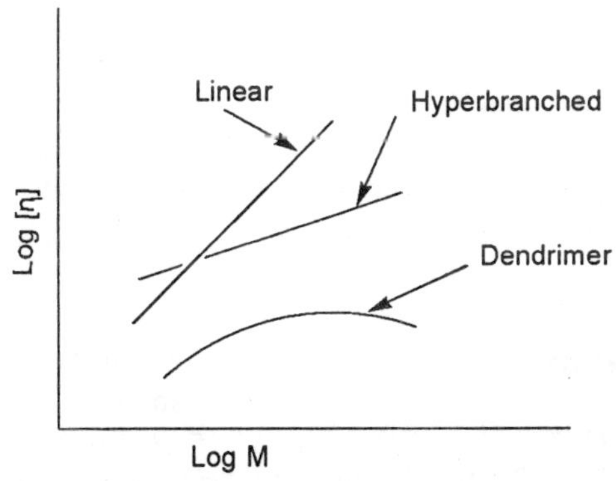

FIG.1.9 The plot of log [η] vs. [M] for dendritic and linear polymers

1.4.1.3 Density

The structure of the dendrimer is such that the density of this macromolecule increases from the core to the periphery. That is, they become densely packed structure at the periphery with the increase of generation number. In general, the density of the dendrimer varies just reverse order as intrinsic viscosity with generation number [71]. The reason is similar to that of viscosity that is the mass increases exponentially, whereas volume increases cubically. However, the individual polymer density depends on the chemical composition and morphology of the structure of the dendritic polymers just like linear polymers.

The densely packed structure on the surface of the dendritic polymers as a function of generation can be easily shown mathematically by the following formula

$$a_s = a_d/n_s \; \alpha \; r^2 / b^n f$$

where a_s = surface area per terminal group, a_d = dendrimer total surface area, n_s = number of surface groups per generation, r = radius of the dendrimer, and n = number of generation, f = number of functionalities or multiplicity of the central core and b = number of functionalities or multiplicity of the building blocks. From this relationship, it can be seen that with increase of 'n' i.e. generation, the surface area per terminal group will continuously decrease and ultimately it will reach a limiting value that is the van der Waal dimension of the surface groups. In this limiting generation the dendritic polymers are densely packed and after that ideal dendritic structure formation is not possible. This critical situation off-course is a function of branch segment length directly related with radius of the dendritic polymer, radius of the central core, building block multiplicity and the steric dimension of the terminal groups.

1.4.1.4 Crystallinity

The dendritic polymers display a tree like architecture and their high branched nature generally make their solid state structures without any 'long range order' [145]. Even, in case of highly dense-surfaced spherical dendrimers where the possibility of some degree of structural orderness appeared, there also some degree of disorder within the generations are present and hence crystal structure can't be formed. There are only a few examples of crystalline dendrimers [146], but that are also confined to lower generations and rigid hindered structures only. The HB polymers are most often reported to be amorphous, which can be anticipated from their highly branched architecture, even though there are some exceptions also.

Although, it is very unlikely to display liquid crystalline (LC) behavior by dendritic polymers, because of their tree-like architecture, but in few cases even this novel material showed LC behavior [147-155]. The dendritic polymers may have the mesogenic groups in each branching unit [147-148] or located on surface of the dendritic scaffold [149-150] or it may be without any mesogenic group also [151-155]. These LC dendritic polymers may display thermotropic, nematic, sematic, etc. mesophase. They are very useful for electro-optic devices, thermochromic and fluid flow sensors for imaging devices, and for information storage.

1.4.1.5 Optical Properties

The dedritic polymers having chiral moiety in their structures, display optical properties. The chiral unit may present in the core unit, or branching blocks, or surface groups or formed by constitutionally different segments anchored to a central core. There are number of research articles [156-160] and reviews [161-163] for this type of dendritic polymers. They have potential applications in the field of chiral chromatographic materials, catalyst in assymmetric synthesis. The chiroptical properties of the dendrimers like optical activity decreases with the increase of generation number and at very high generation, it is not possible to detect also, e.g. PPI dendrimer with 'Boc' protected amino acids as terminal groups, the optical activity is hardly detectable in the fifth generation [156, 162]. This effect is not due to racimization, concentration, temperature or solvent influence but may due to steric effects of the amino acid moieties that lie closely together. The optical rotation of the dendrimer is also very sensitive to the local surroundings. If there is no steric influence in the structure then the optical activity remains almost constant for every generation [162]. Furthermore, the position, number of chiral moieties and the flexibility of the segments strongly influence the resulting macroscopic optical properties of the dendrimers.

The dendrimer containing photoactive components are particularly interesting because of a) luminescence signals offer a handle to better understand the dendritic structures and superstructures, b) co-operation among the photoactive components can allow the dendrimer to perform useful functions, and c) changes in photophysical properties can be exploited for sensing purpose with signal amplification.

1.4.2 Mechanical and Rheological Properties

The correct mechanical and rheological behavior is very much essential for newly developed polymeric materials in order to commercialize them. The tree like architecture with no entanglement and low hydrodynamic volume of the dendritic polymers lead to Newtonian rheological behavior and lower melt viscosity at a given molecular weight compared to linear analogs. The rheological properties for HB polymers are also similar in molten state and no shear thing or thickening was observed [164]. However, this non-entangled state imposes rather poor mechanical properties of the dendritic polymers. This has also limited their uses as a load bearing or structural materials, even they have much more superior properties in other field of applications. The melt behavior has been shown to be greatly affected by the structure of the end groups where an increase in the polarity of the end groups can raise the viscosity by several order of magnitudes [165]. Another interesting feature is that at the critical molecular weight (M_c), the variation of melt viscosity of the dendritic polymers does not follow the same trend as linear polymers. Rather, the increase in viscosity with molecular weight is less pronounced for dendritic polymers and levels off at higher molecular weight [164].

Ordinarily, the conventional linear chain polymers exhibit good mechanical (load bearing) properties. This property increases with the increase of molecular weight. This is due to chain entanglements of the linear polymers. Dendritic polymers, on the other hand, do not possess linear structures rather they are globular. Their high molecular weights are due to increase of branches in the structures from a central point. The situation will be more cleared when we compare with a linear starch polymer, amylose with highly branched bushy type structure, amylopectein. The amylose shows high mechanical properties than even the higher molecular weight amylopectein. Thus amylopectein appears to be some sort of transition molecule from conventional linear polymer to high molecular weight dendritic polymer.

1.4.3 Chemical Properties

The spherical shape, the presence of large number of functional groups at the periphery, low viscosity and high solubility, make the dendritic polymers easily accessible to the other substrates. Thus the dendritic polymers are much more reactive by multiple interactions at a time compared to linear analog polymers. This high reactivity with the presence of large number of reactive groups at the surface may be utilized for curing of different resins.

1.4.4 Thermal Properties

The thermal properties of the polymers are extremely important for their end uses, as these properties are highly sensitive to change in temperature. The important features responsible for thermal behavior of a polymer are bond strength, intermolecular forces, resonance stability, crystallinity, imperfection in structure, molecular weight, etc. As all these factors more or less are similar both for dendritic and linear analog polymers, so there is as such no difference in their thermal properties. The polyester dendrimer, which showed large difference in solubility behavior compared to its linear analog, exhibits almost similar thermogravimetric behavior [3]. Miller *et al.* reported [137-138, 166] high thermal stability (95 % of mass retention up to 500°C) for polyester liquid crystal dendritic polyethers and hydrocarbon dendrimers [137]. In general, the T_g and T_m increase with the increase of molecular weight, rigidity, inter- and intra-molecular interactions, etc. so also the molar enthalpies and entropies of fusion. A study on aliphatic polyesters showed that dendritic polymers with dense structures exhibited smaller thermal expansion coefficient and lower compressibility compared to their linear analogs. The variation of glass transition temperature, within a dendritic series is almost similar to their analog linear polymers. However, the nature of end groups has marked affect on the T_gs of the dendritic polymers. For example, T_g shifted about 100°C, (250 to150)^{o}C on going from carboxylic acid to acetate end groups. For a highly branched macromolecules [167-170] like dendritic polymers when compared with a linear polymer of the same molecular weight, the additional chain ends in the branched molecules act to decrease the $T_{g,}$ while the restriction in mobility caused by the junction point acts to increase T_g. Also with the increase of chain ends polarity, the T_g increased.

1.4.5 Electrical Properties

The dendrimers may be electrically conducting depending on their structural characteristic, just like linear polymers. The naphthalene diimide anion radicals modified PAMAM [171] dendrimer exhibited high electrical conductivity. Actually, these anion radicals aggregated into a π–stacked net-work, which provides the pathway for conductivity. The conductivity of the dendrimers also varies with the humidity, temperature of the experiment, frequency applied, temperature of film casting, etc. just like linear polymers. The modified PAMAM dendrimer [171], cast films at 60^oC showed conductivity, $\sigma = 2X\ 10^{-3}\ scm^{-1}$, ten times higher than that for the film cast at 120^oC. This may due to improvement of stacking of π–network at lower temperture, which facilitates conductivity. The fully reduced dendritic film increased conductivity to $0.12\ scm^{-1}$ (under high humid condition), whereas partially reduced one showed conductivity as high as $11\ scm^{-1}$.

1.4.6 Amphiphilic Behavior

The dendritic polymers may have one polar functionality in their interior and non-polar functionality on the surface and they may behave as micelles. The dendritic polymers are comparable [68] to certain extent with micelles, which are recognized as two concentric spheres. These have an inner core of close-packed solvent incompatible sphere and an outer surface of solvent compatible moieties. The dimensions (10-200A°), spherical shape and number of surface or head groups as a function of generation in dendritic polymers are similar to micelles. However, the head group multiplicity in case of dendrimer is determined by synthesis, while this is determined by free energy minimization [172] and/or head to tail packing ratios [173]. The dendritic micelles are capable not only to molecular inclusion but also can solubilize hydrophobic molecules in aqueous solution to an equal extent than can traditional micelles, while not displaying any critical micelle concentration (CMC). Also very interestingly, the solubilization ability of the dendritic micelles is a sensitive function to the ionic strength of the aqueous solution and the internal chemistry of the dendrimer. Furthermore, the molecular size of the dendritic micelles changes with pH [174], the hydrodynamic radius being the largest at neutral pH and the smallest at acidic pH.

1.4.7 Host-Guest Interaction Chemistry

Vogtle *et al.* [63] first reported on the cascade molecules and other theoretical reports [50-51, 175] including the highly branched molecules as ideal molecular containers, showing the challenges in host-guest interactions of dendritic molecules. After 15 years of real journey of dendritic polymers by Newkome [66] and Tomalia [16], this field of host-guest properties of dendritic polymers has sparked significantly into a special area of supramolecular chemistry [176-178]. Supramolecular chemistry is generally described as the chemistry beyond the covalent bond and takes into account specific molecular interactions and the relationship between the geometrical structure and binding sites.

Analog to the active sites of enzymes, the dendritic polymers have the specific property like localized microenvironment or internal cavities in their structures, which is the basis of this host-guest chemistry. In addition, dendrimers also contain three topologically different regions viz. central core, propagating unit and surface moieties, as discussed earlier. Each of these can exhibit functional properties modulated by the dendrimer as a whole [179]. In dendritic host-guest systems, the interactions are mainly hydrogen bonding, electrostatic or hydrophobic through either central core or branching point or at the periphery.

Host-guest chemistry involves the binding of a substrate molecule, that is the guest in a receptor molecule which is the host. The enhance selectivity and strength of binding sites through the above interaction will be the determining factors in the development of more efficient host-guest system.

The tendency of relatively apolar molecules to assembly in aqueous solution is explained by hydrophobic interactions [180]. This interaction plays a vital role in surfactant aggregation, the assembly of lipid in bio-membranes, and enzyme-substrate interactions. The concept of trapping guest molecules by dendritic host molecules with a spherical structure was first suggested by Maciejeweski [175]. Compared to the relatively open structures of lower generation (< 3rd) dendrimers, the higher generations (> 3rd) trend to adopt an extended conformation with spherical surface containing pockets of space in the interior, which are capable of guest inclusion.

From the studies of structural analysis of particular dendritic polymers, it is suggested that the dendritic structure possesses a flexible core and a dense shell that will have internal cavities available for guest molecules. Actually, the voids in the dendritic interior depend strongly on the exact dendritic structure, secondary interactions between end groups, etc. Depending on the conditions used, the cavities can be filled up by end groups, solvents or other guest molecules. In principle, dendrimers can accommodate different kinds of guests, such as anionic species [181], metal ions [182], and neutral substrates [183], either within the porous interior or at the exterior of the multifunctional periphery.

The dendritic polymers of lower generation are not suitable for encapsulation of guest molecules, as the shell is not densely enough to capture the guests. But at higher generation, a shell is constructed in the final step, so it is possible to perform the coupling reaction in the presence of guest molecules [7]. A large variety of guest molecules such as 3-carboxy-PROXYL, Bengal rose, Eriochrom Black-T, etc. have been encapsulated. Thus this host-guest chemistry opens up a new era of interesting chemical and biological applications.

1.5 APPLICATIONS

The unique structural features, like nanoscopic size, spheroidal surface, high branching, cavernous interior, etc. and exciting properties, like low viscosity, high solubility, high reactivity, no-entanglement, in combination with the high functionalities of the dendritic polymers suggested that they have wide number of potential applications in different fields. These included medicinal and diagnosis applications [184-185], adhesive and coatings [186], gene therapy and chemical sensors [187-188], drug delivery systems [79-81], light harvesting materials [189], catalysts [190-191], electronic applications [192], separating agents

[140], and many more [193-194]. The HB polymers may be exploited as functional crosslinkers, polymeric additives, rheology modifiers or process aids, as well as components in adhesives, advanced coatings, structural hydrogels novel dental composites, etc. Furthermore, the HB polymers might also be of interest in nanotechnology as building block for nanoscale reaction compartments [195] as templates for nanoscale porous materials with low dielectric constant or for the fabrication of defined hybrid particles, along with the other fields where dendrimer might be used and PDI not necessarily to be unity. A general overview of some of these applications of the dendritic polymers in nut-shell is presented here, the details applications in each subclass dendritic polymers are discussed in the subsequent chapters.

1.5.1 Biological and Medicinal

A potential advantage of dendritic polymers is their biomedical applications. This is due to their structural properties, like defined size, inside cavities, and a variety of equivalent peripheral groups, dendritic polymers are ideal synthetic analog to proteins, enzymes and viruses. In addition to these, they can be easily functionalized as already stated earlier. These areas include dendrimer based diagnostic imaging contrast agents, gene-therapy agents [1, 79-81, 196], etc.

Roy *et al.* [196] showed that lysine dendrons, which were substituted with upto 16-α-sialinic acid moieties are highly active in inhibiting the hemaglutination of human erythrocytes by influenza viruses. Gd(III) diethylenetriamine penta acid (Gd-DTPA) attached dendrimers used as effective magnetic resonance imaging (MRI) contrast agent for image target organs, or to detect changes in blood flow in the brain or in the pool of an organ [197-200]. The dendrimers were used as therapeutics in boron neutron capture therapy [201-202] by which the desired malignant cells were selectively destroyed in the presence of normal cells. Boronated dendrimer-monoclonal antibody conjugates [203-204] are used successfully as protein probes in electron spectroscopic imaging.

The attachment of a biologically active molecule to the surface of the dendrimers often increases the efficiency of the compound. Maltews and Holen described [205] the antiviral properties of PAMAM and lysin based dendrimers substituted with a variety of cationic and anionic end groups, significantly inhibit HIVI, HIVL, human cytomegalovirus (CMV) and various herpes simplex viruses (HSV). In this connection, it is worthy to mention that the studies [206] on the effect of size of PAMAM dendrimer showed that the larger dendrimer (M_w ~ 87000 Da) wère passed into the urine and excreted within two days, whereas the smaller dendrimer (M_w ~ 5000 Da) mostly accumulated in the liver, kidney and spleen with no urine excretion, when used inside the human body. Thus it might be noted that application of HB polymers, which are generally polydisperse might create problem for *in-vivo* uses.

1.5.2 Catalysis

There are large number of examples, where dendritic polymers have been used as catalysts, which are discussed in the subsequent chapters where ever it is applicable. There are mainly two important advantages of using dendritic catalyst in any catalytic reaction. Firstly, there is a possibility of creating large dendrimers with many active sites. This type of catalyst is intermediate between heterogeneous and homogeneous catalysts, and can be readily separated by filtration [207]. Secondly, there is a possibility of encapsulating a single catalytic site whose activity selectively becomes modulated or enhanced by the dendritic superstructure [208]. If the dendritic catalytic site also provides for the reversible substrate binding prior to reaction, supramolecular enzyme like catalyst will be formed.

The 'dendrizymes' i.e. diphenyl phosphine ligands substituted with chiral surfaced dendrons were the first dedritic catalysts to model enzymes to act as enantioselective transition metal catalyst [208]. However, very low enantioselectivities have been reported for the several dendrizymes catalyzed reactions including (i) hydrogenation of acetamidocinnamic acid, (ii) hydrosilylation of acetophenone with diphenyl silane and (iii) cyclopropanation of styrene with ethyl diazoacetate.

Zn(II) or Ti(IV) of α,α,α'α'-tetraaryl-1,3-dioxalane-4,5-dimethanol (TADDOL) terminated dendrimers have been used [209] as homogeneous, enantioselective catalysts for (i) nucleophilic addition of diethylzinc to benzaldehyde and (ii) (3+2) cycloaddition of 3-crotoxyloxazolidione to a nitrone with same catalytic activity as parent TADDOL-ate complexes. But advantage here, the catalyst can be reclaimed in greater than 90 % yield.

The addition of diethyl-Zinc to benzaldehyde has also been catalyzed by HB chiral catalysts [190]. Even catalytic dendrimers can be used to create microenvironment on polymerization reactions [210-211]. Frechet polyethers type dendrimers have been found to have a significant effect on the degree of polymerization and polydispersivity indexes of the polymer obtained [210-211]. This may be due to the fact that the growing end of the polymer chain can reside inside the cavity of the dendrimer and thus reduce the rate of termination.

1.5.3 Energy and Electron Transfer

Natural photosynthetic systems collect sunlight by using a vast energy to a single reaction center. The accurate positioning of the chromophore at their central core, focal point (for dendron), periphery, or even at each branching points of the dendrimers may result a similar light harvesting dendritic polymer. However, not only the appropriate placement of chromophores in the dendritic structures is very important, but also the judicious choices must be made both for the selection of the chromophores themselves and for the design of each successive layer in order to maximize favorable energy transfer interactions. It also depends on the rigidity, conjugation, size and polarity of the dendritic backbone i.e. on the type of branched monomer. Thus, the choice of monomer will affect the overall properties and the energy transfer mechanism of the dendritic system. The tree like architecture of a dendritic wedge, naturally suggests itself as a molecular antennae suitable for transferring energy from the surface to the focal point. The rigid polyaromatic ethyne linked dendrimers with a fluorescent perylene chromophore at the focal point were used [212] as the light harvesting materials. The effective energy transfer with high quantum yield (upto 98%) was observed for this system. However, energy transferring ability increases with generation of the dendrimer. The first multi-chromophoric dendrimers undergoing intramolecular energy transfer were reported by Balzani *et al.* [213-214]. The polyaromatic dendrons have also been used to transfer energy and electrons [215-216]. Aida and his coworkers [217-218] used Frechet type poly(benzyl ether) dendrons with porphyrin core unit for efficient energy harvesting system. The dendrimers can also be used for electron transfer to and from a redox-active core moiety with applicability in single molecule binary information storage schemes [219-220]. PPI dendrimers functionalized with ferrocenyl units are used as redox active dendrimers, i.e. they can transfer electron in the system [221]. The redox behavior depends on the generation number and the rate of electron transfer and the diffusion co-efficient decreased with the increase of generation number [222].

1.5.4 Encapsulation

Because of unique geometry of the dendritic structure, they can be used as encapsulating agents for different guest molecules [28]. The host–guest interactions in most of the cases are purely secondary interactions like hydrogen bonding, hydrophobic effect, polar or van der Waal interactions, etc. A large number of guest molecules are being encapsulated into the dendrimer center and interior. Meijer and his coworkers [156, 223-224] reported a so-called 'dendritic box', in which the various molecules were encapsulated on PPI dendrimers and subsequently released only when sterically bulky, hydrogen bonded peripheral groups were cleaved. This type of physical encapsulation of guest molecules within the dendritic structure has potential applicability in catalysts [208,225-226] and drug delivery systems [227-228]. Paleos *et al.* [227] reported the pH-controlled inclusion and the liberation of pyrene in PPI dendrimers. At higher pH it can be encapsulated and at lower pH of the aqueous solution, the pyrene molecules are released, as more and more internal amino groups of the dendrimer are protonated. Solvent molecules are the simplest guests. It is also commonly observed that with the increase of generation it becomes more difficult to remove solvents. Once solvent molecules have been removed, the conformation of the dendritic polymers is supposed to be changed as it is observed that dried polymers required longer time to redissolve.

1.5.5 Unimolecular Micelles

The amphiphilic dendrimers are also known as unimolecular micelles [174] having mainly carboxylate anions or ammonium cations as end groups and a hydrophobic core. As catalysts, dendritic quaternary ammonium ions are active even at very low concentration and high ionic strength [229]. They are also used to solubilize lipophilic compounds in aqueous solutions [67, 174, 230-232] and therefore find applications in molecular encapsulation [224], drug delivery and nanoscopic transport [231]. Other amphiphilic dendritic polymers different from that resemble conventional micelles, also are attractive for such applications.

1.5.6 Liquid Crystal Display (LCD)

The dendritic polymers utilized in a large area of displaying technology such as electro-optical switching, etc. Baars *et al.* [232] developed an electro-optic switch based on the addition of dendrimers to a continuous liquid crystalline matrix. A 5th generation cyanobiphenyl and palmitoyl functionalized PPI dendrimers were used as additive for this purpose.

1.5.7 Surface Coatings

The peculiarities of dendritic structures offer new possibilities to manipulate the properties of surface coatings. HB polyesters [233], a combination of enhanced reactivity and reduced viscosity for alkyd resins was obtained which showed improved properties on coating application. The use of HB methylacrylated polyester in photo-polymerizations of films and fiber reinforced polymer composites enhance the curing rates and lower viscosity compared to linea⁻ unsaturated polyesters [234-235].

1.5.8 Multipurpose Polymeric Additives and Others

Hyperbranched polymers were used as polymeric additives in many linear commercial polymers as compatibilizing agents, process and strength improvers, etc. The hyperbranched aromatic polyamines and polyethers with s-triazine unit were used as multipurpose polymeric additives in industrial linear polymers like LDPE and plasticized PVC [42d, e]. These dendritic polymers act as antidegradants, flame retardants, process aids, semireinforcing agents, etc. One of these hyperbranched polyamines was also used as a crosslinker and performance improver for a vegetable oil based poly(ester-amide) resin [42g]. These hyperbranched polyamines also exhibit fluorescence behavior in the range of 400-650 nm [42h]. They are also used as matrices for preparation of nanoparticles such as Ag and Cu by in-situ reductive technique [42i]. These nano-composite's colloidal solutions have potential antibacterial activity.

1.6 SUMMARY

A brief discussion on the fundamental of dendritic polymers is presented in this chapter. The dendritic polymers have been considered as the 21st century's highly branched macromolecules. They are not only highly branched with large number of active functionalities but most of them are also monodisperse macromolecules. However, the high molecular weight containing dendritic polymers are having more defective structures and hence polydisperse, even though their PDI values are not very large. A good number of synthetic methodologies are available for the synthesis of these novel macromolecules among that accelerated approach for dendrons and dendrimers and single step for HB polymers are very promising. Because of their unique structural features like no entanglement, high surface functionalities, etc. they possess a large variety of unusual properties, like low viscosity, high reactivity, interior cavity, etc. The chemical modification or functionalization of the dendrimers resulted a wide range of variation in physical and chemical properties and hence their applications. They have broad spectra of potential applications including medicinal to polymeric additives. They have the important roles in medicinal and diagnosis applications, adhesive and coatings, gene therapy and chemical sensor, drug delivery systems, light harvesting materials, catalysis, electronic applications, separating agents, and many more. The dendritic polymers of unit ratio of branch sites to repeat units have emerged as an important new concept in macromolecular design.

But even then, in order to realize their full potential outside the academic world, the methodology of synthesis must be developed further, which allows faster and precise synthesis to a level of an industrial scale. As the synthetic method improves, better understanding on their structural aspect and properties will come forward, the number of available dendritic polymers will expand and their potential applications will become prevalent. However, to get the highest output from these novel classes of materials, a multidisciplinary research by collaboration of synthetic polymer chemists, physicists, biologists and engineers is very much essential. Fortunately, off course now a days many research groups started to realize to work together in multidisciplinary way and thereby getting much improved results.

Finally, it is one of the most exciting young field of researches need to be maturated, so there are a lot of scopes to start or more involvement in this field of works.

REFERENCES

[1] J.R.Granja, R.A.Milligan, D.E.McRee, N.Khazanovich and M.R.Ghadiri, *Nature* **366**(1993)324

[2] G.R.Newkome (ed.), 'Advances in Dendritic Macromolecules" Vol. 1-5, CAI press Inc., Greenwich, CT, 1994-1999

[3] G.R.Newkome, C.N.Moorefield and F.Vogtle, 'Dendritic Molecules: Concepts, Syntheses, Perspectives', VCH, Weinheim, Germany, p.37, 1996

[4] J.M.J. Frechet and D.A.Tomalia (ed.), 'Dendrimers and Other Dendritic Polymers', John Wiley & Sons, Chichester, 2002

[5] G.R.Newkome, C.N.Moorefield and F.Vogtle, 'Dendrimers and Dendrons: Concepts, Syntheses, Applications', John Wiley & Sons, Chichester, 2001

[6] C.J Hawker and J.M.J Frechet, 'Three-Dimensional Dendritic Macromolecules: Design, Synthesis and Properties' in New Methods of Polymer Synthesis, edited by J.R.Ebdon and G.C.Eastmond, Chapman and Hall, New York, pp.290-330, 1995

[7] M.H.P.van Genderen, E.M.M. de Brabander van den Berg and E.W.Meijer, 'Dendritic Architectures' in Supramolecular Technology, edited by D.A.Reinhoudt, John Wiley, Chichester, Chap. 2, 1999

[8] S.Gesterman, R.Hesse, B.Windisch and F.Vogtle, 'Dendritic Architectures', in Stimulating Concept in Chemistry, edited by F.Vogtle, J.F.Stoddart and M.Shibasaki, pp.187-198, 2000

[9] G.R.Newkome and C.N.Moorefield, 'Comprehensive Supramolecular Chemistry', Vol.10 edited by N.Reinhoudt, Pregamon press, Oxford, pp.777-832, 1996

[10] D.A.Tomalia, A.M.Naylor and W.A.Goddar III, *Angew Chem.Int.Ed.* **29**(1990)138

[11] N.Karak and S.Maiti, *J.Polym.Mater.* **14**(1997)107

[12] M.Grayson and J.M.J.Frechet, *Chem.Rev.* **101**(2001)3819

[13] A.Archut and F.Vogtle, *Chem.Soc.Rev.* **27**(1998)233

[14] B.I.Voit, *Acta Polym.* **46**(1995)87

[15] J.Issberner, R.Moors and F.Vogtle, *Angew.Chem.Int.Ed.* **33**(1994)2413

[16] D.A.Tomalia, H.Baker, J.R.Dewald, M.Hall, G.Kallos, S.Martin, J.Roeck, J.Ryder and P.Smith, *Polym.J.* **17**(1985)117

[17] R.Engel, *Polym.News.* **17**(1992)301

[18] S.Aoshima, J.M.J.Frechet, R.B.Grubbs, M.Hemmi and M.Leduc, *Polym.Prepr.* **36**(1995)531

[19] J.M.Desimone, *Science* **269**(1995)1060

[20] M.Fischer and F.Vogtle, *Angew Chem.Int.Ed.* **38**(1999)884

[21] F.Vogtle, S.Gestermann, R.Hesse, H.Schwierz and B.Windisch, *Prog.Polym.Sci.* **25**(2000)987

[22] H.Frey, C.Lach and K.Lorenz, *Adv.Mater.* **10**(1998)279

[23] D.K.Smith and F.Diederich, *Chem.Eur.J.* **4**(1998)1353

[24] M.Sciler, *Chem.Eng.Tech.* **25**(2002)237

[25] M.W.P.L.Baars and E.W.Meijer, *Top.Cur.Chem.* **210**(2000)131

[26] A.W.Bosman, H.M.Janssen and E.W.Meijer, *Chem.Rev.* **99**(1999)1665

[27] O.A.Mathews, A.N.Shipway and J.F.Stoddart, *Prog.Polym.Sci.* **23**(1998)1

[28] F.Zeng and S.C.Zimmerman, *Chem.Rev.* **97**(1997)1681

[29] G.E. Oosteron, J.N.H.Reek, P.C.J.Kamer and P.W.N.M.van Leeuwen, *Angew.Chem.Int. Ed.* **40**(2001)1828

[30] D.Astruc and F.Chardac, *Chem.Rev.* **101**(2001)2991

[31] S.E.Stiriba, H.Frey and R.Haag, *Angew.Chem.Int.Ed.* **41**(2002)1329

[32] R.Mezzenga, L.Boogh and J.A.E.Manson, *Comp.Sci.Technol.* **61**(2001)787

[33] M.Jikei and M.Kakimoto, *Prog.Polym.Sci.* **25**(2000)1171

[34] B.Voit, *J.Polym.Sci.Part A Polym.Chem.* **38**(2000)2606

[35] Y.H.Kim, *J. Polym.Sci. Part A Polym.Chem.* **36**(1998)1685

[36] A.Sunder, J.Heinemann and H.Frey, *Chem.Eur.J.* **6**(2000)2499

[37] D.A.Tomalia, H.Baker, J.R.Dewald, M.Hall, G.Kallos, S.Martin, J.Roeck, J.Ryder and P.Smith, *Macromolecules* **19**(1986)2466

[38] Y.H.Kim and O.W.Webster, *Macromolecules* **25**(1992)5561

[39] S.R.Turner, B.I.Voit and T.H.Mourey, *Macromolecules* **26**(1993)4617

[40] E.Malmstrom, M.Johansson and A.Hult, *Macromolecules* **28**(1995)1698

[41] J.Hao, M.Jikei and M.Kakimoto, *Macromolecules* **36**(2003)3519

[42] S.S.Mahapatra and N.Karak, *J.Appl.Polym.Sci.* **106**(2007)95

[42a] J.Borah and N.Karak, *Polym.Int.* **52**(2004)2026

[42b] S.S.Mahapatra and N.Karak, *Polym.Degrd.Stab.* **92**(2007)947

[42c] J.Borah, S.S.Mahapatra, D.Saikia and N.Karak, *Polym.Degrd.Stab.* **91**(2006)2911

[42d] J.Borah and N.Karak, *Int.J.Plastic Technol.* **10**(2006)598

[42e] S.S.Mahapatra and N.Karak, *Euas.Chem Technol. J.* **9**(2007)29

[42f] J.Borah and N.Karak, *J.Appl.Polym.Sci.* **104**(2007)648

[42g] S.S.Mahapatra and N.Karak, *Prog.Org.Coat.* **60**(2007)328

[42h] S.S.Mahapatra and N.Karak, (2007, submitted)

[42i] S.S.Mahapatra and N.Karak, (2007, submitted)

[43] R.Dawkins, *New Scientist*, **25**(1986)24

[44] R.E.Merrifield and H.E.Simmons (eds.), 'Topological Methods in Chemistry', Wiley New York, 1989

[45] B.Stark, B.Stuhn, H.Frey, C.Lach, K.Lorenz and B.Frick, *Macromolecules* **31**(1998)5415

[46] L.L.Miller, T.Hashimoto, I.Tabakovic, D.R.Swanson and D.A.Tomalia, *Chem.Mater.* **7**(1995) 9

[47] A.W.Bosman, M.J.Bruining, H.Kooijman, A.L.Spek, R.A.J.Jansen and E.W.Meijer, *J.Am.Chem.Soc.* **102**(1998)8547

[48] M.W.P.L.Barrs, S.H.M.Sontjens, H.M.Fischer, H.W.I.Peerling and E.W.Meijer, *Chem.Eur.J.* **4**(1998)2456

[49] P.G.de Gennes and H.J.Hervet, *Phys.Lett.* (Fr.) **44**(1983)L351

[50] J.S.Moore and Z.Xu, *Macromolecules*, **24**(1991)5893

[51] Y.Wang, F.Zeng and S.C.Zimmerman, *Tetrahedron Lett.* **38** (1997)5459

[52] R.L.Lescanec and M.Muthukumar, *Macromolecules*, **23**(1990)2280

[53] M.L.Mansfield and L.I.Klushin, *Macromolecules* **26**(1993)4262

[54] M.Murat and G.S.Grest, *Macromolecules* **29**(1996)1278

[55] D.Boris and M.Rubinstein, *Macromolecules* **29**(1996)7251

[56] D.A.Tomalia, *Polym.J.* **12**(1985)117

[57] P.J.Flory, *J.Am.Chem.Soc.* **63**(1941)3083

[58] B.Zimm and W.H.Stockmayer, *J.Chem.Phys.* **17**(1949)1301

[59] P.J.Flory and J.Rehner, *J.Chem.Phys.* **11**(1943)512

[60] W.W.Graessley, *Macromolecules* **8**(1975)865

[61] S.Erlander and D.French, *J.Polym.Sci.* **20**(1956)7

[62] W.Burchard, I.Kranztz and B.Pfannnenmuller, *Makromol.Chem.* **150**(1971)63

[63] E.Buhlein, W.Wehner and F.Vogtle, *Synthesis* (1978)155

[64] C.Worner and R.Mulhaupt, *Angew.Chem.Int.Ed.* **32**(1993)1306

[65] E.M.M. de Brabander van den Berg and E.W.Meijer, *Angew Chem.Int.Ed.* **32**(1993)1308

[66] G.R.Newkome, Z.Q.Yao, G.R.Baker and V.K.Gupta, *J.Org.Chem.* **50**(1985)2003

[67] C.J.Hawker and J.M.J.Frechet, *Chem.Commun.* (1990)1010

[68] D.A.Tomalia, V. Bery, M.Hall and D.M.Hedstrand, *Macromolecules* **20**(1987)1164

[69] H.Hall, A.Padias, R.McConnell and D.A.Tomalia, *J.Org.Chem.* **52**(1987)5305

[70] D.A.Tomalia and J.R.Dewald, US Pat.4737550 (1988)

[71] D.A.Tomalia, M.Hall and D.M.Hedstrand *J.Am.Chem.Soc.* **109**(1987)1601

[72] D.A.Tomalia, US Pat. 4857599 (1989)

[73] R.G.Denkewalter, J.Kole and W.J.Lukasavage, US Pat. 4,410,688 (1983)

[74] R.G.Denkewalter, J.Kole and W.J.Lukasavage, US Pat. 4,289,872 (1981)

[75] R.G.Denkewalter, J.Kole and W.J.Lukasavage, US Pat. 4,360,646 (1982)

[76] S.M.Aharoni, C.R.Crosby III and E.K.Walsh. *Macromolecules*, **15**(1982)1093

[77] H.R.Kricheldorf, Q.Z.Zang and G.Schwarz, *Polymer* **23**(1982)1821

[78] Y.H.Kim and O.W.Webster, *J.Am.Chem.Soc.* **112**(1990)4592

[79] R.Langer, *Chem.Eng.Sci.* **50**(1995)4109

[80] R.Duncan and J.Kopecek, *Adv.Polym.Sci.* **57**(1984)51

[81] P.A.Brady and E.G.Levy, *Chem.Ind.* (London) **18**(1995)21

[82] M.M.Schreiber, B.S.Sharha, D.Trimnell and M.D.White, 'Method of Applying Herbicides. Controlled Release Herbicides', WSS, A monograph, Weed Society of America, 1987, P.177

[83] N.Karak, *J.Polym. Mater.* **16**(1999)309

[84] A.J.Domb, A.Marlinsky, M.Maniar and I.Teomin, *J.Am.Mosqito Control.Assoc.* **11**(1995)29

[85] C.Z.Chen, N.C.BeckTan and S.L.Cooper, *Chem.Commun.* (1999)1585

[86] C.J.Hawker, E.E.Malmstrom, C.W.Frank and J.P.J.Kampf, *J.Am.Chem.Soc.* **119**(1997)9903

[87] E.J.Goethals (ed.), 'Telechic Polymers: Synthesis and Applications', CRC Press, Boca Raton, FL, USA, 1989

[88] N.M.Bikales, 'Encyclopedia of Polymer Science and Engineering', 2nd Edn. Vol.10, Wiley, New York, p.191, 1987

[89] E.S.Wilks, *J.Chem.Inf.Comput.Sci.* **37**(1997)171

[90] G.R.Newkome, G.R.Baker, J.K.Young and J.G.Traynham, *J.Polym.Sci. Part A: Polym.Chem.* **31**(1993)641

[91] G.R.Newkome and G.R.Baker, *Polym.Prepr.* **35**(1994)6

[92] G.R.Baker and J.K.Young, 'A Systematic Nomenclature for Cascade (Dendritic) Polymers', in 'Advances in Dendritic Macromolecules" edited by G.R.Newkome, Vol. 1, Chap. 5, CAI press Inc., Greenwich, CT, 1994

[93] E.W.Kwock, T.X.Neenan and T.M.Miller, *Chem.Mater.***3**(1991)775

[94] C.J.Hawker and J.M.J.Frechet, *J.Am.Chem.Soc.* **112** (1990)7638

[95] E.S.Wilks, *Polym.Prepr.* **37**(1996)6

[96] J.L.Schultz and E.S.Wilks, *J.Chem.Inf.Comput.Sci.* **38**(1998)85

[97] E.S.Wilks, *J.Chem.Inf.Comput.Sci.* **37**(1997)209

[98] M.Trollsas and J.L.Hedrick, , *J.Am.Chem.Soc.* **120**(1998)4644

[99] K.L.Wooley, C.J.Hawker and J.M.J.Frechet, *J.Am.Chem.Soc.* **113**(1991)4252

[100] K.L.Wooley, C.J.Hawker and J.M.J.Frechet, *Angew Chem.Int.Ed.* **33**(1994)82

[101] T.Kawaguchi, K.L.Walker, C.l.Wilkins and J.S.Moore, *J.Am.Chem.Soc.* **117**(1995)2159

[102] T.Nagasaki, M.Ukon, S.Arimori and S.Shinkai, *J.Chem.Soc. Chem.Commun.* (1992)698

[103] T.Nagasaki, O.Kimura, M.Ukon, S.Arimori, I.Mamachi and S.Shinkai, *J.Chem.Soc.Perkin Trans.1* (1994)75

[104] J.M.J.Frechet, I.Gitsov, R.B.Grubbs, C.J.Hawker, M.Hemni, M.Leduc, E.Sanford and K.Yui, Proceedings of the Am.Chem.Soc.Div. of Polym.Mater.Sci.Eng. 73(1995)271

[105] H.T.Cheang, C.T,Chen, T.Kondo, G.Siuzdak, K.B.Sharpless, *Angew.Chem.Int.Ed.* **35**(1996)182

[106] D.Zanini and R.Roy, *J.Org.Chem.* **61**(1996)7348

[107] M.N.Bochkarer and M.A.Kakara, *Usp. Khim.* **64**(1995)1106; CA **124**(1996)R203129

[108] E.M.M. de Brabander van den Berg, J.Brackman, M.Muremak, H.De Man, M.Hogeweg, J.Keulen, R.Scherreuberg, B.Coussens, Y.Mengerink, S.van der wal, *Macromol. Symp.* **102**(1996)9

[109] B.Lambert, E.Barro, N.Ging, S.H.Lim and J.L.J. Pflug, *Organomet.Chem.* **554**(1995)113

[110] M.Chai, Z.Pi, C.Tessier and P.L.Rinaldi, *J.Am.Chem.Soc.* **12**(1999)273

[111] T.J.Norwood, *Chem.Soc.Rev.* **23**(1994)53

[112] M.Chai, Y.Niu, W.J.Youngs and P.L.Rinaldi, *Macromolecules* **33**(2000)5395

[113] T.D.Spitzer, G.E.Martin, R.C.Cronch, J.P.Shockcor and B.T.Farmer II, *J.Magn.Reson.* **99**(1992)433

[114] C.J.Hawker R.Lee and J.M.J.Frechet, *J.Am.Chem.Soc.* **113**(1991)4583

[115] D.Holter, A.Burgath and H.Frey, *Acta Polym.* **48** (1997)30

[116] P.Kambouris and C.J.Hawker, *J.Chem.Soc.Perkin Trans 1.* (1993)2717

[117] C.J.Hawker and F.Chu, *Macromolecules* **29**(1996)4370

[118] R.D.Hester and P.H.Mitchell, *J.Polym.Sci.Chem.Ed.* **18**(1980)1727

[119] R.Scherrenberg, B.Coussens, P.van Vliet, G.Edouard, J.Brackman, E. De Brabander and K.Mortensen, *Macromolecules* **31**(1998)456

[120] P.L.Dubin, S.L.Edwards, J.I.Kaplan, M.S.Mehta, D.A.Tomalia and J.Xia, *Anal.Chem.* **64**(1992)2344

[121] J.W.Leon and J.M.J.Frechet, *Polym.Bull.* **35**(1995)449

[122] J.W.Leon and J.M.J.Frechet, *J.Am.Soc.Mass Spectrom.* **35**(1994)731

[123]P.R.Dvonic and D.A.Tomalia, *Macromol.Sym..* **98**(1995)403

[124] B.L.Schwartz, A.L.Rockwood, R.D.Smith, D.A.Tomalia and R.Spinder, *Rapid Commun.* **9**(1995)1552

[125] J.C.Hummelen, J.L.J.van Dongen and E.W.Meijer, *Chem.Eur.J.* **3**(1997)1489

[126] D.Seebach, J.M.Lappiere, K.Skobridis and G.Greiveldinger, *Angew.Chem.Int.Ed.* **33**(1994)440

[127] T.J.Prosa, B.J.Bauer, E.J.Amis, D.A.Tomalia and R.Scherrenberg, *J.Polym.Sci.Part B: Polym.Phys.* **35**(1997)2913

[128] M.F.Ottaviani, F.Montalti, M.Romanelli, N.J.Turro and D.A.Tomalia, *J.Phys.Chem.* **100**(1996)11033

[129] M.F.Ottaviani, N.J.Turro, S.Jockusch and D.A.Tomalia, *J.Phys.Chem.* **100**(1996)13675

[130] B.A.Herman, U.Hubler, P.Jess, H.P.Lang, H.J.Guntherodt, G.Greiveldinger, P.B.Rheiner, P.Murer, T.Sifferlen and D.Seebach, *Surf.Interface Anal.* **27**(1999)507

[131] L.Merz, J.Hitz, U.Hubler, P.Weyermann, F.Diederich, P.Murer, D.Seebach, I.Widmer, M.Stohr, H.J.Guntherodt and B.A.Hermann, *Single Mol.* **3**(2002)295

[132] D.A.Tomalia, H.Baker, J.Dewald, M.Hall, G.Kallos, S.Martin, J.Roeck, J.Ryder and P.Smith, *Macromolecules* **19**(1986)2466

[133] D.A.Tomalia, V. Bery, M,Hall and D.M.Hedstrand, *Macromolecules* **20**(1987)1601

[134] G.R.Newkome, Z.Yao, G.R.Baker, V.K.Gupta, P.S.Russo, M.J.Saunders, *J.Am.Chem.Soc.***108**(1986)849

[135] J.C.M.van Hest, D.A.P.Delnoye, M.W.P.L.Baars, M.H.P.van Genderen and E.W.Meijer, *Science* **268**(1995)1592

[136] G.R.Newkome, G.R.Baker, M.J.Saunders, P.S.Russo, V.K.Gupta, Z.Yao, J.E.Miller and K.Bouillion, *J.Chem.Soc.Chem.Commun* .**108**(1986)849

[137] T.M.Miller, T.X.Neenan, R.Zayas and H.E.Bair, *J.Am.Chem.Soc.* **114** (1992)1018

[138] T.M.Miller, E.W.Kwock and T.X.Neenan, *Macromolecules* **25**(1992)3143

[139] K.Yamanaka, M.Jikei and M.Kakimoto, *Macromolecules*, **25**(2000)1111

[140] J.M.J.Frechet, *Science* **263**(1994)1710

[141] K.L.Wooley, J.M.J.Frechet and C.J.Hawker, *Polymer* **35** (1994)4489

[142] T.H.Mourey, S.R.Turner, M.Rubenctein, J.M.J.Frechet, C.J.Hawker and K.L.Wooley, *Macromolecules* **25**(1992)2401

[143] C.J.Hawker, K.L.Wooley, and J.M.J.Frechet, *J.Am.Chem.Soc.* **115**(1993)4375

[144] G.Caminati, M.F.Ottaviniani, K.Gopidas, S.Jockusch, N.J.Turro, D.A.Tomalia, Proceedings of the Am.Chem.Soc.Div. of Polym.Mater.Sci.&Eng. 78(1995)80

[145] A.M.Naylor, W.A.Goddard III, G.E.Kiefer and D.A.Tomalia, *J.Am.Chem.Soc.* **111**(1989)2339

[146] A.B.Padias, H.K.Hall Jr., D.A.Tomalia and J.R.McConnel, *J.Org.Chem.***52**(1987)5305

[147] V.Percec, P.Chu, G.Ungar and J.Zhou, *J.Am.Chem.Soc.* **117**(1995)11441

[148] S.Bauer, H.Fischer and H.Ringsdorf, *Angew.Chem.Int.Ed.* **32**(1993)1589

[149] I.M.Saez, J.W.Goodby and R.M.Richardson, *Chem.Eur.J.* **7**(2001)27588

[150] I.M.Saez and J.W.Goodby, *J.Mater.Chem.* **11**(2001)2845

[151] K.Lorenz, H.Frey, B.Stuhn and R.Mulhaupt, *Macromolecules* **30**(1997)6860

[152] H.Meier, M.Lehmann and U.Kolb, *Chem.Eur.J.* **6**(2000)2462

[153] D.J.Peesak and J.S.Moore, *Angew.Chem.Int.Ed.* **36**(1997)1636

[154] V.Percec, W.D.Cho, G.Ungar and D.J.P.Yeardley, *Angew.Chem.Int.Ed.* **39**(2000)1597

[155] J.H.Careron, A.Facher, G.Lattermann and S.Diele, *Adv.Mater.* **9**(1997)398

[156] J.F.G.A.Jansen, E.M.M. de Brabander van den Berg and E.W.Meijer, *Science,* **266**(1994)1226

[157] J.F.G.A.Jansen, H.W.I.Peerlings, E.M.M. de Brabander van den Berg and E.W.Meijer, *Angew Chem.Int.Ed.* **34**(1995)1206

[158] H.T.Chang, C.T.Chen, T.Kondo, G.Siuzdak and K.B.Sharpless, *Angew.Chem.Int.Ed.* **35**(1996)182

[159] H.F.Chow and C.C.Mak, *Pure Appl.Chem.***69**(1997)483

[160] P.K.Murer and D.Seebach, *Angew.Chem.Int.Ed.* **34**(1994)2116

[161] C.W.Thomas and Y.Tor, *Chirality* **10**(1998)53

[162] H.W.I.Peerlings and E.W.Meijer, *Chem.Eur.J.* **3**(1997)1563

[163] D.Seebach, P.B.Rheiner, G.Greiveldinger, T.Butz and H.Sellner, *Top.Curr.Chem.* **197**(1998)125

[164] A.Hult, E.Malmstrom and M.Johansson, 'HB Aliphatic Polyesters', in The Polymeric Materials Encyclopedia: Synthesis, Properties and Application, edited by J.C.Salamone, CRC Press, Florida, 1996

[165] H.Ihre, M.Johanson, E.Malmstrom and A.Hult, "Dendrimers and HB aliphatic Polyester Based on 2,2'-bis(hydroxy methyl) propionic acid (Bis-MPA), in Advances in Dendritic Macromolecules, edited by G.R.Newkome, vol.3, JAI Press, London, p.1, 1995

[166] T.M.Miller, T.X.Neenan, E.W.Kwock and S.M.Stein, *J.Am.Chem.Soc.* **115**(1993)356

[167] Y.H.Kim and O.W.Webster, *Macromolecules* **25**(1992)556

[168] Y.H.Kim and R.Beckerbauer, *Macromolecules* **27**(1994)1968

[169] K.L.Wooley, C.J.Hawker, J.M.Pochan and J.M.J.Frechet, *Macromolecules* **26**(1993)1514

[170] H.Stutz, *J.Polym.Sci. Part B: Polym.Phys.* **33**(1995)333

[171] R.G.Duan, L.L.Miller and D.A.Tomalia, *J.Am.Chem.Sci.* **117**(1995)10783

[172] C.Tanford, 'The Hydrophobic Effect: Formulation of Micells and Biological Membrances', 2nd Edn., Wiley-interscienc, New York, 1980

[173] D.J.Mitchell and B.W.Ninham, *J.Chem.Soc. Faraday Trans. 2* (1981)601

[174] G.R.Newkome, C.N.Moorefield, G.R.Baker, M.J.Saunders and S.H.Grossman, *Angew.Chem.Int.Ed.* **30**(1991)1178

[175] M.Maciejewski, *Macromol. Sci.Chem.* **17A**(1982)689

[176] S.P.Newton, J.F.Stoddart and W.Hayes, *Supramol.Sci.* **3**(1996)221

[177] F.Vogtle, 'Supramolecular Chemistry,an introduction' Wiley, Chichester, 1991

[178] D.N.Reinhoudt, 'Supramolecular Materials and Technologies, Prospective Supramolecular Chemistry, vol.4, Wiley, Chichester, 1999

[179] D.A.Tomalia, *Sci.Am.* **272**(1995)42

[180] H.J.Scneider, *Angew. Chem.Int.Ed.* **30**(1991)1417

[181] A.W.Kleij, R.van de Coevering, R.J.M.K. Gebbink, A.M.Noordman, A.L.Spek and G.van Koten, *Chem.Eur.J.* **7**(2001)181

[182] S.M.Cohen, S.Petound and N.Raymond, *Chem.Eur.J.* **7**(2001)272

[183] S.Mattei, P.Wallimann, B.Kenda, W.Amrein and F.Diederich, *Helv.Chim.Acta* **80**(1997)2391

[184] R.F.Service, *Science* **267**(1995)458

[185] L.J.Twyman, A.E.Beezer, R.Esfand, M.J.Hardy and J.C.Mitchell, *Tetrahedron Lett.* **40**(1999)1743

[186] D.C.Tully and J.M.J.Frechet, *Chem. Commun.* (2000)1229

[187] M.Wells and R.M.Crooks, *J.Am.Chem.Soc.* **16**(1996)3988

[188] L.L.Miller, Y.Kunugi, A.Canavesi, S.Rigant, C.N.Moorefield and G.R.Newkome, *Chem.Mater.* **10**(1998)1751

[189] A.Adronov and J.M.J.Frechet, *Chem.Commun.* (2000)1701

[190] C.Bolm, N.Derrien and A.Seger, *Syn.Lett.* (1996)387

[191] P.Bhyrappa, J.K.Young, J.S.Moore and K.S.Suslick, *J.Am.Chem.Soc.* **118**(1996)5

[192] S.Hecht and J.M.J.Frechet, *Angew.Chem.Int.Ed.* **40**(2001)74

[193] Y.H.Kim and O.W.Webster, *Macromolecules* **25**(1992)5561

[194] Y.H.Kim and O.W.Webster, *Macromolecules* **27**(1994)1968

[195] A.Sunder, K.Kramer, R.Hanselmann, R.Mulhaupt and H.Frey, *Angew Chem.Int.Ed.* **38**(1999)3552

[196] R.Roy, D.Zanini, S.J.Meuier and A.Romanowska, *J.Chem.Soc.Chem.Commun.* (1993)1869

[197] E.C.Wiener, M.W.Brechbiel, H.Brothers, R.L.Magin, O.A.Gansow, D.A.Tomalia and P.C.Lauterbur, *Magn.Reson.Med.* **31**(1994)1

[198] E.C.Wiener, F.P.Auteri, J.W.Chen, M.W.Brechbiel, O.A.Gansow, D.S.Schneider, R.L.Belford, P.B.Clarkson, and P.C.Lauterbur, *J.Am.Chem.Soc.* **118**(1996)7774

[199] M.W.Bourne, L.Margerun, N.Hylton, B.Canpion, J.J.Lai, Derugin and C.B.Higgins, *J.Magn.Reson.Imag.* **6**(1996)305

[200] G.Adam, J.Neurerburg, E.Spuntrup, A.Muhler, K.Scherer and R.W.Gunther, *J.Magn.Reson.Imag.* **4**(1994)462

[201] M.F.Hawthorne, *Angew.Chem.Int.Ed.* **32**(1993)950

[202] A.H.Soloway, W.Tjarks, B.A.Barnum, F.G.Rong, R.F.Barth, I.M.Codogni and J.G.Wilson, *Chem.Rev.* **98**(1998)1515

[203] B.Qualmann, M.M.Kessels, F.Klobasa, P.W.Jungblut and W.D.Sierralta, *J.Microscopy* **183**(1996)69

[204] B.Qualmann, M.M.Kessels, H.J.Musid, W.D.Sierralta, P.W.Jungblut and L.Moroder, *Angew. Chem.Int.Ed.* **35**(1996)909

[205] B.R.Mattews and G.Holan, WO Pat. 95/34595 (1995)

[206] J.Roberts, M.Bhalgat and R.Zera, *J.Biomed.Mater.Res.* **30**(1996)52

[207] J.W.J.Knapen, A.W.van der Made, J.C.de Wilde, P.W.N.M. van Leeuwen, P.Wijkens, D.M.Grove and G.van Koten, *Nature* **372**(1994)659

[208] H.Brunner, *J.Organomet. Chem.* **500**(1995)39

[209] D.Seebach, R.E.Marti and T.Hintermann, *Helv.Chim.Acta* **79**(1996)1710

[210] K.Matyjaszewski, T.Shigemoto, J.M.J.Frechet and M.Leduc, *Macromolecules* **29**(1996)4167

[211] I.Gitsov, P.T.Ivanova and J.M.J.Frechet, *Macromol.Chem.Rapid Commun.* **15**(1994)387

[212] C.Devadoss, P.Bharati and J.S.Moore, *J.Am.Chem.Soc.* **118**(1996)9635

[213] S.Campagna, G.Denti, S.Serroni, M.Ciano, A.Juris and V.Balzani, *Inorg.Chem.* **31**(1992)2982

[214] G.Denti, S.Campagna, S.Serroni, M.Ciano, and V.Balzani, *J.Am.Chem.Soc.* **114**(1992)2944

[215] R.Sadamoto, N.Tomika and T.Aida, *J.Am.Chem.Soc.* **118**(1996)3978

[216] G.M.Stewart and M.A.Fox, *J.Am.Chem.Soc.* **118**(1996)4354

[217] D.L.Jiang and T.Aida, *Nature* **388**(1997)454

[218] D.M.Junge and D.V.McGrath, *Chem.Commun.* (1997)857

[219] C.B.Gorman, J.C.Smith, M.W.Hager, B.L.Parkhurst, H.Sierzputowska-Gracz and C.A.Haney, *J.Am.Chem.Soc.* **121**(1999)9958

[220] C.B.Gorman, *Adv.Mater.* **9**(1997)1117

[221] K.Takada, D.J.Diaz, H.D.Abruna, I.Cuadrado, B.Casado, M.Moran and J.Losada, *J.Am.Chem.Soc.* **119**(1997)10763

[222] C.M.Cardona and A.E.Kaifer, *J.Am.Chem.Soc.* **120**(1998)4023

[223] J.Jansen, E.W.Meijer and E.M.M. de Brabander van den Berg, *J.Am.Chem.Soc.* **117**(1995)4417

[224] J.Jansen, R.A. J.Janssen, E.M.M. de Brabander van den Berg and E.W.Meijer, *Adv.Mater.* **7**(1995)561

[225] R.A.Gossage, J.T.B.H.Jastrzebski, J.van Ameijde, S.J.E.Mulders, A.J.Brouwer, R.M.J.Liskamp and G.van Koten, *Tetrahedron Lett.* **40**(1999)1413

[226] A.W.Kleij, R.A.Gossage, J.T.B.H.Jastrzebski, J.Boersma and G.van Koten, *Angew.Chem.Int.Ed.* **39**(2000)176

[227] G.Pistolis, A.Malliaris, D.Tsiourvas and C.M.Paleos, *Chem.Eur.J.* **5**(1999)1440

[228] M.Jliu, K.Kono and J.M.J.Frechet, *J.Polym.Sci. Part A Polym.Chem.* **37**(1999)3492

[229] J.J.Lee and W.T.Ford, *Macromolecules* **27**(1994)4632

[230] S.A.Kuzdzal, C.A.Monnig, G.R.Newkome and C.N.Moorefield, *J.Chem.Soc.Chem.Commun.* (1994)2139

[231] Y.Sayed-Sweet, D.M.Hedstrand, R.Spindler and D.A.Tomalia *J.Mater.Chem.* **7**(1997)1199

[232] M.W.P.L.Barrs, M.C.W.van Boxtel, C.W.M.Bastiaansen, D.J.Broer, S.H.M.Sontjens and E.W.Meijer, *Adv. Mater.* **12**(2000)715

[233] B.Peterson and K.Sorensen, Proceedings of the 21st Water Borne, Higher Solids and Powder Coating Symposium, New Orleans, Louisiana, p. 753, 1994

[234] W.F.Shi and B.Ranby, *J.Appl.Polym.Sci.* **59**(1996)1937

[235] W.F.Shi and B.Ranby, *J.Appl.Polym.Sci.* **59**(1996)1951

CHAPTER 2

DENDRITIC POLYHYDROCARBONS

2.1 INTRODUCTION

The linear hydrocarbon polymers are one of the most important and a useful class of polymers, whose consumption in the modern society is the highest. So, it is expected that the hydrocarbon dendritic polymers also offer an important class of dendritic polymers for different applications. Indeed, the dendritic hydrocarbons show interesting photochemical and photophysical properties, which make them potential materials for various applications in materials science. However, the hydrocarbon dendritic polymers are synthesized mainly aromatic in nature and hence they are rigid and causing solubility problem, even though they have good stability. So different types of modifications are being reported by which their solubility characteristic has been enhanced. A large number of hydrocarbon dendritic polymers are reported, among these polyphenylenes [1-35], stilbenoids [36-47], poly(phenyl acetylene)s [47-60], etc. are very important and useful as light harvesting materials, catalysts, etc.

2.2 SYNTHESIS, CHARACTERIZATION AND PROPERTIES OF DIFFERENT TYPES OF HYDROCARBON DENDRIMERS

In this section synthesis, characterization and properties of different types of hydrocarbon dendrimers such as polyphenylenes, stilbeniods, phenylacetylenes, phenylcarbazoles, etc. are discussed.

2.2.1 Polyphenylenes

The first synthesis and characterization of polyhydrocarbon monodisperse dendrimers based on 1,3,5-trisubstituted benzenes by the convergent approach upto 4th generation was reported in 1990 [1-2]. They also synthesized another series of polyhydrocarbons of similar type upto 2nd generation in which the outer phenyl rings were fluorinated. The dendrimers were synthesized by palladium catalyzed coupling of arylboronic acids to 3,5-dibromo-1-(trimethylsilyl)benzene and subsequently converted to arylboronic acid by the reaction with borontribromide followed by hydrolysis. The repetition of the above steps resulted higher generation dendrons. Finally, this aryl boronic acid was coupled with 1,3,5-tribromobenzene or 1,3,5-*tris* (3,5-dibromophenyl) benzene to get the desired polyhydrocarbon dendrimers. The whole sequence of reaction is shown in scheme 2.1. For synthesizing polyhydrocarbon dendrimers with perfluorinated benzene end groups upto 3rd generation, the following synthetic strategy is used [1]. First, pentafluoro bromobenzene coupled to 3,5-dibromo-1-(trimethylsilyl)benzene by Cu catalyzed reaction. The resulted trimethylsilyl compound was transformed to aryl-boronic acid by reaction with borontribromide at 180°C in CCl_4 followed by hydrolysis. The repetition of the above steps resulted higher generation dendron, but because of unstability of the higher boronic acid derivative, the product was not possible to isolate. Finally, the coupling of aryl boronic acid with 1,3,5-tribromobenzene in the presence of $Pd(PPh_3)_4$ catalyst, offered the desired dendrimers. The whole sequence of reactions is shown in scheme 2.2. The yield of each step was very high (80–90)%, except the final coupling step in each case, where the yield was relatively low (42–68)%. All dendrimers except lower generation perfluoinated dendrimer are moderate to highly soluble in organic solvents like THF, toluene, CH_2Cl_2, $CHCl_3$, etc. The dendrimers were characterized by ^{1}HNMR, ^{13}CNMR spectroscopy, SEC, mass spectrometry and elemental analysis. The thermal characteristics of the dendrimers were also studied by TGA-MS and DSC analysis. The dendrimers are stable upto 500°C without any decomposition, because of their thermostable aromatic structures. The DSC study revealed that the melting temperature and glass transition temperature of the dendrimers increased with the increase of generation in case of purely hydrocarbon dendrimers, while for the perfluorinated benzene surface groups containing dendrimers, the reverse order was found for melting temperature. The reason was not clearly understood. Also the fluorinated dendrimers melt from 15 to 140°C higher and the molar heat of fusion are 1.4 to 3 times larger than those of their hydrocarbon analogs. As these dendrimers are pure and monodisperse, so they may be used as size standards in GPC analysis. Mullen and his coworkers [3-12] also synthesized, characterized and studied the properties of polyphenylene based dendrimers. The synthesis of polyphenylene dendrimers starts from core unit of different structures (Fig. 2.1) and proceeds by a cascade of Diels-Alder type cycloaddition leading to structurally quite different 3D dendrimeric architectures. The (2+4) cycloaddition of the novel A_4B building block 2,3,4,5-tetrakis(4-triisopropyl silylethynyl phenyl)cyclopenta-2,4-dienone with 3,3',5,5'-tetraethynylbiphenyl allows the rapid synthesis of a spherical polypropylene nanoparticle with 102 benzene rings [11]. This is the first example of an A_4B building block used in dendrimer synthesis. The dendritic polyphenylenes were synthesized by the representative Diels-Alder reaction deprotection sequence of reactions, starting from 3,3,'5,5,'-tetraethynylbiphenyl core (Fig.2.1) and ethynyl and propynyl substituted tetraphenyl cyclopentadienone building blocks [8].

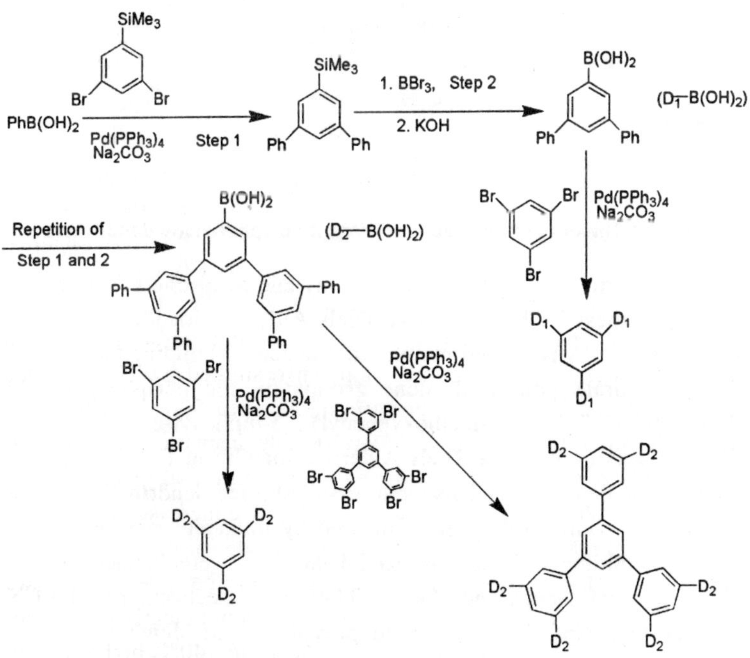

SCHEME 2.1 Synthesis of polyphenylene dendrimers

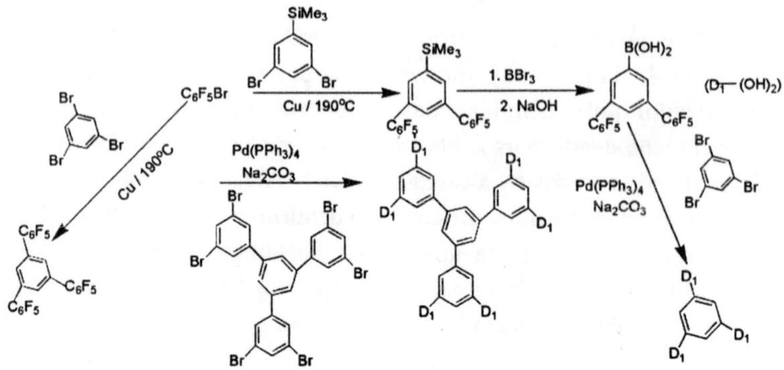

SCHEME 2.2 Synthesis of polyphenylene dendrimers with perfluorinated benzene surface groups

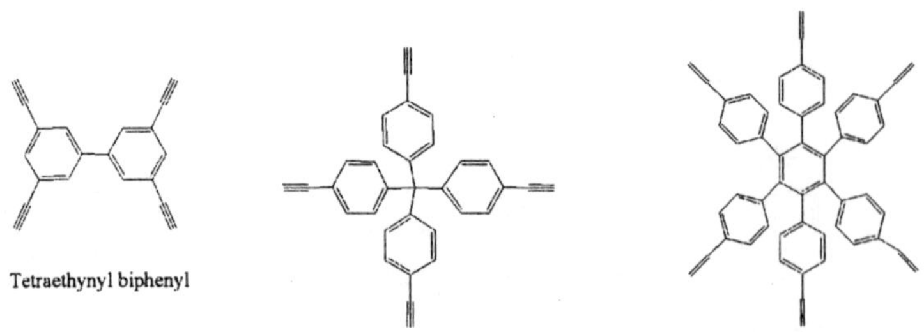

FIG. 2.1 The core structures for different polyphenylene dendrimers

The dithiolane substituted polyphenylene dendrimers were divergently synthesized by reiterative Diels-Alder cycloaddition starting from tetra(4-ethynylphenyl)methane as the central building block [3-4]. To incorporate pentyldithiolene groups on the periphery of the dendrimers 3,4-*bis*(4-methoxyphenyl)-2,5-diphenyl-cyclopenta-2,4-dienone was used in the Diels-Alder cycloaddition forming the last layer of the dendrimers. This methoxy-substituted dendrimer was then treated with boron tribromide followed by thioacetic acid to get the desired dendrimers. The synthesized dendrimers were characterized by ^{1}HNMR spectroscopy and MALDI-TOF mass spectrometry. The CN, COOH and N=CPh_2 substituted polyphenylene dendrimers were synthesized by using correspondingly functionalized cyclopentadienone derivatives and the yield obtained by this approach was excellent [12].

The polyphenylene dendrimers were synthesized [10] by the convergent approach (Scheme 2.3), similar to the divergent (4+2) cycloaddition route [4]. At first, tetra substituted cyclopentadienone and dialkynyl monomer were treated under Diels-Alder reaction condition to yield the corresponding dione. Then Knoevenagel condition of this dione with 1,3-diphenylacetone resulted substituted cyclopentadienone, which offered 2nd generation dendritic wedges by repetition of Diels-Alder reaction. The whole sequence of reaction is shown in scheme 2.3.

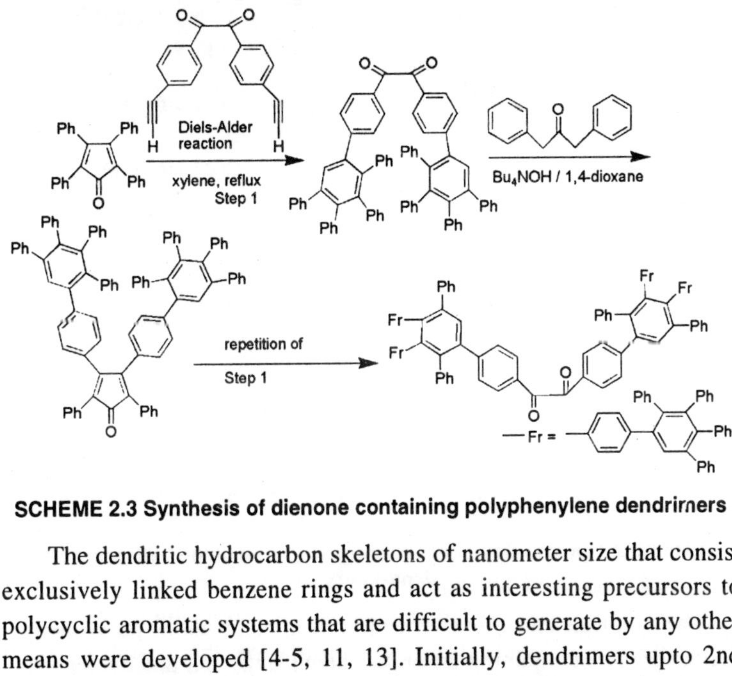

SCHEME 2.3 Synthesis of dienone containing polyphenylene dendrimers

The dendritic hydrocarbon skeletons of nanometer size that consist exclusively linked benzene rings and act as interesting precursors to polycyclic aromatic systems that are difficult to generate by any other means were developed [4-5, 11, 13]. Initially, dendrimers upto 2nd generation and then upto 3rd generation with 142 benzene rings were obtained [5]. The aromatic hydrocarbon dendrimers were also synthesized by using a cyclopentadienone derivative as a tetravalent branch juncture (Scheme 2.4). Here, even the 2nd generation dendrimer containing 102 benzene rings and possesses rigidly spherical shape. Its thermal stability is remarkably high, even in atmospheric oxygen, where it decomposes at temperature above 580°C. The polyphenylene dendrimers with pentaphenylbenzene repeating units were also synthesized [14] by both divergent and convergent approaches utilizing the Diels-Alder cycloaddition reaction between tetraphenyl cyclopentadienones and aryl acetylenes.

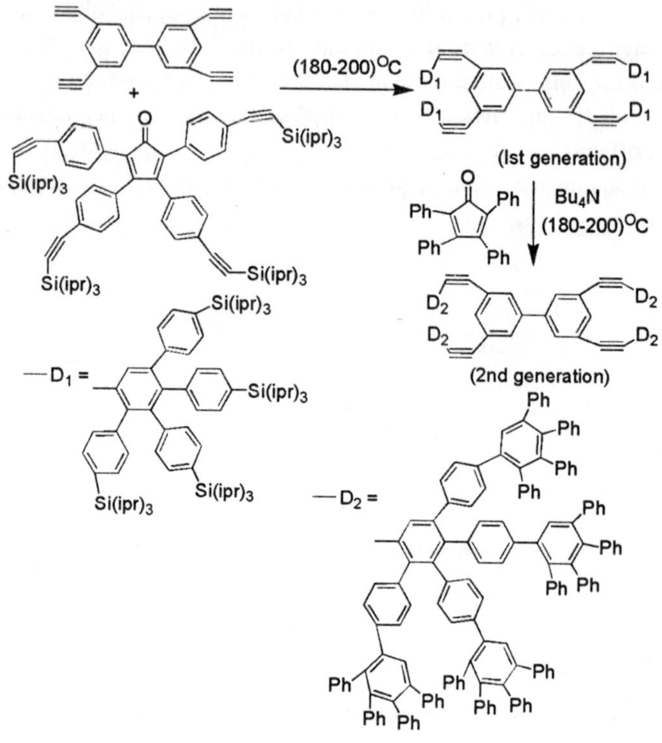

SCHEME 2.4 Synthesis of only benzene ring containing hydrocarbon dendrimers

The synthesis and characterization of organometallic poly-phenylene dendrimers containing 24 or 44 phenyl rings and one cyclopentadienyl cobalt unit were reported [15]. The dendrimers were synthesized by the convergent $CpCo(CO)_2$ mediated dimerization of di- or tetra-ethynyltolanes followed by a divergent core extension utilizing tetraphenyl cyclopentadienone. The synthesized dendrimers were characterized by GPC, scanning pulse voltammetry, etc. techniques. The scanning pulse voltammetry in solution of the dendrimers shows that the oxidation potential ranges from 0.8 V to 0.83 V. Higher the generation, the higher is the oxidation potential i.e. with the increase of steric hindrance the oxidation became more difficult. The effect of terminal groups on the electronic structures of 2nd generation dendritic polyacetylenic oligomers bearing different terminal groups was theoretically investigated [16]. The theoretical quantum mechanical calculation showed that the steric and electronic factors affect the electronic properties of the dendrimers. While the steric hindrance is independent on the nature of the terminal group and is tend to increase band gap and ionization potential but decreased the electron affinity. The interaction of lone pair electrons and electrons of the terminal groups with the rest of the molecules produce specific changes in the electronic structure of each particular group.

The novel light harvesting polyphenylene dendrimers with two as well as three different chromophores were synthesized [17]. The synthesis of these novel multichromophoric dendrimers was based on both convergent and divergent approaches. A tetraethynylterrylene tetracarboxydiimide chromophore (TDI) was used as the core unit. The Diels-Alder reaction of this TDI core with a cyclopentadienone derivative containing perylene dicarboxymonoimide (PMI) chromophore resulted the first generation dendrimer with very high yield. The second generation dendrimer was obtained by the same type of reaction of the core with 10 equivalents of PMI substituted dendron. The synthesis of the dendrimer with three different chromophores was carried out by a complex multi-step route using cyclopentadienone containing a PMI chromophore as well as a protected ethynyl group. The building unit was synthesized by means of a six-step reaction sequence [18]. The Diels-Alder reaction of the TDI core unit with 5 equivalents of above cyclopentadienone was followed by deprotection of the resulted product by tetrabutyl ammonium chloride (TBAC), which was then treated with naphthalene dicarboxmoimide chromophore containing dendron to get the desired dendrimers. The whole sequence of reaction is shown in scheme 2.5.

The hydrocarbon dendrimers of polyphenyl core with phenyleneimide chromophores at the rim were characterized by single-molecule spectroscopy [19-20] and time resolved spectroscopy [21-23]. All these dendrimers have a biphenyl core that leads to a conical disc-shaped structure to get better control over the distribution of chromophores and their interaction in the excited state. A series of first

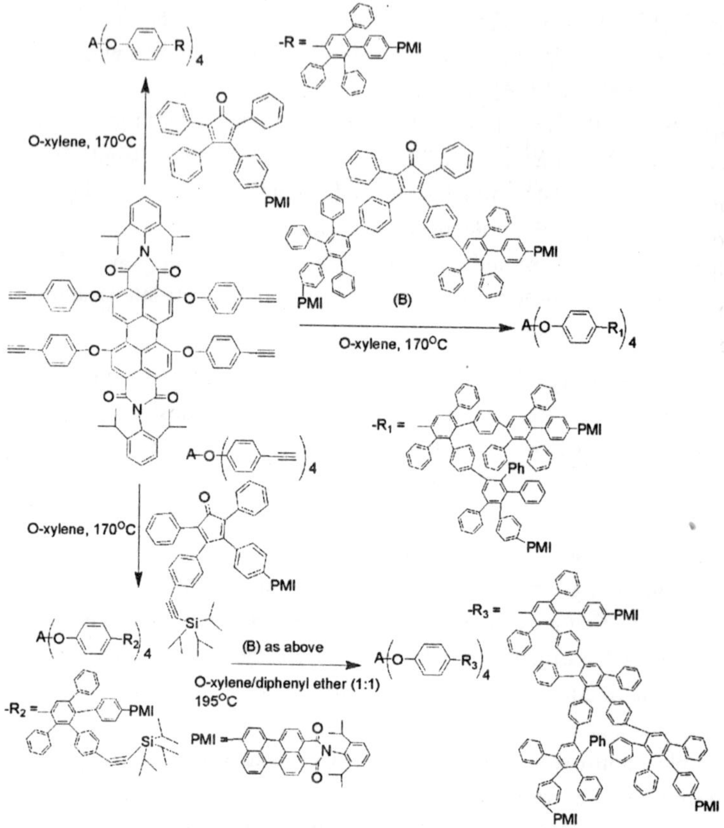

SCHEME 2.5 Synthesis of multichromophore contain polyphenylene dendrimers

generation phenyleneimide dendrimers with a rigid tetrahedral core (Fig.2.2) was synthesized [24]. These dendrimers possess pentaphenyl benzene unit and as a result of the 'inter locking' of twisted phenyl rings, these are persistent of shapes and more closely packed. Also this synthetic process allows one to control the number of PMI chromophores attached to the surface of the nanoparticles from one to four [8, 25]. The interaction of the chromophores with varying number was studied by time resolved fluorescence up conversion with 250 fs resolution for the above dendrimers [26]. It has been observed that a singlet-singlet annihilation occurs in the multi chromophoric dendrimers. The extremely large polycyclic aromatic hydrocarbon based on a cyclohydrogenation method for polyphenylene dendrimer was reported [25]. These polyphenylene dendrimers were synthesized by repeated Diels-Alder reactions of ethynyl substituted tetraphenylcyclopentadienones to the tetraethyhyl biphenyl core. The chemical and physical properties of the dendrimers were controlled by varying cyclopentadienone used in the last growth step [25].

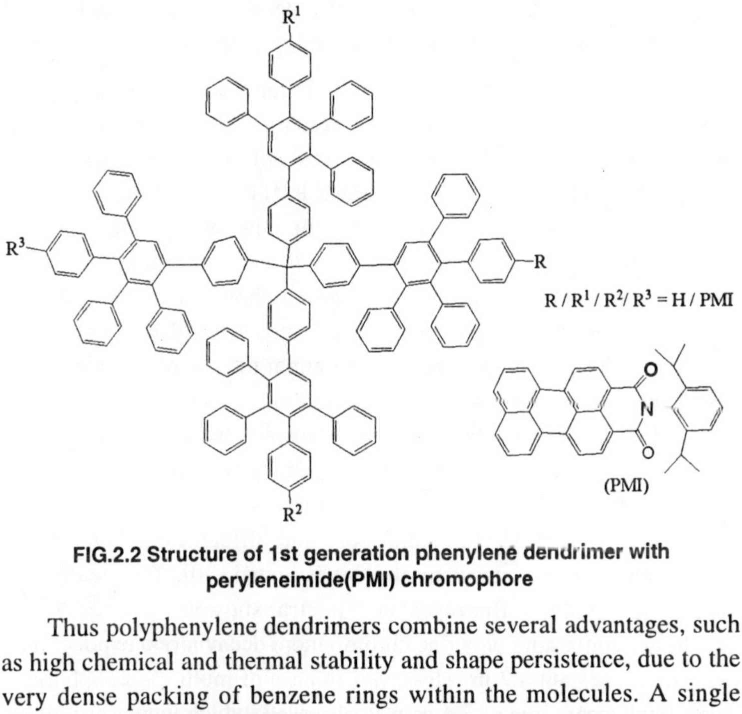

FIG.2.2 Structure of 1st generation phenylene dendrimer with peryleneimide(PMI) chromophore

Thus polyphenylene dendrimers combine several advantages, such as high chemical and thermal stability and shape persistence, due to the very dense packing of benzene rings within the molecules. A single molecule spectroscopic study of a dendrimer based on host-guest system consisting of a 2nd generation polyphenylene dendrimer and the cyamine dye, pinacyanol was reported [27]. The study revealed that at low dye to dendrimer ratios the system was present as a one-to-one complex, while at higher ratios an ion-pair system was formed. Due to the difference in local polarizability, the changes in the spectral properties were observed. The difference of the triplet lifetimes of single free dye molecules and of associated ones was interpreted as deriving from a larger free volume for the dye molecules in the dendritic host relative to the rigid polymer matrix.

A new synthetic approach leading to asymmetrically substituted polyphenylene dendrimers upto 2nd generation with an increasing number of chromophores at the periphery was reported [28]. The synthesis of polyphenylene dendrimers bearing three donor chromophores and one acceptor chromophore, especially is very important as it can act as a light harvesting material through the antennae effect. The intermolecular energy transfer within this dendrimer molecule was also demonstrated by applying absorption and fluorescence measurements.

Maus *et al.* [29] further studied the intramolecular energy hopping and energy trapping by Forester type excitation energy transfer (FTEE) process on a series of first generation polyphenylene dendrimers substituted with spatially well-separated PMI chromophores and a TDI energy trapping chromophore at the rim. The steady state and time resolved fluorescence spectroscopic studies revealed that the energy hopping processes among the PMI chromophores are consistent with FTEE rate constant (K_{hop} = 4.6 ns^{-1}) with anisotropy decay times 50-80 ps. When a TDI chromophore was present at the rim of the dendrimers along with three peryleneimide chromophores, more than 95% of the energy harvested by the PMI chromophores was transferred and trapped in the TDI chromophore. The photophysical properties and the excitation energy transfer (EET) process for a series of 1st generation dendrimers with rigid tetrahedral core and different numbers of PMI chromophore at the rim were also investigated [30]. The steady state and time resolved fluorescence spectra showed that the model dendrimer containing only one chromophore decay mono exponentially with time constant 4.2 ns. However, in case of multi chromophore, an additional decay time of 7.4 ns was observed, which increases with the increase of number of chromophores. The time resolved polarization measurements resulted in a 'ns' relaxation component due to the rotation of the entire dendritic structures. However, for the mutlichromophore dendrimers, an additional fast anisotropy decay time of 100-200 ps was observed, which corresponds to Forester type EET among the chromophores.

A time-reserved fluorescence up conversion on a polyphenylene dendrimer, with eight PMI chromophores on the surface was also investigated [31]. The time dependent fluorescence spectra of the dendrimer showed that the initial excitation results in a locally excited chromophore. The study also indicates the existence of a decay channel that leads to the excited state interaction between chromophores in one dendrimer, which takes place on a 5 ps time scale.

The synthesis and photochemical properties of polyphenylene dendrimers with photoactive stilbene core were reported [32-34]. The dendrimers exhibited fluorescence emission with high quantum efficiency and also underwent *cis-trans* photochemical isomerization. The dendrimers upto 3rd generation based on 4,4'-*bis* (diphenylamino)stilbene repeating units showed ps fluorescence decay [33]. Also the rate of decay showed to be depended on detection wave length and the generation number of the dendrimers. As the generation number increases the energy of accepting states gradually decreases. The synthesis of the dendrimers was carried out by Diels–Alder reaction, as shown in scheme 2.6. The structures of the dendrimers were confirmed by ^{1}HNMR spectroscopy and MALDI-TOF mass spectrometry. The molecular weight of *trans* 2 was 4700 Da. The absorption edges of both the dendrimers are almost same (360 nm) so also the quantum yield of fluorescence emission (0.69 and 0.64 for *trans* 1 and 2, respectively). The dendrimers are also capable to undergo *trans-cis* isomerization.

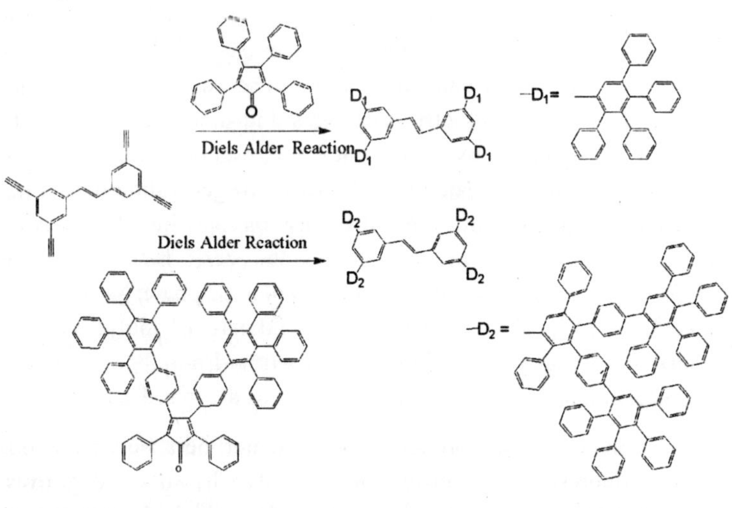

SCHEME 2.6 Synthesis of polyphenylene dendrimers with stilbene core unit

A series of first generation polyphenylene dendrimers based on three different cores (biphenyl, tetraphenylene and hexaphenyl benzene) was synthesized by Diels-Alder cycloaddition reaction and their single crystal structures were determined [35] by X-ray crystallographic analysis. The study showed that in the three crystal structures the solvent molecules are included without specific interaction within the framework of the dendritic molecules and fill the voids. The dendrimers consisting exclusively of interlocked, twisted phenyl rings. These polyphenylene nano-structures have exciting structural and dynamic properties.

2.2.2 Stilbeniods

The stilbenoid dendrimers upto 3rd generation were synthesized [36] by a convergent approach. The benzyl trisphosphonate as the basic core unit and 3,4,5-*tris*(dodecyloxy)benzaldehyde and aldehyde protected 3,5-*bis*(phosphonate) benzene aldehyde were used as the building blocks for the synthesis of the stilbenoid dendrimers. The whole sequence of reaction for this synthetic approach is shown in scheme 2.7. The synthesized dendrimers were characterized by ^{1}HNMR, ^{13}CNMR spectroscopy and MALDI-TOF mass spectrometry. NMR spectra indicate the presence of three-fold symmetry in the structures of dendrimers with all *trans* configuration of the double bonds. Mass spectra for the highest generation dendrimer show molecular ion peak at 8858.6, same as of calculated value. The stilbenoid dendrimers showed three fold symmetric structures and their liquid crystalline behavior was studied by polarization microscope, DSC and X-ray analysis. The study revealed that the first two generation dendrimers show two LC phases. But the 3rd generation does not show any LC behavior, may be due to the fact that the steric hindrance in the periphery probably too great for formation of a LC phase. Meier and his coworkers [37] also reported the synthesis of polyarylalkene dendrimers by using the Horner Wadsworth–Emmons coupling of aldehyde with the *bis*(phosphate) monomer, as described earlier [36].

The photochemical properties of the dendrimers were studied by UV and fluorescence spectroscopies. In CH_2Cl_2, all the dendrimers show an UV absorption at λ_{max} = 329 nm, which decreases upon

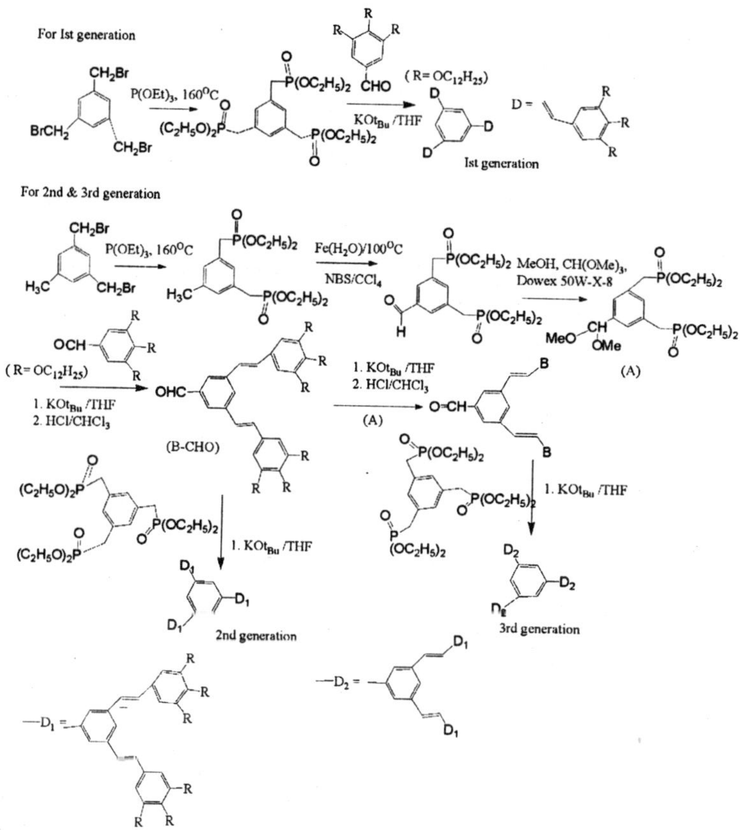

SCHEME 2.7 Synthesis of stilbenoid dendrimers with all trans configuration of the double bonds

irradiation. The fluorescence maximum of the dendrimers also depends on the temperature and concentration of the solution. The synthesis, characterization and properties of the same type of stilbenoid monodisperse dendrimers upto 5th generation were also reported [38] by a combined coupled and a convergent approach. The core unit, building blocks and the synthetic strategy were the same as described earlier [36]. Here, three peripheral alkoxy groups with different chain length viz. 3,4,5-*tris*(dodecyloxy)benzene, 3,4,5-*tris*(propyloxy)benzene and 3,4,5-*tris*(hexyloxy) benzene were used. The last two steps were further repeated to get the highest generation (4th generation) dendron. The hexyloxy containing dendron further treated with the same reagents to get the 5th generation dendron. The final coupling reaction and characterization were same as described earlier. The solubility of these stiff dendrimers, which have primarily a disk-like shape and form columnar LC phases by the higher number alkoxy chain but the shorter (propyloxy) chains are too short to permit such an intermolecular stacking for the first and second generations and a cylindrical shape was observed for the 3rd, 4th, and 5th generations, is governed by the alkoxy chains that are attached to the peripheral benzene rings. The number of stilbene units increases with the generations by the formula $3(2^n-1)$, where n is the generation number. The liquid crystalline behavior is same as described earlier [36].

A novel family of nitrogen core distyrylbenzene dendrimers upto 3rd generation (Fig.2.3) were used as model compounds to investigate the effect of intermolecular interactions on the photophysical and transport properties [39]. The dendrimers containing three distyrylbenzene chromophores around the nitrogen core form a non-planar arrangement. In the UV-absorption spectra two peaks were observed at 410 nm and 320 nm corresponding to the absorption of core chromophore. It was a red shift from the absorption of distyrylbenzene (360 nm), which is due to the lone pair electrons of the nitrogen atom. The absorption moves to the red shift by 10 nm as the generation increased from zero to three. The electroluminescence (EL) and photoluminescence (PL) spectra of the films of dendrimers were comparable and showed a similar trend with generations indicating that the same emissive species were present in both the cases. The emission spectra of the film and solution photoluminescence were independent of generation. Although the solution PL and film EL were almost identical for 2nd generation but for lower generation, the film EL was significantly broader than the solution PL and showed a strong red tail emission, may be due to better interaction between the cores of the dendrimers. The third generation shows a small blue shift in EL spectrum with respect to solution may be due to steric effects in film formation. The current voltage characteristics of the dendrimers indicated that the films become more insulating as the turn-on field for current flow more than doubles from zero to third generation. Thus the dendrimer generation controls the degree of chromophores interaction, which is manifested in the red excimer component in EL, charge carrier mobility and rise time of EL [39]. Lupton *et al.* [40] also studied the effect of dendron branching on light emission and charge transport for the propeller like conjugated dendrimers based on stilbene dendrimer with conjugated core, where the electron delocalization was broken between the core and the dendron by the metal linkages (Fig.2.3) [39]. In these materials, light emission occurs from the core and is independent of generation. The study showed that the mobility was reduced by two orders of magnitude as the size of the dendrimer became double by the increase of generation or by branching. Thus the degree of branching allows a unique control of mobility of the dendrimers. Also an increase in chromophore separation results in a reduction of intermolecular interactions, which reduces the red emission tail in film photo luminescence.

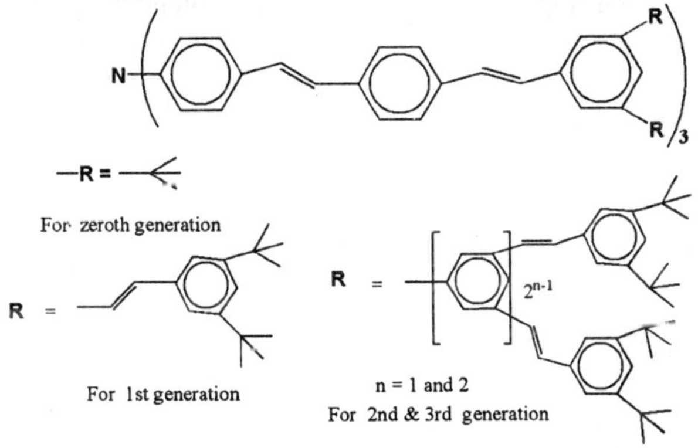

FIG.2.3 Structure of the nitrogen core distyryl benzene dendrimers

The fullerene (C_{60}) functionalized oligophenylene-vinylene (OPV) terminated hydrocarbon dendrons were synthesized [41] as light harvesting materials. OPV units were synthesized first from *p*-ketal benzaldehyde by the following sequences of reactions, as shown in scheme 2.8, 1st part. The double and quartet OPV dendrons were synthesized from ketal protected 3,5-dibenzyl alcohol by the sequence of reactions as shown in scheme 2.8, 2nd part. The synthesized dendrons were characterized by ^{1}HNMR, ^{13}CNMR spectroscopies, mass spectrometry and elemental analysis. The electronic absorption and fluorescence studies of the fluoro dendrimers indicate the presence of both OPV moiety and fullerene unit in the spectra. Again upon excitation at the OPV band maximum (λ_{exe}= 394 nm), dramatic quenching of OPV fluorescence was observed for all the fullerodendrimers. This is attributable to an OPV-C_{60} singlet-singlet energy transfer process as infered by excitation spectra recorded at λem= 730 nm. The quantum yield for the antennae effect increases with the increase of the generation number. It was also observed that OPV-terminated dendrons are capable for the harvesting of increase in amount of light.

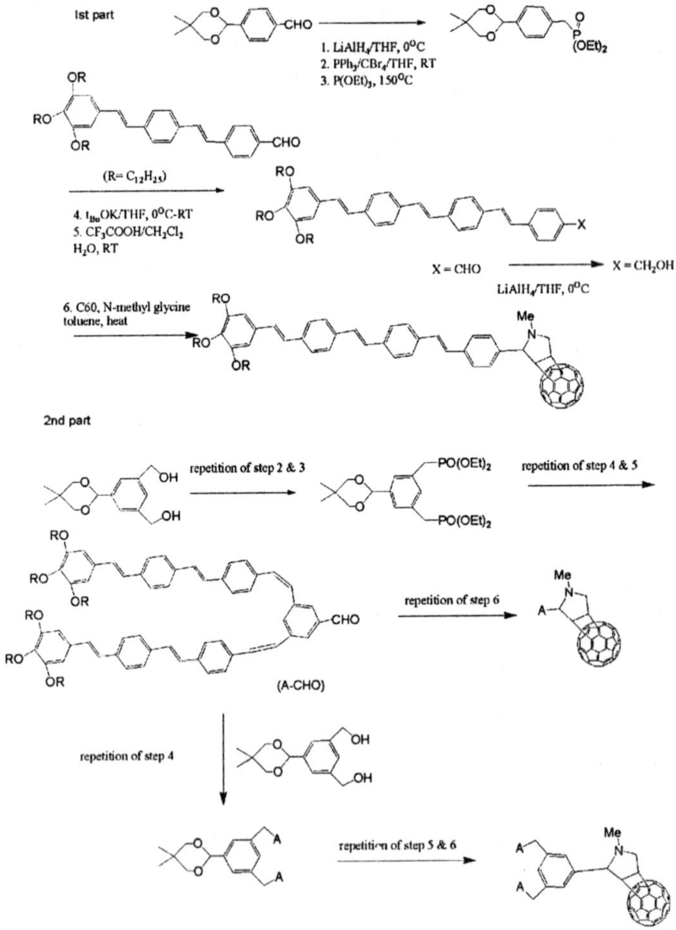

SCHEME 2.8 Synthesis of fullerene functionalized oligo phenylene vinylene dendrons

The synthesis of almost similar type of hydrocarbon dendrimers upto 3rd generation was also reported [42-43] by using the Heck coupling reaction of a derivatized styrene with the 3,5-dibromobenzaldelyde, followed by a Wittig reaction with methyl triphenylphosphonium iodide to produce the activated dendritic styrene. Whereas the chemically homogeneous hydrocarbon dendrons with identical alkenyl linkages in each generation were also synthesized [44]. The repetitive Horner Wadsworth-Emmons and the Heck coupling reactions were used to synthesize poly(aryl alkene) dendrons upto 4th generation, just by 4 steps (Scheme 2.9).

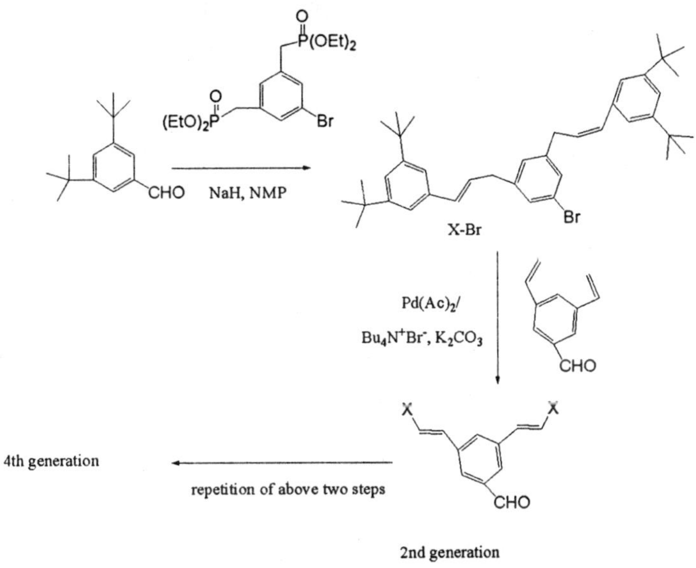

SCHEME 2.9 Synthesis of chemically homogeneous hydrocarbon dendrimer with alkyenyl linkages

The synthesis and energy transfer properties of light harvesting stilbenoid and tolan dendrimers were reported [45-47]. The stilbenoid dendrimers [45] were synthesized by a sequence of reactions starting from methylanthranilate. The synthesized dendrimers were characterized by ^{1}HNMR and ^{13}CNMR spectroscopy. The solution optical properties of the anthracene containing dendrimers indicated energy transfer capability from the periphery to the core unit, as studied by photoluminescence spectrum. The authors [46] also synthesized highly branched anthracene containing stilbenoid dendrimers based on tetraphenylmethan moiety. A cheap dyestuff, New Fuchsin 1 was used as the starting material and by help of three-fold Heck reaction, the desired dendrimers were synthesized (Scheme 2.10). The synthesized dendrimers were characterized by ^{1}HNMR spectroscopy. The absorption and photoluminescence spectra of the dendrimers showed energy transfer from the peripheral stilbene units to the anthracene core. A new Fuchsin dye stuff [47] was utilized to synthesize the peripherally rigid dendritic donor acceptor dyad for energy transfer studies, a tritolan dendron based on a tetraphenyl methane scaffold. The structures of the dendrimers were confirmed by ^{1}HNMR, ^{13}CNMR spectroscopic, and MALDI-TOF mass spectrometric studies. The mass spectrum indicated that the average molecular weight of the dendrimer is 1558 Da. The steady state photophysical studies by UV-absorption, photoluminescence and photoluminescence excitation showed vectorial transfer of excitation energy for one surface tolan units to the anthracene core.

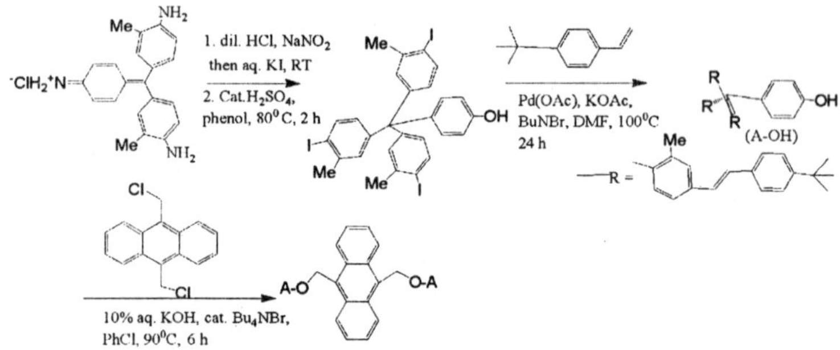

SCHEME 2.10 Synthesis of highly branched anthracene containing stilbenoid dendrimers

The synthesis, characterization and optical properties of dendrimers of 1st and 2nd generations with triphenylbenzene core were also reported [48]. The dendrimers were synthesized from trialdehyde derivative with the phosphonium salt, as shown in scheme 2.11. These dendrimers were fully characterized by ^{1}HNMR and ^{13}CNMR spectroscopies, mass spectrometry and elemental analysis. The photoluminescence studies showed that both the dendrimers exhibited similar photoluminescence properties like emission maximum (438-442 nm), fluorescence quantum yield (0.36-0.37) and life time (0.36-0.60

ns). However, these dendrimers combined wide transparency in the visible range and two photon absorption cross-sections that increased with the generation and thus making these dendrimers promising system for optical power limiting applications.

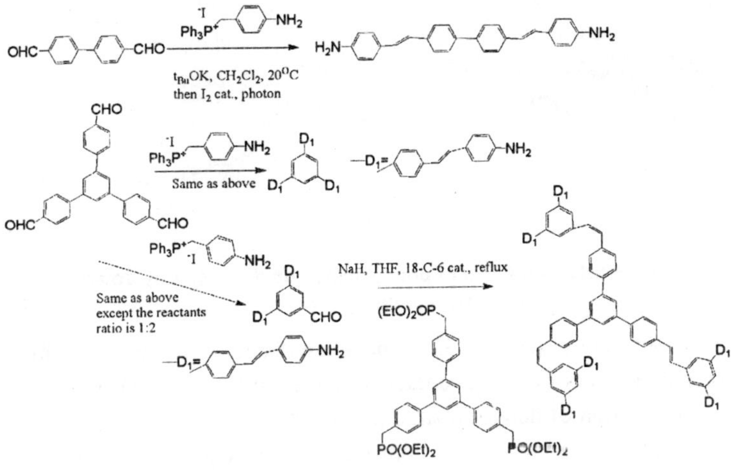

SCHEME 2.11 Synthesis of triphenyl benzene core dendrimers

2.2.3 Poly(phenyl acetylene)s

The poly(phenyl acetylene) dendrimers upto 4th generation were synthesized [49] by 'double exponential growth' strategy. Moore [48] first coined the term 'double exponential'and this concept was the most fundamental breakthrough in the dendrimers synthesis. They employed trimethylsilyl protected triazine monomer and produced 2nd generation dendron, which was utilized to get the 4th generation dendron, as shown in scheme 2.12. The synthetic strategy involved the use of protection chemistry and the palladium catalyzed cross coupling reaction of terminal acetylenes with aryl halides. Xu *et al.* [50] synthesized upto 4th generation poly(phenyl acetylene) dendrimers by an accelerated approach. Chi *et al.* also utilized [51] 'double exponential growth' approach to synthesize a closely related poly(phenyl acetylene) dendrimers.

The synthesis, characterization and properties of different classes of phenyl acetylene based dendrimers were thoroughly studied by Pesak

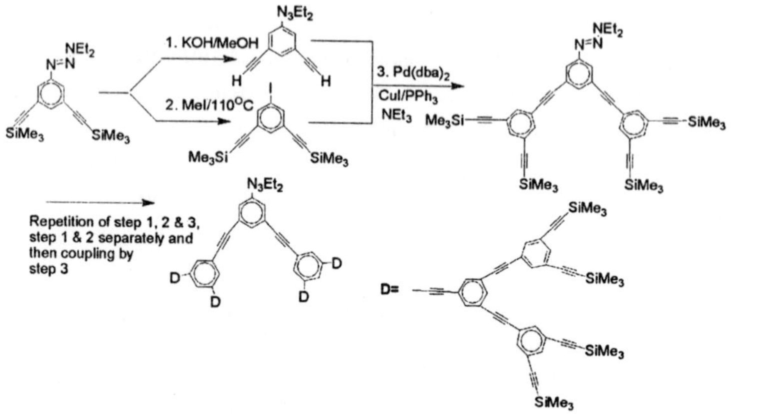

SCHEME 2.12 Synthesis of phenylacetylene dendrons by double exponential growth

in his doctorate research [52]. The phenylacetylene unit is a shift moiety and offers many advantages for their different modifications. The construction of dendritic structures of these categories was carried out by a step-wise convergent approach, which requires a focal point monomer and a peripheral monomer. The coupling of these two units is done by palladium mediated cross-coupling reaction between terminal acetylene and an aryl halide compound. The propagating steps are carried out sequential mutual orthogonal reactions between protective terminal acetylene and aryl halide. Two type focal monomers are generally used to synthesize phenyl acetylene dendrons. 3,5-Dibromo-trimethylsilyl phenyl acetylene, monomer 1 offers high yield mon-odendrons, but higher generation (>4th) could not be synthesized. Whereas the 2nd monomer, 3,5-diacetylenediethyltriazine allows higher generation dendrons with fairly constant yield. The monomer 2 can be easily synthesized from commercially available compound, 2,6-dibromo-4-nitroaniline by the following sequence of reactions (Scheme 2.13). The peripheral monomer, di-butyl-5-iodoisophathalic ester can also be easily obtained from 5-amino isophthalic acid by the following sequence of reactions as shown in scheme 2.14. The advantages of this monomer are better solubility in organic solvents and it can be deprotected the *tert*-butyl groups by simple acid hydrolysis or by heating. The monodendrons upto 3rd generation were synthesized by reiterative coupling of equivalents peripheral monomer with one equivalent focal monomer in the presence of palladium catalyst followed by transformation of focal triazine compound to the corresponding iodo compound (Scheme 2.15). However, in this case also purification was different for higher generation products. Of course, the problem was shorted out by using Frechet double-stage growth strategy. The coupling of tetraacetylene containing focal monomer (Mf_2, Scheme 2.16) with 1st generation iodo monodendron, directly offered 3rd generation monodendron without much purification problem. Same strategy was utilized to get 4th generation monodendron. However, fifth generation was difficult to purify by this technique. To synthesize the phenyl acetylene dendrimers, these monodendrons were coupled with trifunctional core units. There were three different synthetic strategies to get these dendrimers. Route A: Three equivalents of the iodo monodendrons were coupled to 1,3,5-triethynyl benzene to get tridendrons (dendrimers). However, the yield of this route was low, (10-30)%. Route B: Three equivalents of acetylene monodendrons were coupled much more efficiently with 1,3,5 tri iodobenzene to get tridendrons. Route C: An acetylene monodendrons was coupled to an iodo monodendrons of the next or earlier generation to give tridendrons with two different generation monodendrons. This strategy is very useful for lower generation monodendrons but over the entire route, B is the best choice. Alternately, a six number of acetylene containing core unit can be coupled with iodo monodendrons to get the dendrimers. This approach easily offered 5th generation dendrimer. In the synthesis of mono- and tri-dendrons, each step of the reactions resulted very high yield. Interesting only four to five different types of reactions were utilized for the syntheses of all phenyl acetylene dendrons and dendrimers. All the products were purified by chromatographic techniques and characterized by ^{1}HNMR spectroscopy. The mono- and tri-dendron phenyl acetylenes were characterized by ^{1}HNMR, ^{13}CNMR spectroscopy, MALDI/EI/FAB-mass spectrometry, elemental and GPC analyses. The molecular weight of the dendrimers varies from 0.979 to 19 KDa with polydispersity 1.02 to 1.04 for first to 5th generation. The dendrimers were soluble in CH_2Cl_2 and THF for lower generations and in DMSO, aqueous NaOH solution for higher generations. The sodium and tetrabutyl amine salts of the dendrimers were soluble in water also.

However, capillary electrophoresis and light scattering data suggested that the removal of *tert*-butyl ester group by the thermolysis reaction resulted in some sort of crosslinking between dendrimers of higher generation. So, Pesak [52] synthesized 2-(2-(2-methoxy-ethoxy)-ethoxy) ether ester terminated phenyl acetylene dendrimers by using the same strategy of reactions as described above, where such problem was avoided. The monomer was synthesized by hydrolysis of the methyl ester of 3,5-dimethyl ester iodobenzene, followed by DCC medicated esterification with triethylene glycol monoethyl ether. The dendrimers upto 3rd generation were synthesized and characterized by applying the same techniques as described in earlier cases. The hydrolysis of these ester groups in hot aqueous base solution was carried out without any crosslinking. The dendrimers behave as liquid crystalline meso phases over a wide temperature range. The optical microscopy and XRD

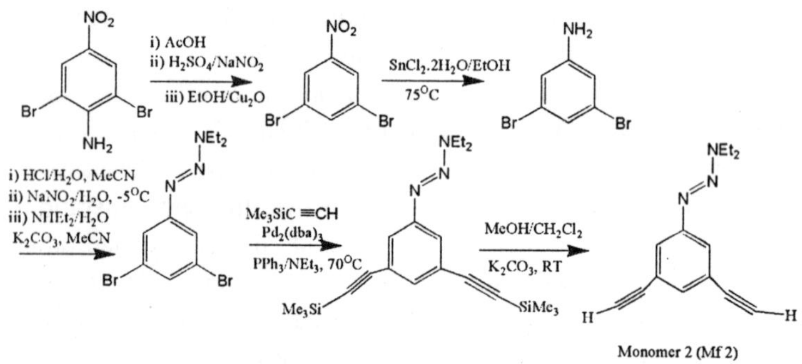

SCHEME 2.13 Synthesis of focal monomer 2 for phenyl acetylene dendrons

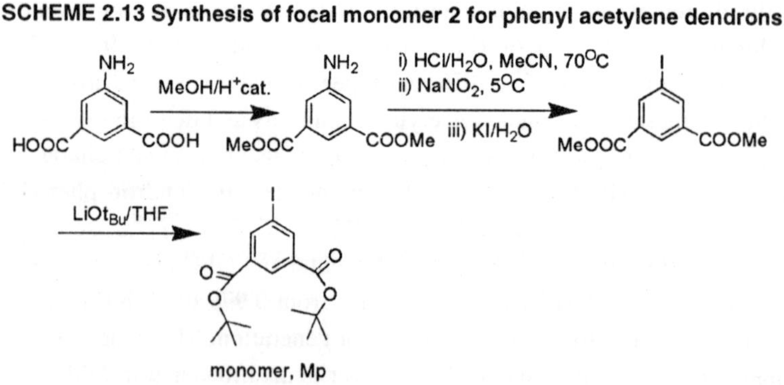

SCHEME 2.14 Synthesis of peripheral monomer (Mp) containing ditert-butyl ester groups

support a hexagonal columnar discotic ordering. The columnar liquid crystals from planar shape persistent dendritic molecules containing a phenyl acetyleneic core were also synthesized [53].

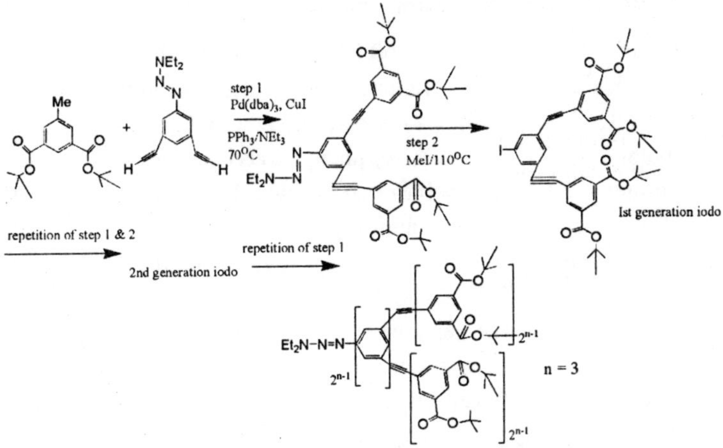

SCHEME 2.15 Synthesis of phenylacetylene monodendrons based on ditert-butyl ester peripheral groups

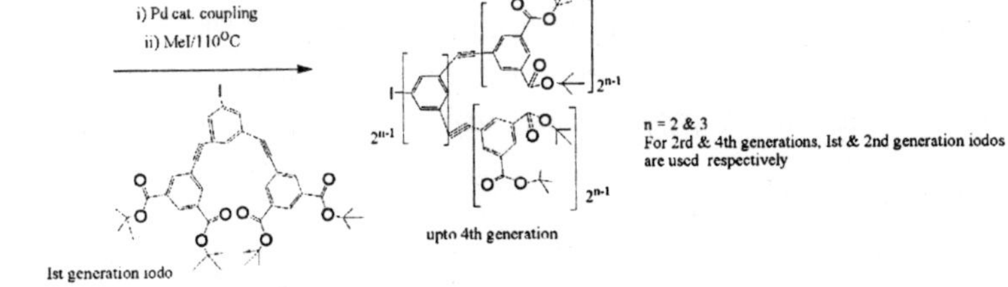

SCHEME 2.16 Synthesis of focal monomer (Mf$_2$)

Kleiman *et al.* [54] measured the ultrafast dynamics of electronic excitations in a light harvesting phenyl acetylene dendrimers (see applications section), a nanostar compound. The degenerate femto second (fs) pump probe spectroscopy study revealed that the nanostar has several components with time scales from hundred of femto seconds to tens of pico seconds (ps). A simple application of the Forester model predicts energy transfer rate constants that are in rough qualitative agreement with the experimental results corresponding to the excitation of the 2-ring chromophores.

The effect of pressure upto 60 Kbar was measured on the luminescence peak location and efficiency for a series of methoxy phenyl acetylene dendrimers (Fig.2.4), where dimethoxy benzene was at the focal point [55]. The neat methoxy phenyl acetylene dendrimers as well as in dilute solutions in poly(*tert*-butylmethylacrylate) were studied. The dimethoxy benzene containing phenyl acetylene dendrimers showed a fluorescence emission that was red shifted relative to the parent dendrimer. This behavior is the characteristic of a fluorescence emission from a charge separated state and suggests that the parent phenylacetylene dendrons are capable of acting as electron acceptors. Also the degree of charge separation increases with the increase of generation. Specially, λ_{max} of the higher generation (63-mer and 127-mer) for the fluorescence emission largely depends on the nature of the solvent used. The methoxy phenyl acetylene dendrimers were synthesized and characterized by Moore *et al* [56]. The fluorescence absorbance (λ_{max}) for the 63-mer in pentane is 421 nm, whereas in cyclohexane it is 380 nm [56]. This hypersensitivity towards solvent may be due to change of molecular shape of the higher generation dendrimers by the polymer-solvent interaction. The study revealed that for the neat dendrimers, when the charge transfer is from the lowest exited state, intermolecular charge transfer occurs with a shift of the emission peak to higher energy as the pressure increases. Whereas in dilute solutions of poly(tertbutyl methylacrylate), π^* was initially the

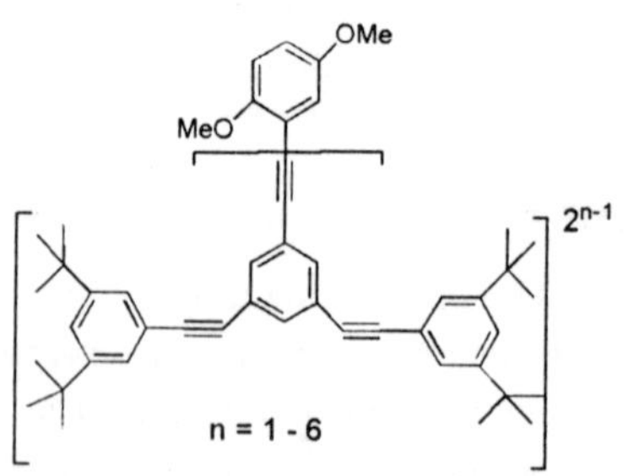

FIG.2.4 Structure of methoxy phenyl acetylene dendrimers

emitting state for the lower generation dendrons, but the charge transfer state was the emitting state for the larger dendrons. However, with the increase of pressure, the inter molecular charge transfer state was rapidly stabilized with respect to the π* state, so that above 40 Kba the emission is from the charge transfer state for all dendrimers.

The effects of electron transfer rate as functions of both molecular weight and dendrimer architecture were studied [57] in the series of ferro-doxin (Fe_4S_4) core phenyl acetylene dendrimers. A significant electron transfer rate attenuation was observed with the increase of molecular weight (generation) of the dendrimers. Gormar and Smith [58] showed that iron-sulfur (ferro-doxin unit) clusters with four arms, in a tetrahedral arrangement with stiff phenyl acetylene dendrimers upto 3rd generation (Fig.2.5) displayed dramatically different electro chemical behavior in films compared to solution. The cyclic voltametry studies showed a large change in redox potential with generation number for the films, while that remains constant in solution. However, in contrast to the large changes in redox potential for the films with generation, little change in the kinetics of charge transfer was observed. Again this behavior contrasts sharply with that observed in solution, where rate of interfacial electron transfer changed greatly with generation. This difference in electro-chemical behavior may be due to the time scale of core motion relative to the electron transfer of the dendrimers cluster.

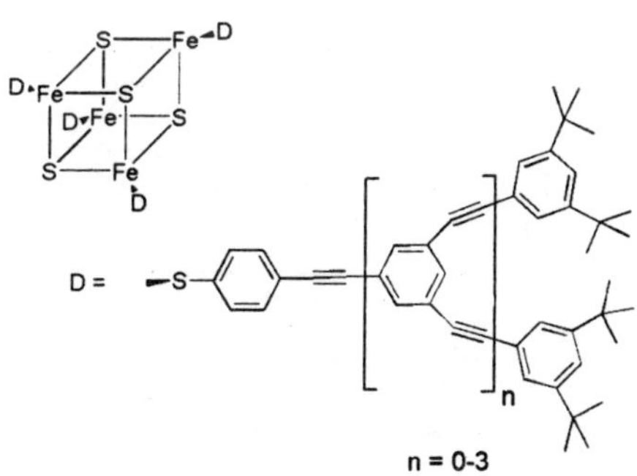

FIG.2.5 Structure of Fe-S cluster core phenyl acetylene dendrimer

A soluble, oxidatively and thermally very stable alkynyl ruthenium dendrimer containing a large two dimensional π-delocalized system was synthesized [59] by the convergent method. At first the dendritic wedge was synthesized from 1,3,5-tribromo benzene followed by coupling of the dendrons with the core unit by the palladium catalyzed reaction offered the desired dendrimers (Scheme 2.17). The dendrimer was characterized by spectroscopic and analytical techniques. The core unit was synthesized by sequence of almost similar type of reactions as described earlier [60]. The core unit can be converted easily to the corresponding phenyl acetylene end capped Ru-complex dendrimers by treatment with PhC≡CH in the presence of $NaPF_6$ and NEt_3. The UV-visible spectra of the higher and lower generations were almost similar. The higher generation dendrimer complex offers absorption multi layers charge transfer band (MLCT) at 403 nm compared to lower generation at 411 nm. This indicates that the increase of the size of dendrimer complex does not reduce the optical transparency. This small blue shift for higher generation may be due to lack of coplanarity through the dendritic structure of higher generation complex. However, interestingly the magnitude of the extinction co-efficient increases four fold for higher generation complex containing nine metal centers compared to lower generation, which contains only three metal centers. The third order nonlinear optical (NLO) property measurement of

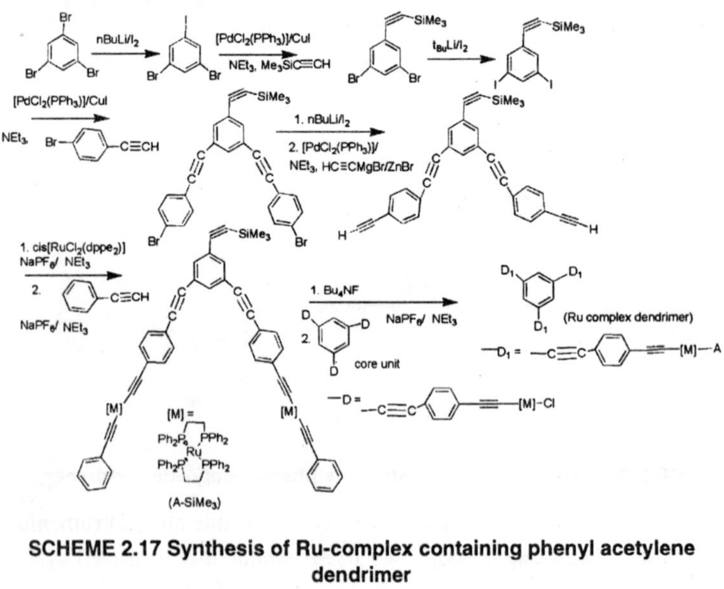

SCHEME 2.17 Synthesis of Ru-complex containing phenyl acetylene dendrimer

Ru-dendrimers showed a significant enhancement of two photon absorption (TPA) upon proceeding from the constituent molecules to the dendritic complex.

A series of arylacetylene dendrons and dendrimers were synthesized [61-62] by the convergent approach. The focal point of these dendrons upto 6th generation was linked to 2,5-dimethoxy benzene (Fig.2.6a) and to a perylene (Fig.2.6b) focal unit [63-64]. The investigation of fluorescence properties of the dendrimers revealed a distinct anomalous solvatochromism for higher generations of

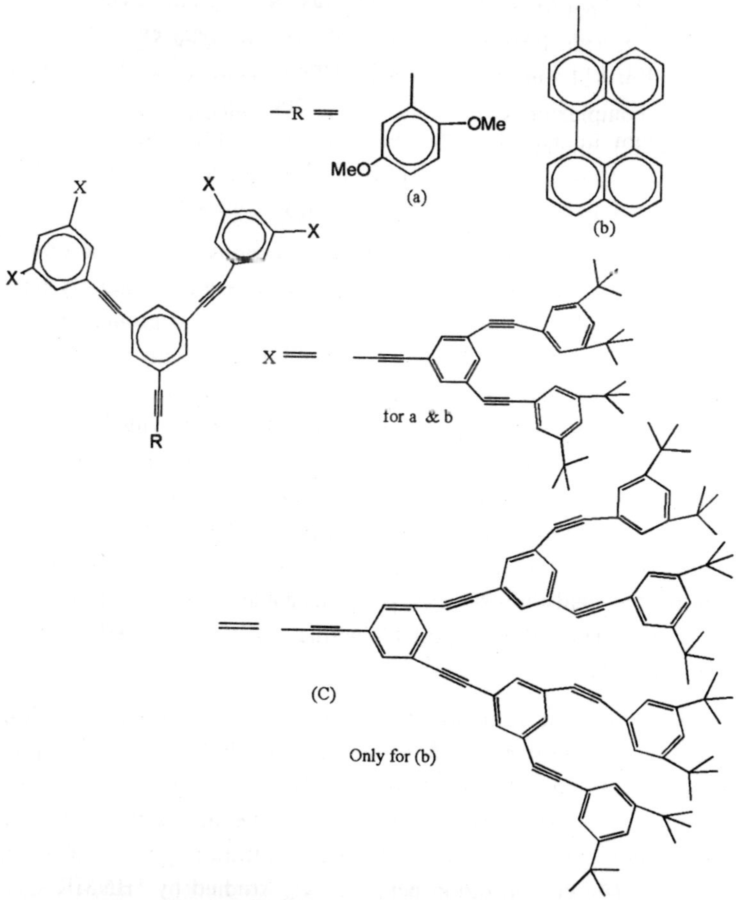

FIG.2.6 Structure of poly(phenyl acetylene) dendrimers

dimethoxybenzene containing dendrimers. This is put down to an electron transfer from the donor dimethoxy benzene to the arylacetylene moieties acting as the acceptor. Whereas for perylene containing dendrimers, the fluorescence intensity of the light emitting perylene group at the focal point increases with growing number of generations, may be due to increase of molar absorption coefficient. Hence they can act as light harvesting materials. However, the efficiency of the arylacetylene units to transfer energy to the perylene group decreases with the increase of generation number [63]. Of course, a significant increase in rate of energy transfer was observed in case of little different type of arylacetylene dendrimers with perylene unit at the focal point (Fig.2.6c). These dendrimers containing additional tolan moieties in the structure, which increase the number of arylacetylene groups from the out side to the inside after each branching unit by one. The results in an energy gradient from the peripheral groups towards the focal point i.e. the system acts as 'energy funnels'

The phenyl acetylene dendrimers with various solubilizing groups were synthesized [65]. As this type dendrimers have generally poor solubility, so they required suitable solubilizing end groups in their structures. The 4-*tert*-butylphenyl peripheral units initially incorporated for this purpose, but it is useful only upto 3rd generation dendrimers. However, 3,5-di-*tert*-butylphenyl peripheral groups provided sufficient solubility upto 4th generation [65]. Later on, the synthesis was optimized to eliminate the necessity of extended monomer units by simply reversing the functionalities on the monomer, using a dialkyltriazene precursor for the focal iodo functionality [66]. In addition to improve the yields, this reverse approach enabled the synthesis upto 5th generation. The solid support technique can also be used for their syntheses [66].

The effect of generation of dendrimers on hydrogen-bonding complexation with guest molecules for *tert*-butyl terminated poly phenyl acetylene dendrimers was also investigated [67]. The dendrimers were synthesized with naphthyridine units at the focal point and capable of H-bonding with benzamidinium guest molecules (Fig.2.7). The complexation behavior was studied by ^{1}HNMR spectroscopy, which indicated that no distinct dendritic environment was present even for the highest generation. That is, the influence of dendrimer is negligible and the structures are highly porous for small guest molecules even for the highest generation. However, for bulkier guests, the binding effect was weaken with the increase of generation may be due to steric effect for complex formation.

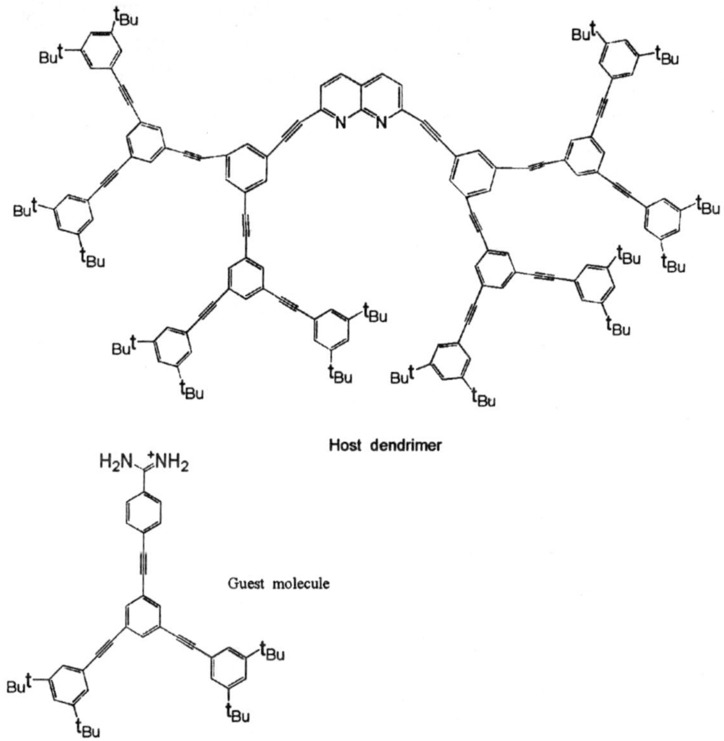

FIG.2.7 Structures of poly(phenyl acetylene) dendrimeric host and guest molecules

A novel hydrocarbon dendrimer possessing platinum-acetylide units and a porphyrin core was synthesized [68] by the convergent approach. At first a tetra(4-ethynylphenyl)porphyrin bridges tetra nuclear platinum acetylide core was synthesized by the reaction of 4-(trimethyl silylethnyl) benzaldelyde and pyrrole in propionic acid and then with $Zn(OAc)_2$ followed by treatment with TBAF to give Zn(II) tetra(4-ethylnylphenyl)porphyrin and then by the reaction of this compound with $Pt(PEt_3)_2Cl_2$ in the presence of CuCl catalyst. A series of organometallic dendritic porphyrins was synthesized by the coupling of the core with Pt-acetylide dendrons upto 3rd generation in the presence of CuCl. The whole sequence of reaction is shown in scheme 2.18. The synthesized core and dendrimers were characterized by ^{1}HNMR, ^{13}CNMR, ^{31}PNMR spectroscopy, GPC analysis, and MALDI-TOF/FSI mass spectrometry. Fluorescence spectral studies showed an intramolecular energy transfer in these dendrimers. Also fluorescence from the porphyrin core at 617 nm decreased with the increase of generation, though 648 nm peak relatively increased. The reason is still

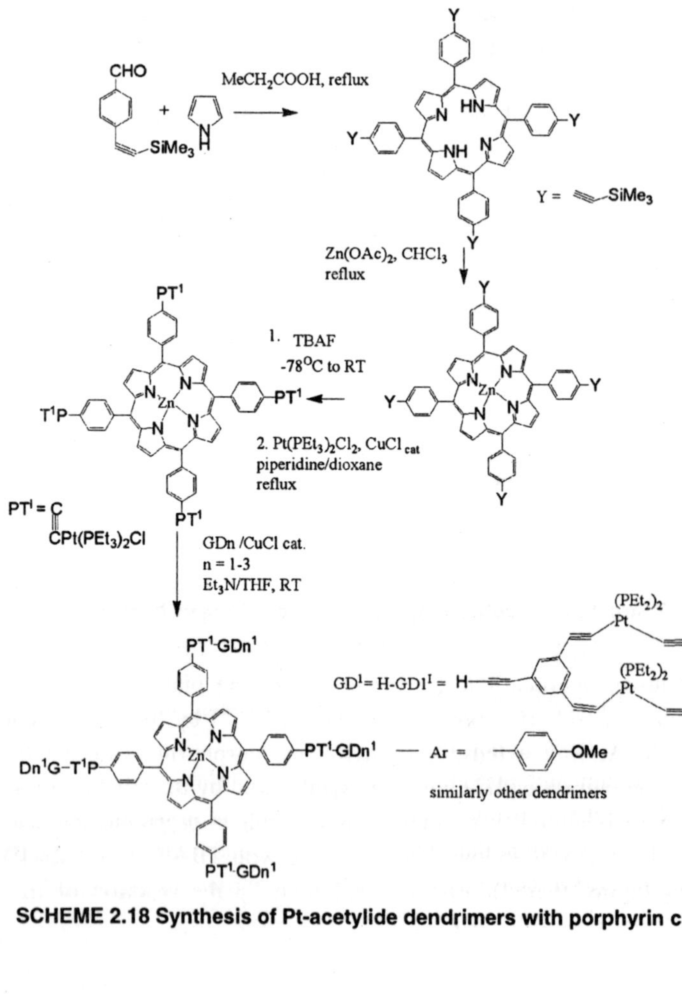

SCHEME 2.18 Synthesis of Pt-acetylide dendrimers with porphyrin core

unknown. The authors [69] also synthesized the giant Pt-acetylides dendrimers upto sixth generation by the divergent technique. The first generation dendrimer was synthesized by the reaction of 2,4,6-triethynylmesitylene with a mononuclear Pt-acetylide complex (I) in the presence of CuCl catalyst. The 2nd generation was obtained by deprotection of 1st generation dendrimer with Bu_4NF followed by repetition of the previous reaction with I. The repetition of these steps generated higher generation. The whole sequence of reactions is shown in scheme 2.19. The synthesized dendrimers were characterized ^{1}HNMR spectroscopy, GPC analysis and MALDI-TOF mass spectrometry. The sixth generation dendrimer contains 189 Pt atoms per molecule. The molecular weight of the dendrimers ranges from 2800 to 40200 Da, as determined by GPC by using polystyrene as the standard. The molecular size of the highest generation dendrimers was 56.8 ± 0.2A° diameter as measured by SANS.

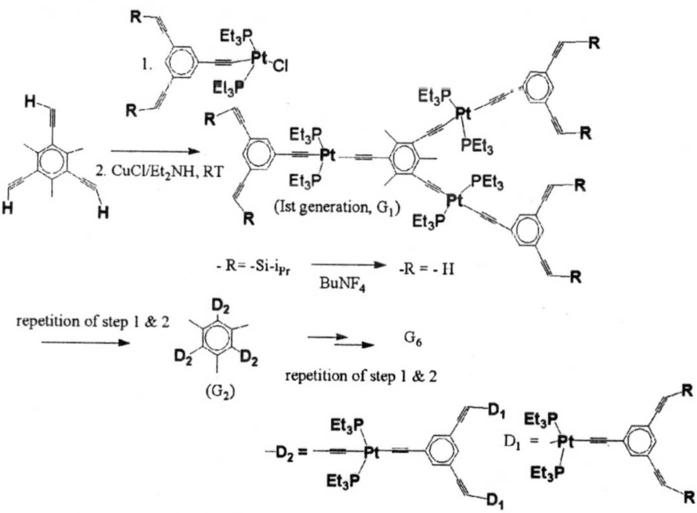

SCHEME 2.19 Synthesis of Pt-acetylide containing dendrimers

Two and three dimensional organanonetallic dendritic polymers with palladium and platinum metals, and acetylene linkages were synthesized [70] by using O-diethynyl benzene as a bridging ligand. The platinum acetylide dendrimers upto 3rd generation were synthesized by the convergent method, where platinum atoms are linked by 1,3,5-triethynyl derivatives.

The synthesis and electrochemical behavior of 61,61-*bis*[4-{3,5-*bis* (3,5-di-*tert*-butylphenylethynyl)phenylethynyl}-phenyl]-1,2-dihydro-1, 2-methanol-60 fullerene and 61,61-*bis*[4-{3,5-*bis*[3,5-bis(3,5-di-*tert*-butyl (phenylethynyl) phenylethynyl] phenylethynyl}phenyl]-1,2-dihydro-1,2-methanol-60-fullerene with first and second generation phenyl acetylene dendrimers were reported [71]. The fullerene coupled phenyl acetylene dendrimers were synthesized by the reaction of the dendrimers substituted tosylhydrazones with fullerene in the presence of sodium hydride. The ultimate products were obtained after a number of steps, as shown in scheme 2.20 [71]. The synthesized dendrimers were characterized by FT-IR, ^{1}HNMR, ^{13}CNMR and UV-vis. spectroscopy and MALDI-TOF mass spectrometry. All these spectral analysis confirmed the structure of the dendrimers. The cyclic voltammetric study showed that no significant interaction occurs in the ground state between the fullerene and the dendrimers. The fullerene derivative retains the typical redox stability of 6,6-closed derivatives in both cyclic votammetary and bulk electrolysis time scales.

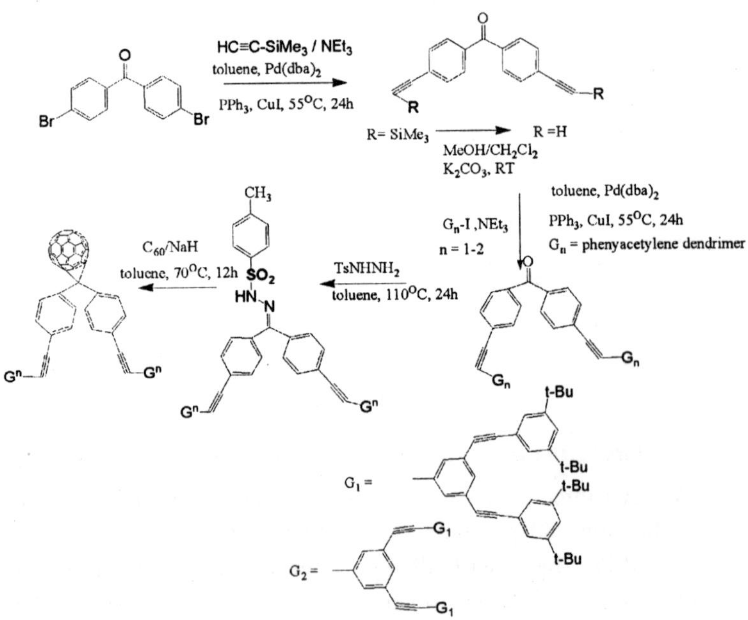

SCHEME 2.20 Synthesis of fullerene coupled phenylacetylene

2.2.4 Dendronized Polyhydrocarbons

The synthesis of dendronized polystyrene upto 3rd generation was reported [72] by the convergent growth technique. All macromonomers were carrying triphenylmethyl (trityl) or trimethylsilylethoxycarbonyl (teoc) protected amine functions in the periphery. The amine groups were chosen, so that it can be modified as per necessity. The synthesis of the macromonomers and polymers were carried out by starting from allyl amine and allyl 3,5-dibromobenzene. The dendronized polymers can be further modified to get different products [72]. The synthesized dendrons and dendronized polymers were fully characterized by ^{1}HNMR, ^{13}CNMR spectroscopy, mass spectroscopy (EI), GPC and elemental analysis. However, GPC analysis indicated the decrease in

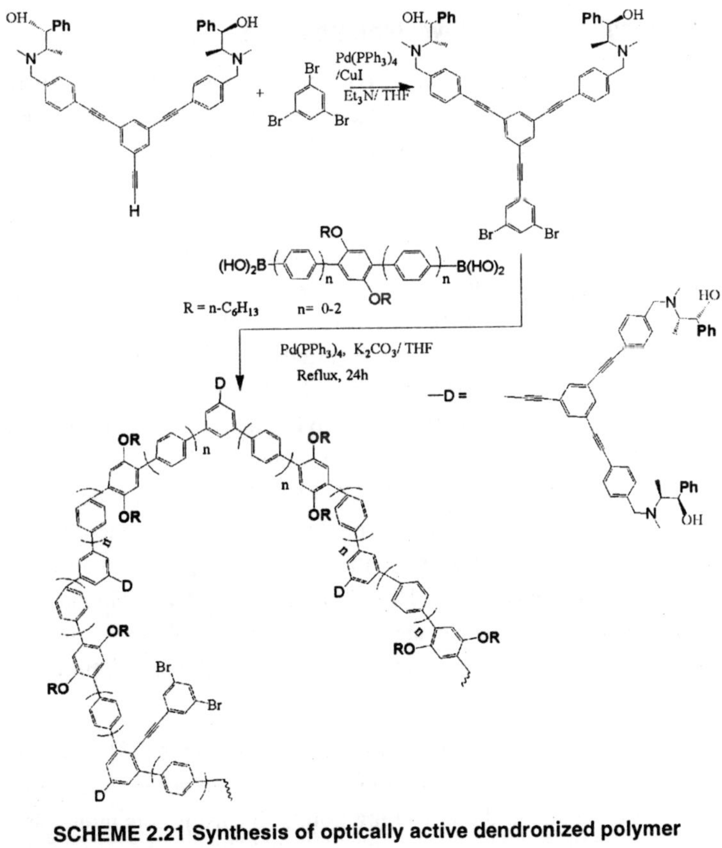

SCHEME 2.21 Synthesis of optically active dendronized polymer

molecular weights with the increase of dendronization (for 1st generation Mn = 50,000 Da and 3rd generation Mn = 37,000 Da). This may be an erroneous result, as polystyrene was used as the standard for the molecular weight determination. These deprotected polymers constitute a new class of polyelectrolytes with an unusually high number of charges per repeating unit. The optically active dendronized polymers were synthesized by the Suzuki coupling polymerization of optically active dendronized monomer with diboronic acids (Scheme 2.21). They are soluble in common organic solvents such as THF, toluene, CH_2Cl_2, etc. The synthesized polymers were characterized by ^{1}HNMR spectroscopy, GPC analysis and by optical rotation measurement. The molecular weight of the polymer was $1.4X10^5$ Da with polydispersity 2.80.

2.2.5 Phenylcarbazole Ethynylene Dendrons

A series of 9-phenylcarbazole ethynylene monodendrons [73] was synthesized by the palladium catalyzed coupling reactions and characterized by ^{1}HNMR spectroscopy, HPLC, GPC analysis and MALDI mass spectrometry. At first the monomer, 3,6-diethynyl-7-(4-(3,3-diethyltriazenyl)-phenyl)carbazole was synthesized from the 9-(4-nitrophenyl)-carbazole. The peripheral unit, 3,6-di-*tert*-butyl-9-(4-iodophenyl)carbazole was also synthesized from the same starting material. First, 9-(4-nitrophenyl)carbazole alkylated with *tert*-butylchloride in the presence of $AlCl_3$ (Friedel-Craft reaction) followed by the same reduction, diazotization and amination reactions to get 3,6-di-*tert*-butyl-9-(4-(3,3-diethyltriazenyl) phenyl)carbazole. This carbazole compound was then treated with methyliodide at 110°C in a sealed tube to get the desired peripheral unit. Lastly, the monodendrons were synthesized by alternatively repeating the coupling reaction and the transformation of the triazene to the corresponding iodo compound. The products were purified by HPLC and high yield was obtained in each case. However, higher generation was not possible to synthesize due to solubility and formation of high amount of dimer and polymer products.

Different type monodendrons were also synthesized [73] by using the monomer, 3-hydroxy-3-methyl-but-1-ynyl group at the focal point and two iodo groups at the branching points. By using this monomer, the synthesis became easier, as purification was easy. This is due to the fact that the resulting monodendrons have higher polarity so as to facilitate separation of any self-coupled by-product and thus higher generation can be easily obtained. The whole sequence of reactions for this synthetic process is shown in scheme 2.22. The iodo peripheral unit and the 1st and 2nd generation iodo monodendrons can be easily transformed into the corresponding acetylene derivative by coupling these iodo compounds with trimethyl silyl acetylene and subsequent removal of trimethyl silyl groups. These acetylene functionalized compounds can be coupled with aryl halides to give other functionalized monodendrons or coupled with monomer (M_2) to get the higher generation (Scheme 2.23). The highest generation monodendron shown to be soluble in solvents of high polarizability (CS_2). Also the elec-

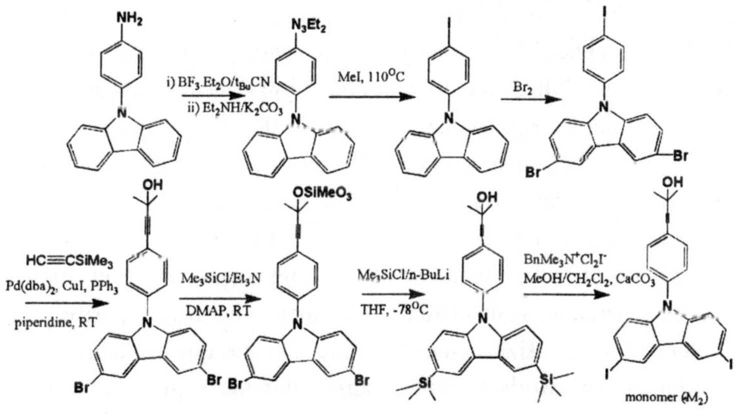

SCHEME 2.22 Synthesis of 9-phenylene carbazole based monomer 2 (M_2)

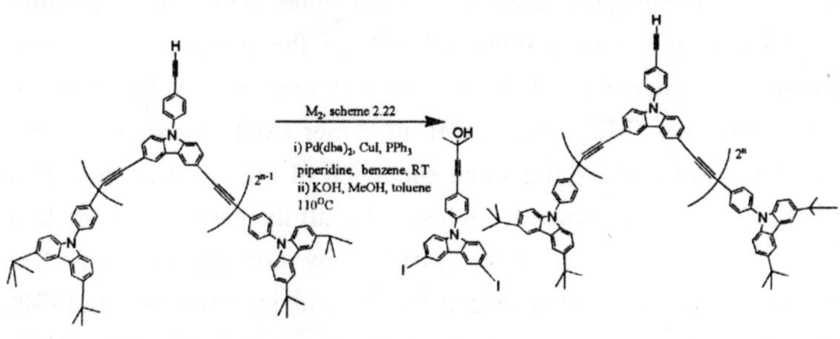

SCHEME 2.23 Synthesis of 9-phenyl carbazole based higher generation monodendron

tronic absorption spectra showed that there is limited delocalization between 9-phenylcarbozole units with the increase of generation number. The fluorescence efficiency increases rapidly from monomer to trimer and upto 7-mer monodendron, where it appears to reach a limit. This is possibly because of the through-space interactions between the 9-phenyl carbazole units, become more significant with the increase of molecular size, providing additional fluorescence quenching pathways.

The 9-phenyl carbazole ethynylene monodendrons upto 3rd generation with 1,1,3,3-tetramethylbutyl and 3,5-di-*tert*-butyl-4-(2-methoxyethoxy)phenylene group as the peripheral units, were synthesized [74] to get enhance solubility than the earlier discussed dendrimers. The first peripheral monomer, 9-(4-ethynyl phenyl)-3,6-*bis*(1,1,3,3-tetramethylbutyl)carbazole was synthesized from carbazole by a long sequence of reactions [74]. The synthesis of focal monomer, 3,6-diiodo-9-(4-(3-hydroxy-3-methyl-but-1-ynyl)phenyl)carbazole (Mf) was carried out by exactly the same way as described earlier. Finally, the coupling of focal monomer (Mf) with peripheral monomer was done also by the same way, as described earlier (Scheme 2.23), but in these repetition of processes resulted up to 3rd generation dendron (Fig. 2.8) because of better solubility of the compounds. However, higher than 3rd generation, again difficult to synthesize because of purification problem of larger size dendrons. This may be due to decrease in retention on the adsorbents during purification techniques. To increase further solubility as well as adsorption retention to facilitate purification, 3,5-di-*tert*-butyl-4-(2-methoxyethoxy)phenylene, at 3,6-position of carbazole unit was synthesized [52] as the peripheral group by following sequence of reactions as shown in scheme 2.24. The coupling of this monomer (MP_2) with focal monomer (Mf) affords the trimer monodendron by using the same sequence of reactions as described earlier. The yield obtained in each step for all the cases was very high and all the compounds were purified by GPC/HPLC, etc. The monodendrons were characterized by ^{1}HNMR spectroscopy and GPC technique. They were soluble in common organic solvents such as benzene, THF, $CHCl_3$,CH_2Cl_2, etc.

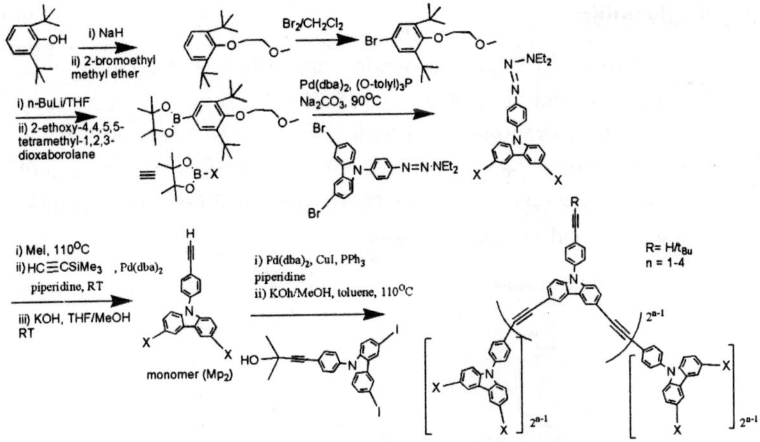

SCHEME 2.24 Synthesis of monodendron trimer with *tert*-butyl methoxy-ethane peripheral groups

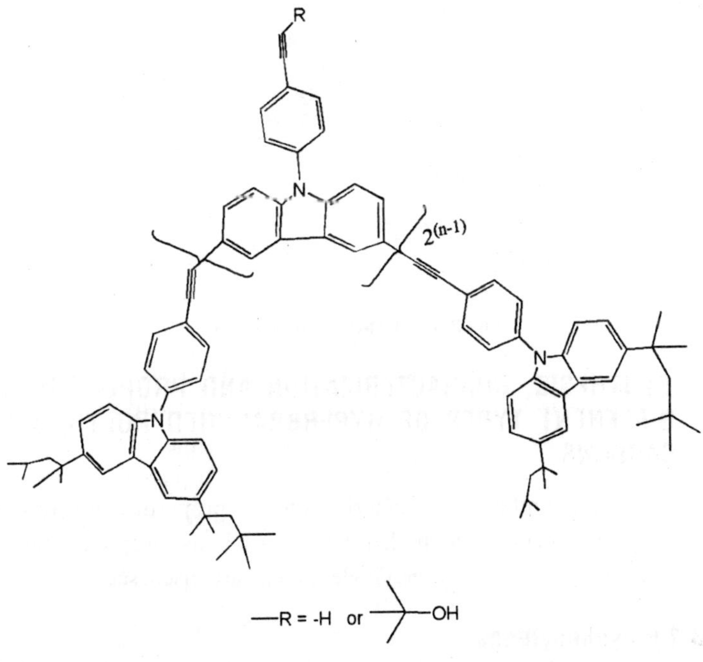

FIG.2.8 Structures of monodendrons with 9-phenylcarbazole containing tetramethylbutyl group at the periphery

2.2.6 Iptycenes

A unique class of hydrocarbon dendrimers called as 'Iptycenes' was reported by Hart and his coworkers [75-76]. These architectures may represent the first example of all hydrocarbon dendrimers prototypes (Fig.2.9). The initiator core used was either benzene or triptycene and the dendrimers were synthesized by Diels-Alder reaction. The structural analysis was carried by XRD analysis.

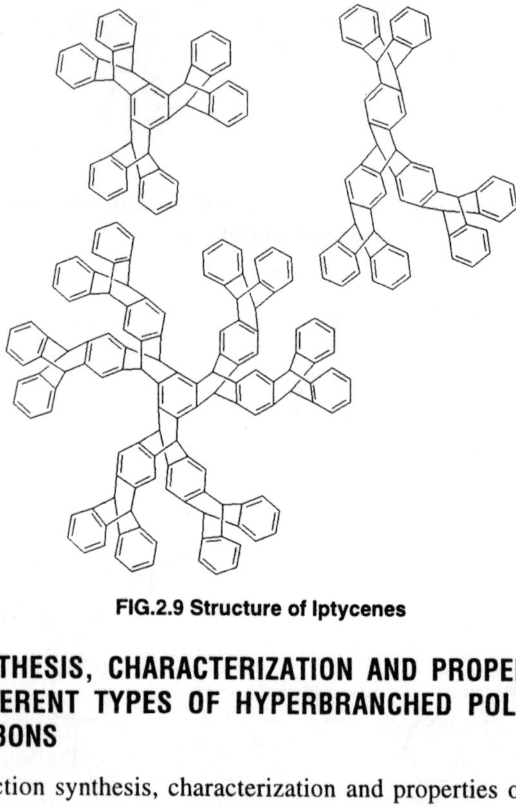

FIG.2.9 Structure of Iptycenes

2.3 SYNTHESIS, CHARACTERIZATION AND PROPERTIES OF DIFFERENT TYPES OF HYPERBRANCHED POLYHYDROCARBONS

In this section synthesis, characterization and properties of different types hyperbranched polyhydrocarbons such as polyphenylenes, polyethynylenes, polyphenylenevinylenes, etc. are discussed.

2.3.1 Polyphenylenes

Many kinds of HB hydrocarbon polymers were synthesized by the one-step polycondensation of AB_2 type monomers, since Kim and Webster reported [77] the synthesis of HB polyphenylenes from AB_2 monomer in 1990. HB polyphenylene polymers were synthesized by either aryl-aryl coupling reactions or Diels-Alder reaction. 1,3,5-tribromo- and trichloro-benzene were used as the starting materials for aryl coupling reactions to get the desired HB polyphenylenes [77-78] as shown in scheme 2.25. The polyphenylenes were brittle and did not form self-standing films when casted from solution. The synthesized HB polyphenylenes were characterized by ^{1}HNMR spectroscopy and GPC analysis. In the ^{1}HNMR spectra, peaks corresponding to aromatic protons became broad, which suggested that the polymer contained many isomeric structures. The degree of branching for Ni catalyzed polymerization was 0.25, while degree of polymerization was 24. Whereas the degree of branching for Pd-catalyzed reaction was much higher (0.7) indicating the presence of more dendritic units [35]. The average molecular weight of the HB polyphenylene was 32000 Da with polydispersity 1.13 as measured by GPC for the polymer obtained by Pt-catalyzed polycondensition reactions. This HB polyphenylene was soluble in organic solvents such as O-dichlorobenzene, tetra-chloroethane, THF, etc. even though the main chain of the polymers consisting only rigid aromatic moieties. The thermal properties and solubility characteristics of the HB polymers depend on nature of terminal functional groups as well as backbone structure. The polyphenylene even became water-soluble by transforming bromide end groups to carboxylate groups. The polymer was stable upto 550°C as measured by TGA. Thus the chemical modification of the terminal functional groups allows to control over the properties of the HB polymers. Kim *et al.* modified [77, 79] the HB polyphenylenes by various functional groups upto 30-80%, applying lithiation and subsequent reactions with various electrophiles. The modified polymers were soluble in diethylether when non-polar functional groups such as methyl and trimethylsilyl were introduced. On the other hand, the modified polymer became soluble in water (1mg/1mL) by the formation of micelles when CO_2 was used as an electrophile to convert the terminal groups into lithium carboxylate. The polymer modified by non-polar functions also have lower T_gs. Bromine terminated polymers having T_gs of 238°C as determined by DSC, whereas trimethylsilyl end groups having only 152°C. The water-soluble automatic HB polyphenylenes [77] with carboxylate end groups complexed with *p*-toluidine. The methyl signal of *p*-toluidine shifted in ^{1}HNMR spectrum, on addition of this HB polymer (pH~10) indicates the complex formation. The poorly defined HB structure and the possibility of multiple complexes with *p*-toluidine (with host to guest ratio low, maximum usable ratio was 2.5) hampers an accurate determination of the equilibrium constant, which was estimated at (510 ± 150) M^{-1}. The HB polymer is further more capable of dissolving naphthalene in higher concentration in aqueous medium, and enhances solubility of methyl red (30 times) and methyl orange (twice) in 0.1 M K_2HPO_4 solution.

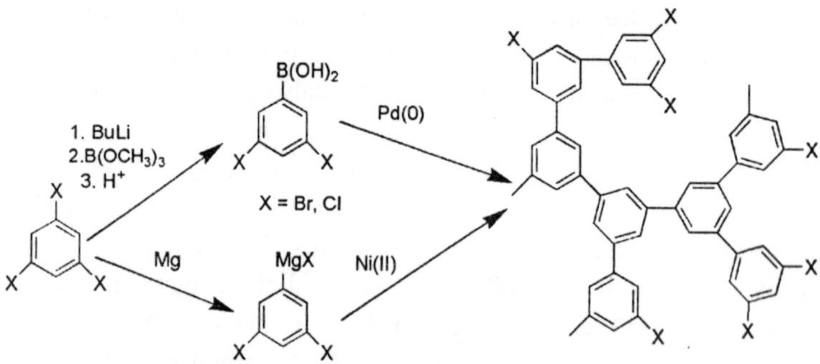

SCHEME 2.25 Synthesis of HB polyphenylene

The HB polyphenylenes were synthesized [10] by Diels-Alder cycloaddition of the AB_2 type of monomers with one cyclopentadienone and two triple bonds as a diene and dienophiles, respectively. The

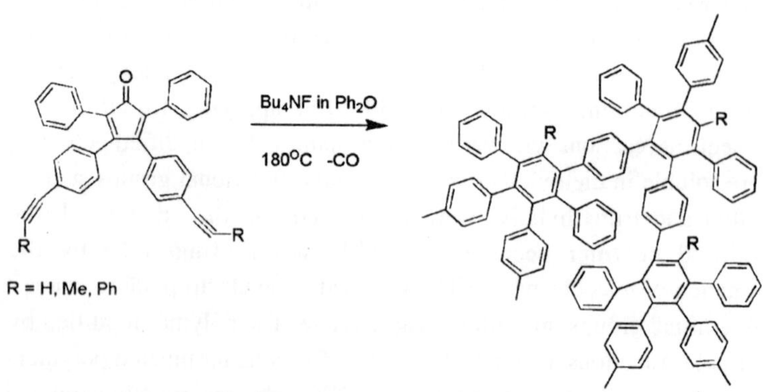

SCHEME 2.26 Synthesis of HB polyphenylene by Diels-Alder reaction

polymerization reaction was carried out in diphenyl ether at 180°C in the presence of tetrabutyl ammonium fluoride, as shown in schene 2.26. The synthesized polymers were characterized by SEC and MALDI-TOF mass spectrometry. The SEC measured the weight average molecular weight of HB polyphenylenes, ranging from 3000 to 107500 Da. The resulting polymers were soluble in toluene and benzene.

2.3.2 Polyethynylenes

The synthesis of HB polyethynylenes by palladium catalyzed carbon-carbon coupling reactions of AB_2 type monomers was reported [80]. The monomer possesses one aryl-ethynyl and two aryliodide units. The HB polymers were produced by self-polycondensation of the monomer in the presence of $Pd(PPh_3)_2Cl_2$, CuI and Et_3N at 0°C in THF, as shown in scheme 2.27. The synthesized HB polymers were characterized by spectroscopic techniques. The GPC analysis indicates the weight average molecular weight for the polymers varies from (6-10)X10^5 Da with polydispersity 1.6 to 3.0. The HB polyethynylenes were soluble in organic solvents. The spectroscopic measurements indicated that the polymers contained diacetylenic moieties (ca. 25%) formed by the oxidative coupling of two aryl-ethynyl units in addition to the desired aryl-ethynylene units.

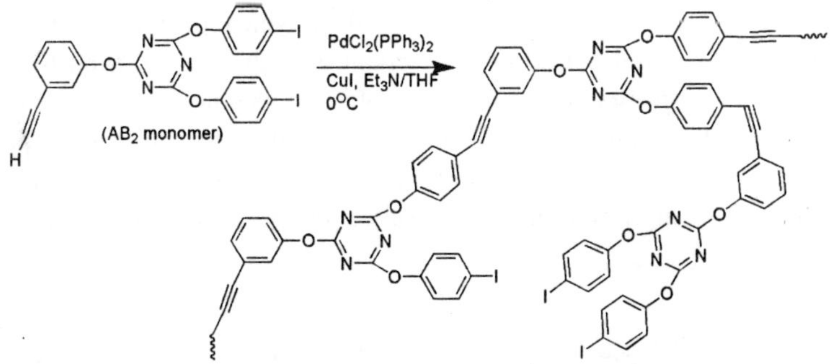

SCHEME 2.27 Synthesis of HB polyethynylene

2.3.3 Poly(phenylene vinylene)s

A HB poly(phenylene vinylene) was synthesized [81] by self-polycondensation of an AB_2 type of monomer. The monomer containing aryl halogen unit, a derivative of 2,6-dibromostyrene, which was polymerized in the presence palladium catalyst in DMF, as shown in scheme 2.28. The resulted HB polymer was soluble in organic solvents. The weight average molecular weight for the polymer was 1.3X10^4 Da with polydispersity 1.5. Again, the deprotection of the acetyl ester and subsequent oxidation of the phenoxyl groups resulted polyphenoxyl radicals, which is a high spin organic macromolecule.

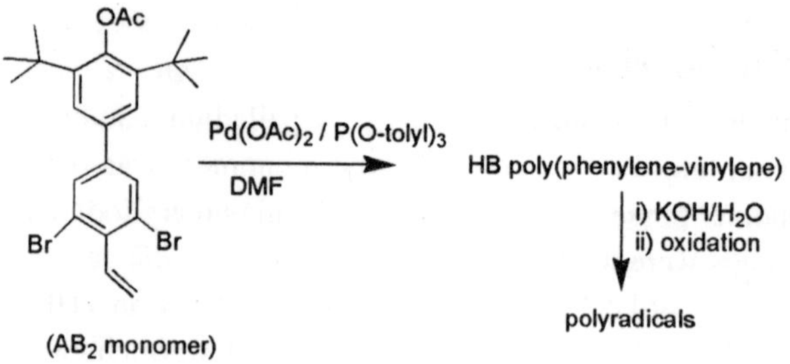

SCHEME 2.28 Synthesis of high spin organic macromolecules

2.3.4 Polystyrenes

The synthesis of HB polystyrene was reported [82-83] by cationic polymerization of 3-(1-chloroethyl)styrene in the presence of $SnCl_4$ and TBAB, as shown in scheme 2.29. The polymerization was performed in dry CH_2Cl_2 under nitrogen at (15 to 20)^{o}C. This self-condensation vinyl polymerization (SCVP) proceeds like a typical polycondensation reaction, an initial period of slow claim growth was followed by a period of exponential increase of the molecular weight. The addition of approximately equimolar amounts of the $SnCl_4$ to the AB* type monomer, where B* is a group capable of initiatives the polymerization of relatively high concentration resulted in the formation of high

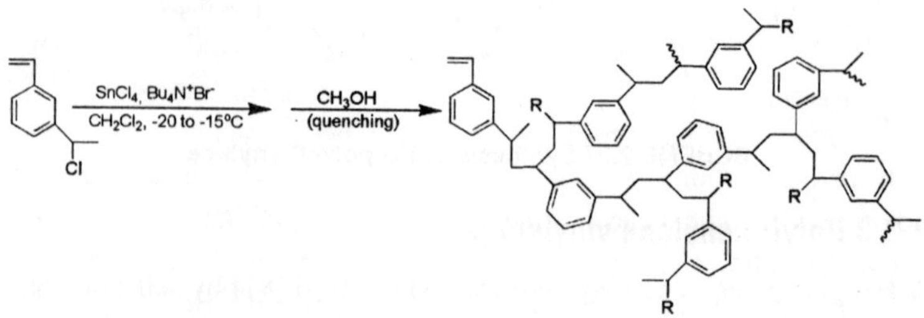

SCHEME 2.29 Synthesis of HB polystyrene derivative by SCVP technique

molecular weight polymers. GPC analysis indicates a wide range of molecular weights from $6.6X10^5$ to $1.7X10^4$ Da with high polydispersity, ranging from 2.9 to 9.8.

HB polystyrenes were also synthesized through the nitroxide mediated living free radical polymerization technique [84]. In this case, the AB* monomer used contains a polymerizable styrene group and an initiating/propagating moiety consisting of a nitroxide linked to a substituted benzylic carbon atom. The polymerization reaction was carried out by using bulk polymerization technique at 130°C under the inert argon atmosphere (Scheme 2.30). The thermal degradation of the AB* monomer generates a propagating benzyl radical and an inactive TEMPO radical, is a reversible system. No insoluble or crosslinked material was observed during the polymerization. The weight average molecular weight for the polymer was found to be 6000 Da with polydispersity 1.4 by SEC analysis. This HB polystyrene can be used as a multi functional initiator for living free radical polymerization to prepare multiarm star polymers.

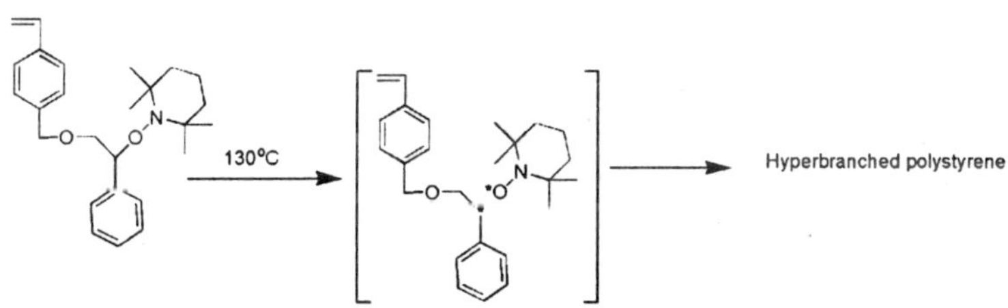

SCHEME 2.30 Synthesis of HB polystyrene by living free radical polymerization process

The synthesis of HB polystyrene by transition metal mediated living free radical polymerization process was also reported [85]. The polymerization of 4-(chloromethyl)styrene in the presence of Cu(I) and 2,2'-bipyridyl resulted HB polystyrene via atom transfer radial polymerization (ATRP) reaction (Scheme 2.31). The chlorine atom at the benzylic position is reversibly substracted by Cu(I) to form Cu(II)Cl and a benzylic radical capable of initiating the polymerization. Both the bulk and solution polymerization techniques were used to polymerize AB* monomer to get the desired HB polystyrene with high yield. The ^{1}HNMR spectrum indicates the presence of double bond in the polymer and used to determine the number average molecular weight, which is ranging from 1900-6280 Da. An almost linear polymer was obtained for the polymerization with a catalyst to monomer ratio of 0.01, while a high catalyst to monomer ratio resulted the formation of HB structure [86]. Gelation or crosslinking product was also formed through radical coupling reaction, when the concentration of the radicals was relatively high. There was a report on stepwise growth of HB polyhydrocarbon from silica surface. The reactions were monitored by XPS and ellipsometry. After reaction with divinylbenzene (DVB), the silicon surface thickness increased by 20A^{O}, much higher than the single monolayer addition (7A^{O}). This suggested that DVB polymerized from the surface by the radical polymerization process. Polymerization at the surface provides for a more complete surface coverage since the long chains fill the void region at the surface, but it loose the control over the film composition.

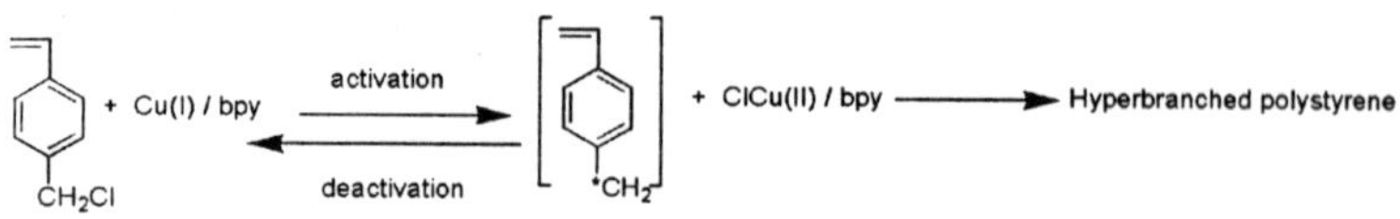

SCHEME 2.31 Synthesis of HB polystyrene by atom transfer radical polymerization technique

The synthesis of HB polystyrene by the anionic polymerization technique was reported [87]. At first, anionically produced linear polystyrene with DP 50 and polydispersity 1.1, was chloromethylated in a statistical process to approximately 30% of the pendent phenyl rings. That chloromethylated polystyrene reacted with living linear polystyrene to get first generation HB polystyrene. Repetition of the choloromethylation and subsequent reaction with living polystyrene resulted higher generation HB polymers. One of the most important advantages of this method is that the polymeric nature of the building blocks and the large number of branching points per chain lead to a rapid increase in molecular weight. The molecular weight obtained was about $4X10^4$ Da just after 2nd generation growth with polydispersity 1.29. However, the hydrodynamic volume of this HB polymer is 27 times lower than the corresponding linear polymer, indicates a compact structure.

2.3.5 Others

Weber and his coworkers [88-90] synthesized several HB polyhydrocarbons based on different acetophenone derivative by ruthenium hydride complex, $[(Ph_3P)_3RuH_2CO]$ and its coordinately unsaturated complex prepared by the hydrogenation of styrene. The acetyl group of the acetophenone derivatives (Fig.2.10) activates both the ortho C-H bonds for the addition reaction to olefins, catalyzed by ruthenium complexes to get the corresponding HB polymers (Scheme 2.32). Both Markovnikov and anti-Markovnikov orientations were observed as the component of the HB polymers and the ratio of composition was dependent on the reaction conditions and monomer structures. The HB polymer prepared from 4-acetylstyrene was soluble in THF and toluene but insoluble in methanol.

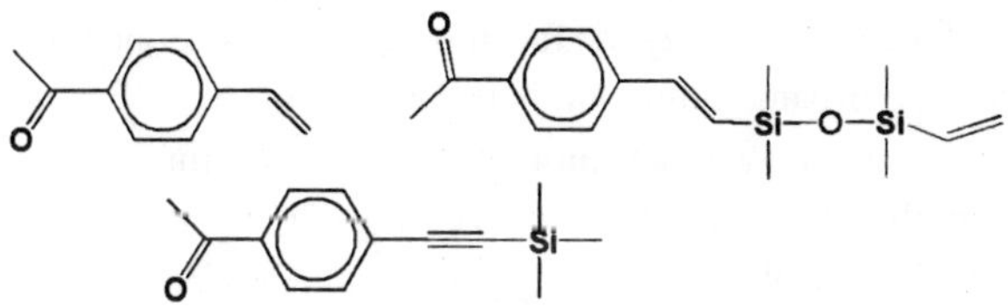

FIG.2.10 Structures of AB_2 monomers based on acetophenone derivatives

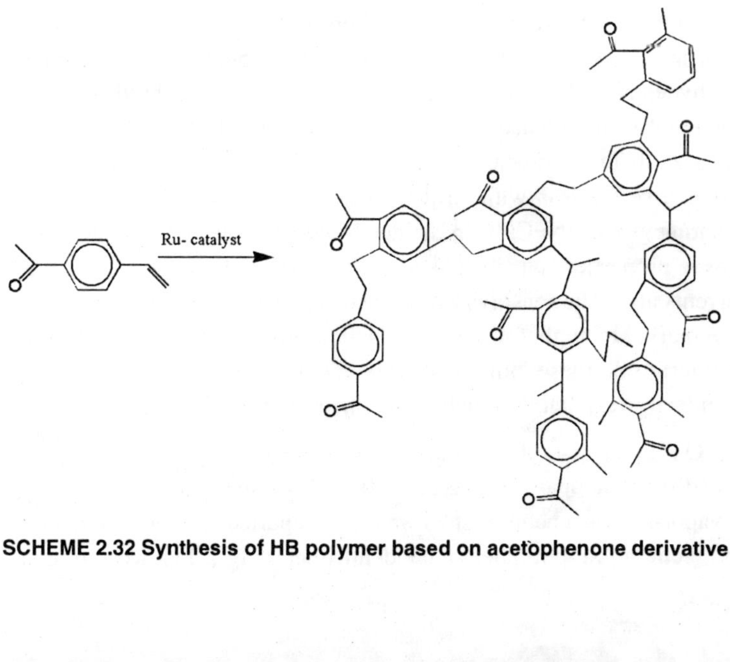

SCHEME 2.32 Synthesis of HB polymer based on acetophenone derivative

2.4 APPLICATIONS OF DENDRITIC POLYHYDRO-CARBONS

The possible applications of polyphenylene dendrimers were described [12]. Due to the presence of large number of benzene rings, this type of dendrimer distinguishes itself by high stiffness, shape persistence, and nanometer dimensions. This allows their use as chemically and thermally inert nano supports for chromophores, catalysis, also as hosts for guest molecules due to large stable cavities. The proper functionalization of the dendrimers opens up new avenues for their applications like redox systems, support for catalysts, drug delivery and biological recognition systems, etc.

2.4.1 Sensors

The polyphenylene dendrimers were utilized [91] as sensitive and selective QCMB sensor layers for monitoring the concentration of various volatile organic compounds (VOC) in different environments. Polyphenylene dendrimers are monodisperse containing internal voids because of their rigid framework, which are useful for this purpose. The 2nd generation unsubstituted polyphenylene dendrimer allows the selective detection with high sensitivity and accuracy of polar aromatic target molecules [91], such as acetophenone, aniline, benzaldehyde, benzonitrile, fluorobenzene, nitrobenzene, etc. However, neither chlorinated nor unsubstituted aliphatic hydrocarbons, alcohols, amines, aldehydes or carbonyl compounds could be monitored by this sensor. The number of included guest molecules depends both on the structure and size of the hydrocarbon dendrimers. Incorporation of functional groups like electron withdrawing groups, CN and COOH or electron donating groups, $N{=}CPh_2$ in the dendrimers enables fine tuning of the sensor properties, which makes it possible to detect even aliphatic solvents also. The sensitivity of these dendrimers coated QCMB sensor is remarkably high for aromatic and acetophenone (5 ppm level). Furthermore, these polyphenylene dendrimers are thermally and chemically very stable, which ensure their long service life.

On the layer-by-layer assembly of mechanically reinforced, cross linked gold nanoparticle/polyphenylene dendrimer films and their use as vapor-sensing chemiresistors were also reported [92]. A hydrophobic 2nd generation polyphenylene dendrimer (Fig.2.11) was used as crosslinking agent for assembling the gold nanoparticle dendrimer films to enhance the sensitivity to VOCs and to suppress undesired cross sensitivity to humidity. During the film deposition, the dendrimer served to crosslink the nanoparticles and thereby enabled the stepwise assembly of homogenous and mechanically reinforced films with precise control over the film thickness. When this film was used as a chemical resistor sensor device, high sensitivity to vapors of toluene and tetrachloroethylene was observed but its sensitivity to humidity was negligible. This high sensitivity to VOCs and low sensitivity to humidity make the material a promising candidate for sensor applications.

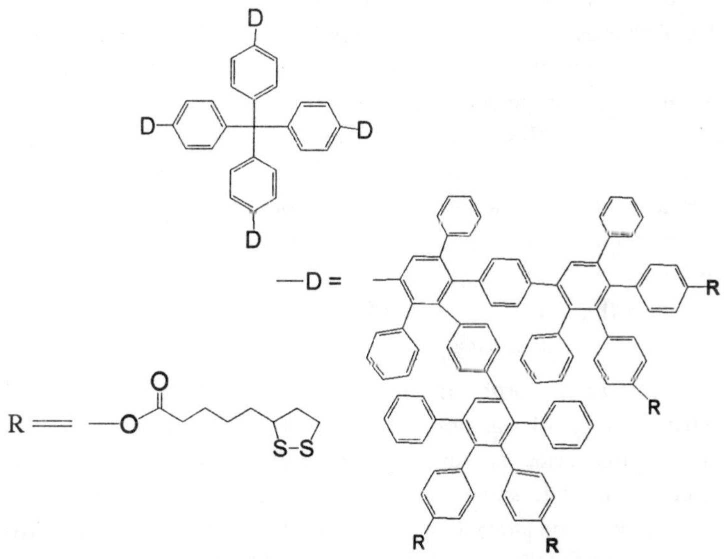

FIG.2.11 Structure of hydrophobic polyphenylene dendrimer

2.4.2 Nanocomposites

The polyphenylene dendrimers as the nanoscopic building blocks were reported [93]. They fabricated multilayer architecture by employing this dendrimers with chargeable groups at the surface and by using the polyelectrolyte deposition technique. Surface plasmon field enhanced fluorescence spectroscopy was used to monitor the luminescent properties of dendrimers with a phthalocyanine core integrated into the multilayer assembly. AFM and SEM micrographs demonstrated that the use of surface functionalized dendrimers in combination with Au-nanoparticles for the controlled assembly of hybrid aggregates as nanoscopic function devices.

2.4.3 Electronics

1st and 2nd generation *fac*-tris(2-phenylpyridine)iridium core based polyphenylene dendrimers with 2-ethylhexyloxy surface groups doped into a wide-gap 4,4'-*bis*(N-carbazole)biphenyl host were used [94] as a single-layer organic light emitting diode (OLED). A very efficient to green phosphorescence with a peak external quantum efficiency of 8.1% at a brightness of 3450 Cd/m^2 and a current density of 13.1 mA/cm^2 was observed from the above spin coated materials. A peak power–efficiency of 6.9 Im/W was measured at 1475 Cd/m^2 brightness and 5 mA/cm^2 current density. However, OLED based on neat film of the dendrimers displayed external quantum efficiencies lower than that of the blend device. This exceptionally high quantum efficiency for a single layer device to the excellent film forming properties and high photoluminescence quantum yield of the dendrimers blend along with efficient injection of charge into the emissive layer suggest that these dendrimers (Fig.2.12) can be used for producing efficient phosphorescent devices by spin coating.

A platinum porphyrin based phosphorescent hydrocarbon dendrimer (Fig.2.4) was used [19] for a triplet harvesting dopant in OLED for EL device. *Tris*(distryrylbenzenyl)amine core of 3rd generation dendrimer with stilbene units containing dendrons and *tert*-butyl surface groups, and another 3rd generation 1,3,5,-*tris*(distyrylbenzenyl) benzene cored dendrimer with same building blocks and surface groups were used as host materials (Fig.2.13). The common branching architecture around the emissive chromophore units of both guest and host materials help to achieve excellent miscibility and phases compatibility. The blends exhibit a large difference between EL and PL spectra. The films of the *tris*(distyrylbenzyl)amine core dendrimer emits green light, whereas other host emits blue light. Also the relative contribution of guest to host emission was found to be depended strongly on the energy level offsets of the two blend materials. Indicating strong trapping processes. The charge carriers are preferen-

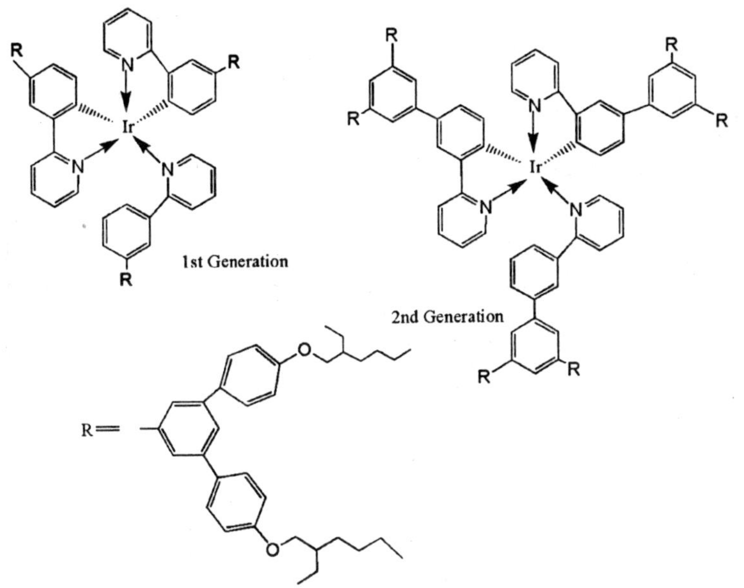

FIG.2.12 Structure of iridium phenylpyridine core phenylene dendrimer

tially trapped on the lower energy dopant site, which affects the ratio of EL emission from each component in the blend. The long phosphorescence lifetime of the guest porphyrin dendrimer gives rise to a

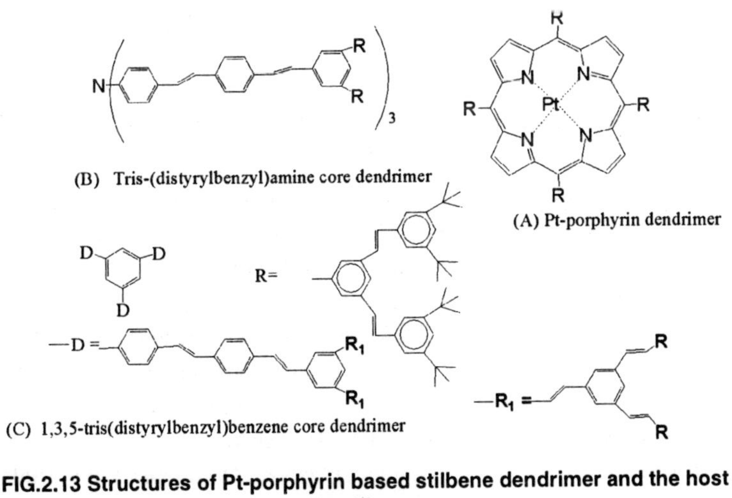

FIG.2.13 Structures of Pt-porphyrin based stilbene dendrimer and the host dendrimers

frequency dependent quenching process, which is independent of the host material. The use of conjugated dendrimers thus provides an elegant way of controlling the phosphorescence, trapping, and quenching of excitations in organic LEDs.

An attempt [95] to mimic natural photosynthetic antennae complexes for light harvesting material by using 1st generation aryl dendrimer with free base or Zn-porphyrin core and boron dipyrrin chromophore at the periphery (Fig.2.14) was made. The energy transfer process occurs with efficiency higher than 85% and time constants 2 and 20 ps. The study showed that the presence of two different conformations at excited state of the dendrimer through different in planarity of the chromophore units and its orientation with respect to the aryl rings.

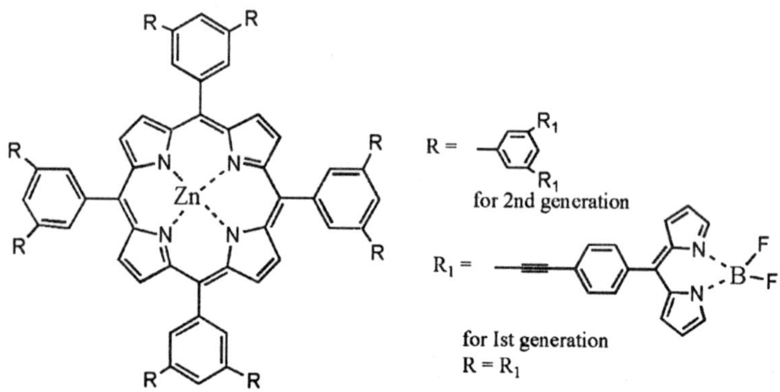

FIG.2.14 Structure of aryl dendrimer with Zn-porphyrin core and boron dipyridine surface groups

It has been found that the dendrimers based on phenyl acetylene chromophores exhibit highly efficient and unidirectional energy transfer and thus act as light harvesting materials [63, 96-98]. The nanostar is an example of a dendrimer with an energy gradient [63, 97] macromolecule (Fig.2.15). This nanostar has a phenyl acetylene backbone with a perylene trap located at core. The energy gradient was engineered by linking chromophore units of increasing length (2, 3 and 4 rings) in the direction of dendrimers periphery to the perylene trap. However, substitution at the meta position on the benzene rings suppresses the π-electron conjugation and leads to localize excitations within the different chromophore segment [96, 99]. The electronic excitations at the periphery then migrate down the gradient and funnel into the perylene trap. When these dendrons were irradiated at 310 nm (absorption band of the dendritic branching), the perylene chromophore emitted at 489 nm and 518 nm with quantum yield upto 98%. This indicates that singlet-singlet energy transfer occurs in the wedges with the increase of number of phenylacetylene moieties, the number of energy collecting sites increases and hence they can act as efficient light harvesting materials. However, the light harvesting ability increased with the increase of generation, while the efficiency of the energy transfer decreased with the same. This dendron has the specialty as it contained an in-build energy gradient through phenylacetylene spacers, which increases by one repeat unit as proceeding from the periphery to the core. That is, the levels of the localized electronic states decrease smoothly in energy from the exterior to the interior, creating a directional energy flow.

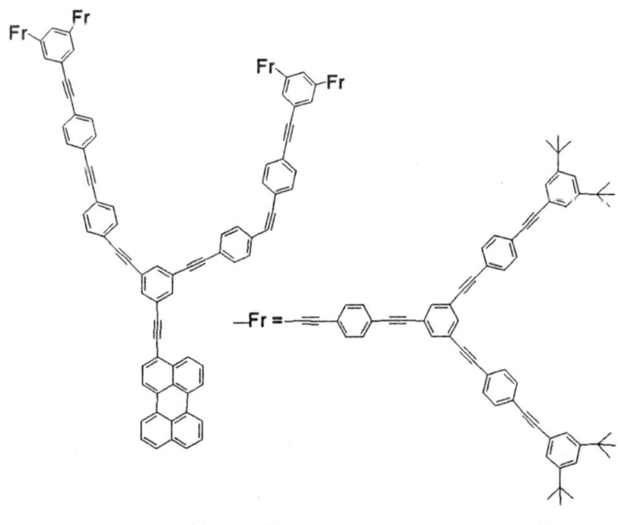

FIG.2.15 Structure of the nanostar phenyl acetylene dendrimer with perylene trapping core

A verity of di-, tri-, tetra-, and penta-meric porphyrin rings containing dendrimers [100-102] were used as light harvesting materials (Fig.2.16a). The dendrimers were synthesized through ethyne linkage between aryl groups on adjacent tetraarylporphyrin macrocycles. This arrangement resulted a semi rigid architecture that limits direct chromophoric interactions. In this arry of five porphyrins efficient singlet-singlet energy transfer from the peripheral Zn-containing units to the free base porphyrin core was observed. A photophysical property studies [101] of this dendrimer revealed that (i) singlet excited state energy transfer from the Zn porphyrin to the free base porphyrin was extremely efficient (95-99%), (ii) competitive electron transfer reactions were observed, (iii) the mechanism of energy transfer predominantly involved through–bond communication via the ethyne linkage, and (iv) energy transfer between two isoenergetic Zn-porphyrins was very fast.

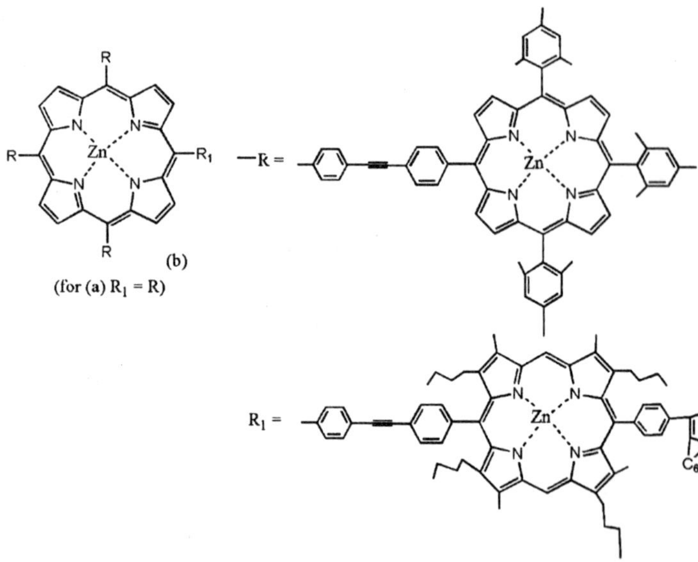

FIG.2.16 Structure of pentameric light harvesting porphyrin dendrimers

An almost similar type of structure of porphyrin containing aryl dendrimer with one fullerene unit attached with one peripheral free base porphyrin unit to form the $(PZn)_4$-(PH_2)-C_{60} hexad structure (Fig.2.16b) was used as a light harvesting material [103]. Here, the four Zn-porphyrin (PZn) units constitute a light harvesting antennae and the free base porphyrin [(PH_2)-C_{60}]-fullerene moiety plays the role of a reaction center for charge separation. The singlet-singlet energy transfer efficiency was very fast (K=3 X 10^{11} s^{-1}) with quantum yield 0.70. Again the charge separated state then return to the ground state with time constant 1.1 ns.

Setayesh *et al.* reported [104] no aggregating light emitting polymers based on polyfluorenes with polyphenylene dendron side chains. In this polyfluorene, the bulky polyphenylene dendrimer substituentes suppress the formation of long wavelength emitting aggregates and hence giving only pure blue emission. The absorption and emission spectral studies and molecular modeling confirmed that the bulky dendrimer side chains do not cause extra torson between the fluorene units. As light emitting diode was made from this material and found to produce blue emission with onset voltages below 4V, so this polymer can be used as light emitting material.

The poly(phenyl acetylene) dendrimers [105] around an optically pure diacetate of 4,4',6,6'-tetrabromo-1,1'-bi-2-naphnol (Fig. 2.17) can efficiently conduct energy migration from periphery to the core and are potentially useful in developing an enantioselective fluorescent sensor. The second generation dendrimer is a very useful catalyst for asymmetric alkylation of benzaldehyde with diethylzinc and the catalytic activity is much higher (98.6% conversion) than (S) BINOL

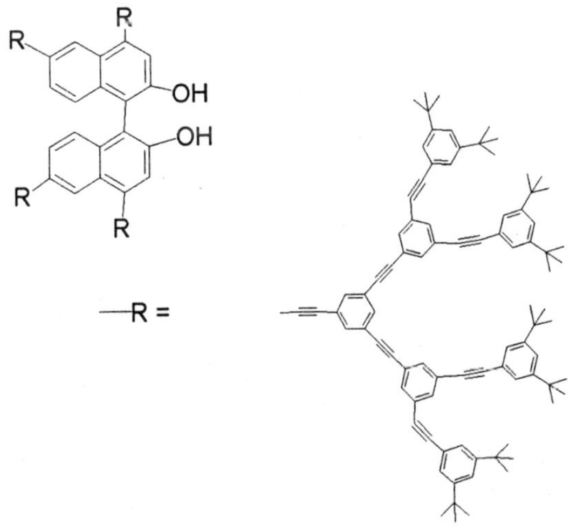

FIG.2.17 Structure of optically pure poly(phenyl acetylene) dendrimer

(37% conversion). Also product is an opposite enantiomer. The advantage of the dendrimer over BINOL is that it can be easily removed from the reaction mixture by precipitation with methanol.

The structurally shape persistent polyphenylene dendrimers with two and three different types of chromophores were used [17] as efficient energy transfer materials with an energy gradient. The dendrimer was bearing a TDI chromophore at the center and PMI as well as naphthalene dicarboxmonoimide chromophores at the periphery (Fig 2.18). The dendrimers containing three different chromophores (triad, A) cover the whole spectrum of visible light and show a stepwise energy transfer over a distant of 30A^{O} from the periphery via the scaffold towards the center of the dendrimers. The high thermal and chemical stability of the novel multi chromophore make these dendrimers attractive candidates as emissive layers for the design of novel photonic devices as well as for solar-cell applications.

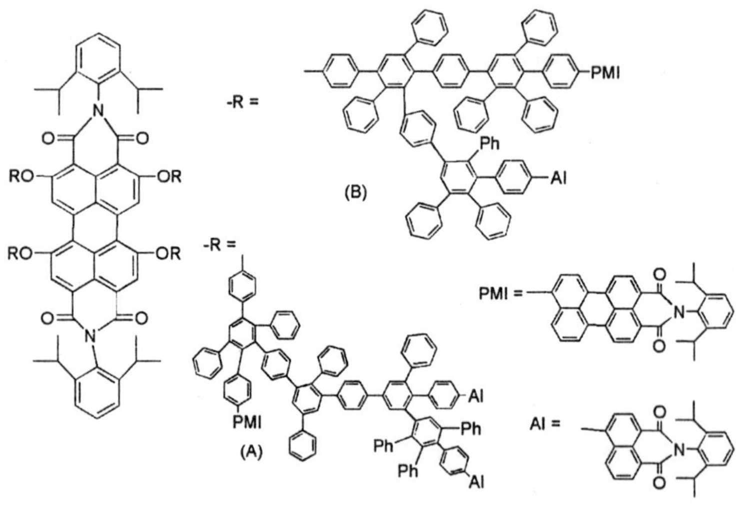

FIG.2.18 Structure of three different chromophore containing polyphenylene dendrimers

2.4.4 Polymeric Additives

The HB polyphenylenes were used as polymeric additives [78] in polystyrene to improve processing and mechanical properties. A mixture of PS and HB polyphenylene (5%) showed lower melt viscosity especially at high temperatures as well as shear rate upto 80% compared to pure PS. It also enhanced the thermal and shear induced stability of PS. However, the improvement of mechanical properties of the blend was not so significant except an initial increase of modulus.

2.4.5 Catalysis

A chiral aminoalcohol terminated polyphenylacetylene dendrimer was used [106] as enantioselective catalyst for the addition reaction of Et_2Zn to N-diphenylphosphinylimines. The enantioselectivity of the dendrimeric compound is same as parent chiral ligand with same rate of reaction. But advantage over the parent ligand is that it helps effective separation of the catalyst by the rigid backbone.

The optically active dendronized polymers were used [107] as a new type of macromolecular chiral catalyst for asymmetric catalytic reaction. These (1R, 2S) ephedrines contain optically active dendronized polymers containing less densely packed, lower generation dendrons with monomeric chiral units in their peripheries have a great potential for asymmetric catalysis. These polymers also overcome the densely packing problem of dendrimers without sacrificing the advantages of easy recovery by either filtration or precipitation. The dendronized polymers were used as chiral catalysts for the asymmetric addition of diethyl zinc to benzadehyde and found to be highly efficient compared to their corresponding linear polymeric and dendritic chiral catalyst.

2.5 SUMMARY

In this chapter the important hydrocarbon dendritic polymers such as polyphenylenes, poly(phenyl acetylene)s, stilbenoids, polyphenylene-vinylenes, etc. are described. They are mainly synthesized by a few important and useful synthetic reactions such as Diels-Alder, Wittig, Heck, Horner, Wadsworth-Emmons, etc. These dendritic polymers are easily characterized by ^{1}HNMR and ^{13}CNMR spectroscopy, as they have mostly chemically equivalent structures. They possess important and very useful properties like photophysical, photochemical, catalytic, etc. and thus they are very promising and useful materials for electronic, catalysis, sensors, etc. applications.

REFERENCES

[1] T.M.Miller, T.X. Neeman, R.Zayas and H.E.Bair, *J.Am.Chem. Soc.* **114**(1992)1018

[2] T.M.Miller and T.X. Neeman, *Chem.Mater.* **2**(1990)346

[3] U.M.Wieler, A.J. Berresheim, F.Morgenroth, G.Lieser and K.Mullen, *Macromolecules* **34**(2001)187

[4] F.Morgenroth, E.Reuther and K.Mullen, *Angew. Chem.Int Ed.* **36**(1997)631

[5] F.Morgenroth, C.Kubel and K. Mullen, *J.Mater.Chem.* **7**(1997)1207

[6] S.Loi, U.M. Wiesler, H.J.Butt and K.Mullen, *Chem.Commun.* (2000)1169

[7] H.Zhang, P.C.Grim, P.Foubert, T.Vosch, P.Vanoppen, U.M.Wiesler, A.J.Berresheim, K.Mullen and F.C.DeSchryver, *Langmuir* **16**(2000)9009

[8] F.Morgenroth and K.Mullen, *Tetrahedron* **53**(1997)15349

[9] A.J.Berresheim, M.Mullen and K.Mullen, *Chem.Rev.* **99**(1999)1747

[10] U.Wiesler and K.Mullen, *Chem.Commun.* (1999)2293

[11] F.Morgenroth, A.J.Berresheim, M.Wagner and K.Mullen, *Chem.Commun.* (1998)1139

[12] U.M.Wiesler, T.Weil and K.Mullen, *Top. Cur.Chem.* **212**(2001)1

[13] V.S.Iyer, K.Yoshimura, V.Enkelmann, R.Epsch, J.P.Rabe and K.Mullen, *Angew.Chem.Int.Ed.* **37**(1998)2696

[14] U.M.Wiesler, A.J.Berresheim, F.Morgenroth, G.Lieser and K.Mullen, *Macromolecules* **34**(2001)187

[15] S.M.Wavbright, K.McAlpine, M.Laskoshi, M.D.Smith and U.H.F.Bunz, *J.Am.Che.Soc.* **124**(2002)8661

[16] L.Fomina, P.Ponce, P.Guadarrama and S.Fomine, *Macromol.Theory Simulation* **8**(1999)403

[17] T.Weil, E.Reuther and K.Mullen, *Angew.Chem.Int.Ed.* **41**(2002)1900

[18] M.Lor, R.De, S.Jordens, G.De Belder, G.Schweitzer, M.Cotlet, J.Hofkens, T.Weil, A.Hermann and K.Mullen, *J.Phys.Chem.A* **106**(2002)2083

[19] J.M.Lupton, I.D.W.Samuel, M.J.Frampton, R.Beavington and P.L.Burn, *Adv.Func.Mater.* **11**(2001)287

[20] T.Gensch, J.Hofkens, A.Herrmann, K.T.Suda, W.Verheijen, T.Vosch, T.Christ, T.Basche, K.Mullen and F.C.DeSchryver, *Angew.Chem.Int.Ed.* **38**(1999)3752

[21] J.Hofkens, M.Maus, T.Gensch, T.Vosch, M.Cotlet, F.Kohn, A.Herrmann, K.Mullen and F.C. DeSchryver, *J.Am Chem.Soc.***122**(2000)9278

[22] J.Hofkens, L.Latterini, G.De Belder, T.Gensch, M.Maus, T.Vosch, Y.Karni, G.SchWeitzer, F.C.DeSchryver, A.Hermann and X.Mullen, *Chem.Phys.Lett.***304**(1999)1

[23] Y.Karni, S.Jordens, G.De Belder, J.Hofkens, G.Schweitzer and F.C.DeSchryver, *J.Phys.Chem.*B **103**(1999)9378

[24] A.Herrmann, T.Weil, V.Sinigersky, U.M.Weisler, T.Vosch, J.Hofkens, F.C.DeSchryver and K.Mullen *Chem.Eur.J.* **7**(2001)4844

[25] F.Morgenroth, C.Kubel, M.Moller, U.M.Wiesler, A.J.Berresheim M.Wagner and K.Mullen, *Carbon* **36**(1998)833

[26] G.DeBelder, G.Schweitzer, S.Jordens, M.Lor, S.Mitra, J.Hofkens, S.DeFeyter, M.Vander Auweraer, A.Herrmann, T.Weil, K.Mullen and F.C.DeSchryver, *Chem.Phys.Chem.* (2001)49

[27] F.Koehn, J.Hofkens, U.M.Wiesler, M.Cotlet, M.vander Auweraer, K.Mullen and F.C.DeSchryver, *Chem.Eur.J.* **7**(2001)4126

[28] T.Weil, U.M.Wiesler, A.Herrman, R.Bauer, J.Hofkens, F.C.DeSchryver and K.Mullen, *J.Am.Chem.Soc.* **123**(2001)8101

[29] M.Maus, R.De, M.Lor, T.Weil, S.Mitra, U.M.Wiesler, A.Herrmaan, J.Hofkens, T.Vosch, K.Mullen and F.C.DeSchryver, *J.Am.Chem.Soc.***123**(2001)7668

[30] M.Mans, S.Mitra, M.Lor, J.Hofkens, T.Weil, A.Herrman, K.Mullen and F.C.DeSchryver, *J.Phys.Chem.A* **105**(2001)3961

[31] Y.Karni, S.Jordens, G.De.Belder, G.Schweitzer, J.Hofkens, T.Gensch, M.Mans, F.C.DeSchryver, A.Herrmann and K.Mullen, *Chem.Phys.Lett.***310**(1999)73

[32] M.Imai and T.Arai, *Tetrahedron Lett.* **43**(2002)5265

[33] M.Drobizhev, A.Rebane, C.Sigel, E.H.Elandaloussi and C.W.Spangler, *Chem.Phys.Lett.* **325**(2000)375

[34] M.Iomai and T.Arai, *Tetrahedron Lett.* **43**(2002) 5265

[35] R.E.Bauer, V.Enkelmann, U.M.Wiesler, A.J.Berresheim and K.Mullen, *Chem.Eur.J.* **8**(2002)3858

[36] H.Meier and M.Lehmann, *Angew.Chem.Int.Ed.* **37**(1998)643

[37] M.Lehmann, B.Schartel, M.Hennecke and H.Meier, *Tetrahedron* **55**(1999)13377

[38] H.Meier, M.Lehmann and U.Kolb, *Chem.Eur.J.* **6**(2000)2462

[39] J.M.Lupton, I.D.W.Samuel, R.Beavington, P.L.Burn and H.Bassler, *Adv.Mater.* **13**(2001)258

[40] J.M.Lupton, J.D.W.Samuel, R.Beavington, M.J.Frampton, P.L.Burss and H.Bassler, *Phys.Rev.B* **63**(2003)155206

[41] G.Accorsi, N.Armaroli, J.F.Eckert and J.F.Nierengarten, *Tetrahedron Lett.* **43**(2002)65

[42] J.N.G.Pillow, M.Halim, J.M.Lupton, P.L.Burn and I.D.W.Samuel, *Macromolecules* **32**(1999)5985

[43] J.N.G.Pillow, P.L.Burn, I.D.W.Samuel and M.Halim, *Synthetic Met*als**102** (1999)1468

[44] S.K.Deb, T.M.Maddux and L.Yu, *J.Am.Chem.Soc.* **119**(1997)9079

[45] S.Sengupta, S.K.Sadhukhan, R.S.Singh and N.Pal, *Tetrahedron Lett.* **43**(2002)1117

[46] S.Sengupta and N.Pal, *Tetrahedron Lett.* **43**(2002)3517

[47] S.Sengupta and P.Purkayastha, *Org.Biomol.Chem.* **1**(2003)436

[48] O.Mongin, J.Brunel, L.Porres and M.Blanchard-Sesce, *Tetrahedron Lett.* **44**(2003)2813

[49] T.Kawaguchi, K.L.Walker, C.L.Wilkins and J.S.Moore, *J.Am.Chem.Soc.* **117**(1995)2159

[50] Z.Xu, M.Khar, K.L.Walker, C.L.Wilkins and J.S.Moore, *J.Am.Chem.Soc.* **116**(1994)4537

[51] C.Chi, J.Wu, X-Wang, X-Zhao, J.Li and F.Wang, *Tetrahedron Lett.* **42**(2001)2181

[52] D.Pesak, Ph.D.Theses, University of Illinois, 2002

[53] D.J.Pasak and J.S.Moore, *Angew.Chem.Int.Ed.* **36**(1997)1636

[54] V.D.Kleiman, J.S. Melinger and D.Mc Morrow, *J.Phys.Chem.B* **105**(2001)

[55] A.Zhu, P.Bharathi, H.G.Drickamer and J.S.Moore, *J.Polym.Sci.Part A Polym.Chem.* **39**(2001)2859

[56] C.Devadoss, P.Bharathi and J.S. Moore, *Angew.Chem.Int.Ed.* **36**(1997)1633

[57] C.B.Gorman, J.C.Smith, M.W.Hager, B.L.Parkhurst, H.Sierzputowska–Gracz and C.A.Haney, *J.Am.Chem.Soc. 121*(1999) 9958

[58] C.B.Gorman and J.C.Smith, *J.Am.Chem.Soc.* **122**(2000)9342

[59] A.M.McDonagh, M.G.Humphrey, M.Samacr and B.Luther-Davies, *Organometallics* **18**(1999)5195

[60] A.M.McDonagh, M.G.Humphrey, M.Samor and B.Luther-Davies, S.Houbrechts, T.Wada, H.Sasabe and A.Persoons, *J.Am.Chem.Soc.* **121**(1999)1405

[61] J.S.Moore and Z.Xu, *Macromolecules* **24**(1991)5893

[62] J.S.Moore and Z.Xu, *Angew Chem.Int.Ed.* **32**(1993)1354

[63] C.Devadoss, P.Bharathi and J.S. Moore, *J.Am.Chem.Soc.* **118**(1996)9635

[64] C.Devadoss, P.Bharathi and J.S.Moore, *Angew.Chem.Int.Ed.* **36**(1997)1709

[65] Z.Xu and J.S.Moore, *Angew.Chem.Int.Ed.***32**(1993)246

[66] P.Bharathi, U.Patel, T.Kawaguchi, D.J.Pesak and J.S.Moore, *Macromolecule* **28**(1995)5955

[67] S.C.Zimmerman, Y.Wang, P.Bharathi and J.S.Moore, *J.Am.Chem.Soc.* **120**(1998)2172

[68] K.Onitsuka, H.Kitajima, M.Fujimoto, A.Iuchi, F.Takei and S.Takahashi, *Chem.Commun.* (2002)2576

[69] K.Onitsuka, A.Shimizu and S.Takahashi, *Chem.Commun.* (2003)280

[70] S.Takahashi, K.Onitsuka and F.Takei, *Macromol.Symp.* **156** (2000)69

[71] A.G.Avent, P.R.Birkett, F.Paolucci, S.Roffia, R.Taylor and N.K.Wachter, *J.Chem.Soc.Perkin Trans.2* (2000)1409

[72] L.Shu, A.Schafer and A.D.Schluter, *Macromolecules*, **33**(2000)4321

[73] Z.Zhu and J.S.Moore, *J.Org.Chem.* **65**(2000)116

[74] Z.Zhu and J.S. Moore, *Macromolecules* **33**(2000)801

[75] A.Bashir-Hasheimi, H.Hart and D.L.Wart, *J.Am.Chem.Soc.* **108**(1986)6675

[76] H.Hart, A.Bashir-Hashemi, J.Luo and M.Meador, *Tetrahedron* **42**(1986)1641

[77] Y.H.Kim and O.W. Webster, *J.Am.Chem.Soc.* **112**(1990)4592

[78] Y.H.Kim and O.W. Webster, *Macromolecules* **25**(1992)5561

[79] Y.H.Kim and R.Backerbauer, *Macromolecules* 27(1994)1968

[80] C.Kim, Y.Chang and J.S.Kim, *Macromolecules* **29**(1996)6353

[81] H.Nishide, M.Nambo and E.Tushida, *Polym.Prepr.Jpn.* 47(1998)1670

[82] J.M.J.Frechet, M.Henmi,I.Gitsov, S. Aoshima, M.R.Leduc and R.B.Grubbs, *Science* **269**(1995)1080

[83] S.Aoshima, J.M.J. Frechet, R.B. Grubbs, M. Henmi and M.R.Leduc, *Polym.Prepr.* **36**(1995) 531

[84] C.T. Hawker, J.M.J. Frechet, R.B. Grubbs and J. Dao, *J.Am.Chem.Soc.***117**(1995)10763

[85] S.G.Gaynor, S.Edalman and K.Matyjaszewski, *Macromolecules* **29**(1996)1079

[86] M.W.Weimer, J.M.J.Frechet, and J.Gitsov, *J.Polym.Sci. PartA Polym.Chem.* **36**(1998)955

[87] M.Gauthier and M.Moller, *Macromolecules* **24**(1991)4548

[88] P.Lu, J.K.Paulasaari and W.P Weber, *Macromolecules* **29**(1996)8583

[89] T.M.Londergan and W.P.Weber, *Polym.Bull.* **40**(199815

[90] T.M.Longlergan, Y.You, M.E.Thompson and W.P.Weber, *Macromolecules* **31**(1998)2784

[91] M.Schlupp, T.Weil, A.J.Berresheim, U.M.Wiesler, J.Bargonand and K.Mullen, *Angew.Chem.Int.Ed.* **40**(2001)4011

[92] T.Vossmeyer, B.Guse, I.Bernard, R.E.Bauer, K.Mullen and A. Yasuda, *Adv.Mater.* **14**(2002)238

[93] J.L.Hernander–Lopez, R.E.Bauer, W.S.Chang, G.Glasser, D.Grebel-Koehler, M.K.Lapper, M.Kreiter, J.Leclaire, J.P.Majoral, S.Mittler, K.Mullen, K.Vasilev, T.Weil, J.Wu, T.Zhu and W.Knoll, *Mater.Sci.Eng.C* **23**(2003)267

[94] J.P.J.Markham, S-C.Lo, S.W.Magenuis, P.L.Burn and I.D.W Samuel, *Appl.PhysLett.* **80**(2002)2645

[95] F.Li, S.I.Yang, Y.Ciringh, J.Seth, C.H.Martin III, D.L.Singh, D.Kim, R.R.Birge, D.F.Bocian, D.Holten and J.S.Lindsey, *J.Am.Chem.Soc.* **120**(1995)10001

[96] R.Kopelman, M.Shortreed, Z.Y.Shi, W.Tan, Z.Xu, J.S.Moore, A.Bar–Haim and J.Klafter, *Phys.Rev.Lett.* **78**(1997)1239

[97] M.R.Shortreed, S.F.Swallen, Z.X.Shi, W.Tan, Z.Xu, C.Devadoss, J.S.Moore and R.Kopelman, *J.Phys.Chem. B* **101**(1997)6318

[98] Z.Xu and J.S.Moore, *Acta Polym.* **45**(1994)83

[99] S.Tretiak, V.Chernyak and S.Mukamel, *J.Phys.Chem.B* **102**(1998)3310

[100] R.W.Wanger, T.E.Johnson, and J.S.Lindsey, *J.Am.Chem.Soc.* **118**(1996)11166

[101] J.S.Hsiao, B.P.Krueger, R.W.Wanger, T.E.Johnson, J.K.Delaney, D.C.Mauzerall, G.R.Fleming, J.S.Lindsey, D.F.Bocian and R.J.Donohoe, *J.Am.Chem.Soc.* **118**(1996)11181

[102] P.G.van Pattern, A.P.Shreve, J.S.Lindsey and R.J.Donohoe, *J.Phys.Chem.B.* **102**(1998) 4209

[103] D.Kuciauskas, P.A.Liddell, T.E.Johnson, S.J.Weghorn, J.S.Lindsey, A.L.Moore, T.A.Moore and A.Gust, *J.Am.Chem.Soc.* **121**(1999)8604

[104] S.Setayesh, A.C.Grimsdale, T.Weil, V.Enkelmann, K.Mullen, F.Meghdedi, E.J.W.List and G.Leising, *J.Am.Chem.Soc.* **123**(2001)946

[105] Q.S.Hu, V.Pugh, M.Sabat and L.Pu, *J.Org.Chem.* **64**(1999)7528

[106] I.Sato, R.Kodaka, T.Shibata, Y.Hirokawa, N.Shirai, K.Ohtake, and K.Soai, *Tetrahedron Assymm.* **11**(2000)2271

[107] Q.S.Hu, C.Sun and C.E.Monaglan, *Tetrahedron Lett.* **43**(2002)927

CHAPTER 3

DENDRITIC POLYAMIDES

3.1 INTRODUCTION

Dendritic polyamides are similar with the biological polypeptide system, as both are formed through amide bonds. So, there is resemblance in synthetic methods and in properties. In fact, in 1981 Denkewalter [1] employed the same techniques of peptide chemistry i.e. activation and protection techniques to synthesize highly branched polylysine based first polyamide dendritic polymers, which having chiral centers in each branching unit. After wards many different types of polyamide dendritic polymers have been synthesized, which possess many synonymous properties to biological peptide molecules. By incorporation of different building blocks and core units, many unusual properties like liquid crystalline, redoxactive, photoactive, catalytic, etc. are also obtained. The most important advantage of the synthesis of this dendritic polymer is that the formation of amide bonds has been an extermelly well studied reaction of the past decades and there is a rich collection of suitable coupling reagents. Also they are especially important for their biological and medicinal applications, along with their other applications. For example, polyamide dendrimers such as branched peptide dendrimers also known as multiple antigen peptides (MAPs) have been synthesized to mimic proteins for applications like synthetic vaccines, serodiagnostics, peptide inhibitors, intracellular delivery vehicles, etc.

3.2 SYNTHESIS, CHARACTERIZATION AND PROPERTIES OF DIFFERENT TYPES OF POLYAMIDE DENDRIMERS

In the following section the synthesis, characterization and properties of different subclasses of polyamide dendrimers are discussed.

3.2.1 Polypeptides

These polypeptides dendrimers are formed mainly through peptide synthetic methods from amino acids or their derivatives. They are subdivided into three different categories, viz. poly(amino acid)s, poly(nucleic acid)s and other type polypeptides.

3.2.1.1 Poly(amino acid)s

The synthesis of poly(L-lysine) was patented [1-3] by Denkewalter and his coworkers. The poly(L-lysine) was synthesized by the stepwise repetitive protection-deprotection technique that was used for standard peptide coupling chemistry. Activated *p*-nitrophenyl ester of *bis*(Boc)-L-lysine as the propagating unit and benzhydrylamine as the initiator core were used for the synthesis. The synthetic strategy of polylysine dendrimer is shown in scheme 3.1. The amino protected N,N'-*bis*(Boc)-L-lysine nitrophenyl ester reacts with the initiator in the presence of DCC, followed by deprotection by acid catalyst, that is the dendrimer surface initially possesses hydrophobic 'Boc' units and then, after deprotection that are converted to hydrophilic amine moieties by this way they synthesized upto 10th generation dendrimer with molecular weight $\approx$ 50,000 g/mole. Although the dendrimeric structure is not

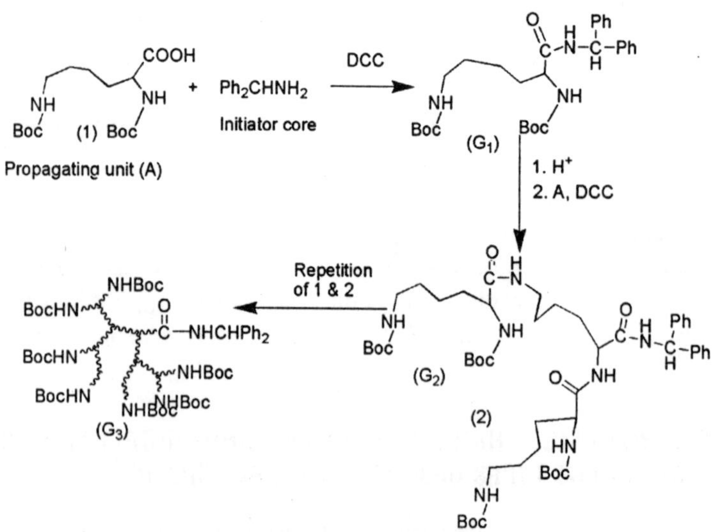

SCHEME 3.1 Synthesis of Boc-protected poly(L-lysine) dendrimers

rigorously verified, but subsequent work by Aharoni *et al.* [4-5] showed that the polymers are monodisperse and behave as 'nonaraing' spheres. Unlike the ideal dendritic polymers the branch junctures in this system are constructed from unequal lengths of branch segment. However, the physical properties of the branched poly(L-lysine) were found to be rather unremarkable [5]. The behavior observed is actually resemble to HB macromolecules [6] and either points to a large number of defects occurring during the synthesis and leading to a distorted ellipsoid shape.

These dendrimers were synthesized [7] on a Merrifield resin (solid support) followed by surface functionalization with various antigenic polypeptide sequences. The selected peptide antigen was then added to the resin bound lysine core matrix to create the MAP (multiple antigen peptides) dendron. However, this was not turned out to be a suitable procedure for obtaining dendritic macromolecular products with high purity. Tam *et al.* developed [8] more efficient approaches by using unprotected peptides as building blocks and chemoselective 'ligation' orthogonal coupling method. These included conjugation through thioalkylation, thio-disulfide exchange, thioester and oxime formation, thiozolidine and oxazolidine ring formation, hydrazone reverse proteolysis and fragment coupling (Scheme 3.2). The circular dichroism spectra indicated a stronger helical structure of the dendritic peptides compared with the peptide monomer VAZO [8].

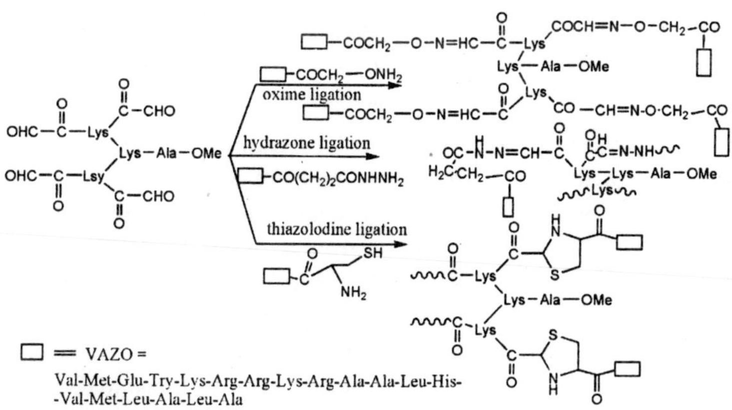

SCHEME 3.2 Lysine core based peptide coupled with model peptide VAZO by various ligations method

PEG was used as a polymer supports for the synthesis of Denkewelter type dendritic lysine upto 8th generation [9-10]. The characters of the dendrimers were changed completely by deprotection of hydrophilic NH_2 groups with trichoroacetic acid. The PEG molecule can also be cleaved easily from the dendrimer by saponification of the glycine ester bond. Polypeptide dendrimers of 1st and 2nd generation were synthesized [11] by the convergent growth method using L-glutamic acid, protected by Boc groups and activated as *bis*(succinyl) ester, as the branching units. The dendrimers were synthesized by treatment of the diester with L-diethyl-glutamate and subsequently deprotecting the N-termini of the resulting 1st generation dendron. Both the dendrimers were characterized fully by NMR, MS and elemental analysis and they are found to be monodisperse.

A solid phase synthesis of poly(L-lysine) was also reported [12]. The poly(L-lysine) dendrimer with PEG block copolymer of AB-type [13] and ABA-type [14] were synthesized by the liquid phase peptide synthetic method with a stepwise approach upto 4th generation. The triblock copolymer of poly(L-lysine) dendrimer-block-poly(ethylene-glycol)-block-poly(L-lysine) dendrimer was synthesized by using the following sequence of reactions as shown in scheme 3.3. Amine terminated PEG was used as polymeric supporter and poly(L-lysine) dendrimer was synthesized by the liquid phase peptide synthetic method by using fluoren-9-yl methoxycarbonyl (Fmoc) chemistry [15]. The coupling and deprotection reactions were performed 3 and 4 times for 3rd and 4th generation dendrimers. The products were characterized by ^{1}HNMR and MALDI-TOF MS analysis.

The synthesis of two dendritic polypeptides containing branched lysine core with upto 8 azobenzene moieties at the periphery was reported [16] on solid support by employing the ω-amino acid (aminomethyl phenylazobenzoic acid). The water solubility of the dendrimer was achieved by an introduction of peptidic tail consisting of an oligolysine portion. Thus the *cis/trans* isomerizations of the dendritic azobenzene compounds in both organic and aqueous media were studied. The *cis/trans* isomerization was found to occur in a reversible manner, at higher dilution even to extent that should allow photo modulation of molecular recognition processes. The synthesis of

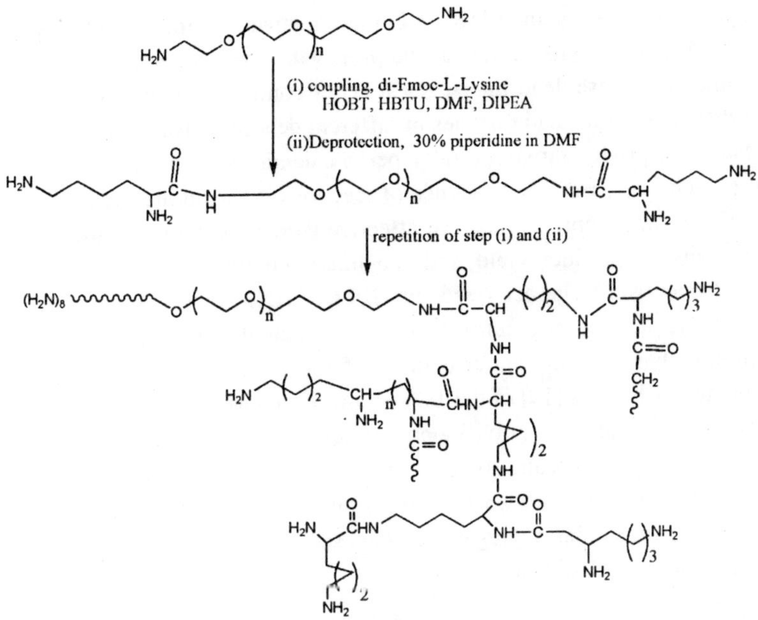

SCHEME 3.3 Synthesis of triblock poly(L-lysine) dendrimers by solid phase technique

these peptide dendrimers was performed on chlorotrityl resin adopting the Fmoc/t_{Bu} strategy. The products were formed with good yield and satisfactory homogeneity under optimized conditions without the need of non-trival purification. The dendrimers were characterized by UV, ^{1}HNMR, ^{13}CNMR spectroscopy and MALDI-TOF mass spectrometry. The polyamide dendrimers with polydispersity of 1.1-1.4 were successfully synthesized [17] from trimesic acid, 5-aminoisphthalic acid and 4-aminophenylpropionic acid as a core molecule, a dendron and a spacer unit respective by using a novel one pot reaction procedure. This synthesis technique involves successive activation of end carbonyl groups with the condensing agent, DBOP followed by condensation of active amide with 5-aminoisophthalic acid or 4-aminophenyl propionic acid and finally end capping with *p*-aniside. The synthesized dendrimers were characterized by ^{1}HNMR spectroscopy.

Four β-alanine based dendrimers with Boc-carbamate surface functionality were synthesized and characterized by the solution phase peptide coupling method [18]. 1,2-diaminoethane as the interior core and 3,5-diaminobenzoic acid as the propagating unit were used for the synthesis of these dendrimers. For a better control over the placement of different amino acid residues in different dendritic layer and a better chance of producing structurally perfect dendrimers, the convergent approach was used. The sequence of reactions is shown in scheme 3.4. This synthetic approach is more affective than the solid phase method in terms of product yield and the homogeneity. The synthesized dendrimer were characterized by HPLC, ^{1}HNMR, ^{13}C-NMR spectroscopy, mass spectrometry, GPC and elemental analysis. The NMR studies characterized the structures of the dendrimers and *syn/anti* rotamers behaviors [19] of the 1st generation dendrimer. ^{1}HNMR study also reveals that these poly(β-alanine) based dendrimers have an open structure in polar aprotic solvents, which allows the free interaction of the interior functionalities with solvent molecules. The dendrimers also form intermolecular aggregates of nanoscopic size in both non-polar and polar aprotic solvents.

The formation of the amide generations in the aromatic polyamide dendrimers required a highly efficient coupling method. The coupling of carboxylic acids with amines by using DCC [20-21], DCC-HOBT [22-24] and BOP-Cl [9, 25] as activating reagents has been used widely

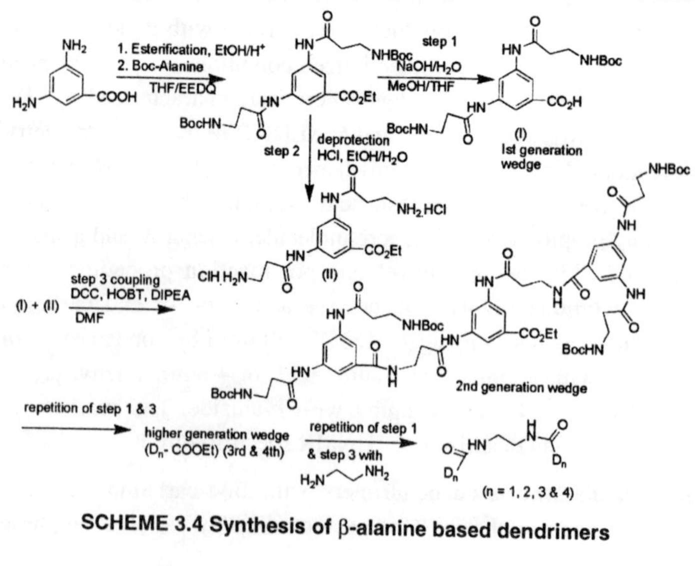

SCHEME 3.4 Synthesis of β-alanine based dendrimers

for synthesis of aliphatic polyamide dendrimers. The coupling of aromatic polyamide dendrimers using DCC-HOBT, BOP-Cl and DPPA [26] was studied, but DPPA was proved to be the most useful for this purpose. The polyamide dendrimers [27] upto 5th generation, were synthesized by the convergent method. The amino acid building blocks were derived from hydroxy benzoic acid derivatives and amino alcohol derivatives. The synthesis of polyamide dendrimers starting from 3,5-dihydroxybenzoic acid or chiral alcohol N-Boc-phenyl alaninol is described here (Scheme 3.5). The authors used different types of acids and amines (Fig.3.1) for synthesizing varieties of polyamide

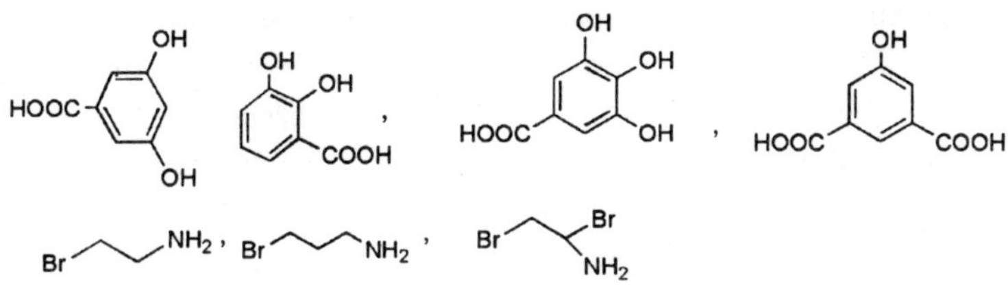

FIG.3.1 Structures of acids and amines used for synthesizing poly(amino acid) dendrimers

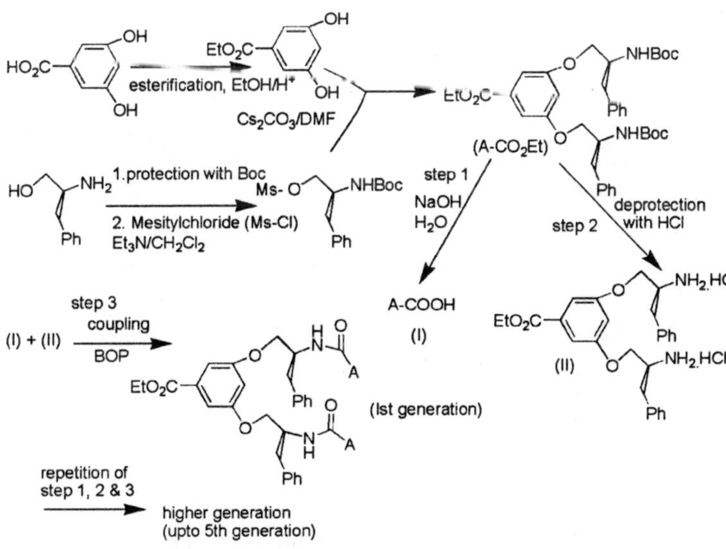

SCHEME 3.5 Synthesis of poly(amino acid) dendrimers based on 3,5-dihydroxy benzoic acid core

dendrimers. The products were purified by column chromatography and characterized by ^{1}HNMR, ^{13}CNMR, GPC analysis, MS and elemental analysis. The variation of aromatic residue and/or amino alcohol part of the dendrimers building blocks lead to considerable diversity in dendrimer, which can be easily extended to the synthesis of libraries by using combinatorial approaches. These dendrimers can also be synthesized by the solid phase divergent technique [27].

3.2.1.2 Poly(nucleic acid)s

The synthesis of poly(nucleic acid) dendrimers upto 3rd generation with 87 nucleotides were reported [28] by the convergent growth approach. The strategy employed was the solid phase synthetic method for oligo-nucleotide synthesis. A fully protected adenosine 2′,3′-O–*bis*(phosphoramidite) derivative was coupled with neighboring polymer bound oligomer to form the branching point. The synthesis starts from thymidine anchored to solid support long chain alkylamine controlled pore glass (LCAA-CPG) and assembled on the surface by the successive addition of standard deoxy thymidine phosphoramidity reagent in an automated chemical process. The branching junction was introduced in a sequence of two reactions involving branch formation by adenosine 2′,3′-O–*bis*(phosphoramidite) reagent with triazole and methyl cyanide followed by oxidation (I_2/H_2O) of the vicinal phosphite triester linkages to the most stable phosphate triester. The capping was performed after each monomer addition and branching step. The chain extension and branching steps were repeated to form successive generation each with twice as many chain ends as the previous generation. The oligomers were cleaved from the support and deprotected by the standard treatment with 29% NH_4OH. The steps of whole synthetic process are shown in scheme 3.6. The synthesized dendrimers were fully characterized by the conventional enzymes technique after purification by HPLC, Further a variety of biological activity, labeling in DNA blots [29], pharmaceuticals [30], signal processing, etc. were also reported for the dendrimers. The signals of blots that were probed using digonucleotides with dendrimers showed amplification of over 100 folds, when compounded to identical blots probed with specific oligonucleotide alone [29].

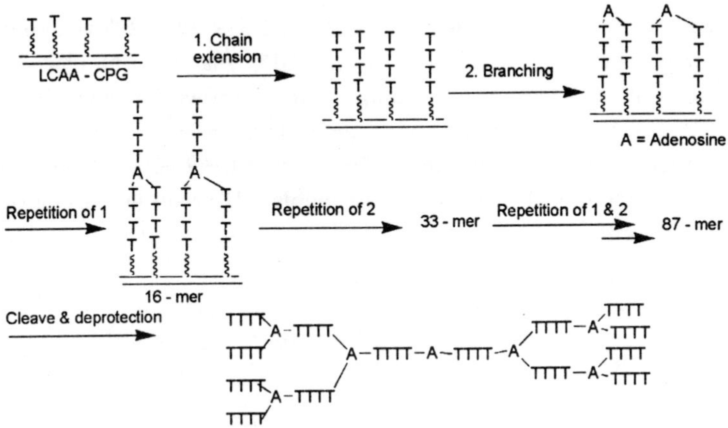

SCHEME 3.6 Synthesis of poly(nucleic acid) dendrimers

3.2.1.3 Other Polypeptides

The aliphatic polyamide dendrimers upto 3rd generation, were synthesized [31] by the convergent method using the selective reaction of aliphatic imidazolides. In this approach the steric problem was overcome by the increase of the size of the triamine branch unit and by using a bifunctional core molecule. The molecular weight of 3rd generation dendrimer was 4000 Da. The synthesis of dendritic polyamides by the convergent approach using N-(tert-butoxycarbonyl)-iminodipropionic acid and a protected amine was described by Uhrich and Frechet [32]. The dendrimers were synthesized by classical peptide chemistry with sequential coupling–deprotection steps to afford polyamides with molecular weight upto 5000 Da. The final step of the growth was carried out by condensation of the pre-final step growth dendrimers with a trifunctional acid chloride core to get the desired dendrimers. The resulted dendrimers were characterized by a variety of chromotographic and spectroscopic techniques [32].

Two series of dendritically modified [32] tryptophane derivatives were synthesized and the emission data measured in a range of different solvents. In the first series the NH groups of the indole ring available for H-bonding, whilst in the second series, NH groups were converted to NMe. The targeted dendritic branches were synthesized by the convergent coupling strategy as shown in scheme 3.7. All the products were successfully characterized by IR, ^{1}HNMR and ^{13}CNMR spectroscopy, mass spectrometry, TLC and GPC analysis. The emission spectra in non-hydrogen bonding solvents showed a red shift of hydrogen-bondable tryptophane dendrimer in contrasts to non-hydrogen bondable dendrimer. This proved the role of H-bonding between the branched shell and the indole NH group in causing the dendritic effect.

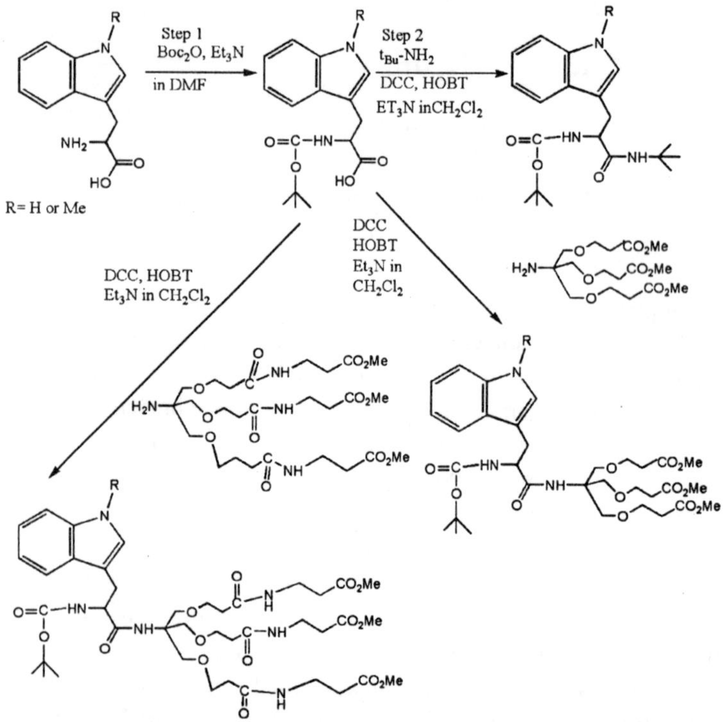

SCHEME 3.7 Synthesis of polypeptide dendrimers based on tryptophane derivatives

The synthesis of amino functionalized dendritic building block, i.e. dendron of the 1st and 2nd generations with differently protected (Boc and Cbz) amino groups in the periphery was reported [34]. These dendrons were used to synthesize the corresponding acrylic and methacrylic macromonomers. The first generation dendronized polymer was decorated at their peripheral amino group with the chiral amino acid L-phenyl alanine by attach to approach. The authors also reported the convergent synthesis of a series of low generation dendrons containing amino groups at the peripheral and carboxylic ester at the focal point [35-37].

The viscoelastic and other properties of t-butyl and methyl terminated amide based dendrimers (Fig. 3.2) upto 3rd generation were

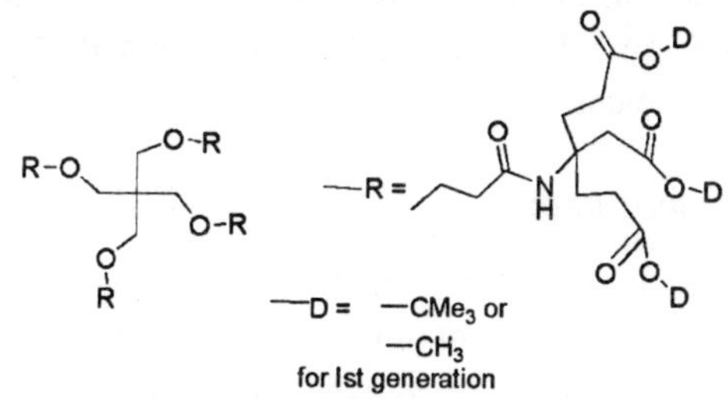

FIG.3.2 Structures of t-butyl and methyl terminated polyamide dendrimers

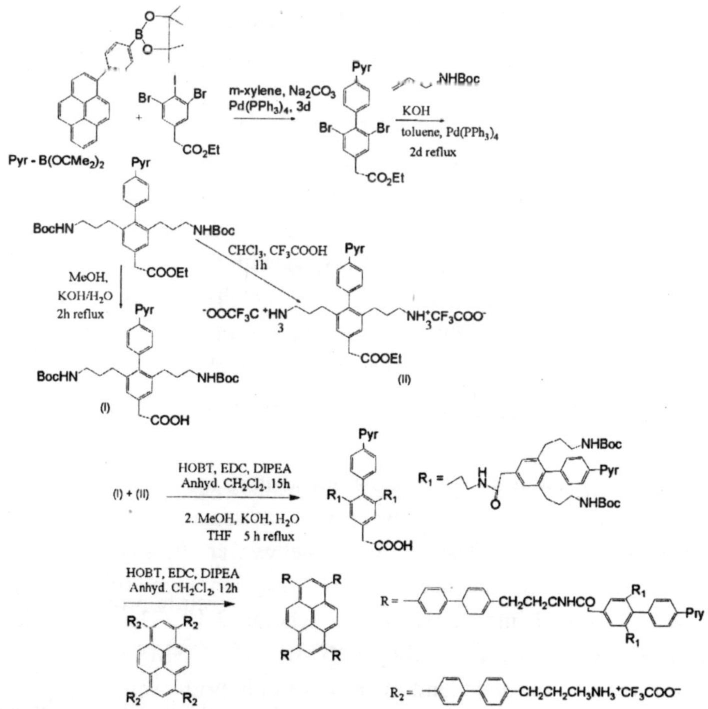

SCHEME 3.8 Synthesis of pyrene based polyamide dendrimers

studied [38] by MALDI-TOF MS, thermal analysis (TGA and DSC), XRD and dielectric analysis. This study revealed that the chemical nature i.e. functionality and branching complexity of the molecular structures influenced the ability of these macromolecules to assemble into crystalline lattices. A novel pyrene containing building block was efficiently synthesized to construct amide base dendrimers [39]. The sequence of reactions involved in this synthesis is shown in scheme 3.8. The products were fully characterized by ^{1}HNMR, ^{13}CNMR, mass spectrometry and elemental analysis.

3.2.2 Glycodendrimers

The synthesis of cluster galactosides constructed on β-CD (cyclodextrin) scaffolds containing 7 β-lactoses through different spacer arms, as well as three first generation dendrimers with same β-CD core, containing 14 β-D-galactosides, β-lactoses and β-melibioses were reported [40]. The synthesis of dendrimers were carried out by nucleophilic displacement reaction using compounds 1 and 4 as nucleophile precursors, which in combination with CD electrophiles 5 and 6 allow to generate the series of heptavalent lactosyl CDs, as shown in scheme 3.9. Similarly, other HB *β*-CDs were synthesized by another series of reactions, starting from 3,5'-*bis*(N,N'-benzyloxycarbonyl)-3,3'-*bis*(propylamine)imine. Finally, the glycodendrimers were obtained by reactions of these HB *β*-CDs with per-6-deoxy-6-iodo-β-CD (5) followed by removal of acetyl groups. These macromolecules were used to bind different guest molecules [40]. These HB β-CDs were characterized by NMR spectroscopic techniques with COSY, HMQC and selective TOSCY experiments and MALDI-TOF mass spectrometry.

The preparation of chitosan-sialodendrimer hybrid by using gallic acid as focal point and tri(ethyleneglycol) as spacer arm was reported by Sashiwa *et al.* [41]. The dendron moieties were first synthesized by using the following sequence of reactions, which were utilized then to get the sailodendrimer as shown in scheme 3.10. The synthesized dendrimers were characterized by ^{1}HNMR and ^{13}CNMR spectroscopic studies. Finally, the sialodendrimer-chitosan hybrids were prepared by efficient reductive amination of trivalents sialodendrimer aldehydes and

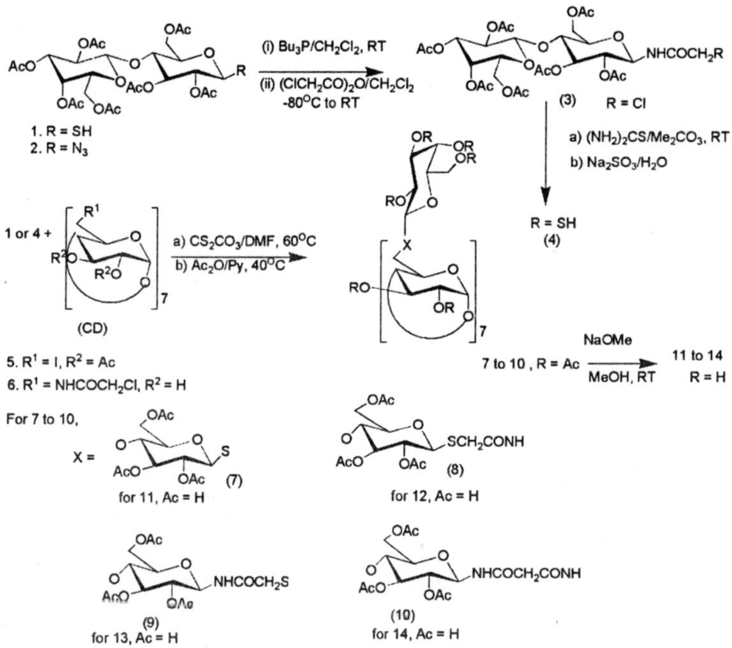

SCHEME 3.9 Synthesis of heptakis lactose β–cyclodextrin derivatives

chitosan with $NaCN.BH_3/AcOH$, as shown in scheme 3.11. The degree of substitution of this hybrid was about 0.13 as measured by NMR spectroscopy, which indicates that 87% of aldehyde-dendrimer (D_1)

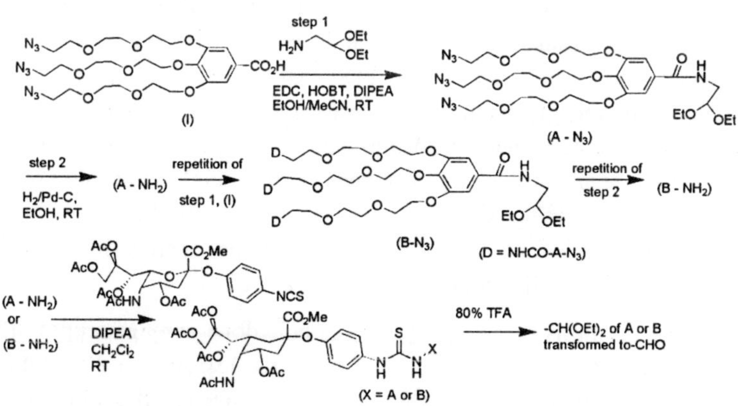

SCHEME 3.10 Synthesis of sialodendrimers

reacted to the primary amino group of chitosan. However, both the hybrids after deprotection by hydrolysis were only slightly soluble in neutral water and thus would not be useful in biological evaluation. Of course after succinylation, they become freely water-soluble.

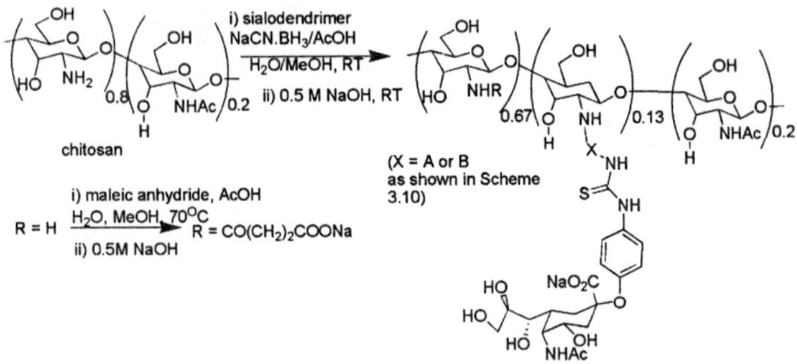

SCHEME 3.11 Synthesis of sialodendrimers-chitosan hybrids

Vrasidas *et al.* synthesized [42] lactose functionalized dendrimers and bound them to the cholera toxin B subunit (CTB). The dendrimers of 1st, 2nd and 3rd generations containing 2, 4, and 8 end groups, based on 3,5-di(2-aminoethoxy) benzoic acid as repeating unit. These dendrimers were coupled to lactose isothiocyanate to generate thiourea-linked glycodendrimers. The glycodendrimers were characterized by ^{13}CNMR spectroscopy and ESI-mass spectrometry. The binding affinities of glycodendrimers towards the cholera toxin B subunit were studied by fluorescence spectroscopy. CTB contains one tryptophane residue, the fluorescence of which is effected by the binding of carbohydrate ligands. The results indicated that the multivalent sugar moieties of the glycodendrimers bridges between toxin molecules. However, for the larger glycodendrimers, binding to a single molecule would not result in cooperativity because this single molecule could satisfy all the binding sites of one molecules of CTB, thus no sites of higher affinity would be available. Further no aggregation of the glycodendrimer-bridge-toxic was found in any case.

The synthesis of poly(L-lysine)-carbohydrate conjugate upto 4th generation was reported for the first time [12, 43]. At first the doubly branched poly(L-lysine) dendrimer were synthesized by the solid phase technique. Then the terminal amino groups were transformed into electrophilic N-chloroacetyl groups by treatment with N-chloroacetyl glycine hydroxy benzotriazol ester to give the corresponding chlorides. Then they were functionalized with peracetylated glycosyl derivatives bearing thiol groups and in a final step they were cleaved from the resin (Wang) support with trifluoroacetic acid. After deprotection the dendritic glycosyl compounds (Fig.3.3) were obtained carrying α-thiosialoside, A [44], β-D-lactoside, B, N-acetyl lactosamines, C [45], 1-thio-β-D-N-actylglycosamines, D [46], α-D-mannosides, E [47] or

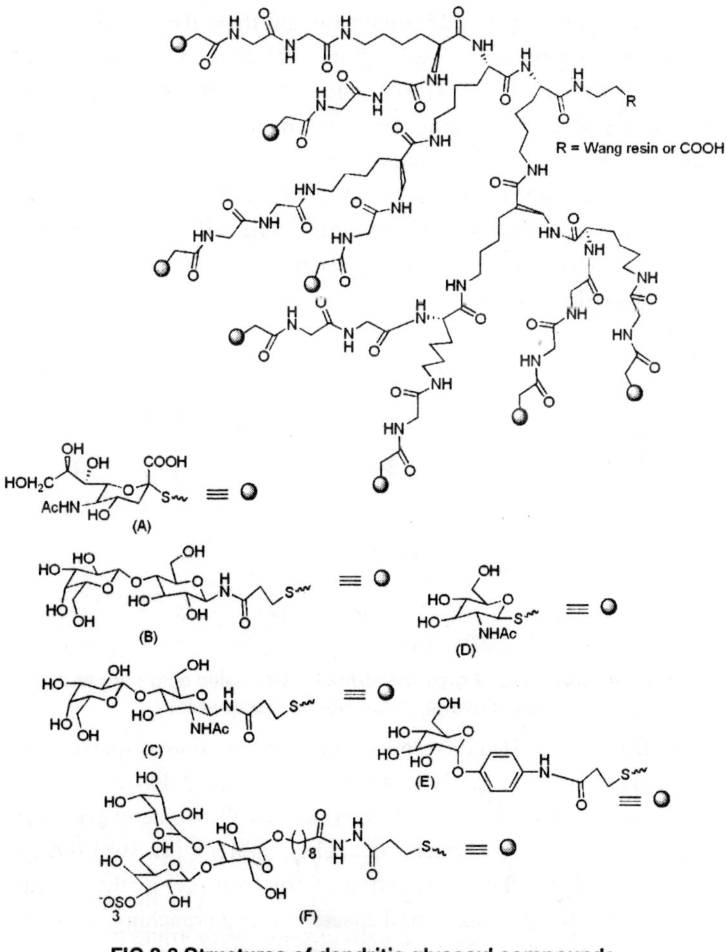

FIG.3.3 Structures of dendritic glycosyl compounds

3'-sulfo Lewis^x glycosides, F [48]. All the products were fully characterized mainly by ^{1}HNMR and mass spectrometry. The glycodendrimer C was also synthesized [45] by using an enzymatic approach starting from peracetylated 1-S-acetyl--thio-β-D-N-acetylglycosamine. The interesting result was obtained from all the lysine based dendrimer when they were tested in enzyme-linked lectin-assays (ELLA) and showed strong biological activities. The compound F containing 3'-sulfo-Lewis^x-epitope was found especially to be very active and showing an IC_{50} (concentration causing 50% inhibition) value towards L-selectin of ca.~1μM, 625 times higher than the corresponding monovalent compound. The authors [49] also synthesized triply branched dendrimer (Fig.3. 4) with gallic acid as core and tetraethylene glycol amine spacers by using the above-mentioned functionalization with thiolated glycosyl derivatives. Similarly, the compounds built from phosphotriester backbones were also synthesized [50] and they were found to be inhibitors of *vicia villosa* binding to asialoglycophorin. The results indicated a 3 to 10 folds enhanced affinity of the dendrimers.

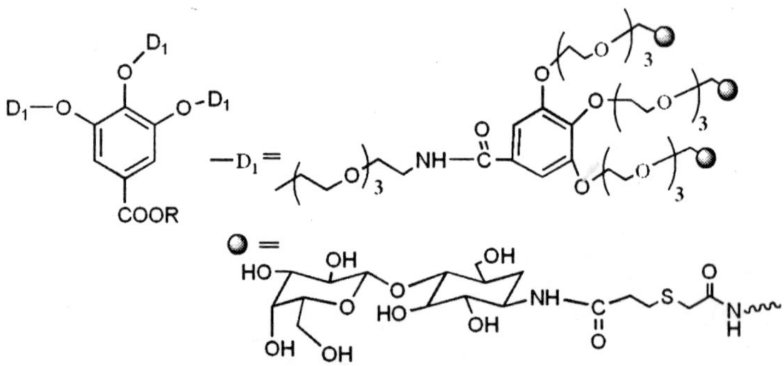

FIG.3.4 Structures of glycodendrimers with gallic acid core and tetraethylene glycol amine spacer unit

Stoddart *et al.* [51] convergently synthesized monodisperse glycodendrimers by using 1,3,5-benzenetricarboxylic acid derivative as the core, 3,3'-iminodipropionic acid derivative as the branching unit and glycine derivative as the spacer unit. Their strategy involved the synthesis of a triglycosylated derivative of *tris*(hydroxyl methyl) methylamine (TRIS), the introduction of spacer unit and branching units on to the *tris*-derivative by amide bond formation subsequently coupling of these dendrons with the central core unit offered the desired dendrimer. (Fig.3.5 for 18-mer).

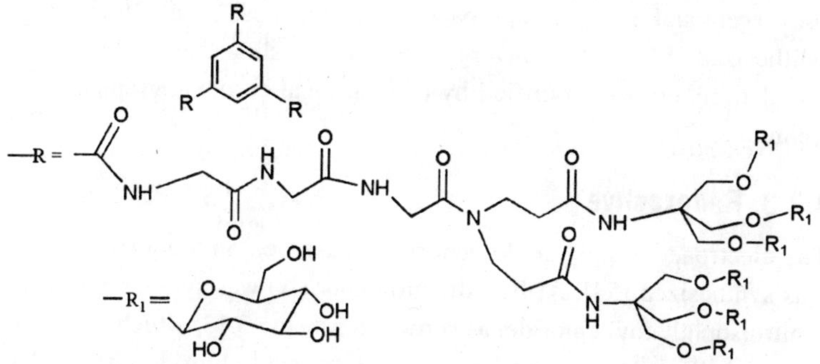

FIG.3.5 Structure of 18-mer glycodendrimer with 1,3,5-benzene tricarboxylic acid core unit

There are two limiting designs under the broad heading of glycodendrimers [52]. The first comprises a non-carbohydrate dendritic scaffold with saccharides present only at the periphery and the second is a fully carbohydrate dendrimer in which the internal saccharides form the branching points of the polymer. Although there are many possible variations between these two extremes, most of the glycoden

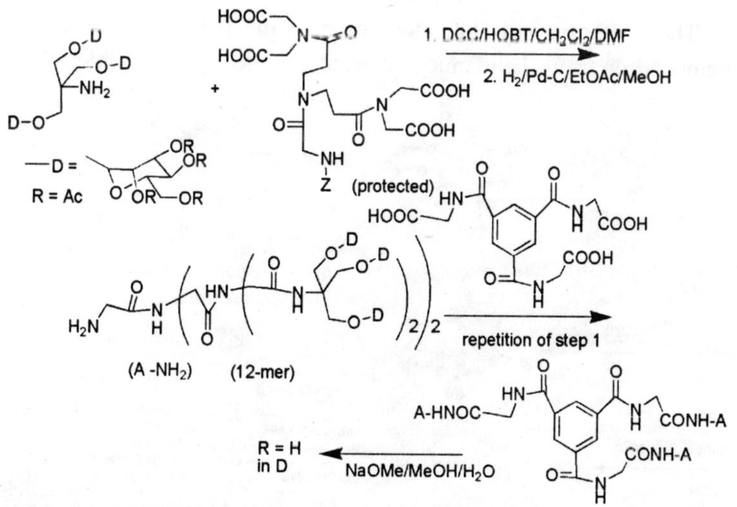

SCHEME 3.12 Synthesis of mannose based glycodendrimers (36-mer) by the convergent method

drimers researches reported on the synthesis of carbohydrate coated dendrimers. Ashton reported a number glycodendrimers by both convergent and divergent approaches. The mannose based 36-mer was synthesized [53] in the convergent fashion as shown in scheme 3.12. The dendrimers were purified by conventional chromatographic GPC technique.

3.2.3. Redoxactive

The electroactive polyamide dendrimer based on anthraquinone units was synthesized [54]. At first the monomer (A) was synthesized from 5-nitroisophthaloyl chloride, as shown in scheme 3.13, which was then treated with the four-directional core, tetra acid chloride (B), in dry THF to get the desired dendrimers (Scheme 3.13). All the products in each step were characterized by IR, ^{1}HNMR, ^{13}CNMR spectroscopy, elemental analysis and MALDI-TOF MS. The electrochemical properties of the dendrimer were studied and it was suggested that anthraquinone moieties can undergo intramolecular or intermolecular chemical reactions. The consequence of this, an induced excitation signal is relevant for the design and synthesis of molecular device and catalyst.

The synthesis and characterization of the H-bonded polyamide monodendron from chelidamic acid were reported [55]. 4-Chloropyri-

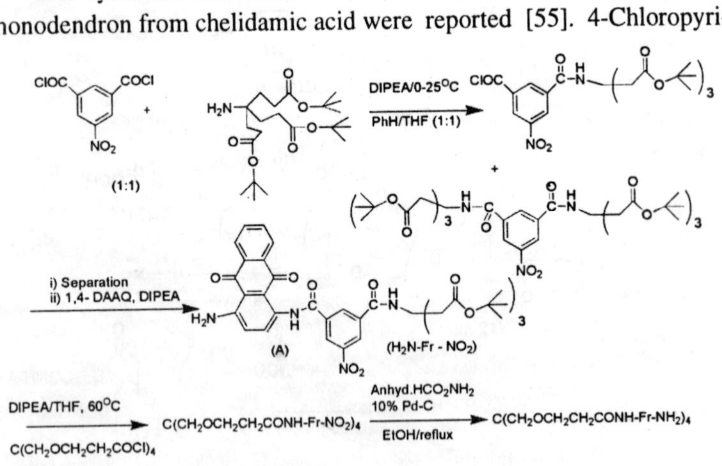

SCHEME 3.13 Synthesis of anthraquinone based electroactive polyamide dendrimers

dine-2-6-dicarbonyl chloride (A) was used as the branching monomer and from this monomer dendrons were synthesized by the convergent approach using amide bond formation reaction with the acid chlorides. The focal point activation occurs by NaN_3 displacement of the 4-chloro group followed by hydrogenation reaction. The synthesized monodisperse dendrons were characterized by NMR spectroscopy, SEC analysis and FAB or MALDI-TOF mass spectrometry. Both the dendrons having hydrogen bonding between the amide N-H and the pyridine N, with PyN----H bond distances ranging from 2.13 to 2.23 A°. Further more, two of the four terminal ester carbonyl oxygen atoms participate in additional H-bonding with the outer amide NH's. This network of H-bonds serves to restrict the branches of each monodendron to a cisoid conformation. Thus these hydrogen bonded dendritic networks offer a new scaffold for designing intramolecularly ordered macromolecules. The synthesis of Co(III) based self-assembling polyamide metallodendrimers by using the convergent approach were also reported [56]. At first a *tert*-butyl ester terminated diamine was synthesized, which was then used to synthesize a six-arm ethylene diamine backbone based dendron. The two-arm to the six-arm dendrons were used as bidentate ligands self-assembled around a Co(III) center to get the desired metallodendrimers. The whole sequence of reactions

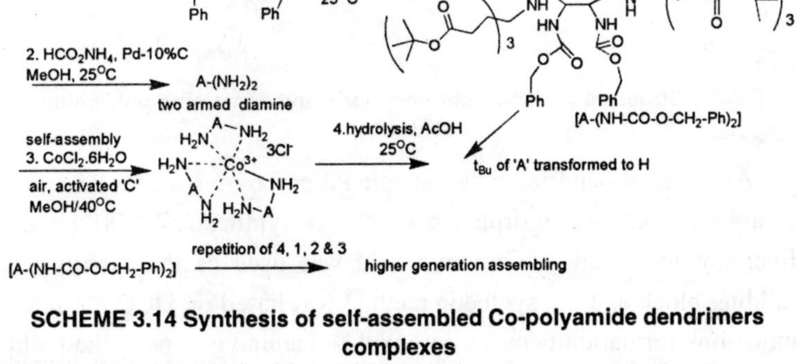

SCHEME 3.14 Synthesis of self-assembled Co-polyamide dendrimers complexes

is shown in scheme 3.14. The products were characterized after purification through column chromatography by ^{1}HNMR, ^{13}CNMR, FT-IR, UV-visible spectroscopy, elemental analysis and FAB or ESI mass spectrometry.

3.2.4. Photoactive

The complexation of the luminescent Nd^{3+}, Eu^{3+}, Gd^{3+}, Tb^{3+}, Er^{3+} and Yb^{3+} ions by a polylysin dendrimer containing 21 amide groups in the interior and in the periphery 2,4-chromophoric dansyl units (Fig.3.6), which showed an intense fluorescence band in the visible region were investigated [57]. These complexes were characterized by very different energy level patterns and redox properties, and capable of exhibiting their own luminescence. The formation of metal complexes were accompanied by quenching of the fluorescent excited state of the dansyl units. The study revealed that quenching effect was very large for Nd^{3+} and Eu^{3+}, moderate for Er^{3+} and Yb^{3+}, small for Tb^{3+} and very small for Gd^{3+}.

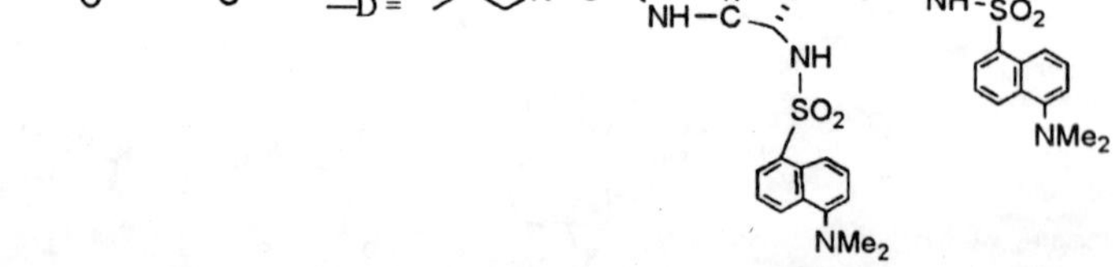

FIG.3.6 Structure of dansyl chromophoric unit containing polyamide dendrimer

As series of dendritic polyglutamic Pd-porphyrins with a Pd-*meso*-tetra-4-carboxyphenylporphyrin core was synthesized [58] by the divergent approach. L-Glutamic acid was used as the monomeric building block and the synthetic method was based on DCC promoted amide link formation between carboxyl and amino groups of the building block. The Pd-*meso*-tetra-4-carboxyphenyl prophyrin (PdTCPP) was used as the core. The synthesis of the ester and acid terminated dendrimers upto generation 4 was performed by sequential repetition of steps 1 and 2 as shown in scheme 3.15. The 1st generation dendrimer was also end capped with 11-amino undecanoic acid by a DCC coupling reaction in order to increase the flexibility of the shell surrounding the porphyrin core. The synthesized dendrimers were characterized by ^{1}HNMR and ^{13}CNMR spectroscopy and MALDI-TOF mass spectrometry. However, the adequate characterization of the 4th generation dendrimer was found to be difficult. All these dendrimers exhibit phosphorescence strongly at λ_{max} = 690 nm with lifetimes in the range of 0.5-1ms in deoxygenated solution. The time resolved phosphorometry study demonstrated that polyglutamic dendritic layers partially protected the porphyrin central core, even when the surrounding shells are not yet fully closed in topological sense. However, the changes in oxygen quenching constants for the Pd-porphyrin dendrimers of different sizes are significantly dependent on the solvent.

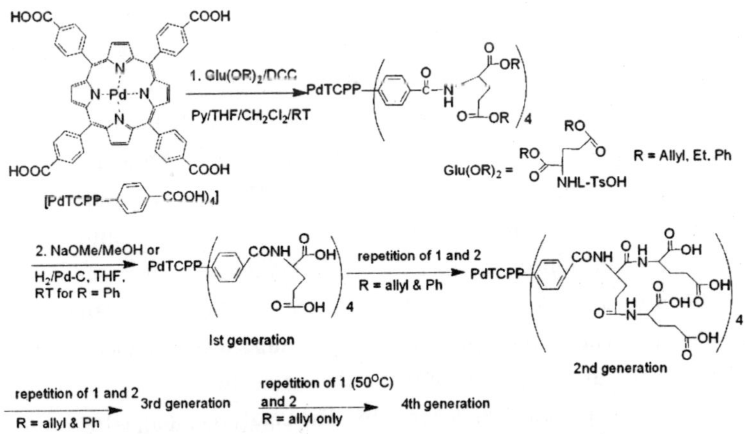

SCHEME 3.15 Synthesis of dendritic polyglutamic porphyrins

Another series of polyglutamic free base *meso*-tetra-4-carboxyphenyl porphyrin (H_2TCPP) dendrimers upto 3rd generation were also synthesized [59] by using the same synthetic strategy as described earlier. The synthesized dendrimers were characterized by using ^{1}HNMR and ^{13}CNMR spectrocopies, and MALDI mass spectrometry. The free acid terminated dendrimers were found to be highly soluble in water and their absorption and fluorescence spectra were dependent on pH. At pH values less than 3.5 noticeable changes in fluorescence intensity and quantum yield even for the highly soluble dendrimer were found. The results [59] suggested that interactions between individual dendritic molecules in solution be favored by full protonation of the peripheral glutamic carboxyls. An interesting result of the first protonation constant of 3rd generation dendrimer was 6.98, which was appropriate for board pH sensing in biological systems. This combined with almost unlimited water solubility, make this through the biological membranes, and the high quantum yields of fluorescence make these dendrimer a promising new pH indicator.

3.2.5 Chiral

The syntheses of a family of polyamide chiral dendrimers upto 2nd generation by the convergent technique were reported [60-61]. The dendrimers were synthesized by peptide coupling technique using a series of organic reactions [62-65] with good yields as shown in scheme 3.16 (only one series) . The chiroptical studies of the protected THF and $CHCl_3$ soluble dendrimers indicate that the solution shapes of these dendrimers were decongested with little steric interaction between different parts of the dendritic structures. The amine groups were protected by Boc and acid groups by methyl ester formation. The selective deprotection of either free carboxylic acid or free amines or both can be unmarked the structures, at the surface or throughout the dendrimers. The chiroptical property i.e. circular dichroism and optical rotation are useful parameters for the determination of the gross solution structure of chiral dendrimers. The catalyst employed was Rh'-(S,S)-Et-DuPHOS benzene, TFFH, and DIPEA complex known to produce amino acids of the (S)-configuration with very high steroselectivity [66]. The dendritic wedge was then synthesized from the monomer as shown in scheme 3.16.

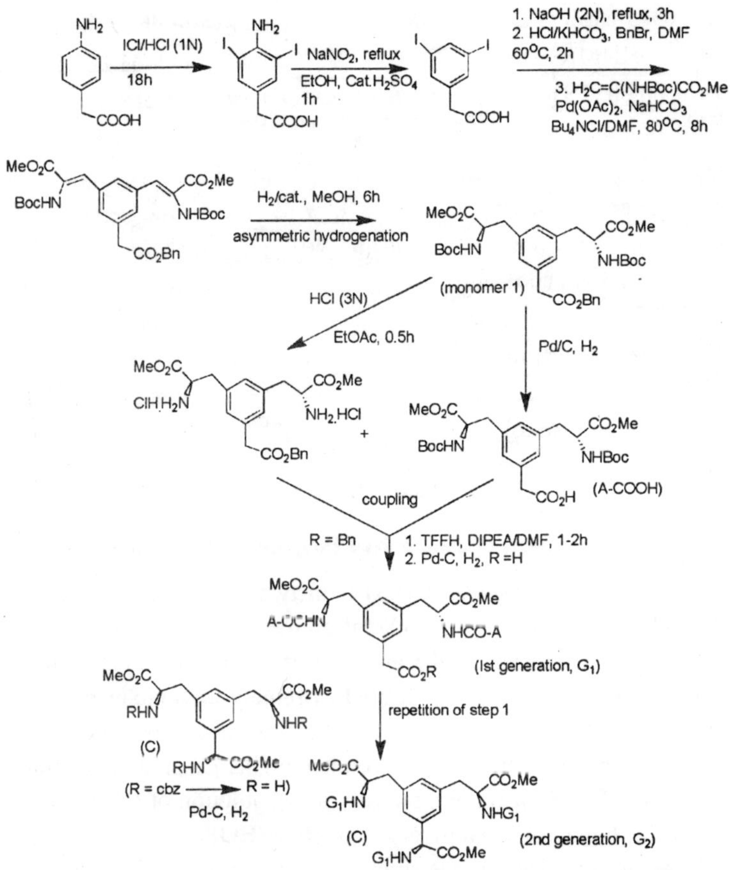

SCHEME 3.16 Synthesis of chiral polyamide dendrimers starting from *p*-aminobenzoic acid

3.2.6 Dendronized

Amine terminated 1st and 2nd generation dendronized macromonomers based on methyl acrylate unit were synthesized very efficiently [67]. The synthesis of 1st generation dendron was carried out by sequence of reactions with good yields starting from 3-bromo-4-hydroxy benzaldehyde as shown in scheme 3.17.

The synthesis of 4th generation surface functionalized dendronized polystyrene was reported [68] by the divergent approach. The 1st

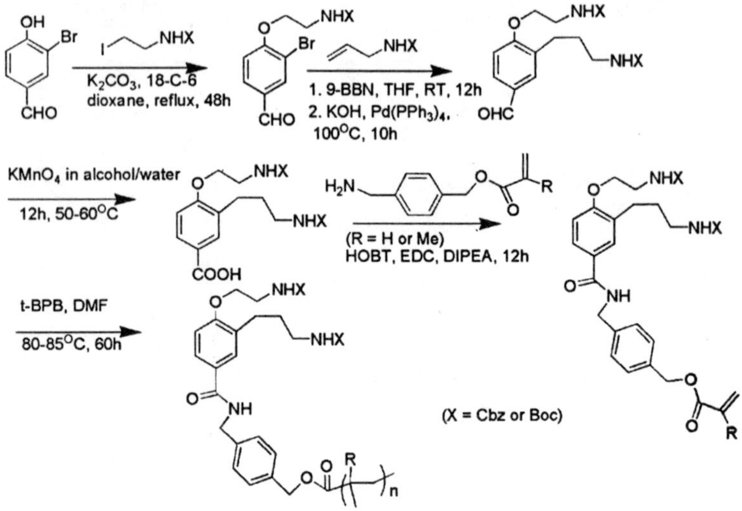

SCHEME 3.17 Synthesis of dendronized 1st generation polyamide dendron

generation dendronized PS (G_1PS) with two trimethylsilyl ethyloxy carboxyl (Teoc) protected peripheral amine groups was used as the starting material and activated ester of 1st generation dendrons A and B as propagation unit to get the desired extremely high molar mass 4th generation dendronized PS. The iterative deprotection of dendronized PS followed by amidation reaction resulted higher generation, as shown in scheme 3.18. Spin coating for very dilute solution of (G_1-G_4)PS in THF on the highly oriented pyrolytic graphite (HOPG) was used to take the SFM images. The images showed that the molecules adsorb with the molecular backbones parallel to the surface, but clearly not in a conformation, which is given by a random walk in two dimensions. The homologous series of (G_1-G_4)PS were decorated with two dodecyl chains per amine groups by treating the respective deprotected dendronized polymers with an active ester to increase the adsorption energy on graphite.

The complex formation of DNA molecules with cylindrically shaped dendronized polymers were also reported [69]. The amine terminated dendronized polystyrene of Ist and higher generations (upto 4th generation) were synthesized by 'attach-to-approach' [35, 70]. The dendronized polymers had approximately the same chain length and

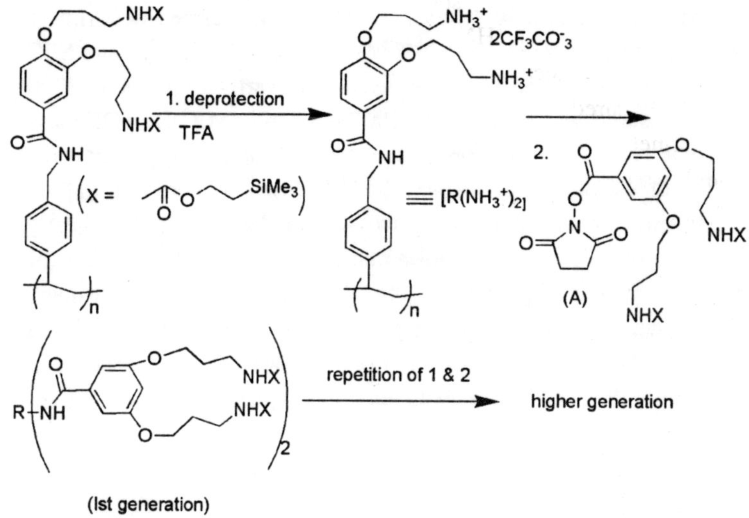

SCHEME 3.18 Synthesis of dendronized polystyrene through amidation reaction

polydispersity but varying numbers of charges as well as radii. The obtained complexes were adsorbed on bare, negatively charged mica or on mica coated with a positively charged polymer and then imaged by SFM. The analysis of SFM data indicates that DNA wrapped around the dendronized polymers. The complex with 2nd generation showed negatively charged and the electrostatic energy is larger than that in the complex, while for the DNA-4th generation complex, those two terms are almost equal. This result indicates that the overall charge of the complex might be selected by using polyelectrolyte with different flexibilities and linear charge densities. Thus these complexes may be of interest for the development of non-viral gene delivery system.

3.2.7 Poly(ether amide)s

The synthesis and characterization of novel poly(ether-amide) dendrimers upto 3rd generation were reported [71] by the convergent approach. The dendrons based on 5-(2-aminoethoxy)isophthalic acid were synthesized by using sequential activation and protection strategy like the peptide synthesis. Finally, these dendrons were coupled with different central cores. The syntheses of core molecules, dendrons and dendrimers are shown in scheme 3.19. The synthesized dendrimers were characterized by ^{1}HNMR, ^{13}CNMR, MALDI-TOF MS and SEC analyses. The molecular weight of dendrimers varies from 900 to 3300 Da, as measured by SEC and 917 Da 2793 Da by MALDI-TOF mass spectrometry. The thermal properties were studied by DSC analysis. The T_gs of the dendrimers were 82-83°C based on core unit (1) are independent of molar masses, whereas other dendrimers T_gs greatly depend on the generation number and increase with the increase of generation number. T_gs varies from 70°C to 114°C for dendrimers based on core (2). Also at lower generations T_gs decreased with the increase of the chain length of alkyl units, however higher generation T_gs are not influenced by the chain length of aliphatic moiety.

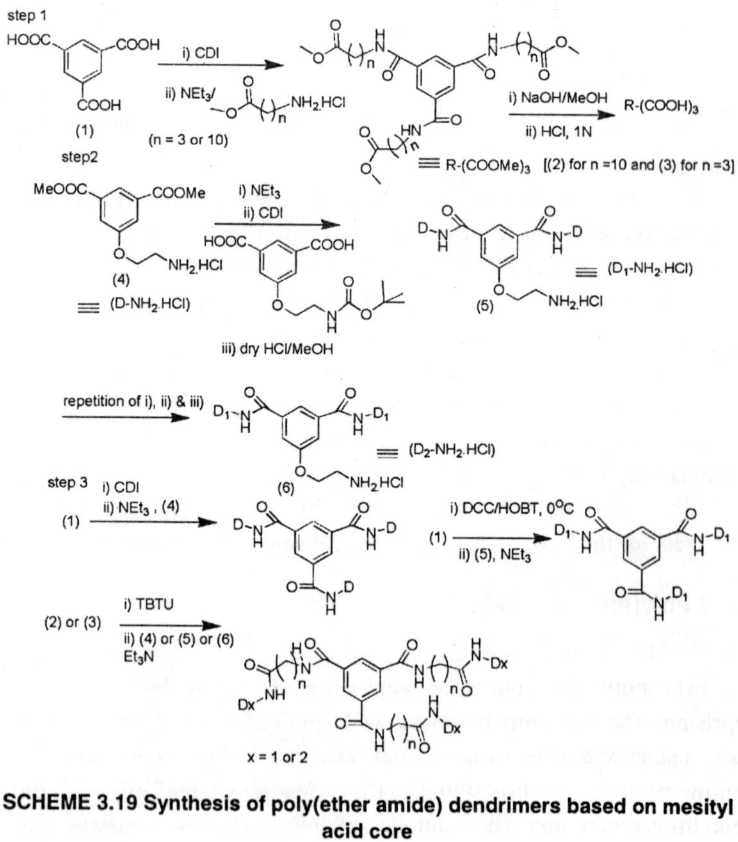

SCHEME 3.19 Synthesis of poly(ether amide) dendrimers based on mesityl acid core

The synthesis and characterization of a new family of polyamide dendrimers have been reported [23]. The dendrimers were synthesized by using 5-hydroxy isophthalic acid and diethanolamine as the sources of the branching units. By applying well-known protection-deprotection techniques and the amide bond forming growth reaction dendrimers up to 4th generation were synthesized. The family consists of one, two and three dimensioned second and fourth generation dendrimers and a mixture of sixth generation dendritic products. A series of reactions involve for these syntheses as shown in scheme 3.20. The yield obtained in each step is considerably high. The synthesized dendrimers were characterized by ^{1}HNMR, ^{13}CNMR spectroscopy, FAB-mass spectrometry and GPC analysis. The GPC analysis indicated that the 4th generation dendrimer adopted a globular shape, while the variable temperature NMR spectroscopic investigations demonstrated that the periphery of the 4th generation dendrimer is less mobile than either of its internal regions or the analog position of the second generation dendrimer.

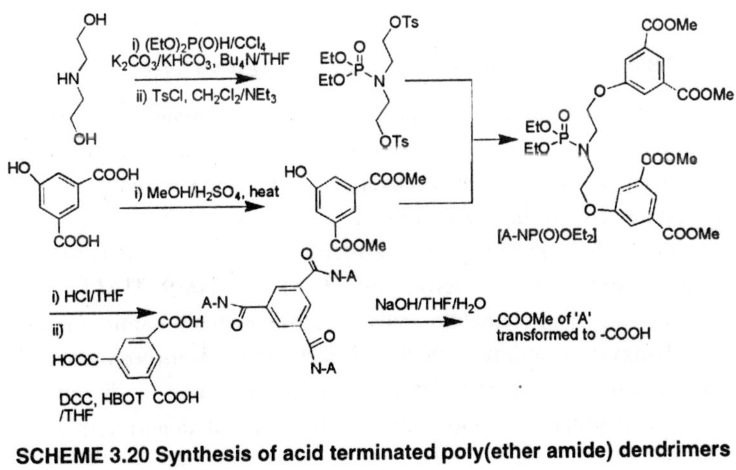

SCHEME 3.20 Synthesis of acid terminated poly(ether amide) dendrimers

The poly(ether-amide) dendrimers were functionalized [72-73] with non-racemic amino acid moieties at the surface. Pentaerythritol was used as a four direction core and *tris*(carboxyethoxymethyl) aminomethane as branching units. The DCC peptide coupling conditions were applied to get 1st generation, which after hydrolysis again coupled with the branching units. This 2nd generation was functionalized with trytophane methyl ester to get the modified dendrimer [3] (Fig. 3.7).

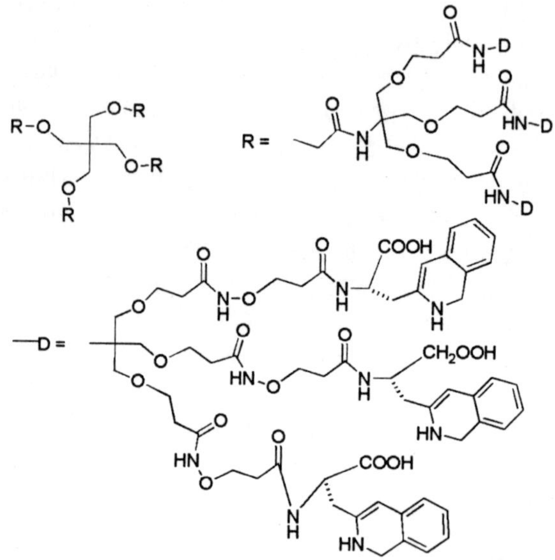

FIG.3.7 Structure of 2nd generation modified poly(ether amide) dendrimer

A series of ester and amide terminated aromatic polyamide dendrimers containing alternating ether and amide linkages were synthesized [74] by the convergent approach by using Williamson ether synthesis and coupling of carbonylic acids with amine as the main reactions. The synthesized dendrimers were purified by chromatography and characterized by ^{1}HNMR, ^{13}CNMR, MALDI-TOF MS or ESI MS and elemental analysis. 3,5-*Bis* (methyl aminomethyl) phenylbenzyloxy methyl ether and *tert*-butyl-3,5-*bis*(bromomethyl) benzoate were used as the building blocks and dibromo- or diamino-*p*-xylene as the initiator core to synthesize the targeted dendrimers.

3.3 SYNTHESIS, CHARACTERIZATION AND PROPERTIES OF HYPERBRANCHED POLYAMIDES

The hyperbranched polyamides have been synthesized by polycondensation of amines and acid chlorides or acids of AB_2, AB_4, and AB_8 type monomers and the condensation of bifunctional (A_2) and trifunctional (B_3) compounds. The synthesis of HB aromatic polyamides was first reported [75] by self-polycondensation of II and I at low temperature (Fig.3.8) in an amide solvent. The polymers were characterized by GPC, solubility and liquid crystalline behavior studies. These HB polymers have relatively high molecular weights ranging from $2.4X10^4$ to $4.6X10^4$ Da with PDIs 2-3.2. Both the carboxyl and amino terminated polymers were soluble in DMF, DMAc, NMP, etc. amide solvents, whereas amino terminated polymers were also soluble in water. Also these carboxyl terminated HB polymers exhibited a nematic liquid crystalline behavior in NMP with more than 40% weight polymer content, though no such behavior was observed for amino terminated polymers.

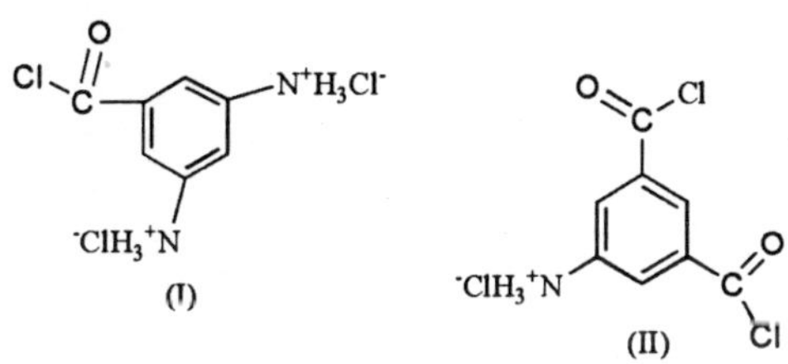

FIG.3.8 Structures of AB_2 and A_2B monomers used to synthesize aromatic HB Polyamides

The HB polyamides were synthesized [76] by direct polycondensation of AB_2, AB_4, and AB_8 monomers (Fig.3.9) with various degrees of branching (DB). The DB was 0.32 for AB_2, 0.72 for AB_4 and 0.84 for AB_8 type monomers as determined by ^{1}HNMR spectroscopy. The solubility and thermal properties of the polymers were also studied. The molecular weights of the resulting polymers were $1.0X10^5$ to $2.0X10^5$ Da with polydispersity ~3. The monomers AB_2 and AB_4 were also melt polymerized to get HB polymers without solvents and catalyst with molecular weight $2.0X10^3$ Da with PDI ~2. The melt polycondensation of 3,5-*bis*(4-aminophenyl) benzoic acid at 235°C produced a glassy solid, which transferred into a HB aromatic polyamide (Scheme 3.21) [77] after pouring DMF solution of the solid into LiCl containing MeOH solvent. The molecular weight was ≈ 7.5 $X10^4$ Da and PDI was 2.6 as determined by GPC-MALLS. The direct polycondensation reactions were carried out by using a triphenylphosphite (TPP)/pyridine system and a (2,3-dihydrodioxo-3-benzoxazolyl) phosphoric acid diphenyl ester (DBOP) system. At first the monomers were synthesized as shown in scheme 3.22. These monomers were used to get the desired HB polymers. The synthesized HB polyamides were characterized by ^{1}HNMR spectroscopy and GPC analysis. The polymers were soluble in aprotic polar solvents like DMF, DMSO, DMAc, NMP, etc. Copolymerization of AB_2 and AB_4 monomers at different molar ratios allowed the control of DB of the resulting copolymer [76]. The end capping of the primary amine terminated polymers were carried out by benzoyl chloride. T_gs of the end capped

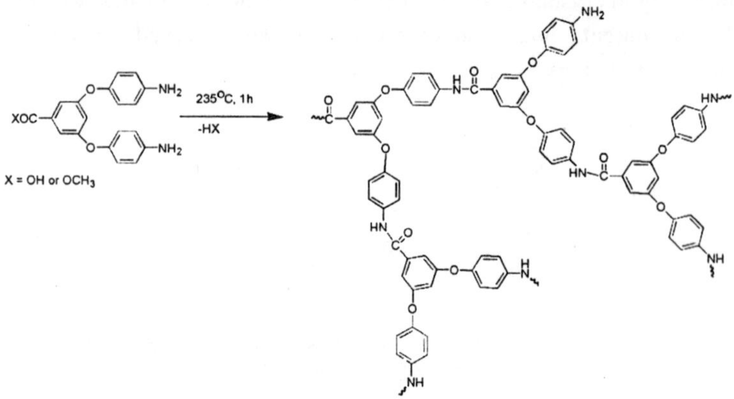

SCHEME 3.21 Synthesis of aromatic HB polyamide

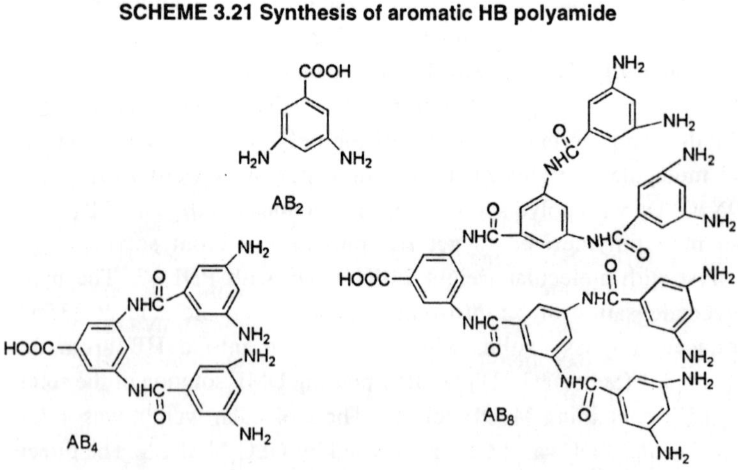

FIG.3.9 Structures of AB_2, AB_4 and AB_8 monomers used to synthesize aromatic HB Polyamides

polymer were almost independent of DB (305°C for DB 0.34 and 309°C for DB 0.72) of the original polymers, and higher than that of the 5th generation polyamide dendron composed of the same units.

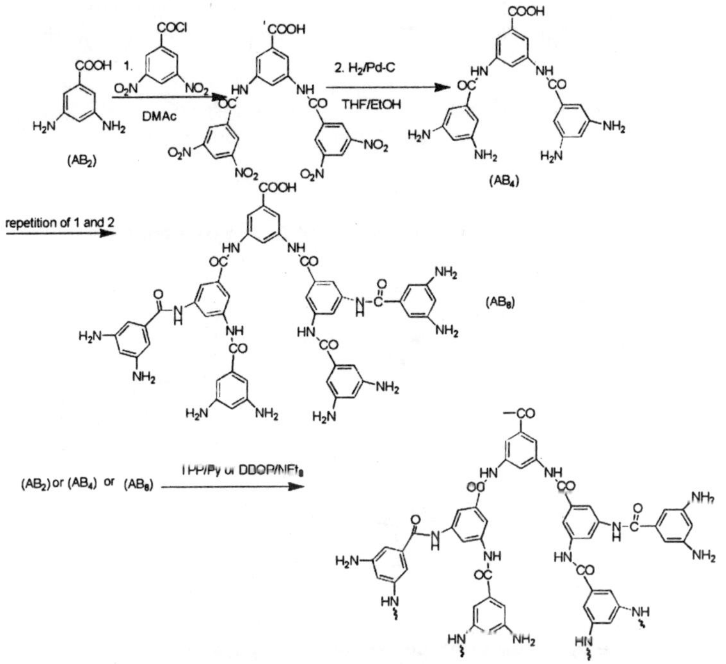

SCHEME 3.22 Synthesis of monomers and HB polyamides

The authors [78] also synthesized HB polyamides by condensation of aromatic diamine (A_2) with trimesic acid (B_3) as shown in scheme 3.23. The reactions of *p*-phenylene diamine with trimesic acid at 80°C in the presence of triphenylphosphate and pyridine proceeded without gelation to form a HB aromatic polyamide. The product was soluble in aprotic polar solvents and contained large amounts of unreacted carboxylic acid groups. The nucleophilic ring opening polymerization of 2-(3,5-dihydroxyphenyl)-1,3-oxazoline in N-methylcaprolectam produced a HB poly(ether amide) [17] (Scheme 3.24). The resulted polymer was characterized by GPC, MALDI-TOF mass spectrometry and thermal studies. These studies indicate a high molecular weight of the polymer (4.7 X 10^4 Da) with Tg 176°C and decomposition temperature 330°C.

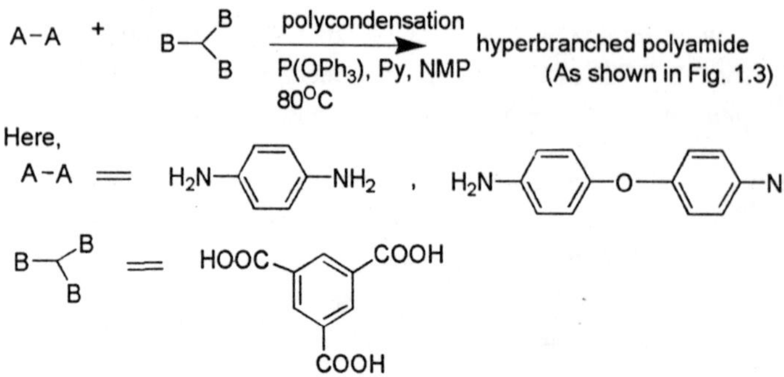

SCHEME 3.23 Polycondensation of A_2 and B_3 monomers

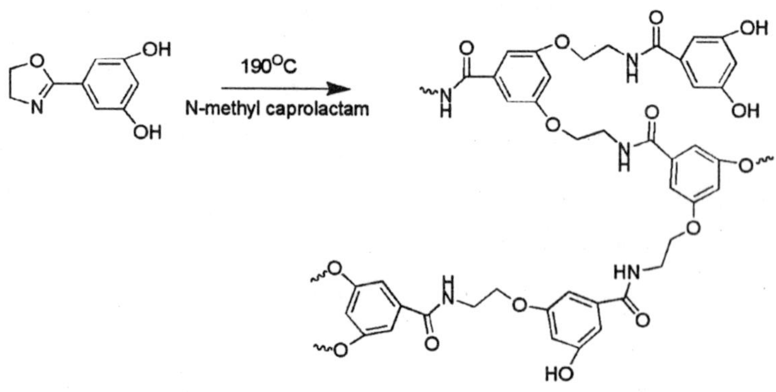

SCHEME 3.24 Synthesis of HB polyamide by ring opening polymerization of oxazoline monomer

The synthesis, characterization and properties evaluations of HB aromatic polyamide copolymers were reported [79]. The copolymers were synthesized by direct polycondensation of 3-(4-aminophenoxy) benzoic acid (AB monomer) and 3,5,-*bis*(4-aminophenoxy)benzoic acid (AB_2 monomer) in the presence of triphenylphosphate and pyridine as condensing agents. The monomers were first synthesized from 3,5-dihydroxybenxoic acid (for AB_2), as shown in scheme 3.25. Then the homo and copolymers were synthesized by direct polycondensation of the above monomers (Scheme 3.25). The structures of the synthesized HB copolyamides were characterized by IR, ^{1}HNMR and ^{13}CNMR spectroscopy. ^{1}HNMR spectra confirmed the copolymer composition, which was almost identical to the feed ratio of the monomers. ^{13}CNMR spectroscopic studies indicated that the copolyamides were composed of five kinds of repeating units in the structures. The transparent copolyamide was soluble in aprotic polar solvents such as DMF, NMP, and DMSO. The thermal study by TGA indicated only 5% weight loss of the copolymer over 400°C and degradation behavior was independent of ratio of the monomer used. However, the ratio of the monomers affected the T_g as determined by DSC and softening points (T_m) by TMA. The Young's modulus decreased from 2.4 to 1.6 GPa with the increase of the AB_2 monomer in the range 0-60% and then became constant over a 60% of the AB_2 monomers as determined by tensile test. The decrease may be caused by less entangled nature of the HB polymers synthesized from AB_2 type monomer.

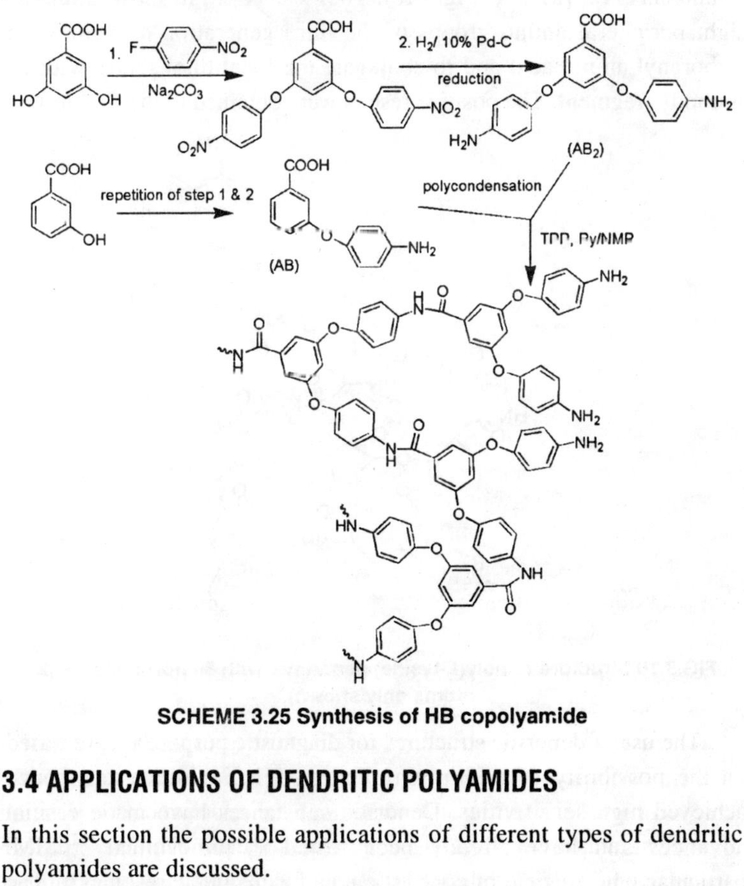

SCHEME 3.25 Synthesis of HB copolyamide

3.4 APPLICATIONS OF DENDRITIC POLYAMIDES

In this section the possible applications of different types of dendritic polyamides are discussed.

3.4.1 Biomedical

The polyamide dendrimers have been used as therapeutics in boron neutron capture therapy [80]. This therapy represents a method for the treatment of presently incurable form of cancer. The exceptionally high neutron capture cross section of the ^{10}B isotope for thermal neutrons is used to produce a radiation energy that is lethal to the surrounding cells. The problem is that a compound as boron rich as possible is required, which can be selectively coupled to molecules that recognize tumor cells. The first lysine dendron that specifically couples to proteins and carries 80 terminal boron atoms was prepared by Qualmann *et al.* [81-82] (Fig.3.10). They succeeded in substituting the eight peripheral amino groups of the third generation dendron by a carboranyl amino acid and to conjugate the focal thiol group with an antibody fragment. The positive results were obtained in the first test.

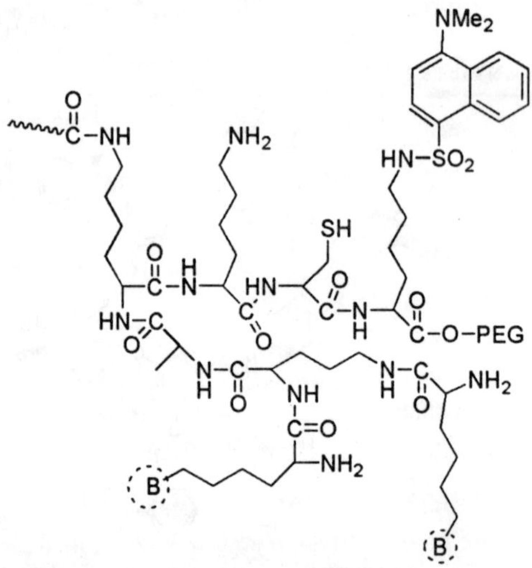

FIG.3.10 Structure of poly(L-lysine) dendrimer with 80 boron atoms (2 atoms only shown).

The use of dendritic structures for diagnostic purpose is also based on the possibility of multiplying certain functionalities and hence achieved high sensitivities. Dendritic substances have made crucial advances and have already been tested in pre-clinical studies, particularly in the field of contrast media for magnetic resonance. The magnetic resonance imaging (MRI) method [83] serves to picture organs, blood vessels or tissues inside the human body by generating defined inhomogeneous magnetic fields that allow the nuclear resonance signal of water to be assigned to its place of origin and then to covert it into pictures. The relaxation time of the water protons are significantly shortened by application of the contrast media (paramagnetic metal cations) into the organ that is to be visualized and leads to a picture with a very good signed to noise ratio. However, clinically used contrast media such as Mgnevist (gadolinium salt of DTPA) [84] reveal a disadvantage. This low molecular compound easily diffused into the extravenous areas. The higher molecular several gadolinium complexes also show same problems like bad renal elimination of the compounds. The Schering AG group [12] developed a lysine based 2nd generation dendron Gd–complex with trimesinic acid as the central building block. The animal tests showed that the quantitative renal elimination and the higher intravascular retention time, which ensures an excellent signal to noise ratio, turn out to be the outstanding properties of this new contrast medium.

L-lysine based polyamide-glycodendrimers [12, 85] showed enhanced binding properties in a direct enzyme linked lectin assay (ELLA) by using horseradish peroxidase labeled wheat germ agglutinin (WGA). Also L-lysine dendrimers with N-acetyl glucosamine or N-acetyllactosamine terminal resides showed enhanced inhibition of lectin binding to porcine stomach mucin. The inhabitation of binding of yeast mannan to concanavaline A and to pea lectins by mannosylated L-lysine dendrimers displays maximum for the tetra- and octa-meric dendrimers respectively [46].

A dendritic core based on lysine unit was developed [7, 86-88] for the construction of MAP, which are carrying antigens at their periphery. It also developed the polyamide dendrimers that carry multiple cyclic antigen peptides at their surface [89-90].

An amphiphilic novel biomaterial, α-GalNAc containing *p-tert* butyl calix-4-arene with upto 16 carbohydrate units, which can bind lectin in aqueous environment can be directly adsorbed on to the lipophilic surface of polystyrene microtiter plates and thus might be useful in bioanalytical devices [91].

Dendritic poly(L-lysine) of the 6th generation showed high transfection efficiency into several cultivated cells with low cyto-toxicity [92]. The antuors [93] also reported the synthesis of monodisperse poly(L-lysine) dendrimers with hexamethylene diamine, as the core and the dendrimer showed high transfection efficiency into several cultivated cells with low cyto-toxicity. Interestingly, the DNA-complex of this dendrimer also showed high transfection efficiency even in the presence of 50% serum.

3.4.2 Encapsulation

A series of L-lysine based dendrimers were functionalized [94] at the focal point with benzo-18-crown-6, which were utilized for controlled release application of encapsulated template molecules. The dendrimers can bind with potassium and benzyl ammonium cations by assembling dendritic branches around the cation and then encapsulating it within a supramolecular dendritic shell in polar MeOH solution. Again this supramolecular dendrimer was smoothly disassembled by the addition of K^+ ions to the solution and thereby releasing the template molecule.

The dendritic polypeptides from L-lysine building blocks upto 3rd generation with carboxylic acid at the focal point of the branched structure specifically solubilized hydrochloride salt of proflavine, hydrophilic aromatic dye containing amine groups [95] in CH_2Cl_2. These results clearly showed that like spherical dendrimer the individual dendritic branches with suitable funtionalization at the focal point can non-convalently solubilized a hydrophilic dye. Also the optical properties of the encapsulated dye indicate that it experiences a unique microenvironment as a consequence of supramolecular encapsulation.

The extent of dendritic branch (generation number) controlled the extent of dye uptake as well as optical properties of the dyes [96]. Further more, it was shown that suitable functionalized dendritic branches can transport hydrophilic dyes through an apolar phase and delivered them continuously into an aqueous medium. It was suggested that proflavine dye is solubilized through the complementary interactions between COOH groups the dendritic branches and NH_2 groups of the dye, whereas aurin tricarboxylic acid dye is solubilized through the interaction of NH_2 groups of dendritic branches with COOH groups of dyes. In the latter case, the dendritic branches containing esterified focal carboxylic groups. These dendritic branched structures also form a two-component supramolecular organogel [97] with linear aliphatic diamines like diaminododecane in non-hydrogen bonding solvent, CH_2Cl_2. It showed that the gelation of 2nd generation dendritic structure with diaminododecane occurred very rapidly in CH_3CN solvent.

Two series of first to third generation dendrophanes (Fig.3.11) (cyclophane type) of molecular weight upto ≈ 18000 Da with poly(ether amide) branching and 12, 36 and 108 terminal carboxylic groups were used [98-100] as the first water-soluble dendrimers with defined active recognition site at the central core. In the first series, the central cyclophane core binding cavity complementary in size to benzene and naphthalene guests, whereas in the 2nd series the larger steroidal substrate can be accommodated. The cyclophane cores formed complexes with the guest molecules by π-π stacking and C-H--π interactions. As a consequence, dendritic cyclophanes mimic apolar-binding sites buried within globular protein superstructures. The enlargement of the cyclophane core is used as a tool to complex larger steroid molecules [101]. With the increase of dendritic generation, the micropolarity around the core cyclophone is significantly reduced and is comparable to that of ethanol. Also the host guest exchange kinetics are very fast (10^2-10^3 s^{-1}). This result is contrast to Meijer and his coworker's findings [102] for PPI dendrimers of similar experiments. This is due to the difference in initiator core size in the above two cases. In PPI dendrimer, a tight densely packed superstructure diverging from a small initiator core, whereas in dendrophanes, the form dendritic wedges are attached to large, nanometer sized cyclophone cores, which produces apertures through which substrates can rapidly enter or leave the cavity. Further more, the PPI dendrimers possess strongly and structurally encumbering surface groups that generate tight substrate encapsulation at the interior, whereas the carboxylate at the dendrophone surface will not densely pack for electrostatic reasons. This suggests that these water-soluble dendrophanes are attractive targets for catalytically active mimics of globular enzymes.

The polyamide dendrimers [103] upto 36 end groups with four 2,6-diamidopyridine units in the interior (Fig. 3.12) can associate with

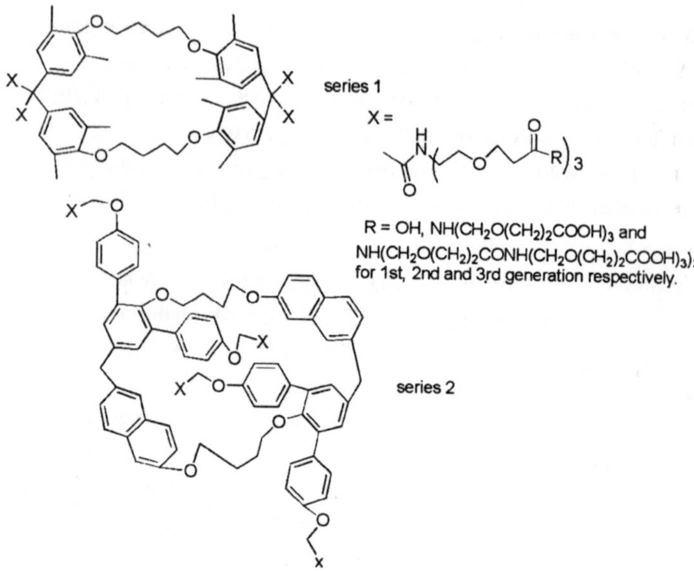

FIG.3.11 Structure of a dendrophane

complementary guests [21] such as glutarimide, barbituric acid and 3'-arido-3'-deoxythymidine. However, the system suffers from same drawbacks like self-association between the different generations and inefficient donor–acceptor alignment.

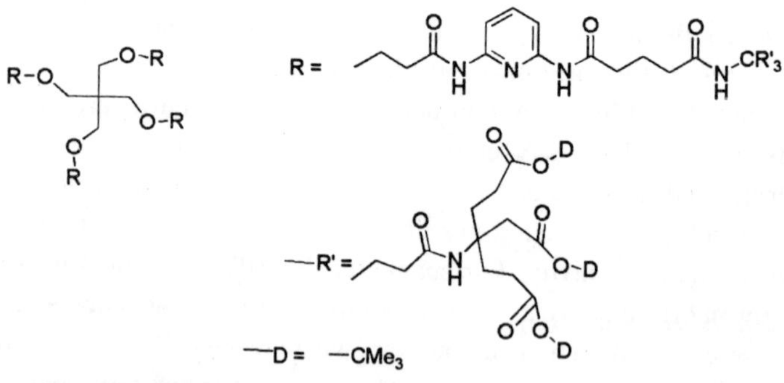

FIG.3.12 Dendrimer with four 2,6-diamino pyridine as H-bonding unit

3.4.3 Self-assembly

Dendrophanes have also been successfully applied to dendritic self-assembly [104-105]. Large multinanometer sized assemblies with molecular weight exceeding 14,000 Da were obtained in aqueous methanol (pH=8.7) by threading two dendrophanes (3rd generation, 2nd series) onto a ~5.5 nm long molecular rod in which two testosterone termini are attached by rigid oligo(phenyl acetylene) spacer to a central phenyl ring bearing quaternary ammonium side chains [106]. The formation of these aggregates in which two dendrophanes preferentially encapsulate the steroidal termini is driven by a combination of a polar forces, hydrophobic desolvation and ion pairing, and depends strongly on the correct length of the rigid oligo(phenyl acetylene) spacers.

3.4.4 Others

In the β-alanine based dendritic, β-peptides [33] due to a cooperative effect of a large number of carbamate and amide groups, these dendrimers exhibit an unusual strong binding ability towards protic solvents and behave as H-bonded sponges. These peptide dendrimers also exhibit a strong tendency to form nanoscopic aggregates in nonpolar or polar aprotic solvents through intermolecular H-bonding interactions. The dendrimer consists of a benzene core branched in the 1, 3 and 5 positions. Each branch starts with a dialkylcarbonamide type moiety and carriers: a) six aliphatic amide groups and b) eight fluorescent 5-dimethylamino-1-naphthalene sulphoamido (dansyl) units, used [8] as a light harvesting ligand for Nd^{3+} luminescence. The dendrimer is capable of coordinating lanthoxide ions and the energy transfer from the 24 light harvesting dansyl units to the encapsulated metal ion is efficient. This is an interesting result for construction of novel light converting materials for optical communication systems.

The dendrons possessing multiple 2,6-di(acylamino)pyridinyl sites, each of which was capable of molecular recognition was used to develop new polymeric supports of an activated agarose matrix by surface modification [107].

3.5 SUMMARY

In this chapter mainly different subclasses of polyamide dendritic polymers with respect to their synthesis, properties and applications are described. They have been generally synthesized by protection/ deprotection (activation) strategy like the peptide synthesis. The dendritic polyamides possess many unusual properties along with their resembling with biological peptide chemistry depending on the structures. They have tremendous potential in biological and medicinal applications along with encapsulation and self-assembly applications, etc.

REFERENCES

[1] R.G.Denkewalter, J.F.Kole and W.J.Lukasavage, U.S.Pat. 4,289,872 (1981)

[2] R.G.Denkewalter, J.F.Kole and W.J.Lukasavage, US Pat. 4,360,646 (1982)

[3] R.G.Denkewalter, J.F.Kole and W.J.Lukasavage, US Pat. 4,410,688 (1983)

[4] S.M.Aharoni, C.R.Crosby III and E.K.Walfh, *Macromolecules* **15**(1982)1093

[5] S.M.Aharoni and N.S.Murthym, *Polym.Commun.* **29**(1983)132

[6] S.R Turner, B.I.Voit and T.H.Mourey, *Macromolecules* **26**(1993)4617

[7] D.N.Posnett, H.McGrath and J.P.Tam, *J.Biol Chem*, **264**(1988)1719

[8] S.Jun and J.P.Tam, *J.Am.Chem.Soc.* **117**(1995)3894

[9] T.M.Chapman, G.L.Hillyer, E.J.Mahan and K.A.Shaffer, *J.Am.Chem.Soc.* **116**(1994)11195

[10] T.M.Chapman and E.J.Mahan, *Polym.Mater.Sci.Eng.* **77**(1997)103

[11] L.J.Twyman, A.E.Beerer and J.C. Mitchell, *Tetrahedron Lett.* **35**(1994)4423

[12] R.Roy, D.Zanini, S.J.Meunier and A.Romanovska, *J.Chem.Soc. Chem.Commun.* (1993)1869

[13] J.S.Choi, E.J.Lee, Y.H.Choi, Y.J.Jeong and J.S.Park, *Bioconjugate Chem.* **10**(1999)62

[14] J.S.Choie, D.K.Joo, C.H.Kim, K.Kim and J.S.Park, *J.Am.Chem.Soc.* **122**(2000)474

[15] E.Bayer and M.Mutter, 'The Liquid Phase Method for Peptide Synthesis' in 'The Peptide edited by E.Gross and J.Meienhofer, Academic press, New York, Vol.2, pp. 285-332, 1979

[16] A.Cattani-Scholz, C.Renner, D.Oesterhelt and L.Moroder, *Chembiochem.* **2**(2001)542

[17] Y.Yamakawa, M.Ueda, K.Takachi and M.Asai, *J.Polym.Sci. Part A.Polym.Chem.* **37**(1999)3638

[18] T.K.K.Mong, A. Niu, H.F.Chow, C. Wu, L.Li and R.Chen, *Chem.Eur.J.* **7**(2001)686

[19] D.Marcovici-Mizrabi, H.F.Gottlieb, V.Marks and A.Nudelman, *J.Org.Chem.* **61**(1996)8402

[20] J.Jones, 'The Chemical Synthesis of Peptides', Oxford University Press, Oxford, 1994

[21] T. Shiori and S.Yamada, *Chem.Pharm.Bull.* **22**(1974)849

[22] S.J.E Mulders, A.J. Bronwer and P.G.J. Liskamp, *Tetrahedron Lett.* **38**(1997)3

[23] P.R.Ashton, D.N.Andrson, C.L.Brown, A.N.Shipway, J.F.Stoddart and M.S.Tolley, *Chem.Eur.J.* **4**(1998)781

[24] P.R.Ashton, S.E.Boyd, C.L.Brown, N.Jagaraman, S.A.Nepoaediev and J.F.Stoddart, *Angew.Chem.Int.Ed.* **36**(1997)732

[25] J.K.Young, G.R.Baker, G.R.Newkome, K.F.Morris and C.S.Johson, *Macromolecules* **27**(1994)3464

[26] T.Shioiri, K.Ninomiya and S.Yamada, *J.Am.Chem.Soc.* 94(1972)6203

[27] A.J.Brouwer, S.J.E.Mulders and R.M.J.Liskamp, *Eur.J.Org.Chem.* (2001)1903

[28] R.H.E. Hudson and M.J.Damha, *J.Am.Chem.Soc.* **115**(1993)2119

[29] T.W.Nilsen and W.Prensky, US Pat. 5,175,270 (1986)

[30] R.Haner and K. Skobridis, PCT Pat. WO 96/19240 (1996)

[31] S.Rannard, N.Davis and H.McFarland, *Polym. Int.* **49**(2000)1002

[32] K.E.Uhrich and J.M.J. Frechet, *J.Chem.Soc.Perkin Trans. 1* **13**(1992)1623

[33] S.Koenig, L.Muller and D.K.Smith, *Chem.Eur.J.* **7**(2001)979

[34] S.Velter, S.Koch and A.D.Schluter, *J.Polym.Sci.Part A Polym.Chem.* **39**(2001)1940

[35] A.D. Schluter and J.P.Rabe, *Angew.Chem.Int.Ed.* **39**(2000)864

[36] I.Neubert and A.D.Schluter, *Macromolecules* **31**(1998)9372

[37] R.Klopsch, S.Koch and A.D.Schluter, *Eur.J.Org.Chem.* (1998)1275

[38] S.K.Emran, G.R.Newkome, C.M.Weis and J.P.Harmon, *J.Polym.Sci. Part B Polym. Phys.* **37**(1999)2025

[39] C.Modrakowski, S.C.Flores, M. Beinholf and A.D.Schluter, *Synthesis* (2001)2143

[40] A.Vargas-Berenguel, F.Ortega-Caballero, F.Santoyo-Gonzaler, J.J.Garcia-Lopez, J.J.Gimenez-Martinez, L.Garcia-Fuentes and E.Oriz-Salmenon, *Chem.Eur.J.* **8**(2002)812

[41] H.Sashiwa, Y.Shigemasa and R.Roy, *Macromolecules* **34**(2001)3905

[42] I.Vrasidas, N.J.de Mol, R.M.J.Liskamp and R.J.Pieters, *Eur.J.Org.Chem.* (2001)4685

[43] R.Roy, D.Zanini, S.J.Meunier and A.Romanovska, *ACS Symp.* **560**(1994)104

[44] D.Zanini and R.Roy, *J.Am.Chem.Soc.* 119(1997)2089

[45] D.Zanini, W.K.C.Park and R.Roy, *Tetrahedron Lett.* **36**(1995)7383

[46] D. Zanini and R.Roy, *Bioconjugate.Chem.* **8**(1997)187

[47] D.Pagel, D.Zanini and R.Roy, *Bioorg.Med.Chem.* **4**(1996)1949

[48] R.Roy, W.K.C.Park, D.Zanini, C.Foxall and O.P.Srivastava, *Carbolydrate Lett.* **2**(1997)259

[49] R.Roy, W.K.C.Park, Q.Wu and S.N.Wang, *Tetrahedron* **36**(1995)4377

[50] W.K.C.Park, B.Kratzer, D.Zanini, Q.Wu, S.J.Meunier and R. Roy, *Glycoconjugate J.* **12**(1995)456

[51] P.R.Ashton, S.E.Boyd., C.L.Brown, N.Jayamaran, S.A.Nepogolier and J.F.Stoddart, *Chem.Eur.J.* **2**(1996)1115

[52] N. Jayaran, S.A.Nepogodiev and J.F.Stoddart, *Chem.Eur.J.* **3**(1997)1193

[53] P.R.Ashton, E.F.Hounsell, N.Jayaraman, T.M.Nilsen, N.Spencer, J.F.Stoddart and M.Young, *J.Org. Chem.* **63**(1998)3429

[54] G.R. Newkome, V.V.Narayanan and L.A.Godinez, *J.Org.Chem.* **65**(2000)1643

[55] B.Huang and J.R.Parquette, *Org.Lett.* **2**(2000)239

[56] V.V.Narayanan and E.C.Wiener, *Macromolecules* **33**(2000)3944

[57] V.Vicinelli, P.Ceroni, M.Maestri, V.Balzani, M.Gorka and F.Vogtle, *J.Am.Chem.Soc.* **124**(2002)6461

[58] S.A.Vinogrador, L.W.Lo and D.F. Wilson, *Chem.Eur.J.* **5**(1999)1338

[59] S.A. Vinogrador and D.F. Wilson, *Chem.Eur.J.* **6**(2000)2456

[60] A. Ritzen and T. Frejd, *Eur.J.Org.Chem.* (2000)3771

[61] A. Ritzen and T. Frejd , *Chem.Commun.* (1999)207

[62] T.W.Barnett, F.A.Robinson and B.M.Wilson, *J.Chem.Soc.Chem.Commun.* (1947)202

[63] A. S. Carlstrom and T. Frejd, *Synthesis* (1989)414

[64] E.W. Cranlall and L. Harris, *Org.Prep.Proceed.* **1**(1969)147

[65] U. Schmidt, A.Lierberknecht and J.Wild, *Synthesis* (1984)53

[66] M.J.Burk, J.E.Feaster, W.A.Nugent and R.L.Harlow, *J.Am.Chem.Soc.* **115**(1993)10125

[67] A.Zhang, S.Vetter and A.D.Schluter, *Macromol.Chem.Phys.* **202**(2001)3301

[68] L.Shu, A.D.Schluter, C.Ecker, N.Severin and J.P.Rabe, *Angew.Chem.Int.Ed.* **40**(2001)4666

[69] I.Gossl, L.Shu, A.D.Schluter and J.P. Rabe, *J.Am.Chem.Soc*. **124** (2002)6860

[70] L. Shu, A.Schafter and A.D.Schluter, *Macromolecules* **33**(2000)4321

[71] D.Appelhans, H.Komber, D.Voigt, L.Haussler and B.I.Voit, *Macromolecules* **33**(2000)9494

[72] G.R.Newkome, X.Lin and C.D.Weis, *Tetrahedron Asym.* **2**(1991)952

[73] G.R.Newkome and X.Lin, *Macromolecules* **24**(1991)1943

[74] Y. Pan and W.T.Food, *J.Polym.Sci. Part A Polym.Chem.* **38**(2000)1533

[75] Y.H.Kim, *J.Am.Chem.Soc.* **114**(1992)4947

[76] Y. Ishida, A.C.F. Sun, M.Jikei and M.Kakimoto, *Macromolecules* **33**(2001)2832

[77] G.Yang, M.Jikei and M.Kakimoto, *Macromolecules* 31(1998)5964

[78] M.Jikei, S.H.Chon, M.Kakimoto, S.Kawauchi, T.Imase and J.Watanable, *Macromolecules* 32(1999)2061

[79] M.Jikei, K.Fujii, G.Yang and M.Kakimoto, *Macromolecules* **33**(2000)6228

[80] A.H.Soloway, W.Tjarks, B.A.Barnum, E.G.Rong, R.F.Barth, I.M.Godogni and J.G.Wilson , *Chem.Rev.* **98**(1998)1515

[81] B.Qualmann, M.M.Kessels, H.J.Musiol, W.D.Sierralta, P.W.Jungblut and L.Moroder, *Angew.Chem.Int.Ed.* **35**(1996)909

[82] B.Qualmann, M.M.Kessels, F.Klobasa, P.W.Junglbat, and W.D.Sierralta, *J.Microse.* **183**(1996)69

[83] R.B.Lauffer, *Chem.Rev.* **87**(1987)901

[84] H.J.Weinmann, W.Ebert. B.Misselwitz, B.Raduchel, H.Schmitt-Willich and J. Platzek, *Eur.Radiol.* **7**(1997)196

[85] R. Roy, *Polym.News* **21**(1996)226

[86] J.P.Tam, *Proc.Natl.Acad.Sci.* **85**(1998)5409

[87] C.Rao and J.P.Tam, *J.Am.Chem.Soc.* **116**(1994)6975

[88] K.J.A.Chang, W.Pugh, S.G.Blanchard, J.McDermed and J.P.Tam, *Proc.Natl.Acad.Sci.* **85**(1988)4929

[89] J.C.Spetrler and J.P.Tam, *Peptide Res.* **9**(1996)290

[90] L. Zhang and J.P.Tam, *J.Am.Chem.Soc.* **119**(1997)2363

[91] R.Roy and J. M.Kim, *Angew.Chem.Int.Ed.* **38**(1999)369

[92] T.Okuda, S.Kidaki, M.Ohsaki, Y.Koyama, K.Yoshikawa, T.Niidome and H.Aoyagi, *Org.Biomol.Chem.* **1**(2003)1270

[93] M.Ohsaki, T.Okuda, A.Wada, T.Hirayama , T.Niidome and H.Aoyagi, *Bioconjugate Chem.* **13**(2002)510

[94] G.M.Dykes, D.K.Smith and G.J.Seeley, *Angew.Chem.Int.Ed.* **41**(2002)3254

[95] D.K.Smith, *Chem.Commun.* (1999)1685

[96] G.M.Dykes, L.J.Brierley, D.K.Smith, P.T.Mcgrail and G.J.Seeley, *Chem.Eur.J.* **7**(2001)4730

[97] K.S.Partridge, D.K.Smith, G.M.Dykes and P.T.McGrail, *Chem.Commun.* (2001)319

[98] S.Mattei, P.Seiler, F. Diederich and V.Gramlich, *Helv.Chim.Acta* **78**(1995)1904

[99] P.Wallimann, P.Seiler and F.Diederich, *Helv.Chim.Acta* **79**(1996)779

[100] S. Mattei, P.Wallimann, B.Kenda, W.Amrein and F.Diederich, *Helv.Chim.Acta* **80**(1997)2391

[101] P.Willimann, S. Mattei, P.Seiler and F. Diederich, *Helv.Chim.Acta* **80**(1997)2368

[102] J.F.G.A Janson, E.M.M. de Brabander-van den Berg and E.W.Meijer, *Science* **266**(1994)1226

[103] G.R. Newkome, B.D. Woosley, E. He, C.N.Moorefield, R. Guther, G.R. Baker, G.H Escamilla, J.Merrill and H. Luffmann, *Chem.Commun.* (1996)2737

[104] S.C.Zimmerman, F.Zeng, D.E.C.Reichert and S.V.Kototuchin, *Science* **271**(1996)1095

[105] N.Tomioka, D.Takasu, T.Takashashi and T.Aida, *Angew.Chem.Int.Ed.* **37**(1998)1531

[106] B. Kenda and F. Diederich, *Angew.Chem.Int.Ed.* **37**(1998)3154

[107] M.C.Strumia, A.Halabi, P.A.Pucii, G.R.Newkome, C.N.Moorefield and J.D.Eppereson. *J.Polym.Sci. Part A Polym.Chem.* **38**(2000)2779

CHAPTER 4

DENDRITIC POLY(AMIDO AMINE)S

4.1. INTRODUCTION

Poly (amido amine)s are the most important and widely studied dendritic polymers available commercially as 'starburst'® dendrons and dendrimers by Dow Chemical and Dendritech Inc. USA. This dendrimer is popularly known as PAMAM dendrimer. The general structure of this subclass of dendrimers is shown in Fig.4.1. PAMAM dendrimers possessing different types of core units, branching units and terminal groups are discussed in this chapter. In short hand notation, like poly(propylene imine), it is generally written as core-dendr-$(NZ)_x$, where x is the number of terminal groups, e.g. EDA-dendr-$(NH_2)_{32}$, indicates ethylene diamine core 3rd generation PAMAM dendrimer with 32 terimal primary amine groups. For the sake of simplicity, in this chapter, this notation is used. It is noteworthy to mention here that when a dendrimer terminates with an amine function it is termed as full generation (G = n.0.), whereas if the dendrimer terminates with a carboxylate function, it is termed as a half (G =n.5) generation.

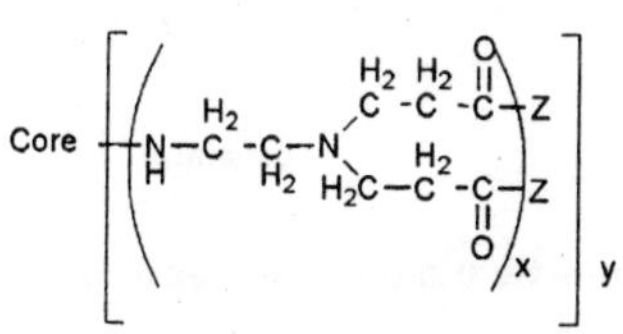

No. of terminal groups, $x = (N^n_b - 1)/(N_b - 1)$
where Nb = branch multiplicity and n = generation number
Z = surface group

FIG. 4.1 General structure of PAMAM dendrimer

This class of dendrimers has a wide range of enormous applications such as standard materials for SEC, viscosity modifiers, materials for drug delivery system, gene therapy, NMR contrast agents, synthetic vaccines, surfactants, cross-linking agents, chemical sensors, processing aids, etc.

4.2 SYNTHESIS, CHARACTERIZATION AND PROPERTIES OF POLY(AMIDO AMINE) DENDRIMERS

In this section the synthesis, characterization and properties of poly(amido amine) dendrimers are discussed.

4.2.1 Synthesis

Poly(amido amine) was the first dendritic polymer synthesized successfully by using the divergent approach. Tomalia *et al.* [1] synthesized PAMAM by a two-step process, involving (i) exhaustive Michael addition (similar to Vogtle [2]) of a suitable amine initiator core such as ammonia and ethylene diamine (EDA) to methylacrylate and (ii) exhaustive amidation of the resulting ester with a large excess of EDA, as shown in scheme 4.1, instead of troublesome reduction step as previously attempted in case of polyamine [2]. Both the Michael addition and the amidation reactions were conducted in methanolic solvent at temperature below 50°C with a slight excess of methylacrylate in step one and a large excess of EDA in step two to prevent bridging and gelation of dendrimer. The excess reagents and solvent were removed under vacuum at temperature below 55°C. The product was obtained as viscous syrup or glassy liquid with very high yield. Now PAMAM dendrimers of 0 to 10th generation are produced on a kilogram scale and these are commercially available from a number of sources [3].

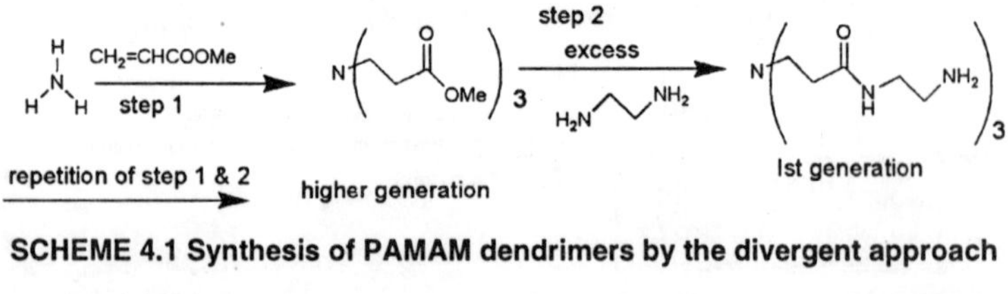

SCHEME 4.1 Synthesis of PAMAM dendrimers by the divergent approach

The molecular weight of the dendrimers varies from 517 to 935000 Da and they contain 4 to 1096 terminal amino groups. The PAMAM dendrimer increases its diameter by ≈10A° per generation evolving from a dish-like shape (zero to second generation) to an oblate spheroid (generation 3 to 4) to a nearly symmetrical spheroid at generation 5 and higher. Now a days initiator cores have been tethered to Merrifield resin solid support and by using the protein synthetic procedures to avoid work up difficulties due to use of large excess of reagents, one can easily control the dendrimer growth. Synthesis of a variety of mono-, di-, tri-, tetra- and poly-dendron PAMAM dendrimers were reported [4-8]. The high aspect ratio, rod like dendrimers upto 4th generation were synthesized [9] using linear polyethylenimines initiator core with average multiplicity 300-400 by repetitive Michael addition to methacrylate, followed by amidation with ethylenediamine (Scheme 4.2). The length and diameter of the dendrimers were determined by the degree of polymerization of initiator core and number of generation respectively. PAMAM dendrimers were also synthesized [4-8] using variable alkyl chain length of diamino central core by the same synthetic procedure (Fig. 4.2).

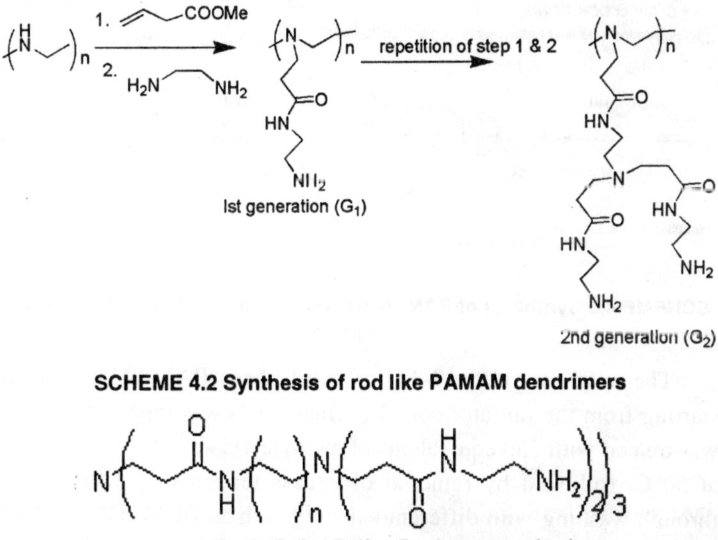

SCHEME 4.2 Synthesis of rod like PAMAM dendrimers

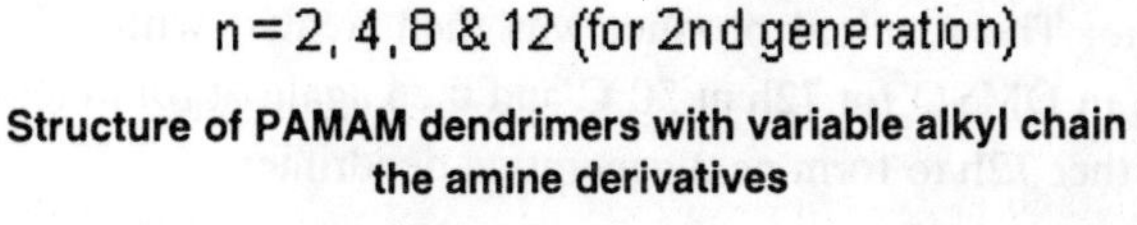

n = 2, 4, 8 & 12 (for 2nd generation)

FIG. 4.2 Structure of PAMAM dendrimers with variable alkyl chain length of the amine derivatives

The synthesis of PAMAM dendrimers on solid phase, starting from Tenta Gel^{TM} resin was described [10] in 1997. An acid labile linker attached to a polyamine scaffold [11] was used to allow cleavage of the dendrimer from the resin in order to monitor the synthesis. The synthesis of the dendrimer was carried out by cyclic treatment with methylacrylate and 1,3-diaminopropane (Scheme 4.3) in methanol. The 3rd generation dendrimer linked Tenta Gel^{TM} was found to have a loading of ≈2.3 mmole NH_2 g^{-1} of resin, with each bead having 5 to 6 nanomoles of free amino groups. This was converted to a 4th generation dendrimer with a resin loading of 2-8 mmole/g and a single-bead loading of approximately 9.6 mmole/bead. The 2nd generation dendrimer was also synthesized on large polystyrene beads (250-300 μm size, 1.02 mmol/g loading). A small inert PEG spacer was first introduced to allow efficient PAMAM dendrimer synthesis [12].

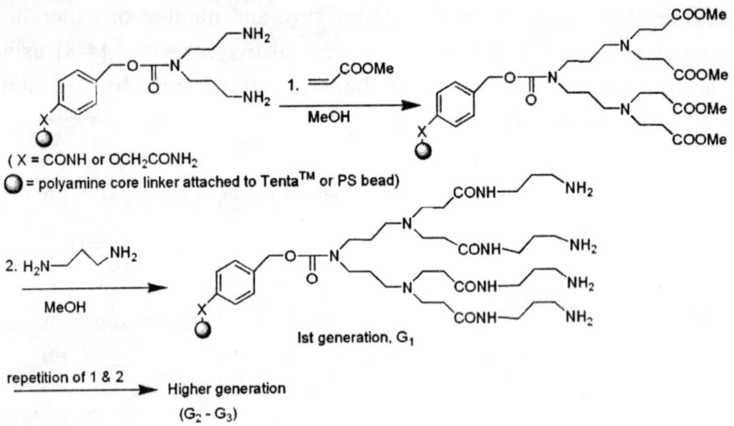

SCHEME 4.3 Synthesis of PAMAM dendrimers on Tenta Gel resin as solid support

The synthesis [10, 13-14] of solid phase PAMAM dendrimers starting from the initiator core 1 (Scheme 4.4) was reported. The core was treated with 250 equivalents of methylacrylate in methanol for 96h at 50°C, followed by removal of excess reagents by filtration and through washing with different solvents such as DCM, DMF, MeOH, Et_2O, etc. The resulted product was then treated with 1,3-diaminopropane in DMSO for 72h at 50°C and then again at room temperature for a further 72h to form next generation dendrimer.

PAMAM dendrimers were also efficiently synthesized by the solid phase methodology and characterized by ^{1}HNMR and ^{13}CNMR spectroscopy and electrospray mass spectrometry [13]. These dendrimers were used as a bead loading amplification tool in the synthesis of a hexapeptide library to allow single bead analysis as well as efficient synthesis of dendrimer conjugates.

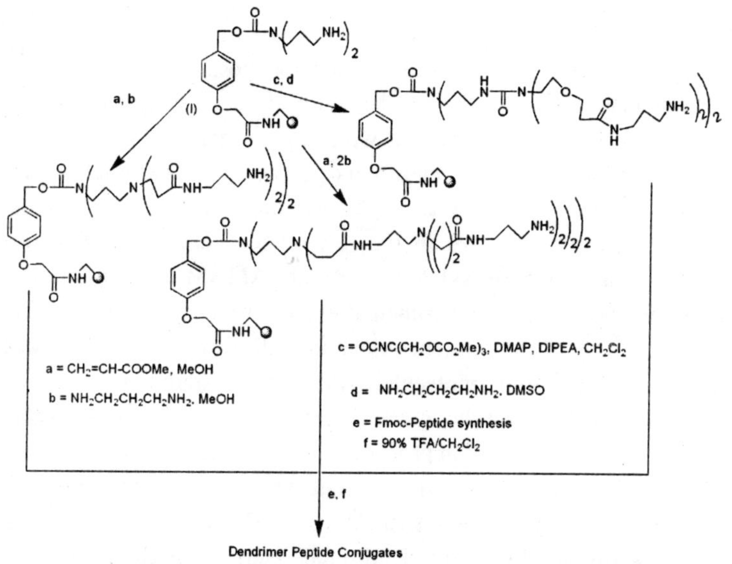

SCHEME 4.4 Synthesis of PAMAM dendrimers by solid phase technique

^{14}C labeled 2.5th and 5.5th generation PAMAM dendrimers were also synthesized [15] by Michael addition of amines with ^{14}C-labelled methylacrylate. The radio-labeled dendrimers had the specific activities between 30 and 45 mCi/mmole in very high radiochemical purity. The AFM was used to characterize the higher generation PAMAM dendrimers. This technique was employed [16] to characterize 9th generation PAMAM dendrimer packed on a mica surface under various conditions. The study showed that the dendrimer deposited on mica surface is of well ordered two dimensional hexagonal arrays with diameter of 11.4 nm size particles. This may be one of the smallest regular monolayer arrays of PAMAM dendrimer. These arrays were formed by the interaction of dendrimers molecules on mica surface, though the other factors like molecular interpenetration and the rigidity of the branch structure obviously play an important role.

4.2.2 Characterization

PAMAM dendrimers are the first starburst dendrimers characterized in details by a number of studies. A series of 10 different 'half generations, n.5" (n = generation number) PAMAM starburst dendrimers, which differ systematically in molecular weight, size and surface charge density were studied [17] by fluorescence spectroscopy. It characterizes the structures of the dendrimers, behavior of the carboxylated anionic dendrimers and their interaction with ionic surfactants. This unique class of anionic starburst dendrimers possesses' an external anionic surface. Pyrene was used as a photoluminescence probe to sense various hydrophobic sites in the microheterogeneous architecture offered by sodium-carboxylate terminated PAMAM dendrimers. This probe method provides experimental evidence for a structural surface tension between different generations as well as the interactions with small ionic amphiphilic molecules. Although the starburst dendrimers do not noticeably affect the micellization process of anionic surfactant such as SDS, the association process with a cationic surfactant (DTAB) leads to the formation of two different types of starburst dendrimer template surfactant aggregates (SBD-TSA). The addition of DTAB to aqueous solution of starburst dendrimers leads to formation of SBD-TSAs that result from non-cooperative, random condensation of surfactant molecules on the anionic dendrimer surface for lower generations (0.5 to 3.5). Whereas for higher generations (4.5th to 9.5th), SBD-TSA resulted from initial non-cooperative, random condensation followed by cooperative condensation of surfactant molecules on the anionic dendrimer surface. These results [17] are consistent with the theoretical generalization of change in the morphology of starburst dendrimers from an open branched structure for generations 0.5 to 3.5 to a closed, increasingly compact surface for generations 4.5 to 9.5. Again, the fluorescence spectroscopy along with laser flash photolysis were utilized to study [18] the supramolecular dynamics of triplet excited states of the 2-naphthol-EDA core PAMAM dendrimer non-covalently bound complex. By varying the acidity and dendrimer size, PAMAM dendrimers were shown to possess both nonspecific and specific binding properties. 2-Naphthol binds preferentially to the tertiary amine groups within the dendrimers interior and can be released by lowering the pH of the solution. Again, as the dendrimer interior becomes increasing charged, the spacing of the branches may expand, thereby leading to increase dynamic behavior of the guest molecules. Also the amount of protonation of tertiary amine groups varies with the varying size of the dendrimers, 12%, 10% and

<1% for 2nd, 4th and 6th generation dendrimers respectively. The computer aided continuous wave (CW) and pulsed EPR, negative-staining TEM, and DLS techniques were also used [19] to characterize the starburst dendrimer, vesicle solutions and to study their interactions. Different sizes (generations) and level of protonations of the surface amino groups of the PAMAM dendrimers and solution of dimyristoyl phosphatidylcholine (DMPC) vesicles were analyzed, both separately as well as in the mixture of the two components by the above mentioned instruments. TEM indicated that the vesicles are of multilumellar structure of spherical shape with diameter of 0.2 to 1.2 mm, whereas DLS measurement provided the mean diameter 400 nm and 7 nm for the vesicle and 6th generation dendrimer respectively. However, EPR measurements at room temperature were not very informative as (i) a fraction of the dendrimers was not interacting with the vesicles and (ii) the labels that were interacting with the vesicles were rotating quickly around the main axis. It is noteworthy to mention here, that the starburst dendrimers were labeled with a nitroxide radical for EPR study. But the interactions between the dendrimers and the vesicles were observed and quantified by EPR study below the freezing transition of a portion of the solution. The fraction of the dendrimers interacting with the vesicles underwent a glass transition also. Dendrimer-vesicle interactions modified the direction of the fast rotation axis of the radical, and the interaction was more effective for protonated dendrimers of higher generation (larger size). The interaction between the nitroxide label dendrimer and the vesicle partially compressed the hydration layer of the N-O group, and/or the hydration layer of the vesicle head groups was compressed onto the unpaired-electron site. This study is very useful, as it provides the information on the possible utilization of starburst dendrimers as gene carriers.

Cu(II) was also used as a probe to investigate [20] the structure and ion binding capacity of the different generation PAMAM dendrimers. The computer aided analysis of the EPR spectra provided information on the formation of Cu complexes in various internal or external locations of the dendrimers as well as the structures of the dendrimers as a function of size of the dendrimers, pH, temperature, Cu(II) concentration and aging of the samples. At low pH, Cu^{2+} ion competes with protons for binding with the external amino groups, which became available for complexation at pH>3.5. The $Cu(H_2O)_6^{2+}$ complexes were localized in proximity to the surface of the dendrimers. About 20% of $Cu(H_2O)_6^{2+}$ residues in the water pools in the open structures of the lower generation dendrimers (upto 3rd generation) and were capable of undergoing a freezing transition. The EPR spectral studies showed the superposition of three components corresponding to (i) a $Cu(H_2O)_6^{2+}$ complex, (ii) a complex with two surface NH_2 groups and (iii) a complex with two surface NH_2 groups and two internal NR_3 groups. The amount of 3rd component increases with the increase of generation and pH of the medium. For example, at pH 6 that was the only component and located only at the external layers of the dendrimers. At higher pH, Cu^{2+} migrates to the internal dendrimer structure and the complexes show higher mobilities than the previous cases. The dendrimers decomposed upon aging and decomposition was almost completed for the earlier generation at pH>5.5 and for 60 days after preparation [20].

The utilization of ultrafast time resolved photoluminescence in a PAMAM dendrimer-Au and Ag nanocomposite systems was reported [21]. High-resolution TEM revealed that the guest-metal domains in these nanocomposties can form clusters with sizes distribute around 15 nm [22]. The fluorescence anisotropy measurements showed that the metal emission shows a depolarized character, while the dendrimer host shows a relatively polarized emission. The isotopic fluorescence decay was found much slower than the depolarization rate. There is a fast (70 fs) fluorescence decay component associated with the metal nanocomposties. Anisotropic measurements showed that the fast component is depolarized. The enhanced emission is suggestive, due to local field enhancement in the elongated metal dendrimer nanoparticles.

The absorption of EDA based amine terminated PAMAM dendrimers from anhydrous ethanolic or deionized aqueous solution (pH = 6.5-7) onto gold was studied *in-situ* by a QCMB technique [23] at room temperature (21°C). The study revealed that the growth of equilibrium surface coverage was exponential with generation upto 6th in aqueous solution, while in ethanolic solution growth is much weaker and having an approximate linear relation with generation. However, this strongly increase of trend of surface coverage in aqueous medium was broken at 7th generation and a drastic drop was observed, but the increase of trend again resumes then upto 9th generation. In ethanol, the equilibrium frequency shift was proportional to the equilibrium surface coverage, correspond to about a monolayer formation and the absorption is likely driven by weak favorable interaction between the primary amine end groups of PAMAM and gold. While in case of aqueous medium, amine end groups of PAMAM trend to protonate at low pH and thus PAMAM became positively charged. The measurement of thickness of absorption layer indicated that multilayers formed on the clear gold surface with exponential increase of number of layers upto 6th generation. The drop in surface coverage from 6th to 7th generation may be due to surface crowding effect. The formation of multilayers could be explained by a favorable electrostatic interaction between the positively charged dendrimers in aqueous solution and the gold substrate. Thus one can manipulate the surface coverage of PAMAM dendrimers on gold significantly from monolayer to multilayers, just by altering the solution conditions to manipulate the role of electrostatic forces.

The nature of the cavities and voids inside the dendritic structures is of great importance and also how it is affected by dendrimer size, constitution and solvent, is of paramount importance in relation to their potential applications. It is therefore, necessary to discover the ways of characterizing the microenvironments so that it might be understood, engineered and utilized. Many investigators have made use of functional probe in order to study dendritic microenvironments. These probes can either be attached covalently to the dendritic structure, like photochemical and chiral moieties or they can be introduced as guest species. Spectroscopic methods have also produced information about dendrimers microenvironments.

Positively charged nitroxide radicals attached to carbon chains of different lengths were found to be valuable probes for identifying hydrophilic and hydrophobic interacting and binding sites at the surface of the dendrimers [24]. For longer chain length containing nitroxide probes, different supramolecular structures were formed between half generation PAMAM dendrimers and surfactant aggregates [25-26].

Further, when the nitroxide group was attached to a phenanthroline ligand of $Ru(phen)_3^{2+}$, interaction with the surface of sodium alkyl sulfate micelles was much stronger than the interactions with the half generation PAMAM dendrimers [27], though the half generations are resemble to amine micelles [28]. However, as nitroxide groups is too large to enter the internal dendrimer structures, the EPR techniques in these cases monitor the interactions between polar nitroxide groups and polar or charged terminal groups of the PAMAM dendrimers, whereas paramagnetic metal ions may enter into the dendrimer structure and bind to both internal and external ligand groups [20, 29-30].

The EPR spectra of Cu(II) complexes of anionic half generation PAMAM dendrimers in aqueous solution at room temperature showed [29] a distinction between lower (generation <3) and higher (generation >3) generations and thus confirmed the change in dendrimer structures of different generations. The lower generations have more open structures as indicated by greater mobility of the Cu complexes.

It was also found that Cu^{2+} ions formed different complexes with carboxylate and amino groups as a function of generations, pH and temperature. Thus Cu^{2+} ions were distributed both inside and outside of the structure of half generation PAMAM dendrimers. These structures of Cu(II) dendrimers complexes are equivalent to the complexes with biomacromolecules such as proteins or enzymes of similar structures. By utilizing photochemical and spectroscopic probes the surface properties of ammonia core and ethylenediamines core PAMAM dendrimers were also studied [31]. Photoinduced electron transfer quenching of *tris*(2,2′-bipyridyl) ruthenium(II) chloride by methyl viologen in the presence of dendrimers was found to be enhanced when the donor and acceptor are adsorbed on the higher generation dendrimers. The ability of aggregation formation with different dyes is similar to the earlier studies in both the cases.

The structure and properties of several PAMAM dendrimers have been studied extensively. In general, these studies reported that PAMAM dendrimers are highly ordered arrays with an open structure for earlier generations and dense spheroid topology for the latter generation [32].

The interaction between PAMAM dendrimers and vesicles consisting of DMPC and its phosphalidylcholate sodium salt (DMPA-Na) in various relative ratios from 0% to 40% were studied [33] by EPR spin label and spin probe techniques supported by TEM. The probes used were the hydrophobic 5-doxylstearic acid (5DSA) and positively charged nitroxide CAT16, a cationic surfactant with a C16 chain to monitor the formation of dendrimer mixed vesicle (DMPC/ DMPC-Na) complexes by electrostatic interaction. The computer aided EPR spectra gave the information about the mobility and structural parameters of the probes and labels. The spin labels at the vesicles-dendrimer interface modified the structural and mobility parameters upon complexation with the dendrimer. At higher level of protonation, dendrimers showed a larger interaction with the vesicles, especially for the mixed vesicles compared to dendrimers at lower level of protonation. TEM micrograph results are well agreed to EPR results and showed that the addition of DMPC-Na to DMPC modifies the vesicle shape from spherical to rod like. Again, the larger dendrimers are more affected by the structural variation of the bilayer with respect to the small dendrimer, but at the same time the smaller dendrimers are more sensitive to the electrostatic interaction between the charge groups of the dendrimers and the vesicles. These results suggested different interacting models between the small and the large dendrimers and the vesicle. Though the authors [34] previously shown that the uranyl ions are completely sequestered by the dendrimers and the vesicles become invisible by TEM in the presence of the dendrimer. As a consequence, the interaction between the dendrimer and the vesicles could not be analyzed by TEM and only vesicles without dendrimers provide clear TEM features.

Computer aided EPR techniques were used to investigate [35] the supramolecular structures formed by the probe technique between anionic PAMAM dendrimers and varieties of cationic surfactants of different structures. Polarities change the solubilities and self-aggregating abilities. The results obtained in this study are as follows. (i) The critical surfactant aggregation concentration was much lower than the critical micelle concentration in the presence of dendrimers. (ii) Although these aggregates formation does not depend on ionic strength of the medium but it depends on the dendrimer size and concentration.

The aggregation was favored in the presence of higher generation dendrimers at low concentration. (iii) The primary interactions of the probe (CAT16) with the dendrimer surface and the secondary interactions of the probe belonging to the surfactant aggregate formed on the dendrimer surface compete with one another. Also the ratio of primary binding to secondary binding depends on the size of the supramolecular structure, which was formed by surfactant aggregates in the presence of dendrimers. Though the longer dendrimers should form too large supramolecular structure with the surfactant aggregates but because of steric reasons their function becomes unfavorable. Thus a large number of fractions of the surfactant that aggregated and the probes give rise to primary noncorporactive interaction with the hydrophilic and hydrophobic site of the latter generation dendrimers.

However, if the core of PAMAM dendrimers was modified by using a long (10-12) carbon chain containing diamine in place of NH_3 or short carbon chain containing diaminoethane core, then that modification may dramatically change the morphology and properties of these dendrimers. This could be performed by (i) incorporating a hydrophobic site at the core of a hydrophilic periphery, (ii) making the overall morphology less compact and (iii) allowing access to the interamines even at higher generation. This type of modified dendrimer may have potential application in host-guest interactions, which could be utilized in membrane transport and drug delivery system. As PAMAM dendrimers have the characteristic hydrophilic external periphery for solubility in aqueous medium along with hydrophobic internal core of large molecules. However, the small central core like NH_3 or diaminoethane containing PAMAM dendrimers only incorporate smaller organic molecules, though the drugs are generally larger in size. So, to encapsulate larger drug molecules the long chain diamino central core based PAMAM dendrimer may be useful.

In nearly, all earlier studies photochemical and spectroscopic probe methods of PAMAM dendrimers with ammonia core showed that the absorption behavior of the dyes on the dendrimer surfaces allowed to classify the dendrimer into two different classes: (i) lower or earlier generation ($n \leq 2.5$) and (ii) higher or latter generation ($n \geq 3.5$). This difference in adsorption behavior is due to the fact that the earlier generations have more open 'starfish' surface structure and the latter generations have more closed and nearby spherical external structures. This model structure of the PAMAM dendrimers has been proved repeatedly for ammonia core PAMAM dendrimers by various techniques including fluorescence and optical absorption probes; pyrene [17], metal complexes [36-37] adsorption and aggregation of organic dyes [38-39] and EPR techniques [27, 29].

The only computer aided EPR spectroscopy was used in several studies [20, 24-27, 29-30] to characterize the structures and interacting (binding) abilities of this novel class of anionic half generation and also full generation PAMAM dendrimers. EPR studies have provided the information about the binding and structural properties of host molecules and surfaces in the case of various spin labels and spin probes as guest molecules. In these studies for the half generation of the PAMAM dendrimers the probe used were nitroxide radicals [24-25, 27] or paramagnetic metal ions like Cu(II), Mn(II), etc. whereas for full generation paramagnetic metal ions Cu(II) and Mn(II) were only used. Thus the study [24] suggested that the increase of size of the dendrimers lead to an increasingly hydrophobic cavity, favoring the encapsulation of the more hydrophobic nitroxide radicals. It can also be found that the electrostatic interactions between the positively charged nitroxide radicals and the negatively charged dendrimer surface increased as the size of the dendrimer increases. Whereas Mn(II) was also used as a probe to investigate the interacting abilities of the various ligands sites at the surface of PAMAM dendrimers at different protonation conditions and as a functions of generation, for both half and full generations. The computer aided EPR analysis showed that Mn(II) did not interact at all with the full generation dendrimers and only interacted probably with 2nd- 3rd salvation shells with the surface carboxylate groups of half generations. The spectral analysis showed two main components: (i) a minor component was due to Mn(II) that was localized in the salvation shells of the internal and external dendrimer surface and was in fast exchange with the bulk solution. At high pH, mainly for the more open surface earlier generation dendrimer (n<4.5), the Mn-aqueous complexes coverage is close to internal dendrimer sites of low polarity. On the other hand, the dense packing of the external layer of the latter generation (n>4.5) and the high concentration of the R_3N^+H groups at the internal layers at low pH, prevented the entrance of $Mn(H_2O)_6^{2+}$ inside the dendrimer structure and (ii) a major component arising due to complexation of Mn(II) with carboxylate groups of dendrimers at the 2nd salvation shell of the ions. The interaction was greater for latter generation than the lower generation due to the larger charge density in the former. The complexation of Mn(II) was favored for higher generations at high pH and low Mn(II) concentration. On the basis of this analysis, the maximum carboxylate groups involved in direct complexation of $Mn(H_2O)_6^{2+}$ was 0.05-0.1 %. Thus Mn(II) did not showed any interaction with full generation PAMAM dendrimers. Whereas Cu(II) ions proved to be good probes to study the structure and dynamics of full generation PAMAM dendrimers. The study also showed that the carboxylate groups at the surface of the half generation dendrimers formed complexes with both Cu^{2+} and Mn^{2+} ions over a large range of pH [29-30]. Whereas the amino groups at the surface of the full generation PAMAM dendrimer interact with Mn^{2+} ions [30], but Cu^{2+} ions formed complexes with the amino groups though at low pH (>3.5) only [20].

ESI-MS has been used to characterize the imperfections in PAMAM dendrimers [40-43]. This study indicates PAMAM dendrimers showed imperfection and these defect structures arise from retro-Michael additions and intramolecular lactam formations. The molecular dynamics simulation study on PAMAM dendrimers showed [44] back folding of the branches in the dendritic structure. The conformational behavior of the PAMAM dendrimers has been studied with several techniques. SEC in combination with intrinsic viscosity measurements has been utilized to obtain the hydrodynamic radii of the PAMAM dendrimers [45-47]. The studies showed that the PAMAM dendrimers have a core and a densely packed outer layer, which is agreed well to the de-Gennes theoretical model. ^{13}CNMR relaxation studies on PAMAM dendrimers [48] indicated that there is no dramatic change in chain dynamics upto tenth generation. The chain motion is most rapid near the surface and is slower in the core of the dendrimers. These studies concluded that back folding to some extent on the dendritic surface is generated to relieve the steric strain. This conclusion was supported by ^{2}DNMR study [49] on various generation of deuterium labeled PAMAM dendrimers. SAXS measurements on PAMAM dendrimers have not given any clear-cut results [50].

Near IR studies of the surface functionalized PAMAM dendrimer with naphthalene diimide anion radicals showed that there are strong π-π stacking interactions between the modified end groups [51-52]. In these cases, the back folding is presumably prohibited by inter end group interactions. The formation of unimolecular anionic micelle of the sodium salt of carboxylated surface functionalized half-generation PAMAM dendrimers was supported by various probe techniques. Each of these techniques has revealed that the binding (adsorption) of small cations to the surface of anionic dendrimers is qualitatively different from earlier generations (2.5 SBD or less) relative to the latter generations (3.5 SBD or greater).

The amine terminated PAMAM dendrimers were found [53] to form a tight nonpolar association with the majority of the dyes within the interior of the dendrimer, as observed by fluorescent, solvatochromic probe phenol blue in aqueous solution of the dendrimers. The absorption and steady state fluorescence emission spectra showed a decrease of polarity with the increase of generation of the dendrimers (upto 3rd generation). However, for 4th to 8th generation dendrimers, the microenvironment possesses equal polarity. Fluorescence anisotropy values for phenol blue in the dendrimers demonstrate the sensitivity of the dye changing the molecular volumes of the dendrimers. This study thus confirmed that the phenol blue is associated with only inside of the dendrimers but it does not interact with the dendrimer surface groups.

The computer simulations of EPR spectra of vanadyl chelate complexes of PAMAM dendrimers were reported [54]. The chelate 2-(4-isothiocyanatobenzyl)-6-methyl diethylene triaminopenta acetic acid was covalently attached to ammonia core PAMAM dendrimers via a thiourea linkage.

The carboxylate terminated PAMAM dendrimers were characterized [55] by excitation-emission matrices fluorescence and lifetime fluorescence techniques. Excitation-emission matrices fluorescence study showed a broad peak with an excitation and emission maximum of 380 nm and 440 nm respectively for all the dendrimers but an overall increase in relative fluorescence emission with the increase of generation was observed. Two life times (1.3-2.5) ns and (4.2–7.1) ns for the dendrimers and a shorter lifetime, (0.2–0.4) ns for the solvent were observed in lifetime measurement fluorescence study. The lifetime increased with the increase of generation, for example, both the lifetimes became almost double by shifting 2.5th generation to 7.5th generation dendrimer. Thus the dendrimer became densely packed with the increase of generation. This fluorescence may due to an n-π^* transition from amido groups through out the dendritic structure.

The 5th generation PAMAM dendrimer with different surface groups was characterized [56] with respect to their dispersity, size and self-interaction by several physico-chemical techniques. Amine terminated PAMAM dendrimers form oligomeric aggregates in aqueous solution and also in the presence 0.6 M sodium phosphate solution. Whereas the hydroxyl terminated dendrimer behaves as a single homogenous species at low concentrations. The measurement of density increment and the sedimentation and diffusion co-efficient of hydroxyl terminated dendrimers suggest a more swollen, porous structure than a globular protein of comparable mass. However, the measurement of the concentration dependence of sedimentation equilibrium of G_5-OH at pH 7.2 phosphate buffer indicates the presence of significant electrostatic repulsion overlaid on weakly attractive interaction, leading to the formation of non-specific aggregates at sufficiently higher dendrimer concentration.

4.2.3. Properties

The dense starburst PAMAM dendrimers exhibit greater and more uniform reactivity than their corresponding conventional star polymers because of higher density of the terminal groups of the former than the latter. Thus such dense dendrimers are useful [57] as emulsifier for oil/water emulsions, wet strength agents in the manufacture of paper and for modifying viscosity in aqueous formulation such as paints.

The amine terminated PAMAM dendrimers upto 5th generation are found to be much faster [58] in aminolysis reaction in water at pH=8-5 compared to an equivalent amine concentration of ethylenediamine [58] or N-acetylethylene diamine. The fourth generation dendrimer gives the maximum rate of enhancement. This enhanced rate of aminolysis may be due to (i) hydrophobic binding of the substrate on the surface of the dendrimers making them close proximate to the reactive amine groups and (ii) stabilization of the transition state of the reactant by the internal amide groups within the dendrimers.

An unusual pH dependent polarity changes in PAMAM dendrimers was observed [59]. In order to investigate the effect of pH, the sixth generation PAMAM dendrimer was bound with polarity responsive probe, 5-(dimethylamino)-1-naphthalenesulfonic acid (DNS) at a fixed pH. The studies on polymer solutions have shown directly that DNS probes exhibit an increase of emission with the increase of polymer concentration. The decrease in polarity between pH 10 and 8.3 by DNS bound dendrimer most likely to indicate that the probe is sampling more density packed binding sites at the lower pH. Two models can be utilized to understand the conformational changes due to pH variation. A 'denser core' model, if the interior of the dendrimer is less polar than the surface, then protonation of the inward-folded termini would result in the observed polarity decrease as the pH is lowered to pH 8.3. Also further lowering the pH would have to cause molecular expansion, in order to account for the polarity increase below pH 8.3. The 'denser shell' model, if protonation of the dendritic termini is able to drive than to the molecular surface, a less polar surface region could result. Further, the lowering of pH would then begin to protonate the interior, more polar tertiary amines.

The adsorption and aggregation of methylene blue (MB) on anionic half-generations PAMAM starburst dendrimers were studied [38] by using UV-vis. and fluorescence spectroscopy. Although, MB did not showed any significant dimerization or aggregation in 10^{-5} M aqueous solution, however, dimers and aggregates of MB was observed on the surface of the dendrimers by UV-vis. spectroscopy. The absorption of MB decreases with the increase of concentration of the dendrimers, and the aggregation between the dye molecules depends strongly on the generation of the dendrimers. The higher generations formed aggregations more easily than the lower generations. As the inside cavities of the dendrimers are too small to host the aggregates of dye molecules, so aggregation occurs only on the surface of the dendrimers. Also for higher generations, because of a closed surface structure, the dyes are prevented to enter into the interior of the dendrimers. Fluorescence spectroscopic studies also showed similar results. In addition to MB, thionine and thiopyronine were also tested [38] and found analog results. However, MB was found to be the most suitable dye, because of its strong tendency to form aggregates on the dendrimers and also because of the large differences in optical absorption of the different forms of MB aggregates.

The random complexation of host molecules in dendritic structure was reported [44] by monitoring the change in guest ^{13}C spin lattice relaxation times. The methylester terminated PAMAM dendrimers used as host with aspirin and 2,4-dichlorophenoxy acetic acid as guest molecules. The relaxation time of the guest molecules in $CDCl_3$ changed in the presence of dendrimers which depends on the generation. The values decreased as the generation number increased form 0.5 to 3.5, but remained constant for higher generations. The maximum ratio of the guests was ~3:1, based on a molar comparison of the carboxylic guest and the interior tertiary amines, suggesting an acid-base interaction between the host and the guest. The interaction of acetylsalicylic acid and 2,4-dichlorophenoxy acetic acid with PAMAM dendrimers was studied by ^{13}CNMR spin lattice relaxation times. The results of these experiments indicated that the guest molecules are encapsulated within or at the surface of 4th to 6th generation PAMAM dendrimers.

With the increasing ability to synthesize perfectly branched polymers, the efforts to understand their solution and bulk properties also increased. The study [60] on rheological behavior of PAMAM dendrimers upto 7th generation showed that all the generations of the dendrimer solutions exhibited Newtonian flow behavior as manifested by the direct proportionality of share stress to share rate. Again, Brownian dynamics was used to simulate the intrinsic viscosity and various statistical properties of dendrimers under the influence of share flow [61]. The hydrodynamic and excluded volume interactions were taken into account explicitly. The onset of shear thinning was observed to occur at lower share rates for higher generation dendrimers. As the generation increases, the zero shear rate intrinsic viscosity reached a maximum and begins to fall. The fact was also observed experimentally for PAMAM dendrimers [62]. The calculation of intrinsic viscosity of PAMAM dendrimers showed [63] that the maximum dependence of intrinsic viscosity against generation number was observed only when the characteristic size of the excluded volume interactions of the monomer is smaller than its hydrodynamic radius. It was also observed [61] that shear flow leads to a monotonically increase in the radius of gyration of dendrimers in contrast to a linear chain. Further, a radius of gyration based only on terminal units that is larger than a radius of gyration based on all monomers of a dendrimer does not automatically imply that the terminal units are located on the periphery of the dendrimer.

A lattice based thermodynamic framework for the vapor-liquid equilibrium (VLE) of PAMAM dendrimer/methanol system has been suggested [64]. It accounts for the effects of highly branched structures and end groups at the external surface of the dendrimers and the specific interaction among the end groups and the solvent molecules. They showed that the specific interaction between the solvent molecules is much smaller than that between the solvent and the end groups of PAMAM dendrimers. Thus the solvent-end group interaction dominates VLE of the dendrimer in solvent, and hence the type of end groups is one of the major factors determining VLE in dendrimer/solvent system.

The preparation and characterization of PAMAM dendrimer monolayer and dendrimer alkanethiol mixed monolayers adsorbed to gold were described [65]. The single component monolayers were prepared by exposing an Au substrate to ethanolic solutions of amine or hydroxy terminated PAMAM dendrimers. The resulting monolayers were highly stable and nearly close packed for dendrimer generation ranging from 4 and 8. Electrochemical ac-impedance measurements indicate that the dendrimer surface is very porous towards the electroactive redox couple $Fe(CN)_6^{3-/4-}$. Exposure of higher generation dendrimer monolayers to ethanolic solutions of hexadecanethiol results in a dramatic compression of the dendrimer, and causes them to reorient on the surface from oblate configuration. The physical and chemical properties of the monolayers were evaluated by reflection infrared spectroscopy, ellipsommetry, contact-angle measurements, ac-impedance spectroscopy, cyclic voltammetry and surface acoustic wave base analyte-dosing experiments.

The structure of a series of PAMAM dendrimers up to 6th generation with diamino dodecane core was investigated [66] by using photophysical properties of an external hydrophobic phenoxazone-9 Nile red dye as probe in aqueous solution. The binding ability of diaminododecane, NH_3 and diamino ethylene core dendrimers were studied with the Nile red dye and found that long chain dodecane core containing dendrimers having strong tendency to interact with the dye. Also the emission of the Nile red dye in aqueous medium was significantly enhanced in the presence of dodecane core dendrimers but not for NH_3 or ethane core dendrimers. The supramolecular assemblies were generated by the interactions of these dendrimers with anionic surfactants in aqueous medium. The emission spectra also indicated a strong evidence for hydrophobic environment to the dye for the longest chain (C=12) but not for the shorter chains (C= 4 or 8). Again only the lower generation (n≤2) dendrimer blocks are open to access the Nile red dye but not the higher generation blocks. The higher generations are effectively close and densely packed, and hence only surface is accessible to the dye. These dendrimers are sterically less capable of hosting hydrophilic organic molecules than those of 2nd generation. Again, in comparison with NH_3 or diaminoethane core dendrimer, diaminododcane core dendrimers exhibited 10 times enhanced emission of the hydrophobic Nile red dye. Since the number of exterior groups for electrostatic interaction is the same for all above cases, so it is definitely the more open core structure of long alkyl chain PAMAM dendrimer is probable reason for the difference in emission spectra. This open structure also provided anionic surfactants to access the interior tertiary amines, which is not possible for the other two type dendrimers [66] (Fig.4.3). Further, the enhanced emission at lower pH values suggests that pH is also a useful tool to adjust the macromolecules structures of the dodecane core dendrimers.

The emission intensity and life times of $Ru(4,7-(SO_3C_6H_5)_2$-

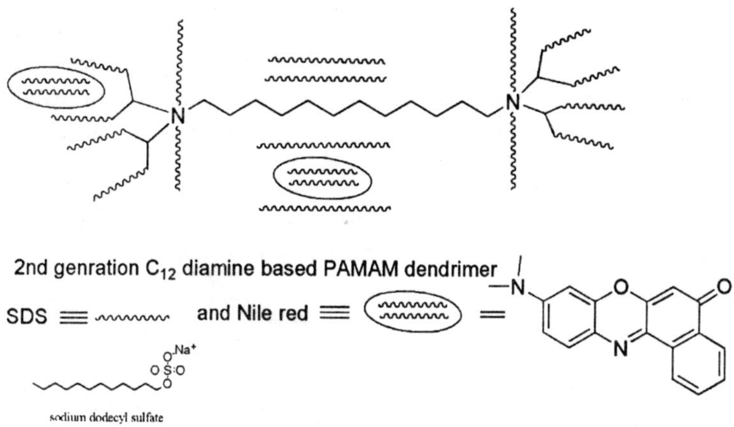

FIG. 4.3 Structure of PAMAM dendrimers with nile dye

phen)$_3^{4-}$ ions in solutions containing various cationic PAMAM dendrimers in the presence and the absence of potential quenchers have been investigated [67]. The anionic Ru complex binds strongly to the positive surface of the full generation dendrimers by electrostatic interaction, and the life time and emission intensity of the probe were enhanced due to decrease of O_2 concentration on the dendrimer surface. The branches at the external surface are terminated at an amine stage, which was protonated in an aqueous environment to provide a cationic surface. The quencher of like charges is strongly repelled the external surface of this dendrimers when methylviologen (MV^{2+}) quencher reacts with Ru complexes in the presence of the dendrimers. However, for like charge probe and quencher molecules [$Fe(CN)_6^{4-}$ and $Fe(CN)_6^{3-}$], it was found that the kind of quenching reaction depends on the dendrimer concentration. If the ratio is ≥ 1 for [probe]/[dendrimer] and [quencher]/[dendrimer] then the quenching rate constant is bimolecular compared to unimolecular reaction rate at lower ratio [67].

The H-bonding interactions between dendrimers and guests have been achieved by incorporation of coordinating sites that are complementary to the guest at the focal point or in the core of the dendrimers. Pyridine formed complexes with external and internal sites of the amine functionalized dendrimers but not with the surface groups of ester terminated dendrimer, though overall weak interaction was found. Trimethadione and nicotine showed no significant interaction with any type of dendrimers. However, the interaction of quinazoline and quinoline with the ester-terminated dendrimers proved to be highly dependent on the chain length of the methyl ester [68].

In the presence of carboxylate terminated PAMAM dendrimer, $CaCO_3$ crystallized to spherical vaterite crystal but rhombohedral calcite crystals were formed in the absence of the dendrimer [69].

PAMAM dendrimers of 2nd to 6th generation with aromatic rings, dodecyl chains, methyl ester groups or perfluoroalkyl chains and two series of benzyl ether linear polymers that are analogs to the dendrimers were used to study [70] the vapor-liquid equilibrium at 35-89°C in different solutions of the dendrimers. The solvents used were acetone, acetonitrile, chloroform, cyclohexane, methanol, n-pentane, n-propylamine, tetrahydrofuran and toluene. The amount of solvent absorbed by the dendrimers depends on the type of end functionalities. The relation between solvent absorption and generation number, or molecular weight of the dendrimers also depends on the solvent-dendrimer system and on the temperature. However, the solvent absorbed by the linear polymers was lower than that for the analog dendrimers, may be due to the presence of crystalinity in the structure of linear polymers.

The structural distortion of high generation PAMAM dendrimer at the liquid/solid interface was investigated [71] *in-situ* by using tapping mode AFM. The Feature size in AFM image indicates that the surface adsorption of the 9th generation dendrimer on mica substrate is accompanied by a significant compression along normal of the surface but only moderate lateral expansion. The more easily disturbed, web like inter connected features found at the water-graphite interface. However, AFM images of lower generation dendrimers at the water-graphite interface revealed that a large three-dimensional aggregate formation.

4.3 MODIFICATION OF POLY(AMIDO AMINE) DENDRIMERS

In this section synthesis, characterization and properties of modified PAMAM dendrimers are discussed.

4.3.1. Photoactive

The ester terminated PAMAM dendrimers formed deep blue complexes with $CuSO_4$ solution, where NH_2 terminated homologues produced deep purple solution [72]. The amide groups of PAMAM also form coordination complexes with different metal ions [73], which are seems to be exciting areas for further research. The most facile entry into surface complexation involved hydrolysis of ester terminated dendrimers with stoichiometric amounts of alkali metal hydroxides of Na, K, Cs, Rb, etc. (Scheme 4.5). Gennes from his theoretical studies, hypothesized that the starburst PAMAM dendritic structures are flexible in the lower generation but became more rigid and constrained, as one approach to the starburst dense packed limit. This may due to the effect of branching angle of the branch cell as well as their rational angles.

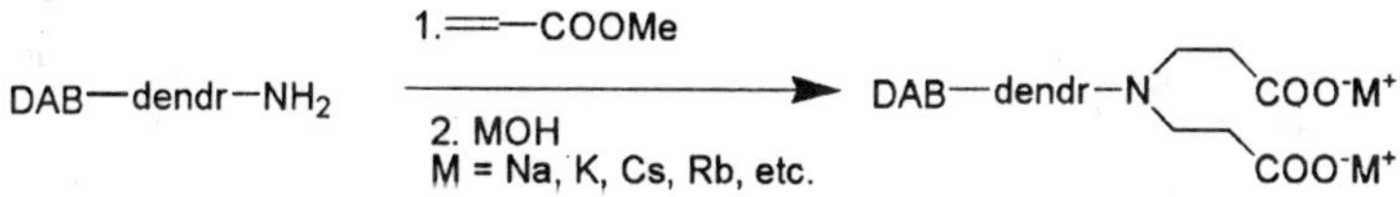

SCHEME 4.5 Modification of PAMAM dendrimers with alkali metal hydroxides and methyl acrylate

It is also possible to incorporate other metal ions such as Fe^{3+}, Gd^{3+}, Mn^{2+}, Pr^{3+} and Yr^{3+} by using methyl bromoacetate instead of methyl acrylate to obtain ester terminated dendrimers followed by hydrolysis. In general, the $N(CH_2CO_2^-M^+)_2$ complexes were more stable than the $N(CH_2CH_2CO_2^-M^+)$ complexes. The sub-stoichiometric yttrium complexes displayed intramolecular exchange reactions as found by ^{13}CNMR spectroscopy and ^{90}Y-trace experiments [23].

The synthesis of chromoionophores endcapped PAMAM was reported [74] in 1996. The 2nd generation of amine terminated PAMAM reacted with anthracene aldehyde and subsequently reduced it to the secondary amine. This mono functionalized dendrimer was then further reacted with a bronic acid derivative to get the desired dendrimer (Scheme 4.6). The resulting dendrimer acts as a sensor for saccharides and shows varying fluorescence intensity depending on the number and kind of the sugar molecules present. By the complexation of D-galactose and D-fructose, respectively, in the form of an intramolecular 2:1 boronic acid-saccharide complex, enhanced fluorescence intensity is caused by a photo-induced electron transfer. Actually, this bronic acid-saccharide complex is more stable than flexible diboronic acids or boronic acids themselves. This enhanced bonding is probably due to the high local concentration of receptor units. When the monosaccharide interacts with one boronic acid moiety, it can intramolecularly bind to a second boronic acid of one of the other seven receptor-units.

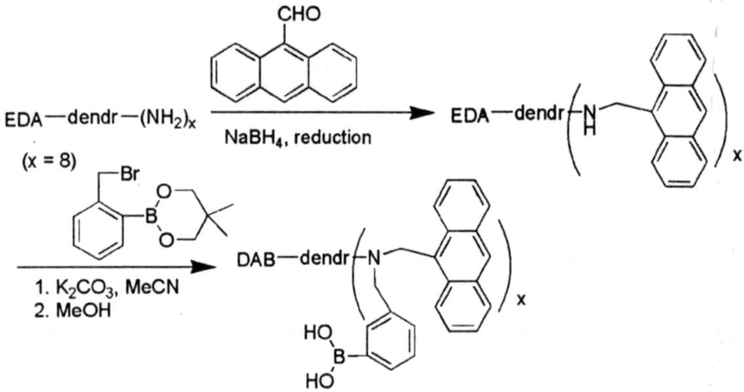

SCHEME 4.6 Synthesis of chromoionophore modified PAMAM dendrimers

PAMAM dendrimers with a cinnamoyl shell were synthesized [75] by reaction of the dendrimers with 2-chloroethanol and cinnamonylchoride, which resulted in densely packed polymerizable unsaturated groups on the periphery. The cinnamoyl shell of the dendrimers dimerized when irradiated under the UV light by using 5-nitroacenaphthylene as an initiator in dilute DMF solution. The modified dendrimers were characterized by UV-visible, FTIR, ^{1}HNMR spectroscopy, SEC and viscosity measurement. The FTIR spectra indicate that the photocyloaddition of the cinnamonyl shell of the dendrimers took place within the molecules with disappearance of the double bonds. The authors also found that the cinnamonyl shell-modified dendrimers possessed fluorescence property and the fluorescence intensity became stronger when the shell was photocyclized under UV-irradiation.

The photoinduced electron transfer reactions were used [37] to probe the structure of starburst dendrimers as a function of their generations. The luminescence from triplet metal-to-ligand charge transfer excited state of *tris*(2,2′-bipyridine) ruthenium(II), [(Ru(bpy)$_3$$^{2+}$] was quenched by electron transfer to the 1,1′,-dimethyl-4,4′-bipyridinium ion (methyl viologen, MV^{2+}) and was monitored with Pekin-Elmer LS-5-spectrofluorometer in the presence of sodium carboxylate terminated half generation PAMAM dendrimer. In the absence of dendrimers as well as in the presence of lower generation dendrimers (0.5 to 2.5 generation), a bimolecular quenching rate constant was observed, whereas for the higher generation, a strong interaction between the dendrimers and the ruthenium complexes was also observed by enhanced luminescence measurement.

Ghosh and Banthia [76] synthesized and characterized a photoresponsive PAMAM dendrimer with an azobenzene central linker. The dendrimer undergoes reversible *cis-trans* isomerization upon exposure to UV-light.

The photophysics of PAMAM modified upto 4th generation with $(Ru(tpy)_2]^{2+}$ and $[Ru(bpy)_3]^{2+}$ were studies in both fluid solution at 25°C and frozen glasses at -196°C [77]. It was shown that the photophysical behavior of these modified dendrimers is complex, as evident form the dendrimer generation dependent variations in the emission wave lengths, the exited state lifetimes, and the solvent dependencies.

The non-radioactive decay rates of the dendrimer complexes in solution weakly depend on the size of the dendrimers. In case of terpyridine complexes, the rate decreases by about 8% from [Ru(tpy)(mtpy)] to dendr-4-Ru(tpy)$_2$ and then slows by an additional 32% as the size increased from 1st to 4th generation. Whereas in case of the bipyridine complexes, the rate increased by 3% from $[Ru(bpy)_2$-$(dmbpy)]^{2+}$ to first generation dendrimer-bipyridine complex and then increases by ≈35% from 1st to 4th generation. Thus the behavior of these dendrimer complexes exhibited by discrete molecular analogs with the bipyridine derivatives exhibiting longer lifetimes and higher quantum yields when compared to the corresponding terpyridine complex.

The photophysical and photochemical properties of the PAMAM dendrimers upto 3.5th generation with anthracene core were studied

[78] by using steady state and transient spectroscopic method in solution. The fluorescence life time of the anthracene moiety in the higher generation is shorter than that of the lower generation. The photodimerization of 3.5th generation dendrimer was observed through chromophoric anthracene unit on irradiation at 355 nm laser light in deaerated solution. Again on irradiation at 266 nm laser light the dimer dissociated to monomeric anthracene. However, the 3.5th generation dendron showed much slower rate of dissociation and lower yield compared to the lower generation dendrimer (0.5th generation). These results suggested that the unlocking reaction is slightly dependent on the dendron generation. Thus several types of dye/dendrimer complexes exist for PAMAM dendrimers. These are (i) penetration of ET-30 beyond the surface groups into the outermost region or interfacial region of the dendrimers, (ii) aggregation of smaller dendrimer around ET-30 molecules and (iii) aggregation of the dye-dendrimer complexes into larger suspensions with limited solubility. However, the sample preparation conditions highly influence the nature of dye-dendrimer association. For example, sample stirring with and without heat results in further association of dye-dendrimer complexes into a suspension of higher order species, whereas sonication prevents the formation of such aggregates.

In contrast to previous fluorescent probe studies [17, 79] the fluorescence emission of phenol blue was not quenched by PAMAM-AT's tertiary amine groups. The authors [80] also investigated the interaction between Reichardt's ET-30 dye (2,6-diphenyl-4-(2,4,6-triphenyl-1-pyridino) phenoxide and amine– and carboxylate-terminated PAMAM dendrimers by spectroscopic study. The absorption spectra illustrated that the solvatochromic, micro-environmental polarity probe associated with PAMAM dendrimer is less polar region with a polarity comparable to 1-decanol. For the most generations, their association results in the increased solubilization of the dye, above its solubility limit in water. It was also found that the ET-30 dye interacts both interfacial region and the surface groups. The dye-dendrimer aggregates were also observed under certain experimental conditions [80]. There was no significant change in the absorbance spectrum between pH 7 to 11.

4.3.2 Redoxactive

It has been shown [73] that the photoinduced electron transfer from *tris*(2,2′-bipyridine)-ruthenium(II) complexes to 1,1′-dimethyl-4,4′-bipyridirium ions occurs much more easily in the presence of carboxylate terminated PAMAM dendrimers of higher generations. In the absence of dendrimer as well as in presence of lower generations, the electron transfer was found to have a bimolecular mechanism. Under the addition of higher generation PAMAM dendrimers, the electron transfer proceeds as a first order reaction. Hence, both cationic species were meant to be electrostatically bound to the dendrimer surface, which acts as a polyelectrolyte.

The synthesis and characterization of viologen functionalized PAMAM dendrimers were reported [81]. An amidation reaction between the succinimide ester of 1-ethyl-1'-(3-propionic acid)-4,4'-bipyridylium dibromide and terminal amino groups of 2nd, 4th and 6th PAMAM dendrimers was employed to get 13-34 % end group functionalized desired dendrimers (Fig. 4.4). The modified dendrimers were characterized by ^{1}HNMR, UV-vis. spectroscopy, MALDI-TOF MS and DLS. DLS measurement indicates that the oxidized viologenated dendrimers were predominately unaggregated in aqueous solutions of electrolytes. The reversible voltammetry study of the modified dendrimers indicates that incomplete electrolysis of the dendrimer-viologen moieties, though the presence of a single wave showed that all electrochemically addressable groups were equivalent. The adsorption of a stable film of oxidized viologented–dendrimer on Au electrode was also demonstrated by cyclic voltammetry.

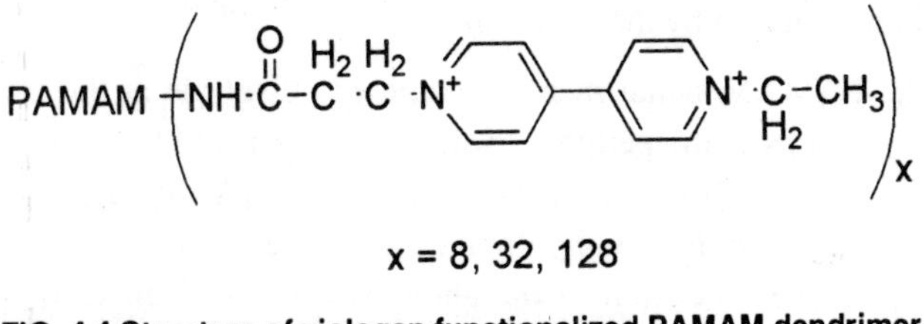

x = 8, 32, 128

FIG. 4.4 Structure of viologen functionalized PAMAM dendrimer

A carboxy terminated PAMAM dendrimer formed an electrostatic interaction with Ru(phen)$_3^{2+}$ complexes [28, 36]. In case of negatively charged half-generation dendrimers and micelles, it was found that quenching of the luminescence of $Ru(phen)_3^{2+}$ by $[Fe(CN)_6]^{4-}$ is diffusion controlled and depends on the size of the micelles and the generation of dendrimers [28]. The negatively charged ferrocyanide ions are strongly repelled by like charges of the dendrimer molecules by electrostatic forces. Thus the effective quenching only takes place when the probe molecules are capable of exit from the host molecules during their lifetime and are quenched in bulk aqueous phase.

The PAMAM dendrimer was modified [82] at the periphery with terpyridine ligands and then it was complexed with Fe(II). The Fe(II) complex of ligand substituted PAMAM dendrimer was characterized by cyclic voltammetry. The cyclic voltammetric studies indicated that the external iron centers on the outer surface of the dendrimer are electrochemically equivalent.

The durable, polyfunctional and highly cross-linked HB dendrimer/ polyanhydride composite thin films were prepared from either amine or hydroxyl terminated PAMAM dendrimers and copolymer of poly(maleic anhydride)–co-poly(metheyl-vinyl-ether) [83]. The dendrimers serve two roles. They act as building blocks, which permeate into, and covalently crosslinked with the copolymer. Additionally, internal α-amino groups within the dendrimers undergo retro Michael reactions. Subsequently, a combination of inter-dendrimer Michael reactions and inter-dendrimer imidization reactions lead to additional crosslinks in the heated film. These films were prepared on Au, Si and Al. Further, when a hydrophobic octadecyl layer was attached as a final step of these composites on Al, Al was resistant against alkaline solution or form pitting in neutral chloride containing solution by these new materials.

The PAMAM dendrimer with diphenyl disulfide at the core and hydroxy groups at the periphery was synthesized [84] by the treatment of *bis*(aminomethylphenyl)disulfide with MA (methylacrylate), followed by reaction with ethylene diamine, using the divergent technique. The repetition of the above two steps resulted the higher generation (Scheme 4.7). The structure of the synthesized dendrimer was confirmed by ^{1}HNMR, ^{13}CNMR spectroscopies and MALDI-TOF mass spectrometry. The molecular weight obtained from mass spectrometry for the dendrimer was 3489 Da. The synthesized dendrimer was used as host for encapsulation of phenol blue guest. The dendrimer also showed a reversible redox behavior between dendron and dendrimer in chemical reactions.

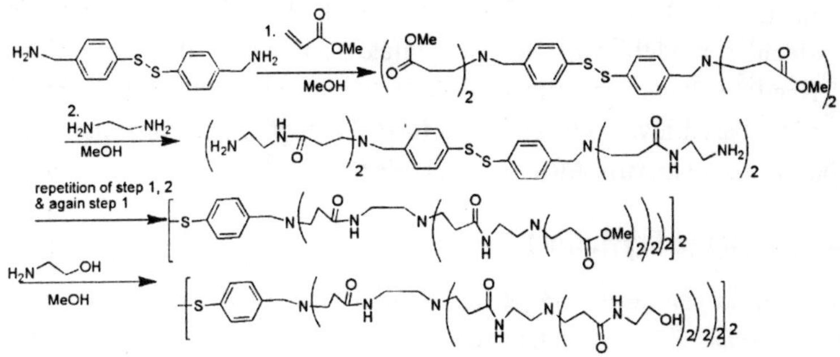

SCHEME 4.7 Synthesis of PAMAM dendrimers with diphenyldisulfide core and hydroxyl terminal groups

Tripyridyl cobalt complexes reacted with terminal functionalization of PAMAM dendrimer upto 3rd generations in order to get the metallodendrimers. The modified dendrimers [85] were characterized by EQCM. EQCM showed that the dendrimers were deposited onto Pt electrode following the Co(II/I) reductive process and the resulting films prevent further deposition upon successive potential scanning. The effects of electronic self-exchange were observed for Co(I/II) redox processes for adsorbed layers of the dendrimers. These dendrimers adsorbed upto a monolayer equivalent onto Pt electrodes at 0.20 V Ag/AgCl. Where the Co site has +2 charges. Further, the kinetics of adsorption is activation controlled and the rate constant is larger for the higher generation metallodendrimers.

A combination of 1-(3-dimethylamino)propyle-3-ethylcarbodiimide hydrochloride (EDC) and N-hydroxysuciimide (NHS) coupling reagents were used to form the amide between the dendrimer and the viologen precursor [86].

The interfacial reactions of 4, 8 and 32 terpyridyl-pendant PAMAM dendrimers and 1,4-*bis*[4,4''-*bis*(1,1-dimethylethyl)-2,2':6',2''-tetrapyridine-4'-yl]benzene (DDTB) *bis*-terpyridine containing bridging ligands in CH_2Cl_2 with aqueous Fe^{2+} or Co^{2+} ions gave rise to film formation on HOPG and Pt(III) single crystal surfaces [87-88]. Molecularly resolved STM images revealed that these films form highly ordered 2D trigonal [87] and hexagonal [88] arrays. The formation of ordered domains appears to be controlled by the rigidity of the molecules containing the terpyridine groups. The dimensions obtained from the STM images are consistent with the estimated size from molecular modeling of the molecules from which the films were derived. In all cases the films are electrochemically active and exhibited a metal based reversible wave at a formal potential that corresponds to that for the respective $M(tpy)_2^{2+}$ complex.

4.3.3. Liquid Crystalline

The liquid crystalline properties of some PAMAM dendrimers have been reported [89-90] after functionalization of the terminal amino groups by promesogenic units. Liquid crystalline polymers were synthesized [89] by functionalization terminal amino groups of PAMAM dendrimers by one-chain promesogenic calamitic units. The series consists of five new modified dendrimers containing 4, 8, 16, 32 and 64 peripheral mesogenic ester units and exhibited liquid crystalline smectic. A mesophase by parallel aggregation of the mesogenic units. The mesogenic functionalized dendrimers also exhibited a high degree of plasticity and tended to adopt the most convenient shape to afford liquid crystalline behavior. In these dendrimers the weak interactions between the mesogenic units make the molecules to adopt a cylindrical geometry, which resembles a heavy thick rod like structure that induced smectic phases. Whereas Macros *et al.* [90] obtained a series of homologous PAMAM dendrimers that carry two decyloxy chains in the terminal promesogenic units at 3- and 4-position of the peripheral aromatic ring (Scheme 4.8). The functionalization of the dendrimers was performed by the condensation of 4-(3',4'-didecyloxybenzoloxy)-salicylaldehyde with the terminal amino groups of the corresponding PAMAM dendrimer of zero to 4th generations. ^{1}HNMR, ^{13}CNMR and IR spectroscopy, FAB-mass spectrometry, GPC and elemental analysis were characterized the functionalized dendrimers. The results are comparable to those obtained for the one terminal chain calamitic homolog LC dedrimers [89].

The liquid crystalline properties and thermal stability of the functionalized dendrimers were studied by optical polarized microscopy,

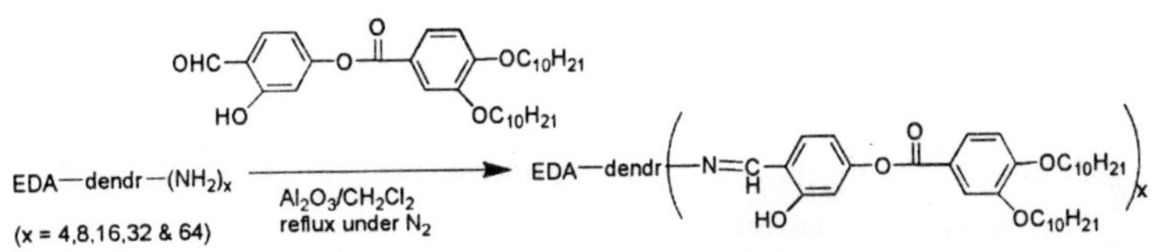

SCHEME 4.8 Funtionalization of PAMAM dendrimers by didecyloxybenzoloxy salicylaldehyde derivative

DSC, TGA and XRD analyses. All the dendrimers showed liquid crystalline mesophase, as analyzed by optical microscopy and DSC. The XRD data indicate the presence of typical hexagonal columnar mesophase. The T_gs and T_ms of the dendrimers are (32, 45, 47, 29, 30)°C and (61.9 & 80.1, 64.2 & 112.8, 58.3 & 121.1, 63.8, 53.5)°C, respectively for Ist to 5th generation, as indicated by DSC studies. The weight losses of the dendrimers started at (175, 200, 210, 200 and 210)°C, for the Ist to 5th generation respectively, obtained by TG analysis. These columnar mesophases formation occurred through hydrogen bonding between the amide groups to generate a rather rigid internal core.

The LC PAMAM dendrimers of zeroth to 5th generation [89-91] were obtained by end capping salicylaldehyde derivatives bearing one, two or three terminal aliphatic promesogenic units to the terminal amino groups as shown in scheme 4.9. The synthesized dendrimers were characterized by ^{1}HNMR, ^{13}CNMR, IR spectroscopy, FAB-MS and GPC analyses. The liquid crystalline behavior of the functionalized dendrimers was studied by polarized optical microscopy (POM), DSC and XRD analyses. The nature of the LC behavior of the dendrimers was directly correlated to the number of terminal chains grafted on the peripheral anisotropic groups. Also, the presence of one chain per mesogenic unit favors their parallel arrangement, thus forming smectic mesomorphism. Whereas with the increase of the number of aliphatic chains favors their radial disposition, leading to the formation of columnar structures. It was also found that the dendrimers with one or two chains per mesogenic unit, T_g does not vary significantly, and the temperature ranges of the mesophases for the higher generations were comparable.

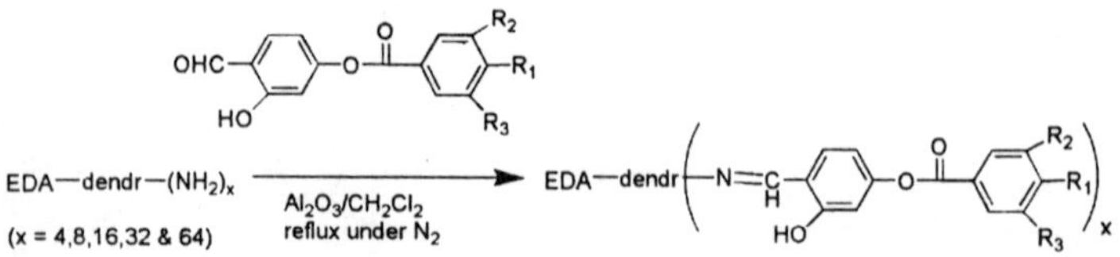

i) When $R_1 = OC_{10}H_{21}$ then $R_2 = R_3 = H$, ii) When $R_1 = R_2 = OC_{10}H_{21}$ then $R_3 = H$ and iii) for x = 16, $R_1 = R_2 = R_3 = OC_{10}H_{21}$

SCHEME 4.9 Functionalization of PAMAM dendrimers at the surface by salicylaldehyde derivative to get LC dendrimers

4.3.4. Glycodendrimers

PAMAM dendrimer was also utilized to synthesize [92] well-defined artificial glycoconjugates of O-(D-glucopyranosyl)-L-serine-persubstituated dendrimer and O-(2-acetamido-2-deoxy-D-glucopyranosyl)-L-serine-persubstituted dendrimer. The glycoconjugates were synthesized by the reactions of PAMAM dendrimers with the sugar substituted amino acid N-carboxyanhydrides that is O-(tertra-O-D-glucopyranosyl)-L-serine N-carboxy anhydride and O-(2-acetermido-3,4,6,-tri-O-acetyl-2-deoxy-D-glucopyrosyl)-L-serine-N-car oxyanhydride respectively, followed by deacetylation with hydrazine monohydrate.

Okada and his co-workers [93] functionalized commercially available 3rd and 4th generation PAMAM dendrimers bearing 24 or 48 amino groups at the surface, with disaccharide lactones of lactose (O-β-D-galactopyranosyl-(1,4)-D-glucono-1,5-lactone) via amide bond formation to obtain so called "sugar ball" (Scheme 4.10). The resulting dendrimers showed strong interactions with the lactin concanavalin-A. The synthesis of a new class of oligoglycopeptide-type 'Sugar-ball' by the 'radical growth polymerization' process using α-amino acid N-carboxyanhydride with PAMAM dendrimers of different generations was also reported [72].

Tetra-, hexa- and octa-valent PAMAM dendrimer derivatives were synthesized [94] by reaction of amine terminated dendrimer with peracetylated glycosylisothiocyanates of β-D-glucose, α-D-glucose, α-D-mannose, β-D-galactose, β-cellobiose and β-lactose. The synthesis of smaller dendrimers using the same coupling reaction that does not rely on the use of protecting groups was also reported [95].

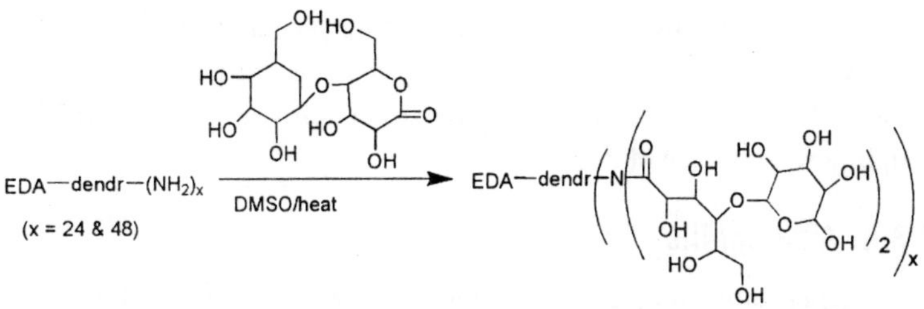

SCHEME 4.10 Functionalization of PAMAM dendrimers at the surface by disaccharide derivative

PAMAM dendrimers of 1st to 4th generation were functionalized by T-antigen derivative [96]. T-antigen (Thomson–Friedenrich antigen) has been reported as a cancer-related epitope and as an important antigen for the detection and immunotherapy of carcinoma particularly relevant in breast cancer patients [97-98]. The T-antigen–glycoPAMAM dendrimers were synthesized with high yield by coupling reactions between T-antigen acid and NH_2–terminated PAMAM dendrimers in the presenceof TBTU and DIPEA (Scheme 4.11). The T-antigen acid (A) was prepared from α-D-N-acid-glucosamine (GalNAc) by sequential protecting groups strategy C-4 epimerization method [99]. The resulting reaction mixtures were lyophilized to dryness and the residue was purified by either GPC or dialysis technique. The functionalized dendrimers were characterized by ^{1}HNMR and ^{13}CNMR spectroscopies. The glyco-PAMAM dendrimers having T-antigen residues with 4, 8, 16 and 32 valences and higher valence conjugate generated stronger binding interactions

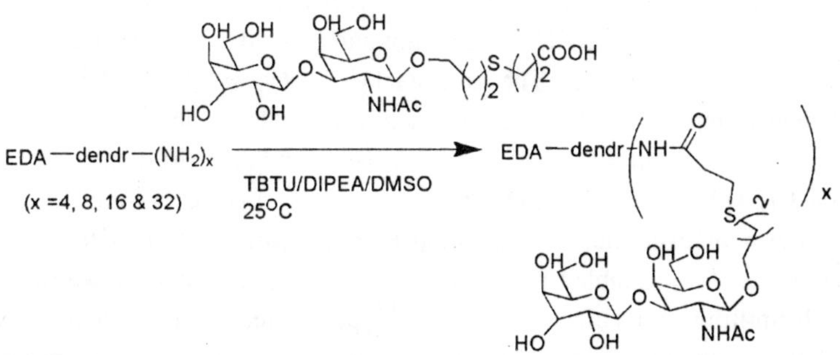

SCHEME 4.11 Functionalization of PAMAM dendrimers at the surface by T-antigen derivative

indicate a cluster effect [96]. The inhabiting potential of these glyco-PAMAM conjugates toward antibody coating antigen interactions enhanced upto 3800 times over that of T-antigen residue without coupling with PAMAM dendrimer.

4.3.5. Amphiphilic

A polymerizable amphiphilic dendrimer was synthesized by end capping of amine terminated PAMAM dendrimer with 10,12-pentacosadiynoic acid (PDA) [100]. The PDA succinimidyl ester and 3rd generation PAMAM were taken in CH_2Cl_2 and stirred at room temperature for 24h. After solution was concentrated under vacuum, NaOH solution was added over the solid and suspension was maintained at 60°C for 6h. The disserted product was obtained with relatively high yield (66 %) after separation of solid followed by purification. The modified dendrimer (Fig.4.5) was characterized by ^{1}HNMR, IR spectroscopy and MALDI-TOF mass spectrometry. The amphiphilic property of the modified dendrimer was studied. The measurement of surface pressure and surface dipole moment area isotherm of PDA-PAMAM by Whilhelmy method using an UV-vis. spectrophotometer showed that the modified dendrimer formed a stable monolayer at the air water interface with a limiting molecular area of 460 Å^2 per molecule. The compressed monolayer can be readily polymerized upon UV irradiation, as like other PDA derivatives. A good correlation between size and shape of monolayer domains was observed by the authors using Brewster angle microscope (BAM) and environmental SEM. The modified dendrimer was readily polymerized within a few seconds by UV irradiation and formed a colloidal particles in chloroform solution. The authors [101] also synthesized a disk-shaped amphiphilic dendrimer by reacting 12-hydroxy dodecanoic acid-N-hydroxysuccinimide ester with 4th generation PAMAM dendrimer in CH_2Cl_2 (Fig.4.6). The surface pressure and surface potential area isotherm measurement of the modified dendrimer showed that the dendrimer forms stable Langmuir monolayers at the water-air interface with limiting molecular area of 160 Å^2 per molecule. This small area relative to the huge size of the dendrimer indicates that the dendrimer molecules form an edge on disk-shaped structure at the air-water interface. The topological characteristics of the modified dendrimer are similar, both under BAM images of Langmuir films at air-water interface and ESEM images of deposited Langmuir-Blodgett (L-B) film. In both the cases, circular and rectangular regions were observed within the same dimension range. However, one important observation of this modified PAMAM dendrimer was that the monolayer becomes more uniform and regular sized strip like topography with increased surface pressure [100].

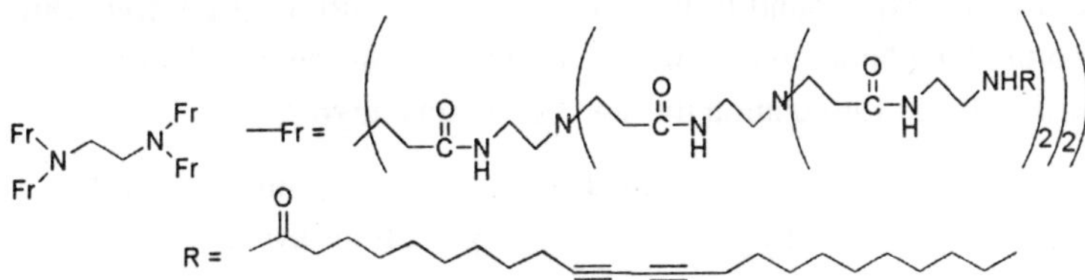

FIG. 4.5 Structure of PDA end capped PAMAM dendrimer

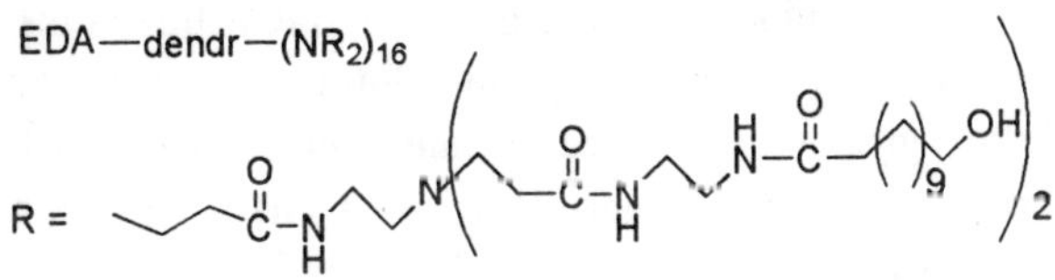

FIG. 4.6 Structure of dish-shaped amphiphilic PAMAM dendrimer

The fatty acids were co-ordinated [102] into the surface of amino functionalized PAMAM dendrimers, which were formed electrostatic ion-pair by protonation of all terminal amino groups and thus solubilized in non-polar solvents by formation of inverse micelles. These inverse micelles allow phase transfer of hydrophilic guests inserted in the dendrimer interior as well as of dye molecules or palladium nanoparticles.

The, ethylene diamine based PAMAM dendrimers were modified [103] by grafting with fatty acid and thereby converting them into unimolecular micelle systems. The modified dendrimers were characterized by IR, ^{1}HNMR and ^{13}CNMR spectroscopic studies. The drug 5-fluorouracil was entrapped in the grafted dendrimers. The effects of various solvents such as ethanol, CH_2Cl_2, THF, etc., pH and ionic strength on solubilization of 5-fluorouracil were determined. An over-coating on the grafted dendrimer was given with phospholipids. The product was lyophilized as yellowish white powder. The maximum loading of the drug was ≈53 % by weight and it was stable for 1 month at room temperature as well as at 40°C. The average particle size ≈375 nm as determined by Malvern's Masersizer 4. The release of the drug was found satisfactory across cellulose tubing in PBS. *In-vivo* studies were performed in albino rats and pharmacokinetic parameters and bioavailability were determined from the plasma profile of 5-fluorouracil. The phospholipid coated dendrimer grafted with 5-fluorouracil was found to be more effective orally than the free drug, also the lymphatic uptake increased, which indicates the absorption of the developed formulation through the lymphatic route.

A 4th generation PAMAM dendrimer was modified [104] to amphiphilic dendrimers by end capping with an average of 7 to 46 hydrophobic chains of varying lengths, C_{10} to C_{14}. The synthesis was carried out by the following three sequential steps. (i) Protection of a desired number of amines with Boc groups. (ii) Reaction of the remaining amine groups with long chain acid chlorides and (iii) deprotection from the Boc groups to produce the amphiphilic dendrimer as HCl salts. The modified dendrimers were well characterized by pulse gradient spin-echo (FT-PGSE) NMR spectroscopy and DLS techniques. The NMR and DLS studies indicated that the modified dendrimers behave as unimolecular entities with coiled chain within the dendrimer periphery. However, when the number of chains becomes very large, they can promote hydrophobic induced self-assembly.

The salicylate, catecholate and hydroxypyridinonate bidentate chelators were coupled [105] to the terminal amino groups of PAMAM dendrimers upto 5th generation. The products were purified by high quality chromatographic technique. The ^{1}HNMR and ESI-MS studies confirmed the exhaustive coupling of these chetalors to the surface amino groups of the dendrimers. Also this study showed that ferric ions binding to these ligands are localized to the chelating end groups.

4.3.6. Copolymers and Polymer Blends

PAMAM dendrimers were blended with linear high molecular weight polymers, poly(vinyl chloride) (PVC) and poly(vinyl acetate) (PVAc) [106]. The tensile properties measurements, Xenon NMR spectroscopy and dynamic mechanical analysis were carried but to determine the interaction between the blend components. The results indicate that the better interactions between PVAc-PAMAM blend compared to that of PVC-PAMAM blend.

The grafting of PAMAM dendrimer on ultrafine silica surface was also reported [107]. The amine-functionalized silica was obtained by treatment of silica with γ-aminopropyl triethoxysilane. Then the PAMAM dendrimer was synthesized on surface of silica/carbon black by repetitive reactions of (i) Michael addition of methylacrylate to amine group of silica and (ii) amidation of the resulting ester with ethylene diamine (EDA) or hexamethylene diamine (HMDA). Although authors tried to propagate the dendrimer upto 10th generation but the amount of grafting was considerably lower, 96.2% with EDA and 60.9% with HMDA for 10th generation, than that of theoretical value 13994.6% EDA and 20869.2% for HMDA for 10th generation. This may due to steric reason of highly branched dendrimer and also the heterogeneity of the reaction system. However, the grafted silica dispersed better in solvent with higher stability for higher grafting. Similarly, the carbon black surface was grafted with PAMAM upto 10th generation, the amount of grafting in this case is even much lower (96.1%) [108]. The amino groups were introduced on the surface of carbon black by nitration of aromatic groups, followed by reduction [109]. HB PAMAM dendrimer grafted carbon black offered a stable dispersion in a good solvent for PAMAM. The grafting of HB PAMAM on amino containing polystyrene bead was also carried out by the similar procedure [108]. The amino groups were introduced by nitration of aromatic moieties of crosslinked polystyrene followed by reduction by $SnCl_2/HCl$ systems [110]. Silica grafted PAMAM dendrimers were utilized as effective curing agents for resin and it was found that the curing activity increased with the increase of amount of grafting [111]. The thermal resistance of the cured resin with PAMAM–modified silica was superior compared to that of using only ethylenediamine.

Three-dimensional covalently cross-linked nanodomained networks of radially layered poly(amidoamine-organosilcon) copolymeric dendrimers containing hydrophilic PAMAM interiors and reactive hydrophobic organosilicon (OS) exteriors were described [112]. These networks can be prepared by cross-linking of alkoxysilyl and/or vinylsilyl functionalized PAMAM-OS dendrimers by using sol-gel type of hydrolysis of alkoxysilyl terminated dendrimers, followed by a silanol condensation reaction or irradiation or heat induced free radical double bond opening of vinylsilyl or alkyl silyl-terminated dendrimers, followed by free radical coupling into carbosilane bridges. These dendrimers may be prepared [113] from amine-terminated PAMAM dendrimers either by Michael addition to silicon containing acrylate or by haloalkylation with chloro-or iodo-alkylsilanes. These networks (Fig.4.7) were optically clear, colorless, highly transparent nanostructured materials that showed smoothness at the nanoscopic level and low surface energy and exhibit useful mechanical properties. The glass transition temperature varies with the generation number of PAMAM dendrimers, the type of organosilicon reagent used, the type and molecular density of the PAMAM-OS dendrimer end groups, the type of cross-linking chemistry, initiator, catalyst, reaction condition applied, the type and relative amount of the cross-linking core agents and any other additive(s) used. These dendrimer-based on networks also exhibited nucleophilicity in PAMAM domains, which provide pronounced affinity to complex inorganic, organic, and/or organometallic electrophiles. This can be utilized as nanotemplates for organizing

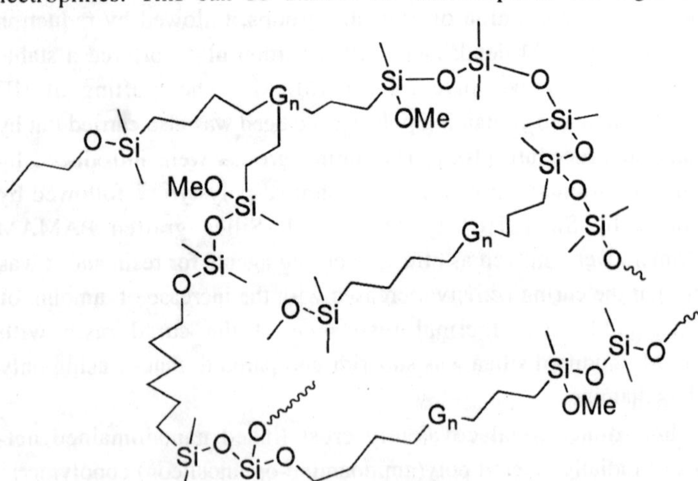

FIG. 4.7 Schematic representation of PAMAM-OS dendrimer network structure

nanoparticles of metals, metal oxides, metal sulfides, metal selenides, as well as organic dyes, drugs, catalysts, indicators, etc.

4.3.7. Conducting

PAMAM dendrimers were functionalized [51] with diimide moieties at the periphery, which upon reduction with dithionite in water offered soluble products with upto 192 anion radicals per molecule. The near-IR studies indicated the π-stacks formation in aqueous solution for all 2nd to 6th generations. This π-stacks are interesting because the π-electrons are delocalized among the molecules in the stacks sometimes, which may lead to the formation of electroconducting materials. All the generations offered same type of π-stacks, which confirmed the fact that the structure of PAMAM dendrimers had only little effect on the aggregation. The diimide modified 3rd generation PAMAM dendrimer cast film from formalin on glass surface under reduced pressure at elevated temperatures [114]. The films showed isotopic characteristic in conductivity and optical properties. Thus these macromolecules have 3-dimensional interaction between the stacks in contrast to small molecular films. A mixed valence film (0.55 electron/diimide) exhibits conductivity of 11 scm^{-1} at 95% relative humidity and showed that the conductivity depends on humidity. Also partially reduced mixed valence film had stronger optical conduction bands and was more conducting than that of fully reduced one.

The surface of the amine terminated PAMAM dendrimers is nucleophilic enough to enter Michael addition with the electrophilic center of a acetylenic monomer and thus acts as a multifunctional macroinitiator for the ionic polymerization of N-methyl-2-ethnylpyridinium trifluoromethanesulfonate (2E PyMeTf) [115]. The 3rd and 4th generations aliphatic PAMAM dendrimers were used as the primary amine terminated macromolecular semiflexible core units for the PAMAM-PA (polyacetylene) product (Scheme 4.12). This copolymer may be described as nanoscopic sea Urchin in which the soft interior is provided by the aliphatic dendrimer and the conjugate rigid ionic polyacetylene chains represent the 'spikes'. The synthesized copolymers were characterized by FTIR, UV-visible, ^{1}HNMR and ^{13}CNMR spectroscopy, TEM and SEC analyses. The color of the bulk copolymers changed from orange-yellow to dark brown with the increase of degree of polymerization of acetylene monomer, indicating an increase in conjugation length. The thermal stability of the PAMAM-PA copolymers increased with the increase of PA content in the copolymers. The decomposition temperature under the nitrogen atmosphere increased from ≈100°C to 240°C. This may be due to the decreased mobility of the semiflexible branches of PAMAM dendrimers and the strong ionic interaction between the ionic oligoacteylene rods on the dendrimer surface. The observed T_g values of the copolymers were about 10-15°C higher than that reported for pure PAMAM dendrimers.

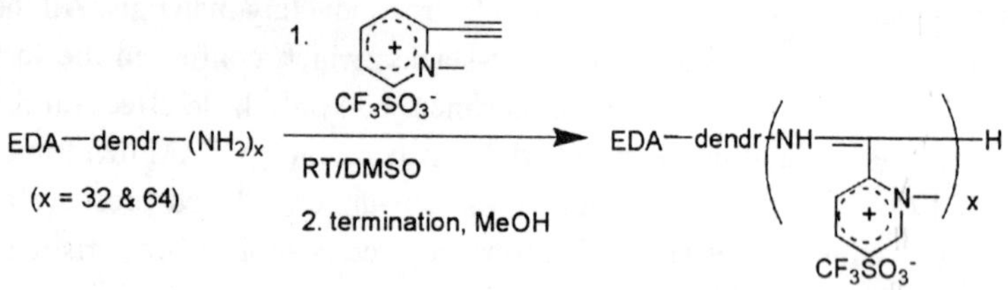

SCHEME 4.12 Synthesis of novel copolymer of PAMAM and PA as conducting dendrimers

The conductivity of the undoped polymers was less than 10^{-7} scm^{-1}, while iodine doping brought the conductivity of the copolymers 2-3 $X10^{-5}$ scm^{-1}. However, the conductivity with $NOPF_6$ doping was $4X10^{-4}$ scm^{-1}. This value is less than the conductivity of doped ionic homopolyacetylene (10^{-3}-10^{-2} scm^{-1}) [116]. This is due to the fact that the conjugation length is shorter and the effective cross section of the layer is smaller than homopolyacetylene. Also this conductivity results indicate that the PA chains are in interaction with each other [115].

A series of electrically conducting dendrimers was synthesized [52] by surface modification of PAMAM dendrimers upto 5th generation, with cationically substituted naphthalene dimers. Reduction with sodium dithionite in water or formalin formed anion radicals on the diimide moieties, which on evaporation of solvent under an argon atmosphere produced electrically-conducting powder of the reduced dendrimers. The films were casted from DMF solvent of reduced dendrimers. The conductivity of the half reduced (0.55 electron/ diimide) dendrimers ($\sigma = 10^{-2}$ s/cm) was more than the corresponding fully reduced one ($\sigma = 10^{-3}$ s/cm) under ambient conditions. The π-stacks formation with medium absorbance beyond 2000 nm was confirmed by near-IR studies. The conductivity largely depends on humidity and increases with the increase of humidity ($\sigma = 18$ s/cm at 90% relative humidity). The nature of the conductivity was electronic and isotropic. The studies of near IR spectroscopy, XRD and QCMB measurements suggested that the amorphous film absorb water from air and at high humidity, they were plasticized allowing faster stack-to-stack electron hopping and thereby increasing conductivity.

4.3.8 Bioactive

A novel water-soluble peptide shelled dendrimer containing a poly(L-glutamic acid) segment was synthesized [117] from the third generation PAMAM dendrimer. The secondary structural properties and interaction with amino acids (Trp, Phe, and Tyr) were studied by spectroscopic measurements. At low pH, the peptide-dendrimers formed almost 100% helix structure from the 3D aggregation of the segment. The interaction with amino acids proceeds with positive cooperatively and the D-isomers were preferentially bound to the helical segments relative to the L-isomers. The bound amino acids were not released into the water phase but were transferred into the inner core of the dendrimer, where they remained stable. Even though a conformational change of the helix segment was caused by pH variation.

Amphiphilic α-helix peptides (4, 8, 16, 32 and 64 segments) were introduced at the end groups of PAMAM dendrimers and then Fe(III) or Zn(II) mesoporphyrin was coordinated between the 2α-helix peptides to get a multi Zn mesoporphyrin assembly [118-119]. These nano-covalent assemblies of Zn-mesophorphyrin have the potential as photoinduced electron transfer agents. The electron transfer properties were more prominent with the growth of the dendrimer generation. The synthesis of the peptide-dendrimers was performed by a domain ligation strategy [120]. The chloroacetyl group was introduced at each amino terminal group of PAMAM dendrimers upto 4th generation by treatment with N-ethoxy carbonyl-2-ethoxy-1,2-dihydroquinoline and chloroacetic acid in MeOH. The chromatographic purified perchloroacetylated PAMAM dendrimers were then coupled with thiol side chain of cystein in α-helix peptide. The α-helix peptide was synthesized by the solid phase method using 'Fmoc' strategy. The peptide dendrimers were purified by SEC and reverse phase HPLC, and characterized by MALDI-TOF mass spectrometry or ultra centrifuge technique and circular dichroism (CD) studies. The binding ability of Zn^{2+}-mesoporphyrin with the above peptide-dendrimers was studied by UV-vis. absorption spectrophotometry and found to be efficient. The electron transfer properties of multi-Zn-mesoporphyrin conjugated peptide-dendrimers were examined by fluorescence quenching studies after addition of a negatively charged electron acceptor naphthalene sulfonate and a positively charged methyl viologen. The study showed [119] that the electron transfer ability was more with the growth of the dendrimer generations.

The formation of supramolecular structures between calf thymus DNA and starburst PAMAM dendrimers of 2nd and 6th generations by various physico-chemical techniques such as EPR, CD, UV and melting profiles was studied [121]. The studies showed that depending on the ratio of dendrimer (surface group) to DNA (base pairs), the structures and the interactions of the starburst dendrimer with the DNA were varied. At low ratios ($0<r<1$) both the generations of the dendrimer offered strong electrostatic interactions with the DNA molecules. At intermediate ratios, ($1<r<200$) the neutralization of DNA negative charge by the dendrimers leads to precipitation and subsequent opalescence of the solution. The interaction was also the highest, as detected by EPR, CD and UV techniques and thus forming supramolecular structures constituted by aggregates of DNA and dendrimer molecules. However, at high ratios ($r>200$), the dendrimers solublized DNA completely due to a 'salting in-effect'. Both the dendrimers showed the occurrence of a further broad maximum of their interaction with DNA in EPR spectra, though CD, UV and melting profiles indicate variations but reversible of the DNA structure and thereby forming supramolecular structure between the dendrimer and DNA. However, there are some differences in the interaction between 2nd and 6th generations of dendrimers, which may be due to their great difference in size. The small size 2nd generation is more fluidic structure and thus better fits in the DNA conformation. In nut-shell, therefore, the study confirmed that the PAMAM dendrimers could bind calf thymus DNA molecules by strong electrostatic interaction between the positively charged groups of terminal amines and negatively charged phosphate groups of DNA molecules.

In another report [122] the structural modification of vesicle, DMPC was investigated by interaction with PAMAM dendrimers using computer aided CW-pulse-EPR and electrospin-echo (ESE) techniques. Mainly mobility and polarity parameters were obtained by CW-EPR and ESE techniques. The vesicle structure became more rigid and ordered due to the interaction with the surface of dendrimers. The enhanced rigidity of the structure also caused the filling of the chains of about 50° with respect to the surface line. Again, in the presence of dendrimer the ordering decreased from the vicinity of the polar heads to the end of the chain. The mobility of the probe, doxyl-functionalized stearic acid (DXSA) at different position of the steric chain (5DXSA, 12DXSA and 16DXSA) increased in the region close to the polar heads, which may be due to the compression of the water layer at the interface caused increased water permeability in that region. The rotational mobility in the rest of the chain was essentially unchanged. The permeability of water increased in the vicinity of the polar heads with addition of the dendrimers, due to the same reason. However, the packing and the ordering of the vesicle chains diminished, if protonated dendrimers were used. This may be due to swelling of the bilayer. Thus the bilayer of vesicle is only partially perturbed by addition of dendrimer and the integrity of the vesicle is preserved, which further encourages the use of PAMAM dendrimers as gene and drug carriers. The half generations carboxy terminated 1.5 to 7.5 PAMAM dendrimers were characterized by EPR techniques by using positively charged nitroxide radicals of variable chain length as probes. Analysis of the EPR spectra as a function of pH demonstrated that the electrostatic interactions promote bonding at the external dendrimer-water interface.

There are a few studies on the interactions of DNA with PAMAM dendrimers. The interactions between plasmid DNA and the dendrimers were reported [123] by examining their electromicroscopic structures, nuclease sensitivity and transcriptional activity. In another study, Chen *et al.* investigated [124] the interactions between double–stranded DNA and amine terminated ethylenediamine core PAMAM dendrimers of generation 2, 4 and 7 by fluorescent studies using ethidium bromide as the probe. The results showed that there is mainly one binding mode between ethidium and DNA-dendrimer complexes and that dendrimers bind DNA too strongly to be displaced by ethidium. The binding ability of ethidium bromide to DNA decreases with the increase of dendrimer concentration until a saturation dendrimer concentration is reached. Although, DNA can be wrapped by 7th generation dendrimer but the wrapping of DNA was not possible for 2nd and 4th generation dendrimers. The DNA-dendrimer interaction at pH 6.5 is stronger than that at pH 7-5. This may be due to higher degree of protonation of the surface amine groups of the dendrimers. Thus the binding is electrostatic in nature. Further the binding between DNA and dendrimer became weaker at higher ionic strength, consistent with the importance of the electrostatic interactions in the DNA dendrimer binding.

This solid-phase immobilized dendrimeric scaffold was then coupled with a range of biologically active ligands [125], such as LDV and EILDVPST. The couplings were performed with 5 equivalents of HOBT and amino acid in DMF. The peptides were synthesized on the dendrimer resins by using the standard Fmoc solid phase peptide chemistry. The dendrimer conjugates were purified by reverse phase HPLC. An ELISA was used to screen the peptide dendrimer conjugates for their ability to compete with a biotinylated peptide component of the CS1 region of fibranectin. The inhibition potential of the different generation peptide-dendrimer conjugate was found to increase with higher generations.

4.3.9. Fullerodendrimers

The synthesis and characterization of a new class of water-soluble PAMAM grafted fullerene (C_{60}) were described [126-127]. The fullerodendrimer was synthesized by a reversible Diels-Alder reaction between C_{60} and anthracene derivative of PAMAM (Scheme 4.13). The yield obtained by this process (21%) was much higher compared to thermal process (12%) by heating 45°C for 4 days. The product was purified by HPLC and characterized by ^{1}HNMR, ^{13}CNMR, UV-vis spectroscopy and MALDI-TOF mass spectrometry. Similar to other C_{60}-anthracene compounds, this product also undergoes retro Diels-Alder reaction under heating at 45°C in methanol for 12h and return back to the component molecules, C_{60} and anthracene derivatives of PAMAM with 89% and 87% yields respectively. The fullerodendrimer was readily soluble in acidic aqueous solutions to different extent depending on the pH of the medium (20.5 mmole at pH 1.50, 2.19 mmole at pH 5.46 and 0.038 mmole at pH 6.51). The decay of the absorption peak at 700 nm in transition spectra was much higher in O_2-saturated water ($1.3X10^{-9}$ $mol^{-1}dm^3s^{-1}$) compared to deaerated water [127]. The potential applications of this grafted dendrimer are as an agent to remove solubilized group for fullerenes and as a biologically active agent that release C_{60} intracellularly.

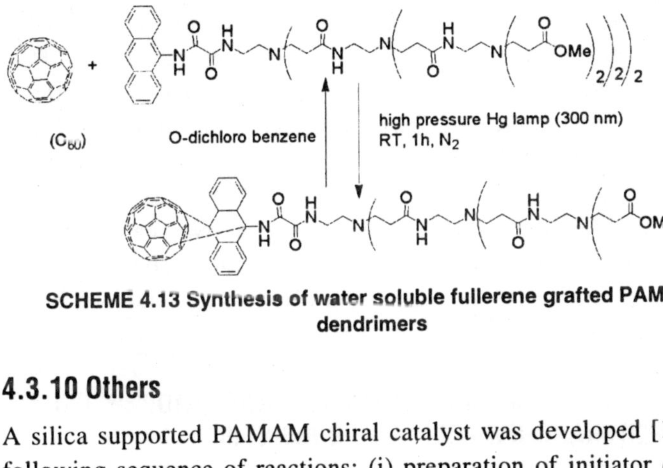

SCHEME 4.13 Synthesis of water soluble fullerene grafted PAMAM dendrimers

4.3.10 Others

A silica supported PAMAM chiral catalyst was developed [128] by following sequence of reactions: (i) preparation of initiator core by treatment of surface silanol groups of silica with 3-aminopropyl triethoxy silane under refluxed xylene, (ii) propagation of PAMAM dendrimer on amine functionalized silica by the standard two step procedure and (iii) immobilization in chiral auxiliaries at the periphery of the dendrimers.

PAMAM dendrimers upto 4th generation, were constructed [129] on the surface of amino propyl-silica by using the divergent technique. By using long chain diamine from EDA to 1,4-diamine butane to 1,6-diamine hexane and diamine dodecane as spacer units, the problems of isolation of dendrimers became critical. The long chain diamine clogged the filtering system, the filtration and washing took upto 3 days for diaminododocane linkers. The silica supported PAMAM dendrimer ligand of Rh (Rh-PPh_2-PAMAM-SiO_2) was a highly active and regioselective catalyst for the hydroformylation of aryl olefins and vinylester had proved earlier [130]. It was found that the lower generation dendrimers were more active than the higher generation (3rd and 4th) dendrimers as catalysts. This may be due to incomplete phosphonation reactions because of steric crowding on the surface of the higher generation dendrimers. Thus by using longer chain length of the chain extender of each generation the steric crowding would reduced and thus it allows increased catalyst loading at higher generation.

The dendrimers on silica surface were then phosphonated by reaction with diphenyl phosphino methanol, prepared *in-situ* from paraformaldehyde and diphenylphosphine; followed by complex formation with chloro(dicarboxyl) rhodium(I) dimer in distilled and degassed hexane at 40°C for 12h under the argon atmosphere. These targeted Rh complex of silica supported PAMAM dendrimers (Rh-PPh_2-PAMAM-SiO_2] were utilized as effective catalysts for hydroformylation of styrene or vinyl acetate [129]. Also the catalytic activity and recyability increased significantly with increased chain length of the spacer unit.

4.4 SYNTHESIS, CHARACTERIZATION AND PROPERTIES OF HYPERBRANCHED POLY(AMIDO AMINE)S

Amine functionalized ultra fine silica filler was grafted with PAMAM in solvent free dry system [131]. The PAMAM was grown on the surface of amine terminated silica by repetition of (i) Michael addition of methacrylate (MA) to amino groups of silica and (ii) exhaustive amidalion of terminal ester group with EDA. MA was sprayed onto silica having amino groups and the mixture was agitated at 300 rpm at 50°C. After the removal of unreacted MA under vacuum, same procedure was followed by using EDA. After eight times repetition of these processes, 14.1% grafting on silica surface was achieved. However, this value is far from the theoretical value. This result shows that the exact

PAMAM dendrimer was not possible to form on silica surface, rather a HB PAMAM was grafted onto silica surface may be due to steric hindrance of the grafted polymer. The product was characterized by IR spectroscopy, element analysis and determination of terminal amino groups. The amine terminated silica i.e. the initiator core was obtained first by the reaction of surface silanol groups of silica with α-amino propyl triethoxy silane in a solvent free dry system. The silicas used for the scale-up investigation as an initiator core containing 0.12, 0.19, 0.33, 0.45 and 0.62 mmole/g of amino groups. The authors showed that the percentage of grafting was higher for lower amine containing silicas than that having higher amino groups content. With the increase of temperature and time of reaction, the amount of grafting increases but at higher temperature (above 70°C) and time, decreases the percentage of grafting and increases the amount of ungrafted polymer respectively [131]. The authors also investigated the effects of solvents like methanol and water, on the percentage of grafting compound to the dry system and found that Michael addition reaction was hardly affected by the presence of solvent after the first MA addition. This may be due to the fact that the silanol groups of silica surface help in MA reaction by catalytic effect. Thus the method may be utilized for the production of modified silica. In fact, the Tsubokawa and coworkers [111] were reported the use of HB PAMAM grafted ultrafine silica for curing of epoxy resin. It was found that this modified silica (80.2 % grafting) successfully and effectively cured epoxy resin with gel content 77% at 170°C for 48h heating. Although with the increase of reaction time and temperature, the curing reaction of epoxy resin with terminal amino groups of HB PAMAM grafted silica was accelerated but at lower temperature such as 120°C was insufficient to cure epoxy resin even after 48h. The gel content increases with the increase of amino content of the grafted silica surface and also the curing ability of the silica increased by complexation of the terminal amino groups with boron trifluoride diethylether [132]. This BF_3-amine complex system cured epoxy resin at lower temperature. The heat resistance property of the cured resin was superior compared EDA cured system, though T_g (≈105°c) in both the cases are the same.

However, the modulus of elasticity (10^{10} dyne/cm^2) was lower in the former case than the latter. This result suggested that the terminal amino groups of HB modified silica act as effective curing sites and the silicas were uniformly distributed throughout the resin matrix. It may also be speculated that the density of network of the cured resin is high in the vicinity of the silica surface and decreases as distance increased from the surface. This may be due to the fact that the curing reaction proceeds from the surface amino groups in a heterogeneous system.

The grafting of HB PAMAM on the amino group of ultrafine silica, chitosan powder, glass fiber and carbon black filler surface was also reported [133-134] by using the divergent synthetic approach. The silica supported PAMAM dendrimers upto 4th generation were prepared [135] by using aminopropyl silica as the initiator core followed by the standard Michael addition and amidation reaction. The terminal amino groups were phosphonated by diphenylphosphino methanol, which were coordinated with $PdCl_2$ to get effective catalysts. The functionalization of terminal amino groups of 3rd and 4th generation PAMAM dendrimer was carried out by ring opening polymerization of α-amino acid-N-carboxy anhydride [72]. Nanoclusters formation of PAMAM with copper [136-137] and CdS [138] were also reported.

4.5 APPLICATIONS OF DENDRITIC POLY(AMIDO AMINE)S

The dendritic PAMAM is one of the most important and widely used dendritic polymers. They have the following uses in different fields of applications as described below.

4.5.1 Catalysis

PAMAM dendrimer supported complexes [135], Pd-PPh_2-PAMAM-SiO_2 are new and highly attractive catalysts for the 'Heck reaction' of a variety of bromoarenes with styrene and butyl acrylate (Scheme 4.14). This heterogeneous Pd catalyst can also be recycled and reused with only moderate reduction in activity. The authors also used silica supported PAMAM dendrimer ligand Pd-complexes as highly active, regioselective and reusable catalysts for hydroformylations of aryl olefins and vinyl esters [129-130].

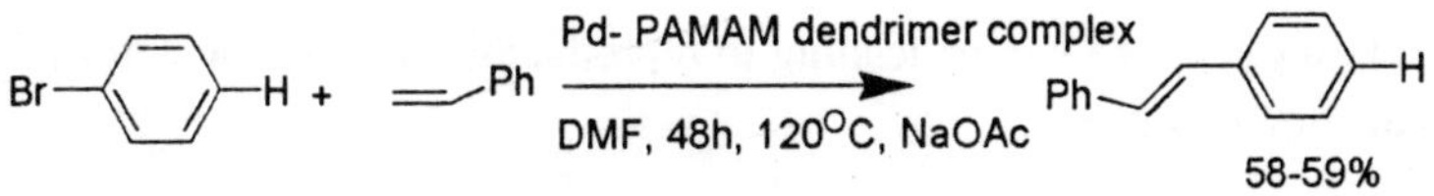

SCHEME 4.14 Use of Pd-PAMAM dendrimer complex as catalyst for Heck reaction

Chung and Rhee [128] developed PAMAM based dendritic auxiliaries on silica, as a new heterogeneous catalyst for the enantioselective addition of diethylzinc to benzaldehyde. The conversion, selectively and enantioselectivity of the reaction primarily depends on the amino groups content of initiator sites and the number of generation. The catalyst is also recyclable and can be reused. The Ist and 2nd generation PAMAM dendrimers were modified with (1R, 2S)-ephedrine moieties, to get an enantioselective catalyst for addition of diethylzinc to N-diphenylphosphinylimines [139]. The use of bifunctional ephedrine ligands results in an addition with a high enantiomeric excess (ee= 92%), whereas the use of dendritic ligands induces substantially lower stereoslectivities (ee = 43% and 39% for 1st and 2nd generation respectively).

Hydroxy terminated PAMAM dendrimers of higher generations (4th, 6th and 8th) encapsulated catalysts (DEC) consisting of Pd-nanoparticles were used for hydrogenation of alkenes (allyl alcohol in a 4:1 MeOH/ water mixture) [139-140]. In this case, an intrinsically non-selective catalyst is rendered selectivity by addition of dendritic 'nonfilter'. These DECs challenged with alkyl alcohol and form other α-substituted allylic alcohols (methyl, ethyl, dimethyl and methyl-ethyl) [139] and found that only smaller alcohol gave higher hydrogenated products, especially for higher sized dendrimeric catalysts. This result indicated that steric crowding on the dendrimer periphery, which increases with dendrimer generation, could act as an adjustable mesh nanofilter. Thus by controlling the packing density on the dendrimer periphery, it is possible to control access of substrates to the encapsulated catalytic nanoparticle. The modified dendrimer [141] was reported to act as a good chiral ligand for the reduction of various prochiral ketones by $NaBH_4$, both in heterogeneous (THF) and homogenous (water) conditions. Under heterogeneous conditions, this amphiphilic dendrimer can be recovered by filtration regenerated and recycled upto

10 times and also can be leading to reproducible results in asymmetric reduction of ketones.

4.5.2. Biological and Medicinal

Therapeutic genes, antisense oligonucleotides or ribozymer could be delivered to cells by various methods or viral and non-viral categories of delivery systems. Cationic PAMAM dendrimers have recently emerged as a novel synthetic carrier for various nucleic acids, e.g. the expression plasmids, single strand DNA oligoneucliotides or RNA transfer [142-146]. These dendrimers are highly efficient for *in-vitro* transfection and appeared to be non-cytotoxic in the concentrations, relevant for gene transfer. Also these dendrimers are not immunogenic or carcinogenic, which is enhancing their potential as vectors for an *in-vivo* gene transfer system [147]. The possible applications of PAMAM dendrimers of 5th, 7th and 9th generations as the DNA and PLGA biocompatible membrane to facilitate transfection of dermal cells *in-vitro* and *in-vivo* was studied [148]. The dendrimer-DNA complex containing membrane can also be desiccated with retention of transfection activity following rehydration.

The appropriate surface modified PAMAM dendrimers have been used for imaging. Fluorescent markers, such as fluorescein, or chelated metal ions, such as Gd^{3+} for MRI are used to generate a measurable signal for imaging [148-149], radiolabled compounds such ^{3}H, ^{14}C, ^{90}Y, ^{125}I, ^{153}Gd, ^{212}Bi, etc. were used in biodistribution studies [150].

This particular function of PAMAM dendrimers relies on their ability to bind all forms of charged nucleic acids on the electrostatic attractions. The resulting dendrimer-DNA complexes enhance DNA transfer, protect against intra- and extra-cellular degradation and result in the transgene expression in a variety of mammalian cell lines *in-vitro* as well as *in-vivo* in variety of animal tissues [145, 147, 151-152]. Formation of DNA–PAMAM dendrimer complexes seems to be based entirely on electrostatic interaction between negatively charged phosphate groups of the nucleic acid and positively charged (by protonation) amino groups of the PAMAM dedrimers [152-153]. In the majority of cells, the transfection can also be enhanced with the addition of chloroquine indicates involvement of endosomes in trafficking of the complexes.

PAMAM dendrimers have also been used to target tumor cells either specific or non-specific ways. Further, receptor specific targeting and optimization of surface properties are also possible. Targeting groups can be conjugated to the host dendrimers surface [154] to allow the imaging agent to bind selectively to specific sites such as receptors on tumor cells to improve detection. The PAMAM dendrimers can be conjugated to a number of cell surface receptors [155] that are known to be over-expressed on prostatic cancer cells, including epidermal growth factor receptor (EGFR) and the lanrinin receptor VLA-6. Whereas for a non-specific targeting, it has been shown that appropriately sized dendrimers and dendrimer-based metal (Pt) complexes became trapped in tumor vasculature which is due to the enhanced permeability and retention (EPR) effect [156].

The 3rd and 4th generation cationic PAMAM dendrimers and 2.5th and 3.5th anionic PAMAM dendrimers were modified [157] to incorporate, on average a single pendant amino group and radioiodinated by using the Bolton and Hemter reagent. ^{125}I labeled dendrimers were incubated with everted sacs *in-vitro* and the transfer of radioactivity into the tissue and serosal fluid was followed with time. The serosal transfer rate was observed for all anionic generations were extremely high with Endocytic Indices (EI) in the range of 3.4 - 4.4.mL/mg protein per hour. This faster serosal transfer rate of the anionic dendrimers may have potential for further development as oral drug delivery systems. Although the ^{125}I–labeled cationic PAMAM dendrimer displayed a tissue uptake (EI= 3.3 - 4.8 mL/mg protein/h) higher than the rate of serosal transfer (EI= 2.3 -2.7 mL/ mg protein/h), which may be due to non–specific adsorption of cationic dendrimer to the mucosal surface.

A general approach for reversible affinity of biomolecules at immunosensing surfaces was described [158], which was based on the biospecific association and displacement reaction between functionalized antigen, ligands and antibody molecules. The under laying template for the affinity surfaces were constructed with 64 PAMAM dendrimer monolayers on (self assemble monolayer) SAM/Au electrode; whose dendritic surface groups were double–functionalized with biotinyls and ferrocenyls for the bio-specific recognition and electron transfer medium reactions. The functionalized monolayer provide a platform for biospecific recognization with monoclonal anti-biotin immunoglobulin G (IgG) and the bound antibody molecules were dissociated by the addition of free biotin in solution. By the displacement reaction using free biotin bound antibody molecules were dissociated from the surface and the bioelectrocatalytic signal was restored. The association of antibody molecules and renewal of the affinity surface were tracked with electrochemical method by using glucose oxidase (GO_x) as a free diffusion signal generator and ferrocenyl–tethered dendrimers (Fc-D) as an electron transferring mediator in the reaction medium. The shielding of affinity surface by biospecifically bound antibody molecules caused hindrance in electron transfer resulting in reduced signal from cyclic voltammetry. Thus the off-on (association-dissociation) reaction cycles were observed by blockage of electrode surface with bound antibody molecules and subsequent hindrance in electron transfer by freely diffusing GO_x and electron transferring mediator and *vice versa*. Therefore, this method may be applied to the biomolecular switches.

There has been a tremendous demand for efficient analytical tools for immunoassays in the fields of clinical analysis and biochemical studies. Recently a new scientific field in bioassay has been emerging, linking bioanaytical techniques with microelectronic technology. Specially, the immune or affinity sensing techniques, registering biospecific interactions such as antigen-antibody, ligand-receptor and protein-protein recognition reactions, etc. are very important for miniaturization and automation. In addition, recent completion of the human genome project opened a new research field of proteomics, making the development of immunoassay techniques more important. For biosensor construction and its miniaturization for high throughput analysis, the surface on which biological entities are immobilized and biospecific interactions take place is very important.

A strategy of signal generation was developed [158] for immunosensors that transduces biospecific affinity recognition reactions into electrochemical signals. The immunosensing interface was developed onto the dendrimer-activated self-assembled monolayers, as the base template for the functionalization of the antigen moiety and signal generation. It was shown that the dendrimer activated self-assembled monolayers (SAMS) are advantages over plain modified thiolate SAMS for sensing surface with respect to sensing performance and the analytical characteristics of the resulting immunosensor.

The PAMAM dendrimer modified surfaces could effectively be made onto electrode surfaces and functionalized with chemical and biological ligand groups for the purpose of affinity and catalytic biosensing [159-161]. They developed an affinity biosensor system [159] based on avidin-biotin interaction on a gold electrode. For this purpose a 4th generation PAMAM dendrimer with partial ferrocenyl–tethered surface groups was used as the building block of an affinity sensing monolayer. In another report [161] a 4th generation PAMAM dendrimer was functionalized with redox–active ferrocenecarbonaldehyde and deposited alternately, layer by layer of partial ferrocenyl-tethered dendrimers with periodate-oxidize glucose oxidase (GO_x) on Au surface. Thus an electrochemically and enzymetically active multilayered were assembled, which were used as reagentless biosensor.

A new strategy of developing a repeatedly renewable affinity-sensing surface based on the reversible association-dissociation reaction between avidin and biotin analogs was reported [160]. A fourth generation PAMAM dendrimer was functionalized with desthiobiotin or desthiobiotin-amidocaproate and used to construct the monolayer on 11-mercapto undecanoic acid self-assembled monolayer on gold surface for the affinity-sensing surface. This affinity sensing surface exhibited effective biospecific association/dissociation reaction cycles, promising for the reversible affinity biosensor.

The silver complexes of PAMAM dendrimers as well as different Ag-PAMAM dendrimer nanocomposite solutions were used [162] as antimicrobial agents, *in-vitro* against *Staphylococcus aureus, Pseudomonas aerugiosa,* and *Escherichia coli* bacteria. Both the PAMAM-Ag salts and nano-composites displayed considerable antimicrobial activity without the loss of solubility and activity, even in the presence of sulfate or chloride ions, as measured by standard agar over-lay diffusion method. This may be due to the atomic/molecular level dispersion of the guest in a dendrimer host.

Besides viruses and liposomes, which were used hitherto, dendrimers also function as artificial gene-transfection reagents. It turned out however that structurally imperfect dendrimers, as they might obtained by thermal treatment, are clearly more efficient gene-transfer reactors [163]. Such dendrimers are now commercially available as transfection reagent for *in-vitro* application. The ability of dendrimers to transport genetic material through cell membranes was extended to the tunneling of oligonucleotides, which inhibit the transfection of m-RNA in proteins. It emerged that the dendrimers not only enable the transport of these oligonucleotides, but also significantly stabilize them inside the cell [164].

The 3rd generation PAMAM dendrimer (Mn=6909 Da) was formed complex [165] with phosphorothioate oligonucleotides with positive, negative or neutral in charge depending on stoichiometry as observed by agarose gel electrophoresis. The complexes were stable in 50% serum to variations in pH (3, 5 and 10) and ionic strength (0-500 mμ). The flow cytometry and confocal microscopic studies revealed that the dendrimer resulted in a 50-folds enhancement in cell uptake of oligonucleotide, and enhanced cytosolic and nuclear availability respectively. Thus PAMAM dendrimer can be used as a potential delivery vehicle for oligonucleotides.

The anionic PAMAM dendrimer conjugated [166] with ibuprofen through amide bond formation by reaction of terminal amino groups of the dendrimer with acid groups of the drug in the presence of carbodiimide linker. A maximal loading of 13.4 weight 1.0% and molar ratio of 1.0 dendrimer to 4 ibuprofen molecules was obtained (30.2% efficiency) and the amount of free ibuprofen in the conjugate was 0.08% of the total ibuprofen bounded.

The use of PAMAM dendrimers as drug delivery system is of great interest. This is mainly due to factor: (i) the drugs molecules can be attached to the end groups of dendrimer through covalent bonds or by electrostatic forces and (ii) the dendrimer is capable of carrying drugs at high density due to presence of large number of terminal groups.

Klarjnert and Bryszewska [167] coupled two fluorescent compounds, tryptophan and 1-anilinonaphthalend-8-sulfonic acid with PAMAM dendrimers containing different types of surface groups (-COOH, $-NH_2$, -OH, etc.). The fluorescence of these molecules was highly sensitive to the nature of the microenvironment. Highly effective immuno assays based on antibodies covalently attached to fifth generation PAMAM dendrimers was reported [168-169]. The dendrimer-antibody conjugates have unique characteristics like high solubility, high charge density, high affinity, specificity, etc. Because of these properties they can be immobilized on the glass fiber matrix, which enables radial partition immunoassayed (RPIA). This procedure has much improved analytical sensitivity compared to any established system, for creating kinase MB isoenzyme (CKMB), thyrotropin and myoglobin, and reduced instrumental assay time for the CKMB assay.

The combination of properties like size, internal cavities and surface groups make PAMAM dendrimers ideal synthetic analogs of proteins, enzymes and viruses. The biocompatibility and pharmacokinetics of dendrimers are very important factors when their *in-vivo* applications are considered. For above mentioned reasons and also due to easy functionalization, the PAMAM dendrimers are used indifferent biological and medicinal areas, such as gene therapy [142-143], drug delivery agents [170-172] antigen conjugates [168-169] NMR contrast agents [173] and synthetic vaccines [174-175].

A sensing film containing a 4th generation hydroxy terminated PAMAM dendrimer and SYTOX green fluorescent nucleic acid stain was used to detect live pseudomonas aeruginosa bacteria [176]. SYTOX green is generally considered a dead cell stain because it is able to cross the membranes of live cells. In the presence of PAMAM dendrimer in water, the bacterial cell becomes permeable to the SYTOX dye and the fluorescence is significantly enhanced. The fluorescence increases with the bacteria concentration and the intensity 3-4 times higher than the liquid controls without PAMAM dendrimer. Further, the dendrimer also stabilized the sensing film.

Avidin–G_6-$(dB4M\text{-}Gd)_{254}$ [Av-G_6Gd] was synthesized [177] from 6th generation PAMAM dendrimer, biotin, avidin and 2-(p-isothiocyanatobenzyl)-6-methyl-diethylene triaminepenta acetic acid (1B4M).

The Av-G_6Gd was radiolabeled with Gd(III) doped with ^{153}Gd. An in vitro internalization study showed that Av-G_6Gd accumulated and was internalization into SHIN3 cells (a human ovarian cancer) 50-and 3.5 folds great than Gd-DTPA (Magne-visit) and G_6-(1B4M-Gd)$_{256}$(G_6Gd). In addition, accumulation of Gd(III) in the cells was detected by the increased signal on relaxation time-weighted MRI. Thus Av-G_6Gd may be a promising agent for Gd neutron capture therapy of peritoneal carcinomatosis. This reagent also has the potential to permit monitoring of its pharmacokinetic progress with MRI.

The artificial photosynthetic system with hydrogen evolution function was successfully constructed [178] by using peptide bound PAMAM dendrimer with multi Zn(II) mesoporphyrins (MP). The hydrogen evolution function was employed more effectively by PAMAM dendrimer architecture. The electron transfer according to the dynamic interaction between the peptide dendrimer-multi Zn-MP and viologen in the ground state was favorable for the function. Thus this approach is important in the development of an artificial photosynthesis and a novel photodevice. The peptide PAMAM dendrimers bonded Fe(III) or Zn(II) mesoporphyrin IX per two α-helices, afforded [118-119] a multimetalloporphyrin assembly similar to the natural light harvesting antennae in photosynthetic bacteria.

By attaching the Gd(III)-chelate to 2nd and 6th generation PAMAM dendrimers through thiourea (TU) lingkage, its efficiency increased [173] upto 6 folds, as larger molecules increased the ion-relaxivity of the contrast agent. Each dendrimeric macromolecular contrast agent binds upto 170 Gd(III) ions and enhances significantly conventional magnetic resonance and 3D-TOF magnetic resonance angiography (MRA) images. Also the half-lives of these PAMAM–TU-DTPA (Fig.4.8) contrast agents are much higher (40-300 mins.) compared to the non-dendrimer bound analog (20-28 min). In fact, the sixth generation PAMAM-TU-DTPA is a more effective contrast agent than most of the other currently available macromolecular chelate conjugates, such as those formed between albumin polysin– and dextran- Gd(III)-DTPA monoamide. In order to produce more effective PAMAM dendrimer based contract agents, the molecular dynamics of the 2nd and 6th generation vanalsyt PAMAM-TU-DTPA chelate complexes were also studied. From this study, it was found that the use of lower molecular weight dendrimers with denser interiors and ellipsoidal, flattened or elongated shaped may result in improved dendritic MRI contrast agents. There are several other reports [144-145, 149, 179-183] also on the dendrimer-based Gd(III) chelates can be used in 3D TOF MRA and as blood pool contrast agents. The prospect of monoclonal antibody guided radiotherapy and imaging of 2nd generation PAMAM dendrimer bound monoclonal antibody was demonstrated [149].

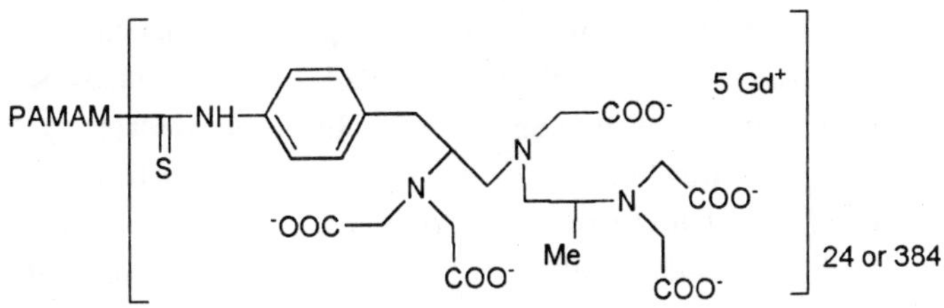

FIG. 4.8 Structure of PAMAM dendrimer based MRI contrast agent

The transfer of luiferase and galactosidase genes into a variety of cell types by using different types of PAMAM dendrimers have been reported by different authors [1'42, 144]. It was found that the transfection efficiency depends on the type of dendrimers and the cells, which are being transfected. In certain instances, the efficiency of transfection is 10-100 folds that of the commercially available cationic lipids. The transfection of antisense oligonucleotides into a variety of cell line has been carried out *in-vitro* by using PAMAM dendrimers [142, 145, 181]. These antisense oligonucleotides complementary to specific target sequences on cellular nucleic acid are able to modulate *in-vitro* and *in-vivo* gene expression [142]. The lack of toxicity, high transfection efficiency and the stability of DNA-dendrimers complexes suggest that PAMAM dendrimers may promote the *in-vivo* transfer of genetic material into cells. One of the most obvious applications of such a system in the treatment of chromosomal detects in human hereditary diseases, such as cystic fibrosis.

The antiviral properties of PAMAM dendrimers substituted with a variety of cationic and anionic end groups showed [182] significant inhibitions against a variety of enveloped viruses, including HIV1, HIV2, human cytomegalovirus (CMV) and various herpes simplex viruses (HSV).

The combination of PAMAM dendrimers with different carbohydrate moieties has broadened their potential application to biological systems [72, 183-184]. A variety of T-antigen bound glycol-PAMAM dendrimers with 4,8,16 and 32 active sites were utilized to bind protein [96]. The inhibiting potential of these glyco-PAMAM conjugates towards antibody containing antigen interactions was enhanced up to 3800 times over that of the monomeric T-antigen reside. By the coordination of viruses, which are blocked, similarly to cell surfaces occurs and an infection can be prevented [183]. In boron neutron capture therapy, the generation of cytotoxic and energetic products from nuclear fission reactions of low energy neutrons and ^{10}B nuclei are used to destroy malignant cells. Barth *et al.* [185] studied boronated PAMAM dendrimers attached to a monoclonal antibody for this purpose.

Two novel dendrimer conjugates, PEG_{1-2}-G_4D-$(IB4M\text{-}Gd)_2$ were used [186] for MRI contrast agents. Their pharmocokinetic excretion and properties as vascular MRI contrast agents were studied and compared with those of G_4D-$(1B4M\text{-}Gd)_{64}$. PEG_2-G_4D-$(IB4M\text{-}Gd)_2$ remains in the blood significantly longer and accumulated significantly lesser in the liver and kidney compared to G_4D-$(IB4M\text{-}Gd)_{64}$ and PEG-G_4D-$(IB4M\text{-}Gd)_{63}$. Although the blood clearance was slower, PEG_2-G_4D-$(1BM4\text{-}Gd)_{62}$ was excreted more readily without renal retention than that of other two contrast agents. Thus the positive effects of PEG conjugation on a macromolecular MRI contrast agents were found to be prolonged retention in the circulation, increased excretion and decreased accumulation in the organs. The effects of co-injecting lysine to accelerate renal excretion of macromolecular MRI contrast agent, G_4D-$(1B4M\text{-}Gd)_{64}$ were also investigated [187]. 3D-dynamic–micro-MRI with G_4D-$(1B4M\text{-}Gd)_{64}$ was evaluated with and without lysine coinjection and found that when lysine was co-injected urinary excretion of $G_4D(IB4M\text{-}Gd)_{64}$ increased 5.4 folds compared to that without lysine, resulting in decreased renal accumulation of $G_4D(IB4M\text{-}Gd)_{64}$ from 150% to 40% injected dose per gram. On dynamic MRI with G_4D-$(IB4M\text{-}Gd)_{64}$, when lysine was co-injected, the kidney-to-blood SI (serial signal intensity) ratio was significantly lower than that obtained without lysine.

4.5.3 Electronics

Laser-induced optical breakdown (LIOB) has shown great importance in diverse fields, including laser surgery, micro machining, three dimensional optical data storage, solid state physics, etc. A 5th generation amine terminated ethylenediamine core PAMAM dendrimer and gold nanocomposties (DNC) can be used to remarkable change the LIOB threshold of a material, owing to a large enhancement of the local electric field [188]. The DNC lower the breakdown threshold values significantly upto two orders of magnitude lower than that of pure dendrimer or normal tissues [188].

Now a days, there is widespread interest in the physical and optical properties of nanometer sized semiconductor nanoparticles. Lakwicz *et al.* [189] prepared CdS nanoparticles in the presence of 4th generation amine terminated PAMAM dendrimer. These nanoparticles displayed strongly heterogeneous intensity decays with mean decay times of 93 ns. Thus they can provide a new class of luminophores for use in chemical sensing, DNA sequencing, high through put screening and other biotechnological applications.

Micro contact printing offered an interesting alternative for lithographic techniques. A simple method was described [190] for the fabrication of arbitrary submicro meter patterns of cobalt on a non-conducting surface with wet chemical methods and microcontact printing by using a hydroxy terminated 4th generation PAMAM dendrimeras as a host molecule in ink. A selective deposition of cobalt films on the patterned region was realized in the presence of Pd-bound dendrimer complex ink solution. Therefore, PAMAM dendrimer may be the good replacement of chlorosilones for small pattern sizes.

A straight forward parallel approach for the wet chemical fabrication of metal cluster arrays was reported [191]. The combined micro-contact printing is much simpler than the standard lithography techniques. The metal cluster size can be tuned by confining the metal ion in and at a PAMAM dendrimer. The lateral dimensions of the pattern in μm scale and an extension was 100-1000 μm range.

4.5.4 Nanocomposites/Nanoparticles

Over the past few years, metal dendrimer hybrid nanocomposites have gained tremendous interest as inorganic-organic hybrid nanostructred materials. This interest has come from the nanoreactors for the catalysis of selected compounds. Dendrimers can act here as nanotemplates as well as separators between the metal particles. Nanoparticles thus prepared are expected to show advantages for use as electro-cataysts due to prevention of aggregation between particles.

PAMAM dendrimers of 2nd to 10th generation were used [192] as potential nanotemplates for the formation of inorganic–organic hybrid colloids in aqueous solution. The nanotemplates were formed by the addition of $H^+AuCl_4^-$ to a neutral dendrimer with primary and tertiary amine groups, which results a protonated dendrimer with $AuCl_4^-$ counter ions. The gold ions were then reduced to metallic gold without the formation of macroscopic metallic precipitate. The stable brown to red solution of colloidal gold in the range of size, 1-5 nm was formed indicated that the metal colloids were stabilized by the dendrimer. The dendrimers not only stabilized the colloid metal in aqueous solution but also determine the architecture of the forming nanostructures. The lower generations (G_2-G_4) dendrimers behave like low molecular mass colloid stabilizers, i.e. several dendrimer molecules surround the surface of the metal particle formed. Again the higher generations (G_6-G_{10}) dendrimers act as effective 'polymeric' nanotemplates. The metal particles were completely formed inside the individual dendrimer.

The amphiphilic gold/PAMAM dendrimer nanocomposites was prepared [193] by using hydrophobically modified EDA core dendrimers of 2nd and 4th generations as templates. He *et al.* also used [194] PAMAM dendrimer as a polymeric template/stabilizer/reservoir to prepare stable Au-dendrimer nanocomposites by reducing PAMAM-tetrachloroaurate polysalts using hydrazine. The large optical limiting effects were observed [195] form Ag-PAMAM dendrimer nanocomposites. The average size of the metal domains in these nanocomposites was found to be 5-2 nm, as measured by TEM [22], depending on the template used. These novel metal dendrimer-nanocomposites have the potential in optical limiting device and medicine. The photophysical properties of CdS-PAMAM dendrimers aggregates were controlled [137] by the synthetic conditions such as solvent, pH, etc.

The novel dendrimer-silica gel hybrids [196] were also prepared from amino-terminated 3rd generation PAMAM and its partly ester-terminated derivatives with tetraethoxysilane in the presence of a coupling agent 3-glycidoxypropyl trimethoxysilane by *in-situ* sol-gel process. Balogh and Tomalia [137] prepared a PAMAM dendrimer Cu(0) nanocomposite by utilizing the reduction procedure. PAMAM dendrimers were also used as container molecules [32, 44], which allow *in-situ* generated reaction products to be dispersed as amorphous or slightly ordered domains at the atomic or molecular level with in the dendrimers interior, where the branch cells may act as separators (Table 4.1) [197].

Table 4.1 Comparison of PAMAM dendrimer-metal nanocomposites characteristics

No. and nature of terminal groups	*No. of 3°N branches*	*No. of 3°N and 1°N ligands*	*No. of metal sites*	*Possible regions for metal*
64, $-NH(CH_2OH)_3$	62	62	15.5	interior
64,$-NH(CH_2)_2NH_2$	62	126	31.5	Interior and exterior
128, $NH(CH_2)_2NHCO(CH_3)_3$	62	126	31.5	interior

The highly stable, isolated and monodispersed Ag nanoparticles protected by PAMAM dendrimers were prepared [198] via direct chemical reduction. pH had a critical effect on aggregation of the nanoparticles in PAMAM dendrimers and in turn, dendrimer aggregation strongly influences the nanoparticles stability to flocculation. Pd-nanoparticles were also prepared [199] by reduction of metal salts in the presence of three different stabilizers: (i) hydroxy terminated PAMAM dendrimers, (ii) polystyrene-block-poly(sodiumacrylate) and (iii) poly(N-vinyl-2-pyrolidone). The stability of the Pd nanoparticles was measured by the amount of Pd black powder formed after the catalytic reaction. The stability of these Pd-nanoparticles used as catalyst, depends on the type of stabilizer, the reactant and the base used in the reaction system. The 4th generation PAMAM dendrimer was found to be the best stabilizer, though stronger encapsulation of Pd-nanoparticles resulted lower catalytic activity. Thus Pd-nanoparticles stabilized by 3rd generation dendrimer, the block copolymer and poly(N-vinyl-2-pyrrolidone) were found to be effective catalysts for the Suzuki reaction between phenyl boronic acid and iodobenzene.

The preparation, electrochemical and catalytic activity of Ni, Ru, and Ni-Ru oxide nanoparticles were reported [200]. The PAMAM dendrimers act both as templates and stabilizers for the nanoparticles. The electrochemical oxidation of ethane was studied by using the above nanoparticles and found a significant catalytic activity. Further the catalytic activity for hydroxy terminated dendrimer templates is much superior than the amine terminated one, for both Ru and Ni nanoparticles.

4.5.5 Sensor Materials

A thin film for sensor applications was developed [201] by using globular PAMAM dendrimer. A 4th generation amine terminated PAMAM dendrimer was covalently linked to the carboxylic acid groups of a self-assembly monolayer of mercaptoundecanoic acid on a gold surface via amide bond formation, which showed the essential attributes for an ideal chemical sensor. The sensor material also guaranteed a differentiation among acids, alcohols and hydrophobic substances by varying the signal intensity. The response to the test chemicals is very rapid, typically completely reversible and the device had an excellent signal to noise ratio.

In addition to this material, PAMAM dendrimers can be irreversibly attached to gold surfaces even without covalent binding [202-206]. By additional adsorption of hexadecylthiol, the dendrimer layer was compressed and thus distorting the molecules into flattened spheroids. These materials act as reversible switchable gates for redox-active ions, which at pH 6.3 let only $[Fe(CN)_6]^{3-}$ ion pass, and at pH 11 $[Ru(NH_3)_6]^{3+}$ions also to exchange with electron on the gold surface. At low pH, these mixed membranes were permeable exclusively to anions, whereas at high pH only cations can passed.

A thickness tunable multilayered enzyme film was constructed by alternative layer-by-layer depositions of 4th generation PAMAM dendrimers and peroxidate oxidise, glucose oxidase on Au surface and used as a biosensing interface [207]. The response of the electrode was significantly enhanced with multilayer growth and thus the sensitivity was tunable by controlling the film thickness.

4.5.6 Others

A 4th generation PAMAM dendrimer was also utilized to stabilize Au-colloid monolayers on a variety of modified silicon oxide surfaces [208]. These metal colloid films were stable over a period of weeks and were not affected by washing repeatedly with water or ethanol. A number of practical applications were reported for this rough metal surface [209].

Amine terminated PAMAM dendrimers were used as adhesion promoters between vapor-deposited Au films and silica based substrates [210]. This method is relatively simple, requiring only substrate cleaning, dipping and rinsing.

PAMAM dendrimers can serve as high capacity nanoscale containers [211] for toxic metal ions such as Cu(II) in aqueous solution. This chelating agent has distinct advantage over conventional chelating agents. Traditional chelating agents with nitrogen donors, which can typically bind only one Cu(II) ions per molecules, whereas PAMAM dendrimers can bind 7 to 173 Cu(II) ions per molecule depending on generation number. Further just by ultrafiltration, the metal bound dendrimers can be separated and also they can be regenerated by simple protonation. Thus the PAMAM dendrimer has great potential as efficient and cost effective functional materials for polymer supported ultrafiltration (PSUF) treatment of water contaminated by toxic metals.

4.6 SUMMARY

From this chapter it has been found that PAMAM dendrimers are the most critically and thoroughly studied dendrimers. A large variety of modifications of this dendrimer were also found as the amine groups are highly reactive towards different type of reactants. These functionalized dendrimers generate a wide range of different properties and so also the applications. As a consequence, a large numbers of potential applications in different fields, especially in biological and medicinal uses have been successfully developed.

REFERENCES

[1] D.A.Tomalia, H.Baker, J.R.Dewould, M.Hall, G.Kallos, S.Martin, J.Roeck, J.Ryder and P. Smith, *Macromolecules.* **19**(1986)2466

[2] E. Buhleier, W.Wehner and F.Vogtle, *Synthesis* (1978)155

[3] D.A.Tomalia, *Aldrichim. Acta* **26**(1993)91

[4] D.A.Tomalia, H.Baker, J.R.Dewould, M.Hall, G.Kallos, S.Martin, J.Roeck, J.Ryder and P. Smith, *Polym. J.* **17**(1985)117

[5] D.A. Tomalia, V.Berry, M.Hall, D.M.Hedstand, *Macromolecules* **20**(1987)1164

[6] H. Hall, A. Padias, R.McCornal and D.A. Tomalia, *J.Org.Chem.* **52**(1987)53067

[7] D.A. Tomalia and J.R.Dewld, USPat. 4,507,466 (1986) and US Pat. 4,857,599 (1989)

[8] D.A.Tomalia, M.Hall and D.M.Hedstrand, *J.Am.Chem.Soc.* **109**(1987)1601

[9] D.A.Tomalia and P.Dvornic, "Dendritic Polymers, Divergent Synthesis (Starburst Polyamido amine Dendrimers)", Polymeric Materials Encyclopedia, Vol.3, J.C.Salamone (Ed.), CRC Press, New York (1996)

[10] V. Swali, N.J.Wells, G.J.Langley and M.Bradley, *J.Org.Chem.* **62**(1997)4902

[11] I.R.Marsh, H.Smith and M.J.Bradley, *J.Chem.Soc.Chem.Commun.* (1996)941

[12] N.J.Wells, M.Davies and M.J.Bradley, *J.Org.Chem.* **63**(1998)6430

[13] N.J.Wells, A.Basso and M.Bradley; *Biopolymers* **47**(1998)381

[14] I.R.Marish, H.K.Smith, C.Lablanc and M.Bradley, *Molecular Diversity* **2**(1996)165

[15] B.D.Maxwell, H.Fujiwara, S.Habibi-Goudarri, J.P.Ortiz and S.J.Logusch, *J.Labelled Compounds and Radiopharmaceuticals* **41**(1998)935

[16] J.Li, D.Qin, J.R.Baker, Jr. and D.A.Tomalia, *Macromol. Symp.* **167**(2000)257

[17] G.Caminati, N.J.Turro and D.A.Tomalia, *J.Am.Chem.Soc.* **112**(1990)8515

[18] M.H.Kleinman, J.H.Floxy, D.A. Tomalia and N.J.Turro, *J.Phys.Chem.B* **104**(2000)11472

[19] M.F.Ottavani, P.Matteini, M.Brustolon, N.J.Turro, S.Jockusch and D.A. Tomalia, *J.Phys.Chem.B* **102**(1998)6029

[20] M.F.Ottaviani, F.Montalti, N.J.Turro and D.A.Tomalia, *J.Phys.Chem.B* **101**(1997)158

[21] O.Varnavski, R.G.Ispasoiu, L.Balogh, D.A.Tomalia and T.Goodson III, *J.Chem.Phys.* **114**(2001)1962

[22] L.Balogh, R.Valluzzi, G.L.Hagnauer, K.S. Laverdure, S.P.Gido and D.A. Tomalia, *J. Nanoparticle Res.* **1**(1999)353

[23] K.M.A.Rahman, C.J.Durning, N.J.Turro and D.A.Tomalia, *Langmuir* **16**(2000)10154

[24] M.F.Ottaviani, E.Cossa, N.J.Turro and D.A. Tomalia, *J.Am.Chem.Soc.* **117**(1995)4387

[25] M.F.Ottaviani, N.J.Turro, S.Jockusch and D.A. Tomalia, *J.Phy.Chem.* **100**(1996)13675

[26] M.F. Ottaviani, N.J.Turro, S.Jockusch and D.A.Tomalia, *Colloids Surf.* **115**(1996)9

[27] M.F.Ottaviani, C.Turro, N. J.Turro, S.H.Bossmann and D.A.Tomalia, *J.Phys.Chem.* **100**(1996)13667

[28] K.R.Gopidas, A.R.Leheny, G.Caminati, N.J.Turro and D.A.Tomalia, *J.Am.Chem.Soc.* **113**(1991)7335

[29] M.F.Ottaviani, S.Bossmann, N.I.Tarro and D.A.Tomalia, *J.Am.Chem.Soc.* **116**(1994)661

[30] M.F.Ottaviani, F.Montalti, M.Romanelli, N.J.Turro and D.A.Tomalia, *J.Phys.Chem.* **100**(1996)11033

[31] S.Jockusch, J.Ramiver, K.Sanghvi, R.Nociti, N.J.Turro and D.A.Tomalia, *Macromolecules* **32**(1999)4419

[32] D.A.Tomalia, A.M.Naylor and W.A.Goddard III, *Angew.Chem.Int.Ed.* **29**(1990)138

[33] M.F.Ottaviani, P. Favuzza, B.Sacchi, N.J. Turro, S. Jockusch, and D.A.Tomalia, *Langmuir* **18**(2002)2347

[34] M.F.Ottaviani, P.Favuzza, M.Bigazzi, N.J.Turro, S.Jockusch and D.A. Tomalia, *Langmuir* **16**(2000)7368

[35] M.F.Ottaviani, P.Andechaga, N.J.Turro and D.A.Tomalia, *J.Phy.Chem. B* **101**(1997)6057

[36] C.Turro, S.Niu, S.H.Bossmann, D.A.Tomalia and N.J.Turro, *J.Phys.Chem.* **99**(1995)5517

[37] M.C.Moreno-Bondai, G. Orellana, N.J.Turro and D.A. Tomalia, *Macromolecules* **23**(1990)910

[38] S.Jockusch, N.J.Turro and D.A. Tomalia, *Macromolecules* **28**(1995)7416

[39] S.Jockusch, N.J.Turro and D.A.Tomalia, *J.Int. Rec.* **22**(1996)427

[40] G.J.Kallos, D.A. Tomalia, D.M.Hedstrand, S.Lewis and J.Zhou, *Rapid Commun. Mass Spectrom.* **5**(1991)383

[41] B.L.Schwartz, A.L.Rockwood, R.D.Smith, D.A. Tomalia, R.Spindler, *Rapid Commun.Mass.Spectron.* **9**(1995)1552

[42] P.R.Dvornic and D.A.Tomalia, *Macromol. Symp.* **98**(1995)403

[43] P.T.Tolic, G.A.Anderson. R.D.Smith, H.M.Brothers II, R.Spindeler and D.A.Tomalia, *Int. J.Spectrom, Ion. Proc.* **165/166**(1997)405

[44] A.M.Naylor, W.A.Goddard III, G.E.Kiefer and D.A.Tomalia, *J.Am.Chem.Soc.* **111**(1989)2339

[45] D.A. Tomalia, A.M.Naylor and W.A.Goddard, *Angew.Chem.Int.Ed.* **29**(1990)138

[46] D.A.Tomalia, V.B.Hall and D.M.Herdstrand, *Macromolecules* **20**(1987)1167

[47] P.L.Dubin, S.L.Edward, J.I.Kaplan, M.S.Mehta, D.T.Tomalia and J.Xia, *Anal.Chem.* **64**(1992)2344

[48] A.D.Meltzer, D.A.Tirrel, A.A.Jones, P.T.Inglefield, D.M.Hedstrand and D.A.Tomalia, *Macromolecules* **25**(1992)4541

[49] A.D.Meltzer, D.A.Tirrel, A.A.Jones and P.T.Inglefield, *Macromolecules* **25**(1992)4549

[50] T.J.Prosa, B.J.Bauer, E.J.Amis, D.A.Tomalia and S.Scherrenberg, *J.Polym.Sci.B* **35**(1997)2913

[51] L.L.Miller, T.Hashimoto, I.Tabakovic, D.R.Swawson and D.A.Tomalia, *Chem.Mater.* **7**(1995)9

[52] L.L.Miller, R.G.Duan, D.C.Tully and D.A.Tomalia, *J.Am.Chem.Soc.* **119**(1997)1005

[53] D.L.Richter-Egger, J.C.Landry, A.Tesfai and S.A.Tucker, *J.Phys.Chem.* A **105**(2001)6826

[54] E.C.Wiener, F.P.Auteri, J.W.Chen, M.W.Brechbiel, O.A.Gansow, D.S.Schneider, R.L.Belford, R.B.Clarkson and P.C.Lauterbur, *J.Am.Chem.Soc.* **118**(1996)7774

[55] C.L.Larson and A.S.Tucker, *Appl. Spectrosc.* **55**(2001)679

[56] A.Nourse, D.B.Miller and A.P.Minton, *Biopolymers*, **53**(2000)316

[57] D.A.Tomalia and J.R.Dewald, U.S.Pat 4,687,737(1986)

[58] L.J.King, S.H.Amy, I.K.Martin and L.J.Twyman, *Tetrahedron Lett.* **43**(2002)243

[59] W. Chen, D.A.Tomalia and J.L.Thomas, *Macromoleclues* **33**(2000)9169

[60] S.Uppuluri, S.E.Keinath, D.A.Tomalia and P.R.Dvornic, *Macromolecules* **31**(1998)4498

[61] A.V.Lyulin, G.R.Davies and D.B.Adolf, *Macromolecules* **33**(2000)3294

[62] D.A.Tomalia, A.M.Naylon and W.A.Goodard III, *Angew.Chem.* **102**(1990)119

[63] M.L.Mansfield and L.I.Klushin, *J.Phys.Chem.* **96**(1992)3994

[64] J.G. Jang and Y.C.Bae, *Chem.Phys.* **269**(2001)285

[65] H.Tokuhisa, M.Zhao, L.ABaker, V.T.Phan, D.L.Dermody, *J.Am.Chem.Soc.* **120**(1998)4492

[66] D.M.Watkins, Y.Sayed-Sweet, J.W.Klimash, N.J.Turro and D.A.Tomalia. *Langmuir* **13**(1997)3136

[67] P.L.Schwarz, N.J.Turro and D.A.Tomalia, *J.Photochem.Photobio.A Chem.* **112**(1998) 47

[68] M.Santo and M.A.Fox, *J.Phys.Org.Chem.* **12**(1999)293

[69] K.Naka, Y.Tanaka, Y.Chujo and Y.Ito, *Chem.Commun.* (1999)1931

[70] C.Mio, S.Kiritsov, Y.Thio, R.Brattman and J.Praunitz *et al.*, *J.Chem.Eng.Data* **43**(1998)541

[71] T.Muller, D.G.Yablon, R.Karchner, D.Knapp, M.H.Kleinman, H.Fang, C.J.Durning, D.A.Tomalia, N.J.Turro and G. W.Flynn, *Langmmir* **18**(2002)7452

[72] K.Aoi, K.Ito and M.Okada, *Macromolecules* **28**(1995)5391

[73] N.J.Turro, J.K.Barton and D.A.Tomalia, *Acc.Chem.Res.* **24**(1991)332

[74] T.D.James, H.Shinmori, M.Takenchi and S.Shinkai, *Chem.Commun.* (1996)705

[75] J.Wang, X.Jia, H.Zhong, H.Wu, Y.Li, X.Xu, M.Li and Y.Wei, *J.Polm.Sci.partA: Polym.Chem.* **38**(2000)4147

[76] S.Ghosh and A.K.Banthia, *Tetrahedron Lett.* **42**(2001)501

[77] S.Glazier, J.A.Barron, P.L.Houston, and H.D.Abruna, *J.Plys.Clen.B* **106**(2002)9993

[78] M.Fujitsuka, O.Ito, Y.Takaguchi, T.Tajima, K.Ohta, J.Motoyashiya and H.Aoyama, *Bull.Chem.Soc.Jpn.* **76**(2003)743

[79] G.Pistolis, A.Mallaiaris, C.M.Paleos and D. Tsiourvas, *Langmuir* **13**(1997)5870

[80] D.L.Richter-Egger, H.Li and S.A.Tucker, *Appl. Spectrosc.* **54**(2000)1151

[81] W.S.Baker, B.I.Buford and R.M.Crooks, *J.Phys.Chem.B* **105**(2001)8885

[82] M.Kimura, K.Mizuno, T.Muto, K.Hanbusa and H.Shirai, *Macromol. Rapid Commun.* **20**(1999)98

[83] M.Zhao, Y.Liu, R.M.Crooks and D.E.Bergbreiter, *J.Am.Chem.Soc.* **121**(1999)923

[84] Y.Takaguchi, K.Saito, S.Suzuki, K.Hamada, K.Ohta, J.Motoyoshiya and H.Aoyama, *Bull.Chem.Soc.Jpn.* **75**(2002)1347

[85] K.Takada, G.D.Storrier, J.I.Goldsmith and H.D.Abruna, *J.Phys.Chem.* **105**(2001)2404

[86] D. Sehgal and I.K.Vijoy, *Anal.Biochem.* **218**(1994)87

[87] D.J.Diaz, S.Bernhard, G.D.Storrier and H.D.Abruno, *J.Phys.Chem.*B **105**(2001)8746

[88] D.J.Diaz, G.D.Storrien, S.Bernhard, K.Takada and H.D.Abruna *Langmuir* **15**(1999)7351

[89] J.Barbera, M.Macros and J.L.Serrano, *Chem.Eur.J.* **5**(1999)1834

[90] M.Marcos, R.Gimenez, J.L.Serrano, B.Donnio, B.Heinrich and D.Guillon, *Chem.Eur.J.* **7**(2001)1006

[91] B.Donnio, J.Barbera, R.Gimenez, D.Guillon, M.Macros and J.L.Serrano, *Macromolecules* **35** (2002)370

[92] K.Aoi, K.Tsutsumiuchi, A.Yamamoto and M.Okada, *Macromol. Rapid Commun.* **19**(1998)5

[93] K.Aoi, K.Tsutsumiuchi, A.Yamamoto and M.Okada, *Tetrahedron* **53**(1997)15415

[94] T.K.Lindhorst and C.Kieburg, *Angew Chem.Int.Ed.* **35**(1996)1953

[95] C. Kieburg and T.K.Linghorst, *Tetrahedron Lett.* **38**(1997)3885

[96] M.G.Baek and R. Roy, *Bioorganic and Medicinal Chemistry* **10**(2002)11

[97] Y.Chen, R.K. Jain, E.V.Chandrasekaran and K.L.Matta, *Glycoconjugatete J.* **12**(1995)55

[98] Y.S. Hung, M. Madej, R.R.Koanty and B.M.Longenecker, *Cancer* **50**(1990)4308

[99] D.Chaplin, D.H.G.Grout, S.Bornemann, D.W.Huchinson and R.Khan, *J.Chem.Soc. Perkin Trans. 1* (1981)235

[100] G.Sui, M. Micic, Q.Huo and R.M.Leblanc, *Colloids and Surfaces A, Physicochemical and Eng. Aspects* **171**(2000)185

[101] G. Sui, M.Micic, Q.Huo and R.M.Leblanc, *Langmuir* **16**(2000)7847

[102] V.Chechik, M.Zhao and R.M.Crooks, *J.Am.Chem.Soc.* **121**(1991)4910

[103] P.K.Tripathi, A.J.Khopade, S.Nagaich, S.Shrivastava, S. Jain and N.K.Jain, *Pharmazie* **57**(2002) 261

[104] F.M.Menger, A.V.Peresypkin and S.Wu, *J.Phys.Org.Chem.* **14**(2001)392

[105] S.M.Cohen, S.Petoud and K.N.Raymond, *Chem.Eur.J.* (2001)272

[106] C.M.Nunez, A.L.Andrady, R.K.Guo, J.N.Baskir and D.R.Morgan, *J.Polym. part A: Polym.Chem.* **36**(1998)2111

[107] N.Tsubokawa, H.I.Chioka, T.Satoh, S.Hayashi and K.Fuijki, *React.Func.Polym.* **37**(1998)75

[108] N.Tsubokawa, T.Satoh, M.Murota, S.Sato and H. Shimiza, *Polym.Adv.Techol.* **12**(2001)596

[109] N.Tsubokawa, K.Kaboyashi, Y.Soni, *Polym.J.* **19**(1987)1147

[110] A.Skogseid, Norway Pat. 72,587(1947)

[111] M.Okazaki, M.Murota, Y.Kawaguchi and N.Tsubokawa, *J.Appl.Polym.Sci.* **80**(2001)573

[112] P.R.Dvornic, J.Li, A.M.de Leuze-Jallouli, S.D.Reeves and M.J.Owen, *Macromolecules* **35**(2002)9323

[113] P.R.Dvornic, A.M.de Leuzep-Jallouli, M. J.Owen and S.V.Perz, *Macromolecules* **33**(2000)5366

[114] R.G.Duan, L.L.Miller and D.A.Tomalia, *J.Am.Chem.Soc.* **117**(1995)10783

[115] L.Balogh, A.de Leuze-Jallouli, P.Dvornic, Y.Kunugi, A.Blumstein and D.A.Tomalia, *Macromolecules* **32**(1999)1036

[116] S.Subramanyam, M.S.Chetam, A.Blumstein and S.Mil'shtein, *Polym.Mater.Sci.Eng.* **64**(1991)186

[117] N.Higashi, T.Koga and M.Niwa, *ChemBioChem* **3**(2002) 448

[118] M.Sakamoto, A.Ueno and H. Mihara, *Chem.Eur.J.* (2001)2449

[119] M.Sakamoto, A.Ueno and H.Mihara, *Chem.Commun.*(2000)1741

[120] C. Rao and J.P.Tam, *J.Am.Chem.Soc.* **116**(1994)6975

[121] M.F.Ottavani, F.Furini, A.Casini, N.J.Turro, S.Jockusch, D.A.Tomalia and L.Messori, *Macromolecules* **33**(2000)7842

[122] M.F.Ottaviani, R.Daddi, M.Brustokon, N.J.Turro and D.A. Tomalia, *Langmuir* **15**(1999)1973

[123] A.U.Bielinska, J.F.Kukowska-Latallo and J.R.Baker Jr., *Biochem.Biophys.Acta* **1353**(1997)180

[124] W. Chen, N.J.Turro and D.A.Tomalia, *Langmuir* **16**(2000)15

[125] S. Managhan, D.Grifith-Johnson, I. Mathews and M.Bradley, *ARKIVOC* (2001)46

[126] Y.Takaguchi, T.Tajima, K.Ohta, J.Motoyoshiya and H.Aoyama, *Chem.Lett.* (2000)1388

[127] Y.Takaguchi, T.Tajima, K.Ohta, J.Motoyoshiya, H.Aoyama, T.Wakahara, T.Akasaka, M.Fujitsuka and O.Ita, *Angew.Chem.Int.Ed.* **41**(2002)817

[128] Y.M.Chung and H.K.Rhee, *Chem.Commun.* (2002)238

[129] S.C.Bourque, H.Alper, L.E.Manzer and P.Arya, *J.Am.Chem.Soc.* **122**(2000)956

[130] S.C.Bourque, F.Maltasis, W.J.Xiao, O.Tardif, H.Alper, P.Arya and L.E.Manzer, *J.Am.Chem.Soc.* **121**(1999)3035

[131] M.Murota, S.Sato and N.Tsubokawa, *Polym.Adv.Technol.* **13**(2002)144

[132] E.H.Brown and C.A.Krans, *J.Am.Chem.Soc.* 51(1929)2690

[133] N.Tsubokawa and T.Takayama *React.Funct.Polym.* **43**(2000)341

[134] K.Fujiki, M.Sakamoto, T.Satoh and N.Tsubokawa, *J.Macromol.Sci.Pure Appl.Chem.A* **37**(2000)357

[135] H.Alper, P.Arya, S.C.Bourque, G.R.Jefferson and L.E.Manzer. *Can.J.Chem.* **78**(2000)920

[136] M.Zhao and R.M.Crooks, *J.Am.Chem.Soc.***120**(1998)4877

[137] L.Balogh and D.A.Tomalia, *J.Am.Chem.Soc.* **120**(1998)7355

[138] K.Sooklal, L.H.Hanus, H.L.Ploehn and C.J. Murphy, *Adv.Mater.* **10**(1998)1083

[139] Y.Niu, L.K.Yeung and R.M.Crooks, *J.Am.Chem.Soc.* **123** (2001)6840

[140] M.Zhao and R.M.Crooks, *Angew.Chem.Int.Ed.* **38**(1999)364

[141] A.R.Schmitzer, S.Franceschi, E.Perez, I.Rico-Lattes, A.L.Thion and M.V.Erard, *J.Am.Chem.Soc.* **123**(2001)5956

[142] J.F.Kukowska-Latollo, A.U.Bietinska, J.Jonson, R.Spindler and D.A.Tomalia, *Proc. Nalt. Acad. Sci.* **93**(1996)4897

[143] M.X.Tang, C.T.Redmann and F.Szoka, Jr. *Bioconjugate Chem.* **7**(1996)703

[144] J.Haensler and F.C.Szoka Jr. *Bioconjugate Chem.* **4**(1993)372

[145] A.U.Bielinska, J.F.Kukowska- Latallo, J.Johnosn, D.A. Tomalia and J.R.Baker Jr. *Nuel. Acids Res.* **24**(1996)2176

[146] M.X.Tang and F.C.Szoka, *Gene Ther.* **4**(1997)823

[147] J.C.Robert, M.K.Bhalgat and R.T.Zera, *J.Biomed.Mater.Res.* **30**(1996)53

[148] A.U.Bielinksha, A.Yen, H.L.Wu, K.M.Zahos, R.Sun, N.D.Weiner, J.R.Baker Jr. and B.J.Roessler, *Biomaterials* **21**(2000)877

[149] C.Wu, M.W.Brechbiel, R.W.Kozak and O.A.Gansow, *Bioorg.Med.Chem.Lett.* **4**(1994)449

[150] H.Kobayashi, N.Sato, S.Kawamoto, T.Saga, A.Hiraga, T.L.Haque, T.Ishimori, J.Konishi, K.Togashi and M.W.Brechbiel, *Bioconjugate Chem.* **12**(2001)100

[151] L.Qin, D.R.Pahud, Y.Ding, A.U.Bielinska, J.F.Kakowska-Lattollo, J.R.Baker, Jr. and J.S.Bromberg, *Hum.Gen.Ther.* **9**(1998)553

[152] A.U.Bielinska, C.Chen. J.Johnson and J.R.Baker Jr. *Bioconjugate Chem.* 10(1999)843

[153] H.Bonkhoff, *Hum.Pathol.* **24**(1993)243

[154] R.Duncan, N.Malik, S.Richardon and P.Ferruti, *Polym. Prep.* **39**(1998)180

[155] F.M.Muggia, *Clin.Cancer Res.* **5**(1999)7

[156] R.Wiwattanapatapee, B.Carreno-Gomez, N.Malik and R.Duncan, *Pham.Res.* **17**(2000)991

[157] H.C.Yoon, D.Lee and H.S.Kim, *Analytica Acta* **456**(2002)209

[158] H.C.Yoon, M.Y.Hong and H.S.Kim, *Anal.Biochem.* **282**(2000)121

[159] H.C.Yoon, M.Y.Hong and H.S.Kim, *Langmuir* **17**(2001)1234

[160] H.C.Yoon, M.Y.Hong and H.S.Kim, *Anal. Chem.* **72**(2000)4420

[161] H.C.Yoon, M.Y.Hong and Y.T.Kim, *Analyst* **127**(2002)1082

[162] L.Balogh, D.R.Swanson, D.A.Tomalia, G.L.Hagnauer and A.T.McManus, *Nano Lett.* **1**(2001)18

[163] M.Tang, C.T.Redemann, F.C.J.Szoka, *Bioconjugate Chem.* **7**(1996)703

[164] A.U.Bielinka, J.F.Kukowska-Lattallo, J.Johnson, D.A.Tomalia and J.R.J.Baker, *Nucleic Acids Res.* **24**(1996)2176

[165] R.Delong, K.Stephenson, T.Loftus, M.Figcher, S.Alahari, A.Nothing and R.L.Juliano, *J.Pharm.Sci.* **86**(1997)762

[166] R.Wiwattanapatapee and R. Duncan, *Thai. J.Pham.Sci.* **24**(2000)67

[167] B. Klajnert and M.Bryszewska, *Cell Mol.Bio.Lett.* **7**(2002)288

[168] P.Singh, *Bioconjugate Chem.* **9**(1998)54

[169] P.Singh, F.Moll III, S.H.Lin and C.Ferzti, *Clinic.Chemistry* **42**(1996)1567

[170] R.Duncan and N.Malik, *Proceed.Intern.Symp.Control.Rel.Bioact.Mater.* **23**(1996)105

[171] R.Duncan, *Chemistry and Industry* **7**(1997)262

[172] D.A.Tomalia and R.Estand, *Chemistry and Industry* **11**(1997)416

[173] E.C.Wiener, M.W.Brechbiel, H.Brothers, R.L.Magin, O.A.Gansow, D.S.Schneider, R.L.Belford R.B.Clarkson and P.C.Lauterbur, *Magnetic Resonance in Medicine* **31**(1994)1

[174] C.Rao and J.P.Tam, *J.Am Chem.Soc.* **116**(1994)6975

[175] J.P.Defort, B.Nardelli, W.Huang, D.D.Ho and J.P.Tam, *Proc.Natl.Acad.Sci.* **89**(1992)3879

[176] A. Chang, J.B.Gillespie and M.B.Tabacco, *Anal. Chem.* **73**(2001)467

[177] H.Kabayashi, S.Kawamoto, T.Sega, N.Sato, T.Ishimori, J.Konishi, K.Ono, K.Togashi and M.W.Brechbiel, *Bioconjugate Chem.* **12**(2001)587

[178] M.Sakamoto, T.Kamachi, A. Ueno and H.Mihara, *Biopolymers* **59**(2001)

[179] M.W.Bourne, L. Margerun, N.Hylton, B. Campion, J.J.Lai, N.Derugin and C.B.Higgins, *J.Magn. Reson. Imag.* **6**(1996)305

[180] G.Adam, J.Neurerburg, E. Spuntrup, A.Muhler, K.Scherer and R.W.Gumther. *J.Magn.Reson. Imag.* **4**(1994)462

[181] J.A.Hughes, A.I.Aronsohn, A.V.Avrustskaya and R.L.Juliaano, *Pharm.Res.* **13**(1996)404

[182] B.R.Matthews and G. Holan, WO Pat. 95/34595(1995)

[183] D.Page and R.Roy, *Bioconjugate Chem.* **5**(1997)714

[184] D.Zamini and R.Roy, *J.Org.Chem.* **63**(1998)3468

[185] R.F.Barth, D.M.Adams, A.H.Soloway, F.Alam and M.V.Darby, *Bioconjugate Chem.* **5**(1994)58

[186] H. Kobayshi, S.Kawamoto, T.Saga, N.Sato, A.Hiraga, T.Ishimori, J.Konishi, K.Togashi and M.W.Brechbiel, *Magn.Reson.Med.* **46**(2001)781

[187] H.Kobayashi, N.Sato, S.Kawamoto, T.Sega, A. Hiraya, T.Ishimori, J.Konishi, K.Togashi and M.W. Brechbiel, *Magn.Reson.Med.* **46**(2001)457

[188] J.Y.Ye, L.Balogh and T.B.Norris, *Appl.Phys.Lett.* **80**(2002)1713

[189] J.R.Lakowicz, I.Gryczynski, Z.Grycznski and C.J.Murphey, *J.Phys.Chem. B* **103**(1999)7613

[190] X. C.Wu, A.M.Bitterner and K.Kern, *Langmuir* **18**(2002)4984

[191] A.M. Bitterner, X.C.Wu and K. Kern, *Adv.Funct.Mater.* **12**(2002)432

[192] F. Grohn, B.J.Bauer,Y.A.Akapalus, C.L.Jackson and E.J.Amis, *Macromolecules* **33**(2000)6042

[193] Y.S.Seo, K.S. Kim, K.Shin, H.White, M.Rafailovich, J.Sokolov, B.Lin, H.J.Kim, C.Zhang and L.Balogh, *Langmuir* **18**(2002)5927

[194] J.A.He, R. Valluzzi, K.Yang, T.Dolukhanyan, C.Sung, J.Kumar, S.K.Tirpathy, L.Samuelson, L.Balogh and D.A.Tomalia, *Chem.Mater.* **11**(1999)3268

[195] R.G.Ispasoiu, L.Balogh, O.P.Varnavski, D.A.Tomalia and T.Goodson III, *J.Am.Chem.Soc.***122**(2000)11005

[196] A.Gong, Y.Chen, X.Zhang, H.Liu, C.Chen and F.Xi, *J.Appl.Polym.Sci.* **78**(2000)2186

[197] N.Beck Tan, L.Balogh and S.Trenino, *Polm.Mater.Sci.Eng.* **779**(1997)120

[198] J.Zheng, M.S.Stevenson, R.S.Hikida and D.G.van Pattern, *J.Phy.Chem.B* **106**(2002)1252

[199] Y.Li, M.A. El-Sayed, *J.Phys. Chem.B* **105**(2001)8938

[200] J.N.Kim and S-M.Park, *Electrochem. Solid State-Lett.* **3**(2000)385

[201] M.Wells and R.M.Crooks, *J.Am.Chem.Soc.***118**(1996)3988

[202] Y.Liu, M.Zhao, D.E.Berghreiter and R.M.Crooks, *J.Am.Chem.Sci.* **119**(1997)8720

[203] M.Zhao, H.Tokushisa and R.M.Crooks, *Angew.Chem.Int.Ed.* **36**(1997)2596

[204] A.Hierlemann, J.K.Cambell, L.A.Baker, R.M.Crooks and A.J.Ricco, *J.Am.Chem.Soc.* **120**(1998)5323

[205] M.Zhao, Y.Liu, R.M.Crooks, D.E.Bergbreiter, *J.Am.Chem.Soc.* **121**(1999)923

[206] H.Tokuhias, M.Zhao, L.A.Baker, V.T.Phan, D.L.Dermody, M.E.Garcia, R.F.Peez, R.M.Crooks and T.M.Mayer, *J.Am.Chem.Soc.* **120**(1998)4492

[207] H.C.Yoon and H.S.Kim, *Anal.Chem.* **22**(2000)922

[208] G.Bar, S.Rubin, R.W.Cults, T.N.Taylor, T.A.Zawodzinski Jr. *Langmuir* **12**(1996)1172

[209] S.Rubin, G.Bar, T.W.Taylor, R.W.Cults, T.A.Zawodzinski Jr. *J.Vac.Sci.Technol.A* **14**(1996)1870

[210] L.A. Baker, F.P.Zamborini, L.Sun and R.M.Crooks, *Anal Chem.* **71**(1999) 4403

[211]M.Diallo, L.Balogh, A.Shatagati, J.H.Johnson Jr. W.A.Goddard III and D.A.Tomalia, *Envir.Sci.Technol.* **33**(1999)820

CHAPTER 5

DENDRITIC POLY(ALKYLENE IMINE)S

5.1 INTRODUCTION

Poly(alkylene imine) is an important category of dendritic polymers not only for historical reason, as it was the first attempt to synthesize dendritic polymer in 1978 but also it offers many unique properties and application, which are rare in other classes of polymers. This class mainly includes poly(propylene imine) (PPI) and poly(ethyleneimine) (PEI) dendritic polymers along with other polymers of similar type, which are discussed in this chapter. PPI dendrimers are very interesting macromolecules because of their acid-base properties, which were discussed in great details by Kopper *et al* [1]. This dendrimers are now available in commercial scale as 'Astramol' manufactured by DSM, Geleen, The Netherland. Structurally poly(alkylene imine)s consist of alkyl diamine or polyamine core with alkylimine, such as propyleneimine or ethyleneimine branching units. Because of the presence of terminal primary amine groups, they can be easily functionalized to varieties of macromolecules to get desired modified poly(alkylene imine) dendritic polymers.

Through out this chapter the abbreviation system, DAB-dendr-X_n was taken from the literature to simplify this discussion of the functionalized dendrimers. This abbreviation consists of three parts. The first part symbolized the core, the starting diamine, which is abbreviated as DAB for diaminobutane. The second part is the phrase "dendr" implying that each nitrogen from that DAB core molecule out to but not including, the surface of the dendrimer is a tertiary nitrogen having propyl groups as its substitutes. The last part X_n contains the number and type of end groups on the dendrimers surface.

5.2 SYNTHESIS, CHARACTERIZATION AND PROPERTIES OF DIFFERENT TYPES OF POLY(ALKYLENEIMINE) DENDRIMERS

In this section the synthesis, characterization, properties of PPI and PEI dendrimers along with their different modification are discussed. The properties of different modified PPI dendrimers are also presented here.

5.2.1 Poly(propylene imine)

The PPI dendrimers are well known as PPI, but they are also abbreviated as POPAM. The alternative name is poly(1,3-trimetylene imine) dendrimers and trade name by DSM, The Netherlands as Astramol^{TM}. The details of this dendrimers are described in this sub section.

5.2.1.1 Synthesis

The first attempt to synthesize this novel class of macromolecules by Vogtle and his coworkers [2] in 1978 was a nitrile terminated PPI dendrimers using the divergent approach. This historical attempt was to synthesize PPI dendrimers upto 3rd generation with molecular weight 873 Da, by a series of Michael addition and subsequent reduction by Co(II) catalyzed borohydride (Scheme 5.1). However, unfortunately

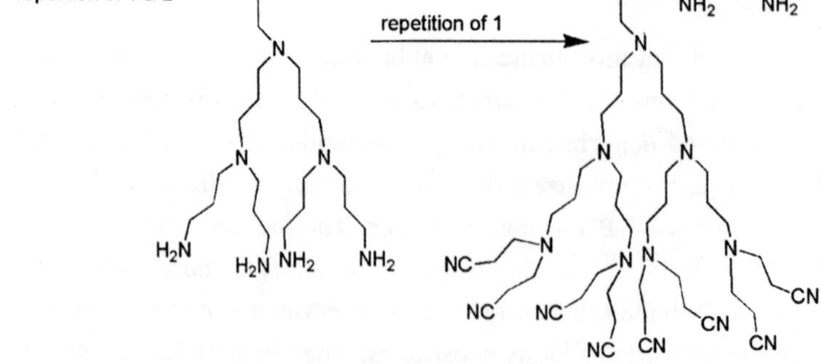

SCHEME 5.1 Synthesis of oligomeric nitrile terminated PPI

due to difficulties in the reduction step and the likely occurance of cyclization side reactions limited the successful synthesis of higher generation dendrimers. As a consequence this oligomeric products did not show any dendritic like properties and hence they cannot be considered as actual dendrimers.

By reapplying Vogtle's initial synthetic strategy [2], the reaction conditions were elaborated that allowed water to be used as the solvent for both the addition and reduction steps. A repetitive cycle of Michael addition with acrylonitrile and catalytic hydrogenation with H_2/Raney-Co produces dendrimers with nitrile end groups. By using 1,4-diaminobutane (DAB) as the central core, DAB-dendr-(CN)x or with amino end groups, DAB dendr-(NH_2)x, where x = 4, 8, 16, 32, 64 and 128, were reported [3] for the successive generations of the PPI dendrimers, as shown in scheme 5.2. This was the first report on synthesis of PPI dendrimer in 1993. The use of Raney-cobalt as the reduction catalyst and accurate control of the hydrogenation condition gave quantitative hydrogenation of the nitrile groups without cyclization. This was also removed the needs of intermediate isolation and purification steps. The production of these PPI dendrimers as 'Astramol' dendrimers has already been started industrially by DSM , The Netherlands. [4].

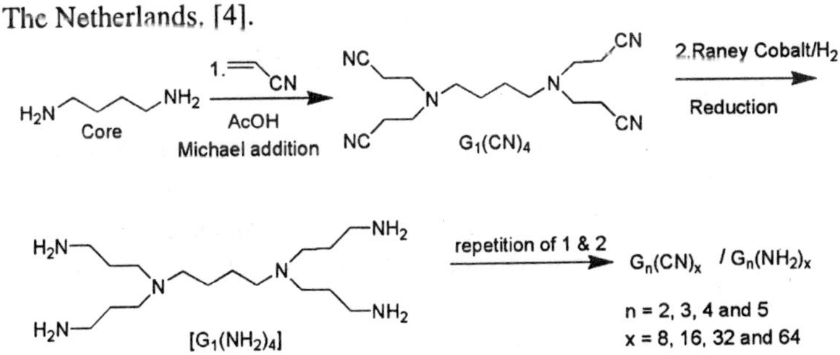

SCHEME 5.2 Synthesis of first true PPI dendrimers by accelerated approach.

5.2.1.2 Characterization

The protonation behavior of the PPI dendrimers and also other related oligoamines has been reported [1] by using ^{15}NNMR spectroscopy. Actually, the natural abundance ^{15}N with inversegated proton decoupling was used for this purpose. Genderen *et al.* also used [5] natural abundance ^{15}NNMR spectroscopy to characterize PPI dendrimers. The inversegated decoupled measurements showed the number and intensities of signals that reflect the generation of the dendrimers. By this technique, one can obtain quantifying spectra in which the individual shell of dendrimers can be identified because all nuclei in one shell are chemically equivalent and the number of nucleus in one shell differs by a factor of 2 from the previous shell.

The structure and conformation of DAB core PPI dendrimers of various generations has been characterized by using NOESY-HSQC 3D NMR and NOESY 2D NMR techniques [6-7] in chloroform and benzene solutions. The high field, 750 MHz, multi dimensional NMR spectroscopy provides the chemical shift dispersion, needed to resolve all of the unique resonance in the dendrimers. The studies of NOE interactions among dendritic chain protons offer information about through space interactions between protons on different parts of the dendrimer chain, which is directly related to the conformation of the dendrimers. The study on effect of concentration on chemical shift revealed that the interactions between solvent and different parts of dendrimer molecules are different. The detail study indicated that folded chain conformation can occur in non-polar solvent such as benzene and extended chain conformation predominant in polar solvent such as chloroform.

The structure of a 4th generation PPI dendrimer was characterized [8] by using SANS and liquid integral equation theory. The structural factor of this flexible dendrimer relates the average shape and observed that lower generations are present fluctuating structures akin to star polymers.

5.2.1.3 Properties

A study on Cu(II) PPI dendrimers (upto 5th generation, with DAB core) complexes was reported [9] for the first time by means of extended X-ray absorption fine structure (EXAFS) and X-ray absorption near edge structure (XANES) spectroscopies. Analysis of XANES spectra, points to an identical environment of the Cu ions in each of the end group for each generation. The DAB-dendr-$(NH_2)_x$ Cu_n(II) (x = 4,

8, 16, 32, 64 and n = x/$_2$) complexes were found to be pentacoordinate with a square pyramidal geometry. Each Cu ion binds to three nitrogen atoms from the dendrimers dipropylene triamine end group (Cu-N distance = 2.03A^O) and an oxygen atom present in the equatorial plane (Cu-O distance = 1.92A^O). Another oxygen atom occupies an axial position at about 5.6A^O. The evidence of nano cluster formation upon reduction with $NaBH_4$ is demonstrated in the XANES dendrimers by a shift of the K-edge threshold of 1.4 eV. The EXAES results indicate a decrease in the size of the clusters with the increase of dendrimer generation and are in agreement with the TEM results. TEM results indicate that it is possible to further decrease the size of the clusters by varying factors such as concentration and the NH_2 to Cu ratio.

The 3rd generation PPI dendrimer was chelated [10] with Cu(II) and Ni (II) ions in aqueous solutions. The formation of the metal-PPI tridecanoic acid self-assembly in toluene was used as a separation and purification steps in the synthetic process. The energy dispersive X-ray and UV-visible spectroscopies were used to determine the binding ability of metal by the dendrimer and found a greater affinity towards chelating of Cu(II) than Ni(II) when both the ions were present in the solution.

Surface induced dissociation of single and multiple protonated PPI dendrimers with DAB core, upto 4th generation was investigated [11]. The ions of single and double charges of 2nd generation, double and triple charges of 3rd generation and triple and quartet charges of 4th generation PPI dendrimers were supposed to be fragmented by charge induced intramolecular nucleophilic substitution (SN^i) reactions. The abundance of different charges were estimated from the coulomb energies and gas-phase basicities of the protonation/fragmentation sites. The laboratory collision energies at which the fragment ion current is 50% of the total ion current were found to increase with size, but to be independent of charge-state of the protonated dendrimers.

The physical and chemical properties of PPI dendrimers were studied [12] by using fluorescent, solvatochromic probe phenol blue. In aqueous solution, two discrete dye populations were observed for all the generations. The PPI dendrimers were shown to form a light, non-polar association with the vast majority of available dyes, within the interior, near the core of the dendrimers. The steady state fluorescence emission spectral studies showed a micro-environment of decreasing polarity with the increase of generation number upto 3rd generation, while the higher generations showed equal polarity. Further, the polarity within the PPI dendrimers micro cavities was shown to be similar to that of n-heptane and is in agreement with the other study [13] of 4th and 5th generation PPI dendrimers using pyrene probe. The studies also showed that the majorities of the dyes did not associate at the surface of dendrimers, but rather within the microcavities of dendrimers, where phenol blue was completely encapsulated near the core. The second dye population consists of a small remaining portion of the dye in the form of an unassociated self-aggregate and was the result of the inability of dendrimers to truly sequester 100% of phenol blue in solution.

Electrospray mass spectrometry studies [14-15] of PPI dendrimers upto 5th generation showed the average globular shape of the dendrimers [14], though an accumulation of statistical defects is observed at higher generations [15]. As there is tremendous difference in fragmentation behavior between single charged and multiple charged dendrimers, so the site of protonation within the dendritic framework has a large impact on the fragments. For singly charged dendrimers, electrostatic interactions are absent and the fragmentation is predominantly governed by the relative difference in the proton affinity between the basic sites present in the dendrimer. As the core nitrogen atoms of butanediamine are the most basic sites, so $[M+H]^+$ ions carry a protonated diaminobutane core in all the cases.

The diaminobutane based 5th generation PPI dendrimer with CN groups at the periphery was used to study [16] the mid frequency Raman scattering data. The results imply that the shape and width of the CN band contour are generated by a vibrational dephasing process in the intermediate modulation regime under inseparable contributions of intra- and inter-molecular oscillator-lattice coupling.

The molecular dynamics (MD) technique was used [17] to examine the encapsulation of Bengal Rose (BR) molecules in the 'Meijer dendritic Box'. This was formed by the addition of *Boc*-L-phenylalanine cap molecules to the 64 terminal primary amines of a fifth generation of PPI dendrimer. The study showed that BR molecules established a concentration dependent equilibrium between the interior and the surface regions of the dendrimers and the solvent (CH_2Cl_2) region outside the dendrimer. The number of BR molecules calculated to associate with the interior of the dendrimer agreed exactly with the experimental value. The MD simulations also showed that the Boc-L-phenylalanine surface is completely impermeable to encapsulate BR molecules, even when an excess is forced inside the box. The molecular dynamics was also used to determine the various structural properties of N-Boc-L-phenylalanine modified PPI dendrimers upto 5th generation [18]. The simulation indicates that the cavities of the dendrimers are partially filled with terminal groups and in these cavities guest molecules can be incorporated.

5.2.1.4 Modifications

In this subsection the synthesis, characterization and properties of different modified PPI dendrimers are discussed.

5.2.1.4.1 Photoactive

The synthesis and property evaluation of functionalized PPI dendrimers up to fifth generation with dansyl unit at the periphery [19] were reported. The dansyl chromophoric group shows intense absorption bands in the near UV and a strong fluorescence in the visible region, is extensively used for sensing or labeling purposes [20]. The dansylated dendrimers were synthesized by the reaction between PPI dendrimers of 1st to 3rd generation with 1,2-diaminoethane core, 4th to 5th generation with DAB central core and dansyl chloride in boiling dichloromethane under the argon atmosphere in the presence of triethylamine as a base. For higher generations high excess of dansyl chloride and longer time (up to 5 days) were used. The products were characterized by ^{1}HNMR and ^{13}CNMR spectroscopy and mass spectrometry. The photophysical properties such as fluorescence spectrum, quantum yield and life time of the dendrimers were also studied [19]. The changes in the absorption spectra and fluorescence properties upon addition of acid show that there is an interesting interplay between the protonation processes of the tertiary amine groups contained in the interior of the dendrimers and the dansyl units appended in the periphery.

The photophysical behavior of the functionalized PPI was also reported [21] when eosin Y was extracted from water. A 4th generation PPI dendrimer functionalized with 32 dansyl units at the periphery showed strong fluorescence. But the fluorescence of all dansyl units is quenched when a Co^{2+} ion is incorporated into dendrimers [22]. The host-guest interactions of dansyl functionalized PPI dendrimers up to 5th generation with dye molecules were investigated [23] by fluorescence study. The study showed that dansyl functionalized dendrimers effectively extract eosin dye molecules from aqueous solution to dichloromethane solvent. The maximum efficiency of hosting dye molecules increases with the increase of generation and maximum 12 eosin molecules can be hosted for each 5th generation dendrimer. The fluorescence properties of the peripheral dansyl units of the dendrimers are quenched via energy transfer by the hosted eosin molecules, whose fluorescence is partially quenched. Analysis also showed that a single eosin guest is sufficient to quench completely the fluorescence of any excited dansyl unit in the host dendrimers. The dye molecules occupied two different sites of the dendritic structures. The behavior of fluorescein and Rose Bengal (RB) is qualitatively similar to that of eosin, whereas naphthofluorescein is not extracted. The maximum number of dye molecules extracted by 4th generation dendrimer was 25 for RB and ~1 for fluorescein, showing that the formation of host-guest species is related to the electronic properties rather than to the size of dye molecules.

The modified PPI dendrimers up to 5th generation with fluorescence dansyl units at the periphery were used to investigate [24] the interactions with Co^{2+} (as a neutral salt). The study showed that the metal ions coordinated at the interior of the dendrimers and quenched the fluorescence of dansyl units by a static mechanism. This coordination process is fully reversible. Also a strong amplification of fluorescence quenching signal was observed with the increase of dendrimer generation. For the large dendrimers (2nd to 5th generation) at very low metal ion concentration, 1:1 metal/dendrimer species were formed, in which the Co^{2+} ion guest quenches each one of the dansyl units that became excited after light absorption.

The PPI dendrimers up to 5th generation were reacted [25] at the periphery with 4'-(dimethylamino)azobenzene-4-sulfonic acid chloride to obtain methyl orange (MO) bearing PPI dendrimers (Scheme 5.3). The products were purified by extraction of organic phase with aqueous Na_2SO_4 followed by thick layer chromatography. The purified dendrimers were used to pH-controlled inclusion and release of oxyanions. The compounds were then applied for liquid–liquid partition of $^{99}TiO^-_4$ and ^{14}C-lable nucleotides AMP, ADP and ATP to characterize binding properties. A remarkable enhancement of pH sensitivity is obtained for dendrimers having MO units at the periphery compared to urea functionalized analogs [26].

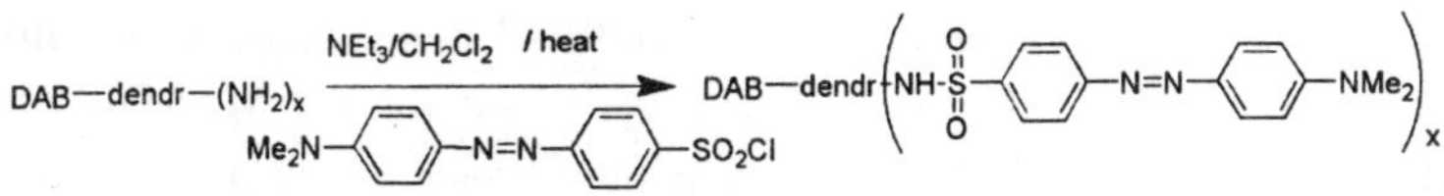

x =4, 8, 16, 32 and 64

SCHEME 5.3 Functionalization of terminal amino groups of PPI dendrimers by methyl orange derivative

The dynamics of electronic energy transfer (EET) for a series of spherical porphyrin array based on different (1st, 3rd and 5th) generations of PPI dendrimers was studied [27] by time resolved fluorescence anisotropy measurement (TRAMS) in a glass environment. The dendrimers depolarized the fluorescence, which indicates that EET takes place between the chromophores within the dendrimers. The experimental time resolved fluorescence anisotropy measurement results were compared to computationally simulated data obtained from the Pauli master equation [27]. Anomalous results were obtained for 3rd generation dendrimers, where EET was confined within a dendron. This may be due to dendron segregation.

PPI dendrimers were functionalized [28] with oligo(*p*-phenylenevinylene) (OPV) at the periphery. The modified dendrimers showed fluorescence properties that depend on the number of end groups. The site-selective and time resolved fluorescence techniques indicate a red shift of the low temperature emission spectrum with the increase of dendrimer generation may be due to rapid migration of electronic excitation energy among OPV units. These interchromophoric interactions are sufficiently strong to induce delocalization of the excitation over more than one chromophoric groups. Because of the flexible nature of the dendrimers an off-diagonal disorder is observed, which leads to a broad distribution of distances between the chromophoric end groups. Also a delay emission component of the dendrimer is attributed to excimer and/or exciplex formation within the dendrimers.

The pH sensitivity and photoactivity of Methyl Orange (MO) functionalized PPI dendrimers upto 5th generation were reported [29]. The modified dendrimers were synthesized (Scheme 5.3) by the reaction of sulfonic acid chloride of MO with amine functionalized PPI dendrimers. The structures of the modified dendrimers were fully characterized by ^{1}HNMR, ^{13}CNMR spectroscopy and MALDI-TOF mass spectrometry. The studies of photophysical properties revealed that the functionalized dendrimers exhibited π-π^* bands at 305 nm and at 445 nm. It was also observed that the extinction coefficient values (ε) of the 445 nm band were proportional to the number of MO units present in the dendrimers for upto 3rd generation. But for 4th and 5th generations, the ε values deviates from Lambert-Beer law, may be due to high local concentration of MO units. Further more, on irradiation of a solution of a MO functionalized dendrimers at (440-448) nm resulted 'E' to 'Z' isomerization, as indicated by the decrease of intensity at 445 nm band. The isomerization process was reversible for these dendrimers and rate of isomerization was extremely high, though it is generation independent. The protonation characteristics of (zeroth and 1st generations) dendrimers were different from the higher generations (2nd to 5th). The lower generations like MO itself, towards CF_3COOH, i.e. on addition of acid causes a decreased in π-π^* transition intensity of 445 nm and increased π-π^* band at 540 nm. However, the higher generations do not behave like MO itself towards protonation. The absorption band of the protonated form has became broaden and less red shifted.

A self-assembled porphyrin sphere with a molecular weight of 15000g/mole was described [30]. This self-assembled porphyrin was synthesized by the following steps. At first, the PPI dendrimers of 3rd generation were functionalized at the surface by activated ester of 4-

pyridine carboxylic acid followed by complexation with oligomeric porphyrin trimer ligand (Scheme 5.4). The UV-visible absorption studies showed that the intensity of the porphyrin Soret band increases as a function of the amount of porphyrin-trimer.

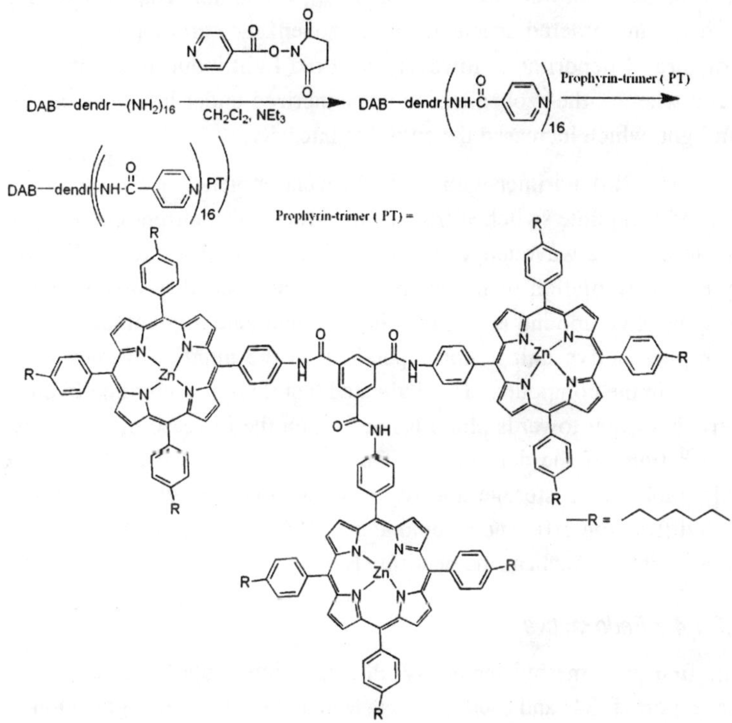

SCHEME 5.4 Synthesis of photoactive PPI dendrimers with porphyrin moiety

Different generations of PPI dendrimers were functionalized [31] with second order nonlinear optical chromophores, 4-dimethyl aminophenyl carbonamide and their non-linear optical properties were studied by the hyper-Rayleigh scattering technique. The study showed that the higher generation (3rd and 4th) dendrimers from globular structure in solution with restricted mobility of the end groups.

The higher generation PPI dendrimers were functionalized [32] by aliphatic chains of palmitoyl and azobenzene moeities that form large micrometer (50 nm to 20 nm) sized spherical objects in aqueous solution below pH 8. The size distribution and the structure of the vesicles (sphere) depend on the pH of the solution and the end groups at the periphery of the dendrimers. The vesicles containing 2 to 3 benzene units showed fluorescence at $\lambda_{max} = 600$ nm. This may be due to dense and ordered arrangement of azobenzene chromophores in the structure of dendrimer. However, infra-red light induced a rearrangement, whereas the azobenzene units isomerized under influence of 420 nm light, which increased the emission intensity.

The PPI dendrimer with 32 azobenzene moieties at the periphery showed reversible switched from E- to Z-form (azobenzene) by light of the appropriate wave length and can be converted back to the E-form by either irradiation or by heating [33]. The speciality of this photo switchable compound is that the photo isomerization quantum yield of each photoactive unit is not dependent on the number of such units present in the compound. This indicated that there is so far no effective steric constrant towards photo isomerism on the increase of dimension (generation) of the dendrimers. This dendrimer was tried to use for holographic data storage and result shows that holographic gratings with diffraction efficiencies upto about 20% can be optically recorded in thin films of azobenzene dendrimer.

5.2.1.4.2 Redoxactive

The first pentamethyl ferrocenyl PPI dendrimers upto 5th generation was reported [34] and used them as clean and selective recognization of $H_2PO^-_4$ anions. The PPI dendrimers were modified on the surface by the treatment of the dendrimers with pentamethyl ferrocenyl chloride in CH_2Cl_2 at room temperature in the presence of triethylamine for over night. The modified dendrimers were characterized by spectroscopic and analytical technique. The cyclic voltammetric study showed that the rotation of the dendrimers was much more rapid than the electrochemical time scale. The dendrimers also showed no chemical irreversibility and much less absorption for recognization of $H_2PO^-_4$ ions. A modest positive dendritic effect, i.e. larger shift of potential in cyclo voltammogramms, by addition of the anion for higher generations was also observed [34] in DMF solution.

The electrochemical properties of 1st generation PPI dendrimers functionalized with ferrocene moieties were reported [35]. The functionalization was carried out by reacting (ferrocenylmethyl)trimethyl ammoniumiodide with PPI tetraamine dendrimers using acetonitrile as the solvent and K_2CO_3 as the catalyst under the reflux condition. These small dendrimers were synthesized with the building block *bis*(ferrocenyl methyl)alkyl amine, which contains the tether $-CH_2$-N(R)-CH_2- linking two ferrocene units.

The redox activity of the cobaltocenium centers of the modified [36] dendrimers was characterized by cyclic voltammetry and EQCM. All the dendrimers exhibited reversible redox chemistry associated with the cobaltocenium-cobaltocene redox couple. The dendrimers also exhibited a tendency to electro deposit onto electrode surface upon reduction, which was more pronounced for the higher generation EQCM measurements and showed that the dendrimers absorbed at open circuit onto Pt surfaces at monolayer coverage. The authors also obtained molecularly resolved images of dendrimer absorbed on a Pt(III) electron by using tapping mode AFM.

A 5th generation primary amine terminated PPI dendrimers was functionalized [37] with ferrocene units in MeCN/CH_2Cl_2 (2:1) in the presence of NEt_3 (Scheme 5.5). The synthesized dendrimer was characterized by IR, ^{1}HNMR ^{13}CNMR spectroscopy and cyclovoltammetry. The complex [Fe^{II}(η^5-C_5H_4-Co)(η^6-C_6Me_6)] was themostable, in addition the metallodendrimer can be reduced to Fe^I by the 19-electron complex [Fe^ICp(η^6-C_6Me_6)] in MeCN at -30°C. This Fe^I dendrimer in acetronitrile solution reacts with C_{60} and found 64-C_{60} attached on each dendrimer. However, in toluene C_{60} could be oxidized Fe^I-dendrimer to Fe^{II}-dendrimer (Scheme 5.6). The [dendr-Fe^{II}]$^+$-C_{60}^- units of the dendrimer being very large, they were most likely located at the periphery as rather tight ion pairs.

The thermodynamics and kinetics of absorption of the redox-active ferrocene containing PPI dendrimers from 2nd to 5th generation were studied [38] onto a Pt-electrode by using cyclic voltammetry and EQCM techniques. The adsorption thermodynamics of the reduced form at an applied voltage of dendrimers in CH_2Cl_2 solution were well represented by the Langmuir adsorption isotherm. The kinetics of

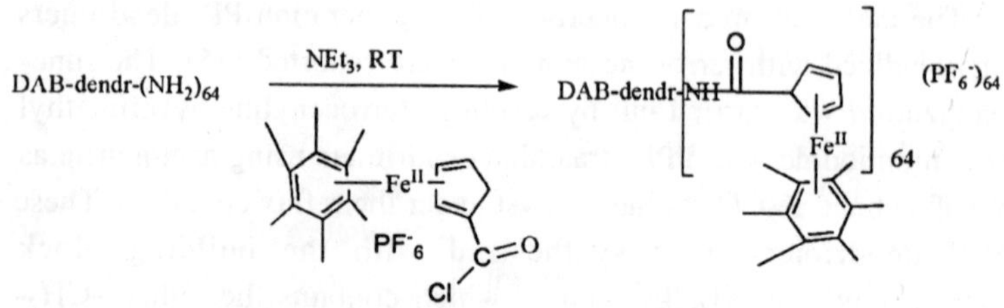

SCHEME 5.5 Functionalization of terminal amino groups of PPI dendrimers by ferrocene derivative

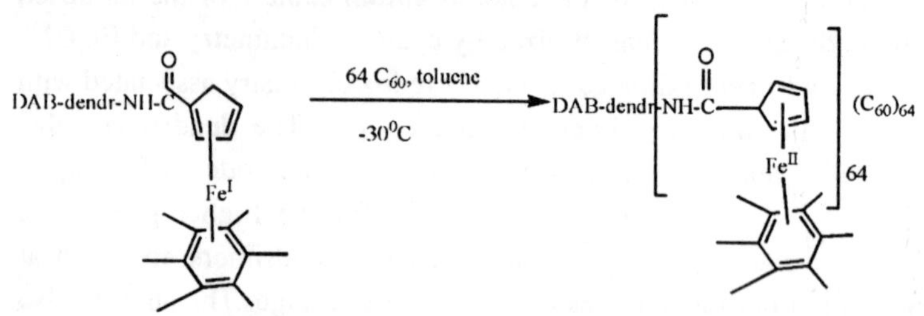

SCHEME 5.6 Attachment of ferrocene modified PPI dendrimers with C_{60} unit

absorption was activation-controlled and the rate constant decreased with the decrease of generation number, that is size of dendrimers. From the analysis of admittance and EQCM measurements showed that the oxidized form of dendrimers deposits on the Pt-electrode. This may be due to low solubility of the salt of oxidized dendrimers and ClO^-_4 anions, while the reduced form of dendrimers easily redissolves except for the first monolayer, which was strongly adsorbed. The AFM studies of ferrocenyl dendrimers absorbed onto a Pt(III) surface revealed that the apparent size of dendrimers absorbed on the surface was significantly larger than estimated values based on calculation, which may be due to flattering of dendrimers upon absorption.

The first four generations of cabaltocenium functionalized, diaminobutane based PPI dendrimers, DAB-dendr–$(Cb_x(PF_6)_x$ (x = 4, 8, 16 and 32; Cb ={Co(η^5-C_5H_4CONH) (η^5-C_5H_5)}were synthesized and characterized [39]. The redox activity of the cobaltocenium centers was characterized by cyclic voltammetry EQCM. All the generations of the dendrimers exhibit reversible redox chemistry associated with cobaltocenium/cobaltocene redox couple upon reduction, the dendrimers exhibit a tendency to electrodeposit onto the electrode surface, which is more pronounced for the higher generations. EQCM measurements show that the dendrimers absorb, at open circuit onto Pt surfaces at monolayer or submonolayer coverage. In general, it was found that the film of first generation is rigid, whereas 4th generation exhibits viscoelastic behavior, and 2nd and 3rd generations behave in between those two. The author [40] found that while the oxidized [Co(III)] dendrimers do not interact with β-cyclodextrin hosts, multiple inclusion complexes are formed upon dendrimers reduction to the cobaltocene [Co(II)] form. A new family PPI dendrimers were modified [41] with phenolthizines as the redox active unit. The resulted dendrimers show redox activity.

5.2.1.4.3 Amphiphilic

The PPI dendrimers of 4th and 5th generations were functionalized with acetyl chloride and deuterated acetyl chloride in DMF at room spectroscopic studies. The functionalized dendrimers were characterized by IR, ^{1}HNMR and ^{13}CNMR spectroscopic studies. The structure and rheological properties of both individual functionalized dendrimers as well as with poly(ethylene oxide) blend were studied in details [42].

A variety of quaternary ammonium ion dendrimers by methylation of amine terminated PPI dendrimers were synthesized [43-45]. The quaternization of the PPI dendrimers were performed: (i) at both chain ends and branching points, (ii) mainly at chain ends but not at branching points and (iii) at branching point only. The PPI dendrimers were quaternized, only at chain ends with iodomethane [46], while dibenzyldiethyl ammonium ion dendrimers were prepared with mesitylene derivatives at branching points [47]. The reaction of glycidyl trimethyl ammonium chloride with PPI dendrimers gave secondary amines at dendrimer ends with quaternary ammonium chloride tails [48]. PPI dendrimers with octyl and triethlene glycol methyl ether chains on tertiary amines were quaternized at both chain ends and the branching points with iodomethane [43]. PPI-polystyrene block copolymers with up to 16 chains ends were methylated at the amino chain ends [49]. The PPI dendrimers, DAB-dendr-$(NH_2)_X$ (X= 8, 32 and 64) were methylated only at chain ends with excess formic acid and formaldehyde to obtain the tertiary amine dendrimers. Then the tertiary amine dendrimers were treated with stoichiometric amount of iodo methane form quaternary ethyl ammonium iodide ion at each chain end. The sequence of reaction is shown in scheme 5.7. The products were characterized by ^{1}HNMR, ^{13}CNMR spectroscopy and element analysis. By using a large excess of iodomethane, instead of stoichiometric amount at high temperature (110°C), methylated PPI dendrimers were produced quaternary methyl ammonium ions both at chain ends and branching points as shown in scheme 5.8. The resulted dendrimers were characterized by ^{1}HNMR, ^{13}CNMR spectroscopy, ESI-MS and elemental analysis. The quaternary ammonium ions of the dendrimers only at

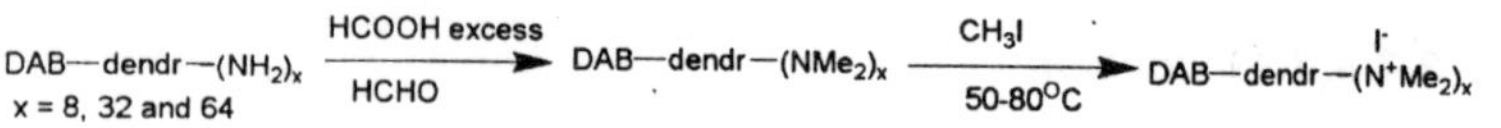

SCHEME 5.7 Quarternization of terminal amino groups of PPI dendrimers by methyl iodide

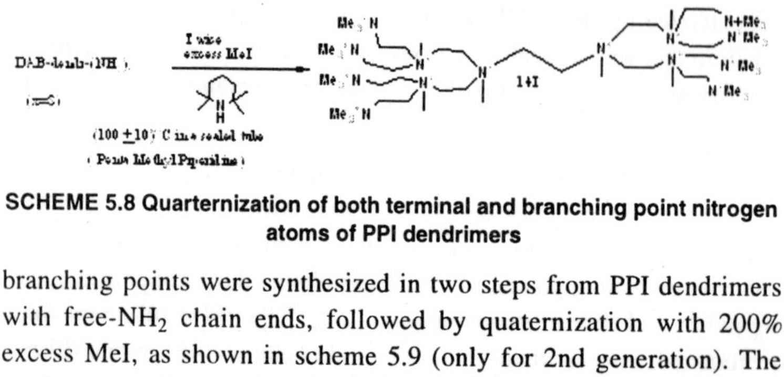

SCHEME 5.8 Quarternization of both terminal and branching point nitrogen atoms of PPI dendrimers

branching points were synthesized in two steps from PPI dendrimers with free-NH$_2$ chain ends, followed by quaternization with 200% excess MeI, as shown in scheme 5.9 (only for 2nd generation). The product was characterized by ^{1}HNMR, ^{13}CNMR spectroscopies and elemental analysis.

The PPI dendrimers, DAB-dendr-(NH$_2$)$_x$, (x = 8, 32 and 64) were fully converted [43] to quaternary ammonium ions at both chain ends and branching points with iodomethane, whereas with less iodomethane partially converted to quaternary ammonium ions mainly at end groups. Again, amidation of the primary amine end groups followed by treatment with iodomethane gave the first dendrimers with quaternary ammonium ion only at branching points. These dendrimers were used as catalyst for decarboxylation of 6-nitrobenz-isoxazole-3-carboxylate

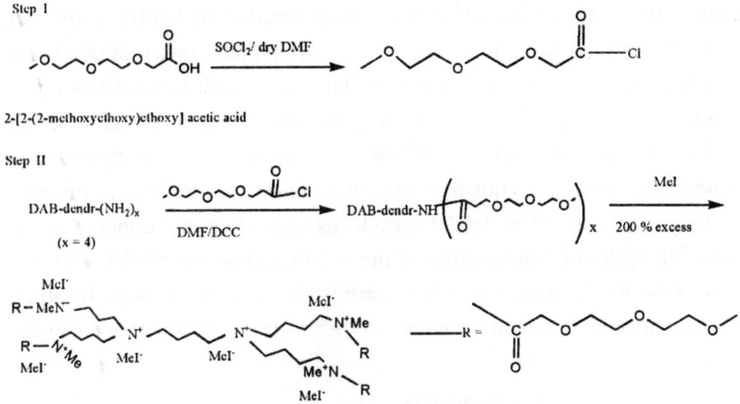

SCHEME 5.9 Quarternization of only branching point nitrogen atoms of PPI dendrimers

ion in aqueous solution, after an exchange of iodide counter ions for chloride. Further, similar quaternary ammonium dendrimers with more hydrophobic interiors or more hydrophobic chains ends are much more active catalysts, for decarboxylation. The polyelectrolyte behavior of Astramol™ PPI dendrimers upto 5th generation [DAB-dendr-$(NH_2)_x$, x = 4, 8, 16, 32 and 64] was investigated by means of potentiometric titration in a salt free water solution and also in the presence of a shielding low molecular weight electrolyte, NaCl. Each dendrimer containing 'x' number of primary amine groups at the periphery and 'x-2' number of inner tertiary amine groups.

The PPI dendrimers were functionalized by both hydrophilic, tri(ethyleneoxymethyl ether) (TEO) and hydrophobic octyl chain on every amine end group. The functionalized dendrimers [43-44] were also converted to methylated quaternary ammonium ion derivatives. The synthesis of quaternized amphiphilic dendrimer required five steps, from the amine end groups of PPI dendrimers, as shown in scheme 5.10. The octyl arms was introduced first as octyl modified dendrimers facilitates solutilization in organic solvents and subsequent reduction by $LiAlH_4$, amidation and purification. The octyl and TEO chain end modified quaternized dendrimers were then fully characterized by FTIR, ^{1}HNMR and ^{13}CNMR spectroscopy, ESI-mass spectrometry and elemental analysis. The ^{13}CNMR spin-lattice relaxation time measurements show that the conformations of the dendrimers depend on solvent used. The amphiphilic dendrimers were soluble in common organic solvents such as toluene, ether, THF, $CHCl_3$, acetone, methanol, etc. and also readily soluble in water. They also solubilized lipophilic compounds, such as pyrene and Reichardt's dye in aqueous solution. However, the limiting solubility corresponds to one pyrene per dendrimer molecule. Thus they behave as unique unimolecular micelles as they retain their colloidal structures regardless of concentration, ionic strength and temperature of the solution. The rate of decarboxylation of 6-nitrobenzene oxazole-3-carboxylic acid in aqueous solutions of the cationic dendrimers were up to 500 times faster than in water alone.

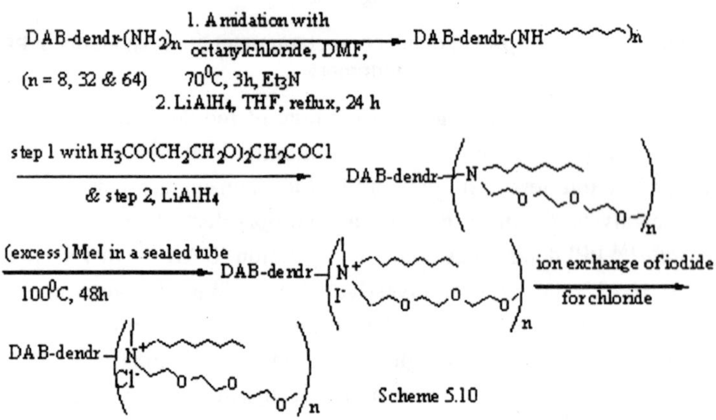

SCHEME 5.10 Synthesis of amphiphilic PPI dendrimers with TEO and octyl chains

In general, the amphiphilic dendrimers have carboxylate anions or ammonium cations as end groups and a hydrophobic core. The dendritic inverse micelles were synthesized by amidation of the primary amine end group of PPI dendrimers such as DAB-dendr-$(NH_2)_{64}$ with fatty acids [50] and by creating alcohol functionality in the core of an aryl benzyl ether dendrimers having tetradecyl chain ends [51].

5.2.1.4.4 Liquid Crystalline

Two series of PPI dendrimers functionalized with pentyloxy and decyloxy cyanobiphenyl mesogens were synthesized and characterized [52]. All the dendritic mesogens were obtained by treating the amine terminated dendrimers with a small excess of pentafluorophenol derivatives, followed by precipitation of the resultant products in MeOH or n-hexane. The sequence of reactions is shown in scheme 5.11. The functionalized dendrimers were characterized by IR, ^{1}HNMR, ^{13}CNMR spectroscopy and mass spectrometry. The FAB-MS and MALDI-TOF-MS were used to examine the defects in the structures of the dendrimers. The properties of the functionalized dendrimers were studied by means of DSC, polarization microscopy and XRD studies. The PPI dendrimers were functionalized [52] with rod like cyanobiphenyl mesogens through various spacer lengths, at the periphery. The modified dendrimers showed smectic liquid crystalline behavior. The resulted dendrimers were characterized by DSC, optical microscopy, and XRD and found that the dendrimers are highly distorted conformation under the influence of the mesogenic end groups.

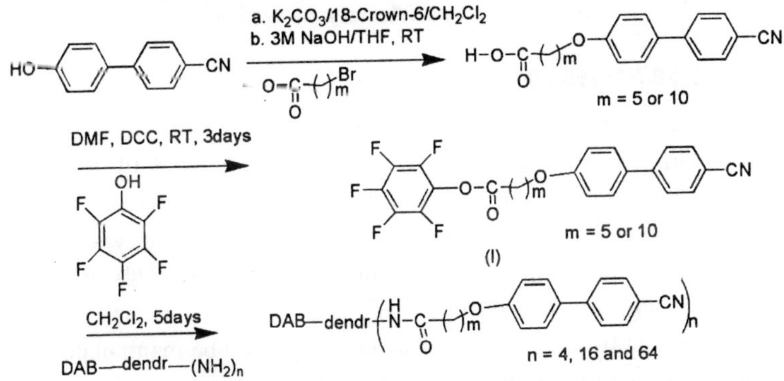

SCHEME 5.11 Synthesis of cyanobiphenyl functionalized PPI dendrimers

The coupling of 3,4-*bis*(decyloxy)benzoyl group onto a PPI dendrimers skeletone induced shape anisotropy, yielding hexagonal columnar mesophases for the 2nd to 4th generation dendrimer [53]. In case of the fifth generation with 64 end groups no mesophase was formed. The 3D geometry of a 5th generation is likely to suppress the formation of a columnar structure.

The first three generations DAB based PPI dendrimers were functionalized [54] with cholesteryl moiety to form liquid crystalline carbamate dendrimers (Scheme 5.12). The liquid crystalline behavior of the resulted dendrimers was studied by optical microscopy, DSC and

XRD analysis. An isotropic phase was observed at a broad thermal range from 150°C upto above 210°C. The lamellar spacing determination suggested that in each layer, the cholesterol moieties were standing almost upright. During first heating run, at temperature below the transition to the smectic phases were observed. However, on cooling from the isotropic phase, the compounds form glass phase, which retains the structural characteristics of the smectic A phase.

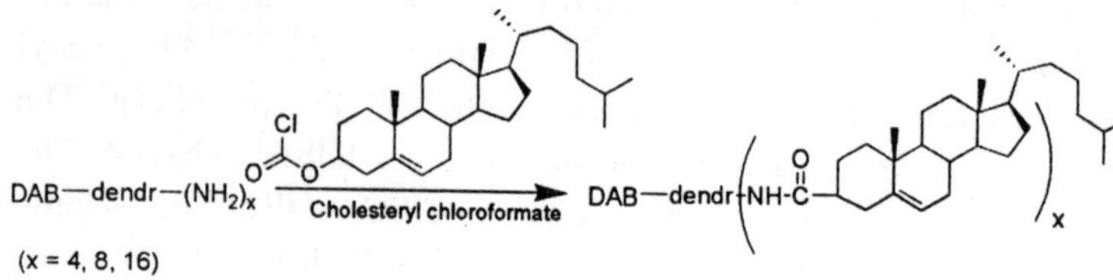

SCHEME 5.12 Functionalization of terminal amino groups of PPI dendrimers by cholesteryl derivative

5.2.1.4.5 Self-assembly

PPI dendrimer of 3rd generation was modified [55] with π-conjugated OPV to form spherical and rod like aggregates (Scheme 5.13) that can be manipulated to optical tweezer. The aggregates behavior of the modified dendrimers were studied by UV spectrophotometry, circular dichroism measurement, TEM and tapping mode AFM. This oligo(*p*-phenylene) terminated dendrimers aggregate in protic polar solvents and for the first time π-conjugated assemblies could be manipulated by the aid of optical tweezers resulting novel architectures. This result showed that micro-manipulation is a new and promising tool to gain control over shape and dimension in π-conjugated assemblies. It was [56] also reported that PPI dendrimers decorated with palmitoyl and azobenzene containing alkyl groups form vesicles in water that could be merged into rod shaped assemblies.

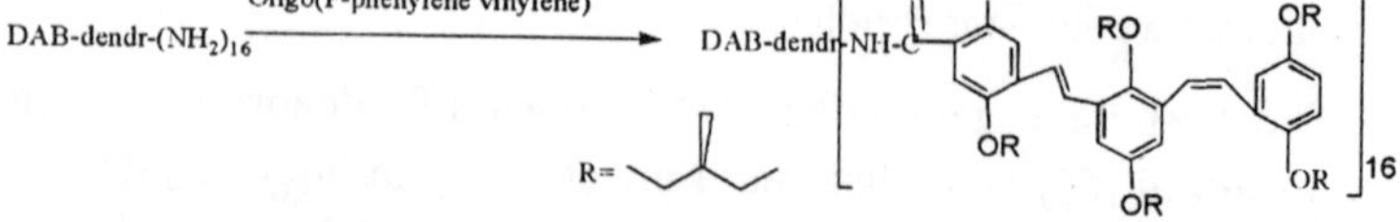

SCHEME 5.13 Functionalization of terminal amino groups of PPI dendrimers by oligo(*p*-phenylene vinylene) derivative

The self-assembly of amphiphilic dendrimers based on PPI dendrimers upto 5th generation were modified [56] with three different types of compounds viz palmitoyl, azobenzene chromophore and adamantane (Scheme 5.14). The modified dendrimers were characterized by ^{1}HNMR, ^{13}CNMR, IR spectroscopy and mass spectrometry. All the modified dendrimers showed good solubility in apolar organic solvents. The langmuir studies at water-air interface suggested that these modified dendrimers were able to arrange themselves as monolayers in which the polar dendrimeric part was in contact with the water phase and non-polar alkyl part was in air, forming a parallel-packed hydrophobic layer. However, in case of adamantane modified dendrimers no well-defined monolayer was found. This may be due to bulkiness of the adamantane group. These amphiphilic dendrimers also formed spherical aggregates in acidic water solution. This may be due to the fact that PPI dendrimers are protonated in an acidic solution, which leads to a more extended conformation. The absorption study by UV spectroscopy of the benzene modified dendrimers in dichloromethane solution showed absorption maximum at 355 nm, which indicated the absence of intramolecular stacking or supramolecular aggregates in organic solution.

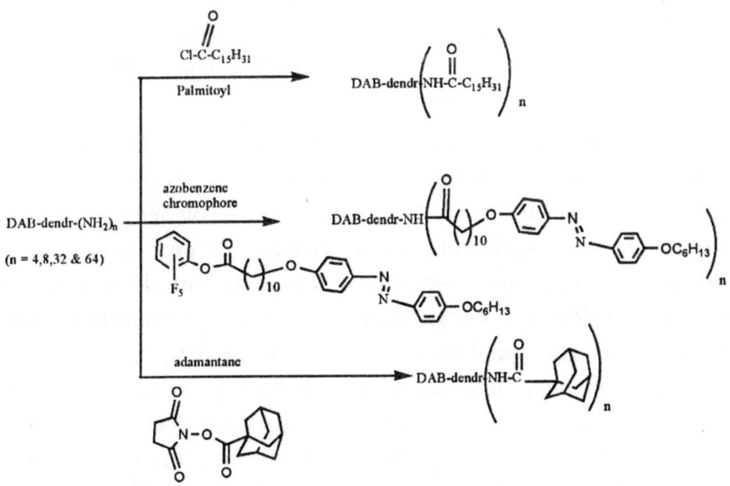

SCHEME 5.14 Functionalization of terminal amino groups of PPI dendrimers by palmitoyl, azobenzene chromophore and adamantane derivative

The formation of well-defined supramolecular assembly builted up by adamantyl-terminated PPI dendrimers upto 5th generation was reported [57]. These dendrimers were dissolved in water in the presence of β-cyclodextrin through strong non-covalent interaction between the β-cyclodextrin and adamantyl groups [57]. The solubilization is maximum at pH = 2, may be due to complete protonation of the tertiary amino groups, which causes the dendrimers to adopt a stretched conformation. All the dendrimers remain in solution upto neutral pH, but precipitated out completely at basic pH, except the first generation. The qualitative fluorescence measurements at pH = 1 by using 8-anilinonaphthalene-1-sulfonate as a probe showed that the assemblies 2-5 with β-cyclodextrin act as suparmolecular hosts in water. The binding of the probe is electrostatic in nature and increases for higher generations. The well-defined assemblies were formed through complexation of all terminal adamantyl units upto 4th generation dendrimers, but for the 5th generation, only 40 β-cyclodextrins complexed instead of 64. This may be due to steric reasons, not cooperative weakening of the adamantyl-β-cyclodextrin interaction.

The modified PPI dendrimers with long chain alkyl groups self-assembled [58] to form well-defined aggregates with unstable calcium carbonate in the presence of octadecyl amine. The aggregates have stability of more than 14 days.

PPI dendrimers were modified [59] with long alkyl chains self-assemble to form well defined aggregates. The geometry and surface chemistry of the dendrimers assemblies can be varied through the addition of surfactants. The use of octadecylamine results in the formation of polyhedral aggregates that became embedded within an amorphous calcium carbonate phase that persists in competition with the thermodynamic product, calcite. In combination with hexadecyl trimethyl ammonium bromide, small spherical aggregates are formed that include the formation of vaterite. The use of the negatively charged surfactant, SDS results in growth retardation by the Ca^{2+} induced agglomeration of dendrimer/surfactant aggregates into giant spherical particles. Eventually these particles became overgrown by rhombohedral calcite.

PPI dendrimer was used [60] as multi(tridentate) ligands to complexate with transition metal ions such as Cu(II), Zn(II) and Ni(II). These strong complexes formed nanoscopic assemblies of defined structures and size by incorporating a distinct number of metal ions from 2 upto 32.

5.2.1.4.6 Glycodendrimers

The PPI dendrimers upto 5th generation were modified with β-thiolactosyl residues and the hydrodynamic properties [61] of the modified dendrimers were studied by velocity sedimentation, translation diffusion, and velocity measurements in 0.168% aqueous NaCl solutions. The apparent molecular weights ranges from 1900 to 45000 Da for first to 5th generation with corresponding hydrodynamic radii ranges from ~10 to 37A^{O}. The investigation of scaling relations of the conformation dependent sedimentation co-efficient translational diffusion co-efficient and intrinsic viscosity with molecular weight indicate a spherical molecules with an inhomogeneous distribution density.

The dibutaneamine based PPI dendrimers of 1st and 5th generations were functionalized to spacer armed glycodendrimers [62], by a reaction between the primary amine end groups of the dendrimers with N-succinimidyl-activated ester of spacer-armed acetyl-protected thio-glucopyranoside units. The linear alkyl chain spacers between the dendrimer surface and the saccharide units increased in length with 1, 5 and 10 carbon atoms. These N-succinimidyl activated esters were obtained by a series of reactions, as shown in step 1 of scheme 5.15, starting from 2,3,4,6-tetra-O-acetyl-α-D-glucopyranosyl bromide. The desired acetylated glucodendrimers were obtained by mild peptide coupling of N-succinimidyl esters to the PPI dendrimers using CH_2Cl_2 as the solvent, as shown in step II of scheme 5.15. The yields of every step were considerably high, may be due to mild reaction conditions and choice of appropriate solvent. The glycodendrimers were characterized by ^{1}HNMR, ^{13}CNMR, IR spectroscopy and MALDI-TOF-MS studies. The solubility behavior of the resulting glycoldendrimers proved to be strongly dependent on the hydrophobic part i.e. the alkyl chain length in the molecules.

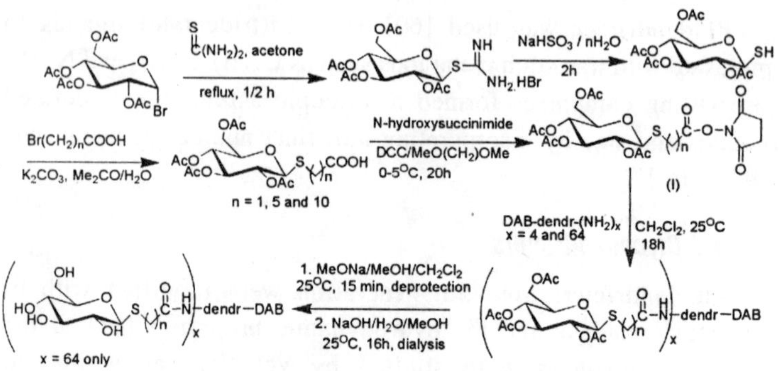

SCHEME 5.15 Functionalization of terminal amino groups of PPI dendrimers by thioglucopyranoside derivative

Ashton *et al.* synthesized [63] 'sugar balls' by modification of PPI dendrimers. The dendrimers were functionalized with the carbonyl derivatives of D-galactose and D-lactose to the amino surface groups of PPI through amide bond formation by using N-hydroxy succinimide coupling method. To avoid the undesirable other coupling products, the acetyl protection was done, which was deacetylated after coupling at the periphery under the standard zemplex deacylation conditions, followed by treatment with an aqueous solution [63]. The functionalized polymers were characterized by ^{13}CNMR spectrocopy.

5.2.1.4.7 Host-Guest Chemistry

Pseudorotaxane-terminated PPI dendrimers upto 4th generation were synthesized and characterized [64]. The modified dendrimers were obtained by reaction of compound A (obtained from mono-Cbz protected DAB) with the dendrimers (Scheme 5.16). The resulted product was treated with HBr-AcOH to protonate diaminobutane units at the termini. Further treatment with CB[6] affords the targeted pseudorotaxane-terminated dendrimers (Scheme 5.16). The resulted pseudorotaxane PPI dendrimers were characterized by ^{1}HNMR and ^{13}CNMR spectroscopic studies. The 5th generation pseudorotoxane terminated dendrimer contains 64 'beads' threaded at the periphery of the dendrimer with molecular weight of ~94,500 Da. An AFM image of the 5th generation dendrimers on a mica surface reveals a flattened sphere with a diameter of 15 nm and height of ~1 nm. This novel pseudorotaxane dendrimers having a number of molecular beads threaded on to the periphery through non-covalent interactions. Further, a large number of terminal pseudorotoxane units can form a rigid shell at the outer surface of the dendrimers, which can potentially inhibit the escape of guest, trapped in the interior. If the beads can be dethreaded to make the exterior flexible once again by treatment with base and thus allow the escape of entrapped guests from the interior, such systems can potentially function as vehicles for translocations guests such as drug molecules.

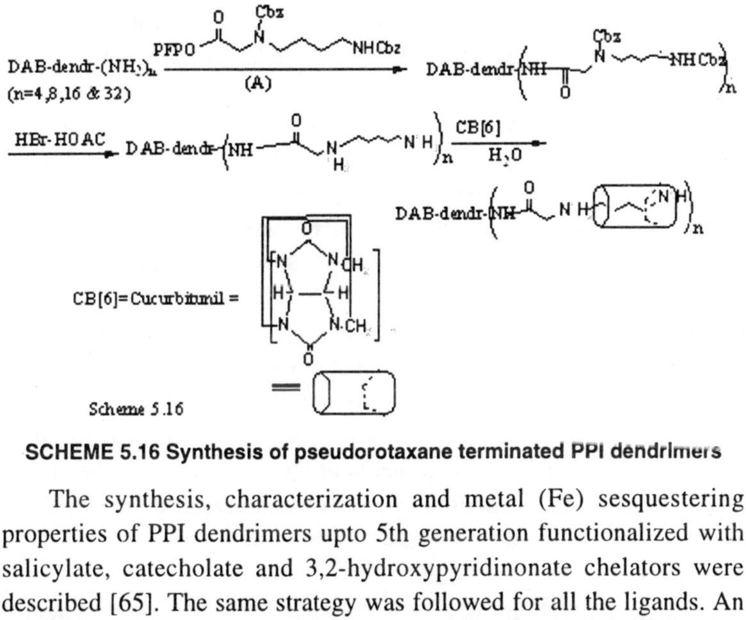

SCHEME 5.16 Synthesis of pseudorotaxane terminated PPI dendrimers

The synthesis, characterization and metal (Fe) sesquestering properties of PPI dendrimers upto 5th generation functionalized with salicylate, catecholate and 3,2-hydroxypyridinonate chelators were described [65]. The same strategy was followed for all the ligands. An activated ester of the chelator was reacted with the dendrimers in CH_2Cl_2 solution for 18h. They purified the products by high quality flash chromatographic techniques. The modified dendrimers were characterized by IR, ^{1}HNMR, ^{13}CNMR spectroscopic and ESI-MS spectrometric studies. One and two dimensional (total correlation spectroscopy, TOCSY) ^{1}HNMR spectroscopic and electrospray ionization mass spectrometric studies confirmed the exhaustive coupling of these bidentate chelators to the terminal primary amine functionalities of the dendrimers. UV-vis. spectrophotometric titration and spectral analysis showed that the ferric ion binding to these ligands are localized to the chelating end groups. Since these novel modified PPI dendrimers have the ability to bind a large numbers of metal ions, so they may have tremendous potential as metal sequestesing agents for waste remediation technology.

PPI dendrimers upto 4th generations were functionalized [66] with benzo-15-crown-5 ether-4-sulfonyl chloride at the primary amine of the periphery (Scheme 5.17). The purified products soften at 74-85°C. The mono functionalized multicrown dendrimers were characterized by ^{1}HNMR, ^{13}CNMR spectroscopy and MALDI-TOF mass spectrometry. They are soluble in most of the common organic solvents but not in water. Because of their high solubility even in low polarity solvent such as CH_2Cl_2, $CHCl_3$, etc. they may be attractive, extractant and carrier molecules for separation and transportation processes. Aqueous solutions of sodium pertechnetate and mercury(II) chloride, a typical contaminant of nuclear waste and a pollutant of chloride, a typical contaminate effluents of incineration processes respectively, were extracted with solutions of multicrown dendrimers in chloroform. The binding ability of the guests was studied by using the radioactive probe such as $^{22}Na^{+}$, $^{99}TiO^{-}_{4}$ and $^{203}Hg^{2+}$. In this study an excellent coalescence behavior and the rapid attainment of extraction equilibria (within a few minutes) was observed for 3rd and 4th generation dendrimers. The study also revealed that the guest molecules are mainly bound in the interior of the polyamine skeleton of the dendrimers, as the crown ether moieties may be interior in terms of their complexation behavior.

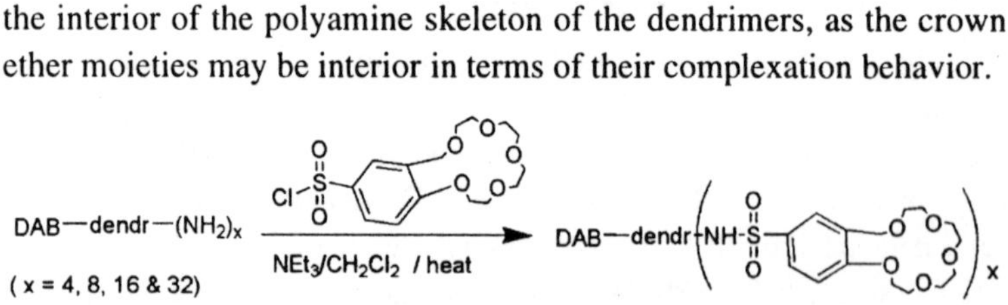

SCHEME 5.17 Functionalization of terminal amino groups of PPI dendrimers by benzo-15-crown ether derivative

The PPI dendrimers upto 5th generations were modified [67] by palmityl and adamantyl end groups via a thiourea linkage. The amine terminated PPI dendrimers were reacted with adamantyl and palmityl isothiocyanates to get the targeted thiourea dendrimers. The resulted dendrimers were characterized by IR and ^{1}HNMR spectroscopy. The interaction of these dendrimers with a number of guest molecules containing a terminal area-glycine unit in organic solvents was studied by 1D and NOESY-NMR spectroscopy. It was found that the guest molecules bind to the dendrimers through thiourea (host)-urea (guest) hydrogen bonding and ionic bonding between the terminal guest carboxylate moiety and the outer shell tertiary amines of the dendrimers.

5.2.1.4.8 Others

The PPI dendrimers were partially modified [68] at the periphery by direct amidation at 150°C with stearic acid. This partial end group modification leads to two types of heterogeneity in the product structure viz. compositional heterogeneity i.e. distribution of degree of substitution and positional heterogeneity i.e. spatial distribution of the substituents over the dendrimer molecule. The HPLC and ES-MS studies revealed that compositional distribution follows a random pattern of binomial distribution, whereas ^{13}CNMR spectroscopy and Cu(II) complexation titration showed that positional distribution can be expressed as dyad distribution of the type AA (two unreacted end groups), AB (one unreacted and one reacted end groups) and BB (both reacted end groups), where a dyad consists of two end groups originating at a common final branching point in the dendrimers. However, the lower generations gave a random distribution of the dyads, whereas higher generations showed a marked preference for single substitution of dyads. This may be due to intramolecular interaction or steric hindrance of higher generations.

Four different generations of oligoethyleneoxy functionalized [69] PPI dendrimers, DAB-dendr-(NHCOEG)n, where EG represents the 3,4,5-*tris*(tetraethyleneoxy)benzoyl unit and n = 4, 16, 32 and 64 were synthesized by reaction of the corresponding penta-fluorophenyl ester with the amine functionalized PPI dendrimers over night in CH_2Cl_2. The functionalized dendrimers (Fig.5.1) were characterized by ^{1}HNMR, ^{13}CNMR spectroscopy and ESI or MALDI–TOF mass spectrometry.

PPI dendrimers of different generations were grafted [70] through selective functionalization with sulfonic acid chloride followed by

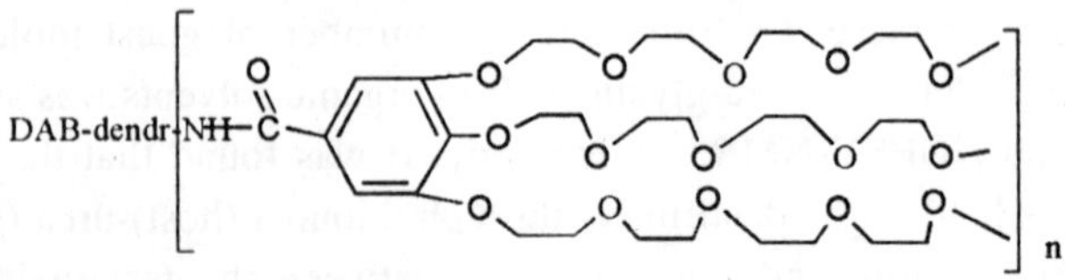

FIG.5.1 Structure of oligoethyleneoxy functionalized PPI dendrimer

substitution of sulfonamide proton of each branch by means of alkyl and benzyl halides.

PPI dendrimers upto 5th generation were functionalized [71] with 3-carboxy-2,2,5,5-tetramethyl-1-pyrrolidinyloxyl radical end groups (Fig.5.2) and characterized by EPR technique. The functionalization was carried out by the treatment of preactivated 3-carboxy-2,2,5,5-tetramethyl-1-pyrrolidinyloxyl (A) with PPI dendrimers viz. DAB-dendr-$(NH_2)_x$ (X=2, 4, 8, 16, 32, 64). The activation was performed by the treatment of 'A' with N-hydroxy succinimide in 1,2-dimethoxy ethane in the presence of DCC. The functionalized dendrimers were characterized by FT-IR and EPR spectroscopy. The EPR spectra reveal directly the number of end groups from the splitting between the hyper-line transitions for the lower generation dendrimers. Due to the exchange interaction of the pendant nitronyl end groups, the higher generations dendrimers narrowed the EPR spectra, which hampers the analysis of exact splitting of spectrum. This strong interaction of the peripheral end groups resulted thermally populated high spin states. Whereas FT-IR spectroscopy revealed that a H-bonding net-work was formed between the amide functionalities of the end groups, consistent with the observed solvent effect on the EPR spectra.

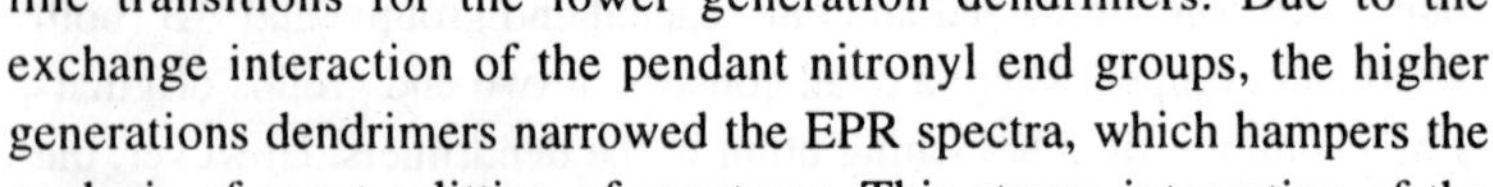

FIG.5.2 Structures of 2nd generation PPI dendrimers with tetramethyl pyrrolidinyloxy radical at the surface

A series of chiral phosphine functionalized PPI dendrimers was synthesized by the reaction of 3,4-*bis*(diphenylphosphino)pyrrolidine (pyrophos) with zeroth to 4th generation PPI dendrimers by EDC/1-hydroxy benzotriazol as a coupling reagent [72]. At first the phosphorous containing compound reacted with glutamic anhydride followed by condensation of the resulted compound with PPI dendrimers offered the modified dendrimers (Scheme 5.18). The functionalized dendrimers were characterized by ^{1}HNMR, ^{13}CNMR, ^{31}PNMR spectroscopy, elemental analysis and FAB or MALDI-TOF mass spectrometry. The molecular weight of the modified dendrimers varies from 1159 to 20632 Da as measured by mass spectrometry. These dendrimers were then treated with $[Rh(Cod)_2]BF_4$ at room temperature in CH_2Cl_2 to get multisite cationic pyrophos-rhodium metalliod dendrimers, which were utilized in the asymmetric hydrogenation of Z-methyl-α-acetamido-cinammate (Scheme 5.19).

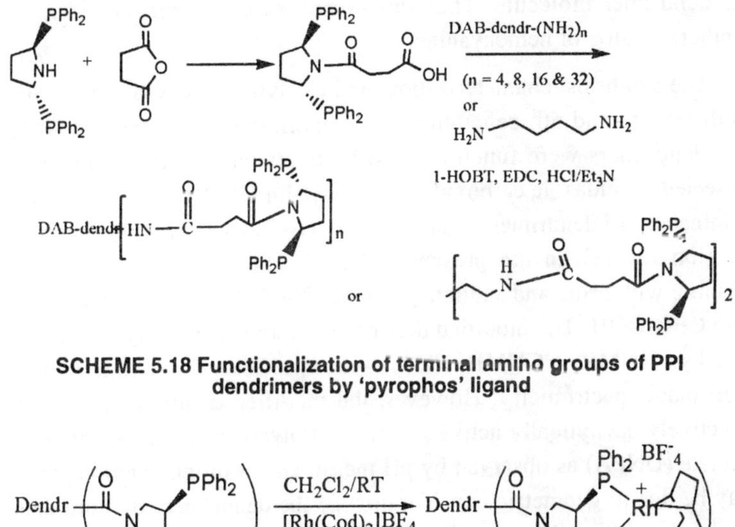

SCHEME 5.18 Functionalization of terminal amino groups of PPI dendrimers by 'pyrophos' ligand

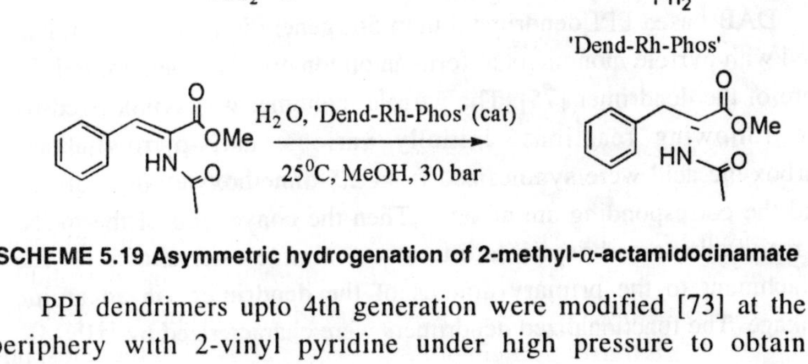

SCHEME 5.19 Asymmetric hydrogenation of 2-methyl-α-actamidocinamate

PPI dendrimers upto 4th generation were modified [73] at the periphery with 2-vinyl pyridine under high pressure to obtain dendrimers with (2-(2-pyridyl)ethyl) amine (PYZ) ligands. They can easily be formed complexes with different metal ions such as Zn(II), Cu(II) and Cu(I), etc. The complexation was studied by ^{1}HNMR, UV-vis. and EPR spectroscopy. The result indicates that one metal ion was present per ligand (PYZ), in all the cases. However, UV-vis. titration of the 4th generation dendrimer in CH_2Cl_2 with $(Cu(I)(CH_3CN)_4)(ClO_4)$ in acetonitrile revealed that approximately 30 Cu(I) ions were present in the structure. Further, low temperature UV-vis. spectroscopy of this complex in CH_2Cl_2 at -85°C in the presence of dioxygen showed that approximately 60-70% of the copper centers can bind dioxygen, corresponding to 10-11 of these molecules per dendrimer molecule. Thus this complex can be considered as a synthetic analog of hemocyanine.

The synthesis, characterization, and catalytic activity of imidazole modified 3rd and 5th generations PPI dendrimers were reported [74]. The dendrimers were functionalized by treatment of triphenylmethyl protected 4-imidazole carboxylic acid (trityl-Im-COOH) with the amine terminated PPI dendrimers in anhydrous CH_2Cl_2 using peptide coupling reaction for 18h in the presence of DCC and HOBT followed by washing with brine and saturated aqueous Na_2CO_3 solution and finally with CF_3COOH. The modified dendrimers were successfully characterized by ^{1}HNMR, ^{13}CNMR, IR, UV-vis. spectroscopies and MALDI-TOF mass spectrometry. However, the modified dendrimers are not effectively catalytically active for the hydrolysis of 2,4-dinitropheny-lacetate (DNPA) as observed by pH measurement of the reaction. This may be due to geometric consideration of the dendrimer structure and weak substrate-catalyst association.

DAB based PPI dendrimers upto 5th generation were functionalized with pyrrole monomers to form an oligomerizable shell around the core of the dendrimer [75]. The pyrrole monomer was synthesized by the following reactions. Initially various, ω-(N-pyrrolyl)alkane carboxylic acid were synthesized from 2,5-dimethoxytetrahydrofuran and the corresponding amino acid. Then the conversion of the ω-(N-pyrrolyl)alkane carboxylic acid to a succinimide ester allows for the attachment to the primary amines of the dendrimer via an amide linkage. The functionalized dendrimers were characterized by ^{1}HNMR,

^{13}CNMR, FTIR spectroscopies and MALDI-MS spectrometric studies. $FeCl_3$ was used to intramolecularly oligomerized the pyrrole monomers surrounding the dendrimers, which was confirmed by FTIR and end group analysis.

A series of carboxy terminated PPI dendrimers [76] upto 5th generation was synthesized from a double Michael addition of amine functionalized PPI dendrimers to methyl acrylate followed by base hydrolysis using LiOH in a water/methanol mixture (Scheme 5.20). The synthesized dendrimers were characterized by ^{1}HNMR, ^{13}CNMR, IR spectroscopy and ESI-MS spectrometry. The main characteristics of these dendrimers are not only consist of acid functionalized co-ordination sites at the periphery but also contain a functional interior of tertiary amines. These type of compounds have also been synthesized previously by acid hydrolysis of nitrile functionalized PPI dendrimers [77] and used in electrokinetic chromatography [78]. However, in this case the synthesized compounds contain a high amount of salts (30 %wt.), hence they are very hygroscopic and brownish in color.

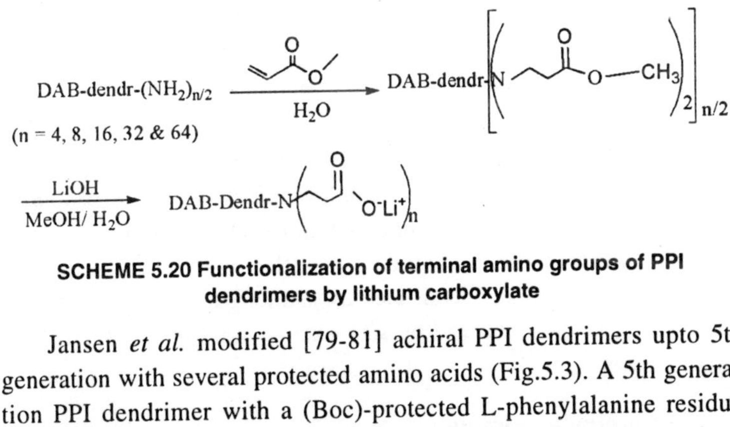

SCHEME 5.20 Functionalization of terminal amino groups of PPI dendrimers by lithium carboxylate

Jansen *et al.* modified [79-81] achiral PPI dendrimers upto 5th generation with several protected amino acids (Fig.5.3). A 5th generation PPI dendrimer with a (Boc)-protected L-phenylalanine residue resulting in a dendrimer with a flexible core/rigid cell structure, coined as 'dendritic box' [79] with molecular weight of ~ 23 kDa. The modified dendrimers showed no constant optical activity per chiral end groups, rather a decrease of optical activity with the increase of generation number was observed. The effect was most pronounced with the aromatic amino acids, like S-tyrosine and S-phenylalanine. Also in the model studies the contribution to the optical activity of the N-Boc-

S-phenylalanine end groups showed to be very sensitive to the local environment. The structures of the modified dendrimers were characterized by IR, UV, ^{1}HNMR and ^{13}CNMR spectroscopic studies. From ^{13}CNMR relaxation time measurements [80] a 'solid phase' behavior of the peripheral groups of higher generations is indicated. According to the authors, the end groups in dense packing seem to adopt 'frozen in' conformations with contributions to the optical rotation that internally cancel each other to give a resulting average of almost zero.

DAB-dendr-$(NR_2)_x$

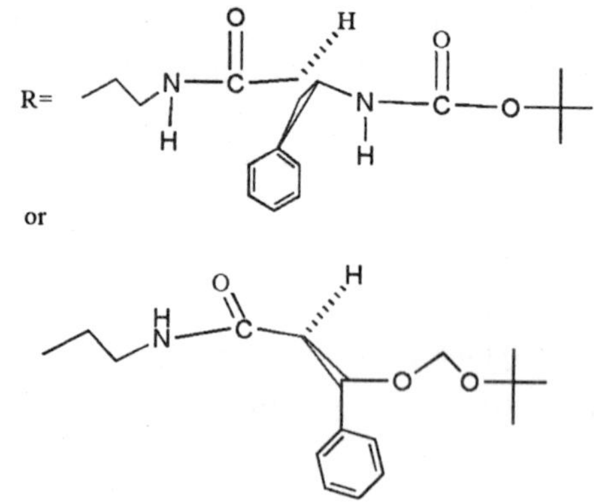

FIG.5.3 Structures of Boc protected L-phenyl alanine derivative functionalized PPI dendrimers

A novel AB_x type dendrimers/linear polymer block copolymer [82], i.e. poly(trimethyleneimine) dendrimers block poly(sarcosine)$_{64}$ was synthesized by the ring opening polymerization of sarcosine N-carboxyanhydride, initiated with 64-NH_2 terminal poly(trimethyleneimine) dendrimer as a micro initiator. The block copolymer was characterized by size exclusion chromatography and SANS studies.

5.2.2 Poly(ethylene imine)

The first example of this type of dendrimers was synthesized [83] by alkylation of diethylene triamine (DETA) with aziridine to form the initiator core, *tris*-(2-amino ethyl) amine. Subsequently exhaustive alkylations of the terminal amino groups with activated N-tosyl- or N-mesyl-aziridine resulted protected first generation poly(ethylene imine) (PEI), which can be deprotected by acid hydrolysis. The repetition of these reactions sequence offered higher generation dendrimers (Scheme 5.21). This technique allows to synthesize upto 4th generation dendrimers with very high yield in each step. The synthesized dendrimers were characterized fully and found that the segmental length is almost half of similar type (PAMAM i.e. 5A° versus 10A° but having high physical and chemical stability. The molecular weight obtained upto 4000g/mole. However, the defect level became progressively higher and differs from ideality. This defect became substantial as one approaches to higher generations, (say 4th generation) and especially for generation five, where defected level makes it's a hyperbranched polymer rather than a dendrimer.

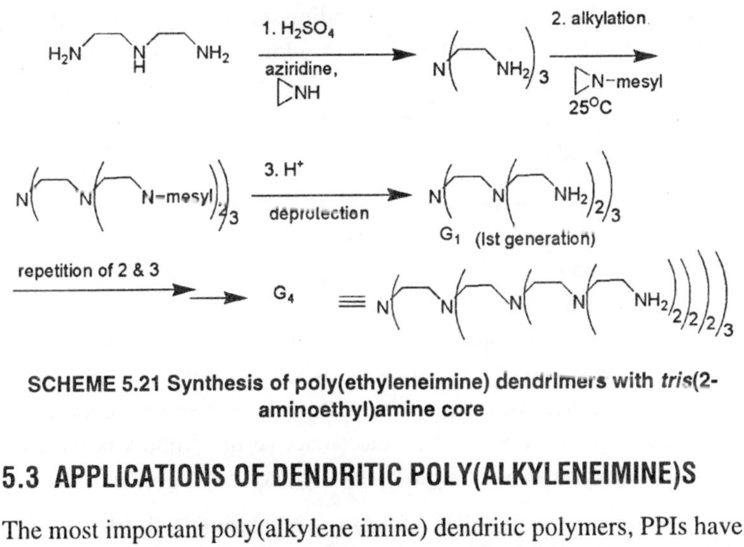

SCHEME 5.21 Synthesis of poly(ethyleneimine) dendrimers with *tris*(2-aminoethyl)amine core

5.3 APPLICATIONS OF DENDRITIC POLY(ALKYLENEIMINE)S

The most important poly(alkylene imine) dendritic polymers, PPIs have various useful applications, especially in electronics and biomedical fields. These are discussed in the following section.

5.3.1 Electronics and Nanoparticles

The novel PPI dendrimer based clusters of clusters were developed [84] as quantum dots by reacting PPI functionalized DAB-dendr-$\{N(CH_2PPh_2)_2\}_{16}$ dendrimers with stoichiometric amount of $[Ru_5C(Co)_{15}]$ or $[(CH_3)_2SAuCl]$. An electro-optical switch based on the addition of modified PPI-dendrimers into a continuous liquid crystalline matrix was also developed [85]. Two types of PPI-dendrimers were used for this purpose; the 5th generation cyanobiphenyl and another palmitoyl functionalized PPI dendrimers. The continuous nematic liquid crystalline matrix is a mixture of cyanobiphenyl and cyanoterphenyl compounds. The cyanobiphenyl functionalized dendrimers showed more effective compatibility at lower concentrations may be due to the presence of mesogenic end groups. It is also observed that the blends have a low transmittance indicates that the visible light is efficiently scattered and thus a static scattering state can be obtained if no voltage is applied. The blend of cyanobiphenyl functionalized dendrimers with nematic liquid crystalline matrix, the switching properties are complex and two regimes can be distinguished depending on the applied voltage. However, the absence of hysteresis and the low voltage required for the switching process of blends containing palmityl functionalized dendrimers offer advantages compared to polymer filled nematics and polymer dispersed liquid crystal systems.

A 3rd generation PPI dendrimers functionalized with OPV was used [86] as a stabilizing agent in the fabrication of CdS nanoparticles. CdS spherical nanoparticles were formed in $CHCl_3$ solution, whereas in $CHCl_3/CH_3OH$ mixture (6:1, v/v) plate like structure evolved. At the air-water interface, nanoparticulate films with a leaf nerve like organization of nanoparticles were formed, whereas vesicles generated in aqueous media became over grown with CdS. The solvent dependent conformational behavior of this electroactive dendrimers is used to generate various organic-inorganic semiconductor composites.

The aqueous assemblies of adamantyl derivatized 4th and 5th generations PPI dendrimers and β-cyclodextrin (β-CD) were used as nano reactors in the preparation of size selective gold and platinum nano particles in water [87]. These nano particles were formed by the reduction of aurate or platinate in the presence of the supramolecularly assembled dendrimer-CD complexes. The persistent shape of the complexes provided a kinetic barrier for the escape of nanoparticles and thus prolonging their lifetime.

5.3.2 Host-guest and Controlled Release Systems

The host-guest interactions between various dyes with 5th generation

PPI dendrimers in aqueous solutions were studied [88] by using a surface plasmon resonance, (SPR) sensor. The study showed strong interactions among the dendrimers with the guest molecules such as RB and erythrosine B. It may be due to the good match in charge and shape between the cavities of the dendrimers and the dye molecules. The effect of pH and ionic strength on affinity of guests to host showed that the maximum absorption of guests occurred at neutral pH.

PPI dendrimers upto 32 azobenzene moieties functionalized at the periphery were potentially used to encapsulate host of eosin (2',4',5',7'-tetrabromofluorescein dianion) [89]. All the E-form azobenzene dendrimers transferred to Z-form by light excitation. Both the forms are quenched eosin fluorescence by a static mechanism. The results obtained from the study indicate that eosin is hosted by the dendrimers and Z-forms are more efficiently hosted than that of E-form.

The covalently attached adamantyl urea end groups of PPI dendrimers up to 5th generation were used as a scaffold to reversibly bind glycinylurea building blocks through strong and directional multiple interactions [90] (Scheme 5.22). This supramolecules were characterized by IR, ^{1}HNMR spectroscopy and ^{1}H-^{1}H NOESY spectroscopic studies. The globular shape of the dendrimer confined the ionic and H-bonded interactions in the curved two dimensional plane of the periphery. The architectures obtained showed increased rigidity at the

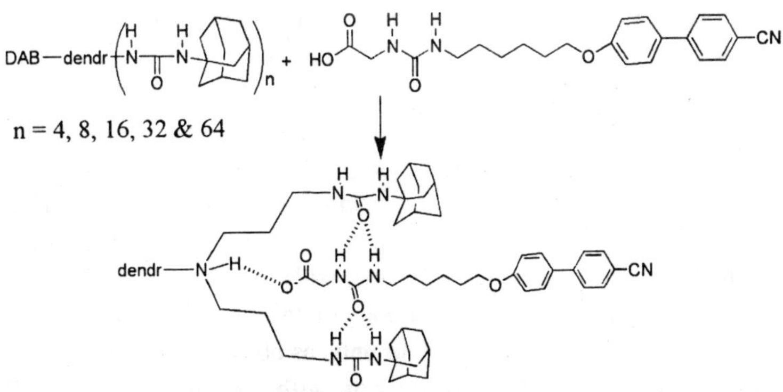

SCHEME 5.22 Binding of glycinylurea unit with PPI dendrimers through multiple interactions

periphery of the core, while local conformational at flexibility remains at the end of the molecules that is selectively bound.

The oligo-pyrrole terminated PPI dendrimers with a DAB core was used to encapsulate guest molecules [91]. The neutral core can be protonated to form cation, which permitted the incorporation of neutral or anionic guests. The encapsulation and release of the guest molecules are controlled through oxidation and reduction of the oligomeric pyrroles.

A 5th generation perfluorooctanoyl functionalized PPI dendrimers was used as extracting agent, in a viewing cell at 40°C under 300 bar pressure to remove MO dye from water to CO_2 phase, whereas the palmitoyl functionalized dendrimers extracted into a tetrachlorocarbon phase [92]. These modified dendrimers are also used to catalyze the following SN^2 reaction in supercritical CO_2 and found to be effective. The higher generations (4th) is less effective than lower 2nd and 3rd generations. Similarly, the hydrophilic guests can be transported from water into organic phases or liquid CO_2 by using inverted unimolecular micelles based on a PPI skeleton with grafted aliphatic and fluorinated chains.

Meijer's 'dendritic box' is an ideal structure with internal cavities available for guest molecules. As the shell is constructed in the final step, so it is possible to perform this coupling reaction in the presence of guest molecules. A large variety of guest molecules have been encapsulated and this open up a plethora of interesting chemical and biochemical applications. The guest molecules used were 3-carboxy-PROXYL, Rose Bengal and Eriochrome Black T (Fig. 5.4). When 3-carboxy-PROXYL is using as the guest, the number of entrapped radicals varied from 0.3 to 6 molecules per dendritic box as determined by ESR spectroscopy [93]. The number of 3-carboxy-PROXYL radicals in the dendritic box does not increase above 6, clearly demonstrated that the maximum attainable number of radicals is restricted by the shape of the cavities of the box. In case of Rose Bengal, the number is limited to four only, as observed in fluorescence spectra in $CHCl_3$. The dendritic box with one molecule of RB encapsulated exhibits an induced circular dichroism spectrum related to the UV spectrum, in which all bands possess a negative Cotton effect. Again, an exciton-coupled spectrum is observed in the case of four guests per box, indicative of the close proximity of chromophores with a certain fixed orientation [94]. Whereas for Eriochrome Black T [95], the diffusion of the encapsulated dye was not observed in the aqueous phase even after prolonged heating, dialysis or sonication indicated that the diffusion of the dye out of the box is measurably slow.

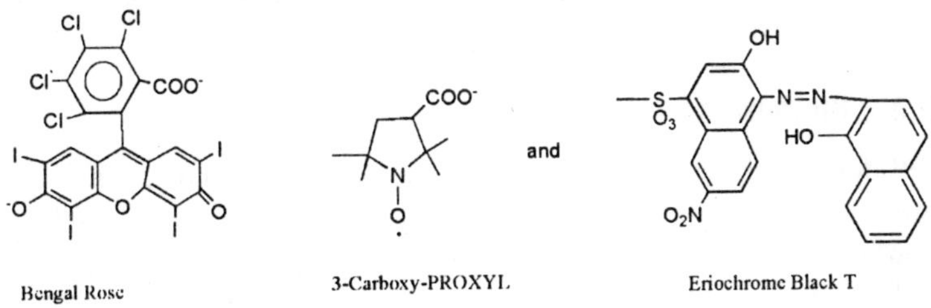

FIG.5.4 Structures of 3-carboxy-PROXYL, Rose Bengal and Eriochrome Black T dye

The PPI dendrimers were modified [96] with adamantyl urea and adamantyl thiourea to act as hosts for N-terminal Boc protected peptides and form chloroform soluble complexes. It was found that the peptide is bound to the dendrimers by ionic interactions between the dendrimer outer shell tertiary amines and the C-terminal carboxylic acid of the peptide, and also through host urea to peptide amide H-bonding. The host-guest interaction of these systems was studied thoroughly by the help of NMR, FTIR, and NOESY spectroscopies. The dendrimer also capable to make multiple peptide dendrimer complexes. The different peptides showed specific preferences and could be released under mild acidic conditions.

PPI dendrimers were used [97] as pH-sensitive controlled released systems. The molecular composite prepared by the solubilization of pyrene in PPI dendrimers. The pyrene incorporation was favored in basic medium, so the release was induced in acidic medium. The interaction of pyrene with the dendrimer was studied by fluorescence and absorption spectroscopy at a range of pH.

OPV functionalized PPI dendrimers were [98] used for effective extraction of anionic dye molecules from water to organic solvents.

These host-guest assemblies of the dyes with dendrimers are very efficient for energy transfer and the emission wavelength can be adjusted by using variety of dye molecules, though only dye/poly(*p*-phenylene vinylene)(PPV) showed a little energy transfer. This may be due to the fact that the modified dendrimers and dye systems mixed very well and thus forming good quality of films in contrast to the films obtained from dye/PPV without dendrimer. Thus OPV units of the dendrimer act as a compatibilizer in these systems and energy transfer is observed from the organic PPV polymer to the dye.

5.3.3 Biomedical

PPI dendrimers were used [99] as the multivalent bioconjugates. The presence of multiple biologically active molecules on a dendritic scaffold may lead to enhance effect. These complexes are made by non-covalent functionalization of dendrimer at the periphery.

The self-assembled ternary complex of DAB based 4th and 5th generation PPI dendrimers, cucurbituril (CB) and DNA was reported [100] as an example of an totally self-assembled gene delivery carrier. The complex was formed in a non-covalent way in which DNA interacts with dendrimers electro-statically and CB with dendrimers through multiple non-covalent interactions. DLS data indicates that the diameter and size distributions of the complexes were dependent upto the sequence of mixing of each component with unimodal distribution ranging from 150.8 to 210.2 nm under favorable conditions. The complex was able to transfect mammalian cells with high efficiency and the cytotoxicity of the PPI-CB complex was relatively low.

A 5th generation amine terminated PPI dendrimer with DAB core (DAB-Am 64) conjugated with a bifunctional conjugate of DTPA, (IB4M) for complexing Gd(II) atoms and used as liver MRI contrast agent [101-102]. This MRI contrast agent was evaluated to accumulate in the liver at a 50% of the injected dose within 15 mins after injection. The dynamic micro-MRI with DAB-Am 64-$(1B4M\text{-}Gd)_{64}$ was able to homogeneously enhanced the normal liver parenchyma and visualized micrometa static tumors of 0.3 mm diameter in the liver of the mice with better contrast than that of (Gd-DTPA). Therefore, this novel liver

MRI contrast agent is potentially useful for diagnosis of micro metastatic in the liver.

5.3.4 Unimolecular Micelles

The PPI amphiphilic dendrimers (in quaternary ammonium chloride form) upto 3rd generation with both octyl and triethylenoxy methyl ether chain ends were utilized [42] to solubilize lipophilic compounds such as pyrene and Reichardt's dye and the solubility limited for pyrene was found to be one pyrene molecule per dendrimer molecule. It was also found that the rates of decarboxylation of 6-nitrobenzioxazole-3-carboxylic acid in aqueous solutions of the cationic dendrimers were upto 500 times faster than in water alone.

PPI dendrimers functionalized with OPV groups act as amphiphilic dendrimers (Fig.5.5). These electroactive dendrimers consists of a polar core and an apolar periphery and consequently also behave as macro molecular surfactants [103-104]. When they are dissolved in an organic solvent, the molecules behave as unimolecular inverted micelles. In water, the molecules adopt a geometry in which the hydrophilic core is in contact with water while the alkyl chains are oriented away from the aqueous phase, resulting in the formation of vesicular aggregates.

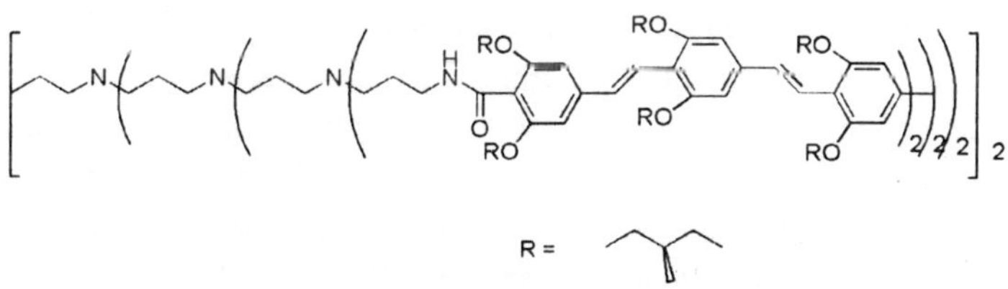

FIG.5.5 Structures of oligo(phenylene vinylene) functionalized PPI dendrimers

The palmitoyl modified PPI dendrimers at terminal groups upto 5th generation form inverted unimolecular dendritic micelle like structure with a polar interior of tertiary amines and an apolar periphery. The extraction properties of these dendrimers were studied [105] by using different anionic xanthene dyes such as 4,5,6,7-tetrachlorofluorescein, Erythrosin B and RB. The study showed that the 5th generation modified dendrimers effectively extracted the dyes from aqueous solution at lower pH with 1:1 mole ratio to organic layer. However, at higher pH no extraction is possible, except for RB. The large hydrophobicity of RB may be the possible reason for this extraction even at higher pH (pH >12). From the results of all generations, it may be concluded that the extraction is directly related to the number of 3° amine present interiors of the dendrimers.

5.3.5 Catalysis

The PPI dendrimers of 3rd to 5th generations were reported [106] to form diamino Cu(II), diamino Zn(II), and tetramino Co(II) complexes confirmed by spectroscopic and titrametric study. These dendrimer–metal ion complexes catalyzed the hydrolysis of *p*-nitrophenyl diphenylphosphate in zwitterionic buffer solutions at pH = 8.1 with relative activities Cu(II)>Zn(II)>Co(III). The rate of hydrolysis was faster with Na-perchlorate than with NaCl to control ionic strength. In NaCl solution, the rate of hydrolysis decreases for Cu(II) complexes with the increase of generation number.

The supramolecular complex of the urea-adamantyl PPI dendrimers with phosphine ligands was used [107] as multidentate ligand system in the palladium catalyzed allylic amination reaction in a batch process and in a continuously flow membrane reactor. The catalytic activity and selectivity of the dendrimeric complex are similar to that of the monomeric complex, hence catalytic centers act as independent sites. As the size of the supramolecular system is sufficiently large and the binding of the guests is strong, so a good separation of the catalyst components from the reaction mixture by using nano filtration technique can be easily done.

DAB–dendr-$(NH_2)_{16}$ dendrimer functionalized with diphenyl phosphane groups at the periphery in formaldehyde [108] and used as catalyst. The use of phosphine-based dendritic catalysts for allylic substitution reaction in a continuous membrane reactor was also reported [109]. The transition metal complexes of Pd, Ir, Rh etc. were prepared in which the dendr–N-$(CH_2PPh_2)_2$ groups act as bidentate ligands. The Pd-complexes of these dendrimers (Fig.5.6) were used as catalysts in the Heck reaction of bromo-benzene and styrene to from stilbene. Again, these dendritic catalysts are reusable with same efficiency.

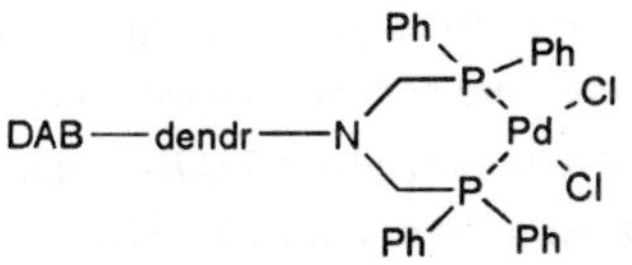

FIG.5.6 Structures of Pd-complexes of PPI dendrimers

5.4 SUMMARY

In this chapter the synthesis, characterization, properties and possible applications of different types of poly(alkylene imine)s, mainly PPI dendritic polymers are discussed. Because of the active amine terminal groups in the dendrimers, they can be functionalized to a great variety of different functional groups containing dendritic polymers, which are very much useful for different potential applications. From the above discussion, it has been found that dendrimers are mainly reported for this category and no hyperbranched polymer was reported so far. Even though, their development took long time but they have important roles in host-guest chemistry, biomedical applications, catalysis, electronic applications, etc.

REFERENCES

[1] G.J.M.Koper, M.H.P.van Genderen, C.Elissen-Roman, M.W.P.L.Baar, E.W.Meijer and M.Borkovec, *J.Am.Chem.Soc.* **119**(1997)6512

[2] E.Buhleier, W.Wehner and F.Vogtle, *Synthesis* (1978)155

[3] E.M.M.de Brabander van den Berg and E.W.Meijer, *Angew.Chew.Int.Ed.* **32**(1993)1308

[4] A.W.Bosman, H.M.Janssen and E.W.Meijer, *Chem.Rev.* **99**(1999)1665

[5] M.H.P.van Genderen, M.W.P.L.Baars, J.C.M.van Hest, E.M.M.Brabander van den Berg and E.W.Meijer, *Recl.Trav.Chem.Pays.Bas.* **113**(1994)573

[6] M.Chai, Y.Niu, W.J.Youngs and P.L.Rinaldi, *J.Am.Chem.Soc.* **123**(2001)4670

[7] M.Chai, Y.Niu, W.J.Youngs and P.L.Rinaldi, *Macromolecules* **33**(2000)5395

[8] C.N.Likos, M.Schmidt, H.Lower, M.Ballauff, D.Potschke and P.Linder, *Macromolecules* **34**(2001)2914

[9] P.N.Floriano, C.O.Noble, M.Schoonmaker, E.D.Poliakoff and R.L.McCarley, *J.Am.Chem.Soc.* **123**(2001)10545

[10] http://www.users.marshall.edu/holley23/index.htm

[11] J.de Maaijer-Gielbert, C.Gu, A.Somogyi, V.H.Wysocki, P.G.Kislemaker and T.L.Weeding, *J.Am.Chem.Soc.Mass.Spectrum.* **10**(1999)414

[12] D.L.Richter-Egger, A.Testai and S.A.Tucker, *Anal.Chem.* **73**(2001)5743

[13] G.Pistolis, A.Malliaris, D.Tsiourvas and C.M.Paleos, *Chem.Eur.J.* **5**(1999)1440

[14] J.W.Weener, J.L.J.van Dongen and E.W.Meijer, *J.Am.Chem.Soc.* **121**(1999) 10346

[15] J.C.Hummalen, J.L.J.van Dongen and E.W.Meijer, *Chem.Eur.J.* **3**(1997)1484

[16] M.Perrot, W.G.Rothschild, and R.M.Cavagnat, *J.Chem.Phys.* **110**(1999)9230

[17] P.Miklis, T.Cagin and W.A.Goddard III, *J.Am.Chem.Soc.* **119**(1997)7458

[18] L.Cavallo and F.Fraternali, *Chem.Eur.J.* **4**(1998)927

[19] F.Vogtle, S.Gestermann, C.Kauffmass, P.Ceroni, V.Vicinalli, L.De Cola and V.Balzani, *J.Am.Chem.Soc.* **121**(1999)12161

[20] L. Prodi, F.Bolletta, M.Montalti and N.Zaccheroni, *Chem.Eur.J.* **5**(1999)445

[21] V.Balzani, P.Ceroni, S.Gestermann, M. Gorka, C. Kauffmann, M. Maestrim and F.Vogtle *Chem.Phy.Chem.* (2000)224

[22] V.Balzani, P.Ceroni, S.Gestermann, C.Kauffmann, M.Gorka and F.Vogtle, *Chem.Commun.* (2000)853

[23] V.Balzani, P.Ceroni, S.Gestermann, M.Gorka, C.Kauffmann and F.Vogtle, *Tetrahedron* 58(2002)629

[24] F.Vogtle, S.Gestermann, C.Kauffmann, P.Ceroni, V.Vicinalli and V.Balzani, *J.Am.Chem.Soc.* **122**(2000)10398

[25] H.Stephan, H.Spies, B.Johannsen, C.Kauffman and F.Vogtle, *Org.Lett.* **2**(2000)2343

[26] H.Stephan, H.Spies, B.Johannsen, L.Klein and F.Vogtle, *Chem.Commun.* (1999)1875

[27] E.K.L.Yeow, K.P.Ghijgine, J.N.H.Reck, M.J.Crossley, A.W.Bosman, A.P.H.J.Schenning and E.W.Meijer, *J.Phys.Chem.B.* **104**(2000)2596

[28] S.C.J.Meskers, M.Bender, J.Huebner, Y.V.Romanovski, M.Oestreich, A.P.H.J.Schenning, EW.Meijer and H.Baessler, *J.Phys.Chem.A* **105**(2001)10220

[29] A.Dirksen, E.Zuidema, R.M.Williams and L.De Cola, C.Kauffman, F.Vogtle, A.Roque and F.Pina, *Macromolecules* **35**(2002)2743

[30] P.Ballester, R.M.Gomila, C.A.Hunter, A.S.H.King and L.J.Twyman, *Chem.Commun* (2003)38

[31] E.J.H.Put, K.Clays, A.Persoons, H.A.M.Biemans, C.P.M.Luijk and E.W.Meijer, *Chem.Phys.Lett.* **260**(1996)136

[32] K.Tsuda, G.C.Dol, T.Gensch, J.Hofkens, L.Latterini, J.W.Weener, E.W.Meijer and F.C.DeSchryver, *J.Am.Chem.Soc.* **122**(2000)3445

[33] A.Archut, F.Vogtle, L.De Cola, G.C.Azzellini and V.Balzani, *Chem.Eur.J.* **4**(1998)699

[34] J.Ruiz, M.J.R.Medel, M.C.Daniel, J.C.Blais and D.Astruc, *Chem.Commun.* (2003)464

[35] J.Alvarez, T.Ren and A.E.Kaifer, *Organometallics* **20**(2001)3543

[36] K.Takada, D.J.Diaz, H.D.Abruna, I.Cuadrado, B.Gonzalez, C.M.Casado, B.Alonso, M.Moran and J.Losada, *Chemistry* 7(2001)1109

[37] J.Ruiz, C.Pradet, F.Varret and D.Astruc, *Chem.Commun.* (2002)1108

[38] K.Takada, D.J.Diaz, H.D.Abruna, I.Cuadrado, C.Casado, B.Alonso, M.Moran and J.Losada, *J.Am.Chem.Soc.* **119**(1997)10763

[39] K.Takada, D.J.Diaz, H.D.Abruna, I.Cuadrado, B.Gonzalez, C.M.Casado, B.Alonso, M.Moran and J.Losada, *Chem.Eur.J.* **7**(2001)109

[40] B.Gonzalez, C.M.Casado, B.Alonso, I.Cuadrado, M.Moran, Y.Wang and A.K.Kaifer, *Chem.Commun.* (1998)2569

[41] J.B.Christensen, M.F.Nielsen, J.A.E.H.van Haare, M.W.P.L.Baars, R.A.J.Janssen and E.W.Meijer, *Eur.J.Org.Chem.* **2001**(2001)2123

[42] I.Bodnar, A.S.Silva, Y.H.Kim and N.J.Wagner, *J.Polym.Sci. PartB Polym.Phys.* **38**(2000)874

[43] Y.Pan and W.T.Ford, *Macromolecules* **32**(1999)5468

[44] Y.Pan and W.T.Ford, *Macromolecules* **33**(2000)3731

[45] J.L.Kreider and W.T.Ford, *J.Polym.Sci.Part A Polym.Chem.* **39**(2001)821

[46] J.J. Lee, W.T. Ford, J.A.Moore and Y.Li, *Macromolecules* **27**(1994)4632

[47] P.R.Ashton, K.Shibata, A.N.Shipway and J.F Stoddart, *Angew.Chem.Int.Ed.* **36**(1998)2781

[48] Z. Sideraton, D. Tsiourvas and C.M. Paleos, *Langmuir* 16(2000)1766

[49] C Elissen-Roman, J.C.M.van Hest, M.W.P.L.Baars, M.H.P.van Genderen and E.W.Meijer, *Polym.Mater.* **77**(1997)145

[50] S.Stevelmans, J.C.M.van Hest, J.F.G.A. Jansen, D.A.F.J.van Boxtel, E.M.M.de Brabander van den Berg and E.W.Meijer, *J.Am.Chem.Soc.* **118**(1996)7398

[51] M.E.Piotti, F.Rivera Jr., R.Bond, C.J.Hawker and J.M.J.Frechet, *J.Am.Chem.Soc.* **121**(1999)9471

[52] M.W.P.L.Baars, S.H.M.Sontjens, H.M.Fischer, H.W.I.Peerlings and E.W.Meijer, *Chem.Eur.J.* **4**(1998)2456

[53] J.H.Cameron, A.Facher, G.Latterman and S.Diele, *Adv.Mater.* **9**(1997)398

[54] D.Tsiouvas, T.Felekis, Z.Sideratou and C.M.Paleos, Poster presentation 19th International Liquid Crystal Conference, Edin Burg, UK, July 2000

[55] A.P.H.J.Schenning, P.Jonkheijm, J.Hofkens, S.D.Feyter, T.Asavei, M.Cotlet, F.C.DeSchryver and E.W.Meijer, *Chem.Commun.* (2002)1264

[56] A.P.H.J.Schenning, C.Elissen-Roman, J.W.Weener, M.W.P.L.Baars, S.J.van der Gaast and E.W.Meijer, *J.Am.Chem.Soc.* **120**(1998)8199

[57] J.J.Michels, M.W.P.L.Baars, E.W.Meijer, J.Huskens and D.N.Reinhoulds, *J.Chem.Soc.Peskin Trans.2* (2000)1914

[58] J.J.J.M.Donners, B.R.Heywood, E.W.Meijer, R.J.M. Nolte, C.Roman, A.P.H.J.Schenning and N.A.J.M Sommer dijk, *Chem.Commun.* (2000)1937

[59] J.J.J.M.Donners, B.R.Heywood, E.W.Meijer, R.J.M.Notle and A.J.M.Sommerdijk, *Chem.Eur.J.* **8**(2000)2561

[60] A.W.Bosman, A.P.H.J.Schenning, R.J.A.Janssen and E.W.Meijer, *Chem.Ber.Recl.* **130**(1997)725

[61] G.M.Pavlov, E.V.Korneeva, K.Junnel, S.E.Harding, E.W.Meijer, H.W.I.Peerling, J.F.Stoddart and S.A.Nepogodiev, *Carboh. Polym.* **38**(1999)195

[62] H.W.I.Peerlings, S.A.Nepogodiev, J.F.Stoddart and E.W.Meijer, *Eur.J.Org.Chem.* (1998)1879

[63] P.R.Ashton, S.E.Boyd, C.L.Brown, S.A.Nepogodiev, E.W.Meijer, H.W.I.Peerling and J.F.Stoddart, *Chem.Eur.J.* **3**(1997)974

[64] J.W.Lee, Y.H.Ko, S.H.Park, K.Yamaguchi and K.Kim, *Angew.Chem.Int.Ed.* **40**(2001)746

[65] S.C.Cohen, S.Pefond and N.Raymond, *Chem.Eur.J.* **7**(2001)272

[66] H.Stephan, H.Spies, B.Johannsen, K.Gloe, M.Gorka and F.Vogtle, *Eur.J.Inorg.Chem.* (2001)2957

[67] U.Boas, A.JKarlsson, B.F.M.de.Waal and E.W.Meijer, *J.Org.Chem.* **66**(2001)2136

[68] P.E.Froehling and H.A.J.Linssen, *Macromol.Chem.Phys.* **199**(1998)1691

[69] M.W.P.L.Baars, R.Kleppinger, M.H.J.Koch, S.-L.Yen and E.W.Meijer, *Angew.Chem.Int.Ed.* **39**(2000)1285

[70] A.Archut, S.Gestermann, R.Hesse, C.Kauffmann and F. Vogtle, *Syn.Lett.* **5**(1998)546

[71] A.W.Bosman, R.A.J.Janssen and E.W.Meijer, *Macromolecules* **30**(1997)3606

[72] G.D.Engel and L.H.Gade, *Chem.Eur.J.* **8**(2002)4319

[73] G.Klein, J.M.Roberters, A.W.Bosman, M.C.Feiters, E.W.Merjer and R.J.M.Nolte, *Chem.Eur.J.* **5**(1999)65

[74] L.A.Baker, L.Sun and R.M.Crooks, *Bull.Korean Chem.Soc.* **23**(2002)647

[75] C.O.Noble IV and R.L.McCarley, *J.Am.Chem.Soc.* **122**(2000)6518

[76] R.C.van Duijvenbode, A.Rajanayagam, G.J.M.Koper, M.W.P.L.Baars, B.F.M.De Waal, E.W.Meijer and M.Borkovec, *Macromolecules* **33**(2000)46

[77] E.M.M.de Brabander van der Berg and E.W.Meijer, *Angew.Chem.Int.Ed.* **32**(1993)1308

[78] P.G.H.M.Muijselaar, H.A.Claessens, C.A.Cramers, J.F.G.A.Jansen, E.W.Meijer, E.M.M.de Brabander van der Berg and S.van der Waal, *J.High.Resolut.Chromat.* **18**(1995)121

[79] J.F.G.A.Jansen, E.M.M.de Brabander van der Berg and E.W.Meijer, *Science* **266**(1994)1226

[80] J.F.G.A.Jansen and E.W.Meijer, *J.Am.Chem.Soc.* **117**(1995)4417

[81] R.A.J.Janssen, E.M.M.de Brabander van der Berg and E.W.Meijer, *Adv.Mater.* **7**(1995)561

[82] K.Aoi, T.Hatanaka, K.Tsutsumiuchi, M.Okada and T.Imae, *Macromol.Rapid Commun.* **20**(1999)378

[83] D.A.Tomalia, A.M.Naylor and W.A.Goddard III, *Angew.Chem.Int.Ed.* **29**(1990)138

[84] N.Feeder, J.Geng, P.G.Goh, B.F.G.Johnson, C.M.Martin, D.S.Shepherd and W.Zhou, *Angew.Chem.Int.Ed.* **39**(2000)1661

[85] M.W.P.L.Baars, M.C.W.van Boxtel, C.W.M.Bastiaansen, D.J.Broer, S.H.M.Sontjens and E.W.Meijer, *Adv.Mater.* **12**(2000)715

[86] J.J.M.Donners, R.Hoogenboom, A.P.H.J.Schenning, P.A.van Hal, R.J.M.Notle, E.W.Meijer and N.A.J.M.Sommerdijk, *Langmuir 18*(2002)2571

[87] J.J.Michels, J.Huskens and D.N.Reinhoudt, *J.Chem.Soc.Perkin.Trans.2* (2002)102

[88] S.Chen, Q.Yu, L.Li, C.L.Boozer, J.Homola, S.S.Yee and S.Jiang, *J.Am.Chem.Soc.* **124**(2002)3395

[89] A.Archut, G.C.Azzellini, V.Balzani, L.D.Cola and F.Vogtle, *J.Am.Chem.Soc* **120**(1998)12187

[90] M.W.P.L.Baars, A.J.Karlsson, V.Sorokin, B.F.W.der Waal and E.W.Meijer, *Angew.Chem.Int.Ed.* **39**(2000)4262

[91] A.D.Morar, C.O.Noble IV and R.L.McCarley, Symp. Delaware Valey Dal Nogare Award, Delaware, March 2000

[92] E.L.V.Goetheer, L.J.P.van den Broeke and J.T.F.Keurenties, hltp://uspt.cf.utwente.nl

[93] J.F.G.A.Jansen, R.A.J.Janssen, E.M.M.de Brabander van der Berg and E.W.Meijer, *Adv.Mater.* **7**(1995)561

[94] J.F.G.A.Jansen, E.M.M.de Brabander-van der Berg and E.W.Meijer, *Recl.Trav.Chim.Pays-Bas* **114**(1995)225

[95] J.F.G.A.Jansen and E.W.Meijer, *Macromol Symp.* **102**(1996)27

[96] U.Boas, S.H.M.Sontjens, K.J.Jansen, J.B.Christensen and E.W.Meijer, *ChemBioChem* **3**(2002)433

[97] G.Pistolis, A.Malliaris, D.Tsiourvas and C.M.Paleos, *Chem.Eur. J.* **5**(1999)1440

[98] A.P.H.J.Schenning, E.Peeters and E.W.Meijer, *J.Am.Chem.Soc.* **122**(2000)4489

[99] M.W.P.L.Baars, A.J.K.Karlsson, V.Sorokin, B.F.W.de Waal, *Angew.Chem.* **11**(2000)4432

[100] Y.B.Lim, T.Kim, J.W.Lee, S.Kim, H.J.Kim, K.Kim and J.S.Park, *Bioconjugate Chem.* **13**(2002)1181

[101] H.Kobayashi, T.Saga, S.Kawamoto, N.Sato, A.Hiraga, T.Ishimoni, J.Konishi, K.Togashi and M.W.Brechbiel, *Cancer Resarch* **61**(2001)4966

[102] H.Kobayashi, S.Kawamoto, T.Saga, N.A.Hiraga, T.Ishimori, Y.Akita, M.H.Mamede, J.Konishi, K.Togashi and M.W.Brechbiel, *Magn.Reson.Med.* **46**(2001)795

[103] A.P.H.J.Schenning, C.Elissen-Roman, J.W.Weener, M.W.P.L.Baars, S.J.van der Gaast and E.W.Meijer, *J.Am.Chem.Soc.* **120**(1998)8199

[104] A.P.H.J.Schenning, E.Peeters and E.W.Meijer, *J.Am.Chem.Soc.* **122**(2000)4489

[105] M.W.P.L.Baars, P.E.Frochling and E.W.Meijer, *Chem.Commun.*(1997)1959

[106] K.Vassilev and W.T.Ford, *J.Polym.Sci.Part A Polym.Chem.* **37**(1999)2727

[107] D.de Grost, B.F.M.de Waal, J.NH.Reek, A.P.H.J.Schenning, P.C.J.Kamer, E.W.Meijer and P.W.N.M.van Leeuwen, *J.Am.Chem.Soc.* **123**(2001)8453

[108] M.T.Reetz, G.Lohmer and R.Schwickardi, *Angew.Chem.Int.Ed* **36**(1996)636

[109] N.Brinkmann, D.Giebel, G.Lohmer, M.T.Reetz and U.Kragl, *J.Catalysis.* **183**(1999)163

CHAPTER 6

DENDRITIC POLYETHERS

6.1 INTRODUCTION

The dendritic polyethers are one of the most important and extensively studied dendritic polymers. This may be due to their variety of applications and easiness of their synthesis. They can be easily synthesized by simple reactions such as Williamson ether synthesis, base catalyzed etherification reaction, etc. Their potential fields of applications included almost everything where ever the macromolecules can be used. As they are mainly synthesized from aromatic or aliphatic hydroxy group containing compounds, so their dendritic structures contain large number of surface hydroxy groups, which are very useful for their different applications and further modifications. They have the characteristic properties like amphiphilic, redoxactive, photoactive, catalytic, assembly, host-guest interaction, etc. which are most important fields of studies in advanced material science and technology.

6.2 SYNTHESIS, CHARACTERIZATION AND PROPERTIES OF DIFFERENT TYPES OF POLYETHER DENDRIMERS

In this section the synthesis, characterization and the properties of different types of polyether dendrimers are discussed.

6.2.1 Aliphatic

The synthesis of aliphatic polyethers upto 4th generation was reported [1] by the convergent method. Methallyl dichloride was used as the building block, as its allylic functionality provides for the facile nucleophilic substitution of its two electrophilic sites, while the double bond itself acts as a masked functionality for the subsequent activation growth steps. Hence, the nucleophilic displacement of the allylic chlorides in a Williamson ether synthesis method was the propagating step for this synthetic approach. The activation step was the derivatization of the olefinic double bond, which may be accomplished either via hydroboration-oxidation to afford a primary alcohol or by ozonolysis followed by reduction to give a secondary alcohol. The whole sequence of reactions is shown in scheme 6.1. The yield obtained in each step was very high (85-95%) and all the products were purified by flash chromatography. The synthesized dendrimers were characterized by ^{1}HNMR spectroscopy. The weight average molecular weights were determined by FAB/MALDI-TOF MS and SEC analyses. The molecular weight of the highest generation dendrimers was 5414 Da (MALDI-MS)/3500Da (SEC), /5409 Da (theoretical value) with polydispersity index 1.01. The alcoholic focal point is readily accessible towards low molecular weight reactive agents such as trimethoxy silane, etc. The hydrosilylation reaction with this reagent was quantitative under typical conditions.

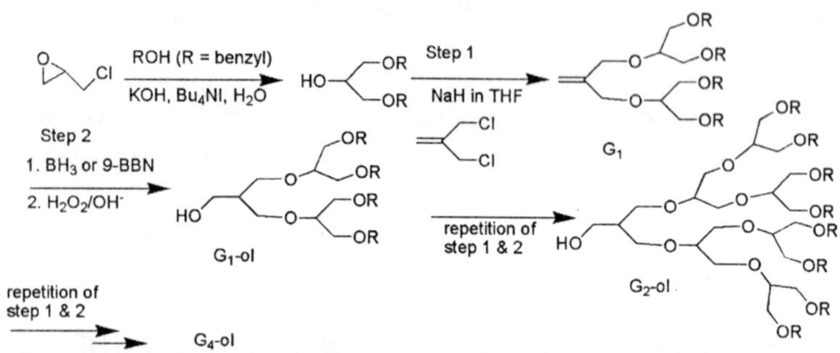

SCHEME 6.1 Synthesis of aliphatic polyether dendrimers based on propane alkane structure

The synthesis of aliphatic polyether dendrons upto 4th generation was also reported [2] by the convergent synthetic approach using the above reaction strategy. Here also methallyl dichloride was used as the monomeric building block along with peripheral benzyl ether and ketal protected hydroxy functionalities, thus enjoyed the above dual facilities. The synthetic process as well as characterization techniques were same as described earlier for benzyl protected dendrons. A second class of aliphatic polyether dendrons were synthesized using symmetrical ketal protected groups to prevent any undesirable stereochemical complication during purification and characterization. Further, even though acetone, pentanone and nonanone were used for protection but nonanone was the most stable in hexane and thereby simplifying flash chromatographic process and hence for actual study. The protected triol was prepared by the reaction of 1,1,1-*tris*(hydroxymethyl)ethane with excess nonanone in the presence of BF_3-etherate catalyst. Then the same sequence of reactions was employed to get the different generation dendrons (upto 4th generation). Both benzylether and ketal protected groups were deprotected by catalytic hydrogenolysis with 10% Pd/C in 50:50 THF/EtOH under 5-10 atmospheric pressure of hydrogen gas and by acid catalyzed hydrolysis with either Dowex acid cation exchange resin and dilute HCl in 1:1 THF/ MeOH respectively, to get desired polyol with very high yield [2]. These 2nd generation polyol dendrons are highly polar and hence soluble in water. Lower generations with alcoholic group at focal points are soluble in water but not alkene focal functional dendrons are. The surface hydroxylic groups as well as focal functionality were also modified with various compounds, which are discussed separately on modification section. The synthesis of hybrid peripheral copolymer dendrons having benzylic and ketal-protected groups was also reported [2].

A new synthetic method for heterogeneous aliphatic polyether dendrimers by the convergent as well as by the divergent approach was described [3], by using triallyl chloride as the core unit. The triallyl chloride core unit was synthesized by the reaction of 1,1,1-*tris*(hydroxymethyl)ethane with an excess of 3-chloro-2-chlormethyl-1-propane in DMF with 20% yield. This can be easily reacted with 2nd to 4th generation benzyl protected dendrons to produce corresponding dendrimers by coupling to the alkoxide dendron created *in-situ* by the reaction of dendron with NaH to the triallyl chloride core. However, the ketal-protected dendrimers obtained similarly as stated above, proved to be more amenable to core functionalization. For example, the treatment of the 2nd generation dendrimer with 9-BBN followed by oxidation with H_2O_2 afforded corresponding triol. These three hydroxy groups can be further grafted with other functionalities such as with polyester dendron by the divergent method. The sequence of all reactions is shown in scheme 6.2, to get the desired dendrimers. The synthesized dendrimers were characterized by spectroscopic and analytical techniques [3].

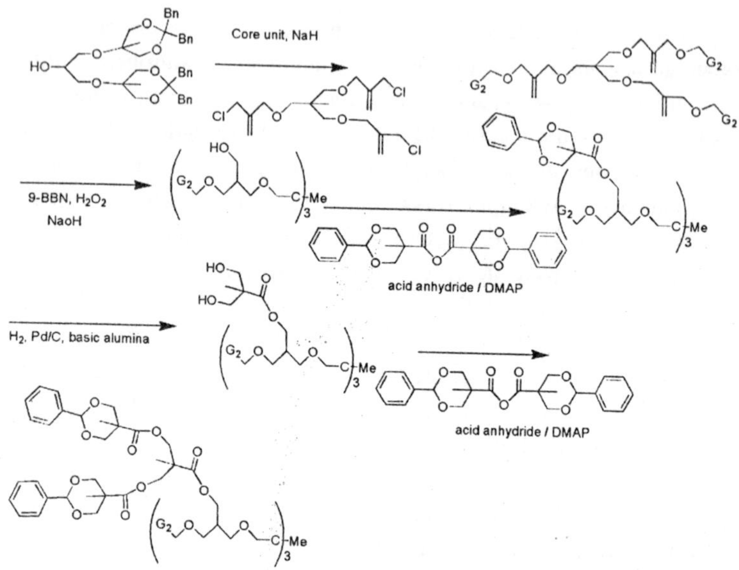

SCHEME 6.2 Synthesis of 2nd generation heterogeneous aliphatic polyether dendrimer

A series of water-soluble hydroxy terminated dendritic polyethers with carborane nucleus covalently attached to the focal point was synthesized [4] by the following sequence of reactions, as shown scheme 6.3. The presence of carborane group allows these materials to

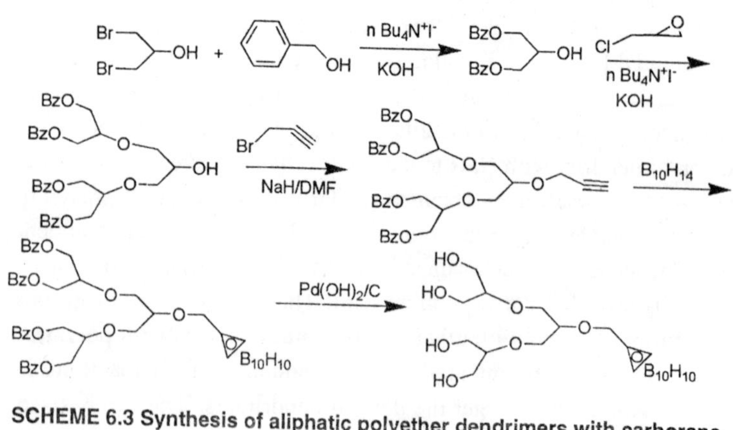

SCHEME 6.3 Synthesis of aliphatic polyether dendrimers with carborane moiety at the focal point

be used for neutron capture therapy and the problem of low solubility of carborane derivatives in aqueous solutions is elegantly overcome by coupling to a water soluble polyether dendrimer. Interestingly, attachment of even a small dendrimer to the carborane is sufficient to increase its water solubility to a level that is comparable to that of trichloroacetic acid and an order of magnitude greater than that of DL-valine.

The aliphatic polyether and polythioether dendrimers, which help to demonstrate dense packing phenomenon for the dendrimers after a certain generation growth were also synthesized [5]. In this divergent approach a tetra functional core, pentaerythritol (PE) tetrabromide and a compact AB_3 type building block, 4-(hydroxymethyl)-2,6,7-trioxalbicyclo-[2,2,2]octane (HTBO) were used. The potassium salt of HTBO was treated with PE-bromide in diglyme followed by deprotection through hydrolysis in methanolic hydrochloric acid to obtain the first generation dendrimer. For the subsequent higher generations (upto 3rd generation), the process was repeated successively. For the regeneration of terminal active bromide groups from deprotected hydroxyl groups it is necessary to treat first with excess TsCl in pyridine, followed by the reaction of tosylate derivative with sodium bromide in DMAc. The yields obtained in various steps are not very satisfactory, which were varying from 52-70%. The whole sequence of reactions is shown in scheme 6.4. As the generation number increases from 1 to 2 to 3, the number of terminal groups also increases from 12 to 36 to 108. Also the very short distance between the branch points (three covalent bonds) leads to a very compact structure, where the growth could not be continued beyond generation three. This may be due to extremely steric congestion in the vicinity of the chain ends.

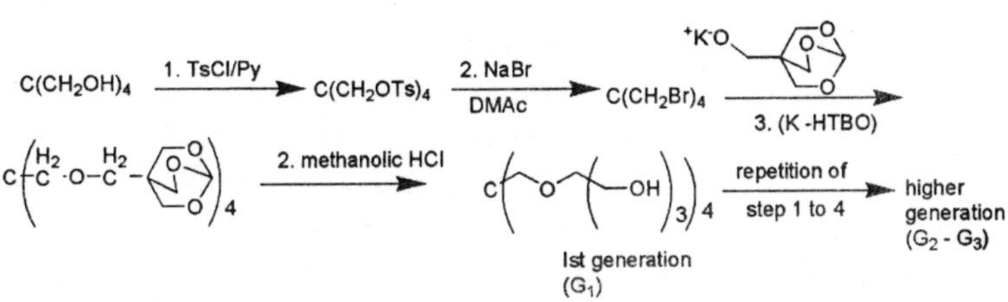

SCHEME 6.4 Synthesis of densely packed aliphatic polyether dendrimers

6.2.2 Aromatic

The convergent method was first demonstrated [6] for synthesis of a number of dendritic polyethers based on 3,5-dihydroxybenzyl alcohol as the monomer unit. The polyether dendrimers were synthesized [6-9] by repetitive alkylation of the monomer unit with either unsubstituted or substituted benzyl bromides followed by bromination i.e. activation of the benzylic alcoholic group to the corresponding bromide (Scheme 6.5). The number and the nature of substitutents (R), viz. single or a large number of cyano or carboxylic groups at the periphery can be placed by using unsubstituted or substituted with cyano or carboxylic group, etc. benzylic bromide in the first step of the growth. The final coupling of these wedges to a polyfunctional core such as 1,1,1-*tris*(4-hydroxyphenyl)ethane or 4,4'-dihydroxybiphenyl gives a series of dendritic polyethers (Scheme 6.5). The yield of the reactions in each steps was very high (~80%) and the products were easily purified by chromatographic technique. This reaction was carried out upto 6th generation to afford almost monodisperse (PDI = 1.03) macromolecule of molecular formula $C_{2687}H_{2304}O_{381}$ with nominal molecular weight 40,689 Da. There was no steric inhibition in this reaction, only a slight retardation and decrease in yield was observed for higher generations (>4th generation). The synthesized dendrimers were fully characterized by SEC, GPC, ^{1}HNMR and ^{13}CNMR spectroscopy and mass spectrometry. Later on, the authors [10-11] used extended core molecules such as bucky minister fullerene, and phenolic terminated dendrimers for synthesizing modified dendritic polyethers. The dendritic polyethers with carboxylate groups at the periphery, behave as unimolecular micelles and their ability to solvate hydrophobic molecule was investigated by UV-visible spectroscopy [9]. A linear relationship between the solubilizing ability and the concentration of the dendrimer was observed, which indicated that these micelles do not have a critical micelle concentration. The dendrimers with high dipole moments were achieved [12] by the placement of electron withdrawing cyano groups and electron donating benzyloxy groups at segmentally opposed regions of the chain ends of the dendrimers (Fig.6.1). The study showed that the dipole moments of the functionalized dendrimers were much larger than those for the symmetrical structures and increased with the increase of molecular weight, i.e. generation number. However, the variation of dipole moment with molecular weight is not linear, may be due to more globular structure of higher generations.

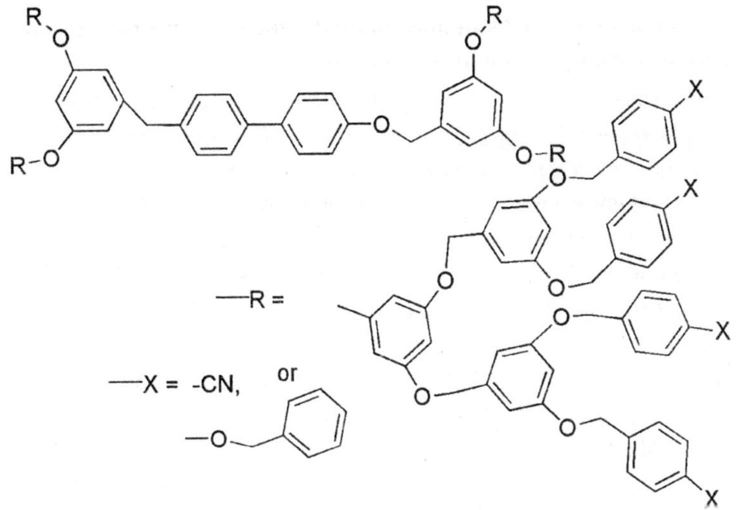

FIG.6.1 Structure of aromatic polyether dendrimer with electron withdrawing cyano and electron donating benzyloxy groups on the surface

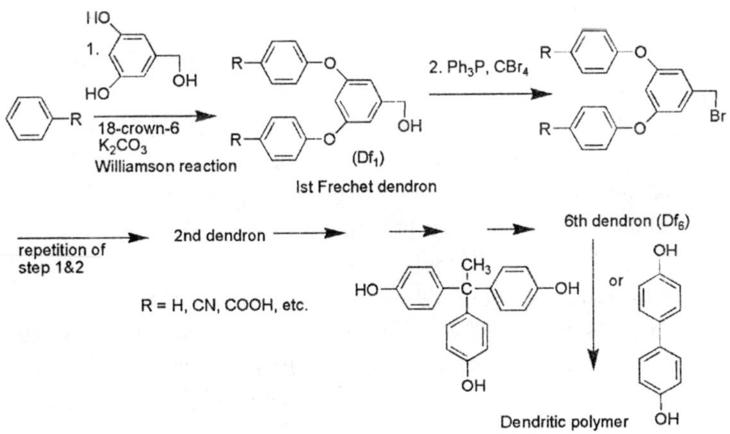

SCHEME 6.5 Synthesis of aliphatic polyether dendrimers based on 3,5-dihydroxybenzyl alcohol

The structures of a series of polybenzyl ether dendrimers of highly dilute solution were derived from SAXS experiments [13]. The study indicates a globular structure for the dendrimers. Electron distributions, calculated from the scattering data point towards an ellipsoidal shape with a rather uniform distribution of the electron density and hence the monomer subunits within the dendrimers.

Similar type of aromatic polyether dendrimers upto generation five with *p*-nitro-N-methylaniline at the focal point were also synthesized [14]. The dendritic branches would be modulated the physical properties at the dendritic core and this focal point is a solvatochromic probe, UV-visible absorption wave length of which is dependent on solvent polarity. Further the increase of generation number and size of the dendritic wedges, the UV-visible absorption band of the central chromophore in CCl_4 shifted to longer wavelengths. As the dendritic wedge expands, the solvatochromic probe is increasingly shielded from the bulk apolar solvent and the effective polarity in its surround is shifted. However, the effect was not observed in polar solvent like DMSO. The authors also pointed out a discontinuity in the absorption band shift between 3rd and 4th generation dendritic wedges, which may be due to the fact that at this point the structure become more globular and encapsulating the probe more effectively.

An alternative synthetic approach for the convergent synthesis of benzene core dendritic polyether via cyclotrimerization was reported [15]. In this approach, 2-butyne-1,4-diol reacts with bromine present at the focal point of different generations of aromatic polyether dendrons (upto 3rd generation) based on 3,5-dihydroxybenzylether by Williamson ether synthesis method. This acetylinic moiety was chosen for it synthetic capability and prevention of occurrence of a Claissen rearrangement reaction. Then the acetylenic moiety was trimerized in reflux toluene by using dicobalt octacarbonyl catalyst by (2+2+2) cycloaddition process to obtain the desired dendrimers. The sequence of reactions is shown in scheme 6.6. The highest genration dendrimer is one of the largest 3rd generation aromatic polyether dendrimer with molecular weight of 10^4 order. However, the trimerization yield decreases with the increase of generation number (80–36%). The dendrimers were fully characterized by NMR, SEC, and MALDI-TOF mass spectrometry. This approach benefits from the ease of purification of the trimerized product and may prove valuable in the highly convergent synthesis of hybrid dendrimers from asymmetric, i.e. different substituted alkyne precursors.

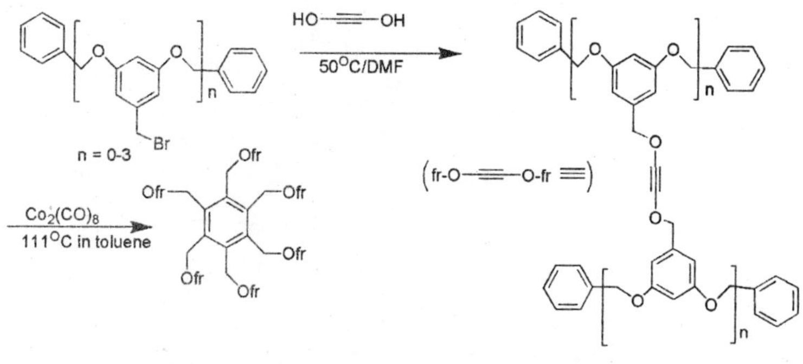

SCHEME 6.6 Synthesis of aromatic polyether dendrimers with benzene core via cyclotrimerization

The synthesis of isophthalate ester terminated dendrimers upto 4th generation based on 3,5-dihydroxybenzyl alcohol building blocks was reported [16] by the convergent approach. The convergent synthesis consists of the construction of diester terminated dendrons containing benzylic bromide functions at the focal point and their subsequent attachment to 4,4'-biphenol core, as shown in scheme 6.7. The dendrons upto 3rd generation were purified by recrystalization. Again, the 3rd and 4th generation bidendron dendrimers were surface functionalized by hydrolysis, ester exchange, and amide-ester interchange reactions. Further, the addition of benzyl alcohol or a first generation 3,5-di(benzyloxy)benzyl alcohol dendron to the terminal groups of the dendrimer increases the generation number by one or two by a double convergent growth approach. The authors [16] also used asymmetric dimethyl 4-(bromomethyl)phthalate as a starting material for synthesis of diester terminated dendrimers upto 4th generation by using the same reaction strategy. The diester-terminated bidendron dendrimers were subject to varieties of transformations at the end groups as shown in scheme 6.8. All the dendrimers and modified dendrimers were fully characterized by FTIR, ^{1}HNMR, ^{13}CNMR spectroscopy, elemental analysis, GPC analysis and MALDI-TOF mass spectrometry. The double growth process increases the molecular weight of the dendrimers from 5644 Da to 14422 Da with high yields.

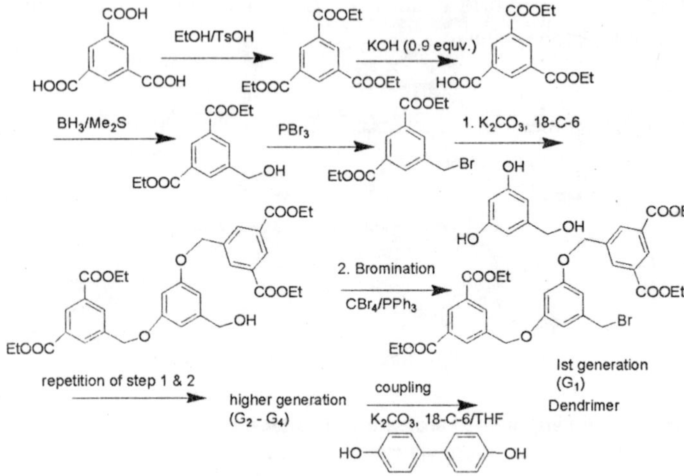

SCHEME 6.7 Synthesis of isophthalate ester terminated aromatic polyether dendrimers

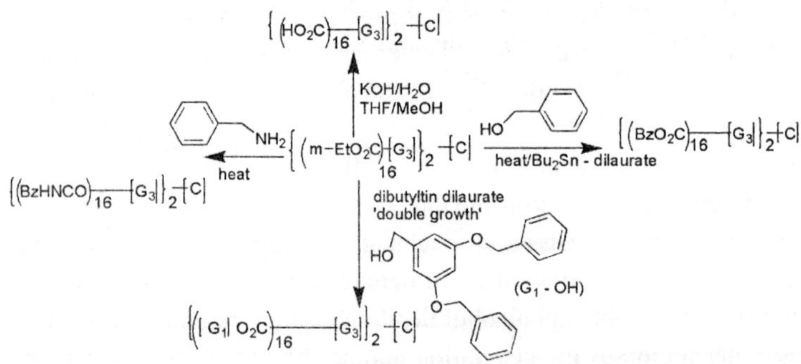

SCHEME 6.8 Surface functionalization of aromatic polyether dendrimers with various compounds

The syntheses of three 4th generation polyether containing macromonomer were reported [17] by three different techniques and their efficiencies were compared for Suzuki polycondensation reaction (SPCR). The SPCR of monomers offers the novel poly(*p*-phenylene)s equipped with 4th generation polyether dendron at every alternative repeating unit. Here the synthetic route was very long, but it was mainly used Williamson ether synthesis, catalytic reduction and bromination reaction as described [18] by Bo *et al.* (see amphiphilic polyether dendrimers) for different compounds. In route one, the synthesis of monomer 1, the Frechet dendron was easy to prepare but it contains benzylic hydroxyl group, which did not offer satisfactory yield in Williamson ether synthesis process and also the monomer is difficult to purify. In route two, for the synthesis of monomer 2, the dendron having reactive phenolic group but the 4th generation dendron has relatively large branching unit. Whereas in route 3, the synthetic process was simple and purification process was easy but the overall yield was low. The polymerization of the monomers was carried out separately by SPCR using exactly the same procedure as described ealier [18]. All the products were characterized by GPC and NMR techniques. The maximum molecular weight of the dendrimer was 639000 Da as obtained by GPC analysis.

A novel parallel monomer combination approach was described [19] to manipulate the dendritic architectural disposition of macromolecules. This technique was used for the synthesis of a family of poly(benzyl ether) dendrimers possessing 1-15 internal allyloxy groups, which are displayed in a highly controlled, layer specific, generational manner. This novel method harness the synthetic speed and power of the double-stage convergent growth approach as well as classical parallel synthetic process to synthesize diverse series of dendrimers that possess a predetermined number and arrangement of internal functional moieties.

The influence of molecular architecture and solvent type, both polar and nonpolar, on the size and structure of poly(benzyl ether) dendrimers was investigated [20] by SANS. The radius of gyration (Rg) of all dendrimers follows a scaling law of $RgM^{1/3}$ (M = molecular weight) consistent with literature values of other chemically different dendrimers. However, the effect of solvent type on size of dendrimers is not very significant.

Difunctionalized benzyl ether dendrimers upto 2nd generation were synthesized [21] from phen-acyl protected compounds by using Williamson ether synthesis and bromination techniques as shown in scheme 6.9. The synthesized dendrimers were characterized by ^{1}HNMR and ^{13}CNMR spectroscopy and MALDI-TOF mass spectrometry. The molecular weight of the dendrimers were found to be 1081 Da and 1929 Da for the 1st and 2nd generations respectively as measured by mass spectrometry.

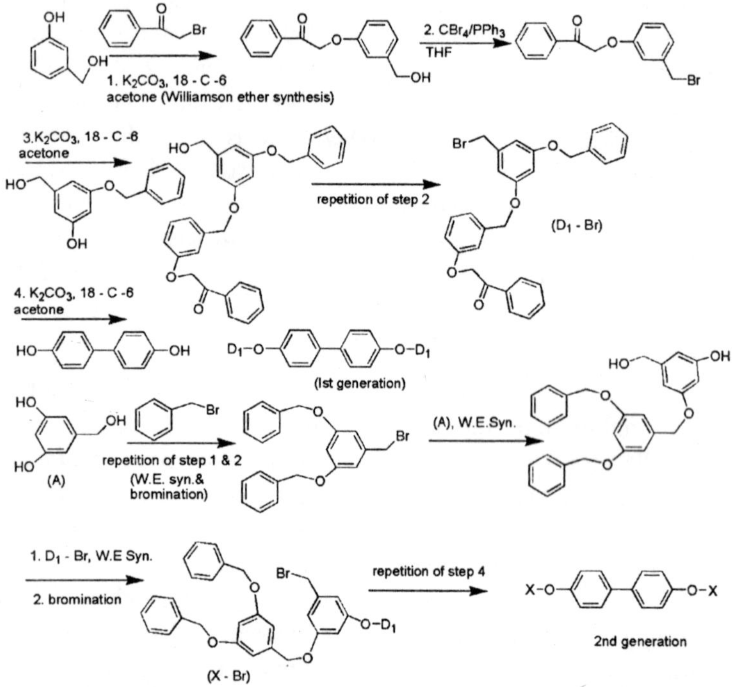

SCHEME 6.9 Synthesis of difunctionalized benzyl polyether dendrimers

Weintraub and Parquette convergently synthesized [22] unsymmetrically branched polyether dendrimer upto 4th generation by using 2,3-dihydroxy benzaldehyde as the branching unit. The propagation reaction of the dendrimer was carried out by O-alkylation of the phenolic hydroxy group and the focal point activation was achieved by a reduction and bromination sequence of reactions on the focal aldehyde group. The whole sequence of reactions is shown in scheme 6.10. The yield obtained in each step is very high. The synthesized dendrimers were characterized by ^{1}HNMR spectroscopy and GPC analysis. The molecular weight of the dendrons varies from 342 to 3574 Da. The theoretical calculation shows that this unsymmetrical branching pattern may force the terminal groups to coil around the focal point of the dendrimers. This coiling effect makes the dendrons more compact globular structure.

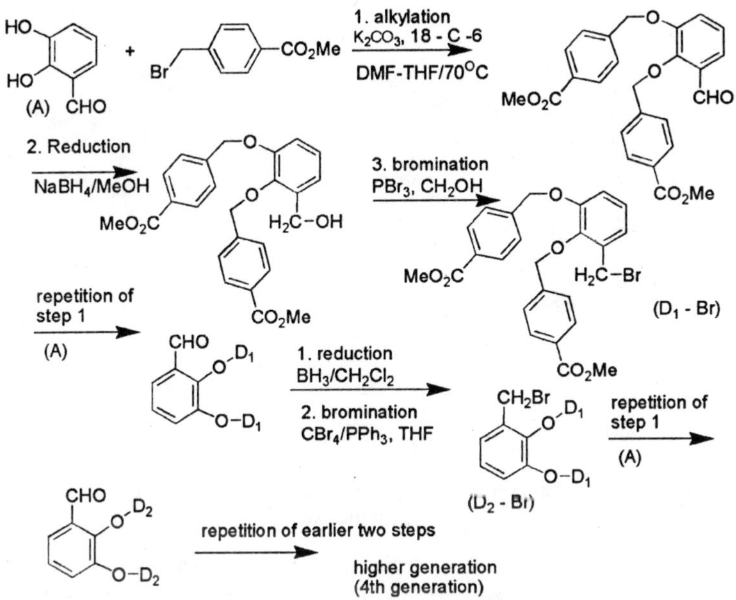

SCHEME 6.10 Synthesis of unsymmetrical aromatic polyether dendrimers

6.2.3 Poly(ether-urethane)s

The dendritic poly(ether-urethane)s upto 3rd generation were synthesized [23] by two step accelerated synthetic approach. In this process, two AB_2 type monomers viz. 3,5-dihydroxybenzyl alcohol and 3,5-diisocyanatobenzyl chloride were used. In the first step of the one-pot multistep approach, the preformed first generation alcohol formed by the reaction of benzyl bromide with 3,5-dihydroxybenzyl alcohol, reacted with 3,5-diisocyanatobenzyl chloride to generate 2nd generation *bis*-carbamate with benzyl chloride at the focal point. The 3rd generation can be obtained by direct treatment of 2nd generation reaction mixture with 3,5-dihydroxybenzyl alcohol, i.e. by Williamson ether synthesis approach. So, there is no need of purification or activation process in this approach. However, by using 3,5-dihydroxy benzoate, the side reactions can be avoided, which were observed in the previous reaction. The whole sequence of reaction is shown in scheme 6.11. The synthesized dendritic polymers were characterized by ^{1}HNMR and IR spectroscopic techniques.

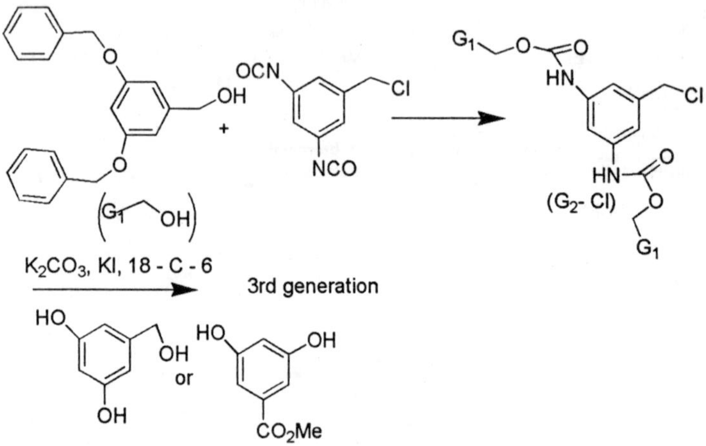

SCHEME 6.11 Synthesis of dendritic poly(ether-urethane)s based on benzyl building blocks

A series of side chain dendritic polymers was synthesized [24] by polymerization of dendritic diols of 2nd to 4th generation with four different semi-rigid and flexible diisocyanates. The polymerization reactions were conducted both in solution and in bulk. The sequence of reactions is shown in scheme 6.12, for the synthesis of the dendrimers. The synthesized polymers were characterized by ^{1}HNMR and ^{13}CNMR spectroscopy, SEC and intrinsic viscosity measurement. The aggregation behavior was studied in solution by interfacial tension measurement and self-assembly in bulk by XRD technique. The molecular

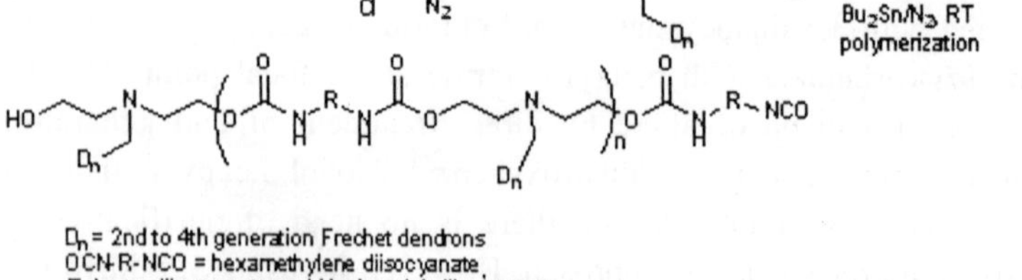

D_n = 2nd to 4th generation Frechet dendrons
OCN-R-NCO = hexamethylene diisocyanate,
Toluene diisocyanate and bisphenol-A diisocyanate

SCHEME 6.12 Synthesis of side chain dendritic poly(ether-urethane)s

weight of the polymers varies from 780 to 112000 Da with PDI 1-5.5. Also it has been found that the flexibility of diisocyanates and size of the dendrons are quite sensitive to the nature of the polymerization technique.

6.2.4 Liquid Crystalline (LC)

The synthesis and characterization of polyether dendrons upto 3rd generation were reported [25]. These monodendrons self-assembly into supramolecular dendrimers, which generate a novel thermotropic LC phase. In this synthetic process, polyhydroxy methylbenzoate was used as the AB_3 type building block. 3,4,5-*Tris*(n-dodecyloxy)benzoate, the monomer was reduced by $LiAlH_4$ to get the corresponding benzyl alcohol followed by activation of the benzyl group to chloride by treatment of $SOCl_2$ with catalytic amount of DMF to get the desired reactive building block for 1st generation. The treatment of this reactive compound with 3,4,5–*tris*(hydroxymethyl)benzoate (methyl-gallate) offered the 1st generation dendrimer. The repetition of the above steps generated the higher generation dendrimers. The yield of the product in each step was very high (~90%). The monodendrons with carboxylic acid group at the focal point was obtained by base hydrolysis followed by protonation with 50% aqueous acetic acid. The synthesized dendrons were characterized by using TLC, GPC, HPLC, ^{1}HNMR and ^{13}CNMR spectroscopy, elemental analysis and MALDI-TOF mass spectrometric analysis. The molecular weight of the 3rd dendron was 19205 Da with narrow polydispersity as measured by GPC analysis. The phase behavior of the monodendrons was analyzed by a combination of thermal OPM and DSC studies and found to display an enantiotropic cubic LC phase. The hemispherical shaped 3rd generation monodendrons with carboxylic acid group at the focal point form a novel thermotropic LC cubic phase. These supramolecular dendrimers generated by self-assembly of the monodendrons contain a poly(benzyl ether) core. The size of the supramolecular dendrimer and of the entire unit cell increases with the increase of generation. However, the average cone angle decreases due to large increase of number of repeat units relative to the number of terminal alkyl chains in the monodendrons.

The retrostructures of a series of benzyl ether based self-assembly monodendrons containing benzo-[15]-crown-5 at the focal point were synthesized and analyzed [26]. These crown ether containing monodendrons are self-assembled and subsequently self-organized into two dimensional hexagonal columnar, semectic B and semectic A, as well as in the three dimensional cubic lattices either in the absence of complexation with NaOTf or mediated by complexation with NaOTf. In other report [27], the authors synthesized and characterized by structural and retrosturctural analysis of three libraries of self-assembled benzyl ether monodendrons. From these libraries, a few were taken for further studies. The monodendrons were first and 2nd generations polyethers based on 3,4- or 3,5-disubstituted benzyl ether of AB_2 type or 3,4,5-trisubstituted AB_3 type monomers with n-dodecanyloxy groups on the periphery and carboxylic acid at the focal point. The desired monodendrons containing benzo-15-crown-5 ether at their focal point were synthesized with high yield by the esterification of the above monodendrons with 4'-hydroxymethylbenzo-15-crown-5 in the presence of DCC/DPTS in CH_2Cl_2 (Scheme 6.13). The

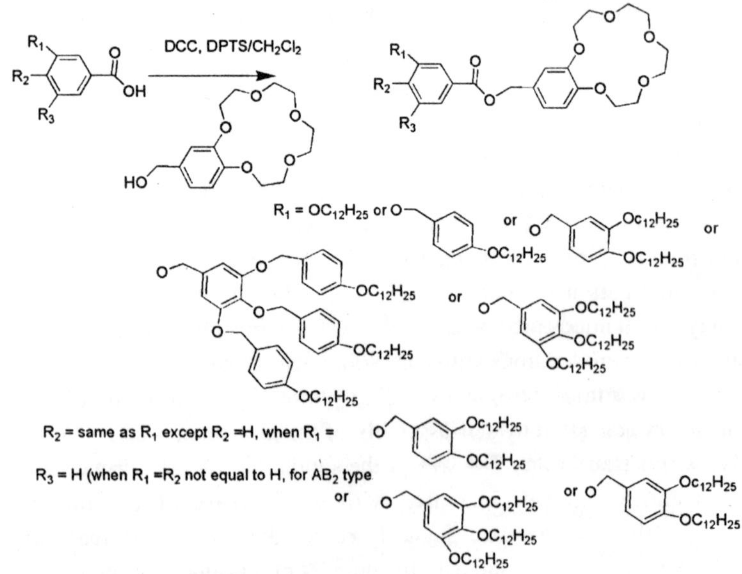

SCHEME 6.13 Synthesis of liquid crystalline polyether monodendrons contain benzo-15-crown-5 ether

products were purified by column chromatographic technique (silica gel) and recrystallization process. The structural characterization of the supramolecular dendrimers and the retrostructural analysis of the corresponding monodendritic crown ether were carried out by thermal OPM, DSC and XRD studies. From these studies, a very delicate dependence between the shape of the dendritic crown ether and the stability of its supramolecular lattice obtained by complexation with NaOTf was demonstrated and mechanisms of complexation mediated self-assembled were suggested. The results showed that the shapes of the supramolecular dendrimers and of their monodendritic building blocks are completely different; 3,4-dibenzyl ether based supramolecular are spherical, whereas 3,5-dibenzyl ether based one is cylindrical.

The polyether dendrimers based on $HO\text{-}C_6H_4C_6H_4C_6H_4CH(CH_2)_{11}$ (Br) $CH_2C_6H_4OH$ monomers were also synthesized [28-29] by using Frechet iterative convergent strategy. Mono- and tri-dendrons upto 4th generation were prepared with 10-undecenyl groups on the periphery and their properties were investigated by using DSC, POM and XRD studies. In contrast to the HB (see LC HB polymer section) polymers, these dendrimers showed sharp phase transition peaks and a low degree of supercooling state in DSC studies. All the dendrimers, (except first generation, which was shown to be monotropic), displayed an enantiotropic nematic mesophase and even a smectic mesophase with increasing phase transition temperature as the generation number increased. The highest generation dendrimer with molecular weight 30,863 Da organized into a nematic texture with a rate comparable to that found with low generation dendrimer, indicating a relatively low viscosity. This nematic phase formation, instead of cubic or discotic phase for all dendrimers suggests that these dendrimers do not exhibit a spherical or disc shape. The structure of the liquid crystal dendrimers (4th generation) suggested that all repeating units adopt a anti conformation instead of gauche conformation so that the overall shape is rod like with a impact structure that minimizes free volume.

The synthesis, characterization and property evaluation of a number of phthalocyanine containing poly(aryl ether) dendrimers upto 3rd generation were reported [30]. Poly(aryl ether) dendrons of 1st to 3rd generation were attached to the phthalocyanine core. The resulted properties of the dendrimers are dominated both by the columnar self-association of the phthalocyanine core and by the glass forming character of the substituent dendrons. The asymmetric phthalocyanines containing a single poly(aryl ether) dendron displayed a columnar mesophase, the structure of which can be frozen into an anisotropic glass at room temperature. By placing the dendrons at the axial sites of silicon phthalocyanine prohibits self-association and gives materials, which can be fabricated robust, isotropic solid solutions of phthalocyanine with high T_g. A single crystal XRD study of one of these dendrimers illustrates the ability of the axial dendrons to prevent cofacial aggregation in the solid state.

6.2.5 Redoxactive

The dendritic polyethers upto 3rd generation with *bis*(terpyridine)iron(II) complex (Fig.6.2) core were synthesized [31] by attachment of dendrimer with the preformed metal complex. The dendritic complexes were characterized by NMR, UV spectroscopy and combustion analysis as well as by the electrochemical technique. The cyclic voltammetry study showed reversible redox chemistry as the generation number increased. The observation may be rationalized by the stabilization of redox products due to an ineffective electron transfer from the electrode surface to the dendritic shell. A series of chiral terpyridines containing Frechet type dendrons upto 3rd generation was reported [32]. The dendrimers were synthesized by coupling of the corresponding dendritic benzylbromide with 4'-[6-(2,2'-dihydroxy-1,1'-binaphthyl)]-2,2',6',2''-terpyridine. These chiral dendrimers were utilized to prepare high molecular weight hybrid metal-organic dendrimers based on the Ru(II)-*bis*(terpy) linkage (Fig.6.3). The synthesized hybrid dendritic complexes were characterized by ^{1}HNMR, ^{13}CNMR, COSY, ESI-MS and MALDI-TOF mass spectrometry. The UV-visible spectra showed a broad band at around 496 nm may be due to metal to ligand charge transfer excitation. The complexes exhibited reversible Ru oxidation process half-wave potential 1.24 to 1.29 in CH_2Cl_2 solution. Further more, the study showed that although the dendrimer-terpyridine ligands exhibited high luminescent at ~430 nm on irradiation at 330 nm light. The chiral metallodendrimers are non-emissive or very weak emissive in solution at room temperature. The chiral metallodendrimer also exhibited circular dichroism signal similar to chiral terpyridine-dendritic ligands, indicating no extra chirality generation upon the formation of metallodendrimers.

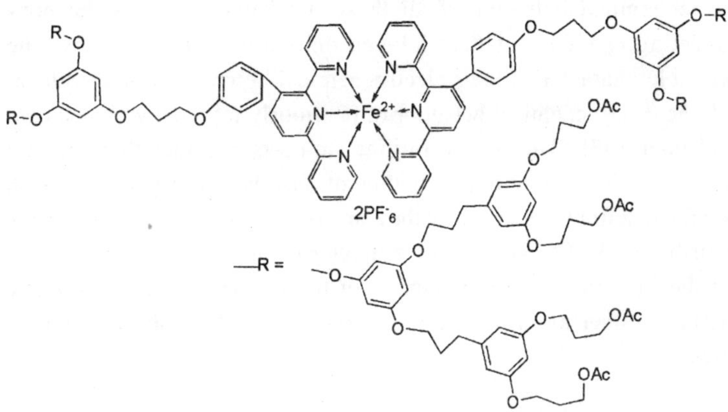

FIG.6.2 Structure of redoxactive dendritic polyethers with *bis*(terpyridine) iron(II) complex core

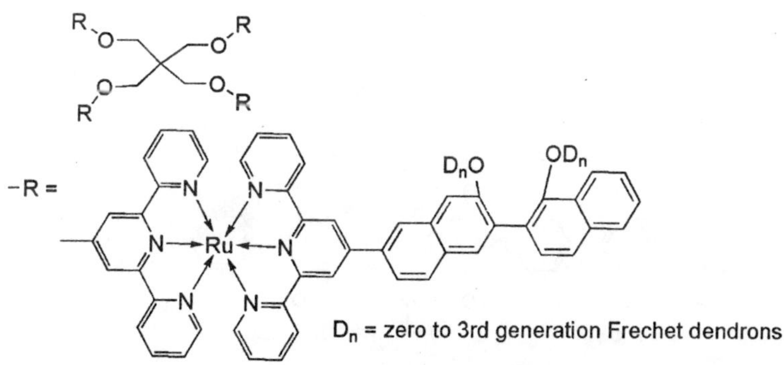

FIG.6.3 Structure of hybrid metal-organic polyether dendrimer based on Ru(II) *bis*(tepy) linkage

A new series of Frechet type dendrimers that contain four central viologen units (4,4'-bipyridinium) covalently attached to a cavitant core was synthesized [33-34] by the convergent method. The cavitant precursor containing four bipyridinium subunits (Fig.6.4) was synthesized by reported procedure [35]. By applying Menschutkin reaction between the previous cavitant subunit and Frechet benzylbromide dendrons [36] (1st to 3rd generation) in DMF at 55°C and after counter ion exchange the dendritic cavitants with the tetrahedral core resulted. The desired dendrimers were then characterized by ^{1}HNMR, ^{13}CNMR spectroscopy, mass spectrometry and elemental analysis. The electrochemical behavior of all these dendrimers reveals the non-interacting character of the viologen units around the cavitant. The reversible character of their electrochemical behavior is in agreement with the electrochemical behavior of previously reported viologen core dendrimer [37]. This study further supports the fact that internal viologen sub-units remain capable of fast heterogeneous electron transfer reactions, even when they are surrounded by a considerable dendritic mass. The voltametric behavior of these compounds reveals that the first one electron reduction of the viologen core is reversible (fast) and rather insensitive to the size of the dendrons through out the series.

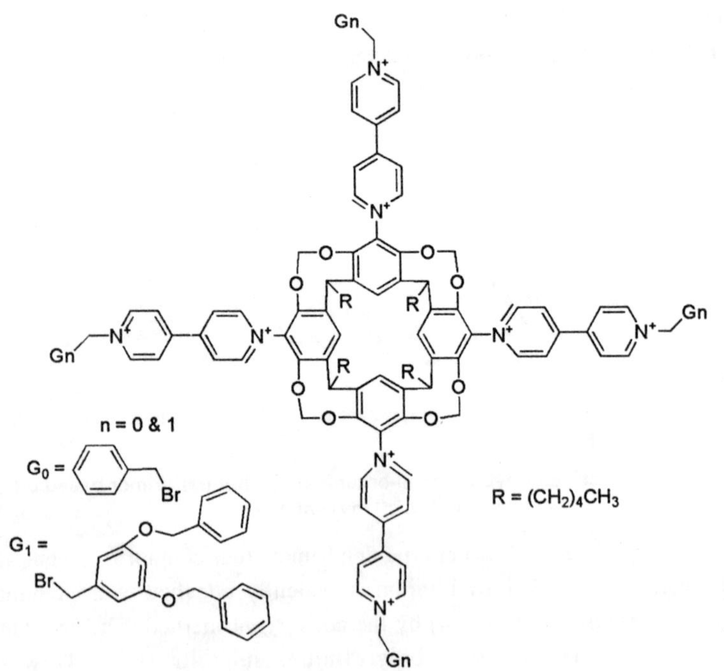

FIG.6.4 Structure of viologen (4,4'-bipyridinium) containing cavitant precursor

The synthesis, characterization and electrochemical properties of a family of polyether dendrimers comprising a 4,4'-bipyridinium core and 1,3-dimethyleneoxy benzene-type dendrons were reported [38]. The dendrimers upto 3rd generation were prepared by using Frechet type of dendron [36] with 4,4'-dibenzylbipyridinium (benzyl viologen) in acetonitrile. The products with bromide counter ion were isolated and converted into PF_6^- salts by ammonium hexafluorophosphate in water free acetonitrile. The electrochemical results showed that there is no significant inhibition of electron transfer involving the dendrimer core by the dendrons. The strong fluorescence of the 1,3-dimethyleneoxy benzene units of the branches is completely quenched as a result of charge transfer interactions with the core. The 4,4'-bipyridinium unit undergoes two distinct and fast one electron reduction process at easily accessible negative potentials, regardless of dendrimer generation, indicating that the site isolation effect on the dendrimer core by the dendrons is not too strong. The bipyridinium unit is reduced at less negative potential in the 3rd generation dendrimer, due to back folding of the dendrons toward the core.

A novel polyether based on 4,4-bis(4-hydroxy phenyl)pentanol dendritic iron-sulfur clusters (Fig.6.5) upto 4th generation having focal arm in tetrahedral arrangement was synthesized [39-41]. As the dendritic generation increased, the core reduction potential in DMF solution became increasingly negative, an effect attributed to the increase of insulating effect of the dendritic shell. The heterogeneous electron transfer rates were obtained by Osteryoung square wave

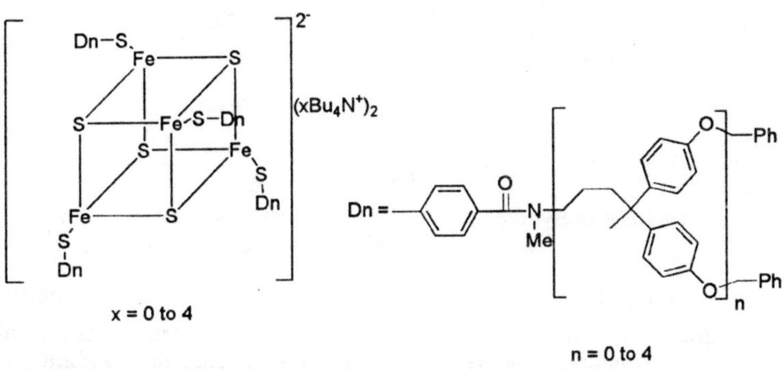

FIG.6.5 Structure of four arms tetrahedral polyether dendritic Fe-S-clusters

voltametric studies [39-41]. The results from conformational searching computations on this flexible architecture suggested that this architecture permitted substantial mobility of the core throughout the structure. These examples clearly demonstrated that the dendritic architecture can modulate physical properties at the core and this effect can be considered as the creation of a dendritic microenvironment.

Thin layer cyclic voltammetry (TLCV) was used as an efficient tool to determine the redox characteristics of polyether dendrimers with an electroactive *bis*(phenanthroline) Cu(I) core [42]. The phenanthroline bound polyether dendron was first prepared from 2,9-dimethyl-1,10-phenanthroline by the following sequence of reactions as shown in scheme 6.14. Finally, the treatment of this resulted dendritic compound with $Cu(CH_3CN)_4BF_4$ in CH_2Cl_2 offered the desired dendritic Cu(I) complexes (Scheme 6.14). The synthesized dendrimers were characterized by ^{1}HNMR spectroscopy, FAB-mass spectrometry and elemental analysis. The UV absorption spectra of the complexes exhibited the intense π-π^* ligand-centered bands and much weaker and broad MLCT bands with a maximum at 456 nm. It also shown a weak electroactivity for the dendritic complexes by TLCV experiment [42].

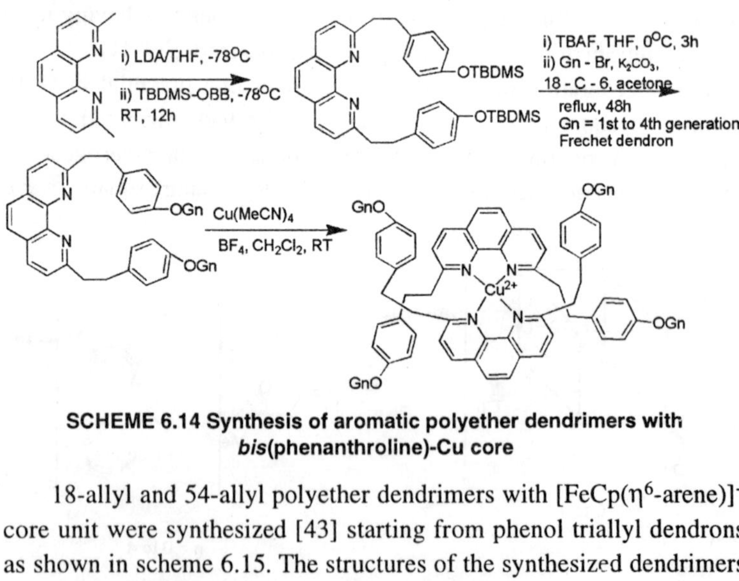

SCHEME 6.14 Synthesis of aromatic polyether dendrimers with *bis*(phenanthroline)-Cu core

18-allyl and 54-allyl polyether dendrimers with [FeCp(η^6-arene)]$^+$ core unit were synthesized [43] starting from phenol triallyl dendrons as shown in scheme 6.15. The structures of the synthesized dendrimers were characterized by ^{1}HNMR and ^{13}CNMR spectroscopy and elemental analysis. The molecular weight of the dendrimers varies from 2061 to 6876 Da as determined by MALDI-TOF mass spectrometry. The cyclic voltammetry was used to study the hetergeneous electron transfer between the metal center of the dendrimers and the electrode. The study showed that the rate of electron transfer decreased with increase of size of the dendrimers, whereas the half-wave potential remain almost constant.

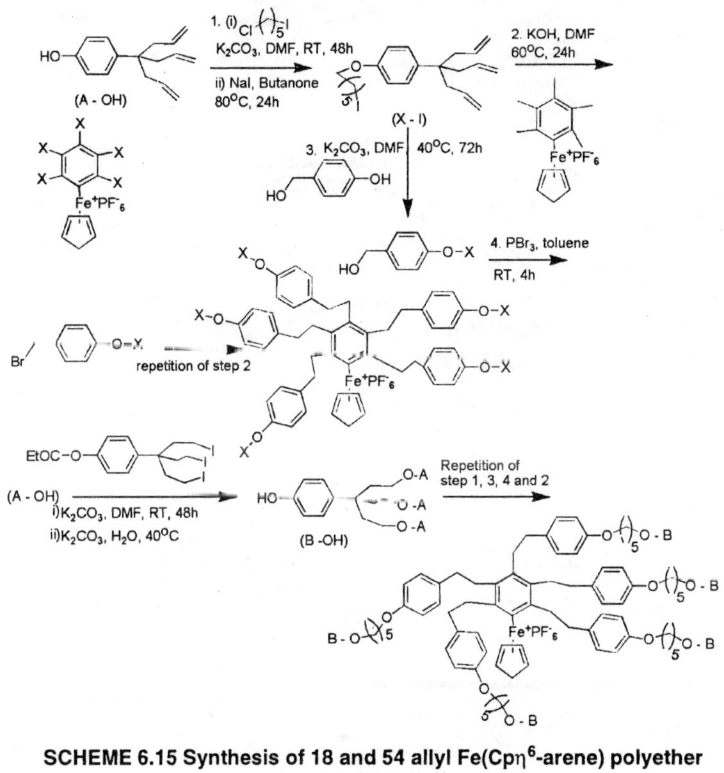

SCHEME 6.15 Synthesis of 18 and 54 allyl Fe(Cpη^6-arene) polyether dendrimers

6.2.6 Amphiphilic

The synthesis of two novel amphiphilic poly(*p*-phenylene)s polyether dendrons bearing both a pendent and a flexible chain at each repeat unit was reported [18]. The total synthetic procedure is very long and critical, so it is divided into five different steps: the synthesis of the (i) unsymmetrically substituted dibromobenzene, as the core part for the final monomer (Scheme 6.16 A), (ii) hydrophilic second generation polyether dendron (Scheme 6.16 B), (iii) final monomer by assembling above two parts with a hydrophobic chain attached with hydroquinone

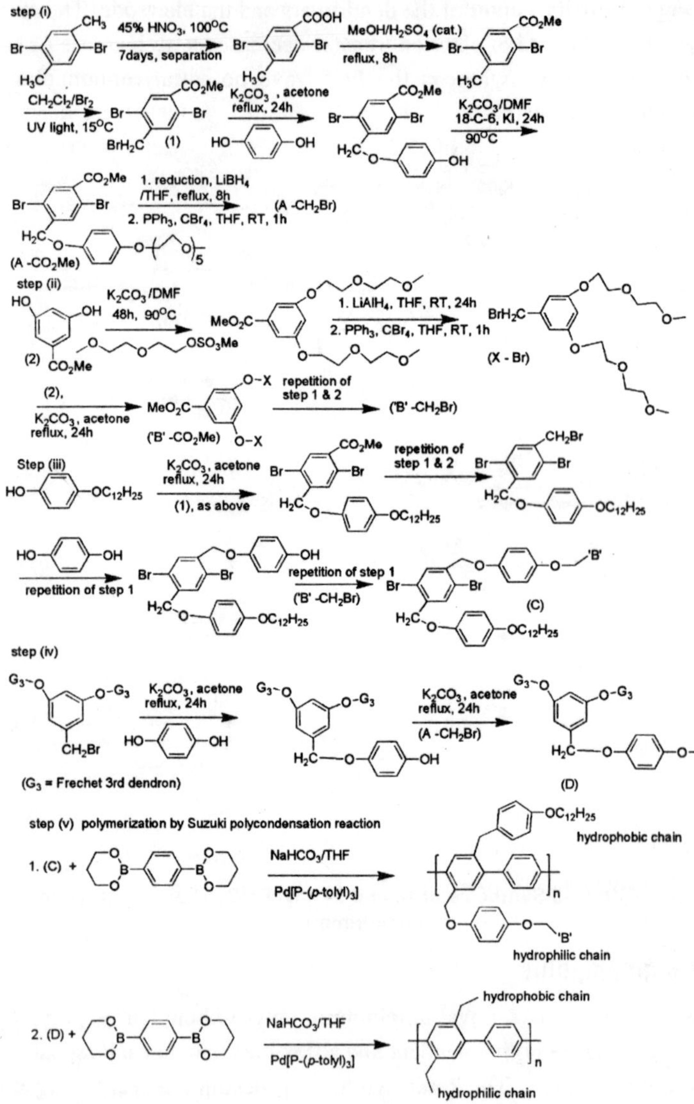

SCHEME 6.16 Synthesis of poly(*p*-phenylene)s bearing amphiphilic polyether dendrons

linker (Scheme 6.16 C), (iv) second monomer by assembling modified unsymmetrical oligoethylenoxy chain containing dibromobenzene and 3rd generation Frechet type of polyether dendrons [36] (Scheme 6.16 D) and (v) finally polymerization of both the monomers with benzene diboronic acid ester by Suzuki polycondensation reaction [44] to get the desired amphiphilic dendrimers (Scheme 6.16 E). The yield obtained in each step was relatively high (60-99%) and the products were purified by chromatographic technique. The synthesized dendrimers were characterized by NMR, SEC, GPC and DSC studies. The T_g for the first generation dendrimer is 130^oC as measured by DSC analysis. Also this amphiphilic dendrimers form stable mono-layers at the air/water interface, which makes it reasonable to assume that it segregates length wise into homophilic domains with the short ethylenoxy chains pointing into the subphase.

The synthesis of amphiphilic hydrogels with highly shape persistent crosslinked junctions was described [45] by using linear blocks such as poly(ethylene glycol), (PEG) and perfectly branched polyether dendrons upto 3rd generation. The synthetic strategy was based on the reaction of PEG with isocyanate or epoxy end groups as the hydrophilic component and hydrophobic dendritic poly(benzyl ether)s with amino groups at the periphery. At first the Frechet type of poly(benzyl ether)s with cyanide terminal groups were synthesized by the convergent method using 3,5-dihydroxy benzyl alcohol as the starting material and 1,1,1-*tris*(4-hydroxyphenyl)ethane as the core. The reduction of CN terminated groups to CH_2NH_2 groups were carried out by $LiAlH_4$. Then the hydrogel was prepared by the interaction of PEG diisocyanate with amino terminated dendrimers. This coupling reaction was done by using two different ratios of the reactive groups, 1:2 corresponding to full crosslinked or 50% crosslinked respectively. This coupling reaction was performed at room temperature in the presence of catalytic amount of triethyl amine. PEG-*bis*(epoxide) were also used to prepare the hydrogel with dendrimers. The synthesized dendrimer and hydrogel were characterized by FTIR and ^{1}HNMR spectroscopic studies. These amphiphilic hydrogels with either isocyanate or epoxide functional groups at both ends formed at faster rate because of better reactivity of the functional groups. They formed loose networks containing highly cross-linked domains weakly connected to one other. The PEG-*bis*(epoxide) formed densely crosslinked structures. Although the studies of swelling behavior and solid state properties of the hydrogels were able to distinguish the extent of crosslink, surface to surface linkage of the multifunctional tridendrons with PEG segment or catenane formation by PEG loops formed on the individual dendritic fragment. The results showed that in the case of PEG-*bis*(epoxide) the physical entanglement plays a little role in the network formation, while in isocyanate gels this mode of crosslinking was definitely present. Further more, the properties like swelling behavior, thermal characteristics, etc. of the hydrogels were influenced by the mode of preparation, degree of crosslinking, nature of the anchoring groups at the surface of the dendrons, the polarity of the medium and the temperature.

6.2.7 Photoactive

A novel poly(aryl ether) dendrimers upto 3rd generation was functionalized [46] with laser dyes both at the periphery and the core by the convergent approach. The targeted dendrimers were functionalized with the laser dyes coumarin 2 (at periphery) and coumarin 343 (at core), so that the energy can transfer from the former to the latter, i.e. the functionalized dendrimers act as light harvesting materials. Similar dye labeled poly(aryl ether) dendrimers as energy transferring materials were also synthesized [47-48] and it was observed that for dendritic antennae molecules, the nucleophilic amino group of coumarin 2 should be on the periphery and the elctrophilic acid group of cumarin 343 should be at the focal point. So, in this study [46] the reverse synthetic strategy was used. In reverse strategy the monomer is protected 3,5-*bis*(bromomethyl)phenol. This monomer was synthesized from 3,5-dimethylanisole and subsequently attached to the coumarin dyes. The whole synthetic process is critical and cumbersome, as given in scheme 6.17 in different steps. The hexa decanesulfonate-protecting group was found to be extremely useful due to solubility, crystalline behavior and ease of purification upto 3rd generation dendrimer. The accelerated synthesis of the laser dye labeled dendrimers was carried out by using compound 'C', as shown in steps 4 and 5 of scheme 6.17. The peculiarity of this hypermonomer 'C' is its extremely high solubility in common organic solvent because of the presence of five lipophilic groups. The products were purified by column chromatography and characterized by FTIR, ^{1}HNMR, ^{13}CNMR spectroscopy,

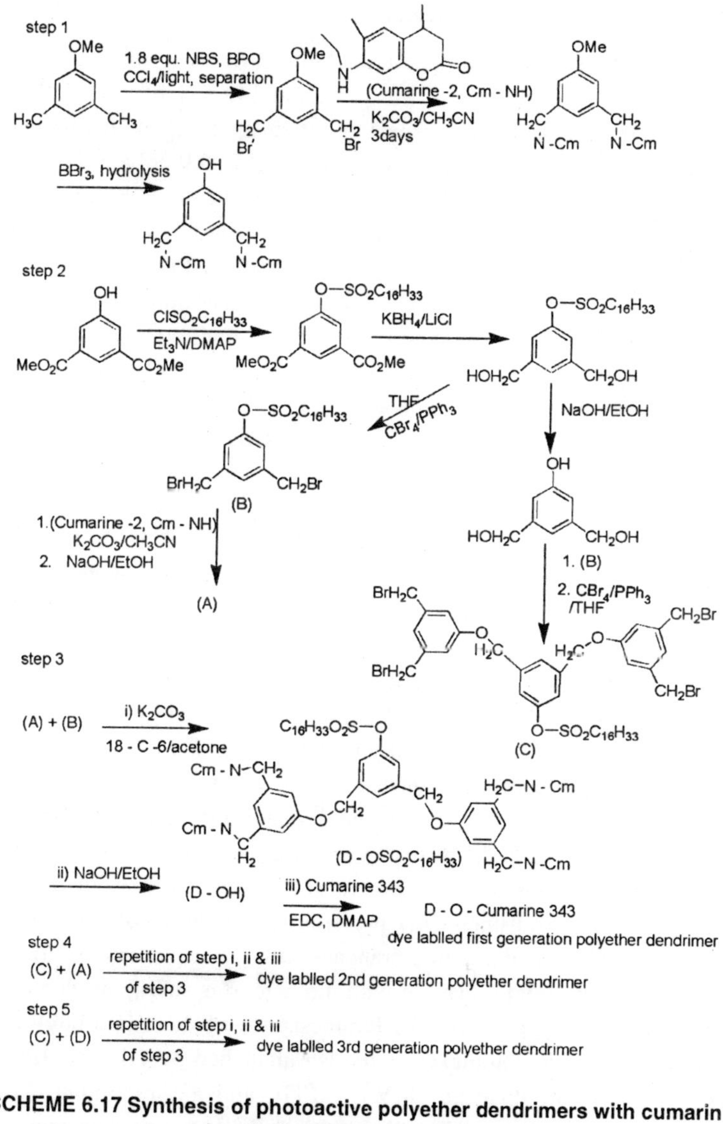

SCHEME 6.17 Synthesis of photoactive polyether dendrimers with cumarin dyes both at core and periphery

MALDI-TOF mass spectrometry and elemental analysis. The preliminary steady state fluorescence measurements indicated the energy transfer from periphery donor cumarin moieties to the core acceptor cumarin is highly efficient (97-90%).

The synthesis of a 4th generation tridendron with 48 ruthenium complexes on the periphery was reported [49-51] by using 3,5-dihydroxy benzyl alcohol as a repeat unit in an approach similar to that of Frechet. Balzane and his coworkers reported [52-53] poly(aryl ether) dendrimers containing $[Ru(bpy)_3]^{2+}$ type of core unit. Since 2,2'-bipyridine can be easily functionalized at 4,4'-positions, so a variety of polyether dendrons were attached in these positions to get the desired $[Ru(bpy)_3]^{2+}$ dendritic complexes. All of these dendrimers exhibit the characteristic luminescence, both in fluid solution at room temperature and in a rigid matrix at -196^{0}C. A small displacements in the wave length of the emission maximum was observed, which may be due to the electronic nature of the groups directly linked to the 'bpy' ligands in the 4,4'-positions and a decrease in the excited state life time in shifting from deaerated to aerated solutions caused by the quenching of the luminescent excited state by dioxygen. The lifetime of the ^{3}MLCT in aerated solution increases with increase of size of the branches. The lifetime of the largest dendrimer was three times longer than that of the smallest one. A long life time of the luminescent excited state in aerated solution is important for immunoassay applications, since the signal of the label can be read after the decay of the back ground fluorescence of the sample, whose life time usually in the nanosecond time scale. The protection of the dendrimer branches on the luminescent ^{3}MLCT level of the Ru(II)-bpy core can be assigned to one or more of the following factors: (i) decrease in the diffusion rate constant with the increase of volume of the compound, (ii) lower solubility of dioxygen in the interior of the dendrimer, (iii) preferential solvation of the metal complex core by the dendrimer branches with consequent difficulty to obtain a suitable orbital overlap with dioxygen for energy or electron transfer quenching process. The luminescence of these $[Ru(bpy)_3]^{2+}$ polyether dendritic complexes were also quenched by 1,1'-dimethyl-4,4'-bipyridinium dication (MV^{2+}), TTF, and anthraquinone-2,6-disulfonate anion (AQ^{2-}). These three quenchers do not possess excited states below the ^{3}MLCT level of the $[Ru(bpy)_3]^{2+}$ chromophoric unit, but are easy to reduce (MV^{2+}, AQ^{2-}) or oxidize (TTF).

In Ru-bpy-dendritic complexes with 2-naphthyl type chromophoric units [52], the inter-chromophoric interactions are weak and the absorption spectra of these type dendritic complexes are substantially equal to the summation of the spectra of the chromophoric groups present in the structure. The strong absorbence at near UV region of the naphthyl group and the strong emission of the core resulted a strong visible emission upon UV excitation even in very dilute solution (10^{-7} M) of the dendrimeric complex [52].

The photophysical properties of polyether dendrimers based on Zn-porphyrin core have been investigated by many authors [54-56]. The absorption spectra of the Zn-porphyrin based dendrimers (upto 3rd generation) exhibited a small red shift of the Soret band upon the increase of generation, which was attributed to an interaction between the dendrons and the core [54]. However, no shift in absorbance band was observed for tetraphenyl porphyrin based dendrimers. This may be due to the fact that the additional phenyl groups increases the distance between the branches and the porphyrin moiety. The volume measurements also indicated that the extra phenyl spacers present in these dendrimers allow more flexibility to the structure.

The synthesis of porphyrin core poly(benzyl ether) dendrimer was reported [57] by two different convergent methods. The first method involves the direct condensation of dendrons having 3,5-disubstituted benzaldehyde focal points with an equivalent amount of pyrrole. This approach requires very mild conditions, which is especially useful for the rapid assembly of small dendrimers but suffers from steric limitations as the size of the dendrons increased above the 4th generation. However, the second method requires the attachment of preformed benzyl bromide dendrons to a functionalized porphyrin through a simple Williamson ether synthesis technique. This route is also very practical but it requires careful purification of the final product. MALDI mass spectrometry proved to be a very useful tool in this case, both for monitoring the formation of the dendritic porphyrins and for their characterization.

The synthesis of anthracene core polyether dendrimers upto 3rd generation was reported [58]. The nucleophilic substitution reaction of 9,10-*bis*(halo[Cl/Br]methyl)anthracene with 3,5-dialkoxy substituted 1st to 3rd generation Frechet type benzylic alcohols [6] in an alkaline medium under phase transfer conditions had provided yields 66 to 30% for the desired dendrimers (Scheme 6.18). The synthesized dendrimers were characterized by ^{1}HNMR and ^{13}CNMR spectroscopies. The 1st and 2nd generation dendrimers were undergone dimerization on exposure to UV radiation. Irradiation in the near-UV region wave length ($\lambda \geq 300$ nm) of an approximately 10^{-3} M solution of the dendrimers in benzene offered quantitative dimerization. However, the products obtained were not the same type as was obtained by dimerization of anthracene, as indicated by ^{1}HNMR and ^{13}CNMR spectroscopic studies. The dimerization occurs through the benzene rings of dendrimers attached with anthracene core. The reaction was reversible by irradiation at shorter wavelength (254 nm) or heating above 100°C.

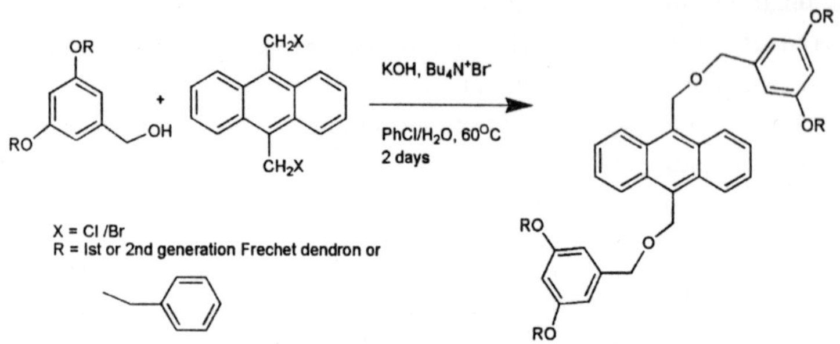

SCHEME 6.18 Synthesis of polyether dendrimers with anthracene core

The synthesis, characterization and photoluminescence properties of anthracene based polyether dendrimer and dendron were reported [59]. This polyaromatic dendrimer composed of six anthracene groups, which showed a monomeric fluorescence feature with a diminished quantum yield. This is due to intramolecular self-quenching process in the dendritic framework. The dendrimer was synthesized from anthrylmethyl chloride and methyl 3,5-dihydroxybenzoate, as shown in scheme 6.19, using the convergent method. The dendrimer and dendron were characterized by IR, ^{1}HNMR, ^{13}CNMR spectroscopies, FAB-mass spectrometry and elemental analysis.

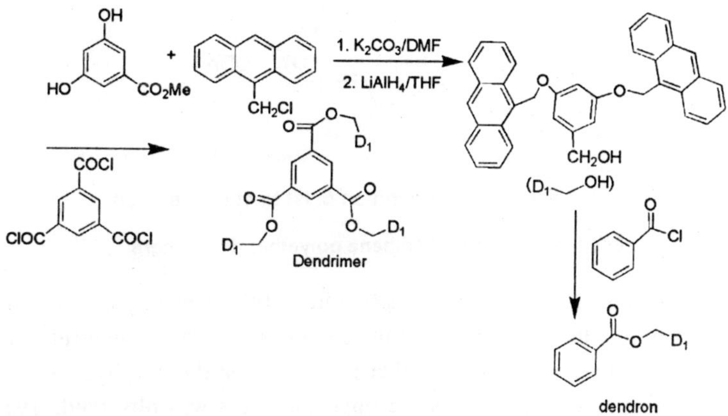

SCHEME 6.19 Synthesis of anthracene based polyether dendrimers and dendrons

A 3rd generation photoresponsive water-soluble stilbene polyether dendrimer undergoes unusual one way *trans* to *cis* isomerization to give 100% of cis isomer at the photostationary state on UV-irradiation in water (2 X 10^{-3}M KOH aqueous solution) at room temperature [60]. The initial absorbance of the *trans* dendrimer is 0.44 at 330 nm, which decreased with irradiation time to give a constant value of 0.08. This result indicated that the *cis* isomer composition at the photostationary state to be as high as 100% on irradiation at 330 nm.

A series of stilbene dendrimers upto 4th generation was synthesized [61] by the Frechet technique [36] and characterized by NMR and mass spectrometric studies. The molecular weight of the highest generation dendrimer was 6548 Da. All dendrimers (Fig.6.6) exhibited photoisomerization from *trans* to *cis* form on irradiation at 330 nm light. The dendrimer exhibited absorption maximum of 310 nm in THF solution is due to stilbene moiety. However, the absorbance at 280 nm increased with the increase of the generation number. The quantum yield of fluorescence emission was same for 1st to 3rd generation (0.6), while for 4th generation, it was 0.71 with emission maximum at 400 nm. This is a first study of photoisomerization of the C=C bond in molecular weight over 6000 Da can take place efficiently within 10 ns time scale.

Polyether dendrimers upto 3rd generation with 3,5-dihydroxybenzene terminated dendrons and Zn-porphyrin core unit (Fig.6.7a) were

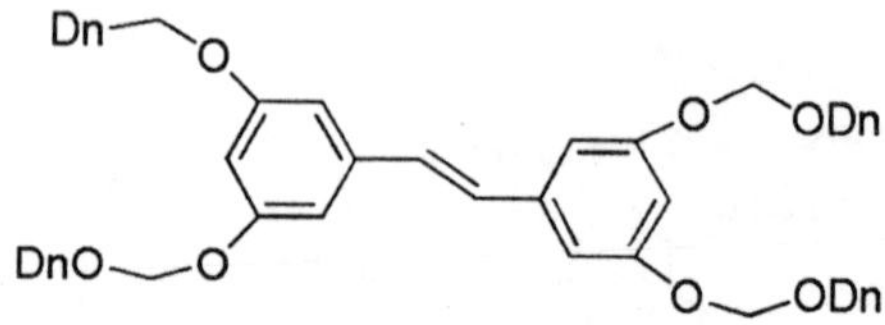

Dn = Frechet dendron of Ist to 4th generation

FIG.6.6 Structure of stilbene polyether dendrimers

synthesized [62]. The dendritic shell reduces the quenching of fluorescence with the increase of size of the dendrons caused by the influence of large electron acceptor quencher molecules on the porphyrin core, Whereas for small quenchers the opposite effect was observed. The decrease of quenching effect in this porphyrin dendrimer may be due to the globular shape filling properties at higher generation and the core shielding induced by it. The enhanced quenching of fluorescence varies with growing dendron generation containing smaller quenchers [63]. In this case, the small quencher molecules may be captured by attractive interactions with the polyether dendrons of the porphyrin dendrimer (Fig. 6.7b). Aida and coworkers [55-56] have synthesized different types of polyether dendrimers upto 3rd generation with porphyrin core unit (Fig.6.7c). A negatively charged water soluble dendrimer [55] Zn-porphyrin with 32 carboxylate ion functionalities on the exterior surface and a positively charged dendrimer [56] Zn-porphyrin with 32 ammonium ion functionalities on its periphery, were synthesized. In the negatively charged dendrimer, when positively charged quenchers were added, they electrostatically bound to the dendrimer surface. Again after the core had been photochemically excited, a long range electron transfer from the quencher to the chromophore can occur, which leads to a quenching of fluorescence. But, if negatively charged quenchers were added, no quenching occurs with higher generation dendrons, since the porphyrin core was effectively shielded by the dendritic shell with like charges [55]. The electrostatic assembly between negatively charged dendrimer free base porphyrin with positively charged Zn-porphyrin dendrimer in phosphate buffer at 20°C showed that transmittance is highly dependent upon the molar ratio of the two dendrimer electrolytes and it dropped sharply at equal concentration of two. The fluorescence of negatively and positively charged

Zn-porphyrin dendrimers at excited state are strong and homogeneous. But in an equimolar ratio of the two, imaged dark and large nanosized fluorescent aggregates were observed. The resulting photochemical excitation enables an intramolecular singlet energy transfer (rate 3 X10^9 s^{-1}) from one dendrimer to an adjacent one. An intramolecular singlet energy transfer for porphyrin dendrimers (Fig.6.7a) shows that the energy transfer is efficient in a globular and dense dendritic shell [64].

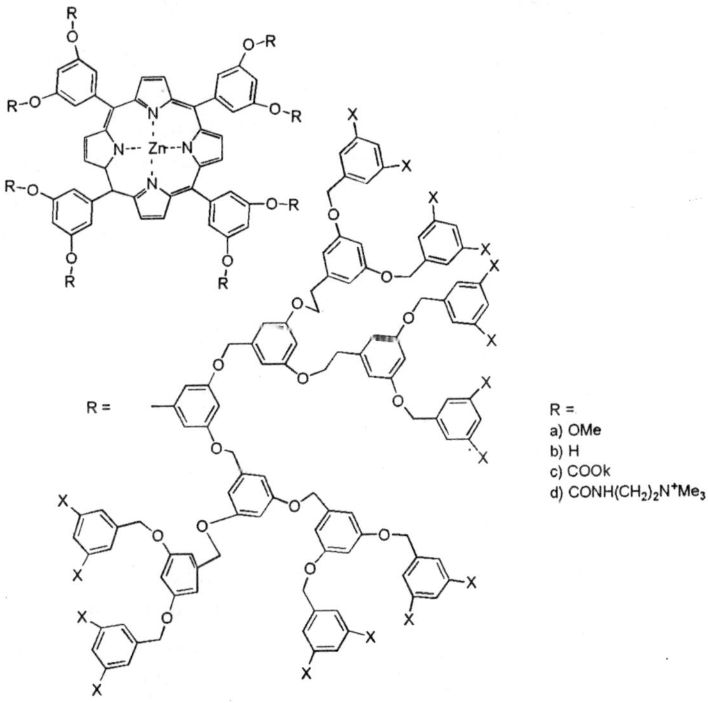

FIG.6.7 Structure of porphyrin containing polyether dendrimers

The dendritic substituents do not influence the photo excitation or any other photophysical process of the porphyrin core of polyether dendrimers. Thus on excitation of such dendrimers (upto 4th generation) neither the fluorescence intensity and its band position nor the lifetime of the excited state changes [62-63, 65]. A shielding effect from oxygen in polyether dendrimers was also observed [65], are when dendritic shells with dendryl substituents. The shielding effect increases from the heteroleptic complexes with only one dendrylated bipyridine moiety and two unsubstituted bipyridines towards the dendritic homoleptic complex [64]. Again, this shielding effect is not restricted to oxygen and the denser dendritic shell protects the central complex from the quenching molecules to a growing extent. The denser complex also has a more positive oxidation potential and less negative potential than non-dendritic analogs and it undergoes incompletely reversible redox processes [65].

A cascade light harvesting system based on a flexible dendrimer scaffold demonstrated that fluorescence resonance energy transfer (FRET) is favorable in a vectorial cascade process, and that shape-persistence is not needed to avoid self-quenching through chromophore aggregation and excimer formation [66]. The targeted dendrimer was synthesized by coupling of fluorol 7GA dye to the protected AB_2 dendritic monomer [46] followed by removal of the hexadecane sulfonyl-protecting group by using ethanolic alkali solution. Similarly, cumarin 2 dye coupled with 2nd generation dendron [46] was attached with the fluorol 7GA dye coupled with *tert*-butyldiphenylsilyl protected AB_2 dendritic monomer, followed by deprotection by CsF in DMF. Then the resulted bichromophoric dendron reacts with succinic anhydride by using a catalytic amount of DMAP in acetone. The sequence of reactions is shown in step 1 of scheme 6.20. The perylene bi(dicarboximide) acceptor core was synthesized as shown in the 2nd step of scheme 6.20. Finally, the donor core and the acceptor core were coupled to get the desired dendrimer, as shown in the 3rd step of scheme 6.20. The steady-state photophysical property measurement of the dendrimer revealed that the energy transfer within this system favors a cascade route, moving from initial chromophore, cumarin 2 through an intermediate chromophore, fluorol 7GA and finally onto a acceptor chromophore, perylene bi(dicarboximide).

The synthesis of spironaphthoxazine functionalized polyether dendrimers upto 3rd generation was reported [67] by coupling photoactive spironaphthoxazine group to the peripheral carboxylic acid of the dendrimers. The surface functionalization of the carboxy terminated polyether dendrimers with spironaphthoxazine unit was carried out in DMF by using DCC and DPTS as catalysts [68]. The yield obtained for

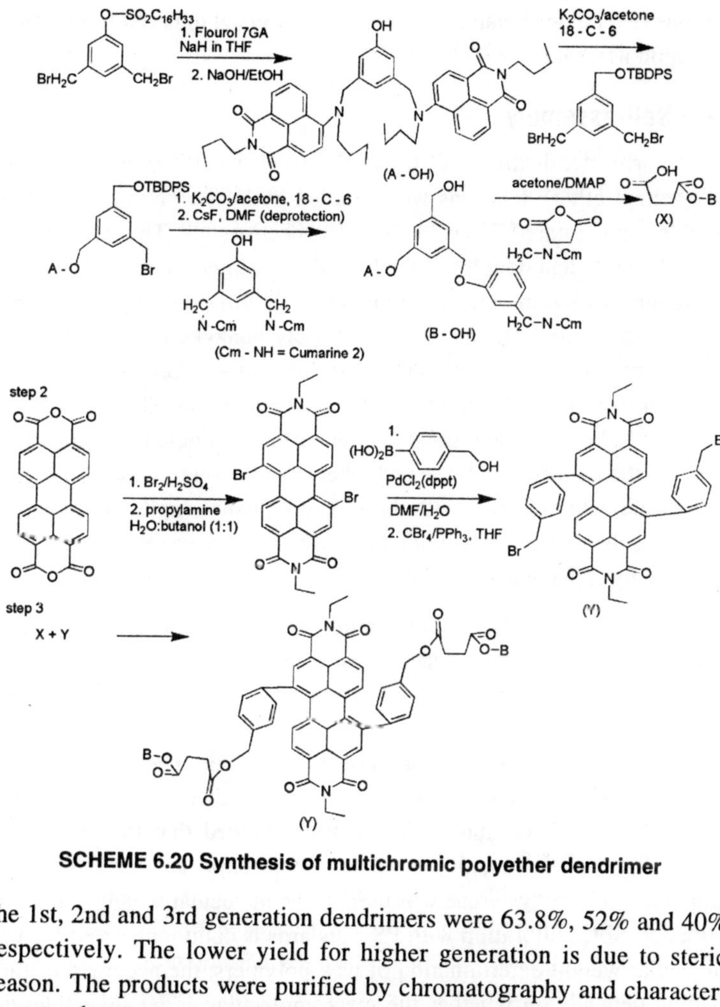

SCHEME 6.20 Synthesis of multichromic polyether dendrimer

the 1st, 2nd and 3rd generation dendrimers were 63.8%, 52% and 40% respectively. The lower yield for higher generation is due to steric reason. The products were purified by chromatography and characterized by ^{1}HNMR spectroscopy and MALDI-TOF mass spectrometric studies. The photoresponsive property of the functionalized dendrimers varies with generation number and showed photo-isomerization on UV irradiation at 340 nm. These dendritic structure in contrast to spironaphthoxazine unit only, can also acts as a carrier of metal ions in a reversible way, which is very important finding in this study. That is, the dendritic structure provides a special microenvironment for spironaphthoxazine groups that can not form a stable complex with metal ions and thus reversible chelating and releasing of metal ions (Ag^+, Zn^+, Al^+, etc.) upon irradiation was obtained.

6.2.8 Self-assembly

The hybrid dendrimers [69] based on oligothiophene core and poly(benzyl ether) dendrons were studied by cyclic voltammetry and variable-temperature UV/visible/near-IR spectroscopy. The redox states (neutral, single and double) of the hybrid dendrimers were measured as a function of oligothiophenes conjugation length (varying from 4 and 17 repeating units). The attachment of poly(benzyl ether) dendrons at the periphery of the oligothiophenes significantly enhances their solubility. In case of undecamer and heptadecamer, the dicationic state consists of two individual polarons, rather than a single bipolaron. The effect of the dendritic poly(benzyl ether) solubilizers on the properties of the redox states varies with the oligothiophene length and dendron size. The dendritic supramolecular assemblies formation were also observed by self-complexation of cation radicals via dimerization.

A novel tetraacid core containing two isophthalic acid units in a *syn*-orientation by a rigid spacer was attached [70] to Frechet type of polyether dendrons upto 4th generation with different periphery and focal point functional groups. The ^{1}HNMR spectroscopic studies suggested the aggregate formation of the core unit. The self-assembly behavior of the core unit was characterized by SEC, laser light scattering (LLS) and VPO studies. The results showed that the molecular weight for each tetraacid was close to that expected for hexameric aggregate when polystyrene was used as the molecular weight standard. Although SEC calibration with PS standards is commonly used for the molecular weight determination of new polymers, the accuracy of this method depends on whether the macromolecular under investigation and its identical molecular weight polystyrene has the same sizes and shapes. Therefore, it would be more exclusive to have a unimolecular analog of the hexameric aggregate for comparison. So a similar size and shape polyether dendrimer was synthesized as the corresponding hexamer of 3rd generation self-assembly structure after computer modeling. The SEC results for both these hexamers were almost identical with slightly broader peak in case of tetraacid core hexamer compared to the dendrimer based on computer modeling. This result further supports the self-assembling of the former dendrimer with well-defined discrete hexameric aggregate. Further studies suggested that the self-assembly behavior of these tetraacids depends on the site of the peripheral group and small peripheral groups containing aggregates are more stable. These dendrimers are self-assembly using hydrogen-bonding interactions. As self-assembly process are usually a balance between enthalpic gain and entropic loss, so stable aggregates require strong H-bonding interaction and preorganized structures. These structures based on self-assembly were verified by means of SANS [71].

The attachment of dibenzo-24-crown-8 (DB24C8) moiety to the focal point of 1st to 3rd generation Frechet polyether dendron to form solution phase self-assembly with a core unit consisting of 1,3,5-tris[*p*-(N-benzyl ammoniomethyl)phenyl] benzene (A) was reported [72]. This assembling is producing the corresponding dendritic pseudorotaxanes structures of nanometer size. The self-assembled dendritic pseudorotaxanes were synthesized by coupling Frechet polyether dendron containing DB24C8 at the focal point with triammonium salts of (A) in $CDCl_3$ for 36 to 48 hours (Scheme 6.21). The assembly formation was confirmed by ^{1}HNMR spectroscopic and MALDI-TOF or FAB mass spectrometric studies.

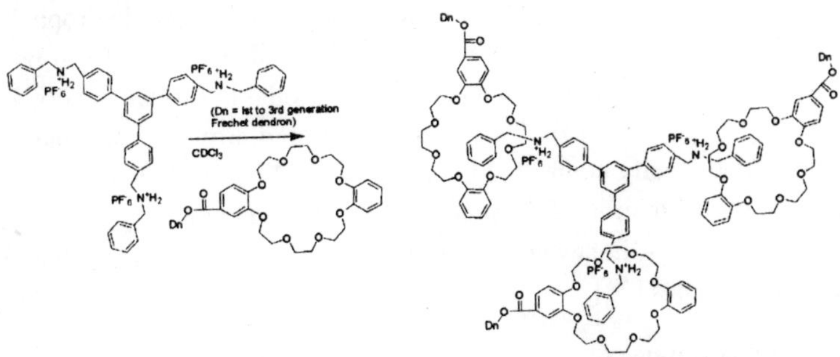

SCHEME 6.21 Synthesis of self-assembled polyether dendrons (pseudorotoxane) with dibenzo-24-crown-8 ether at the focal point

Three generations of self-assembly monodendrons based on the AB_3 building block, methyl-3,4,5-trihydroxy benzoate were synthesized [73] by the convergent approach starting from *p*-hydroxymethylbenzoate. All the synthesized compounds were analyzed by a combination of ^{1}HNMR, ^{13}CNMR spectroscopy, TLC, HPLC and elemental analysis. The phase behavior of the monodendrons was analyzed by a combination of thermal OPM, DSC and XRD studies. The study showed the Ist and 2nd generation monodendrons have disk-like shapes. The 3rd generation monodendron is sixth of a sphere. This system showed a dramatic change in shape of monodendron from hemisphere to a sixth of a sphere. Whereas in the resulting supramolecular dendrimer shape changes from cylindrical to spherical at the transition from generation two to three.

Structurally, very similar surface bound six poly(aryl· ether) dendrons were prepared [74] by the convergent method and used them as building blocks for the preparation of self-assembled monolayers (SAMs) on a gold surface. Surface-bound dendrons with both hydrophobic heptane chains and hydrophilic oligo (ethylene oxide) chains at their periphery were also prepared. The STM and electrochemical behavior studies showed that the precisely tailored structures of surface bound dendrons with different main structures and peripheral substituents allow the formation of some special structures on gold. The symmetric and asymmetric structures with different peripheral substituents resulted different aggregation behavior of the dendrimers.

The 2nd generation polyether dendrons with different hydrophobic and hydrophilic surface groups and alkane thiol site at the focal point formed self-assembled monolayer with controllable surface pattern and wetting properties [75]. The precisely tailored structure of dendron-thiols with locally controlled hydrophobic and hydrophilic peripheries allows the formation of designed surface structures on a gold surface e.g. nano stripes, honey comb and homogeneous structures, as obtained by STM studies [75].

6.2.9 Crosslinked

The first example of a dendrimer with extensively crosslinked peripheral groups but without core was reported [76] by Wendland and Zimmerman. The targeted dendrimers were synthesized from homoallyl alcohol and methyl 3,5-dihydroxybenzoate by using iteratively

Mitsunobu etherification reaction followed by $LiAlH_4$ reduction [77]. Then the desired dendrimers were synthesized by the coupling of trimesic acid with appropriate 3rd generation alcohol (1). The synthesized dendrimers were characterized by ^{1}HNMR spectroscopy, SEC, MALDI-TOF mass spectrometry and elemental analysis. The ring closing metathesis (RCM) reaction of dendrimer 1 with 4 mole % Grubb's ruthenium alkylidene catalyst per dendrimer was found to give exclusively intramolecular product, when carried out in benzene under the nitrogen atmosphere at high dilution (10^{-5} M). The structure of the intramolecular product (2) is still not known, through intramolecularity was established by SEC and MALDI-TOF mass spectrometric studies. The core of this product (2) was then treated with KOH in THF-EtOH-water mixture and the core unit was isolated from the crosslinked product. The ^{1}HNMR spectroscopy and MALDI-TOF mass spectrometric studies confirmed the total loss of the core unit in the product (3). The whole sequence of reactions is shown in scheme 6.22.

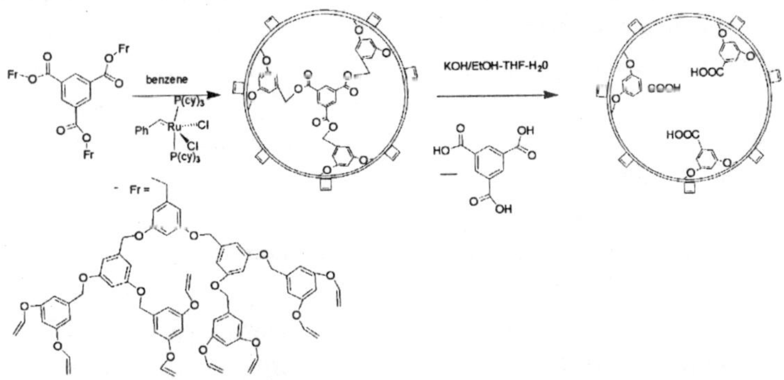

SCHEME 6.22 Formation of polyether dendrimers with crosslinked periphery and without any core unit

Another set of three dendrimers was synthesized [78], where cores of the dendrimers were constructed by ring closing metathesis of homoallyl ether groups using ruthenium benzylidene catalyst. These dendrimers are identical except one cantains an additional dendritic layer on its surface. The dendrimers were synthesized by the convergent approach using either Mitsunobu etherification reaction [79] or phenol-bromide coupling reactions. Another series of dendrimer of similar structures also synthesized [78]. The structure of the dendrimer was such that the ends are conjugated with targeting groups or converted into groups that provide water solubility. The preliminary study on crosslinking of these dendrimers indicated the formation of crosslinks at 10^{-3} M concentration in dichloromethane using 4 mole % of Grubb's catalyst per alkene. The synthesized dendrimers were characterized by ^{1}HNMR spectroscopy, MALDI-TOF mass spectrometry and SEC analysis. Thus these dendrimers can be crosslinked even at higher concentration compared to the earlier reported such type of dendrimers [76].

Methoxy terminated saccharide containing poly(benzyl ether) dendrimers upto 3rd generation were synthesized [80-81] and immobilized by *in-situ* crosslinking. The dendrons were immobilized by the dissolution in $CHCl_3$ using ultrasound technique, followed by addition of 1,3-phenylene diisocyanate and then keeping the solution under the nitrogen atmosphere for 5 days. It was found that 2nd and 3rd generation dendrons formed dendritic aggregates by saccharide-saccharide hydrogen bonding interactions, whereas the first generation tends to form disordered aggregates. It was confirmed from the disappearance of IR peak at 2270 cm^{-1} that 2nd and 3rd generation dendrons crosslinked by 1,3-phenylene diisocyanate, but these are no significant increase of polymerized product formation by the increase of the amount of cross linker. The particle size of the dendrons decreased both in $CHCl_3$ and THF, as measured by DLS method. However, the order of decrease is different in chloroform (1st > 2nd ~ 3rd) compared in THF (2nd ~ 3rd > 1st generation), although change in size in case of 1st generation was very significant but least for 3rd generation. Thus the aggregate of 3rd generation was immobilized as it is by *in-situ* crosslinked with 1,3-phenylene diisocyanate.

6.2.10 Chiral

A series of polyethers based on chiral dendrons, where the chiral subunits were placed in individual generational shells at varying distances from the focal point, was synthesized [82]. The dendrons possess both chiral and achiral branching units. The achiral branching units are 3,5-substituted benzyl aryl ether and chiral units are derivatives of (R, R)-hydrobenzoin [83]. Frechet dendrons were obtained by the similar approach as described earlier [36]. The chiral doubly branched monomers were synthesized by asymmetric dihydroxylation reaction [83]. The yield obtained in each step was high (maximum cases above 85%). The structures of the each series were characterized by ^{1}HNMR, ^{13}CNMR spectroscopy, MALDI-TOF mass spectrometry and combustion analysis. The chiro-optical properties study of the dendrimers revealed that the changes in optical activity with the increase of generation was a result of constitutional effects rather than chiral conformational order. The OH groups on the sterogenic centers in the inner part of the dendrimers are not the branching points, which are used to combine the monomer units by etherification reaction. Rather, they are introduced in the acetonide protected form, and after deprotection they are ready for reaction as well as interaction with each other or with other compounds. The free hydroxy group can give rise to either intramolecular interaction between two branches or to intermolecular interactions between the dendrimer and other molecules [84]. However, the solubility of the deprotected dendrimer (18-OH groups for 2nd generation) in non-polar organic solvent was low. Of course, they may be anchored the metal complexes, which could be used for asymmetric synthesis.

The authors [85] also synthesized another two series of chiral polyether dendritic wedges upto 4th generation to investigate the location of individual chiral units, which affects the chiroptical properties of a single branch. In one series, they attached Frechet type of poly(aryl ether) dendrons [36] of zero to 3rd generation to the chiral acetonide protected hydrobenzoin unit. The chiroptical measurement showed no increase of the molar optical activity of these dendritic wedges with the increase of size of the achiral branches. In the 2nd series, McGrath and his coworkers [85] have synthesized same type of polyether dendritic wedges upto 4th generation, where the chiral acetonide protected hydrobenzoin units were placed in the different shells and at the periphery, in contrast to first series, where the chiral unit was placed only at the center. Interestingly, the dendrons containing chiral units at the interior shell have a large influence on the molar optical activity per chiral unit compared to the dendrons with chiral unit at center or periphery.

Different types of chiral polyether dendrimers of 2nd generation were synthesized [86-88] by the convergent approach using enantiopure threitol derivatives as spacer unit and achiral 1,3,5-benzenetriol derivatives as the branching unit. By combining the spacer derived from (R, R)- or (S, S)-tartaric acid, they were able to construct polyether dendrimer containing only spacers of (S, S)-configuration and layer block dendrimers with combination of spacers of opposite chirality sense, i.e. (R, R) with (S, S). The chiroptical properties study revealed that the molar rotation of the dendrimers is proportional to the excess of (R, R)- or (S, S)- threitol units. That is the chiroptical effect of threitol building blocks of opposite chirality was cancelled out each other. For the same type threitol [either only (R, R)- or (S, S)-threitol] containing dendrimers, an average positive molar optical rotation value of 146 for each (R, R)-threitol unit and –185 for each (S, S)-threitol unit were calculated. However, circular dichroism spectral study indicated that the chiroptical effect of (S, S)-chiral unit in the outer shell of the dendrimer did not compensate that of a (R, R)-chiral unit in the inner shell [89]. Thus the different dendritic layers are chiroptically slightly different [88].

The synthesis of chiral polyether dendrimers was reported [90] by the convergent approach using chiral triol core and achiral Frechet dendrons bromide [6] upto 2nd generation. The chiroptical studies of these series of dendrimers with aromatic spacers and without spacer showed that optical activity decreased with the increase of generation, whereas with aliphatic spacers no optical activity was detected. The molar optical rotation values for the dendrimers with aromatic spacers are constant, whereas for dendrimers with no spacer, the value became double with moving from 1st to 2nd generation. Also the ^{1}HNMR spectroscopic studies indicate the signals for diastereotopic benzylic protons, which can be taken as a measure for the degree of dissimilarity of their environment. The heterotopic benzylic protons become isochronous, when going from the inner to the outer layer of the 2nd generation dendrimers with aliphatic or aromatic spacers. However, dendrimers with no spacer, there was no signal for any of that sets of analog benzylic protons. The 'fully chiral' polyether dendrimers with stereogenic core as well as all-branching units upto 2nd generation were also synthesized [91-92]. At first the stereogenic branches containing benzylic bromide group at focal point was generated by etherification reactions between four molecules of stereogenic building block (Fig. 6.8) of (S)-configuration. In the next step, three such stereogenic branches were coupled with previously mentioned chiral triol core to get the fully chiral polyether dendrimers (Fig. 6.8), with moderate yield. However, when diastereoisomeric branch with (R)-configuration was used, only two such branches were able to couple, even under forcing condition (Fig.6.8). On the other hand, by using enantiomeric triol core, three branches were coupled for both (S)- and (R)–configuration containing branches to offer the desired dendrimers. Other varieties of fully chiral polyether dendrimers were also synthesized [91] by combinations of various diasteroisomeric 2nd generation branch bromides with different core units. The optical activity measurement showed that the dendrimers with all branches of (S)-configuration has a specific rotation and a molecular ellipticity, which deviate from the expected value [91-92], where all other 2nd generation dendrimers have specific rotation that are comparable to those expected by simple addition of appropriate values for their building blocks. The deviation of optical rotation in the first case may be due to the presence of chiral conformational substructures in the dendrimers. The synthesized dendrimers were fully characterized by ^{1}HNMR spectroscopy and MALDI-TOF mass spectrometry. For better characterization of such type dendrimers, the chiral polyether dendrimers upto 2nd generation using fluorine contain (CF_3) stereogenic moiety were synthesized [93]. Since the use of ^{19}FNMR spectroscopy offers a valuable tool to study the conformation of such dendrimers especially with the fluorines attached close to the stereogenic centers. The ^{19}FNMR spectra indicate a clear-cut difference between the inner CF_3 groups and those close to the periphery. From the above discussion, it has been found that with sterically demanding triply branching building blocks no dendrimers larger than 2nd generation could be prepared. So, to get higher generation chiral building blocks i.e. doubly branching blocks can be used. Seebach and his coworkers [91, 94] have used doubly branching building units as the core and synthesized 'fully chiral' polyether dendrimers upto 5th generation by applying the convergent strategy. The dendrimers were obtained from triol with an aromatic elongated group and from double branching units. The exemplified dendrimer contains 93 stereogenic centers (10^{28} possible stereoisomers) and still the dendrimer is monodisperse. ^{13}CNMR spin lattice relaxation time measurements indicated that the segmental mobility of the peripheral units of all such dendrimers upto 4th generation is higher than that of the interior units. The dendrimers were also characterized by SEC and intrinsic viscosity measurement. Further, the polarimetric studies showed that even for dendrimer with so many stereogenic centers, the overall specific rotation could be derived as arithmetic sum of increment of suitable model compounds resembling the building blocks. The chiral polyether dendrimers were also synthesized [95] by attachment of single achiral Frechet type of polyether branches upto the 3rd generation to a chiral pyridyl alcohol.

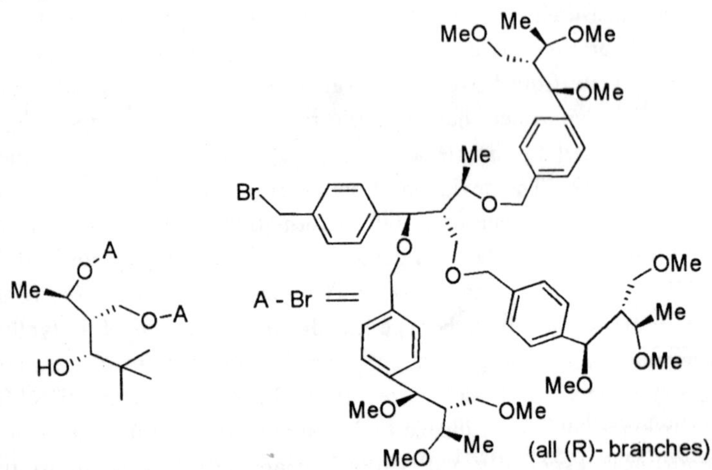

FIG.6.8 Structure of fully chiral polyether dendrimers

The chiral polyether dendrimers upto 3rd generation were synthesized [96] by using Frechet type dendrons. They substituted (S)-1,1'-bi-2-naphthol with Frechet dendrons of first to 3rd generation by etherification reaction (Fig.6.9). The chiroptical property measurement showed that with the increase of generation number the molar optical activity also increases. Even though, the optical activity of the chiral dendrimers with chiral units in the core turned out to drop considerably

[97] in general, but in this case, an enhanced optical molar rotation was observed [96]. This may be due to the steric hindrance of the dendritic arms. Similar results were obtained from the enantiomerically pure axial chiral of the same dendrimers upto 4th generation [98]. Zimmerman and his coworkers [99] also obtained the trimeric dendritic aggregates through H-bonding interaction. They synthesized the dendrimers by functionalization of 3,5-di-*tert*-butyl terminated Frechet type of dendrons with a phthalhydrazide at the focal points. The trimeric disc-like stable aggregate was formed via H-bonding of above three dendrons.

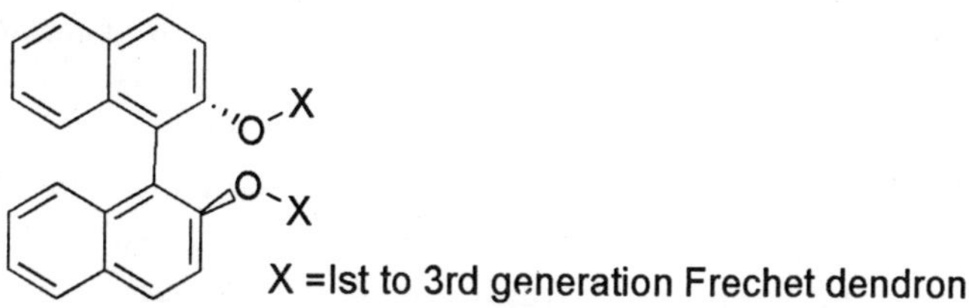

FIG.6.9 Structure of 3rd generation chiral polyether dendrimer with binaphthol moiety

The ester terminated chiral polyether dendrimers upto 3rd generation were synthesized [100] by using (1R, 2S)-2-amino-1-phenyl-1,3-propanediol as the core unit. The chiral properties of the dendrimers were also studied [100]. By applying the Frechet [36] convergent methodology, the polyether dendrons were prepared and attached convergently to the core. The attaching of a phenolic group of the Frechet dendrons to the core simplified by alkylation with the ester-terminated dendrons monobromides by preventing polyalkylation of the amine and *trans*-esterification of the ester terminal with alkoxides generated from the core. The aminodiol core was first acylated with 4-(*tert*-butyl dimethyl siloxy) benzoyl chloride and then deprotected with n-$(C_4H_9)_4$NF. Trialkylation of the resulted product was carried out with the different generation (1st to 3rd) of Frechet polyether dendron monobromides [9] separately, in the presence of K_2CO_3/18-crown-6 in THF/DMF mixture to get the desired chiral polyether dendrimers. The structures of the synthesized dendrimers were characterized by ^{1}HNMR, ^{13}CNMR spectroscopy, electrospray/MALDI-TOF mass spectrometry and GPC analysis. The GPC study revealed the highest molecular weight of the dendrimer was 7178 Da with PDI 1.0032. The chiroptical properties of the dendrimers indicated that the molar rotation of the dendrimers decreased with the increase of generation. The optical rotation of the dendrimers was depend on solvent and temperature. The chiroptical properties of the dendrimers are rationalized by changes occurring in the conformational equilibrium of the chiral, nonracemic central core as a function of generation.

The dendrimers with stereogenic centers generated were synthesized [101] by asymmetric reaction rather than being derived from natural chiral building blocks. The chiral 1,2-diols as branching units through reliable strategies for the efficient construction of the dendrimers with 1,3,5-benzene tricarboxilic acid as the core were synthesized. The chiral polyether dendrimers upto 4th generation were synthesized by the convergent double exponential growth approach, starting from *p*- or *m*–(chloromethyl)phenyl substituted acetonides through enantioselective dihydroxylation reaction. An increasing specific optical rotation was found in dependence of the generation number of the polyether dendrimers. For these dendrimers, the molar optical rotation is directly proportional to the number of chiral moieties and is not influenced by any conformational arrangement of the dendrimer skeleton. The stereogenic centers of the chiral dendrimers were synthesized mostly either by asymmetric synthesis or derived from molecules of the pool of chiral blocks. However, chiral polyether dendrimers consisting of achiral building blocks exclusively were also synthesized [101a]. The dendrimers become chiral due to attachment of four different generations of Fréchet dendrons onto pentaerythritol core unit. Unfortunately, the enantiomers of the racemic product could not be separated. Of course, Meijer and his coworkers [102] synthesized similar type of chiral polyether dendrimers by other route using chiral glycerol derivatives of ketal as the starting material. Even though, all the intermediate products of this synthesis process were optically active but again unfortunately, the final dendrimer was not optically active. The conformational flexibility and the lack of difference of the electronic properties of these polyether branches in the dendrimers may be the possible reasons for zero rotation. So, Meijer's group synthesized [103] a similar dendrimer, which caries a more sterically demanding branch, leading to a more rigid structure. Interestingly, this dendrimer indeed exhibited a very small but measurable optical activity, which underlines the thesis that 'nanoscopic chirality' depends on the rigidity of the investigated structure.

Poly(benzyl ether) dendrimers with TADDOL core unit and vinyl substituted benzyl groups at the periphery were synthesized [104]. The linear benzyl ether and Frechet type of 1st and 2nd generation polyether dendrons (4 equivalents) were coupled with the TADDOL core unit (1 equivalent) by etherification reaction in the presence of K_2CO_3 in acetone at 65^oC. The products were purified by column chromatography with relatively good yield. There were also report [105] on synthesis of TADDOL cross linkers (Fig. 6.10) by coupling of benzyl bromide of flexible spacer moiety (F) and a rigid elongated unit contains a 1,1'-biphenyl unit (R) separately with TADDOL by using the same ether synthesis process. The preparation of polyether dendrimers of 1st to 3rd generation by etherification of 1,1,1-tris(4-hydroxyphenyl) ethane with 4-vinylbenzylchloride or 1st and 2nd generation of vinylbenzyl terminated Frechet type polyether dendrons in the presence of K_2CO_3 and 18-crown-6 was also reported [105] (Scheme 6.23). The dendrimers obtained with very high yield by this process and they were purified by chromatographic technique. The copolymerization of a TADDOL dendrimers with styrene, followed by loading with titanate affords polymer bound Ti-TADDOLate complexes with an unprecedented stability and activity during recycling process. The polymer-bound TADDOL maintained a constant swelling ability during recycling, which might be the reason for their excellent performance. The catalytic activity of dendritic TADDOLates $Ti(Oi_{pr})_2$ is much faster in the Et_2Zn addition to PhCHO than its soluble analog [105]. Whereas copolymerization of vinylbenzyl end capped polyether dendrimers (Scheme 6.23) with styrene in the presence of TADDOL type of compound, followed by addition with $Ti(Oi_{pr})_2$ afford the polymer bound TADDOLate Ti-catalyst, whose catalytic activity for the same type of reactions is much less compared to the dendritic TADDOLate Ti-complexes. These results indicated that a dendritically crosslinked polystyrene support itself is not sufficient to guarantee a long-term stability of the polymer bound catalyst. Also it proved that placing the TADDOL unit in the core of the dendritic crosslinker would be necessary to obtain polymers with an outstanding catalytic activity.

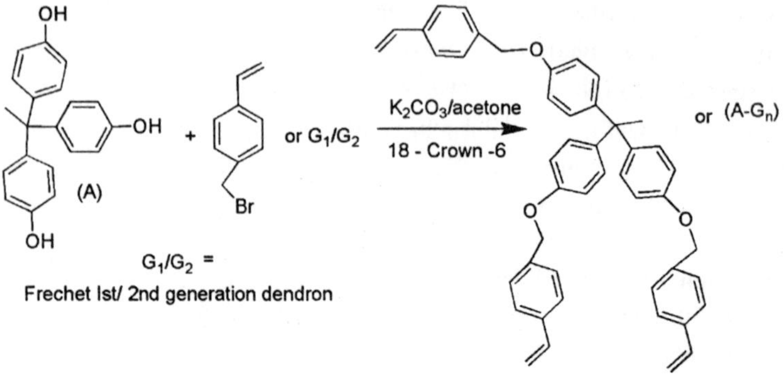

SCHEME 6.23 Synthesis of vinyl benzyl end capped polyether dendrimers

Different types of dendritic polyether-BINOL (1,1'-bi-2-naphthol) derivatives bearing at least two polymerizable styryl group were synthesized and fully characterized [106]. BINOL was first brominated selectively in the 6 and 6' position followed by protection of OH groups by triisopropylsilyl (TIPS) affords TIPS protected brominated BINOL. Subsequent replacement of Br by OH groups resulted the desired core unit. The coupling of 1st and 2nd generation Frechet type of dendrons, (obtained from 3,5-dihydroxy benzoic acid methyl ester and 4-vinyl benzyl chloride) with the BINOL core unit offered the desired dendrimers. The dendrimer obtained were purified by flash column chromatography with 40-50% yields. All the compounds were characterized by ^{1}HNMR, ^{13}CNMR and IR spectoscopic, FAB/MALDI-TOF mass spectrometric studies and elemental analysis. The copolymerization of these dendrimers with styrene by the suspension technique followed by cleavage of the protecting groups and subsequent loading with a Lewis acid afforded catalytically active polystyrene supported BINOLates. The polymer bound dendritic Ti-BINOLates were successfully used for enantioselective addition of Et_2Zn and Me_3SiCN to PhCHO and pivalaldehyde respectively with recyclable capacity.

6.2.11 Fullerene Containing Polyethers

The fullerene containing dendrimers were synthesized [107], where fullerene was functionalized by poly(aryl ether) dendritic branches terminated with peripheral triethylene glycol chains to obtain solubility in a wide range of solvents. Four polyether dendritic branches with methylol at the focal point were synthesized by using the same procedure [108], followed by oxidation with MnO_2 afforded aldehyde group at the focal point. Then these dendritic branches were reacted with C_{60} in the presence of N-methyl glycine. Actually, the 1,3-dipolar cycloaddition of the dendritic azomethine ylides generated *in-situ* from the corresponding aldehydes and N-methylglycine. The whole sequence of reactions is shown in scheme 6.24. The products were purified by column chromatography followed by GPC analysis and the structures of fullerene dendrimers were characterized by ^{1}HNMR, ^{13}CNMR spectroscopy and electronspray mass spectrometry. The photophysical studies of the dendrimer revealed that by enlarging the dendrimer size, changes of the core solvation environment occur, most likely as a consequence of tighter contact between the fullerene unit and the external dendritic wedges due to favorable electronic donor/acceptor interactions [109]. This provides some shielding of the fullerenopyrrolidine chromophore from contact with external molecules such as dioxygen. As a consequence, a reduction of the potent singlet oxygen sensitizing character of the fullerene unit was obtained. The preliminary measurements on the mesoporous silica glasses doped fullerene dendrimers revealed efficient optical properties.

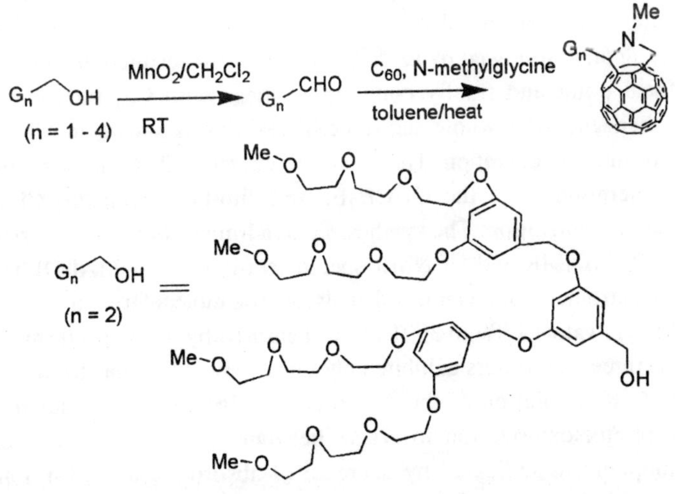

SCHEME 6.24 Functionalization of polyether dendrimer with fullerene moiety at the focal point

The molecular design of novel dendritic polyethers bearing a cyclic porphyrin dimer at the focal point, which form stable supramolecular fullerene/dendrimer hybrid was reported [110]. 3,5-Dibromophenol, protected by *tert*-butyldiphenyl silyl group, was reacted with 3-ethynyl benzaldehyde to get the corresponding dialdehyde, which was further reacted with a Frechet polybenzyl ether bromide dendron (3rd and 4th generation only) to obtain dendritic dialdehyde. Finally, this was undergone condensation reaction with dipyrrolylmethane under acidic condition for 3h to get the corresponding cyclic dimer to tetramer. Then, it was purified by chromatographic technique to get the cyclic dimer, which was treated with C_{60} fullerene to obtain supramolecular fullerenen-dendrimer hybridization. The products in each step were analyzed by mass spectrometry and ^{1}HNMR spectroscopy. The molecular weights of hybrids vary from 3159 to 3998 Da, as measured by ESI-mass spectrometric study. Fluorescence studies showed that on excitation of porphyrin dendrimers at 409 nm in toluene, fluorescence bands at 635.5 nm and 700 nm were emitted, while the hybrids with fullerene those emissive bands were quenched.

The fullerene cored dendrimers upto 3rd generation with photoisomerizable azobenzene groups in the periphery were synthesized [111] by Frechet convergent technique [36] in which repetitive reaction of bromide of azobenzene derivative with the monomer 3,5-dihydroxy benzyl alcohol, followed by azide formation of the resulted dendrons at the focal point and finally coupling of these with C_{60} fullerene in chlorobenzene. The whole sequence of reaction is shown in scheme 6.25 for the 1st generation. The repetition of step 3, 2, 4 and 5 resulted 2nd generation from the G_1-CH_2Br and similarly from G_2-CH_2Br offered 3rd generation. The synthesized dendrimers were characterized by FTIR, ^{1}HNMR and ^{13}CNMR spectroscopy, FAB or MALDI-TOF mass spectrometry and elemental analysis. The molecular weight of the dendrimers varies 1389 to 3991 Da as measured by mass spectrometry. All the three dendrimers exhibited photoresposive behavior. Irradiation of the $CHCl_3$ solution of the dendrimers at 365 nm light resulted in clean photoisomerization to *cis*-azobenzene from *trans*-azobenzene chromophore as indicated by decrease of absorption at 357 nm and increase in absorption at 437 nm. A photostationary state was reached within 150s. POM study suggested that the dendritic films were uniform and isotropic. The dendrimers showed no mesophase upon heating and cooling between 30 to 270°C.

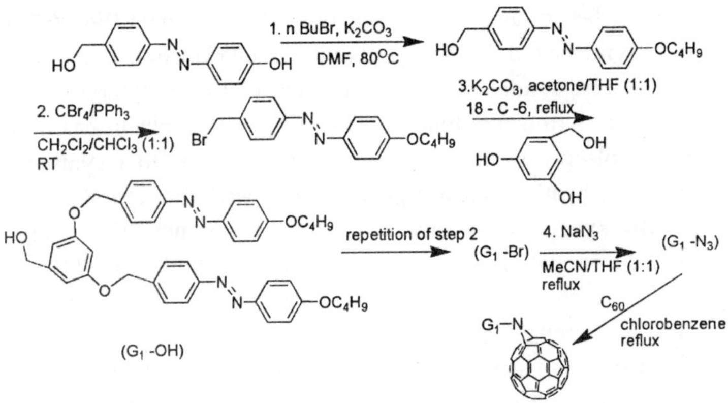

SCHEME 6.25 Synthesis of fullerene core polyether dendrimer with azobenzene surface groups

6.2.12 Copolymers

The synthesis of a novel class of AB and ABA block copolymers was reported [112-113] by using Williamson ether synthesis process. The hydrophobic poly(aryl ether) dendrons of different generations were taken as block A and hydrophilic linear polyether glycol or polyethylene oxide of different molecular weights were the block B. The copolymers were synthesized by the reaction of varying chain length of hydroxyl terminated linear PEG or PEO with different generations (upto 6th) polyether dendrons with benzyl bromide at the focal point in presence of K_2CO_3 / 18-crown-6 in THF. The synthesized block copolymers were characterized ^{1}HNMR spectroscopy, SEC coupled with a viscometric detector and optical microscopy. The collapse and tight packing of either block reduced the mobility of the copolymers, which was detected by ^{1}HNMR spectroscopy from signal broadness and weakened signal intensity. The results suggested that the dendrimers blocks are well shielded by the extended chains of linear block, especially at higher generations. The polymers also formed aggregates in aqueous methanolic solution, in form of large micellar structure. The copolymers with lower generation (upto 2nd) dendrons formed unimolecular micelles, whereas higher generation formed multimolecular micelles, may be due to hydrophobic effect and π-π interaction between the dendritic blocks.

A novel class of amphiphilic star type of copolymers (Fig. 6.10) was also synthesized [114] by using pentaerythritol core. The copolymers were synthesized by attachment of four poly(aryl ether) dendrons on the terminal of four armed linear polyether chains based on pentaerythritol core by using the same Williamson ether synthesis technique. The synthesized star block copolymers were characterized by ^{1}HNMR spectroscopy and SEC/VISC technique. The results suggested the formation of unimolecular micelles in $CHCl_3$, THF or MeOH, but with strikingly different structures. Thus the star copolymers can self-organized into different micellar structures as a function of the environment. A potential application of these 'stimuli responsive wrappers' involves their use as solvent-specific encapsulating agents. The linear-dendritic amphiphilic diblock copolymers of linear poly(acrylic acid) as the hydrophilic block with dendritic poly(bezyl ether) as the hydrophobic block were synthesized [115] by two step reactions. In the first step, a poly(methyl acrylate)-poly(bezyl ether) dendrimer series were synthesized by the ATRP process. Followed by the 2nd step, where hydrolysis of linear PMMA block transformed into poly(acrylic acid) to get the desired amphiphilic

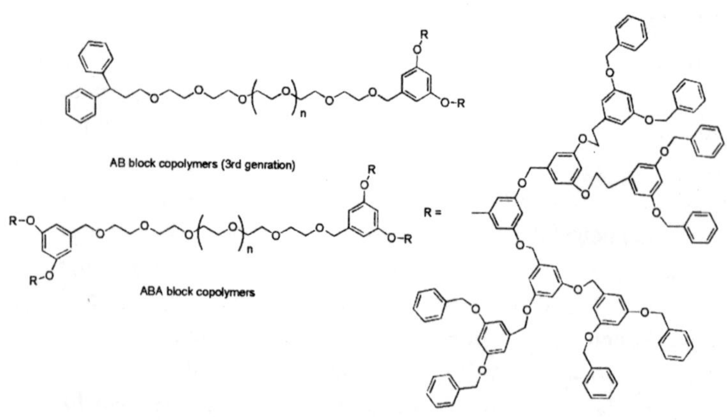

FIG.6.10 Structure of dendritic polyether type AB and ABA block copolymers

block copolymers. The resulted block copolymers were well characterized by ^{1}HNMR, FTIR spectroscopy and SEC technique. A well-defined architectures with low polydispersity of the copolymers was found. The water soluble amphiphiles were obtained from copolymers of 1st to 3rd generation poly(benzyl ether) dendrons with linear poly(acrylic acid) with degree of polymerization greater than 10.

6.2.13 Others

Schluter and his cowokers [116-118] synthesized polystyrene [116] and poly(*p*-phenylene) [117], each of which attached with 3rd generation Frechet polyether dendron [36] at every repeating unit. The dendronized monomers were first synthesized by etherification reaction between styrene and *p*-phenylene monomers with the dendrons separately. This novel class of polyether dendrimers allows to investigate the space demand for dendrons and their sequence. Further, their backbones are dense enough to force the intrinsically more (in the first polymer) or less (in the 2nd polymer) coiled polymeric core into a maximally stretched conformation [118]. The SFM study showed that poly(*p*-phenylene) dendrimer molecules form 'ellipsoidally flattened' cylinders. However, this flattening indicated that they still possess some conformational flexibility, especially in solution and thus it may lead to shape fluctuation and change, which are against the concept of molecular object. However SFM study for polystyrene dendrimer shows a remarkably high degree of order structure, when image was taken as an ultrathin film of the dendrimer on HOPG. There is a very good agreement of the domain size was observed in different study, SFM (5±0.5) nm and MD simulation (4.4±0.3) nm and SANS (6.1±0.5) nm. Thus the cylinderical dendrimer are grouped, in domains in which they are packed to each other with periodicity of (5±0.5) nm. Finally, it was also noticed that the different domains exhibit three molecular orientations at $(60\pm8)^{o}$ to each other. These studies have great importance, as a module for nano materials construction along with other fields such as micelles, membranes, etc.

The synthesis and characterization of a crosslinkable non-linear optical (NLO) polyether type dendrimer, which possessing very large optical non-linearity and excellent thermal stability, were reported

[119]. The dendrimer was synthesized by the Mitsunobu condensation reaction between carboxylic groups of the core with hydroxyl groups of the chromophoric precursor, which has crosslinkable trifluorovinyl ether moiety. The condensation product was then reacted with tetracyanoethylene (TCNE) to activate the phenyl-tetracynobutadienyl (Ph-TCBD) electron acceptor. The purity and structure of the synthesized dendrimer were fully characterized by GPC, MALDI-TOF mass spectrometry, ^{1}HNMR, ^{13}CNMR and ^{19}FNMR spectroscopies. The single, sharp and symmetrical peak in GPC confirmed the purity of the dendrimer. The molecular weight of the dendrimer was 4664 Da. The dendrimer exhibits a combination of large electrooptic coefficient value, 60 pm/V at 1.55 μm and good temporal stability at 85°C, which indicated that the material may be used as NLO material.

The dendritic polypyridines [120-121] based on diethyl-4-hydroxy pyridine-2,6-dicarboxylates were synthesized by the convergent method using 1,3,5-*tris* (bromomethyl)benzene (Scheme 6.26). The synthesized dendrimers were characterized by laser desorption (LD) and MALDI-TOF mass spectrometric studies. The effect of matrix was studied using three differently branched dendrimers with three different matrices at different matrix/sample molar ratios. The study showed that the fragmentation was strongly dependent on the matrix used, the matrix/sample molar ratios and on the molecular structure of the dendrimers.

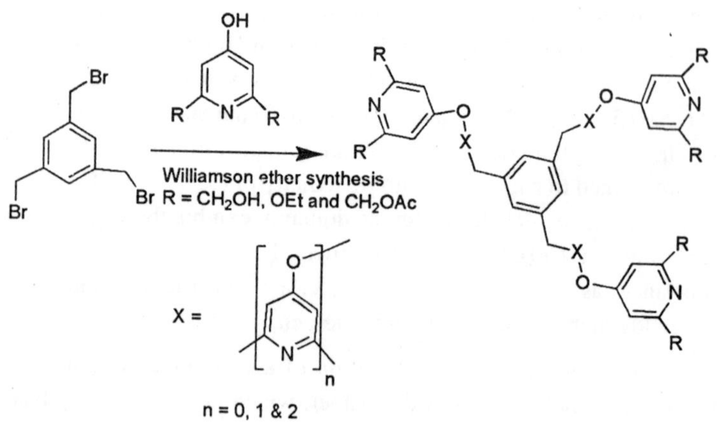

SCHEME 6.26 Synthesis of polyether dendrimer based on pyridine moiety

Polyether dendrimers with phenyl sulfides, selenides and tellurides terminal groups were synthesized [122] and used them as the catalysts for the oxidation of bromide with H_2O_2 to give positive bromine species that can be captured by cyclohexene in two phasic systems. The dendrimers upto 2nd generation were synthesized by using 1,1,1-*tris*(4-hydroxyphenyl)ethane as the core unit and 3,5-dihydroxybenzyl alcohol as the building block by the convergent method. The phenyl chalcogenide groups were introduced through propyloxy spacers. The synthesized dendrimers were characterized by ^{13}CNMR spectroscopy, FAB /MALDI-TOF mass spectrometry and elemental analysis. The study showed that the phenyl sulfide terminated dendrimers exhibited no catalytic activity. The phenyl tellurides terminated dendrimers exhibited increased rate of catalysis with the increase of generation number, while for selenides the increment even much higher for the same, may be due to cooperative affect among the phenyl seleno groups. However, this increment of catalytic activity was lost upon replacing all but one phenylseleno group by phenoxy groups. Further, it was found that the dendrimer with 12 phenyl seleno groups showed a turnover number of >60,000 mole of H_2O_2 consumed per mole of catalyst.

A series of rotaxanes with dendritic stopper was prepared by using a threading approach [123]. The synthesis involves the alkylation of bipyridinium based molecules with Frechet polyether dendritic bromide [36] (3rd generation) in the presence of *bis*(*p*-phenylene-34-crown-10) (BPP34C10). The formation of rotaxanes was confirmed in ^{1}HNMR spectroscopy by observing the upfield shift of both bipyridinium and hydroquinone protons. The presence of dendritic frame works made these rotaxanes soluble in wide range of organic solvents. The ^{1}HNMR studies at variable temperatures in different solvents, suggested that in less polar solvents, the cationic rotaxanes backbone was more surrounded by the dendritic substituent, which led to the restricted molecular motion. Yamaguchi *et al.* synthesized 4-pseudorotaxanes by complexing three Frechet polyether dendrylated dibenzo-24-crown-8 moieties with the tritopic ammonium salts [124]. The dendritic pseudorotaxane was characterized by ^{1}HNMR spectroscopic and FAB/ESI/MALDI-TOF mass spectrometric studies.

A series of polycationic polyether dendrimers was synthesized [125] by a convenient procedure involving quantitative quaternization of arylamine core. The convergent approach was first used to synthesize the series of poly(aryl ether)s. Then the series was expanded to the preparation of the first polycationic, mixed core shell dendrimer. These dendrimers are very unique type as all of the dendrimers consist of an apolar core with a peripheral ionic layer, which is surrounded by a less polar layer of dendritic wedges. These cationic dendrimers were utilized to form assemblies with anionic guest molecules. It was observed that, all the cationic sites throughout the dendrimer core were involved in ion pair formation with anionic guests by help of UV/visible, NMR spectroscopies and MALDI-TOF mass spectrometric studies. The assemblies were formed with predefined anionic guest/host ratios upto 24. The large NMR spectroscopic shifts of resonances corrected with the groupings located in the core of the dendrimers, together with the relaxation time data indicated that the anionic guests are associated with the cationic core of these dendrimers. The x-ray molecular structure of the octacationic, tetraarylsilane model derivative also shows that the iodide counterions are primarily located near the polycationic sphere. Thus this novel polycationic ether dendrimers may be used as phase transfer catalysts and substrate delivering materials over a nanofiltration membrane.

The synthesis of dendritic styryl-substituted salen polyethers for crosslinking suspension copolymerization with styrene and multiple use of the corresponding Mn and Cr complexes in enantioselective epoxidation and hetero-Diels–Alder reactions were reported [126]. The synthetic process involves the etherification reaction of Frechet type of polyether dendron bromides [36] of different generation with 3-*tert*-butyl-5-hydroxy salicylic aldehyde to get the desired polyether dendrimers.

A polyether dendron with acetylene spacer was also synthesized by Sonogashira cross coupling of 3-*tert*-butyl-5-bromosalcylic aldehyde with ethynyltrimethyl-silane followed by deprotection by TBAF in THF and subsequent reaction with 1st generation styrene capped polyether dendron. Now, these dendrons with or without peripheral vinyl groups were coupled with the suitable diamine in ethanol for several hours to get the salen polyether dendrimers [126] (Scheme 6.27). The products were purified by column chromatography and characterized by ^{1}HNMR, ^{13}CNMR spectroscopic studies, MALDI-TOF mass spectrometry and elemental analysis. The salen dendrimers without peripheral vinyl groups were then loaded with Mn to get the desired catalysts, which were used for epoxidation of olefins under homogeneous conditions (see application section). The vinyl terminated salen dendrimers were copolymerized with styrene by suspension polymerization technique to get polymer bound salen ligands, which were subsequently complexed with $Mn(OAc)_2.4H_2O$ in DMF/EtOH medium. These polymer bound Mn-salens were also used for epoxidation of olefins as in earlier cases (see application section). The salen dendrimers of similar type with cyclohexane diamine core were loaded with $CrCl_2$ in THF to get the desired polymer bound Cr-salens, which were used as catalyst for hetero-Diels-Alder reactions (see application section).

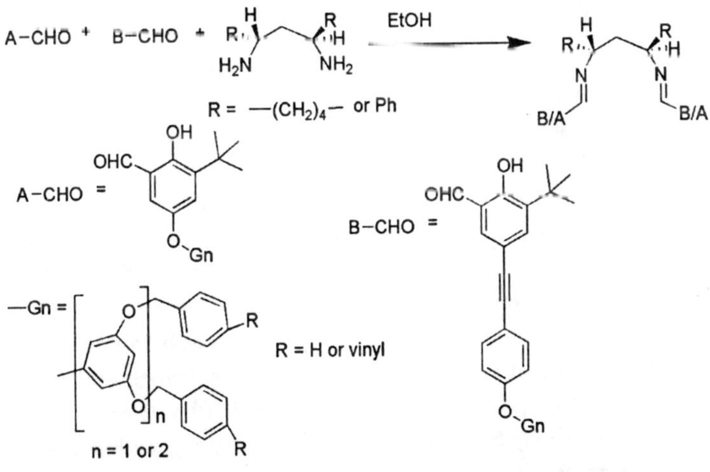

SCHEME 6.27 Synthesis of salen polyether dendrimer with diamine core

The addition of anionic benzylsulfate functionalized polyether dendrons upto 2nd generation to dynamic mixture of Ag^+ and triphosphine ligands to obtain loosely bounded (Fig. 6.11) cage-core dendrimers was reported [127]. These complexes are unusually labile metallodendrimers, which are undergoing reversible dendron exchange for available SbF^-_6. The alcoholic functionalized (at the focal point)

Frechet dendrons [36] were first converted to the corresponding benzylsulfate salts by the reaction with pyridine sulfurtrioxide. Then the titrations of these dendrons into 6:4 $AgSbF_6$/triphos mixture in $CDCl_3$-CD_3CN solution resulted the desired labile coordination dendrimers. The reaction was monitored by ^{31}PNMR spectroscopy.

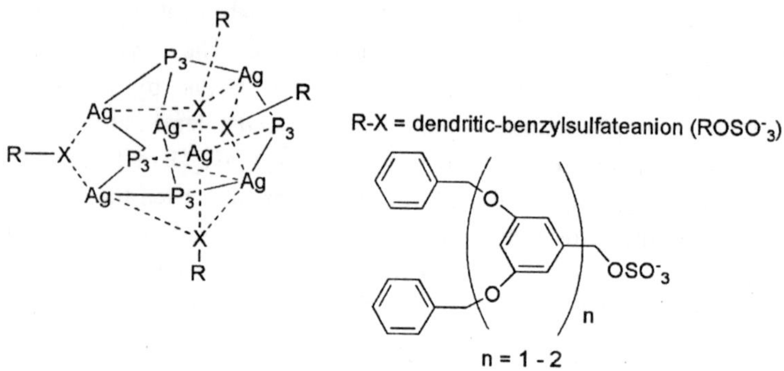

FIG.6.11 Structure of anionic benzylsulfate functionalized polyether dendron

The synthesis and metal complexation chemistry of a novel series of crown ether based cascade molecules were reported [128-129]. The amide linkage was reduced to increase the ion-binding affinity of the crown ether moieties. The dendrimer was able to extract metal ions in a generation independent manner and form a 1:1 complex with Cs^+ unlike the sandwich complexes that Cs^+ forms with polymeric crown ethers. The above results indicate that ether moieties in the dendrimer do not necessarily function in a cooperative manner. Solubilization of myoglobin in organic solvents like DMF through interactions of multiple azacrown ethers with ammonium or carboxylate functions on the peptide is only possible for the lowest generation azacrown. Many such dendrimers may be required to solubilize one myoglobin molecule and higher generations do not allow adequate interactions with protein as they have closed structures.

Uda *et al.* syntehsized and studies [130] the photoisomerization by oxygen for 1,3,5-tristyrylbenzene polyether dendrimers of zeroth, 2nd and 4th generations. The dendrimers were synthesized from the reaction of the Wittig-Horner reagent with 3,5-dimethylbenzaldehyde for zeroth generation and coupling of 1,3,5-*tris* (3,5-dihydroxystyryl)benzene with 2nd and 4th generation of aryl ether dendritic bromide resulted 2nd and 4th generation desired dendrimers, respectively. The synthesized dendrimers were characterized by ^{1}HNMR, UV-visible spectroscopic, MALDI-TOF mass spectrometric studies and GPC analysis. The fluorescence (FL) and fluorescence excitation (FLE) spectra of the aryl ether dendrimers showed almost same nature in FL, whilst in the FLE spectra they are different, especially at shorter wavelength. This difference may be due to singlet energy transfer from the dendron to the core unit. The fluorescence life times of the dendrimers reduced to half (from 18 ns to 4.6-8 ns) under oxygen compared to argon atmosphere. Also on irradiation of 330 nm light, the dendrimers exhibited E to Z isomerization, as indicated by absorption at 300-350 nm with quantum yield 0.07 to 0.08. Thus the efficiency of isomerization does not influenced by the size of the dendrimer, may be due to volume-conserving isomerization mechanism but do affect the mobility of the molecules.

A dendrimer [131] with Frechet type dendrons attached to a dibenzo-24-crown-8 ether unit showed rotaxane like characteristics and assembled under thermodynamic controlled conditions by slippage in CH_2Cl_2. The driving force for the self assembly process was the molecular recognition that exists as a result of [N^+-H---O] and [C-H---O] H-bondings between NH_2^+ centers of the dendron and the core unit. This slippage methodology provides a bridge between kinetically stable, mechanically interlocked compounds and kinetically labile supramolecular assemblies. The dialkylammonium salt of polyether dendron with cyclohexylmethyl group attached through NH_2^+ groups and a *bis*-dendron with dibenzo-24-crown–8 ether unit are assembled under reflux condition in CH_2Cl_2.. The assembly formation was confirmed by ^{1}HNMR spectroscopic and FAB mass spectrometric studies. This assembly formation is a reversible process. So this type of dendritic host can reversibly trap different guest molecules and may be used in controlled released system.

6.3 MODIFICATION OF POLYETHER DENDRIMERS

In this section some of the modified polyether dendrimers are discussed, even though some of these types are already mentioned in the previous section.

6.3.1 Surface

The aliphatic polyether dendrimers upto 4th generation, were modified [2] by acylation of the periphery hydroxyl groups. The polyhydroxy terminated dendrimers were reacted with benzoyl chloride in the presence of pyridine or with tetradecanoic acid in the presence of DCC and DPTS to generate the corresponding polyesters with almost quantitative yield. The modified dendrimers were characterized by MALDI-TOF mass spectrometry. However, these modified dendrimers with alkene functionality at the focal point showed same solubility behavior as before deprotection of the surface groups.

Poly(benzyl ether) dendrons of different generations (upto 5th) with benzyl groups at the periphery were modified [132] by the attachment of oligo(ethylene glycol) chains of varying length from mono to hexakis at the focal point. The langmuir monolayer formation was studied by using these dendrimers. In the case of modified 3rd and 4th generation dendrons, it was found that the hydrophilic oligo(ethylene glycol) chains extended into the water subphase thereby increasing the stability of the monolayer. However, it was concluded that the stability trends for the dendrimer monolayers depended on the relative sizes between the hydrophobic dendrons and the hydrophilic focal point chain. The study also showed that the molecular weight of modified dendrimers possess hexakis(ethylene glycol) chain at the focal point. However, this trend is much smaller than the analog relationship predicted for dendrons assuming a spherical shape, and hence they concluded that the monodendrons assumed an ellipsoidal shape at the air-water interface.

6.3.2 Core

By applying the known synthetic strategy [2] Malenfant and Frechet reported [133] the first conducting polythiophene, having only dendritic solubilizer. The 2nd and 3rd generation benzyl protected polyether dendrons with free hydroxylic group at the focal point was esterified with 2,5-dibromo-3-thiophene acetic acid in the presence of DCC/DPTS. Subsequently, Stille coupling reaction between the esterified dendrons and 5-(trimethylstannyl)–2,2-bithiophene in DMF by using $Pd(PPh_3)_2Cl_2$ catalyst resulted dendron attached to thiophene pentamer with high yield (82%). The treatment of NBS in DMF resulted dibromide of the pentamer. Then the polythiophenes were prepared by A_2 + B_2 polymerization protocol using Stille coupling reaction between the dendrimer bound thiophene dibromide and 2,5-*bis*(trimethylstannyl) thiophene in DMF with $Pd(PPh_3)_2Cl_2$, catalyst. The whole sequence of reactions is shown in scheme 6.28. The other type of such dendron-bound polythiophene was also reported [133]. The modified dendrimers were characterized by SEC, UV-vis. spectroscopy and MALDI-TOF mass spectrometry. Although polythiophene was also synthesized from dendrimer bound monothiophene but they showed no appreciable conductivity by iodide doping. The lack of charge mobility may be due to the steric hindrance by bulky dendrimer that precludes adequate packing of the short polymer repeating unit with one dendrimer for two thiophene units thus preventing charge hopping from chain to chain. However, polythiophene from dendrimers bound pentathiophene exhibit good conductivity (50-100 s/cm) by $NOBF_4$ and I_2 doping. In this case, the ratio of dendrimer to thiophene units is 1:6, thus allowing more space and consequently better packing is expected in the solid state. It is worthy to mention here that the

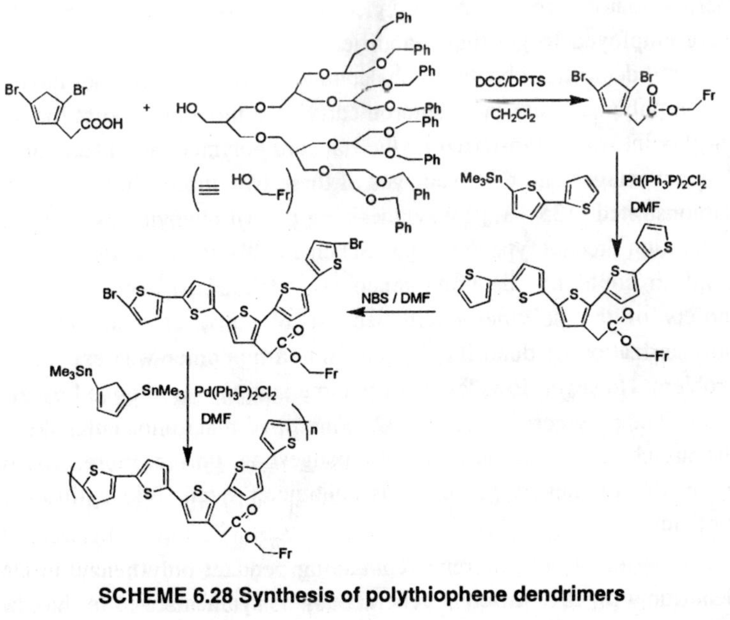

SCHEME 6.28 Synthesis of polythiophene dendrimers

pentathiophene macromonomer synthesis was facilitated by the enhanced solubility and the chemical inertness provided by the aliphatic dendrimer frame work. This allows the preparation of the first conducting polymer functionalized exclusively with dendritic side chains.

Frechet type poly(benzyl ether) dendrons [36] upto 2nd generation were modified [134] by attachment of carbonyl ruthenium(II) *meso*-tetraphenyl-porphyrin in the core. The attachment was carried out by treating the carbonyl ruthenium complex of 5,10,15,20-tetrakis(4'-hydroxyphenyl) porphyrin or 5,10,15,20-tetrakis(3',5'-dihydroxyphenyl)porphyrin with methyl sulfonate dendrons in refluxing dry acetone in the presence of K_2CO_3 and 18-crown-6. The porphyrin-dendrimer complexes were characterized by UV/visible, IR and ^{1}HNMR spectroscopic studies. All these dendritic ruthenium porphyrins are soluble in common organic solvents such as dichloromethane, acetone, etc. They are highly selective catalysts for epoxidation of alkene (see application section).

A new type of macromolecules consist of dendrimers with a polymeric initiator core was reported [135-137]. Two synthetic approaches were employed to get these modified dendrimers. The convergently prepared dendritic substituent was coupled with each monomer moiety of the polymer chain or a dendron carrying a monomer moiety at the focal point was polymerized by the standard polymerization technique. The advantages and disadvantages of these two approaches were also demonstrated [135-137] by synthesizing poly(*p*-phenylene)s with 3rd generation Frechet type dendrons (Scheme 6.29). It was shown that the dendron could not be homogeneously attached to each monomer moiety of the polymer chain because of steric effects, Whereas polymerization of dendritically substituted monomer was caused no problem. However, low degree of polymerization was obtained by this route. The polymeric chains of these modified macromolecules are in the stretched form because of steric congestion. Further more, macromolecules of these types exist as nanometer sized rigid cylindrical particles.

A series of polyfluorenes containing pendant poly(benzyl ether) dendritic wedges of different generations was synthesized [138] via the

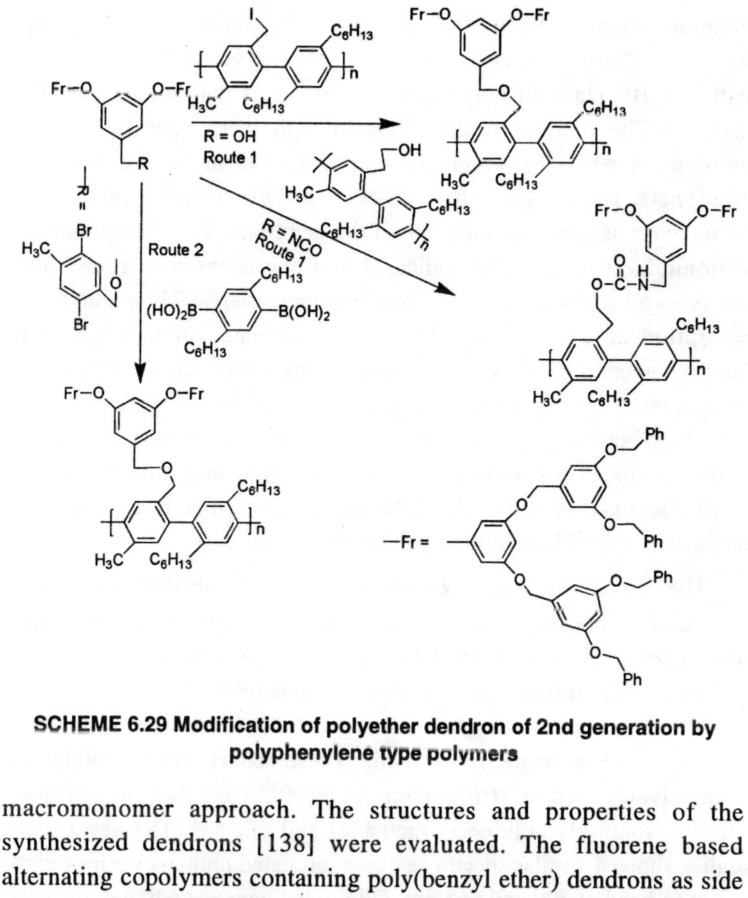

SCHEME 6.29 Modification of polyether dendron of 2nd generation by polyphenylene type polymers

macromonomer approach. The structures and properties of the synthesized dendrons [138] were evaluated. The fluorene based alternating copolymers containing poly(benzyl ether) dendrons as side chains with benzyl focal point attached through a 4-hydroxy phenylene spacer to the C-9 carbon in every alternating fluorene unit. The dendritic monomers were first synthesized for this purpose. The acid catalyzed condensation reaction of 2,7-dibromo-9-fluorenone with excess phenol resulted 9,9-*bis*(4-hydroxyphenyl)-2,7-dibromofluorine. Subsequently, this compound coupled with benzyl bromide functionalized different generation of dendrons. Finally, this macromonomers, individually copolymerized with 2,7-*bis*(4,4,5,5-tetramethyl-1,3,2-dioxaborolan-2-yl)-9,9-dioctylfluorene by Suzuki coupling reaction. The structures of the synthesized copolymers were characterized by ^{1}HNMR and ^{13}CNMR spectroscopy. They are readily soluble in common organic solvents such as THF, chloroform, toluene, xylene, DMF, etc. The molecular weight of the polymer varies from 3.8 X 10^4 to 6.5 X 10^4 Da with polydispersity 1.9-2.5 as determined by GPC analysis. The thermal studies showed high thermostability of the copolymers with 5% weight loss upto 350°C under the nitrogen atmosphere. The T_gs varies from 90-56°C and the declining values with the increase of generation number of the dendrons. The absorption and photoluminescence studies indicated a strong absorption at 388-392 nm. Also all dendron functionalized polymers displayed an additional absorption at 278 nm, at which the absorbance increases as the dendritic generation grew. The highest generation dendrons sufficiently encapsulated the conjugated backbone. This not only prevents self-quenching but also suppresses aggregate/excimer formation, leading to an enhancement in photoluminescence efficiency and stability. These results demonstrated that this alternative copolymer is a promising candidate for the blue emitting polymer LEDs.

The synthesis and aggregation studies of octasubstituted dendritic polyether containing phthalocyanines were reported [139]. The complexes were synthesized by cyclotetramerization of dendritic dialkoxyphthalonitrile. The coupling of diphenolic phthalonitrile with dendritic Frechet polyether bromide (upto 3rd generation) was done by Williamson ether synthesis technique. These dendrimers are soluble in polar solvents such as THF, acetone, chloroform and dichloromethane, whereas sparingly soluble in methanol and ethanol. The absorption studies showed similar spectra as non-aggregated phthalocyanines with a split Q-band at 700 and 665 nm, vibrational bands at 640 nm and 610 nm, and B-band at 340 nm. The band arising from absorption of the benzyl ether dendrons at 250 nm increases appropriately with the increase of generation. The investigation also showed from solution and thin film studies that the ability of the phthalocinine to aggregate decreased as generation number of the polyether dendron increases.

6.4 SYNTHESIS, CHARACTERIZATION AND PROPERTIES OF DIFFERENT TYPES OF HYPERBRANCHED POLYETHERS

In this section synthesis, characterization and properties of different HB polyethers are discussed.

6.4.1 Aliphatic

Glycidal contains one epoxide and one hydroxyl group, and therefore represent a latent AB_2 monomer for ring opening multibranching polymerization reaction (ROMBP). The chiral aliphatic HB polymers and chiral HB dendrons were synthesized [140] by the one pot polymerization of enantiomeric glycidols. The incorporation of precisely one functional group at the core was achieved via an initiator bearing a functional group i.e. a terminal double bond. Trimethylol-propane (TMP) and *bis*(2,3-dihydroxy propyl)-10-undecenylamine were used as initiators for the base catalyzed ROMBP [141] of the enantiomeric glycidols. The core unit was prepared by the reaction of 10-undecenylamine with 2 equivalents of glycidol. The whole sequence of reactions is shown in scheme 6.30. The synthesized polymers were characterized by ^{1}HNMR spectroscopy, SEC and MALDI-TOF mass spectrometry. The molecular weight of the dendrons varies from 1300 to 4800 Da with polydispersity less than 1.5. The DB was in the range of 0.53–0.57, similar as in the case of racemic polyglycerol [141]. The optical molar rotation of the dendron varies from + 5.7 to + 5.1. This value is independent of degree of polymerization. This may be due to the fact that all chiral center are separated by one oxygen and at least one carbon atom, and may also, high have flexibility of polyether structure. The double bond at the focal point can be functionalized [140] further by an excess thiol and cysteamine hydrochloride in the presence of free radicals (AIBN and heat) offers quantitative addition of the thiol to the double bond. The reaction was monitored by NMR spectroscopy and MALDI–TOF mass spectrometry. The aliphatic HB polyethers [141] were also synthesized by the anionic ring opening polymerization of glycidol. 10% Deprotonated 1,1,1-*tris*(hydroxymethyl)propane was used as an initiator in order to control the concentration of the active alkoxides sites during polymerization. Because of the unsymmetrical structure of glycidol, the synthesized HB polymer may have four different types of structural units viz. linear 1,3- and 1,4-dendritic and terminal branches. From the ^{13}CNMR spectra, the DB was found ranging from 0.53 to 0.59 with DP 15 to 83 depending on the monomer/initiator ratio. However, polydispersitivity ranged from 1.13 to 1.47 as obtained from GPC, which indicated narrow polydispersity, which indicates that it is a controlled polymerization reaction.

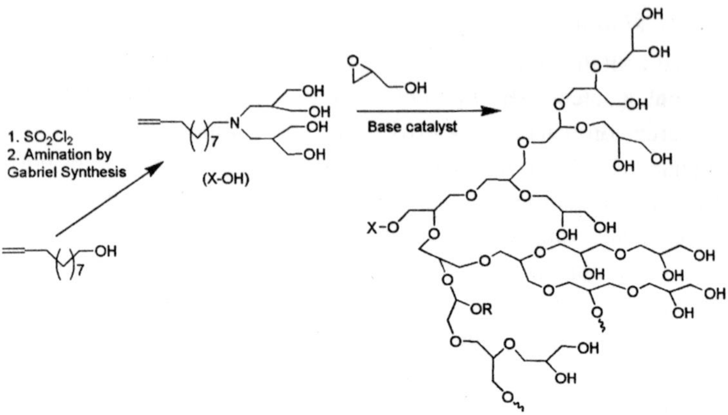

SCHEME 6.30 Synthesis of chiral aliphatic HB polyether based on enantiomeric glycidol

Multiarm block copolymers of glycidol with propylene oxide were synthesized [142] by anionic ROMBP. This is a convenient approach that permits to tailor the polarity of the HB polyglycerol without reducing the functionality and without variation of the basic polyether structure, thus without changing the relative chemical stability of the HB scaffold. Glycidol was first polymerized in the presence of 1,1,1-tri(hydroxymethyl)propane or diethylglycol by the same method as described earlier [141]. Then the polyglycerol was deprotonated by KOH to an extent of ca. 5% per OH group. The resulted polyglycerol was then treated with propylene oxide to get the desired block copolymer. The whole sequence of reaction is shown in scheme 6.31. All HB polyglycerols possess the initiator, which was used as incorporated core functionality. The main benefit of this approach lies in the fact that the molecular weight can be controlled by controlling the monomer/initiator ratio in a range between 1000 to 30000 Da with OH function between 5 to 400 and PDI below 1.8. The synthesized block copolymers were characterized by ^{1}HNMR, ^{13}CNMR spectroscopy, SEC, VPO, GPC, DSC and hydroxy number determination. Four different polyglycerols were prepared and used to get block copolymers after attachment of propylene oxide. The transparent block copolymers solubility behavior completely different from the polar polyglycerol. All hydroxyl groups of the polyglycerol core undergone propagation, which was confirmed from NMR studies. The number average molecular weight of the HB block copolymers varies from 2 X 10^4 to 3.78 X 10^4 Da with PDI 1.19 to 1.66. The hydroxy number decreases with the increase of attachment of propylene oxide (355 to 160). The DSC study measured their glass transition temperature (-36.9 to –70.6 ^{O}C). This indicates the flexibility increases with attachment of the propylene oxide blocks.

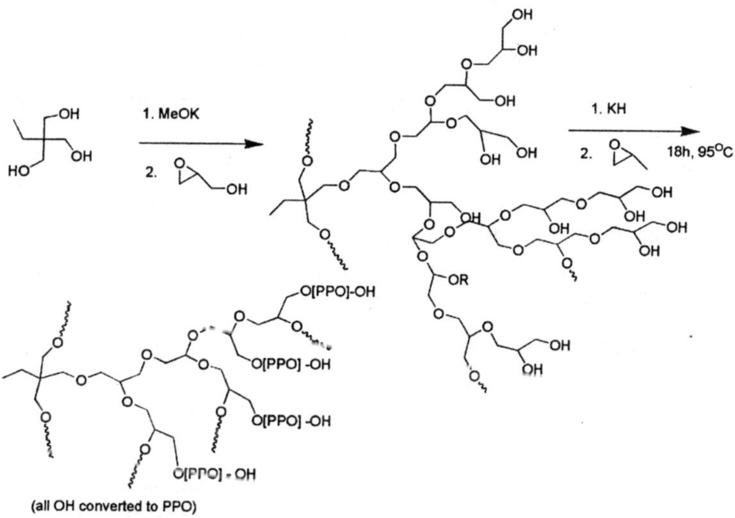

SCHEME 6.31 Synthesis of aliphatic HB polyethers block copolymers of glycidol with propylene oxide by anionic ROMBP

The above block copolymers and HB polyglycerols were also modified [143] by esterification reaction to different extent (25%-100%) with various carboxylic acid viz. benzoic acid, biphenylcarboxylic acid, C_2 to C_{18} acid, etc. The thermal properties like T_g, crystalinity, etc. for all end functionalized HB polyglycerols and hydrophobically modified propoxylated polyglycerols were studied with respect to their molecular weight, nature of the HB scaffold, alkyl substituent attached and degree of substitution. The esterification reactions of HB polyglycerols and propylene oxide end capped polymers were performed by three different ways. These are (i) by using acetic anhydride in the presence of pyridine, (ii) via acyl chlorides with N-methylimidazole catalyst in anhydrous pyridine solvent and (iii) by using free carboxylic acid with TsOH catalyst in toluene. The esterification of polyglycerol with suitable carboxylic acid creates unusual phase transfer properties, i.e. unimolecular inverted micelle-like behavior [144]. Even, broad nematic liquid crystalline phase is observed upon esterification of the end groups with mesogenic acids [145]. The esterified HB polymers were well characterized by ^{1}HNMR, ^{13}CNMR spectroscopy, SEC, GPC, DSC and polarized light microscopy. From the thermal studies, it was found that the T_g decreases on esterification for polyglycerol, whereas it increases on esterification for poly(glycerol-block-propylene oxide) HB polymers. This may be due to the fact that when longer saturated fatty acids are used for esterification, the alkyl chains crystallize and thereby forcing the HB polyether core into a more rigid structure causing an increase of T_g value. Whereas decrease of free OH groups, decrease H-bonding interaction and thereby decreasing T_g. The second effect is much more in case of polyglycerols, whereas for block copolymer, the first effect is more significant.

A marvalous work [146] on HB polyglycerol was reported by Haag *et al.* in which the terminal and linear hydroxyl groups can be distinguished. The study shows that structural inversion of the architecture i.e. apolar periphery with polar core vs. polar periphery with apolar core leads to enormous changes in their physical properties. The inverse architecture was constructed by the following method for HB polyglycerols. First, 1,2-diols of the terminal glycerol unit was selectively protected by acetals or ketal formation. This allows one to distinguish between linear and terminal hydroxy groups. Then the linear hydroxyl groups were reacted with allyl chloride or benzyl chloride under phase transfer conditions to obtain the corresponding HB polyether polyketals. Now selective deprotection of the 1,2-ketals by acidic ion-exchange resin produced linear hydroxy functionalized HB polyether. The whole sequence of reactions is shown in scheme 6.32. All the structures of the polymers were well-characterized by ^{1}HNMR, ^{13}CNMR and IR spectroscopic and DSC studies. HB polyglycerols are soluble in water and found only in water (99%) in chloroform/water mixture. However, on acetylation/ketylation of terminal 1,2-diols, the solubility of the polymer changes to chloroform phase dramatically (99% for the acetal and 95% for ketal). Functionalization of the remaining linear OH groups further decreases the solubility in water. Again, the fully methylated polyglycerol is exclusively present in organic phase. The thermal, DSC studies of the polar/apolar incompatible HB polymer indicates the presence of two glass transition temperatures at around -80^oC and $(-53$ to $-10)^oC$. The first T_g is due to linear polyether moieties, whereas the second one may be due to strong interaction and nanocluster formation of the hydroxyl groups present in

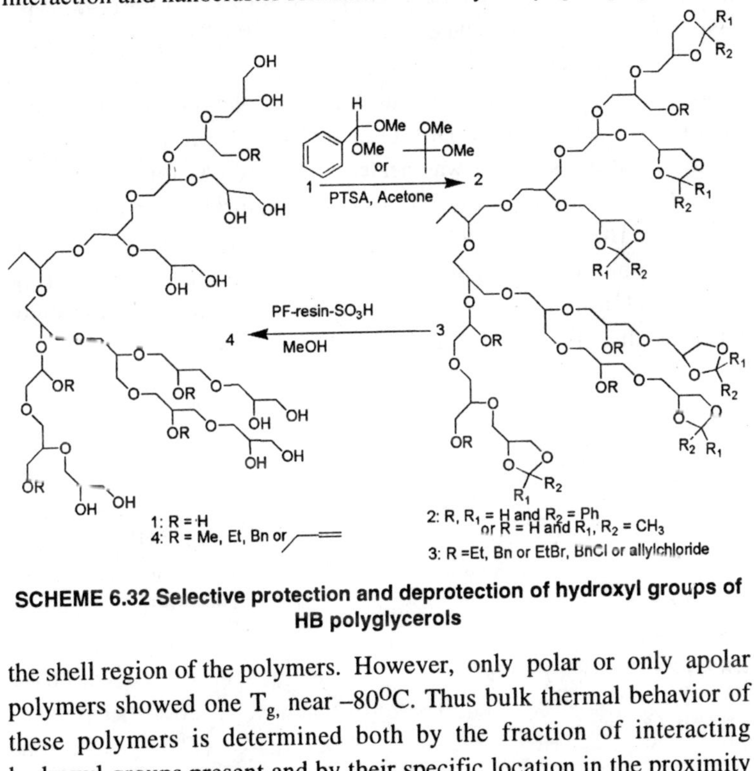

SCHEME 6.32 Selective protection and deprotection of hydroxyl groups of HB polyglycerols

the shell region of the polymers. However, only polar or only apolar polymers showed one $T_{g,}$ near -80^oC. Thus bulk thermal behavior of these polymers is determined both by the fraction of interacting hydroxyl groups present and by their specific location in the proximity of the core or in the shell.

The HB aliphatic polyethers were also synthesized [147] by using cationic ring opening polymerization of oxetane derivatives. The polymerization of 3-ethyl-3-(hydroxymethyl)oxetane was carried out in bulk at 120^oC in the presence of benzyltetramethylene sulfonium hexafluroantimonate as a thermal initiator. The synthesized HB polyether was characterized by ^{1}HNMR, ^{13}CNMR spectroscopy, SEC and MALDI-TOF mass spectrometry. The DB was found 0.41 with number average molecular weight 4170 Da. This HB polyether has promising role as a macro-initiator [147] for ring opening polymerization of ε–caprolactone.

6.4.2 Aromatic

The first HB poly(aryl ether)s were synthesized [148] by self-polycondensation of phenolate monomers in 1993. The polymers were synthesized by the following strategy as shown in scheme 6.33. The monomers in THF were treated with NaH, the solution was filtered and dried under vacuum. Then DMSO or DMAc solvent was added to it and the resulting solution was heated at 100-180°C for 0.5–2h. The product was separated by filtration and neutralized by 10% 1N methanolic HCl. The yields obtained were varying from 50-80%. The resulting polymers recognized as dendritic analog of engineering plastics such as poly(ether ketone) and poly(ether sulfone). SEC study indicates that the weight average molecular weights of the HB

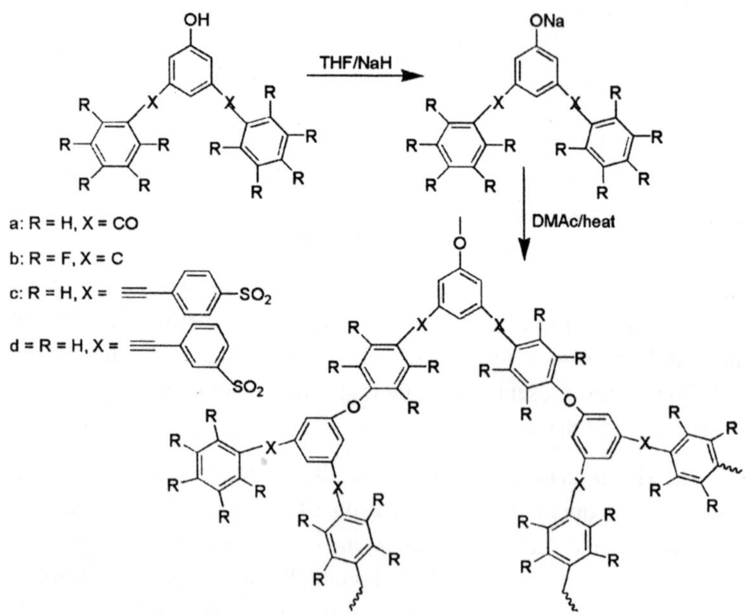

SCHEME 6.33 Synthesis of aromatic HB polyethers by self polycondensation of phenolate monomers

polymers were ranging from 11300 to 13400 Da depending on the structure of the spacer group. However, polydispersity varies from 1.53 to 4.36. The polymers are soluble in common organic solvents such as $CHCl_3$, THF, etc. although the polymers composed of aromatic rings and short spacer units. Thermal study indicates high thermostability of the polymer, 95% of the mass remained upto 500°C under the nitrogen atmosphere. The T_g of the polymers ranges 135 to 231°C but melting point was not reported.

In 1996, Srinivasan *et al.* [149] has used AB_2 type quinoxaline monomers to produce HB poly(aryl ether phenyl quinoxaline)s by aromatic heterocyclic activated nucleophilic substitution reaction. First, the monomers were synthesized by the condensation of a substituted benzil and a fluoro substituted ortho-phenylenediamine. Then the monomers were polymerized by polycondensation reaction at 180-220°C in the presence of K_2CO_3 in aprotic polar solvent (NMP). The synthesized polymers were characterized by solution viscosity measurement, thermal studies and MALDI-TOF mass spectrometric. The intrinsic viscosity of the polymers found to be 0.5 dL/g with T_g of 225°C and 190°C for the polymers. The organic-inorganic hybrid films formation from HB poly(aryl ether phenyl quinoxaline) and polysilsesquinoxane were also reported [150]. The morphology of the film was influenced dramatically by the nature of the end functional groups of the HB polyether. The size scale of phase separation for a film by using triethoxysilyl terminated polyether was less than 50 nm and the film was optically transparent. In contrast, spherically HB 'rich' regions on the order of 1-1.5 μm were observed when hydroxy terminated poly(aryl ether phenyl quinoxaline) was used.

Aromatic aliphatic HB polyethers [151], the main chains of which were composed of benzyl ether linkages were prepared as analog for polyether dendrimers with the same repeating units. 3,5-Dihydroxy benzyl bromide monomer was polymerized in the presence of K_2CO_3 and 18-crown-6 ether in acetone. It was found that the resulting polymers contained C-alkylated methylene [152] upto 30% of the polyether structure in addition to O-alykylated benzylic methylene. The weight average molecular weight of the HB polyethers was > 10^5 Da as obtained by SEC measurement. The T_g could be controlled from 311 to

343°C by chemical modification of the phenolic chain ends. HB polymers containing tetrafluorinated phenylene units [153] were prepared from AB_2 and AB_4 monomers, where A and B denote benzylic hydroxy group and pentafluorobenzene, respectively. The fluorinated HB poly(benzyl ether) was obtained by self-polycondensation of the AB_2 type monomer, which was promoted by the deprotonation of benzylic alcohol with sodium particles in THF (Scheme 6.34). The polymerization of the monomers in the presence of less than 1 mm size sodium particles resulted lower molecular weight polymer with high yield. However, the use of larger sodium particles offered high molecular weight polymer but low yield. Both, the low as well as high molecular weight HB polymers are soluble in common organic solvents, such as THF, DMF,CH_2Cl_2, $CHCl_3$, toluene, acetone, etc. The chemical modification of the pentafluorobenzene ring via nucleophilic displacement reaction changes the properties of the HB poly(benzyl ether). Contact angle measurements of the water and hexadecane on films of a pentafluoro-terminated HB polymer, trifluoroethoxy substituted and 1H,1H,2H,2H-perfluorodecanoxy substituted derivatives indicates a high degree of hydrophobicity and lipophobicity. The 1H,1H,2H,2H-perfluorodecanoxy substituted polymer led to a more than two folds decrease in the coefficient of friction and adhesive force [153]. However, cyclized components were detected by MALDI-TOF mass spectrometry [154] in the polymerization process of the HB polymers.

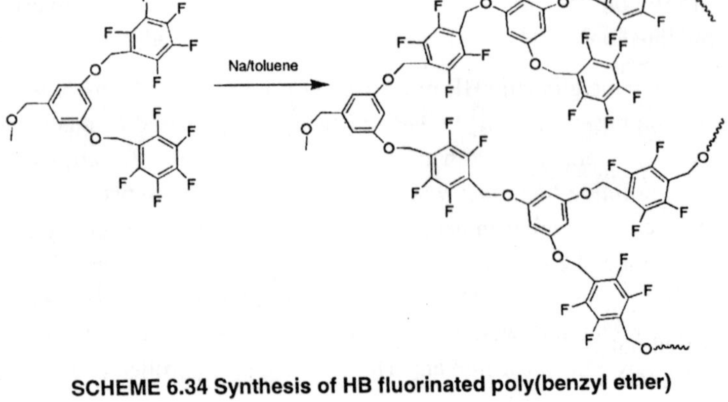

SCHEME 6.34 Synthesis of HB fluorinated poly(benzyl ether)

6.4.3 Liquid Crystalline

The liquid crystalline (LC) HB polyethers [155-157] were obtained from AB_2 type monomers containing two phenolic hydroxyl groups and an alkyl halide and these flexible monomers can exist either in anti or in gauche conformation. The polymerization reaction of the monomers was carried out in O-dichlorobenzene and aqueous NaOH solution (1N) in the presence of tetraammonium hydrogen sulfate under heterogeneous conditions. The resulting polymers were isolated after *in-situ* end capping reactions of phenolic terminated groups with alkyl halides at the end of the polymerization process, which offered soluble HB polyether. Because of the possible intramolecular cyclization reaction, the resulting HB polyethers possessing relatively low molecular weight and high PDIs. Some of these HB polymers exhibited an enantiotropic nematic mesophase derived from anti gauche isomerization. The temperature range for the nematic mesophase was dependent on the structure of the mesogen and the end functional groups. The polymer with terphenylene as a mesogen unit and octyl ether as an end group shows a stable nematic phase [157] in the range of 50-132°C. The DSC studies showed that glass and isotropization temperatures of these HB polyethers were strongly affected by the end groups.

6.4.4 Others

A new approach of proton transfers polymerization technique to synthesize HB poly(hydroxyl ether)s from AB_2 type monomer composed of two epoxides and one phenolic hydroxyl group, was reported [158]. The polymerization reaction was carried out by the addition of a base (KOH) to abstract the phenolic hydrogen and then nucleophilic addition of the phenolate anion to an epoxide ring proceeds to generate a dimer containing a secondary alkoxide. Then the activated dimer undergoes a proton exchange reaction with another AB_2 type monomer to produce the phenolate anion and a neutral dimer. The much lower pKa (~10) of the phenolic group relative to that of the secondary alkoxide enable the first proton. The colorless HB polymer was isolated from the reaction mixture. The whole sequence of reactions is shown in scheme 6.35. The polymerization reaction was carried out at 70-75°C in the THF with KOH as the base catalyst. The molecular weight of the polymer was found to be 2.06 X 10^5 Da as measured by GPC and the PDI was reduced effectively by protion wise addition of the monomer to the polymerization mixture to 2.4.

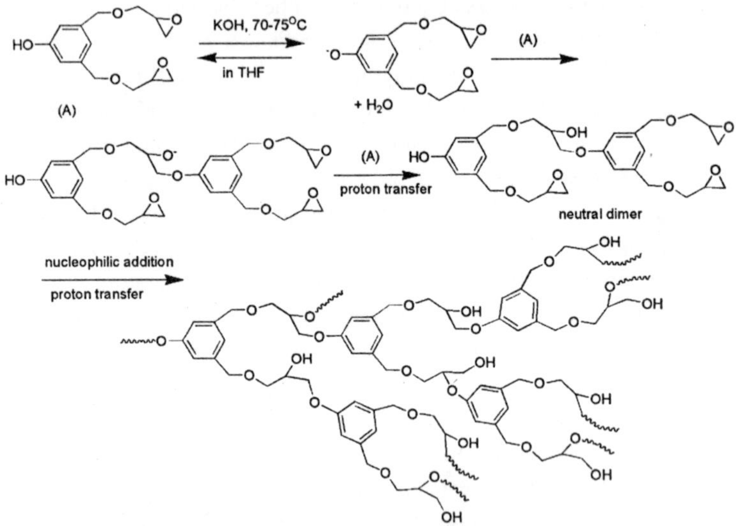

SCHEME 6.35 Synthesis of HB poly(hydroxyether)s by proton transfer polymerization technique

A HB poly(benzyl thioether) was synthesized [159] by self-poly-condensation of an AB_2 type monomer, methyl-3,5-*bis*(phenylthio) phenyl sulfoxide. Poly(aryl sulfonium) cation as a precursor was prepared through electrophilic substitution in trifluoromethane sulfonic acid, and then it was refluxed in pyridine to get the desired HB polythioether [159-160]. The HB polymer was amorphous and soluble in organic solvents, even though linear analog is a crystalline polymer and insoluble in most of the organi: solvent. The synthesis of HB poly(ether ketone)s along with HB poly(aryl ether) were also reported [148] as described earlier. Hawker and Chu [161] systematically studied the synthesis of HB poly(ether ketone)s by aromatic nucleophilic substitution reaction of four different ABx type monomers (Fig. 6.12). The self-condensation polymerization of the monomer was carried out in NMP solvent by using K_2CO_3 as the base catalyst with azeotropic distillation of water by toluene. The polymer is soluble in many organic solvents and yield obtained in the reaction was high. The number average molecular weight varies from 2 X 10^4 to 6.5 X 10^4 Da, whereas DB ranges from 0.15 to 0.49. However, the solubility of the HB poly(ether ketone)s was highly dependent on the terminal units of the polymers. The hydroxy terminated polymer was soluble in aqueous alkaline solutions. The thermal studies of the polymers revealed that the thermal properties of the HB poly(ether ketone)s were independent on the macromolecular architecture i.e. DB, but depend on the nature of terminal groups. The T_gs for these amorphous polymers ranges from 97 to 290°C. The hydroxy and fluoro-terminated HB poly(ether ketone)s are stable upto 480°C under the nitrogen atmosphere and only 10% weight loss was found over 500°C.

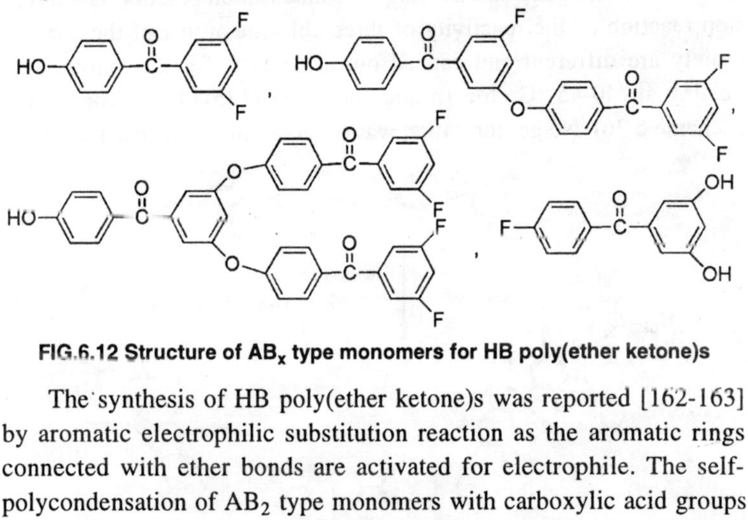

FIG.6.12 Structure of AB_x type monomers for HB poly(ether ketone)s

The synthesis of HB poly(ether ketone)s was reported [162-163] by aromatic electrophilic substitution reaction as the aromatic rings connected with ether bonds are activated for electrophile. The self-polycondensation of AB_2 type monomers with carboxylic acid groups was carried out in methanesulfonic acid by using P_2O_5 as condensing agent. Further, the end groups were functionalized to control the solubility and Tg. For example, the ammonium salt of the HB polymer was soluble in water and behaves as a unimolecular micelle. By applying aromatic nucleophilic substitution reaction a series of poly(ether sulfone) HB polymers was synthesized [164-166]. The AB_2 type monomer possessing aryl fluoride or chloride groups activated by electron withdrawing sulfone group, undergoes easily self-polycondensation reaction in DMAc at 150-170°C by using $CsCO_3$ and $Mg(OH)_2$ as the catalysts. The halogen terminated HB poly(ether sulfone)s were soluble in most of the organic solvents such as THF,

$CHCl_3$, DMAc, etc. and are more stable than hydroxy terminated ones. Depending on the nature of end functional groups, the T_gs varies from 226 to 277°C. The molecular weight was about 10^4 Da as obtained from SEC, even though the formation of cyclic products was indicated by MALDI-TOF mass spectrometry.

The aromatic HB polyethers with s-triazine rings were synthesized from A_2 [sodium salt of aromatic diol e.g. bisphenol-A, tetrabromo-bisphenol-A, 4,4'-sulfonyl diphenol, etc.] and B_3 [2,4,6-trichloro-s-triazine (CYC)] types of monomers by using a single step nucleophilic displacement polymerization technique in a polar solvent (DMAc) by Borah and Karak [166a, b]. This polycondensation reaction is a three-step reaction as the reactivity of three chlorine atoms of the triazine moiety are different and carried out at (0-5)°C for 2h followed by heating at (40-45)°C for 1h and finally at (135-140)°C for 12-14h (Scheme 6.36). No gel formation was observed under the used reaction

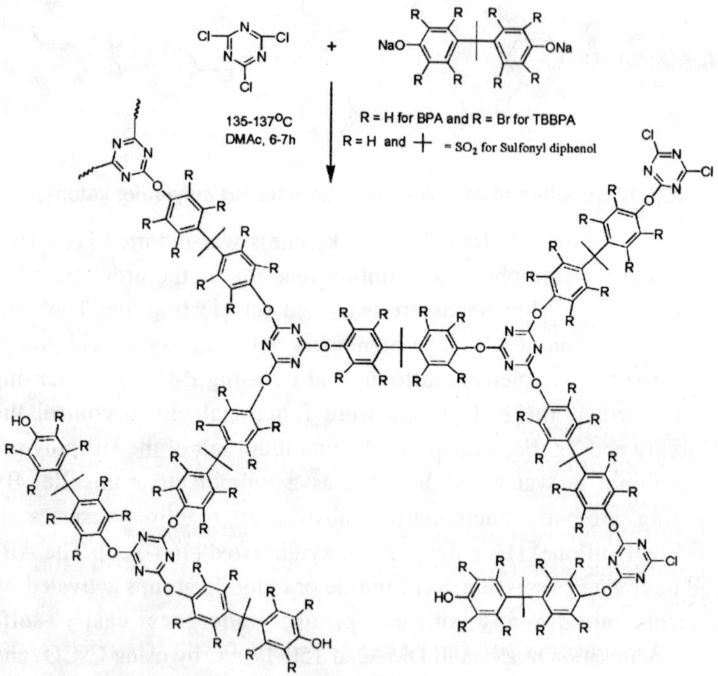

SCHEME 6.36 Synthesis of HB polyethers with s-triazine moiety in the main chain

conditions of this polymerization process, which was confirmed by solubility test in different polar solvents for each case. The HB polymers obtained by this technique have good yield (60-70)%. The synthesized polymers were characterized by solubility test, solubuity parameter measurement, inherent viscosity measurement, FT-IR, UV, ^{1}H-NMR and ^{13}C-NMR spectroscopic, X-ray diffraction, etc. studies. The inherent viscosity of the polymers was found to be in the range of 0.20-0.40 dLg^{-1} in 0.5 % (w/v) DMAc at 27±0.1^{O}C. They are soluble only in highly polar solvents such as DMAc, DMF, DMSO, NMP, etc. The TGA studies showed that the polymers are highly thermostable and modified limiting oxygen index (LOI) measurements show that all the polymers exhibit self-extinguishing characteristic. These dendritic polymers are better soluble in different polar sovent compared to their linear analog polyether and can be used as multipurpose polymeric additives for commercial linear polymers like LDPE and plasticized PVC [166c, d]

6.5 APPLICATIONS OF DENDRITIC POLYETHERS

In this section the applications of different dendritic polyethers are discussed.

6.5.1 Catalysis

The polyether dendrimers with phenylsulfides, selenides and tellurides terminal groups were used [122] as catalysts for the oxidation of bromide with H_2O_2 to give positive bromonium species that can be captured by cyclohexene in two phasic systems. A remarkable increase in catalytic activity and selectivity was observed [167] in the Pauson-Khand reaction for Co-complexes when the catalyst was immobilized on 2nd or 3rd generation polyaryl ether dendronized polystyrene. The polymer supported catalyst was obtained by the reaction of 2 and 4-(diphenyl phosphino) benzoic acid functionalized dendronized polystyrene resin (Wang resin) and $Co_2(CO)_8$ catalyst in dioxane at 75°C.

A polycation poly(aryl ether) dendrimer with tetraphenyl silane core, which forms a well-defined stoichiometric assembly with eight anionic metal complexes through non-covalent bond was successfully used [168] as a Lewis acid catalyst, which functions comparably or even better than unsupported metal complex. Further, it can be recycled and reusable without much loss of activity. This is a novel approach as a non-covalent attachment resulted a reversible assembly from which both the support and the metal complex can be retained separately.

The dendritic polyglycerols were used [169-170] as soluble high loading supports for aldehydes and ketones, and the separation of polymer bound compounds by parallel membrane separation technique. The soluble polyether skeleton allows the parallel synthesis of small libraries on a large scale (1-5 mmole). The developed purification process can separate upto 12 reaction mixtures simultaneously. The terminal 1,2-diol groups of polyglycerol with loading capacity 4.1 mmole diol/g can be directly coupled with carbonyl compounds without additional linker groups at the same time. The polyglycerol support acts as a polymeric ketal protection. The cleavage of the products occurs in high yields (95%) after phase separation or ultrafiltration. Thus the access to large quantities of these well defined polyglycerols and the chemical stability of the scaffold, in addition to the multiple possibilities for functionalization render them as ideal candidates for new soluble polymeric supports in organic synthesis. The polyether dendrimers with redox active core have been proposed as promising materials for miniaturized information storage circuits. Ideally, injecting/extracting electrons into or from the core of the dendrimer would be a simple way to store and/or read a bit of information in a single molecule. However, the redox active core must be surrounded by a considerable amount of inert material to prevent electronic 'crosstalk', i.e., self-exchange electron transfer reaction between neighboring dendrimers that would lead to data losses in a two-dimensional array. However, the isolation of the redox core by growing dendritic branches has always led to significantly decreased rate of heterogeneous electron transfer. Toba *et al.* reported [33, 37] two series of redox active viologen core polyethers dendrimers upto 4th generation, which are promising material for such applications.

The soluble and recyclable polyether based dendritic catalysts were used [171] for the enantioselective transfer hydrogenation of prochiral ketones. The amino functionalized chiral ligand was coupled by condensation with Frechet type polyether dendrons [6] upto 3rd generation (Fig. 6.13). which were used for asymmetric hydrogenation of aromatic ketones, after complex formation with ruthenium. The study showed a slight enhanced reactivity of the dendritic catalyst compared to the monomeric Ru-complex but high enantioselectivity (96%). The most important feature of this dendritic catalyst was their complete maintaining of enantioselectivity with only slight loss of activity in successive uses.

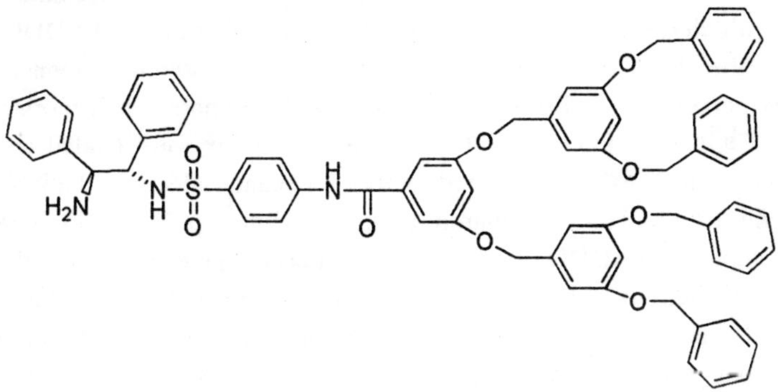

FIG.6.13 Structure of chiral dendritic ligand with 2nd generation polyether dendron

Polyethers based Ru-porphyrin complexes were used [134] as highly selective catalyst for epoxidation of alkene with 2,6-dichloropyridine-N-oxide (Cl_2PyNO). The chemo- or diastereo-selectivity increases with the generation number of the dendrons and the number of dendrons attached with porphyrin core. For example, porphyrin-$[G_2]_8$–complex exhibits remarkable selectivity in catalyzing Cl_2PyNO epoxidation of a variety of alkene substrates such as styrene (Scheme 6.37) *trans/cis* stilbene, 2,2-dimethylchromene, cyclooctene, etc. and unsaturated sterioids such as cholesteryl esters and estratetraene derivative. Whereas the cyclopropanation of styrene and its *para*

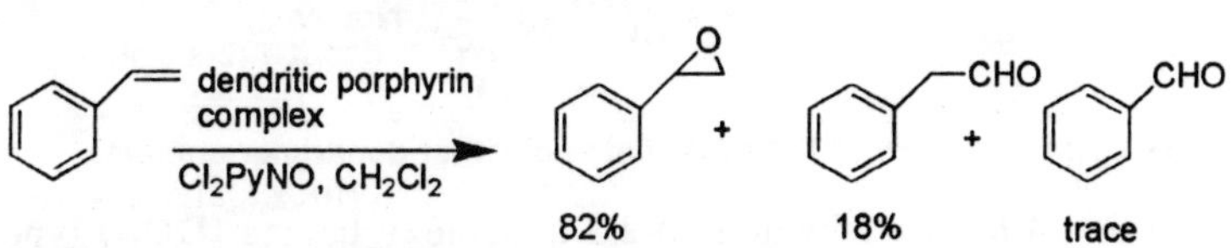

SCHEME 6.37 Epoxidation of styrene in the presence of dendritic polyether-Ru(II) porphyrin complex

substituted derivatives with ethyl diazoacetate catalyzed by the above complex was highly *trans* selective.

The chiral polyether dendrimer-metal complexes of tetraaryl-1,3-dioxolane-4,5-dimethanols (TADDOL) ligands in chiral active sites of the dendrimers were used [172] for a large variety of catalytic and stoichiometry enantioselective transformation. The $Ti(OCHMe_2)_2$ complex of TADDOL moiety in 1st generation polyether dendrimer (Fig. 6.14) catalyzes the enantioselective addition of diethyl zinc to benzaldehyde. They also prepared styryl-substituted TADDOL polyether dendrimers that can act as crosslinkers in styrene suspension polymerizations and thus lead to bead with intimately incorporated TADDOL sites [173-174]. Then these dendrimers were loaded with $Ti(OCHMe_2)_3$ to get the desired catalyst for enantioselective addition of Et_2Zn to PhCHO. Compared to conventional polymer attached insoluble Ti-TADDOLates, the new materials have a much higher catalytic activity, which is close to that of soluble analogs. Employing only 0.2 equivalent of polymer bound dendritic Ti-TADDOLates of 1st and 2nd generations, enantioselectivities upto 98:2 (S:R) were obtained. For the dendrimer crosslinked polymers the selectivity increased from zeroth to 1st generation, whereas with 2nd generation no further improvement of the enantioselectivity was observed.

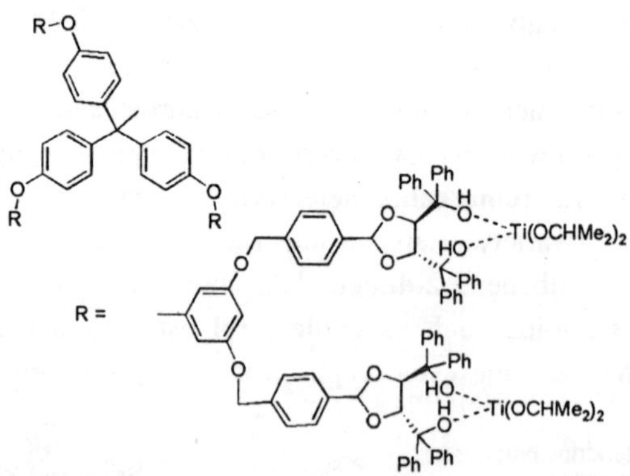

FIG.6.14 Structure of Ti-TADDOLate polyether dendrimer complex

(S,S)-1,4-*bis*(dimethylamino)-2,3-dimethoxy butane (DDB) type core containing polyether dendrimers upto 3800 Da molecular weight were used [175] as dendritic amine catalyst. These dendritic amines were used in several base catalyzed reactions e.g. the addition of ketone to chiral or the Michael addition of thiophenol to cyclohexenone. In this case, the enantioselectivity was not very significant compared to monomeric DDB. However, when dendritic diamine was used to catalyze the enantioselective addition of methanol to methyl phenyl ketone, the selectivity of (R)-methyl-2-propionate product increased by a factor of three. The polyether dendrimeric metal complexes of either Zn(II) or Ti(IV) of TADDOL at the periphery were used as homogeneous, enantioselective catalysts [176]. The catalytic activity was observed [176] for (i) the nucleophilic addition of diethyl zinc to benzaldehyde and (ii) the [3+2] cycloaddition of 3-crotonyloxalidinone to a nitone was approximately the same as that for the non-dendrimeric parent TADDOLate complex. However, the dendrimeric catalyst was reclaimed in >90% yield.

The polymer bound polyether dendritic Ti-BINOLates was used [106] as the catalyst for addition of Et_2Zn to PhCHO. The enantioselectivities (upto 93 %) and the conversions obtained with the polymer bound catalysts were almost samo to their non-polymerized BINOL crosslinker under homogeneous conditions. However, the polymer bound catalysts are reusable and upto 20 consecutive catalytic runs, there is no or only minor loss of selectivity were observed [177]. A dendritically crosslinked BINOL polymer was also used in the Ti-BINOLate mediated cyanosilylation of pivalaldehyde. The enantiopurity of the cyanohydrine was obtained in the first run as high as in the homogeneous reaction (72%). Again, suprisingly the catalytic performance of the supported catalyst increased steadily during the first catalytic cycles to reach 83% enantioselectivity.

The cleft-type dendritic *bis*(oxazoline), metal complexes of 2nd and 3rd generation were used [42, 178] as catalyst for Diels-Alder reaction. The reaction of two differently size dienophiles was investigated and found that a slightly enhanced ratio of initial reaction velocities with reaction of the small dienophile being less hindered by the branching. Actually, upon complexation of the dienophile at the focal point the geometry at the center changes. This result in an increase in steric repulsion between the dendritic wedges, which is more pronounced for the layer systems.

The catalytic film of a polyether dendrimer with phenylseleno group at the periphery [179] can kill algae and bacteria in seawater (Fig.6.15). The selenides catalyze the oxidation of bromide ion with H_2O_2 to give positive bromonium radical [179]. These radicals are very toxic to algae and bacteria and thus dendrimer film on hulk will provide protection against the accumulation of algae.

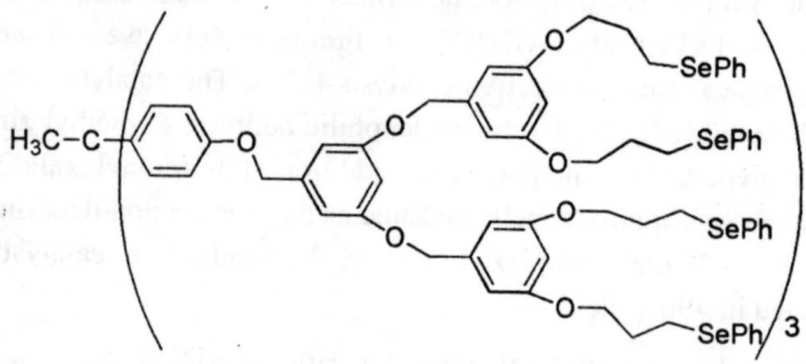

Fig.6.15 Structure of polyetherdeudrimer with selenide groups at the periphery

The salen-polyether dendritic Mn complexes were used [126] for epoxidation of olefins under homogeneous conditions. Also no generation dependent catalytic activity was observed for different generation salen-dendrimers in contrast to TADDOL and BINOL dendrimers. The polymer bound Mn-salens obtained by copolymerization of vinyl terminated salen dendrimers with styrene were used [126] for same epoxidation reaction of olefins. The catalytic activity of the polymer bound Mn-salen dendrimer was found to be same as it is under homogeneous conditions. But the advantage obtained is their reusability in many catalytic cycles. This polymer Cr-salen dendrimer was also used in hetero Diels-Alder reaction of Danishefsky's diene with various aldehydes [174]. The yields obtained were relatively high with very good enantioselectivity.

Different sizes of the core functionalized metallodendrimer containing Frechet type polyether dendritic wedges with anchoring sensor active aryl platinum(II) sites at focal point were used as catalysts and dyes [180]. The strong color of these metallodendrimers in the presence of SO_2 was used to assess the permeability of nanofiltration membrane at ambient pressure. The nanofilter, membrane–capped, immersion vial was used as sensor when loaded with these metallodendrimers with good retention properties. Again, the appropriate substitution of the dye site at the focal point of these metallodendritic wedges by a catalytically active group afforded dendritic catalysts. These catalysts exhibit essentially the same physical properties as the corresponding dye functionalized dendritic wedges, when these homogeneous catalysts are used compartmentalized in membrane capped vials. Further more, such immobilized metallodendritic catalysts can be regenerated and stored for months without losing their activity. Thus they may provide access for the development of novel sustainable homogeneous catalyst.

6.5.2 Macromolecular Initiators

The first use of polyether dendrons as macromolecular initiators for the living/ controlled free radical polymerization of vinyl monomers was reported [181]. The polyether dendrons containing either a single benzylic TEMPO or halide functionality at their focal point have been used for the nitroxide mediated and ATRP respectively (Scheme 6.38). The dendron TEMPO method affords greater controlled over the polymerization at high molecular weight and less prone to chain-transfer than the dendritic halide ATRP process. However, both the techniques offer excellent control over molecular weight and polydispersity index.

The HB polyglycerol was deprotected with diphenyl-methyl potassium and used as polyfunctional initiators for the anionic polymerization of ethylene oxide to prepare poly(ethylene oxide) to multiarm star polymers [182].

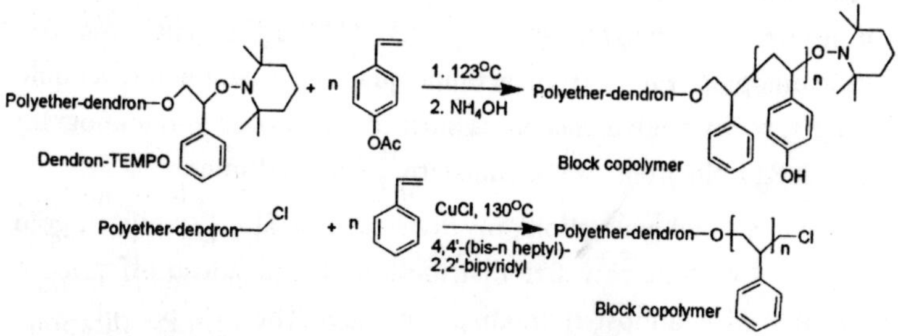

SCHEME 6.38 Use of dendritic polyether as macroinitiator for free radical polymerization of vinyl monomers

The polyether dendrimers were also used to control the degree of polymerization and PDI of the polymers when the polymerization of the said polymers performed in the presence of the Frechet type polyether dendrimers [176, 183-184]. The growing tip of the polymer chain can reside inside the cavity of the dendrimer, thus avoiding termination reaction with another chain. Again, the monomer molecules diffuse into the dendrimer, allowing the polymerization to proceed.

The HB polyglycerols were transformed into polyfunctional macroinitiators [185] for ATRP and polymerized (40%) methyl acrylate.

6.5.3 Electronics /Nanocomposites Materials

The polyether dendrimers with azobenzene core unit may be used as photo-switchable materials [186-187]. It has been found [69] that under irradiation with infrared light at 21°C the isomerization from the Z-form to the E-form takes place 260 times faster than without light. Actually, the polyether dendrons absorbed and accumulated energy and then transfer it to the focal azobenzene unit. The photoswitchable effect was observed [187] for the 1st generation with only two polyether dendrons i.e. without any steric hindrance problem. The photochemical behavior of this dendrimer is analog to monomeric azobenzene moiety. The polyether dendrimers with three photoswitchable azobenzene units attached with the central core unit were utilized [188] as fluorescence materials. These dendrimers are available in four different isomeric states viz. EEF, EEZ, EZZ and ZZZ.

An organometallic arborol serving as a molecular electronic device for multielectron transfer was reported [189-190]. The nonairon dendritic complex (Fig. 6.16) bears nine quasiequivalent and reversible Fe(II)/Fe(I) redox equivalents on its surface. The cyclic voltammogram shows that 8±1 electrons can be transferred per dendrimer.

A novel second generation polyether dendron [177] stabilized gold nanoparticles with narrow size distribution. The dendron offered the best results with almost monodisperse core. The HB polyglycerol nanocapsules can also be employed [191] as nano size templates for the preparation of transition metal nanocomposites.

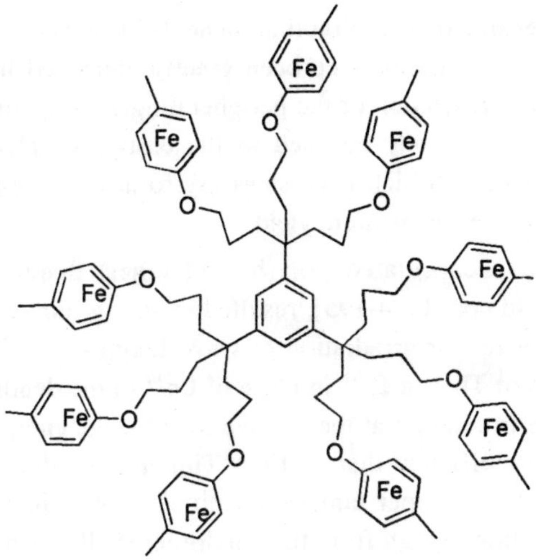

FIG.6.16 Structure of 'nonairon' dendritic complex

6.5.4 Energy Transfer Materials

A series of laser dye labeled poly(aryl ether) dendrimers upto generation four was used [192] as light harvesting and energy transfer materials. They act as extremely efficient light harvesting antennae capable of transferring light energy through space from their periphery (cumarin 2 dye) to their core (cumarin 343 dye). The light harvesting capability of the dendrimers increases with the increase of generation number is due to an increase of number of chromophores at the periphery. But energy transfer efficiency although is quantitative for 1st to 3rd generation, however, it decreases for 4th generation dendrimer [192] (93%). This novel class of dendrimers may have tremendous potential for integral component of molecular photonic devices as they have high extinction co-efficients and fluorescence quantum yields of the chromophores along with the efficient intramolecular energy transfer of the dendritic assemblies. Further the replacement of the cumarin 343 acceptor by other chromophores may result in efficient energy transfer, as long as the absorption spectrum retains good overlap with the donor emission [193]. Oligothiophane chromophores proved to be ideal for this purpose, since it is possible to tune their absorption and emission properties by changing the degree of oligomerization. Again, the emission intensity of the oligothiophene, which has relatively low quantum yield of fluorescence been greatly enhanced by sensitized excitation upon irradiation of the peripheral light-harvesting antennae. Thus, this material may be used in the emission enhancement of photonic devices in which it is necessary to utilize components that have low fluorescence quantum yields.

Ln^{3+} ion encapsulated poly(benzyl ether) dendrimers at the aromatic triacid core [194-195], resulted in strong luminescence from the lanthanide ion on irradiation at wave lengths 280-290 nm. The encapsulation of Tb^{3+} or Er^{3+} in place of Ln^{3+} in the dendrimer, it was found that on irradiation at these wave lengths, the energy transfer to Tb^{3+} was more efficient than to Er^{3+}. This may be due to the better overlap of the dendrimer emission with Tb^{3+} absorption. Thus, the transfer excitation energy from the dendrimer shell to single core unit i.e. antennae effect critically depends on the substitution pattern within the dendritic shell. A variety of polyether dendritic structures [64], differing in the number of dendrons attached to the central porphyrin as well as in the generation number of the dendrons, were used as energy transfer materials. The energy transfer efficiency in both the 4th and 5th generations analogs was calculated to be ~ 80%. However, if the porphyrin was only partially substituted with one, two or three 5th generation dendrons instead of four, the energy transfer efficiency dropped dramatically to 10, 20 and 32% respectively. The temperature dependent effects indicated that increased flexibility and conformational freedom were responsible for the decrease of energy transfer efficiency. However, the highly crowded dendrimer with four 5th generation dendrons retained a constant level of energy transfer, even at high temperature.

The encapsulation of a poly(phenylene-ethynylene) rigid-rod polymer by a poly(benzyl ether) Frechet type dendrimer shell was effective as a steric 'bumper' preventing polymer chains from aggregation and as a light harvesting antennae [196]. The direct excitation of the polymer backbone showed the fluorescence quantum yield remains constant over a wide concentration range only if a large dendrimer envelop (composed of 4th generation dendrons or higher) was utilized. These dendritic polymers may have potential for organic LED device.

Poly(benzyl ether) 1st and 2nd generation dendrons with naphthyl or pyrenyl groups at the periphery and a tertiary amine at the core [197] were used as light harvesting materials. Two dyes labeled polyether dendrimers were also used [198] in two component single layer organic LED. The photoluminescence spectra both in solution and in the solid state, showed that forester energy transfer from the donor chromophores on the dendrimers periphery to the acceptor located at the core is highly efficient, and affords emission exclusively from the core dyes, either cumarin 343 or a benzene-capped penta thiophene. Further, when these dendrimers are doped into single layer LED, the electroluminescence is almost identical to the above photoluminescence, indicating once again exclusive emission from the core chromophores.

6.5.5 Biomedical

The HB polyglycerol and its derivatives represent a versatile and promising class of materials for future biomedical applications [199], may be due to the expected biocompatibility of the polymers. A polyether dendrimer-PEG conjugate, may be used as a model drug carrier system [200]. The parent polyether dendrimer having two different types of chain end functionalities one of which attached with short PEG chain to make it water soluble and the rest type were then used to incorporate model drug molecules of varying polarity such as cholesterol and it's amino acid derivatives.

6.5.6 Micelles

The carboxyl terminated poly(aryl ether) dendrimers [9] act as unimolecular micelles, as the terminal carboxylate groups help the dendrimers to solubilize in basic aqueous solution. A linear relationship between the amount of solubilized pyrene and the dendrimer concentration was observed. The solubilization efficiency of the 4th generation dendrimer is comparable to SDS, although the dendrimer retained its ability to solubilize guest molecules at very low concentration also (5×10^{-7} mole /dm^3), whereas SDS worked only above its CMC (8.1×10^{-3} mole/dm^3). On an average ~0.45 molecule of guest pyrene was dissolved by each dendrimer molecule. Further, this value increased to 1.9 pyrene molecules per dendrimer by the addition of NaCl, which may cause the elimination of some water from the interior of the dendrimer. A possible π-π contribution to the complexation was suggested by using other aromatic guest such as 2,3,6,7-tetranitrofluorenones, 1,4-diamino anthraquinone, etc. The saturation concentration of anthracene increased 58 times, 1,4-diamino anthraquinone 56 times and 2,3,6,7-tetranitro fluorenone 258 times relative to pure water. Again, the whole system can be used as a novel and recyclable solubilization and extraction system. After solubilization of pyrene inside the dendrimer interior, the dendrimer can be precipitated by a decrease in pH of the aqueous medium. Upon extraction of the organic phase with aqueous KOH, the polyether dendrimer migrates to the organic medium, whereas pyrene remains in the organic phase. This solubilization and extraction procedure with no decrease in efficiency, show potentiality of the system.

6.5.7 Encapsulation

The partial esterification of HB polyglycerol with fatty acids such as palmitic or stearic acid offers potential for the preparation of amphiphilic structure with a hydrophobic shell and hydrophilic interior groups [199]. Such 'molecular nanocapsules' do not exhibit aggregation in dilute solution and thus act as inverse unimolecular micelles to irreversibly transfer a well defined number of water soluble guest molecules (dyes) from an aqueous solution to apolar environment. The coordination ability of different dendritic imidazoles to the dendritic porphyrin (Fig. 6.17) was studied [51, 62]. The size selective complexation was observed as the dendritic porphyrin bound preferentially a small vitamin K3 molecule (2-methyl-1,4-naphthaquinone) in the presence of a large porphyrin guest. The dendritic substituents serve as a steric barrier preventing the larger molecule from binding close to the core of the metalloporphyrin [62].

Frechet and his coworkers [201] further extended this solubilization approach by covalently attachment of poly(ethylene oxide) (PEO) oligomer (Mw ~ 2000 Da) to the same dendrimer non-ionic macromolecules. This provides an interesting host-guest system, because it is non-immunogenic and exhibits low toxicity. Studies with pyrene indicate that the guest resides in the dendrimer core and not in the polar

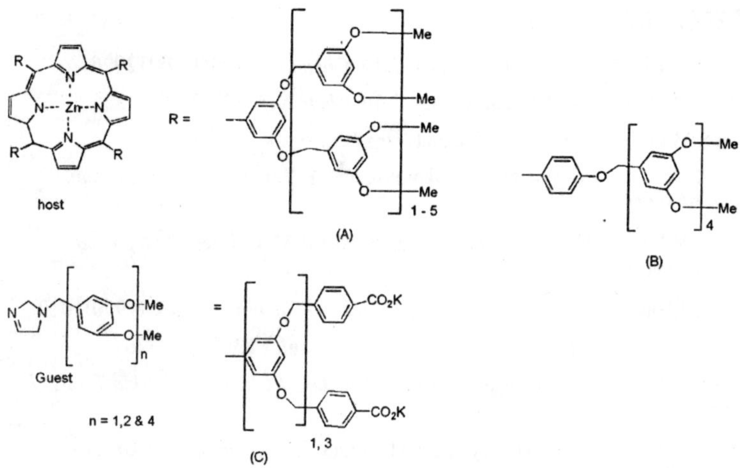

Fig. 6.17. Structure of Zn-porphyrin dendrimers and guest molecule

chains. It has also shown that the polarity inside the dendrimer is similar to that of $CHCl_3$ and that the dissolved guest has a low conformational mobility.

The *tert*-butyl terminated Frechet type polyether dendrimers with naphthyridine or anthypridine units were used [202-203] as host to site specifically complex formation within their interior by guest molecules. The complexation of naphthyridine or anthypridine based dendrimer was through H-bonding interactions with guest benamidinium and pentamidine derivatives.

6.6 SUMMARY

In this chapter one of the most important and extensively studied dendritic polymers are discussed. From this discussion it has been found that this novel class of dendritic polymers has the versatility with respect to their structures, properties, and applications, even though their synthetic methods are quite easy. The iterative reactions of etherification and activation steps result the desired dendritic polymers, in general. They exhibited photoactivity, redoxactivity, catalytic, micelle like characteristics and hence they have potential uses in different fields of application.

REFERENCES

[1] M.Jayaraman and J.M.J.Frechet, *J.Am.Chem.Soc.* **120**(1998)12996

[2] S.M.Grayson and J.M.J.Frechet, *J.Am.Chem.Soc.* **122**(2000)10335

[3] S.M.Grayson and J.M.J.Frechet, *Org.Lett.* **4**(2002)3171

[4] H.Nemoto, J.G.Wilson, H.Nakamure and Y.Yamamoto, *J.Org.Chem.* **57**(1992)435

[5] A.B.Padias, H.K.Hall, D.A.Tomalia and J.R.McConnell, *J.Org.Chem.* **52**(1987)5305

[6] C.J.Hawker and J.M.J.Frechet, *J.Chem.Soc.Chem.Commun.* (1990)1010

[7] C.J.Hawker and J.M.J.Frechet, *Macromolecules* **23**(1990)4726

[8] K.L.Wooley, C.J.Hawker and J.M.J.Frechet, *J.Chem.Soc.Perkin Trans. 1* (1991)1059

[9] C.J.Hawker, K.L.Wooley and J.M.J.Frechet, *J.Chem.Soc.Perkin Trans. 1* (1993)1287

[10] C.J.Hawker, K.L.Wooley and J.M.J.Frechet, *J.Am.Chem.Soc.Chem.Commun.* (1994)925

[11] K.L.Wooley, C.J.Hawker and J.M.J.Frechet, *J.Am.Chem. Soc.* **113**(1991)4252

[12] K.L.Wooley, C.J.Hawker and J.M.J.Frechet, *J.Am.Chem. Soc.* **115**(1993)11496

[13] R.Kleppinger, H.Reynaers, K.Desmedt, B.Forier, W.Dehaen, M.Koch and P.Verhaert, *Macromol. Rapid Commun.* **19**(1998)111

[14] C.J.Hawker, K.L.Wooley and J.M.J.Frechet, *J.Am.Chem. Soc.* **115**(1993)4375

[15] S.Hecht and J.M.J.Frechet, *J.Am.Chem. Soc.* **121**(1999)4084

[16] J.W.Leon, M.Kawa and J.M.J.Frechet, *J.Am.Chem.Soc.* **118**(1996)8847

[17] Z.Bo and A.D.Schluter, *Chem.Eur.J.* **6**(2000)3235

[18] Z.Bo, C.Zhang, N.Serverin, J.P.Rabe and A.D.Schluter, *Macromolecules* **33**(2000)2688

[19] A.W.Freeman, L.A.J.Chrisstoffels and J.M.J.Frechet, *J.Org.Chem.* **65**(2000)7612

[20] G.Evmenenko, B.J.Bauer, R.Kleppinger, B.Forier, W.Dehaenn, E.J.Amis, N.Mischenko and H.Reynaers, *Macromol.Chem.Phys.* **202**(2001)891

[21] R.N.Ganesh, J.Shraberg, P.G.Sheridan and S.Thyumanavan, *Tetrahedron Lett.* **43**(2002)7217

[22] J.G.Weintraub and J.R.Parquette, *J.Org.Chem.* **64**(1999)3796

[23] R.Spindler and J.M.J.Frechet, *J.Chem. Soc.Perkin Trans. 1* (1993)913

[24] S.Jahromi, B.Coussens, N.Meijerink and A.W.M.Braam, *J.Am.Chem.Soc.* **120**(1998)9753

[25] V.S.K.Balagurusamy, G.Ungar, V.Percec and G.Johansson, *J.Am.Chem. Soc.* **119**(1997)1539

[26] V.Percec, W.D.Cho, G.Ungar and D.J.P. Yeardley, *Chem.Eur.J.* **8**(2002)2011

[27] V.Percec, W.D.Cho, G.Ungar and D.J.P. Yeardley, *J.Am.Chem.Soc.* **123**(2001)1302

[28] V.Percec, P.W.Chu, G.Ungar and J.P. Zhou, *J.Am.Chem.Soc.* **117**(1995)11441

[29] J.F.Li, K.A.Crandall, P.W.Chu, V.Percec, R.G.Petschek and C.Rosenblatt, *Macromolecules* **29**(1996)7813

[30] M.Brewis, G.J.Clarkson, M.Helliwell, A.M.Holder and N.B.McKeown, *Chemistry* **6**(2000)4630

[31] H.F.Chow, I.Y.K.Chan, D.J.W.Chan and R.W.M.Kwok, *Chem.Eur.J.* **2**(1996)1085

[32] H.Jiang, S.J.Lee and W.Lin, *Org.Lett.* **4**(2002)2149

[33] R.Toba, J.M.Qiuntela, C.Peinador, E.Roman and A.E.Kaifer, *Chem.Commun.* (2002)1768

[34] A.Jasat and J.C.Sherman, *Chem.Rev.* **99**(1999)931

[35] E.Roman, M.Chas, J.M.Qiuntela, C.Peinador and A.E.Kaifer, *Tetrahedron* **58**(2000)699

[36] C.J.Hawker and J.M.J.Frechet, *J.Am.Chem. Soc.* **112**(1990)7638

[37] R.Toba, J.M.Qiuntela, C.Peinador, E.Roman and A.E.Kaifer, *Chem.Commun.* (2001)857

[38] P.Ceroni, V.Vicinelli, M.Maestri, V.Balzani, W.M.Muller, U.Muller, U.Hahn, F.Osswall and F.Vogtle, *New J.Chem.* **25**(2001)989

[39] C.B.Gorman, *Adv.Mater.* **9**(1997)117

[40] C.B.Gorman, B.L.Parkhurst, W.Y.Su and K.Y.Chen, *J.Am.Chem.Soc.* **119**(1997)1141

[41] K.Y.Chen and C.B.Gorman, *J.Org.Chem.* **61**(1996)9229

[42] Y.Rio, G.Accorsi, N.Armaroli, D.Felder, E.Levillain and J.F.Nierengarten, *Chem.Commun.* (2002)2830

[43] S.Nlate, Y.Nieto, J.C.Blais, J.Ruiz and D.Astruc, *Chem.Eur.J.* **8**(2002)171

[44] Z.Bo and A.D.Schluter, *Macromol. Rapid Commun.* **20**(1999)21

[45] I.Gitsov and C.Zhu, *Macromolecules* **35**(2002)8418

[46] S.L.Gilat, A.Adronov and J.M.J.Frechet, *J.Org.Chem.* **64**(1999)7474

[47] S.L.Gilat, A.Adronov and J.M.J.Frechet, *Polym.Mater.Eng.* **77**(1997)91

[48] S.L.Gilat, A.Adronov and J.M.J.Frechet, *Angew.Chem.Int.Ed.* **38**(1999)1422

[49] Y.H.Liao and J.R.Moss, *J.Chem.Soc.Chem.Commun.* (1993)4922

[50] Y.H.Liao and J.R.Moss, *Organometallic* **14**(1995)2130

[51] J.R.Moss, *Platinum Met.Rev.***39** (1995)33

[52] M.Pleovets, F.Vogtle, L.Decola and V.Balzani, *New J.Chem.* (1999)63

[53] F.Vogtle, M.Pleovets, M.Nieger, G.C.Azzellini, A.Credi, L.Decola, V.DeMarchis, M.Venturi and V.Balzani, *J.Am.Chem.Soc.* **121**(1999)6290

[54] M.S.Matos, J.Hofkens, W.Verheijen, F.C.DeSchryver, S.K.Hecht, W.Pollak, J.M.J.Frechet, B.Forier and W.D.Dehaen, *Macromolecules* **33**(2000)2967

[55] R.Sadamoto, N.Tomioka and T.Aida, *J.Am.Chem.Soc.* **118**(1996)3978

[56] N.Tomioka, D.Takasu, T.Takahashi and T.Aida, *Angew.Chem.Int.Ed.* **37**(1998)1531

[57] K.W.Pollak, E.M.Sanford and J.M.J.Frechet, *J.Mater.Chem..* **8**(1998)519

[58] D.Cao and H.Meier, *Angew.Chem.Int.Ed.* **40**(2001)186

[59] M.Takahashi, T.Odagi, H.Tomita, T.Oshikawa and M.Yamashita, *Tetrahedron Lett.* **44**(2003)2455

[60] J.Hayakawa, A.Momotake and T.Arai, *Chem.Comuun.* (2003)94

[61] T.Mizutani, M.Ikegami, R.Nagahata and T.Arai, *Chem.Lett.* (2001)1014

[62] R.H.Jin, T.Aida and S.Inoue, *Chem.Commun.* (1993)1260

[63] K.W.Pollak, J.W.Leon, J.M.J.Frechet, M.Maskus and H.D.Abruna, *Chem. Mater.* **10**(1998)30

[64] D.L.Jiang and T.Aida, *J.Am.Chem.Soc.* **120**(1998)895

[65] Y.Tomoyose, D.L.Jiang, R.H.Jin, T.Aida, T.Yamashita, K.Horie, *et al., Macromolecules* **29**(1996)5236

[66] J.M.Serin, D.W.Brousmiche and J.M.J.Frechet, *Chem.Commun.* (2002)2605

[67] L.Zhang, F.Huo, L.Wu, Z.Wang and X.Zhang, *Macromol.Chem.Phys.* **202**(2001)1618

[68] B.Osterby, R.D.McKelvey and L.Hill, *J.Chem.Edu.* **47**(1970)176

[69] J.J.Apperloo, R.A.J.Janssen, P.R.L.Malenfant, L.Groenendaal and J.M.J.Frechet, *J.Am.Chem.Soc.* **122**(2000)7042

[70] S.C.Zimmerman, F.W.Zeng, D.E.C.Reichert and S.V.Kolotuchin, *Science* **271**(1996)1095

[71] P.Thiyarajan, F.W.Zeng, C.Y.Ku and S.C.Zimmerman, *J.Mater.Chem.* **7**(1997)1221

[72] H.W.Gibson, L.Hamilton and N.Yamaguchi, *Polym.Adv.Technol.* **11**(2000)791

[73] V.Percec, W.D.Cho, P.E.Mosier, G.Ungar and D.J.P. Yeardley, *J.Am.Chem.Soc.* **120**(1998)11061

[74] B.Dong, F.Huo, L.Zhang, X.Yang, Z.Wang, X.Zhang, S.Gong and J.Li, *Chem.Eur.J.* **9**(2003)2331

[75] L.Zhang, B.Zou, B.Dong, F.Huo, X.Zhang, L.Chi and L.Jiang, *Chem.Commun.* (2001)1906

[76] M.S.Wendland and S.C.Zimmerman, *J.Am.Chem. Soc.* **121**(1999)1389

[77] S.Hoger, *Synthesis* (1997)20

[78] L.G.Schultz, Y.Zhao and S.C.Zimmerman, *Angew.Chem.Int.Ed.* **40**(2001)1962

[79] O.Mitsunobu, *Synthesis* (1981)1

[80] A.Ikeda, M.Numata and S.Shinkai, *Chem.Lett.* (1999)929

[81] M.Numata, A.Ikeda and S.Shinkai, *Chem.Lett.* (2000)370

[82] D.M.Junge, M.J.Wu, J.R.McElhanon and D.V.McGrath, *J.Org.Chem.* **65**(2000)5306

[83] J.R.McElhanon, M.J.Wu, M.Escoban and D.V.McGrath, *Macromolecules* **29**(1996)8979

[84] J.McElhanon and D.V.McGrath, *Polym.Mater.Sci.Eng.* **77**(1997)153

[85] D.M.Junge and D.V.McGrath, *Polym.Mater.Sci.Eng.* **77**(1997)157

[86] C.C.Mak and H.F.Chow, *J.Chem.Soc.Chem.Commun.* (1996)1185

[87] H.F.Chow and C.C.Mak, *J.Chem.Soc.Perkin Trans. 1* (1997)91

[88] H.F.Chow and C.C.Mak, *Pure Appl.Chem.* **69**(1997)483

[89] H.F.Chow and C.C.Mak, *Tetrahedron Lett.* **33**(1996)5935

[90] D.Seebach, J.M.Lapierre, G.Greiveldinger and K.Skobridis, *Helv.Chim.Acta* **77**(1994)1673

[91] P.K.Murer and D.Seebach, *Angew.Chem.Int.Ed.* **34**(1995)2116

[92] P.K.Murer, J.M.Lapierre, G.Greiveldinger and D.Seebach, *Helv.Chim.Acta* **80**(1997)1648

[93] G.Greiveldinger and D.Seebach, *Polym.Mater.Sci.Eng.* **77**(1997)184

[94] P.K.Murer and D.Seebach *Polym.Mater.Sci.Eng.* **77**(1997)69

[95] C.Bolm, N.Derrien and A.Seger, *Syn.Lett.* (1996)387

[96] H.W.I.Peerlings and E.W.Meijer, *Eur.J.Org.Chem.* **1998**(1998)573

[97] H.W.I.Peerlings and E.W.Meijer, *Chem.Eur.J.* **3**(1997)1563

[98] H.W.I.Peerlings, D.C.Trimbach and E.W.Meijer, *Chem.Commun.* (1998)497

[99] M.Suarez, J.M.Lehn, S.C.Zimmerman, A.Skoulios and B.Heinnch, *J.Am.Chem.Soc.* **120**(1998)9526

[100] J.L.Chaumette, M.J.Laufersweiler and J.R.Parquette, *J.Org.Chem.* **63**(1998)9399

[101] H.T.Chang, C.T.Chen, T.Kondo, G.Siuzdak and K.B.Sharpless, *Angew.Chem.Int.Ed.* **35**(1996)182

[101a] J.A.Kremers and E.W.Meijer, *J.Org.Chem.* **59**(1994)4262

[102] M.P.Struijk, H.W.I.Peerlings and E.W.Meijer, *Polym.Prep.* **37**(1996)497

[103] H.W.I.Peerlings, D.C. Trimbach and E.W.Meijer, *Polym.Mater.Sci.Eng.* **77**(1997)73

[104] P.B.Rheiner and D.Seebach, *Chem.Eur.J.* **5**(1999)3221

[105] H.Sellner, P.B.Rheiner and D.Seebach, *Helv.Chim.Acta* **85**(2002)352

[106] H.Sellner, C.Faber, P.B.Rheiner and D.Seebach, *Chem.Eur.J.* **6**(2000)3692

[107] Y.Rio, G.Accorsi, H.Nierengarten, J.L.Rehspringer, B.Honerlage, G.Kopitkovas, A.Chugreev, A.V.Dorsselear, N.Armaroli and J.F.Nierengarten, *New J.Chem.* **26**(2002)1146

[108] M.J.Hannon, P.C.Mayers and P.C.Taylor, *J.Chem.Soc.Perkin Trans. 1* (2000)1881

[109] J.F.Eckert, D.Byrne, J.F.Nicoucl, L.Oswald, J.F.Nierengarten, M.Numata, A.Ikeda, S.Shinkai and N.Armaroli, *New J.Chem.* **24**(2000)749

[110] T.Nishioka, K.Tashiro, T.Aida, J.Y.Zheng, K.Kinbara, K.Saigo, S.Sakamoto and K.Yamaguchi, *Macromolecules* **33**(2000)9182

[111] K.Y.Kay, K.J.Han, Y.Y.Yu and Y.D.Park, *Tetrahedron Lett.* **43**(2000)5053

[112] I.Gitsov, K.L.Wooley and J.M.J.Frechet, *Angew.Chem.Int.Ed.* **104**(1992)1200

[113] I.Gitsov, K.L.Wooley, C.J.Hawker, P.T.Ivanova and J.M.J.Frechet, *Macromolecules* **26**(1993)5621

[114] I.Gitsov and J.M.J.Frechet, *J.Am.Chem. Soc.* **118**(1996)3785

[115] L.Zhu, X.Tong, M.Li and E.Wang, *J.Polym.Sci.Part A Polym.Chem.* **38**(2000)4282

[116] I.Neubert, E.Amoulong-Kristein, A.D.Schluter and H.Dautzenberg, *Macromol. Rapid Commun.* **17**(1996)517

[117] B.Karakaya, W.Claussen, K.Gessler, W.Saengen and A.D.Schluter, *J.Am.Chem.Soc.* **119**(1997)3296

[118] W.Stocker, B.L.Schurmann, J.P.Rabe, S.Foster, P.Lindner, I.Neubert and A.D.Schluter, *Adv.Mater.* **10**(1998)793

[119] H.Ma, B.Chen, T.Sassa, L.R.Dalton and A.K.Y.Jen, *J.Am.Chem.Soc.* **123**(2001)986

[120] G.Chessa and A.Scrivanti, *J.Chem.Soc.Perkin Trans. 1* (1996)307

[121] G.Chessa, A.Scrivanti, R.Seraglia and P.Traldi, *Rapid Commun.Mass Spectrom.* **12**(1998)1533

[122] C.Francavilla, M.D.Drake, F.W.Bright and M.R.Detty, *J.Am.Chem.Soc.* **123**(2001)57

[123] D.B.Amabilino, P.R.Ashton, V.Balzani, C.L.Brown, A.Credi, J.M.J.Frechet, J.W.Leon, F.M.Raymo, N.Spencer, J.F.Stoddart and M.Venturi, *J.Am.Chem.Soc.* **118**(1996)12012

[124] N.Yamaguchi, L.M.Hamilton and H.W.Gibson, *Angew.Chem.Int.Ed.* **37**(1998)3275

[125] A.W.Kleij, R.van de Coevering, R.J.M.K.Gebbink, A.M.Noordman, A.L.Spek and G.van Koten, *Chemistry* **7**(2001)181

[126] H.Sellner, J.K.Karjalainen and D.Seebach, *Chem.Eur.J.* **7**(2001)2873

[127] X.Xu, E.J.Maclean, S.J.Teat, M.Nieuwenhuyzen, M.Chambers and S.L.Janes, *Chem.Commun.* (2002)78

[128] T.Nagasaki, M.Ukon, S.Arimori and S.Shinkai, *J.Chem.Soc.Chem.Commun.* (1992)608

[129] T.Nagasaki, O.Kimura, U.Masakatsu, S.Arimori, I.Hamachi and S.Shinkai, *J.Chem.Soc.Perkin Trans.1* (1994)75

[130] M.Uda, A.Momotake and T.Arai, *Org.Biomol.Chem.* **1**(2003)1635

[131] A.M.Elizarov, T.Chang, S.H.Chiu and J.F.Stoddart, *Org.Lett.* **4**(2002)3565

[132] J.P.Kampt, C.W.Frank, E.E.Malstrom and C.J.Hawker, *Langmuir* **15**(1999)227

[133] P.R.L.Malenfant and J.M.J.Frechet, *Macromolecules* **33**(2000)3634

[134] S.L.Zhang, H.B.Zhou, J.S.Huang, and C.M.Che, *Chem.Eur.J.* **8**(2002)1554

[135] B.Karakaya, W.Claussen, K.Gessler, W.Sacnger, A.D.Schluter, *J.Am.Chem.Soc.* **119**(1997)3296

[136] R.Klopsch, S.Koch and A.D.Schluter, *Eur.J.Org.Chem.* **1998**(1998)1275

[137] W.Stocker, B.Karakaya, B.L.Schurmann, J.P.Rabe and A.D.Schluter, *J.Am.Chem. Soc.* **120**(1998)7691

[138] C.H.Chou and C.F.Shu, *Macromolecules* **35**(2002)9673

[139] C.A.Kernag and D.V.McGrath, *Chem.Commun.* (2003)1048

[140] A.Sunder, R.Mulhaupt, R.Haag and H.Frey, *Macromolecules* **33**(2000)253

[141] A.Sunder, R. Hanselmann, H.Frey and R.Mulhaupt, *Macromolecules* **32**(1999)4240

[142] A.Sunder, R.Mulhaupt and H.Frey, *Macromolecules* **33**(2000)309

[143] A.Sunder, T.Bauer, R.Mulhaupt and H.Frey, *Macromolecules* **33**(2000)1330

[144] A.Sunder, M.Kramer, R.Mulhaupt and H.Frey, *Angew.Chem.Int.Ed.* **38**(1999)3552

[145] A.Sunder, M.F.Quincy, R.Mulhaupt and H.Frey, *Angew.Chem.Int.Ed.* **38**(1999)2928

[146] R.Haag, J.E.Stumbe, A.Sunder, H.Frey and A.Hebel, *Macromolecules* **33**(2000)8158

[147] H.Magnusson, E.Malmstrom and A.Hult, *Macromol. Rapid Commun.* **20**(1999)453

[148] T.M.Miller, T.X.Neenan, E.W.Knock and S.M.Stein, *J.Am.Chem.Soc.* **115**(1993)356

[149] S.Srinivasan, R.Twieg, J.L.Hedrick and C.J.Hawker, *Macromolecules* **29**(1996)8543

[150] J.L.Hedrick, C.J.Hawker, R.D.Miller, R.Twieg, S.Srinivasan and M.Trollsas, *Macromolecules* **30**(1997)7607

[151] K.E.Uhrich, C.J.Hawker, J.M.J.Frechet and S.R.Turner, *Macromolecules* **25**(1992)4583

[152] N.Kornblum, R.Seltzer and P.Haberfield, *J.Am.Chem. Soc.* **85**(1963)1148

[153] A.Mueller, T.Kowalewski, K.L.Wooley, *Macromolecules* **31**(1998)776

[154] J.K.Gooden, M.L.Gross, A.Mueller, A.D.Stefanescu and K.L.Wooley, *J.Am.Chem.Soc.* **120**(1998)10180

[155] V.Percec and M.Kawasumi, *Polym.Prep.* **33**(1992)221

[156] V.Percec and M.Kawasumi, *Macromolecules* **25**(1992)3843

[157] V.Percec, P.Chu and M.Kawasumi, *Macromolecules* **27**(1994)4441

[158] H.T.Chang and J.M.J.Frechet, *J.Am.Chem. Soc.* **121**(1999)2313

[159] E.Tsuchida, E.Shouji and K.Yamamoto, *Macromolecules* **26**(1993)7144

[160] M.Jikei, Z.Hu, M.Kakimoto and Y.Imai, *Macromolecules* **29**(1996)1062

[161] C.J.Hawker and F.Chu, *Macromolecules* **29**(1996)4370

[162] C.F.Shu and C.M.Leu, *Macromolecules* **32**(1999)100

[163] C.F.Shu, C.M.Leu and F.Y.Huang, *Polymer* **40**(1999)6591

[164] C.A.Martinez and A.S.Hay, *J.Polym.Sci.PartA: Polym.Chem.* **35**(1997)2015

[165] C.A.Martinez and A.S.Hay, *J.Polym.Sci.PartA: Polym.Chem.* **35**(1997)1781

[166] C.A.Martinez and A.S.Hay, *J.Macromol.Sci.Pure Appl.Chem.* A **35**(1998)57

[166a] J.Borah and N.Karak, *Polym.Int.* **53**(2004)2026

[166b] J.Borah and N.Karak, *Int.J.Plastic Technol.* **10**(2006)598

[166c] J.Borah, S.S.Mahapatra, D.Saikia and N.Karak, *Polym.Degrd.Stab.* **91**(2006)2911

[166d] J.Borah and N.Karak, *J.Appl.Polym.Sci.* **104**(2007)648

[167] A.Dhan and M.Portnoy, *Chem.Commun.* (2002)

[168] R.van de Coevering, M.Kuil, R.J.M.K.Gebbink, and G.van Koten, *Chem.Commun.* (2002)1636

[169] R.Haag, A.Sunder, A.Hebel and S.Roller, *J.Combinatorial Chem.* (2002)

[170] R.Haag, *Chem.Eur.J.* **7**(2001)327

[171] Y.C.Chen, F.Wu, J.G.Deng, H.Liu, Y.Z.Jiang, M.C.K.Choi and A.S.C.Chan, *Chem.Commun.* (2001)1488

[172] R.Dahinden, A.K.Beck and D.Seebach in L.Paquette (ed.) Encyclopedia of Reagents for Organic Synthesis, Vol.3, John Wiley, Chichester, P.2167

[173] P.B.Rheiner and D.Seebach, *Polym.Mater.Sci.Eng.* **77**(1997)1710

[174] P.B.Rheiner, H.Sellner and D.Seebach, *Helv.Chim.Acta* **80**(1997)2027

[175] T.Butz, P.K.Murer and D.Seebach *Polym.Mater.Sci.Eng.* **77**(1997)132

[176] D.Seebach, R.E.Marti and T.Hintermann, *Helv.Chim.Acta* **79**(1996)1710

[177] M.K.Kim, Y.M.Jeon, H.J.Kim, S.G.Hong, C.G.Park and K.Kim, *Chem.Commun.* (2001)667

[178] H.F.Chew and C.C.Mak, *J.Org.Chem.* **62**(1997)5116

[179] C.Francavilla, F.V.Bright and M.Detty, *Org.Lett.* (1999)1043

[180] M.Albrecht, N.J.Hovestad, J.Boersma and G.van Koten, *Chemistry* **7**(2001)1289

[181] M.R.Leduc, C.J.Hawker, J.Dao and J.M.J.Frechet, *J.Am.Chem.Soc.* **118**(1996)11111

[182] R.Knischka, P.J.Lutz, A.Sunder, R.Mulhaupt and H.Frey, *Macromolecules* **33**(2000)315

[183] I.Gitsov, P.T.Ivanova and J.M.J.Frechet, *Macromol.Chem.Rapid Commun.* **15**(1994)387

[184] K.Matyjaszewski, T.Shigemoto, J.M.J.Frechet and M.Leduc, *Macromolecules* **29**(1996)4167

[185] S.Maier, A.Sunder, H.Frey and R.Mulhaupt, *Macromol. Rapid Commun.* **21**(2000)226

[186] D.L.Jiang and T.Aida, *Nature* **388**(1997)455

[187] D.M.Junge, and D.V.McGrath, *Chem.Commun.* (1997)857

[188] D.M.Junge, and D.V.McGrath, *J.Am.Chem.Soc.* **121**(1999)4912

[189] F.Moulines, L.Djakovitch, R.Boese, B.Gloague, W.Thiel, J.L.Fillaut, M.H.Delville and D.Astruc, *Angew.Chem.Int.Ed.* **32**(1993)1075

[190] J.L.Fillaut, R.Boese and D.Astruc, *Syn.Lett.* (1992)55

[191] S.Mecking, R.Thomann, H.Frey, and A.Sunder, *Macromolcecules* **33**(2000)3958

[192] A.Adronov, S.L.Gilat, J.M.J.Frechet, K.Ohta, F.V.R.Neuwahl and G.R.Fleming, *J.Am.Chem.Soc.* **122**(2000)1175

[193] A.Adronov, P.R.L.Malenfant and J.M.J.Frechet, *Chem.Mater.* **12**(2000)1463

[194] M.Kawa and J.M.J.Frechet, *Chem.Mater.* **10**(1998)286

[195] M.Kawa and J.M.J.Frechet, *Thin Solid Films* **331**(1998)259

[196] T.Sato, D.L.Jiang and T.Aida, *J.Am.Chem.Soc.* **121**(1999)10658

[197] G.M.Stewart and M.A.Fox, *J.Am.Chem.Soc.* **118**(1996)4354

[198] A.Freeman, J.M.J.Frechet, S.Kaene and M.Thompson, *Macromol.Symp.* **154**(2000)163

[199] H.Frey and R.Haag, *Rev.Molecular Biotechnol.* **90**(2002)257

[200] M.Liu, K.Kono and J.M.J.Frechet, *J.Polym.Sci.Part A Polym.Chem.* **37**(1999)3492

[201] K.W.Pollak and J.M.J.Frechet, *Abstr.Pap.Am.Chem.Soc.* **212**(1996)185

[202] S.C.Zimmerman, Y.Wang, P.Bharalhi and J.S.Moore, *J.Am.Chem.Soc.* **120**(1998)2172

[203] Y.Wang, F.Zeng and S.C.Zimmerman, *Tetrahedron Lett.***38**(1997)5459

CHAPTER 7

DENDRITIC POLYESTERS

7.1 INTRODUCTION

Dendritic polyesters are another important class of dendritic polymers like linear polyesters. They have potential applications with higher efficiency than linear analogs as coating materials, binders and crosslinkers for composite materials, catalysts, light harvesting materials, etc. The common feature of all these dendritic polymers is that they have a large number of ester functional groups in their structures, which may easily hydrolyze, aminolyze and undergo ester exchange reactions. They are generally synthesized by condensation of polyhydroxy compounds with polybasic acids or acid chlorides, catalytic ester exchange reactions, ring opening polymerization, etc. The aromatic dendritic polyesters are relatively more thermostable than their linear analogs.

7.2 SYNTHESIS, CHARACTERIZATION AND PROPERTIES OF DIFFERENT TYPES OF POLYESTER DENDRIMERS

In this section the synthesis, characterization and properties of different types of polyester dendrimers are discussed.

7.2.1 Aliphatic

The synthesis of oligonucleotide polyester dendrimers was reported [1] by using a solid phase technique. The polyester dendritic components were synthesized from 2,2-*bis*(hydroxymethyl)propionic acid (*bis*-MPA). At first, two equivalents of isopropylidene protected propionic acid derivative react with one equivalent of DCC to generate the corresponding anhydride. Subsequent acylation of the terminal hydroxy group of benzylic 2,2-*bis*(hydroxymethyl)propionate with the resulted anhydride afforded doubly protected propionic acid derivative. The benzyl ester core was then selectively unmasked under hydrogenolysis conditions to get 2nd generation polyester dendron with carboxylic acid group at the focal point. The oligonucleotide polyester dendron conjugates were then synthesized directly on a controlled pore glass (CGP) solid support containing pre-synthesized oligonucleotide with terminal 5'-dimethoxytrityl (DMT) protecting group. The whole sequence of reactions is shown in scheme 7.1. Finally, the oligonucleotide polyester dendron conjugates were taken out from the solid support and accompanying removal of nucleotide base and phosphate protecting groups by morpholine/MeOH mixture (1:1). The conjugates were then purified by HPLC and characterized by MALDI-TOF mass spectrometry. The self-assembly of dendronized oligonucleotide were then prepared by combining two ketal protected polyesters-dendronized-oligonucleotides with one hydroxy terminated polyester with different amino acids sequences.

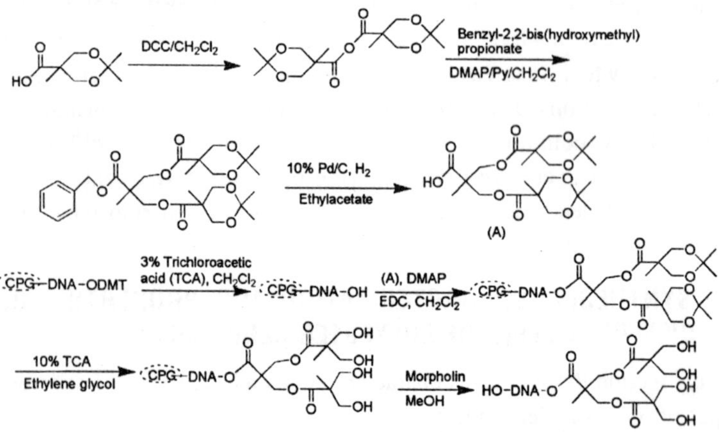

SCHEME 7.1 Synthesis of oligoneucleotide-polyester dendrimers by solid phase technique

A novel architecture of polyester dendrimers were synthesized [2] by the divergent approach. The architecture of these new polymers resembles that of 3D spherical dendrimers as well as classical star polymers. A hexahydroxy functionalized *bis*-MPA derivative was used as the initiator for the living ring opening polymerization of ε–caprolactone in the presence of stanous-2-ethylhexanoate as the catalyst. The polymerization of the monomer generated a six arms polymer. This was then functionalized with a protected *bis*-MPA, followed by deprotection and polymerization generated a 2nd generation 12 arms poly(ε-caprolactone). The repetition of the above steps generated a 3rd generation 24 arms containing star-dendrimer. All dendrimers were fully characterized by ^{1}HNMR, ^{13}CNMR spectroscopy and SEC. The number average molecular weight of the 1st to 3rd generation polyester dendrimer varies from 14300 to 96000 Da with PDI 1.06 to 1.16. These novel star dendrimers are different from the traditional dendrimers as well as defined highly branched polymers can be synthesized by only a few steps. These are also semicrystalline.

The viscoelastic properties and phase behavior of (0–20 %, wt./wt.) 12-*tert*-butylester dendrimer (Fig.7.1) were studied [3] in poly(methyl methacrylate) (PMMA) by refractometry, FTIR, UV-visible spectroscopy, DSC and dielectric analyses. The blends with above dendrimer upto 12 wt.% with PMMA are formed miscible blend, as indicated by single intermediate glass transition temperature in dielectric study of the blends. This miscibility may be due to the hydrogen bonding interaction of C=O groups of PMMA with N-H groups of dendrimer. However, at higher weight percentage of dendrimer in the blends, the T_g values decreased and ultimately two T_g appeared, which indicated phase separation at 15 weight % and onwards, as revealed by Rayleigh scattering in UV-visible spectra and dielectric studies. This is due to the fact that with the increase of

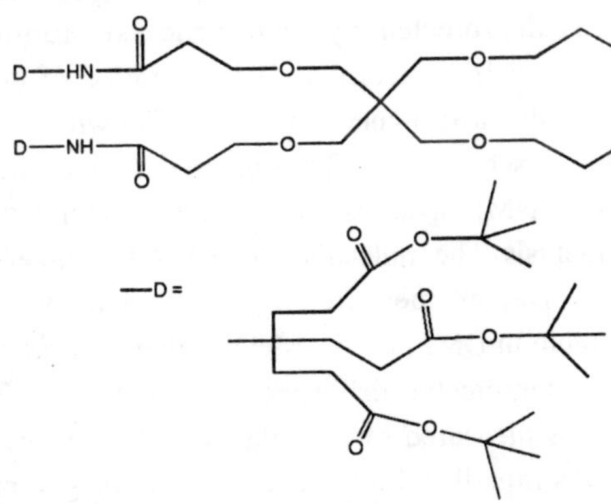

FIG. 7.1 Structures of polyester dendrimer with 12-*tert*-butyl ester surface groups

dendrimer content, the free volume of the system increases and hence the molecular mobility of PMMA chain increases. This resulted decrease in T_g values of the blends with greater than 12 wt.% of the dendrimer.

The monodisperse dendritic aliphatic polyesters upto 6th generation were synthesized [4] by using an acetal protected anhydride derivative of *bis*-MPA as the acylating agent and 1,1,1-*tris*(hydroxyphenyl)ethane as the core unit by a fast and convenient divergent method. This divergent approach is unique in the sense that unlike all others, it only requires simple solvent extraction or precipitation for purification of the products. The linear and star shaped poly(ethylene glycol) derivatives of narrow polydispersity were also used as the core to obtain linear-dendritic hybrid copolymers upto 4th generation. The molecular weight obtained for sixth generation dendritic polyester with *tris*(hydroxyphenyl) core is 30,711 Da and having 192 masked hydroxyl groups at the periphery.

The polyester dendrimers upto 3rd generation from glycerol and lactic acid were synthesized [5] by the divergent approach. At first, a tetra functional core was produced by three-step process. *Cis*-1,3-O-benzylidene glycerol (I) was treated with NaH followed by the addition of 2-bromopropionic acid to afford 2-[(*cis*-1,3-O-benzylidene glycerol)-2-propionic acid] (II). The coupling reaction of this resulted compound (II) with the starting glycerol (I) offered the key protected tetrahydroxy core. This core was deprotected by hydrogenolysis followed by coupling with II, resulted 1st generation. The reiteration of the above two reactions afforded the next higher generation. The whole sequence of reactions is shown in scheme 7.2. The synthesized dendrimers were characterized by FTIR, NMR spectroscopies, MALDI-TOF mass spectrometry and SEC analysis. The molecular weight of dendrimers ranges from 1260 to 4060 Da, as measured by SEC. The deprotected dendrimers were soluble in DMF, EtOH, MeOH, even in water. The T_g of the protected and deprotected dendrimers varies from + 31°C to –34°C respectively, as measured by modulated DSC. Because of the incorporation of biocompatible building blocks (both glycerol and lactic acid), it is expected that the resulted dendrimers will have specific medical and tissue engineering applications.

The synthesis of polyester dendrimers from glycerol and lactic acid or succinic acid were also reported [6] by the divergent approach. The

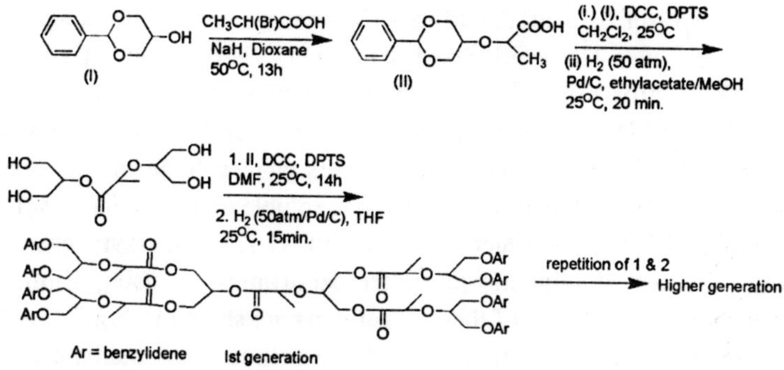

SCHEME 7.2 Synthesis of poly(glycerol-lactic acid) dendrimers

poly(glycerol-succinic acid) dendrimers upto 4th generation were synthesized from a tetrahydroxy core unit (Scheme 7.3). The core unit was synthesized in two steps, as shown in scheme 7.3. Similarly,

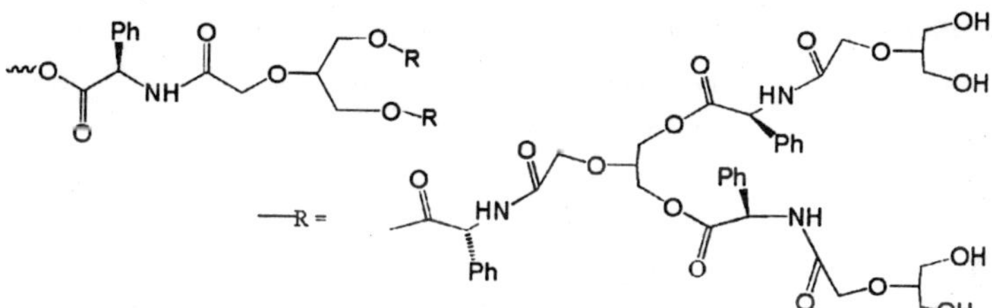

2nd Generation Poly(glycerol-glycolic acid)-phenylamine

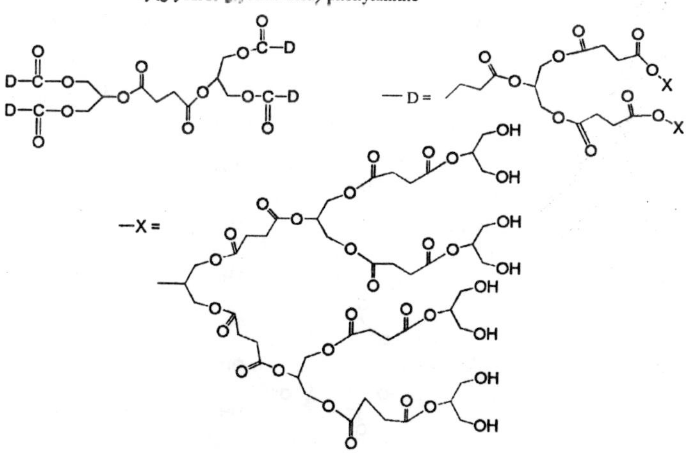

FIG.7.2 Structure of aliphatic polyester-polyether dendritic-linear block based on poly(glycerol-succinic acid)-poly(ethylene glycol)

poly(glycerol-lactic acid) dendrimers [5] were synthesized by using the same technique. The synthesized dendrimers were characterized by ^{1}HNMR spectroscopy, FAB/MALDI-TOF mass spectrometry and SEC analysis. The PDI was ~1.02 and the highest molecular weight obtained was 10,715 Da. The protected dendrimers were soluble in common organic solvents, whereas deprotected one soluble only in polar solvent such as DMF, MeOH, water, etc. The authors [7-8] also synthesized large varieties of aliphatic polyester dendrimers by using similar synthetic strategy. Some of these dendrimers are shown in Fig.7.2. All the above discussed dendrimers are susceptible to both acid/base and enzymatic degradations. A larger dendrimer degrades more slowly than a smaller one i.e. lower generation. Also, poly(glycerol-lactic acid) dendrimers degrade at faster rate than the corresponding poly(glycerol-succinic acid) dendrimers. This may be due to the difference in reactivity to the facile loss of monomer unit from a favorable internal six member, cyclization reaction in the former dendrimers. These

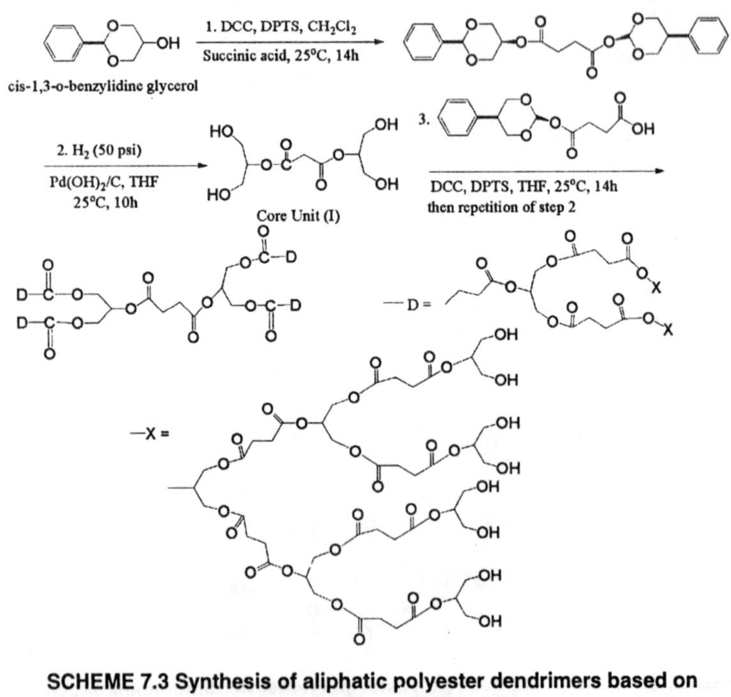

SCHEME 7.3 Synthesis of aliphatic polyester dendrimers based on poly(glycerol-succinic acid)

biodendrimers have potential future for medical applications such as tissue engineering, drug delivery, sensor applications, etc.

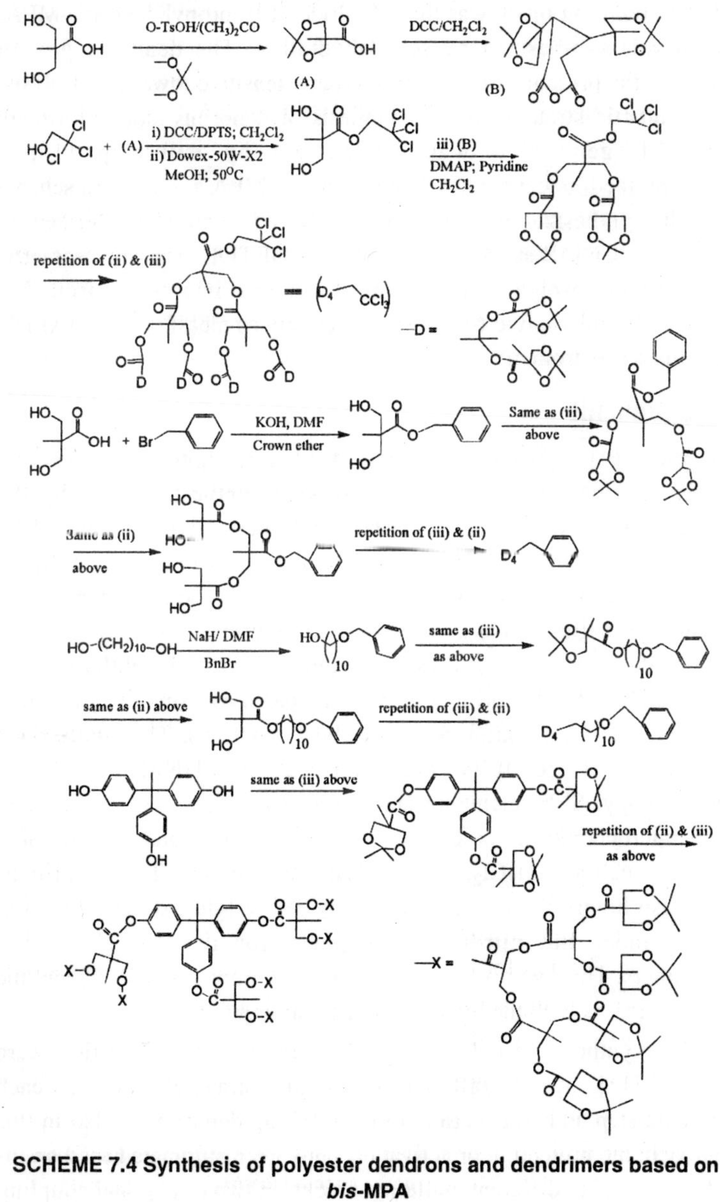

SCHEME 7.4 Synthesis of polyester dendrons and dendrimers based on *bis*-MPA

The aliphatic polyester dendrons and dendrimers based on *bis*-MPA were synthesized [9] by a rapid and efficient divergent approach. The key building block was the anhydride of isopropylidene-*bis*-MPA, which was synthesized by self-dehydration. The dendritic growth occurs in the presence of protecting groups sensitive toward hydrolysis. The synthetic strategy for the synthesis of dendrons and dendrimers upto 4th generation was carried out by multistep protection/deprotection technique starting from *bis*-MPA as shown in scheme 7.4. The synthesized dendrons and dendrimers were characterized by ^{1}HNMR, ^{13}CNMR spectroscopy and MALDI-TOF mass spectrometry. The molecular weight of the dendrons and dendrimer varies from 719 to 2430 Da and 1456 to 6593 Da respectively as measured by MALDI-TOF mass spectrometry.

7.2.2 Aromatic

The monodisperse poly(aryl ester) dendrimers upto 4th generation based on benzene tricarboxylic acid were synthesized [10] by the convergent approach. The polyester dendrimers were synthesized by repetitive coupling and deprotection sequence of reactions, utilizing silyl protected phenols and acyl chlorides. At first the key intermediate in the synthetic process, 5-tert(butyldimethylsiloxy) isophthaloyl dichloride (A), was synthesized from 5-hydroxyisophthalic acid (Scheme 7.5). The dendritic arms were prepared by using the reiterative sequence of esterification and hydrolysis reactions. The synthesized dendrimers were fully characterized by ^{1}HNMR, ^{13}CNMR spectroscopy, SEC, elemental and thermal analyses. The dendrimers were microcrystalline or glassy solids and highly soluble in organic solvents such as CH_2Cl_2, $CHCl_3$, THF, toluene, etc. The dendrimers were also highly thermostable and only 5% weight loss was found at 500°C under the nitrogen atmosphere for the 4th generation dendrimers. The Tgs for the dendrimers vary from 110-141°C, and the value increases with the increase of generation number.

The synthesis of polyester dendrimers upto 6th generation were reported [11] by an orthogonal coupling strategy, where in each synthetic step adds a generation to the exiting dendrimer. Also in this approach, the protection or activation steps were eliminated by sequential use of two different building blocks in two orthogonal coupling

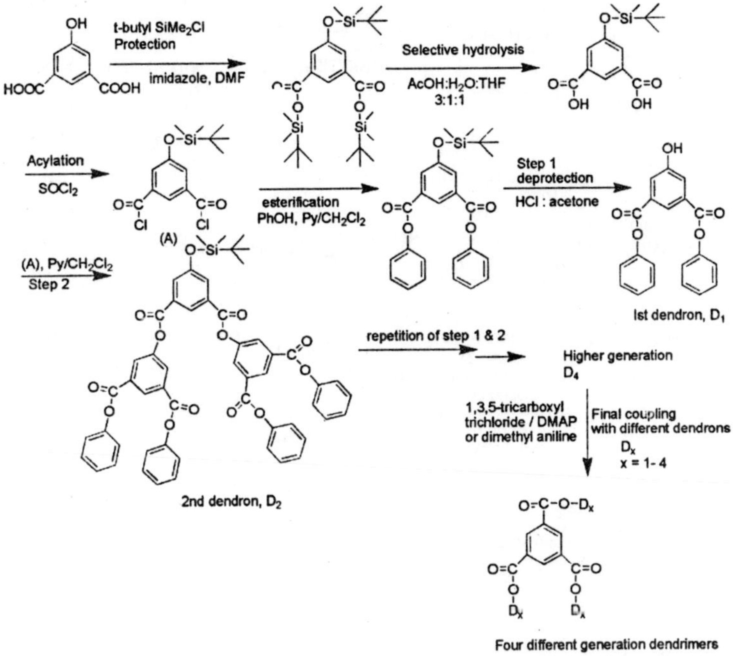

SCHEME 7.5 Synthesis of poly(aryl ester) with tricarboxyl benzene as the core unit

reactions. At first, 5-aminoisophthalic acid was transformed into 5-iodo isophthalic acid, the activated monomer, which was coupled with two hydroxy containing compounds sequentially to get the first generation dendron. Reiteration of previous steps offered higher generation. The efficiency of the strategy was further enhanced by merging this orthogonal coupling with Frechet branched polyether dendrimer approach [12]. By this way, they synthesized tetraacid containing iodo derivative and tetraacetylene containing hydroxy derivative, which were used to get the higher generation dendrons by much lesser number of steps. These dendrons were further coupled with trimesic acid to obtain these desired dendrimers. The whole sequence of reactions is shown in scheme 7.6, in three different steps. All the synthesized dendrimers were characterized by spectral studies such as ^{1}HNMR, ^{13}CNMR spectroscopy, SEC with differential refractometer index (DRI) and dual angle laser light scattering (LLS) detector HPLC, combustion analysis,

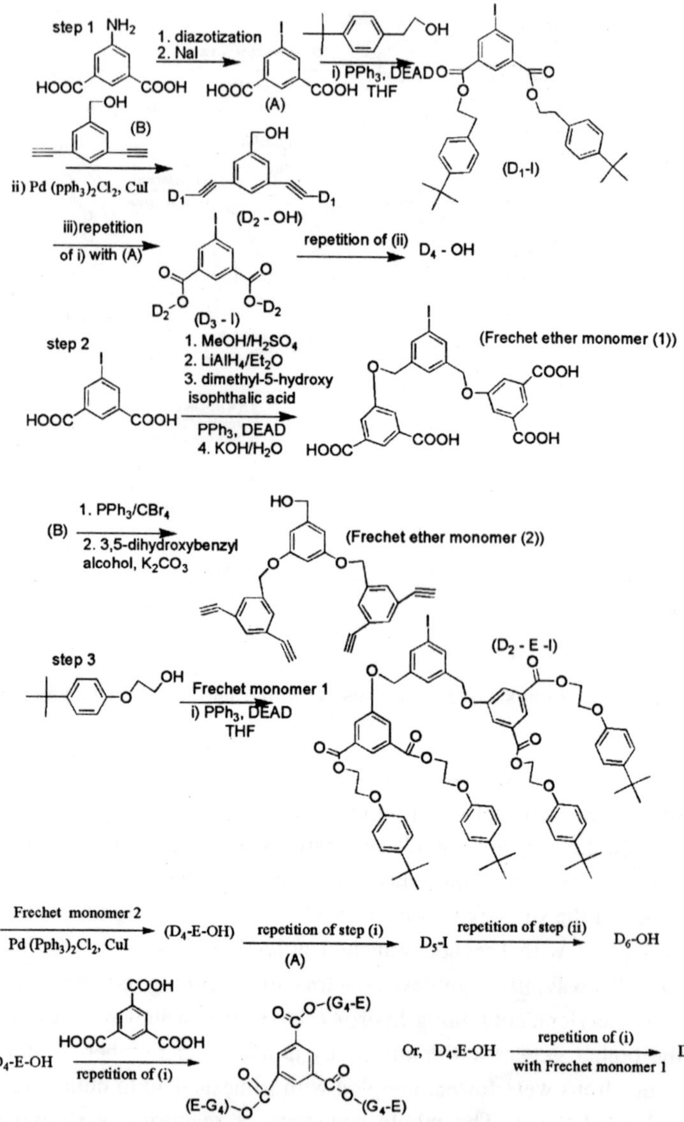

SCHEME 7.6 Synthesis of polyester dendrons and dendrimers by an accelerated synthetic approach

and MALDI-TOF mass spectrometry. The molecular weight of the dendrons and dendrimers were determined by SEC-LLS. The molecular weight of 6th generation dendron was 20896 g/mole and it was synthesized just by three steps using Frechet type ether monomers.

A series of sterically hindered poly(phenylester) dendrimers [13] was synthesized by the convergent approach using 5,10,15,20-*tetrakis*(3',5'-hydroxy phenyl)porphinato-manganese(III) chloride as the core. At first, di-*tert*-butyl benzene terminated polyester monodendrons of 1st and 2nd generations containing carboxylic group at the focal point were prepared by using the Frechet method [14]. Then, they coupled separately with the porphyrin core by DCC coupling reaction using DPTS as the catalyst. The coupling was carried out in two steps. First in THF, which was removed under reduced pressure and then product was redissolution in CH_2Cl_2 far further reaction (Scheme 7.7). The complexes were purified by HPLC and characterized by UV-visible and MALDI-TOF mass spectrometry. The molecular weights of the complexes were found to be 5344 and 10980 g/mole respectively for

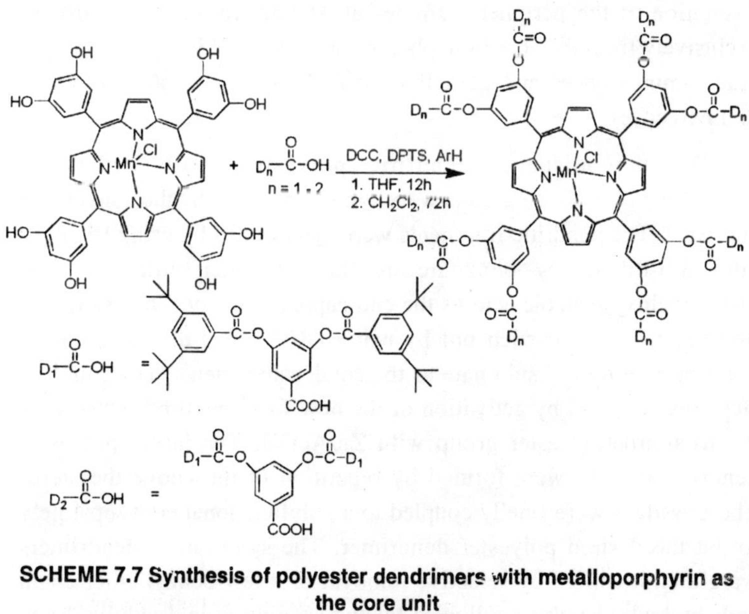

SCHEME 7.7 Synthesis of polyester dendrimers with metalloporphyrin as the core unit

1st and 2nd generations. These complexes were used as regioselective oxidation catalyst for both intra and intermolecular cases (see application section). The dendrimeric complexes were soluble in alkanes but less soluble in more polar solvents. The authors [15] also synthesized, characterized and studied the shape selective ligation of polyester dendrimer-metalloporphyrins from 3,5-di-dendron substituted *meso*-tetraphenylporphyrinato-zinc(II) complex by using the same synthetic strategy [13].

The polyester dendrimers with 12 peripheral redox active triaryl amine groups surrounding a single central dye labeled core synthesized [16] by the convergent method. The central dye was either cumarin C343 or pentathiophene T_5. All the dendrimers were characterized by ^{1}HNMR, ^{13}CNMR spectroscopy and MALDI-TOF mass spectrometry. The photophysical behavior of the dendrimers both in the solution ($CHCl_3$ and CH_3CN) and in the solid state (thin film on glass) photoluminescence and electroluminescence studies revealed that the energy transfer efficiencies in both the cases were essentially quantitative. Excitation of the peripheral amines at 310 nm resulted in emission exclusively from the core lumophores, cumarine C343 (λ_{max} = 480 nm, blue luminescence) and pentathiophene T_5 (λ_{max} = 550 nm, green luminescence).

The monodisperse dendritic polyesters with hydrophobic/hydrophilic surface groups were synthesized [17] by the convergent approach. The dendritic fragments were synthesized by using trichloro ethyl-3,5-dihydroxy benzoate as the monomer unit and 3,5-*bis*(benzyloxy)benzoic acid as the end capping moiety. The esterification reaction was carried out by using DCC and 4-dimethyl amino pyridinium *p*-toluenesulfonate as the condensing agents. The coupling step was followed by activation of the new focal point by removal of the trichloroethyl ester group with Zn/AcOH. The larger polyester dendritic wedges were formed by repetition of the above the steps. These wedges were finally coupled to a polyfunctional core separately to get the desired polyester dendrimer. The synthesized dendrimers were then modified by selective removal of a few benzyl ether chain ends by hydrogenolysis followed by end capping of hydrophilic group containing moiety. Thus, the final polyester dendrimers possesses both hydrophobic and hydrophilic moieties. This modification drastically changes the glass transition behavior of the dendrimers.

The polyester segment of a layer block copolymer [14], (discussed in copolymer subsection) was synthesized from 3,5-dihydroxy benzoic acid using a protection / deprotection strategy. The protection was done through trichloroethyl ester formation, whereas the deprotection reaction was carried out by treatment of protected groups by Zn in glacial acetic acid. The sequence of reactions is shown in scheme 7.8 for upto 3rd generation. The synthesized dendrimer was characterized by ^{1}HNMR, ^{13}CNMR spectroscopy, SEC analysis and mass spectrometry.

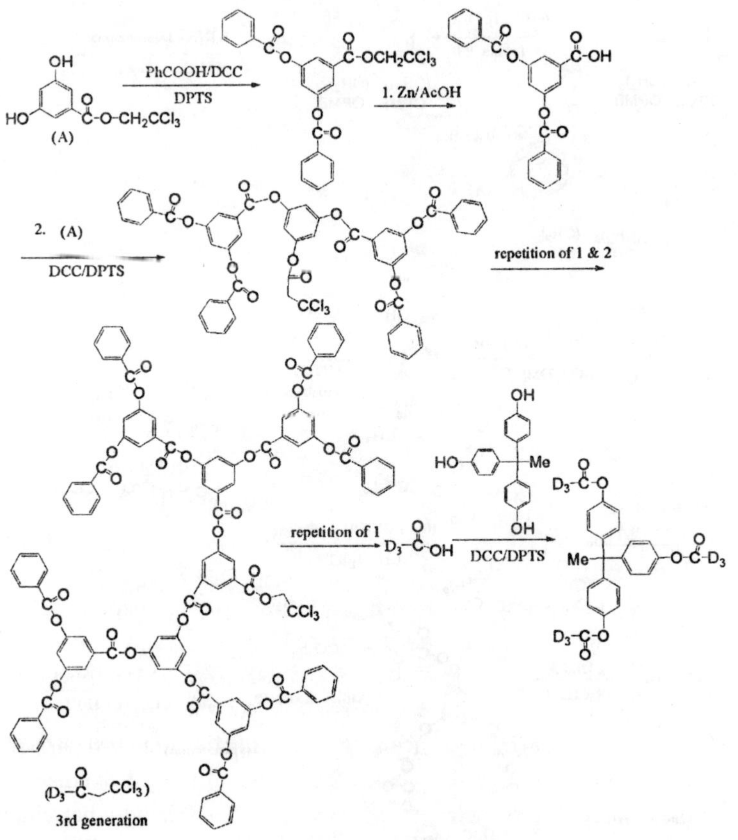

SCHEME 7.8 Synthesis of polyester dendrimers with 3,5-dihydroxy benzoic acid as the core unit

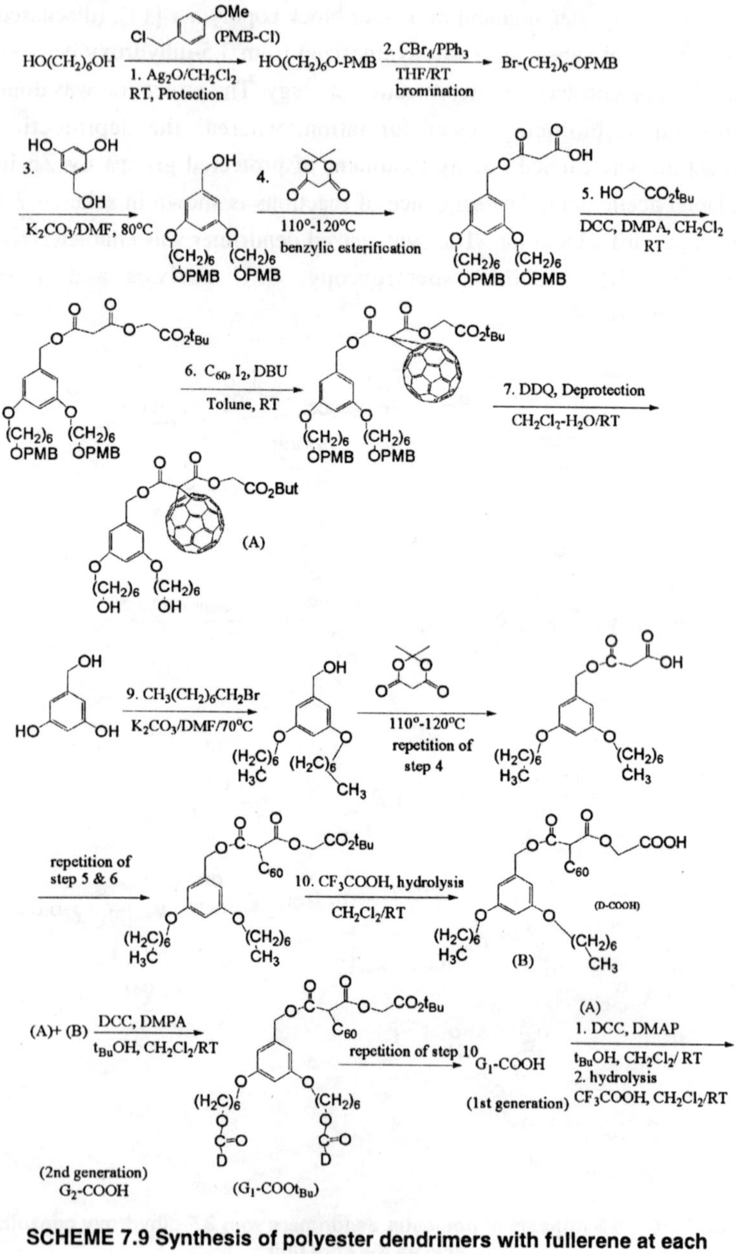

SCHEME 7.9 Synthesis of polyester dendrimers with fullerene at each branching point

Fullerene-functionalized polyester dendritic branches containing C_{60} sphere at the each branching unit were synthesized [18] by the convergent approach using successive esterification and deprotection steps. The main building block, fullerene containing diol was synthesized first, from 1,6-hexanediol by the following sequence of reactions as shown in scheme 7.9. Similarly, a carboxylic group containing fullerene derivative with two long alkyl chains was synthesized from 3,5-dihydroxybenzyl alcohol (Scheme 7.9). Finally, the esterification of the diol with the monocarboxylic acid containing fullerene derivative under DCC coupling conditions in the presence of DMAP and 1-hydroxybenzotriazole, followed by hydrolysis resulted the desired polyester dendrimers (Scheme 7.9 part C). The dendrimers were characterized by ^{1}HNMR, ^{13}CNMR, IR and UV-visible spectroscopies, elemental analysis and ES-mass spectrometry. The studies of electrochemical behavior of the dendrimers suggested that the highest generation behave as independent redox centers. This new ionization strategy of unchanged fullerene derivatives appears to be efficient and offers new perspective in the characterization of complex molecular architectures larger than 10 kDa.

Similar type poly(aryl ester) dendrimers upto 5th generation with *tris*(*p*-hydroxyphenyl)ethane as the core and 3,5-dihydroxy benzoic acid as the building block (Fig.7.3) were also synthesized [19]. The solubility of these polyester dendrimers are extremely high (1.15 g/mL in THF, for 5th generation) compared to their analogs linear polyesters (0.025 g/mL only). This study clearly confirms that the highly branched dendritic structure dramatically affects the solubility. This is due to their branched architecture and also the presence of large number of end groups, which are insignificant in the linear polymers.

The redox active polyester dendrimer based on benzene tricarboxylic acid trichloride was synthesized [20]. The dendrimer was end capped with TTF. The cyclic voltammogram of the dendrimer (Fig.7.4) indicates formation of a multiple positively charged species in which the tetrathiafulvalene units do not interact. These new redox-active species may be useful as organic conductors.

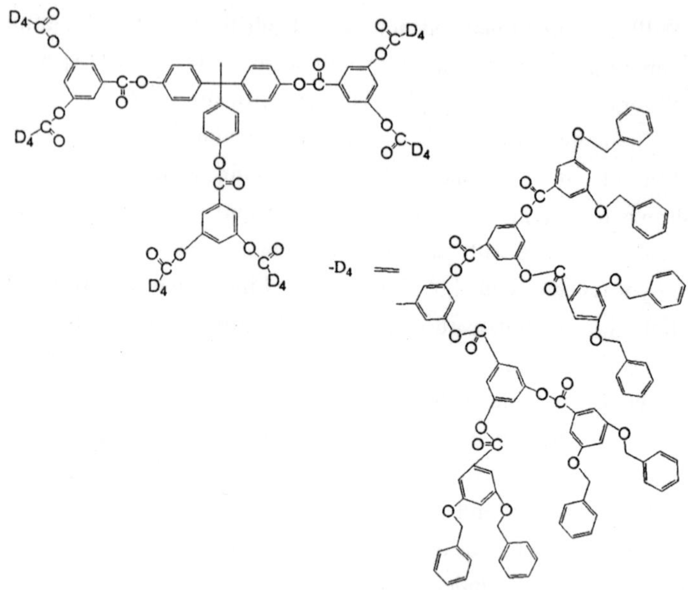

FIG. 7.3 Structures of poly(aryl ester) dendrimer with tris(*p*-hydroxyphenyl)ethane core unit

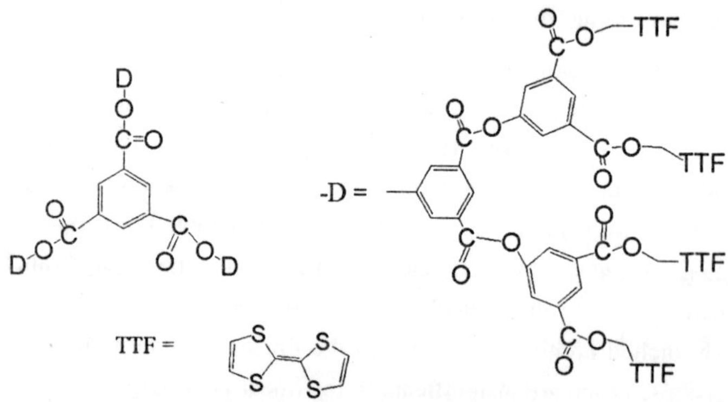

FIG. 7.4 Structures of polyester dendrimer with tetrathiafulvalene units at the periphery

The synthesis of poly(3,4,5-trihydroxy benzoate ester) dendrimers and their chemiluminescence properties were reported [21]. The dendrimers were synthesized by the divergent method using pyrocatechol and 3,4,5-trihydroxy benzene as core molecules and gallic acid units in the periphery, as shown in scheme 7.10 (A and B). The synthesized dendrimers were characterized by ^{1}HNMR spectroscopy, elemental analysis and MALDI-TOF mass spectrometry. The chemiluminescence (CL) intensities of the dendrimers in methanol are 400-600 times stronger than that of gallic acid, as measured by photo counting luminometer.

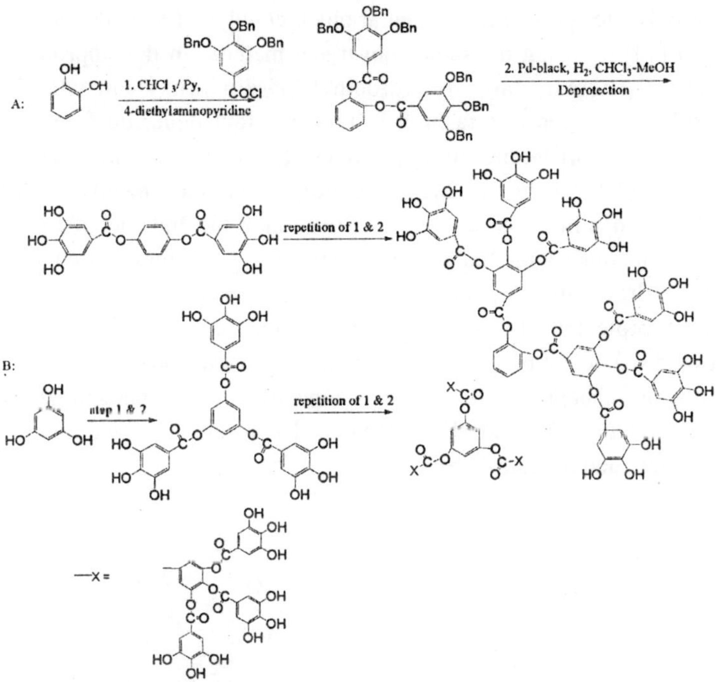

SCHEME 7.10 Synthesis of poly(3,5-trihydroxy benzoate ester) dendrimers

7.2.3 Copolymers

The polyester-polyether layer block copolymers were synthesized [14] by the convergent approach. In this approach, the number and size of each block can be varied by suitable changes in synthetic strategy. A 4th generation of such polymer with ether outer three layers and ester inner two layers was synthesized from 3,5-dihydroxybenzyl alcohol as the monomer unit. However at 3rd generation, a different building block, trichloroethyl-3,5-dihydroxybenzoate was introduced. This allows the coupling reaction to be changed from ether to ester chemistry. The esterification reaction was carried out by using DCC and DPTS as condensing agents, followed by activation of the focal point by deprotection of trichloroethyl group by Zn/AcOH. Then, the synthesis was continued with attachment to a trifunctional core by using the ester synthesis process to get the desired dendrimer (Fig.7.5). The polyester-polyether segmental block copolymer was also synthesized [14] by using the same convergent method. In this approach, a single monomer unit, trichloroethyl-3,5-dihydroxybenzoate was coupled in a stepwise fashion to two different dendritic fragments composed of different building blocks. The monomer unit was first monoalkylated with the 2nd generation benzyl terminated aromatic polyether dendron followed by esterification of the 2nd phenolic group of the monomer with 2nd generation aromatic polyester dendron to get the dendron with polyester and polyether segmental blocks. Then, it was deprotected and coupled with trifunctional core to obtain the desired segmental block dendrimer. The whole sequence of reactions is shown in scheme 7.11. These block copolymers were characterized by NMR spectroscopy and SEC technique. The results indicated that they were pure and monodisperse.

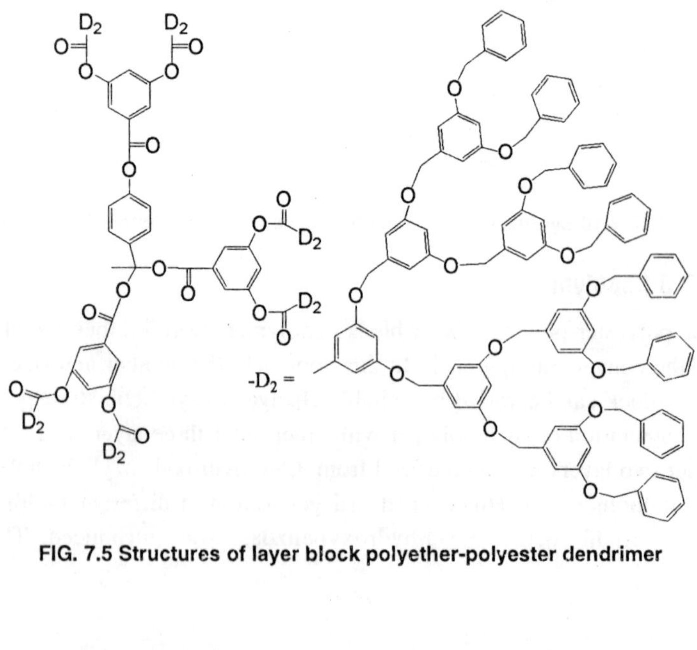

FIG. 7.5 Structures of layer block polyether-polyester dendrimer

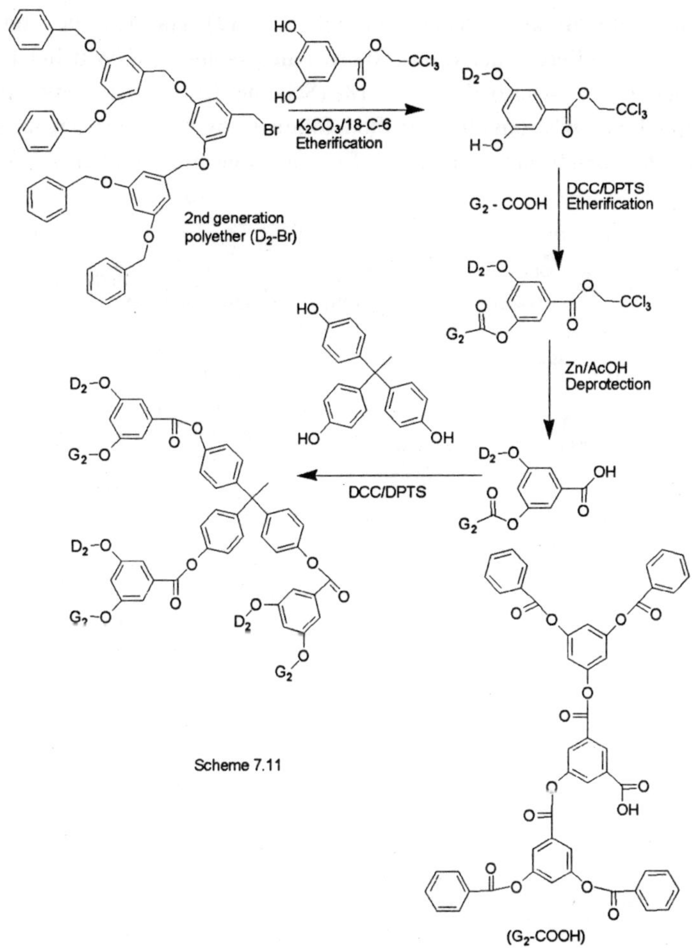

SCHEME 7.11 Synthesis of dendritic polyether-polyester block copolymer

The amphiphilic diblock polyester dendrimers with periphery long alkyl chains and containing five C_{60} fullerene units in the branches were synthesized [22]. Also, the peripheral substitution of the dendrimer with hydrophobic chains on one hemisphere and hydrophilic groups on the other, provides the perfect hydrophobic/hydrophilic balance, allowing the formation of stable Langmuir films. The synthetic method involves the same strategy as described above [18]. The fullerene derivative (A) with carboxylic at focal point was synthesized from 3,5-dihydroxybenzyl alcohol (Scheme 7.12). The aliphatic tetraol containing fullerene derivative (B) was also synthesized by using 1,6-hexanediol as described earlier [18] (Scheme 7.12). Then, they were coupled by DCC coupling method followed by hydrolysis, resulted polyester dendrimer containing five fullerene units (Scheme 7.12).

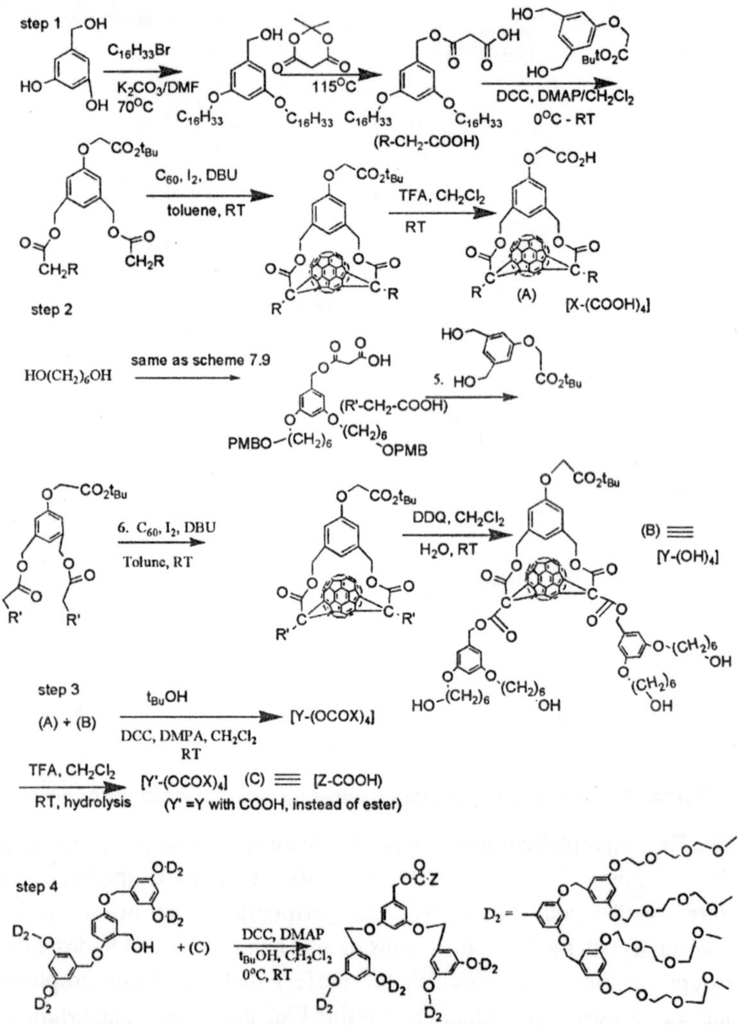

SCHEME 7.12 Synthesis of amphiphilic diblock polyester dendrimers with fullerene moieties and long alkyl chains

Finally, a 3rd generation poly(aryl ether) dendron [1] possessing polyethylene glycol at the periphery and benzylic alcohol at the focal point esterified with the fullerene combining polyester dendron with carboxylic group at the focal point (Scheme 7.12). The dendrons and dendrimers were characterized by NMR spectroscopy and MALDI-TOF mass spectrometry. This polyester-polyether dendrimeric diblock structure allows the efficient transfer of Langmuir films and well ordered multilayered Langmuir-Blodgett film. Further, a perfect reversibility was observed in successive compression/decompression cycles.

The photophysical properties of polyester dendrimers with *bis*(1,10-phenanthroline) copper(I) core and peripheral fullerene units upto 3rd generation (Fig. 7.6) were studied [23]. The luminescent ^{3}MLCT excited state of the metal complex core is completely quenched in CH_2Cl_2 solution at room temperature. The quenching process may take place by (i) energy transfer to the fullerene centered singlet, (ii) energy transfer to the fullerene centered triplet, and (iii) electron transfer yielding a fullerene based anion and an oxidized Cu(II) complex. However, the electron transfer process does not seem to be the prevail phenomenon for two reasons. First, the luminescence takes place even in rigid matrix at -196°C in spite of the low exoergonicity (ΔG = -0.17 eV) of the electron transfer process and secondly, quenching of fullerene fluorescence was not observed.

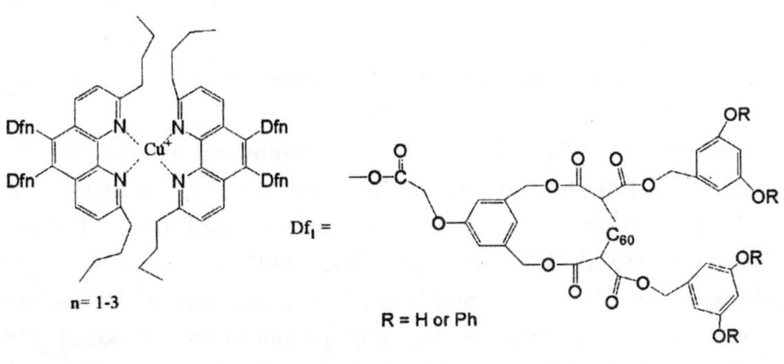

FIG. 7.6 Structures of polyester dendrimers with $[Cu(phen)_2]^+$ core unit and fullerene surface groups

Several chiral polyester dendrimers upto 2nd generation based on (R)-3-hydroxybutanoic acid were synthesized [24-25] by the convergent approach. In this approach, trimesic acid was used as core and the benzyl esters of the dimer and the tetramer of (R)-3-hydroxybutanoic acid as the building blocks (Fig.7.7). The biodegradability of these dendrimers were tested in the presence of poly((R)-3-hydroxybutanoic acid) depolymerase enzyme and found that the benzyl-protected dendrimer with dimeric (R)-3-hydroxybutanoic acid building blocks was not degraded by that enzyme. However, the free acid terminated dendrimer was surprisingly good substrate for the enzyme, even though the simple dimeric (R)-3-hydroxybutanoic acid was not [24]. Furthermore, the rate of the first degradation step was about 100 times faster for tetramer acid containing dendrimers compared to the dimeric one. An esterase, a lipase and a protease were shown to be able to degrade these chiral polyester dendrimers as well [24].

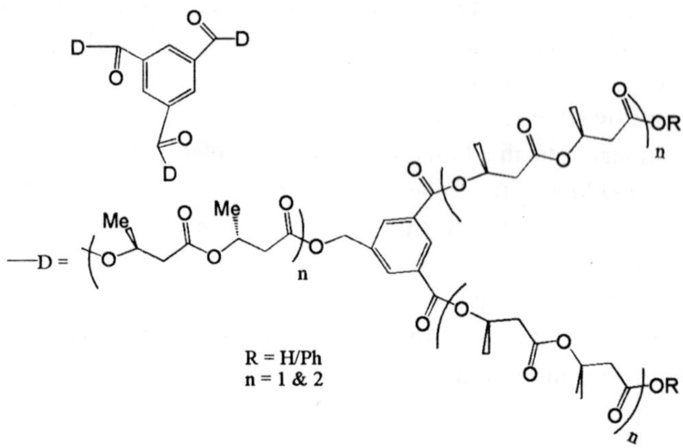

FIG. 7.7 Structures of chiral polyester dendrimers with trimesic acid core unit

The synthesis of dendritic-linear amphiphilic AB macromolecules was reported [26] by the tandem 'core in/core out' approach. The copolymers composed of polyester dendrons and linear PMMA polymerized by ATRP technique. The synthetic strategy to get 'miktoarm' copolymers, a building block containing a site for the initiation of ATRP and the attachment of a dendron was employed. The coupling of this building block, in a 'core in' fashion, to a multifunctional core leads to a multiarms compound with the transformation sites arranged in an alternating sequence. The synthetic steps for the synthesis of these miktoarm copolymers are shown in scheme 7.13. All the products were characterized by ^{1}HNMR and ^{13}CNMR spectroscopic studies. The degree of polymerization varies from 11 to 57, as measured by NMR spectroscopy. The molecular weight was determined by the

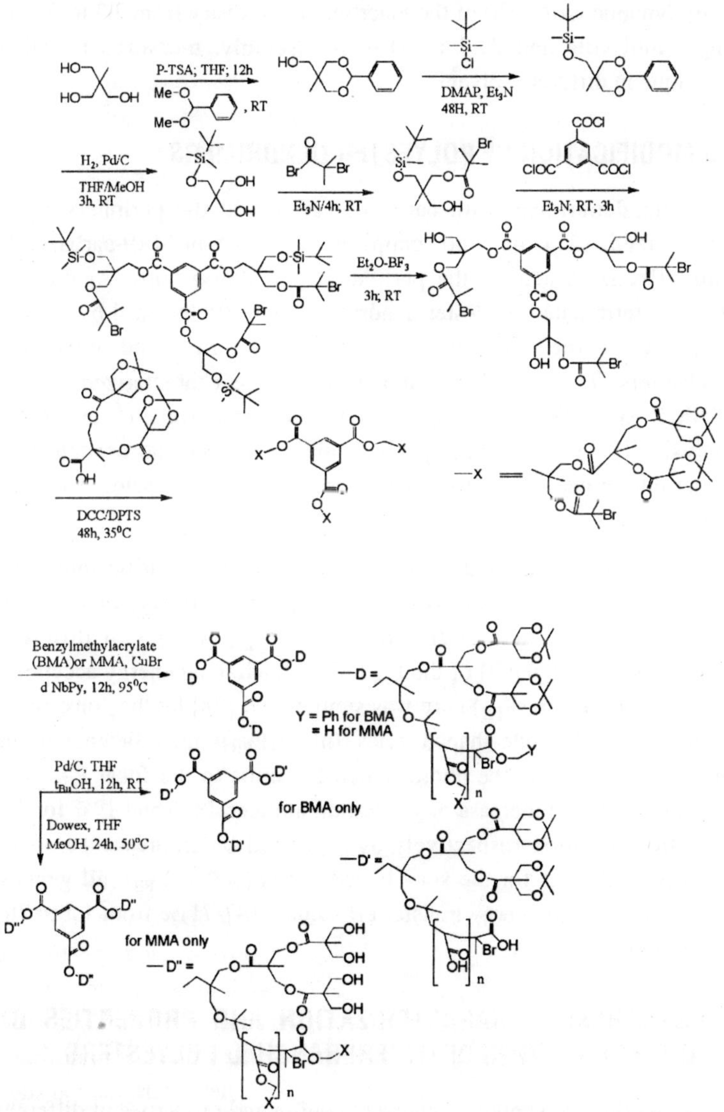

SCHEME 7.13 Synthesis of dendritic-linear block amphiphilic copolymer

end group analysis and varies from 500 to 34700 Da. Narrow polydispersity (PDI = 1–1.17), measured by SEC and good yields were the special characteristic of the above reaction scheme. The hydrodynamic radii (Rh) of the macromolecules vary from 2.3 to 3.0 as single molecule and 29.3 to 91.0 as assembly, measured by DLS technique in different solvents.

7.3 MODIFICATION OF POLYESTER DENDRIMERS

Polyester dendrimers with carbazole groups at the periphery were synthesized [27] by end capping with 3,5-*bis*[4-(9-carbazolyl) butoxyl]benzoic acid to the peripheral phenoxy groups. Again, the phenoxy terminated polyester dendrimers were synthesized by hydrogenation of the corresponding benyloxy-terminated polyester dendrimers. The synthesized dendrimers were characterized by IR, ^{1}HNMR spectroscopy, MALDI-TOF mass spectrometry and GPC analysis. The fluorescence spectral studies showed the existance of strong intramolecular interactions in the carbazole terminated polyester dendrimers.

A series of star- and cone-shaped polyester dendritic multiporphyrin arrays, $(nP_{zn})_4P_{FB}$ and $(nP_{zn})_1P_{FB}$, respectively, that contain energy-donating dendritic zinc porphyrin (P_{zn}) wedges of different numbers (n= 1, 3 and 7) of the P_{zn} units, attach to an energy accepting free base porphyrin (P_{FB}) core was synthesized [28] by the convergent approach. For the cone-shaped series $[(nP_{zn})_1P_{FB})]$, the efficiency of the energy transfer from the photoexcited P_{zn} units to the focal P_{FB} core decreases with the increase of generation number, 86% and 19% for 1st and 7th generation respectively as measured by fluorescence spectroscopy. Whereas for the star-shaped series $[(nP_{zn})_4P_{FB}]$, all generations showed high energy transfer efficiency, (87-71)% from 1st to 7th generation.

7.4 SYNTHESIS, CHARACTERIZATION AND PROPERTIES OF DIFFERENT TYPES OF HYPERBRANCHED POLYESTERS

In this section the synthesis, characterization and properties of different types of HB polyesters are discussed.

7.4.1 ALIPHATIC

In 1995, Malmstrom *et al.* synthesized [29] an aliphatic HB polyester by melt polymerization of *bis*-MPA as an AB_2 type monomer in the presence of 2-ethyl-2-(hydroxymethyl)-1,3-propanediol, as the core molecule (Scheme 7.14). The melt polycondensation reaction was carried out at 140°C in the presence of TsOH as the catalyst. The degree of branching of the polymers was found to be very high, 0.83-0.96, which indicated more symmetrical structure with less defects. The molecular weight of the resulting HB polymers depends on the molar ratio of the AB_2 monomer to the core molecule and varies from ~10^4–10^5 Da. The kinetics of formation of HB polyesters based on *bis*-MPA and radiation curing of these polymers were studied [30-31].

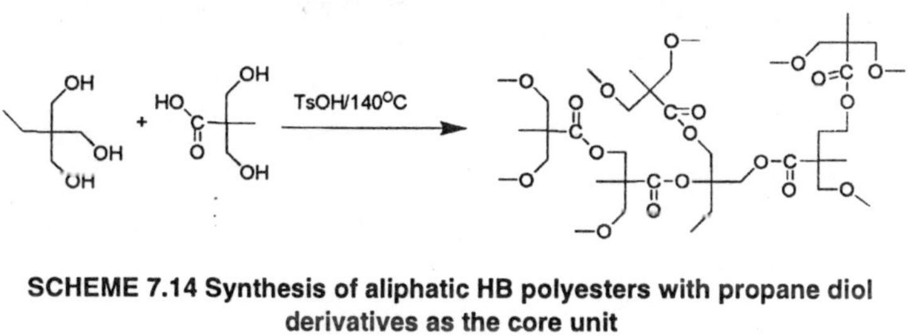

SCHEME 7.14 Synthesis of aliphatic HB polyesters with propane diol derivatives as the core unit

In the same year, HB polyester of poly(ethylene glycol) was synthesized [32] by transesterification reaction of an AB_2 macromonomer (Scheme 7.15). The ethyleneoxide chains of the polymer are not crystallized due to the presence of branching and T_g of the HB polymer depends on the number of ethyleneoxide units present

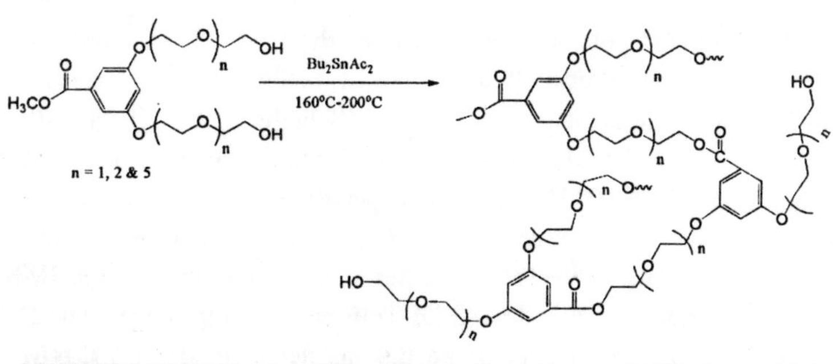

SCHEME 7.15 Synthesis of HB polyesters based on poly(ethyleneoxide) chain

in the structure. It also formed Li-complex through ethyleneoxide chains (n = 5) with $LiClO_4$. The ionic conductivity of these HB polyester-Li complexes was found to be 10^{-5} scm^{-1} at 30°C for a concentration of $LiClO_4$ of 0.62 molecule per repeating unit. The conductivity increases with the increase of temperature and the value is 10 scm^{-1} at 60°C.

HB polyester was synthesized [33-34] by using larger AB_2 type macromonomers, oligo(ε-caprolactone)s and *bis*-MPA as a branching unit (Fig.7.8). At first, the macromonomers were prepared by living ring opening polymerization of ε–caprolactone initiated by aluminum benzyloxide or tin octate, followed by termination with benzylidene-protected *bis*-MPA and subsequent deprotection offered the monodispersed oligo(ε-caprolactone), AB_2 type macromonomers. Then, the macromonomers were polymerized by esterification reaction in the presence of DCC and 4-(dimethylamino)pyridinium-4-toluenesulfonate. The DB of the resulted HB polyesters do not depend on the initial molecular weight of monomer and found to be 0.37 ± 0.03. However, their thermal properties depend on the degree of polymerization of the AB_2 macromonomers.

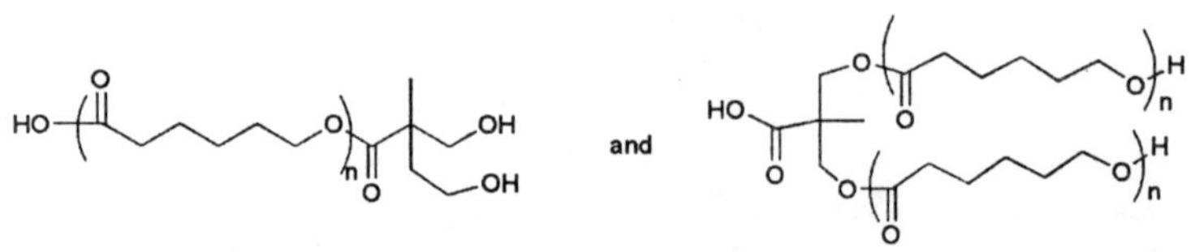

FIG. 7.8 Structures of AB_2 type macromonomers derived from ε–caprolactone

HB aliphatic polyesters were also synthesized [35] by a bulk ring opening polymerization of a lactone at 110°C in the presence of catalytic amount of stannous chloride (Scheme 7.16). The resulted polymer was a clear viscous liquid with weight average molecular weight (6.5-8.5) X 10^4 Da with polydispersity 3.2 as measured by SEC. The molecular weight increases with the decrease of catalyst concentration. The structure of the HB polyester was characterized by ^{1}HNMR, ^{13}CNMR spectroscopy and MALDI-TOF mass spectrometry. The DB of the polymer was found to be 0.5, as determined by ^{13}CNMR spectroscopy. The solubility of the HB polyester highly depends on the nature of the end capping moieties. For example, the hydroxy terminated polymers were soluble in common organic solvents like DMSO, DMF, methanol, but insoluble in THF, CH_2Cl_2 and $CHCl_3$. The HB polyester end capped with acetyl ester became soluble in THF, CH_2Cl_2 and $CHCl_3$, but no longer soluble in methanol.

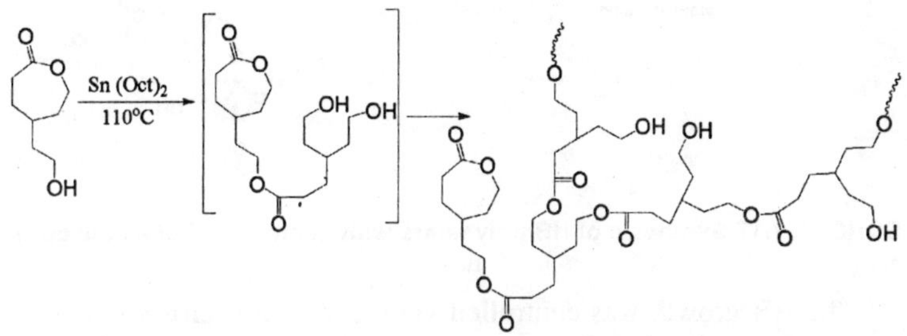

SCHEME 7.16 Synthesis of HB polyesters by ring opening polymerization

The synthesis of HB polyester by ring opening polymerization of a dihydroxymethyl caprolactone derivative (Fig. 7.9) and its copolymerization with ε–caprolactone were reported [36]. The *bis*(hydroxymethyl)caprolactone derivative was obtained from *bis*-MPA. This lactone self-polymerized at 110°C in the presence of $Sn(Oct)_2$ to the corresponding HB aliphatic polyester. The HB polyesters were also obtained by copolymerization of this lactone with ε–caprolactone by using the same procedure. The average molecular weights and polydispersity of both the homo and copolymers vary from 3-8 X 10^3 Da and 2.3–2.8 respectively. The degree of branching was found to be 0.5, as determined by ^{1}HNMR spectroscopy. The rheological properties of the HB polyester [37] upto 5th generation obtained by single step polycondensation reaction of 2,2-dimethylolpropionic acid with pentaerythritol core (Scheme 7.17) were studied in 1-methyl-2-pyrrolidinone solvent. The influence of rheological properties on blending of this HB polyester with a linear poly(2-hydroxy methyl methacrylate) polymer was also studied [37].

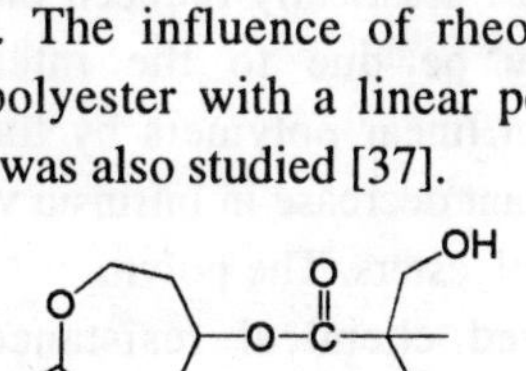

FIG. 7.9 Structure of caprolactone *bis*(hydroxy methyl) propionate as the monomer unit

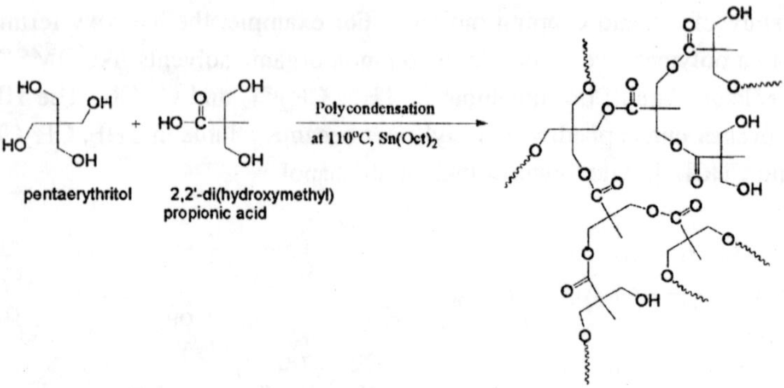

SCHEME 7.17 Synthesis of HB polyesters with pentaerythritol as the core unit

The HB growth was controlled via the stoichiometric ratio of the core to the monomer. The average molecular weight and polydispersity of the HB polyesters vary from 1.75-14.6X10^3 Da and 1.3-2.5 respectively for 2nd to 5th generation. However, the average molecular weight for the linear polymer was 3X10^6 Da. The study of rheological properties revealed that the HB polyesters in NMP solvent upto 50 weight percent solution exhibited Newtonian behavior. This result confirmed the absence of physical entanglements in the structures of the polymers and steady shear viscosities are independent of shear rate. It was also observed [37] that the solution viscosities are only slightly affected by the different generations of the HB polyesters. This polymers have very low intrinsic viscosities and their hydrodynamic radius scales as Rn ~ $M^{0.39}$. This results indicated that the polymers have less packed structure than the corresponding dendrimers. Further, all the polyesters of different generations showed almost same apparent activation energy of flow within the temperature range, 20-40°C. The viscosities of the blends drastically reduced on addition of these HB polyesters. This may be due to the interference of physical entanglements between linear polymers by the globular shaped HB polymers and the constant decrease in intrinsic viscosities of the blends upon addition of HB polyesters. The polymers having better hydrolytic stability and improved chemical resistance compared to their conventional linear polymers may be due to the presence of tertiary carbon in the structures. The HB polyesters were amorphous solids at room temperature with T_g ~ 40-45°C and softening points 70-100°C.

However, they have very good thermal stability and start to degrade only above 350°C. The T_gs are highly dependent on H-bonding interaction of surface hydroxylic groups and decrease dramatically by incorporation of aliphatic groups. Thus, it is possible to get solid to liquid state just by simple end capping of the surface groups with aliphatic groups.

The structure-property relationship was studied [38] for a series of HB polyesters based on dimethoxypropionic acid with ethoxylated pentaerythitol as the core. In DSC studies, a prominent T_g and a small melting endotherm were observed for all HB polyesters. As the T_gs for these polyesters are nearer to the ambient temperature, and small changes in the T_g with pseudo-generation number have a prominent effect on their mechanical properties. They formed monolayer aggregates of spherical shape on deposition from a dilute solution of a good solvent. Also, the aggregates became stable through H-bonding of the terminal hydroxyl groups. However, since the globular structure does not permits the usual processes of orientation and strain hardening, the neck gradually thins until it fractures at an engineering strain above 100%.

HB aliphatic copolyesters were synthesized [39] by using enzymatic route for the first time. The synthetic process involved the combination of ring opening polymerization and polycondensation of ε–caprolactone, as an AB monomer and 2,2'-*bis*(hydroxy-methyl)butyric acid as the AB_2 monomer (Scheme 7.18). By systematic variation of the fraction of the AB_2 comonomer, a series of HB copolyesters with different degree of branching were synthesized. The synthesized polyesters were characterized by ^{1}HNMR and ^{13}CNMR spectroscopies, GPC, MALDI-TOF mass spectrometry and DSC analysis. The yields obtained for different copolymers were very high ~70 - 95%. The weight average molecular weight for the polymers varies from 6600-61500 g/mole. All the HB copolymers having low T_g value (-45 to –59)°C. Thus, this enzymatic process avoids the multistep monomer synthesis and most importantly it avoids the use of heavy metal based catalyst, which might advantages for biomedical applications.

The aliphatic HB polyesters were synthesized [40] from trimethylolpropane (TMP) and *bis*-MPA at 140°C in the presence of TsOH.

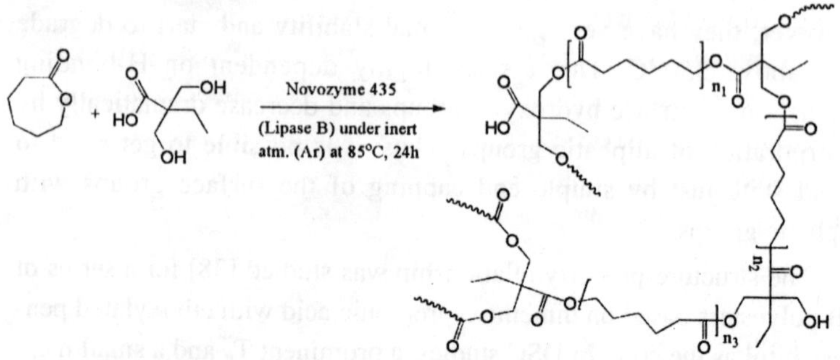

SCHEME 7.18 Synthesis of HB polyesters by enzymatic process

Whereas the aromatic polyester was synthesized [40] by polycondensation of 2,2-*bis*(4-hydroxyphenyl)pivalonic acid (*bis*-PVA) in the presence of DMAP and DCC at 0°C at room temperature for 16h. The synthesized HB polyesters were characterized by ^{1}HNMR and ^{13}CNMR spectroscopies. The DB for aliphatic polyesters was 0.47 with DP 16.5, whereas for the aromatic one, DB was 0.48 with DP 30.5. The molecular weight varies from 1–2 X 10^3 g/mole. The aromatic HB polyester have T_g ~83-116°C. These HB polymers were functionalized with methacryloyl chloride and carboxylic acid chloride in the presence of Et_3N base to obtain isobutyric acid and methacryloyl terminated HB polyesters. These modified polymers have low viscosity (200-36 pa) and exhibit relatively low volume shrinkage (1.9 to 3.6)% during photoinitiated curing. Thus, they can be used for making dental composites (see application section).

HB polyester of alkoxy maleic acid and maleic anhydride were synthesized [41] by the free radical copolymerization technique (Scheme 7.19). The weight average molecular weight of the copolymer was 4.56X10^4 Da with polydispersity 2.7. The synthesized polyester adopts a globular shape in good solvents [42]. The studies on the condensation reaction between this HB polyester and two kinds of diglycidyl ether bisphenol-A type commercial epoxy resins showed a marked difference in their reaction pattern in terms of ability to form crosslinked structures with the HB polyester. The low molecular weight resin does not crosslinked in good solvents such as THF or dioxane but the polymers with degree of epoxide functional group between 0.34–0.5 were formed on heating at 110°C for 5h or 100°C for 12h. Whereas the high molecular weight resin rapidly produced highly crosslinked materials with polyester under the same conditions. This may be due to the fact that the HB polyester is in spherical-shape. Thus, low molecular weight epoxy resins were unable to crosslink the comparatively much bigger spherically shaped HB polyester rather it formed epoxy functionalized HB polyester by chemisorption to the spherical macromolecules. However, longer epoxy molecules increase the linkage possibility on the spherical polyester and thus increase the crosslinking efficiency, greatly.

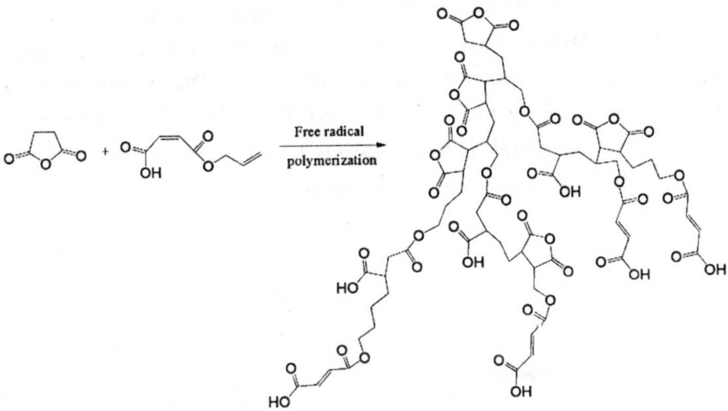

SCHEME 7.19 Synthesis of poly(alkoxy maleic acid-co-maleic anhydride)

The epoxy functionalized polyesters were characterized by ^{1}HNMR, FTIR spectroscopy and SEC. The average molecular weight varies from 1.5 X10^4 to 2.4 X10^4 Da with polydispersity 1.6 to 2.06. These HB polymers with epoxy functional group can be used as an intermediate for further functionalization, in addition to their direct uses.

7.4.2 AROMATIC

The aromatic HB polyesters were mainly synthesized from hydroxy-benzoic acid or 5-hydroxyisophthalic acid as the starting material. Of course, other AB_2 type aromatic monomers were also used. The synthesis of aromatic HB polyesters from an AB_2 type monomer, 3,5-*bis*(trimethyl siloxy)benzyl chloride was first reported [43] in 1982. However, the product was a copolymer of aromatic polyester. Later on in 1991, the HB polyester from trimethyl silane protected AB_2 type monomer (Scheme 7.20) was synthesized [44]. The bulk polymerization of AB_2 monomer was carried out in the absence or presence of catalyst such as trimethylamine hydrochloride and DMF. As the purity of the monomer is very important, so purity has to be high. A small amount of impurity of the monomer gave an insoluble product even if the reaction temperature was lower than 200°C. The trimethyl silane deprotected products were obtained by methanolic hydrolysis. The average molecular weight of the resulted HB polyester was 80,000 Da as measured by GPC-MALLS. The DB was found to be 0.55 to 0.66, as measured by ^{1}HNMR study. The HB polymers was soluble in common organic solvents like acetone, THF, DMF, etc. The polymers were highly thermostable (upto 400 °C) and T_g is 197°C. However, on modification of terminal hydroxy groups with monobenzyl adipic acid ester [45], the Tg dramatically decreased to 6°C.

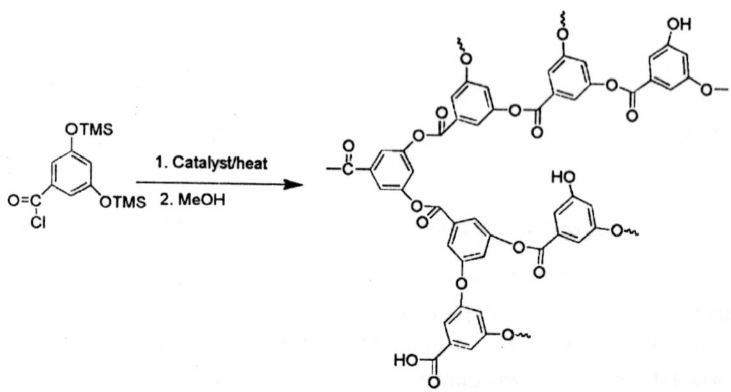

SCHEME 7.20 Synthesis of HB polyesters based on 3,5-bis(trimethyl siloxy)benzyl derivative

The HB polyesters [46] were also synthesized by thermal bulk polymerization of 3,5-diacetoxybenzoic acid, another AB_2 type monomer at 250°C (Scheme 7.21). The resulted polymers were soluble in common organic solvents and molecular weight was found to be 10^6 Da as determined by SEC. The relationship between intrinsic viscosity and molecular weight of the resulting polyester suggested a highly branched structure and the value of α in Mark-Houwink equation is less than 0.5. However, the polymerization of same monomer at above

250°C resulted insoluble product, even though polymerization of trimethyl-silyl-3,5-diacetoxy benzoate at 280°C resulted a HB Polyester, which was soluble in common organic solvents [47] (Scheme 7.21). The melt condensation of AB_3 monomer, 3,4,5-triacetoxybenzoic acid can also be used to synthesize HB polyesters and their copolymers. Several copolyesters such as liquid crystalline polyesters [48], biodegradable polyesters [49], etc. were synthesized by melt polycondensation of AB_3 and AB type monomers.

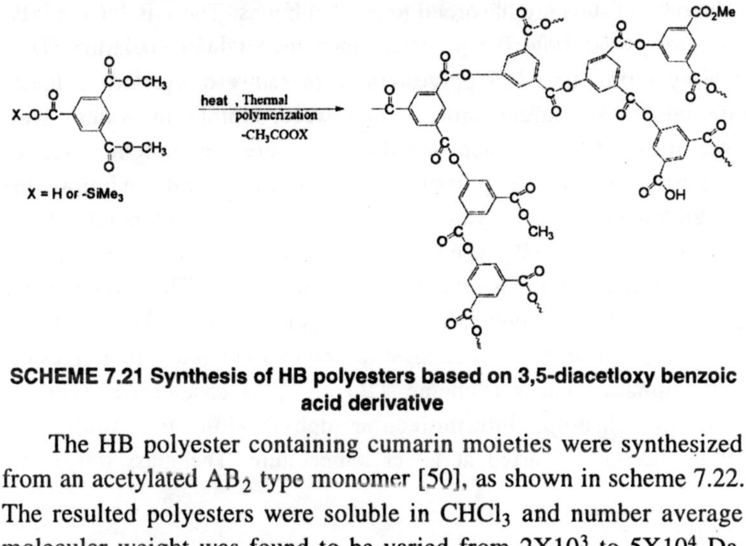

SCHEME 7.21 Synthesis of HB polyesters based on 3,5-diacetloxy benzoic acid derivative

The HB polyester containing cumarin moieties were synthesized from an acetylated AB_2 type monomer [50], as shown in scheme 7.22. The resulted polyesters were soluble in $CHCl_3$ and number average molecular weight was found to be varied from $2X10^3$ to $5X10^4$ Da. This HB polyesters also showed a blue light emission in the range of 450–492 nm, when irradiated at 400 nm.

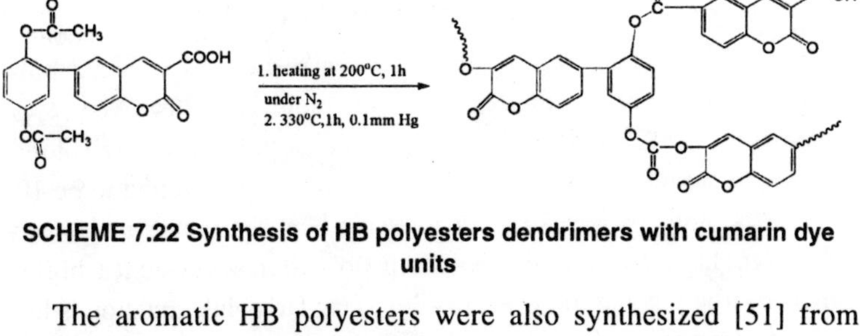

SCHEME 7.22 Synthesis of HB polyesters dendrimers with cumarin dye units

The aromatic HB polyesters were also synthesized [51] from another AB_2 type monomer, 5-acetoxyisophthalic acid (Scheme 7.23) by melt polymerization technique at 250°C. The resulted polymers were insoluble in organic solvents, since intermolecular dehydration between the carboxylic acid groups occurred during the melt polymerization. However, the hydrolysis of polymer resulted soluble polyester with carboxylic acid end groups (Scheme 7.23). The weight average molecular weight and polydispersity of the polymers were found to be 4.29 X 10^4 Da and 3.69 respectively as measured by SEC after protection of carboxylic end groups by trimethyl silane, as free carboxylic groups are absorbed to SEC columns. The DB for the HB polyesters was about 0.5 as determined by ^{1}HNMR spectrum. The carboxy terminated HB polyesters were easily converted to their ammonium or sodium salts, which were soluble in water. The viscosities of the aqueous solutions were dependent on the concentration of the polymer and inorganic salts, similar to conventional polyelectrolytes. The alkyl chains were incorporated [51] in the backbone of AB_2 type monomers, which were melt condensated at lower temperature to get the desired HB polyesters. The monomer, 5-(2-hydroxy ethoxy)isophthalic acid was polymerized at 190°C in the presence of an organic tin catalyst to obtain a HB aromatic polyester with aliphatic spacer (Scheme 7.24). In this case, no anhydride formation through intermolecular dehydration was found as polymerization occurred at lower temperature. The glass transition

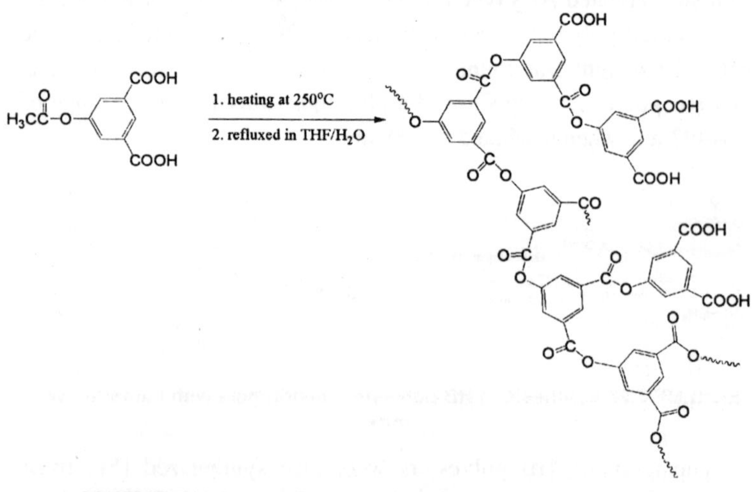

SCHEME 7.23 Synthesis of carboxyl terminated HB polyesters

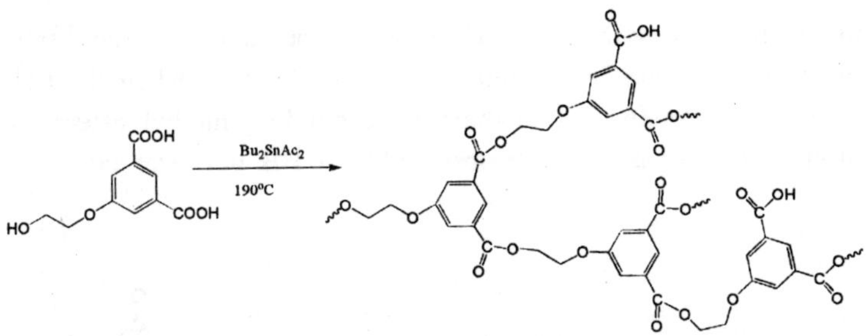

SCHEME 7.24 Synthesis of HB polyesters with aliphatic spacer unit

temperature of the polymers was found to be 150°C, which was much lower than the fully aromatic analog.

HB polyesters [52] were synthesized by a novel CO insertion process. The insertion reaction of CO catalyzed by palladium can be used for the propagation reaction to form linear aromatic polyesters. They polycondensation of AB_2 and A_2B type monomers via the CO insertion reaction resulted HB polyesters. However, the products obtained from AB_2 type monomer, 3,5-dibromophenol were insoluble in organic solvent, which may be due to crosslinking reactions via aryl-aryl coupling. Although, low molecular weights soluble products were obtained by the polymerization of A_2B type monomers viz. 5-bromoresorcinol and 5-iodoresorsinol.

5-(2-hydroxy ethoxy)isophthalic acid dimethyl ester was melt polymerized [53] in the presence of a core molecule, trimethyl-1,3,5-benzene tricarboxylate to get the desired HB polyesters. The synthesis of HB polyesters was reported [54] by polycondensation of 4,4-(4'-hydroxyphenyl)pentanoic acid esters (Scheme 7.25) The determination of DB of the polymers revealed that the value (0.49) is independent of the ester groups in the starting monomer. The extent of cyclization during polymerization process was found to be very less (< 5%), confirmed by ^{13}C labeled NMR experiment [54].

HB aromatic polyesters [55] possessing long chain flexible aliphatic spacer were synthesized by polycondensation of AB_2 type monomers in the presence of thionyl chloride and pyridine (Scheme 7.26). This aromatic–aliphatic HB polymers showed liquid crystalline behavior. The polymers with carboxylic acid end groups formed a nematic phase above their glass transition temperature. It is interesting to note that mesophase formation was not observed when the end groups were transformed to their corresponding methyl esters. A cholesteric mesophase was also observed for a HB polyester contain-

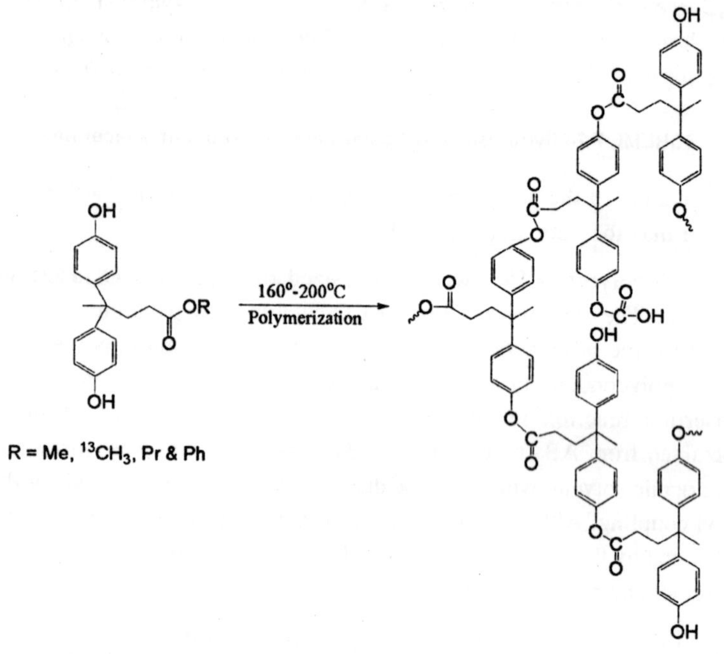

SCHEME 7.25 Synthesis of HB polyesters based on 4,4-(4'-hydroxyphenyl)pentanoic acid derivatives

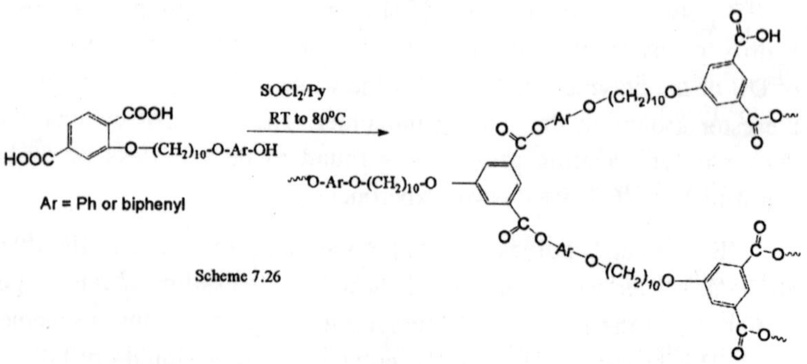

SCHEME 7.26 Synthesis of liquid crystalline HB polyesters

ing biphenyl unit as a mesogen and an aliphatic alkyl chain as a flexible spacer group after the end capping reaction with a chiral group [56].

7.4.3 Copolymers

HB aromatic homo- and co-polymers [57] were synthesized by a random bulk polycondensation and the slow monomer addition strategy. The monomers used here were 3,5-*bis*(trimethyl siloxy)benzoyl chloride (AB_2 type) and 3-(trimethyl siloxy) benzoyl chloride (AB type) with three different core molecules, trimethylol-propane (TMP), 1,3,5-trihydroxybenzene (B3OH) and 1,3,5-*tris*(trimethyl siloxy)benzene. The homo- and co-polyesters were synthesized by the bulk polymerization of the different monomers at about 200°C under the Agron (Ar) atmosphere in the presence of DMF as the catalyst. Whereas in the polymerization under slow monomer addition process the monomer as one weight percent diglyme solution was very slowly added into a highly concentrated solution of the core in dry quinoline, as the catalyst. The reaction was carried out at about 200°C under the Ar atmosphere. The reactions are shown in scheme 7.27 for the homopolymer from AB_2 monomer. The synthesized HB polyesters were characterized by ^{1}HNMR, ^{13}CNMR spectroscopy and SEC-LALLS-VISK [57] technique. The weight average molecular weight of the polyesters varies from 117400 to 505100 Da with polydispersity 1.09 to 1.49 and DB are 0.55 ± 0.02 to 0.66 ± 0.02 for homopolymers whereas for copolymers the values are 37000 to 160000 Da, 1.21 to 2.16 and (0.08–0.64) ± 0.03 respectively, in case of slow monomer addition technique. These results are far better than the results for polyester synthesized by bulk polymerization method. The viscosmetry measurements revealed a linear dependence of the intrinsic viscosity on molecular weight as well as decrease in Mark-Houwink exponent to only 0.18 was obtained for the HB homopolymers prepared under slow monomer addition conditions.

Highly branched radial block copolymers were synthesized [58] by living ring opening polymerization of ε–caprolactone initiated from the numerous chain-end hydroxymethyl groups of analogs dendrimeric and HB polyesters derived from *bis*-MPA. A variety of highly branched radial block copolymers was obtained by controlling the size of the

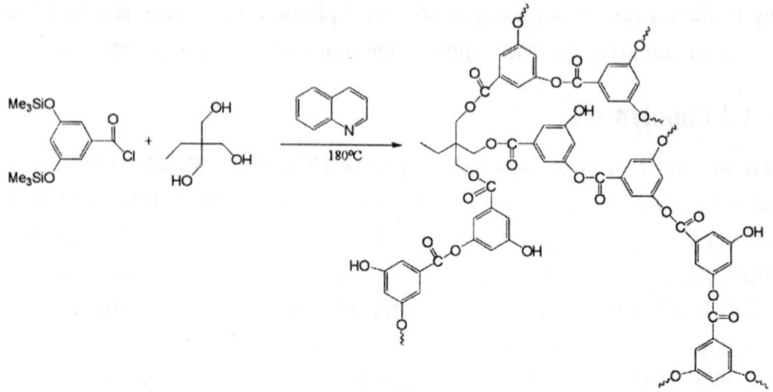

SCHEME 7.27 Synthesis of HB aromatic polyesters by slow addition process

dendritic polymer and the molar ratio of the ε–caprolactone. The study revealed that the reactivity of the chain end hydroxy methyl groups in the dendrimer is significantly greater than in the isomeric HB initiator.

7.5 APPLICATIONS OF DENDRITIC POLYESTERS

In this section different applications of dendritic polyesters are discussed.

7.5.1 Coatings

HB polyester based on *bis*-MPA as a base material was used [31] for radiation curable thermosetting resins. The cure rate and final mechanical properties of the HB polyester cured resins were measured and found much suitable compared to conventional resins. The study showed that HB polyester could be used as versatile scaffolds for various radiation curable resins. Perstop Specialty Chemical AB, Sweden used aliphatic HB polyesters [59-60] based on *bis*-MPA, an AB_2 type monomer and *tris*(hydroxyl methylol)propane as the core unit, as UV curable coating [61] without any photoinitiator. These HB polyesters are now available commercially in large scale [62]. Different grades of these HB polyesters with molecular weight 1750 g/mole to 7500 g/mole and glass transition temperatures (30-40)^{o}C, as amorphous solids are also available. The properties of the commercially available aliphatic polyester, Bolton® H20 (Fig.7.10) are given in table 7.1 [62]. The aliphatic polyester was acrylated with acrylic acid in the presence of Broenstedt acid and free radical inhibitor and the resulted HB acrylated polyester (20 wt.%) was used in an UV curable coating formulation by Perstop Specialty Chemicals AB, Sweden. Because of its high number of functionality and high reactivity, a high crosslink density of the cured resin was obtained. Also, a lower shrinkage with deeper through curing was observed. The chemical resistance and scratch resistance are also very good for the cured coatings.

Table 7.1 Properties of the HB polyester, Bolton® H20

Property	*Value*
Mw (SEC), g/mole	2100
Polydispersity	1.3
Hydroxy value, mgKOH/g	509
Acid value, mgKOH/g	6
Density, g/c.c.	1.29
Tg (DMA), ^{O}C	30
Viscosity, 110°C, 30, S-1, pas	5

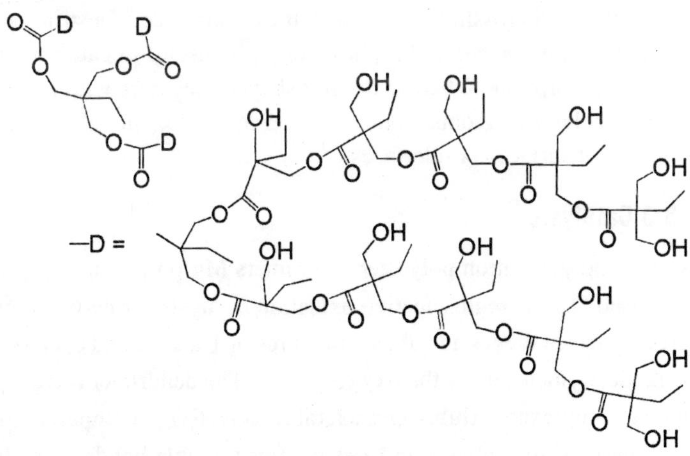

FIG. 7.10 Structure of Bolton ® H20 aliphatic polyester

7.5.2 Binders and Crosslinkers

HB polyesters were used [40] for making dental composites. The dental composites were prepared by mixing this polymerizable resin with 74.3% a bariumalumosilicate glass fiber or 72.8 wt.%, a strontiumalumosodium fluorophosphoro silicate. The composite materials showed compressive strength 158 to 238 MPa and flexural strength of 27 to 64 MPa. An extremely low volume shrinkage (0.5 to 1.55)% was observed for the composites after photochemical crosslinking.

The unsaturated HB polyester, poly(allyoxy maleic acid-co-maleic anhydride) as crosslinker was used [63] in radical copolymerization of vinyl monomers such as vinyl acetate, methyl methacrylate (MMA) and styrene in 50 wt.% solution of THF. The olefinic monomers like styrene and vinyl acetate react in different manner from MMA. Both styrene and vinylacetate afforded interpenetrating polymer network copolymer gels by reactions with the HB polyester. Whereas the copolymerization of polyester with MMA afforded non-crosslinked starbranched copolymers that consisted of a HB polyester core from which long PMMA branches were produced. This is due to the fact that the reactivity ratio of the HB polyester is same to the reactivity ratio of the vinyl acetate and styrene, thus forming alternative copolymers. Whereas the reactivity ratio of MMA is much higher than polyester and thus the homopolymer of MMA was formed, in stead of copolymer. The synthesized crosslinked gels with the coexistence of a strong polar phase and a low or non-polar phase might be useful in catalysis or in selective absorption. Whereas the star shaped polymers with huge and strongly polar end groups could be used in the functionalization and treatment of surfaces or interfaces.

7.5.3 Catalysis

1st and 2nd generation polyester dendrimers Mn-porphyrin complexes were used [13] as regioselective oxidation catalysts for epoxidation of non-conjugated dienes and alkene mixtures of 1-alkene and cyclooctene using idostrylbenzene as the oxygen donor. The dendrimer metalloporphyrin complexes exhibit considerable selectivity compared to its monomeric Mn-porphyrin and less hindered double bonds epoxidized more readily than the more hindered one. The higher selectivity of the dendrimer-metalloporphyrin complex is mainly due to the steric influence of the bulky dendrimers at the periphery of the Mn(II)-porphyrin catalyst rather than any electronic factor, which was the fact for monomeric catalysts.

7.5.4 Encapsulating Materials

Bhyrappa *et al.* showed [15] the shape selectivity of the binding sites of the Zn-polyester dendrimer-porphyrins by using the axial ligation of various nitrogenous bases of different shapes and sizes in toluene. On ligation of bases, the visible absorption spectra of these complexes were red-shifted and showed an increase in the extinction co-efficient of both the Soret and visible bands. The shape–selective ligation was observed with selectivity for the linear and non-linear ligands of more than 10^4.

7.5.5 Electronics

A family of chromophore containing polyester dendrimers (Fig.7.11) was used [16] in single layer light emitting diodes. The dendrimers possess peripheral redox active triarylamine that are responsible for hole transport and are the loci for exciton formation. Since, both the backbone and the core fluorophores have comparatively high redox potentials, charge transportation is expected to occur solely through the hole transporting periphery moieties and the accompanying electron transporting oxadiazole, 2-(4-biphenyl)-5-(4-*tert*-butylphenyl)-1,3,4-oxadiazole. Transfer of energy to the highly luminescent acceptor fluorophores at the core of the dendrimer via a Forster process, subsequently results in emission exclusively from the acceptors. Thus, two component, single layer OLED, consists of polyester dendrimers with either cumarin C343 or pentathiophene T_5 acceptor lumophores core and the electron transporting oxadiazole were prepared. The authors have also tried to get four single layer devices containing mixture of both the dendrimers, i.e. with both acceptor lumophore dyes and observed a degree of site isolation of the central dye afforded by the small surrounding dendrons.

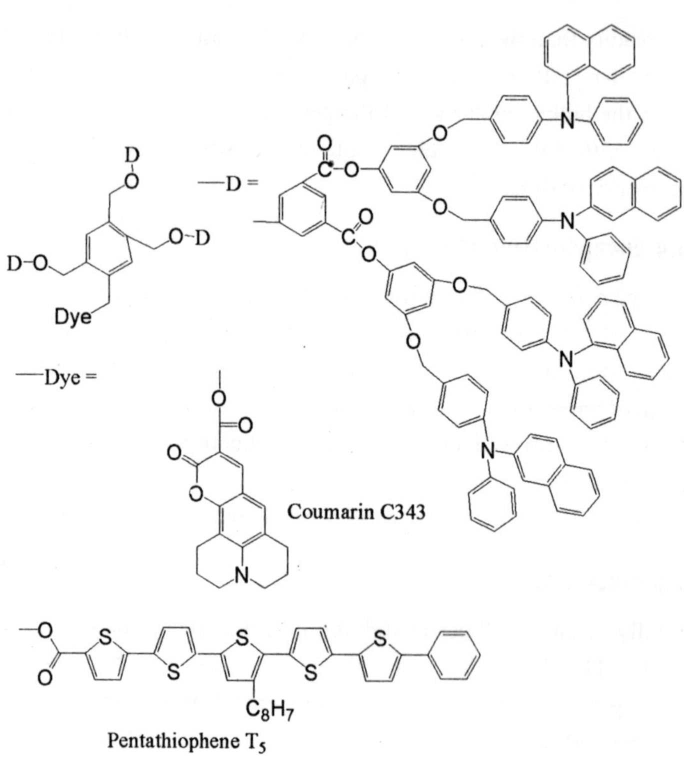

FIG. 7.11 Structures of polyester dendrimers with triaryl amine periphery and cumarin C343 or pentathiophene T_5 as the core unit

7.6 SUMMARY

In this chapter a large variety of different aliphatic and aromatic dendritic polyesters are discussed. They can be easily synthesized like linear polyesters mainly by condensation, ester exchange and ring opening polymerization techniques. These dendritic polymers are also very useful like linear polyesters in different applications, such as surface coatings, catalysts, energy transferring materials, etc.

REFERENCES

[1] S.L.Goh, M.B.Francis and J.M.J.Frechet, *Chem.Commun.* (2002)2954

[2] M.Trollsas and J.L.Hedrick, *J.Am.Chem.Soc.* **120**(1998)4644

[3] S.K.Emran, Y.Liu, G.R.Newkome and J.P.Harmon, *J.Polym.Sci. Part B Polym.Phys.* **39**(2001)1381

[4] H.Ihre, O.L.P.Jesus and J.M.J.Frechet, *J.Am.Chem.Soc.* **123**(2001)5908

[5] M.A.Carnahan and M.W.Grinstaff, *J.Am.Chem.Soc.* **123**(2001)2905

[6] M.A.Carnahan and M.W.Grinstaff, *Macromolecules* **34**(2001)7648

[7] M.T.Morgan and M.W.Grinstaff, Materials Research Society, Boston, MA, Symp. Polymeric Biometrials for Tissue Engineering, 2001

[8] M.A.Carnahan, C.Middleton, J.Kim, T.Kim and M.W.Grinstaff, *J.Am.Chem.Soc.* **124**(2002)5291

[9] M.Malkoch, E.Malmstrom and A.Hult, *Macromolecules* **35**(2002)8307

[10] T.M.Miller, E.W.Kwock and T.X.Neenan, *Macromolecules* **25**(1992)3143

[11] F.Zeng and S.C.Zimmerman, *J.Am.Chem.Soc.* **118**(1996)5326

[12] K.L.Wooley, C.J.Hawker and J.M.J.Frechet, *Angew.Chem.Int.Ed.* **33**(1994)82

[13] P.Bhyrappa, J.K.Young, J.S.Moore and K.S.Suslick, *J.Am.Chem.Soc.* **118**(1996)5708

[14] C.J.Hawker and J.M.J.Frechet, *J.Am.Chem.Soc.* **114**(1992)8405

[15] P.Bhyrappa, G.Vijayanthimala and K.S.Suslick, *J.Am.Chem.Soc.* **121**(1999)262

[16] A.W.Freeman, S.C.Koene, P.R.L.Malenfant, M.E.Thomson and J.M.J.Frechet, *J.Am.Chem.Soc.* **122**(2000)12385

[17] C.J.Hawker and J.M.J.Frechet, *J.Chem.Soc.Perkin Trans.1* (1992)2459

[18] D.Felder, H.Nierengarten, J.P.Gisselbrecht, C.Boudon, E.Leize, J.F.Nicoud, M.Gross, A.van Dorsselaev and J.F.Nierengarten, *New J.Chem.* **24**(2000)687

[19] K.L.Wooley, C.J.Hawker and J.M.J.Frechet, *Polymer* **35**(1994)4489

[20] M.R.Bryce, W.Devonport and A.J.Moore, *Angew.Chem.Int.Ed.* **33**(1994)1761

[21] M.Nakazono, L.Ma and K.Zatisu, *Tetrahedron Lett.* **43**(2002)8185

[22] J.F.Nierengarten, J.F.Eckert, Y.Rio, M.del, P.Carreon, J.L.Gallani and D.Guillon, *J.Am.Chem.Soc.* **123**(2001)9743

[23] N.Armaroli, C.Boudon, D.Felder, J.P.Gisselbrecht, M.Gross, G.Marconi, J.F.Nierengarten and V.Vicinelli, *Angew.Chem.Int.Ed.* **38**(1999)3730

[24] D.Seebach, G.F.Herrmann, U.D.Langweiler, B.M.Bachmann and W.Amrein, *Angew.Chem.Int.Ed.* **35**(1996)2795

[25] D.Seebach, G.F.Herrmann, U.D.Langweiler, and W.Amrein, *Helv.Chim.Acta* **80**(1997)989

[26] T.Glauser, C.M.Stancik, M.Moller, S.Voytek, A.P.Gast and J.L.Hedrick, *Macromolecules* **35**(2002)5774

[27] Z.Bo, W.Zhang, X.Zhang, C.Zhang and J.Shen, *Macromol.Chem.Phys.* **199**(1998)1323

[28] M.S.Choi, T.Aida, T.Yamazaki and I.Yamazaki, *Chem.Eur.J.* **8**(2002)2667

[29] E.Malmstrom, M.Jahnsson and A.Hult, *Macromolecules* **28**(1995)1698

[30] E.Malmstrom and A.Hult, *Macromolecules* **29**(1996)1222

[31] M.Jahnsson, T.Glauser, G.Rospo and A.Hult, *J.Appl.Polym.Sci.* **75**(2000)612

[32] C.J.Hawker, F.Chu, P.J.Pemery and D.J.T.Hill, *Macromolecules* **31**(1998)3439

[33] M.Trollsas, B.Atthoff, H.Claesson and J.L.Hedrick, *Macromolecules* **29**(1996)3439

[34] M.Trollsas and J.L.Hedrick, *Macromolecules* **31**(1998)4390

[35] M.Liu, V.Vladimirov and J.M.J.Frechet, *Macromolecules* **32**(1999)6881

[36] M.Trollsas, P.Lowenhiclm, V.Y.Lee, M.Miller, R.D.Miller and J.L.Hedrick, *Macromolecules* **32**(1999)9062

[37] C.M.Nuner, B.S.Chiou, A.L.Andrady and S.A.Khan, *Macromolecules* **33**(2000)1720

[38] M.Rogunova, T.-Y.S.Lynch, W.Pretzer, M.Kulzick, A.Hiltner and E.Baer, *J.Appl.Polym.Sci.* **77**(2002)1207

[39] S.Skaria, M.Smet and H.Frey, *Macromol.Rapid Commun.* **23**(2002)292

[40] J.E.Klee, C.Schneider, D.Holter, A.Burgath, H.Frey and R.Mulhaupt, *Polym.Adv.Technol.* **12**(2001)346

[41] H.Liu, J.H.Nasman and M.Skrifvars, *J.Polym.Sci.part A Polym.Chem.* **38**(2000)3074

[42] H.Liu, C.E.Wilen and M.Skrifvars, *J.Polym.Sci.part A Polym.Chem.* **38**(2000)4457

[43] H.R.Kricheldorf, Q.Z.Zang and G.Schwarz, *Polymer* **23**(1982)1821

[44] C.J.Hawker, R.Lee and J.M.J.Frechet, *J.Am.Chem.Soc.* **113**(1991)4583

[45] K.L.Wooley, C.J.Hawker, R.Lee and J.M.J.Frechet, *Polym.J.* **26**(1994)187

[46] S.R.Turner, B.I.Voit and T.H.Mourey, *Macromolecules* **26**(1993)4617

[47] H.R.Kricheldorf, O.Stober and D.Lubbers, *Macromolecules* **28**(1995)2118

[48] A.Reina, A.Gerken, U.Zemann and H.R.Kricheldorf, *Macromol.Chem. Phys.* **200**(1999)1784

[49] H.R.Kricheldorf and T.Shukenbrock, *J.Polym.Sci.part A Polym.Chem.* **36**(1998)31

[50] S.Fomine, E.Rivera, L.Fomina, A.Ortiz and T.Ogawa. *Polymer* **39**(1994)3551

[51] S.R.Turner, F.Walter, B.I.Voit and T.H.Mourey, *Macromolecules* **27**(1994)1611

[52] Y.Ishida, M.Jikei and M.Kakimoto, *Kobunshi Ronbunshu* **54**(1997)891

[53] W.J.Feast and N.M.Stainton, *J.Mater.Chem.* **5**(1995)405

[54] P.Kambouris and C.J.Hawker, *J.Chem.Soc.Perkin Trans. 1* (1993)2717

[55] S.W.Hahn, Y.H.Kim and J.I.Jin, *Macromolecules* **31**(1998)6417

[56] S.Bauer, H.H.Fischer and H.Ringsdorf, *Angew.Chem.Int.Ed.* **32**(1993)1589

[57] A.Mock, A.Burgath, R.Hanselmann and H.Frey, *Macromolecules* **34**(2001)7692

[58] M.Trollsas, C.J.Hawker, J.F.Remenar, J.L.Hedrick, M.Jahonsson, H.Ihre and A.Hult, *J.Polym.Sci. Part A Polym.Chem.* **36**(1998)2793

[59] A.Hult, M.Jahnsson, E.Malmstrom and K.Sorensen, Sweden Pat. SE 468771(1992)

[60] B.Pettersson and K.Sorensen, Sweden Pat. SE 503342(1994)

[61] B.Pettersson, 'The use of dendritic polyester Acrylates for photoinitiator-Free UV curable coating', 6th Congress, European Caoting Show, 2001

[62] Bolton® H20, Bolton® H30 and Bolton® H40, Perstorp Specialty Chemicals AB, Sweden, 2001

[63] H.Liu and C.E.Wilen, *J.Polym.Sci.part A Polym.Chem.* **39**(2001)964

CHAPTER 8

SILICON BASED DENDRITIC POLYMERS

8.1 INTRODUCTION

The designing of a polymer with element other than carbon atoms, such as silicon, phosphorous, nitrogen, metal ions, etc. in the main backbone is of great importance because the incorporation of such heteroatoms in the polymer chains is expected to offer some unusual properties. The dendrimers based on heteroatoms resulted peculiar properties, such as variable branching multiplicity, flexibility, electrooptical properties, high thermal properties, etc. The silicon based dendrimers are the most interesting among all of these heteroatoms based dendrimers. This is due to the unique characteristics of silicon chemistry. These are: a) silicon chemistry offers several quantitative (>99% yield) reactions suitable for synthesis of silicon based dendrimers, b) the Si-O-Si bonds are highly polar and very strong due to large difference in electronegativity (Si- 1.90 and O- 3.5), and formation of partial double bond character through pπ–dπ bonds. The main advantage of using silicon chemistry to synthesize dendritic polymers is that the electrophilic silicon can be easily accessed by nucleophilic species. The synthetic methods for preparing silicon containing dendrimers are mainly the use of repetitive hydrosilation, alkenylation and alcolysis reactions [1-17]. The important features for silicon based dendrimers, which is almost common to all, are: a) they are mostly synthesized by the divergent approach, b) they exhibit high flexibility and hence low glass transition temperature, c) the use of Si as the branching point offers tailoring of branching multiplicity, etc. The dendritic polymers based on silicon chemistry have been introduced by Uchida *et al.* [18] and Morikawa *et al* [19]. A few numbers of reviews [20-22] and reports [23-24] have already been published on the synthesis, characterization and properties evaluation of silicon containing dendritic polymers.

8.2 SYNTHESIS, CHARACTERIZATION AND PROPERTIES OF SILICON BASED DIFFERENT TYPES OF DENDRIMERS

The carbosilane and carbosiloxane dendrimers are two main important groups of polymers in this class [21-22]. However, the silicon based dendrimers can be categorized into six different groups, like (i) carbosilane, (ii) carbosiloxane, (iii) siloxane, (iv) silsesquioxane, (v) silazane and (vi) silatrane. The syntheses, characterizations and properties of these silicon based dendrimers are discussed in the subsequent sub-sections.

8.2.1 Carbosilanes

One of the most important silicon based dendrimers is carbosilane dendrimers. This is due to their straight forward syntheses and the possibility to tailor the structures of the dendrimers. Furthermore, they are kinetically as well as thermodynamically very stable owing to similar bond dissociation energy of Si-C bond (306 kJ/mole) with C-C bond (345 kJ/mole) and low difference in electronegativity between Si and C and hence low polarity of Si-C bond. These dendrimers are very useful due to the flexibility of synthetic route, which allows versatile functionalization. Almost, all the carbosilane dendrimers have been synthesized by the divergent method using mainly hydrosilylation and alkenylation reactions.

A branched carbosilane compound to prepare 12 arms star polymer, which may be viewed as a first generation dendrimer with 12 end groups was first reported [25] by Felter and his coworkers. However, the true carbosilane dendrimers of various generations were synthesized and reported by many researches [4, 23, 26-28]. The carbosilane dendrimers upto 5th generation, were synthesized [26-27] by using tetra allylsilane as the core unit and by the reaction of $SiCl_4$ with allylmagnesium chloride. Trichlorosilane as the hydrosilylation reagent and allyl magnesium bromide as the alkenylation reagent were used for this synthesis. The undecenylmagnesium bromide as the alkenylating reagent was also used to obtain less dense dendrimer structure, though the molecular weight and the structural perfection of these dendrimers were not reported with appropriate analytic methods.

A series of carbosilane dendrimers was synthesized by using simple molecule, tetravinyl silane as the central core and dichloromethyl silane as the propagating unit [23]. Two reactions were involved in the synthesis of each generation: (a) simple and clean hydrosilylation of vinyl silane with dichloromethyl silane and (b) nucleophilic replacement of silicon chloride by vinyl magnesium bromide. The sequence of reactions is shown in scheme 8.1. These dendrimers are truly highly branched 3D symmetric structural macromolecules. The product obtained with high yield and minimal side reactions. The structures of the synthesized dendrimers were characterized by ^{1}HNMR and ^{13}CNMR spectroscopy and the molecular weights were determined by VPO and laser light scattering techniques.

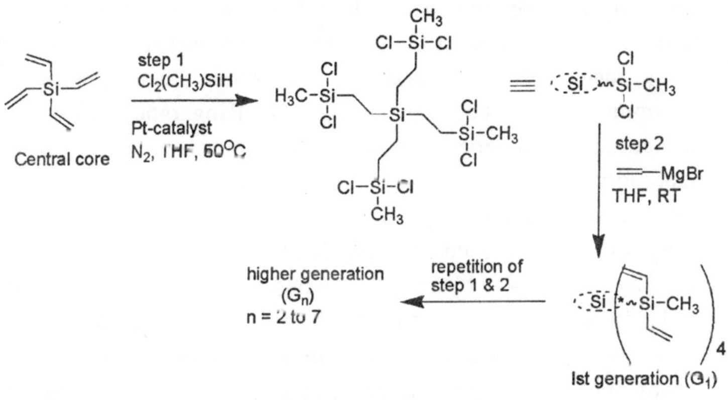

SCHEME 8.1 Synthesis of carbosilane dendrimers with tetravinyl silane as the core unit

Another series of polyallyl carbosilane dendrimers [29] upto 7th generation, were synthesized by repetitive alkylation and hydrosilylation reactions using the divergent technique. $MeSi(OSi(Me)Cl_2)_3$ was used as the core unit, CH_2=$CHCH_2$-MgCl as the alkylating agent and $HSiMeCl_2$ as the hydrosilylation reagent were used for this synthesis (Scheme 8.2). The dendrimer was purified by column chromatography and preparative GPC technique. Thermal studies by DSC and TMA indicated that the Tg of the dendrimers remains constant after 4th generation [29]. Similar type of dendrimers were synthesized by the same technique, using $MeSi(OSiMe_2CH$=$CH_2)_3$ or $MeSi(OSiMe_2CH_2CH$=$CH_2)_3$ as the initiator core [30].

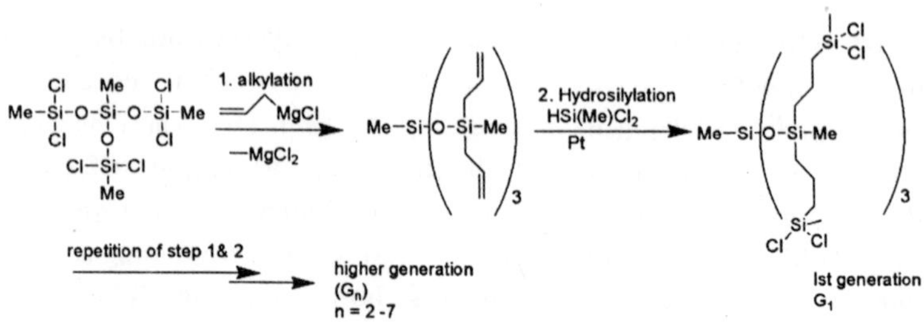

SCHEME 8.2 Synthesis of carbosilane dendrimer with allyl end groups

The carbosilane dendrimers upto 3rd generation were reported [31] by using *tetrakis*(phenyl ethynyl)silane as the core, dichloromethylsilane as hydrosilylating agent and lithium phenyl acetylide as the alkylating agent. The synthesized dendrimers were characterized by NMR spectroscopy and MALDI-TOF mass spectrometric studies. The T_gs of the dendrimers were found to be considerably high compared to other similar type carbosilane dendrimers. Ring type carbosilane dendrimers upto 2nd generation, were also synthesized [7] by using siloxane polymer with SiH bonds. $[Me_3SiO(MeSiOH)_nSiMe_3]$ as the central core and dichlorovinyl silane, dichloromethyl silane, etc. as the building blocks are used in this process. The polymers were synthesized by repetitive hydrosilylation and alkenylation as well as by alkynylation reactions. They also synthesized double-layered dendritic carbosilanes [8] with phenyl ethynyl moieties on the peripheral layer and propyleneoxy and ethynyl moieties in the inner shells. The allyloxy containing dendrimers were synthesized by the reactions of chlorosilylated dendrimers with alkylalcohol in the presence of TMEDA as described above. The dendrimers with ethynyl groups on each terminal arm were obtained by the reaction of the chlorosilyl terminal groups containing dendrimers with lithium phenyl ethynyl. By repetition of hydrosilylation and alkynylation as well as alcoholysis reactions sequence, carbosilane dendrimers were synthesized upto 4th generation with 64 phenyl ethynyl groups on the periphery. IR, UV, and NMR spectroscopies, elemental, SEC, and DSC analyses characterized the synthesized dendrimers.

The carbosilane dendrimers upto 4th generation [32] with 1 to 2 branching system and first generation [33] for 1 to 3 branching system were synthesized. However, by combining both the branching systems, dendrimers upto 4th generation were also synthesized [34]. The synthesis of carbosilane dendrimers upto 4th generation was described [32] from $[MeSiO(CH=CH_2)]_4$ via successive Pt-catalyzed hydrosilylation of vinyl groups of the core unit with $HSiMeCl_2$, alternating with alkenylation of the chlorosilane moieties with $BrMgCH_2CH=CH_2$. These dendrimers were surface functionalized by various moieties [32, 35], as shown in scheme 8.3. Similarly, the corresponding chlorinated dendrimers upto 3rd generation can be functionalized [35].

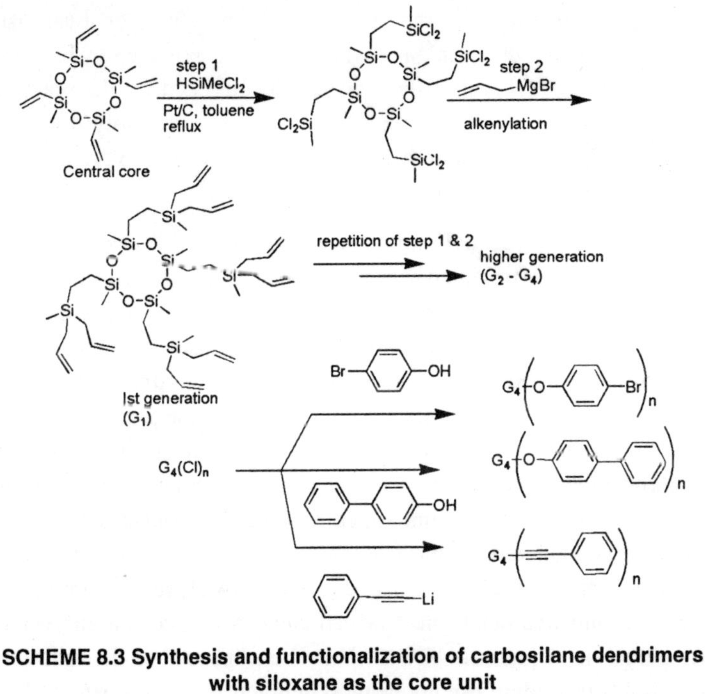

SCHEME 8.3 Synthesis and functionalization of carbosilane dendrimers with siloxane as the core unit

The dendritic carbosilanes containing benzoxazole, quinoline and pyrone derivatives on the periphery were also synthesized [9]. The parent dendrimers were synthesized by the same iterative hydrosilylation and alkenylation reactions, as described earlier. The desired dendrimers, were prepared by the reaction of Si-Cl groups on the periphery with 2-(2-hydroxy phenyl)-benzoxazole, 8-hydroxy quinoline and 3-hydroxy-2-methyl-4-pyrone in the presence of amine catalyst.

The limiting generation of the carbosilane dendrimers based on phenyl ethynyl methyl silanes as the core unit was synthesized [10] by the iterative alkynylation and hydrosilylation reactions cycle of lithium phenylacetylide and dichloromethyl silane. The building blocks of the dendrimers consist of double bonds in the inner shell and triple bonds on the periphery. The synthesized dendrimers were characterized by UV, NMR spectroscopies, elemental analysis, MALDI-mass spectrometry, SEC and DSC analyses.

The preparation of multiarmed star polymers with polylithiated carbosilane dendrimers was reported [36]. The polyfunctional carbosilane dendrimers having terminal allyl groups were extended by hydrosilylation with didecylmethylsilane and partly lithiated by the reaction with butyl lithium. A few starbranched polymers viz. poly(dimethyl siloxane), polystyrene, poly(ethylene oxide), etc. were also prepared by using these dendritic polylithium macromolecules.

A series of ferrocenyl carbosilane dendrimers [37] with 9-, 27-, 81-, and 243-allyl terminated groups was synthesized by the reaction of ferrocenyl dimethylsilane in diethyl ether at $40^{o}C$ using Karstedt's catalyst. The synthesis of carbosilane dendrimers upto 4th generation, was described [38] by iterative reactions of hydrosilylation of tetravinylsilane, as the central core with $HSiCl_3$ in the presence of Pt/C catalyst followed by Grignard reaction with vinyl magnesium bromide. The carbosilane dendrimers with 4, 8 and 12 peripheral zirconocene, hafnocene and titanocene end caped moieties were also synthesized [39] and characterized by spectroscopic techniques. The chlorosilanes of each generation were also reduced [40] by $LiAlH_4$ to the corresponding Si-H terminated dendrimers, which yields SiC as a ceramic residue after pyrolysis in argon at $950^{o}C$. However, because of the laborious process, this procedure can not be used in practice.

A wide variety of carbosilane dendrimers was synthesized [15] with high yield by using repetitive alkyenylation and hydrosilylation reactions cycle. In this process $SiCl_4$ reacts with alkymagnesium bromide (I) in ether to give tetraalkyl silane and then the alkyl group was hydrosilylated by $HSiCl_3$/Pt-catalyst. The product obtained with four $SiCl_3$ functional groups, which were treated with I to get the first generation dendrimer. The repetition of the above reactions sequence resulted higher generations.

The carbosilane based dendrons upto 3rd generation with amine group in the focal point, were also synthesized [41]. Bromopropyltrichlorosilane was used as the core unit for this synthesis and subsequent reaction with ammonia under pressure to generate the focal amine functionality. Coupling of the amine with trimesic acid was employed to obtain hybrid topologies with polar triamide core that may serve as a binding site for the polar guests in the receptor like structure.

A novel organosilane compound, 4-tri(allyl silyl)phenol was used to synthesize functionalizable dendrimers [42] upto 2nd generation by the divergent approach. This organosilane compound was synthesized from 4-bromophenol by a low temperature 1,4-silyl migration reaction. The dendrimers were characterized by NMR, IR spectroscopies, elemental analysis, MALDI-TOF or EI mass spectrometry and GPC analysis. This compound was also utilized [42-43] in convergent synthesis of 1,3,5-*tris*{(4-triallylsilyl)phenyl ester}benzene carbosilane dendrimers. The synthesized dendrimers were characterized by single crystal X-ray studies.

The synthesis of carbosilane dendrimers was reported by the divergent method as shown in scheme 8.4 [44-45]. The synthesized dendrimers were characterized by $^{1}H/^{13}C/^{29}Si$ triple resonance, 3D and pulse gradient NMR techniques.

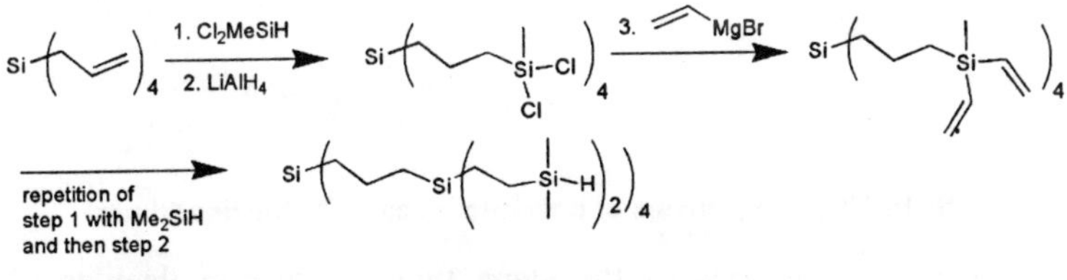

SCHEME 8.4 Synthesis of carbosilane dendrimers by divergent method using tetraallyl silane as the core unit

The first non-coordinating polyanionic carbosilane dendrimers were constructed [46] by means of alternating hydrosilylation and Grignard reactions followed by hydroboration of the resulting alkyl silyl dendrimer with dichloroborane and arylation with penta fluorophenyl lithium (Scheme 8.5). The tetravinylsilane was used as the core unit. The different allyl carbosilane dendrimers were synthesized with high yield by the same route as described earlier [39-40]. The boropentafluoro carbosilane dendrimers were synthesized then by reactions of tetraallylsilane (I), Si-(CH_2-CH_2-Si-(CH_2-CH=CH_2)$_3$)$_4$ (II), and Si-(CH_2-CH_2- Si-(CH_2-CH_2-Si-(CH_2-CH=CH_2)$_3$)$_3$)$_4$ (III), separately with $HBCl_2$, prepared *in-situ* from BCl_3 and $HSiEt_3$ at –60°C, followed by treatment with pentafluorophenyl lithium, prepared *in-situ* by transmetallation of bromopentafluorobenzene with n-BuLi at –70°C; at a temperature below –65°C. Both the reactions were very critical. The tetraallylsilane (I) was synthesized by reaction of $SiCl_4$ with allylmagnesium chloride, II was synthesized by hydrosilylation of tetravinyl silane with $HSiCl_3$ in the presence of Pt/C catalyst, followed by the reaction with allyl magnesium chloride [40]. The carbosilane dendrimer, (III) was obtained by repetition of the above steps. The poly(alkyl dichloroborane)s, products of first step, hydrolyzed rapidly in air but the hydroboration reaction is quantitative and regioselective. In the second step, which is exothermic reaction, if the temperature goes high the pentafluorophenyllithium formed can be decomposed. The products were characterized by ^{1}HNMR, ^{13}CNMR ^{11}BNMR and ^{19}FNMR spectroscopic studies. The stable ammonium salts were prepared by treatment the lithium salt with R-N^{+}Me$_2$H (R = Ph/$C_{11}H_{23}$) and used for the first time as metallocene catalyst.

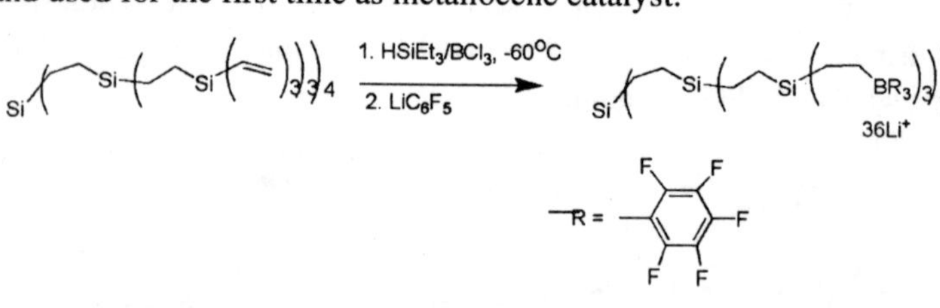

SCHEME 8.5 Synthesis of polyanionic carbosilane dendrimer

A novel synthesis of the same type of macromolecules via hydrosilylation of triallylsilane as well as methyl diallylsilane was also reported [2].

The introduction of perfluoro organosilyl peripheries into carbosilane dendrimers and related core functionalized monodendrons gives rise to anomalous hydrodynamic and viscometric behavior [47]. The intrinsic viscosity is unusually low, resulting in calculated hydrodynamic radii, 6.3A° for the 1st generation and 8.3A° for 2nd generation dendrimers, which imply unusually compact structure. Thus, these low surface energy materials may be very useful for surface coverage to promote non-sticky behavior or low friction coatings.

The, novel block copolymers containing two structurally different blocks, i.e. flexible carbosilane dendrimers and rigid polyisocyanides were reported [48]. In chloroform, these blocks copolymers respond to the addition of silver ions by the generation of nanowires consist of oriented crystalline silver. This process offers interesting possibilities for the generation of conducting nanoscopic architectures.

The carbosilane dendrimers upto 5th generation with 144 phenyl ethynyl groups, were synthesized [49]. For synthesis of these dendritic polycarbosilanes 1,3,5-*tris*(dimethyl vinylsilyl)benzene as the core and *bis*(phenyl ethynyl)methyl silyl groups as the propagating unit were used. The first generation was obtained by reaction of core unit with trichlorosilane followed by reaction of lithium phenyl acetylide. The repetition of the above two reactions offered to get higher generation. But to get less crowded dendrimers, the first generation was treated with dichloromethyl silane and then by repetitive reaction sequence of lithium phenyl acetylide and dichlorosilane to the respective generation allowed to obtain 5th generation carbosilane dendrimer.

A new methodology was developed [50] for the use of carbosilane dendrimers as soluble support in organic reaction. Hydroxy functionalized dendritic carbosilanes, $Si[CH_2CH_2CH_2SiMe_2(C_6H_4CH(R)OH)]_4$ and $Si\text{-}[CH_2CH_2CH_2\text{-}Si\text{-}[CH_2CH_2CH_2Si(Me_2)\text{-}(C_6H_4CH(R)OH)]_3]_4$, where R=H or (S)Me, were synthesized and subsequently converted into the esters $Si\text{-}[CH_2CH_2CH_2SiMe_2(C_6H_4CH(R)OC(O)CH_2Ph)]_4$ and $Si\text{-}[CH_2CH_2CH_2\text{-}Si\text{-}[CH_2CH_2CH_2SiMe_2(C_6H_4CH(R)OC(O)CH_2C_6H_4R')]_3]_4$, where R=H and R'=H or (S)Me or Br, by Suzuki coupling conditions. These ester functionalized dendrimers were converted to the corresponding zinc enolates followed by a condensation reaction with an imine to a β–lactum with an excellent yield and purity. Again, a small combinatorial library of β–lactam could be prepared starting from a carbosilane dendrimer functionalized with different ester moieties. These results showed that carbosilane dendrimers could be used as soluble substrate carriers for the generation of low molecular weight organic molecules. These functionalized carbosilane dendrimers can be easily separated by nanofiltration and also recyclable. At first the Me_2SiCl terminated carbosilane dendrimers were functionalized by the following sequence of reactions, as shown in scheme 8.6.

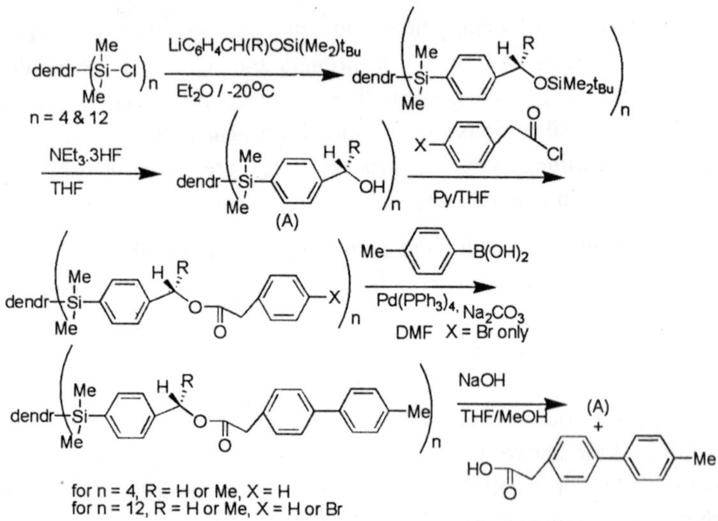

SCHEME 8.6 Functionalization of chloro terminated carbosilane dendrimers and use as soluble support for organic synthesis

The preparation and characterization of nonpolar fluorinated carbosilane dendrimers were reported [51]. The fluorinated dendrimers were obtained with high yield by repetitive hydrosilylation reactions between core hydride terminated carbosilane dendrimers with allyl-1,1-dihydrotrifluoroethyl ether or allyl-1,1-dihydroheptadecafluorononyl ether via the divergent route. The obtained functionalized dendrimers were characterized by atmospheric pressure chemical ionization (APCI) mass spectrometry, SAXS, TGA and elemental analyses.

Silicon based dendrimers upto 4th generation were synthesized [52] by the convergent method using CH_2=$CHCH_2$-CN and (CH_2=$CHCH_2$)(Me)Si($PhSiMe_2OH$)$_2$ as the starting materials and building block respectively. By repetition of hydrosilylation of alkylcyanide with HSiMeCl and amination with Et_2NH yields the different generation of the dendrimer as shown in scheme 8.7. The GPC studies indicated that the dendrimers, were pure with narrow PDI. The

3rd generation dendrimer with only one Et_2N moiety can convergently reacts with $MeSi(Ph\text{-}4\text{-}SiMe_2OH)_3$ central core unit to produce larger spherical dendrimers with 24 CN-end groups [52]. The dendrimers were characterized by NMR spectroscopy and SEC analysis. The dendrimers were highly flexible as indicated by their T_g values, which were arround $-50^{o}C$ or so. Also the T_g value depends on the generation number and increases with the increase of generation number.

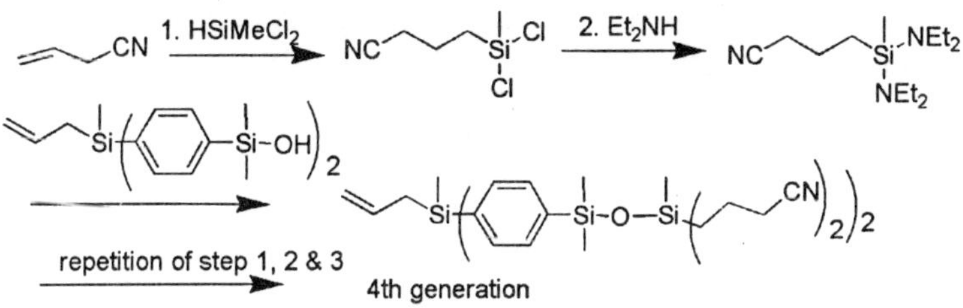

SCHEME 8.7 Synthesis of carbosilane dendrimers from acrylonitrile monomer

The syntheses, characterizations and properties of redox active carbosilane dendrimers containing ferrocenyl groups were reported [53-56]. The first report [53] for synthesis of this interesting class of dendrimers, was published in 1994. These dendrimers were synthesized by hydrosilylation of tetraallylsilane with dichloromethylsilane followed by alkylation with allylmagnesium bromide. Subsequent hydrolysis with chlorodimethyl silane yielded the desired carbosilane dendrimer with 8 Si-Cl groups. Then, these silyl chloride groups react either with ferrocenyllithium or with β–aminoethyl ferrocene resulting direct attachment of ferrocenyl group to the outer most silicon atoms or the same were separated by ethylamino groups, respectively. The cyclic voltametric studies revealed that ferrocenyl units were non-interacting redox centers and undergo 'oxidative precipitation' upon oxidation to polycations. Further investigation [54] shows that the directly attached ferrocenyl dendrimer coated Pt-electrode surface is extremely durable and that the redox response is practically unchanged without loss of electroactive material. The carbosilane dendrimers with electronically communicating ferrocenyl moieties were synthesized [55] by the convergent approach. At first, dichloromethylvinylsilane reacts with ferrocenyllithium to produce diferrocenyl methylvinylsilane. Further growth of this dendron was carried out by hydrosilylation with phenylchlorosilane followed by alkenylation with allyl magnesium-bromide. The coupling of these dendrons with tetra*kis*(dimethylsilyl propyl)silane via hydrosilylation results carbosilane dendrimer with ferrocenyl moieties on the surface. The electrochemical behavior supports the existence of significant interaction between the two ferrocenyl units linked by the bridging silicon atoms. Similar type of carbosilane dendrimers (Fig.8.1) was also synthesized [56] by hydrosilylation of tetra*kis*(dimethyl silylpropyl)silane with the corresponding vinyl ferrocene. The electrochemical behavior of these dendrimers was investigated by CV measurements.

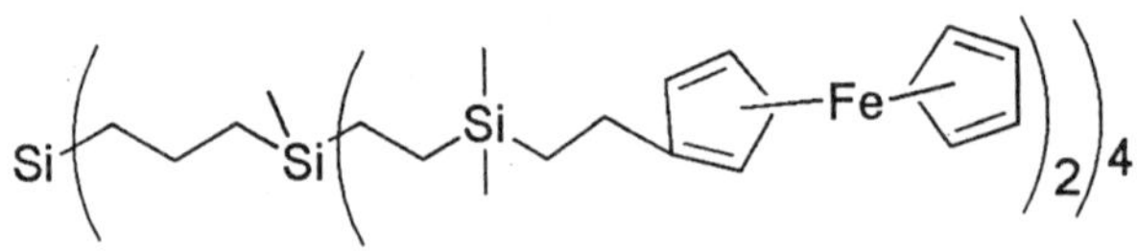

FIG.8.1 Structure of redoxactive carbosilane dendrimer

The synthesis of a core functionalized carbosilane dendrimer was reported [57]. Starting from *p*-bromostyrene, dendrons upto 3rd generation were synthesized by repetitive hydrosilylation with trichlorosilane and alkenylation with allyl magnesiumbromide. The dendrons were coupled with tetraethylferrocene-1,1-diyl*bis*(phosphonite) after lithiation at the bromobenzene moiety. The resulting dendrimers have molecular weight upto 8567 g/mole. Pd-based catalysts were prepared by reactions of different generation dendrimers with $Pd(MeCN)_2Cl_2$, and the catalytic activity was tested. The dendrons were characterized by NMR spectroscopy, elemental analysis, FAB- and MALDI-TOF mass spectrometry.

The carbosilane dendrimers upto 4th generation, were synthesized as nanoscopic building blocks for hybrid inorganic materials and catalyst supports [58]. The dendrimers were synthesized by repetition of general hydrosilylation of terminal olefinic groups with $HSiCl_3$ followed by allyl Grignard addition to the resulting trichlorosilyl functionalities. Finally, the vinyl terminated dendrimers were end capped by direct hydrosilation with triethoxysilane to obtain the desired dendrimers $Si(CH_2CH_2CH_2Si(CH_2CH_2CH_2Si\text{-}(CH_2CH_2CH_2Si\text{-}$

(Me)(CH_2CH_2Si-$OEt_3)_2)_3)_3)_4$. These triethoxy silyl terminated dendrimers were then hydrolyzed with acid catalyst to produce micro- and meso-porous hybrid dendrimer-silica xerogels and aerogels (Scheme 8.8). The acidic conditions employed for these dendrimers to convert Si-OH functionalities, which are useful for using these materials as catalyst supports. The surface area and pore volume of the corresponding gel increased with the increase of generation number of the dendrimer, which are contradictory to general silica based gel system. The authors suggested that this anamoly might due to the higher compressibility of lower generation compared to higher generation as the former structures are less congested. Solid state ^{29}Si cross polarization magic angle spinning (CPMAS) NMR spectroscopy was utilized to determine the degree of condensation in the gels. The gel treated with hot toluene has a high degree of site isolated hydroxyl groups, which is definitely useful as catalyst support. The aerosols were obtained with high surface areas from supercritical CO_2 extraction of hydrolyzed dendrimers.

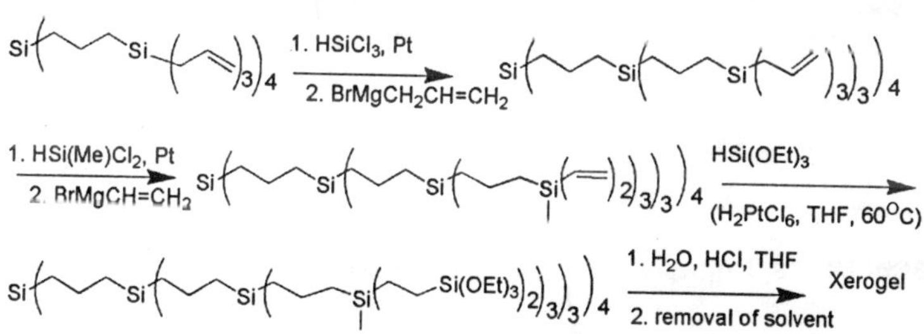

SCHEME 8.8 Synthesis of carbosilane dendrimers for preparation of xerogel

The synthesis and characterization of phosphino carboxylic acid esters terminated carbosilane dendrimers of zeroth and 1st generations were reported [59] as hemilable nanoscale ligands for the Pd-catalyzed co-dimerization of olefins. The ligands were synthesized by the following sequence of reaction as shown in scheme 8.9. The synthesized dendrimers were characterized by ^{1}HNMR and ^{13}CNMR spectroscopy and MALDI-TOF/FAB mass spectrometric studies. These Pd-catalysts were used for the hydrovinylation of styrene in a continuously operated nanofiltration membrane reactor. Under these continuous conditions, the selectivity of the reaction increased considerable, even though they can be used in batch reaction also. However, the complexes have a limited efficiency resulting from a restricted lifetime. The deactivation was noticed during the reactions by the precipitation of Pd-block in the reactor and on the membrane surface.

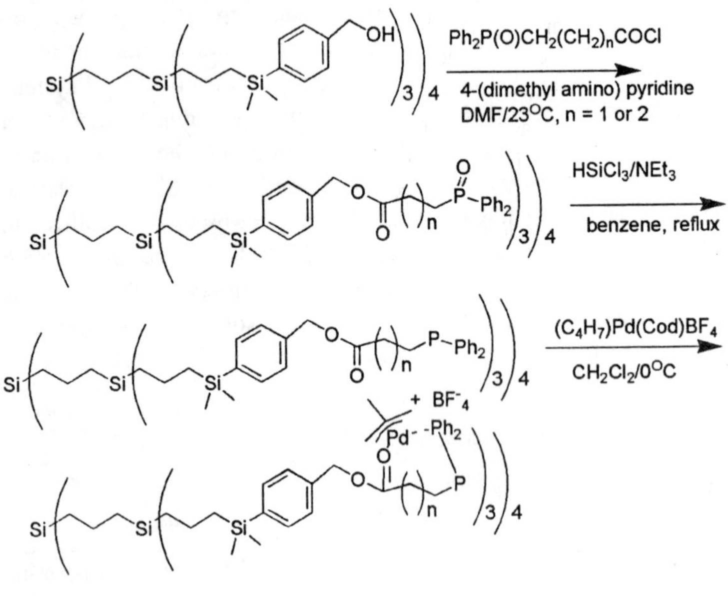

SCHEME 8.9 Synthesis of Pd-phosphino carboxylic acid ester containing carbosilane dendrimer

Modified carbosilanes

Various carbosilane dendrimers were converted into dendritic carbosilane polyols [12, 15] by hydroboration using 9-BBN and subsequent oxidation with $H_2O_2/^-OH$. This transformation of terminal double bonds into primary alcohol is quantitative (Fig.8.2). The functionalized dendrimers were characterized by NMR spectroscopy and MALDI-TOF mass spectrometry. The T_gs of the dendrimers with allyl end groups vary from –100 to –80°C, which indicated their high flexibility. But the T_gs of the dendritic polyols are much higher, -30 to –40°C. This increment of T_g may be due to increase of intermolecular attraction among the branches through H-bonding interaction. Allyl terminal carbosilane dendrimers were functionalized into polyols [60-61] by hydrosilylation with organosilane bearing a trimethylsilyl protected hydroxy group, followed by hydrolysis. IR study showed that the H-bond formation in the structure is temperature and concentration dependent.

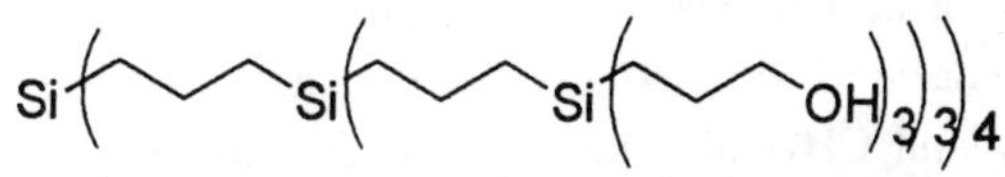

FIG.8.2 Structure of 2nd generation dendritic carbosilane polyol

The other interesting hydroxy terminated carbosilane dendrimers were also reported [62-64]. The water-soluble carbosilane dendrimers were synthesized [62] by nucleophilic reactions between mercapto-substituted amphiphiles and carbosilane dendrimer bearing chloromethyl silyl groups on their terminal branches offered amphiphilic dendrimer with hydrophobic carbosilane and alcohol, dimethylamino, or sodium sulfonate amphiphilic groups at the periphery. The negatively charged sulfonate terminated dendrimers were soluble in water. Again, the positively charged poly ammonium salts were prepared from the dimethylamino terminated derivatives by reaction with methyl iodide, providing quaternary ammonium iodide salts (Fig.8.3). These new amphiphilic dendrimers were characterized by NMR spectroscopy and mass spectrometry. The preliminary studies of the aqueous solution behavior of the 2nd generation, sulfonate terminated dendrimer demonstrated its ability to enhance the solubility of lipophilic alkyl-substituted benzene derivatives, a characteristic property of micelles. An alternative route for synthesis of hydroxy terminated carbosilane dendrimers was also reported [63]. The vinyl terminated carbosilane dendrimers [4] were hydrosilylated with *bis*(6-(2-tetrahydropyranyloxy)hexyl)methylsilane, followed by deprotection of THF protected polyols produced the desired polyols with 4, 8, 16 and 32 hydroxy groups. The functionalized carbosilane dendrimers were characterized by NMR spectroscopy and SEC analysis. A number of carbosilane dendrimers were functionalized by silanol groups and characterized by spectroscopic techniques [64]. The silanol dendrimers were obtained by repetitive hydrolysis/alkenylation reactions and finally, a careful hydrolysis of Si-Cl groups resulted the desired dendrimers. Various molecules with 3, 4 and 8 vinyl groups were used as the starting materials to produce dendrimer of different sizes. Actually, the vinyl end groups of the dendrimers were hydrosilylated with chlorodimethyl silane and subsequently silyl chloride was reduced catalytically with $LiAlH_4$. The catalytic hydrolysis with water over Pd/C yielded the desired silanol-functionalized carbosilane dendrimers. The functionalized dendrimers were characterized by NMR spectroscopy and CHN microanalysis. Although, in general, silanol groups are hydrolytically unstable but these dendrimers are interesting since they may be utilized to mimic silica surface to study the attachment of catalytically active metal to silanol groups. The carbosilane dendrimers with 32, 64 and 128 terminal chloride groups with living poly(butadienyl lithium) chains were modified [3-4] to prepare well-defined multiarmed star polymers. The molecular weight of polybutadiene was (6.4-7.2) X 10^3 Da. The dilute solution properties of these star polymers were investigated by osmometry, viscometry and light scattering techniques. The study showed that the isolated stars behave like hard spheres. A gel like characteristics of these star polymers beyond a certain concentration were also observed [65]. The formation of a microcrystalline order phase with spacing of the order of 100-500A^O was noticed for these polymers.

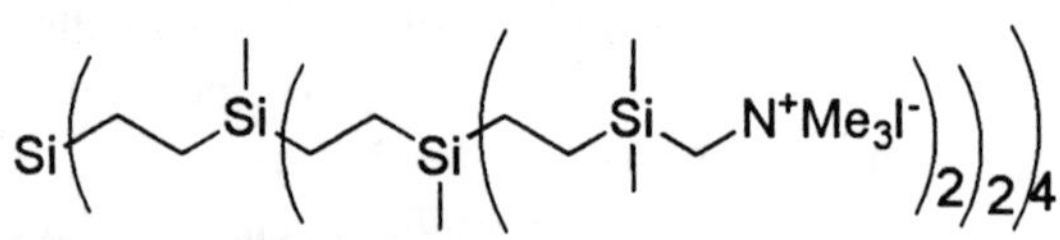

FIG.8.3 Structure of amphiphilic carbosilane dendrimer

A star polymer was also synthesized [66] by the reaction of silyl chloride end groups containing carbosilane dendrimer with poly(styryl lithium) and poly(isopropyl lithium). The carbosilane dendrimers upto 2nd generation were first synthesized from triallyl phenylsilanes with repetitive hydrosilylations with chlorodimethyl silylpropane groups and alkenylation with allyl magnesiumbromide. The modified star shaped 12 arms poly(styrene-block-isoprene) copolymer was characterized by SEC, AFM and TEM studies. SEC indicates narrow PDI and regular order micelle structures in solution as obtained from AFM and TEM studies.

The phosphino-terminated dendrimers [67], Si-(CH_2-CH_2-$SiMe_2CH_2PPh_2$)$_4$ and Si-(CH_2-CH_2-Si(Me)(CH_2PPh_2)$_2$)$_4$ were obtained by the reaction of carbosilane dendrimers, Si-(CH_2-CH_2-$SiMe_{3-n}Cl_n$)$_4$, (n = 1 and 2) with $LiCH_2PPh_2$. The phosphino modified dendrimers were then treated with ClAu(tht) to obtain gold containing dendrimers, Si-(CH_2-CH_2-$SiMe_2CH_2PPh_2AuCl$)$_4$ and Si-(CH_2-CH_2-SiMe($CH_2PPh_2AuCl_2$)$_4$, which were further reacted with $(PPh_4)_2[Fe_3(CO)_{11}]$ to give the first metal cluster containing dendrimers, $(PPh_4)_4$[Si-(CH_2-CH_2-$SiMe_2CH_2PPh_2$){$AuFe_3(CO)_{11}$})$_4$] and $(PPh_3)_4$[Si-(CH_2-CH_2-SiMe(CH_2PPh_2)$_2$) {$AuFe_3(CO)_{11}$})$_4$]. The synthesis of novel block copolymers from amine functionalized carbosilane dendrimers from first to 3rd generation was reported [48] by following techniques. At first, the dendrimers were reacted with nickel complex, Ni(C=N-^{t}Bu)$_4$(ClO_4)$_2$ to form Ni(C=N-^{t}Bu)(C(R')-N(^{t}Bu)(ClO_4)$_2$ (R'= dendr-NH). The desired block copolymers were then obtained by addition of L-Ala-L-Ala-NC to these Ni-complexes in CH_2Cl_2. After precipitation, the resulting block copolymers were isolated as amorphous powder. The products were characterized by IR, ^{1}HNMR, ^{13}CNMR spectroscopy, SEC, TEM, UV-vis. and circular dichroism spectral studies. DLS study revealed that in chloroform at concentration less than 0.5 wt.%, two types of structures were present. One structure has a diameter of 6 nm and the other has a hydrodynamic diameter of ~90 nm. Again, the solubility of the block copolymers is enhanced compared to homopolymer of the L-Ala-L-Ala derived isocyanide.

In 1994, the first homogeneous catalyst based carbosilane dendrimer was reported [68-69]. The first generation of the dendrimer was obtained by hydrosilylation of tetraallylsilane with trichlorosilane followed by alkenylation with allyl magnesiumbromide. The terminal allyl groups were then treated with dimethylchlorosilane to convert into chlorinated terminal groups, which were then connected with 4-amino substituted 2,6-*bis*(dimethylamino)-methyl)-1-bromobenzene (NCN-Br), a potential multidentate monoanionic ligand. To achieve the connection between the chlorine terminated dendrimer and NCN-Br ligands, the amino substituted NCN-Br was reacted with triphosgene to afford the isocyanate derivative, which was then subsequently reacted with an excess of 1,4-butanediol. These modified ligands were then reacted with chlorodimethyl silyl functionalized dendrimers. The targeted homogeneous catalyst based dendrimer (Fig.8.4) was then produced by oxidative addition of the above dendrimer to the zero valent nickel complex. In 1998, the synthetic methodology was further improved [70] and the stability of the homogeneous catalyst based dendrimer was enhanced by direct linkage of ligand moieties with surface Si-atoms of the dendrimer. For this purpose, the chlorotrimethyl silane was treated with lithium derivative of biphosphinoaryl ligand, 3,5-$(Ph_2PCH_2)_2C_6H_3Br$(PCP), a phosphorous analog of the NCN-ligand. The dendritic carbosilanes functionalized with NCN-H surface groups by direct attachment with the dendrimers were also obtained [71-72] via the reaction of zero and a first generation dendrimers bearing chlorodimethyl silyl end groups with 3,5-*bis*[(dimethylamino)methyl] phenyllithium. A polyol carbosilane dendrimer with 12 surface groups was functionalized [73] with iodoarene groups by esterification of terminal –OH groups with 4-iodobenzylchloride. The resulting product was reacted with Pd(dibenzylidene acetone)$_2$ in the presence of N,N,N',N'-tetramethylethylene diamine to obtain a dendrimer with palladium at periphery.

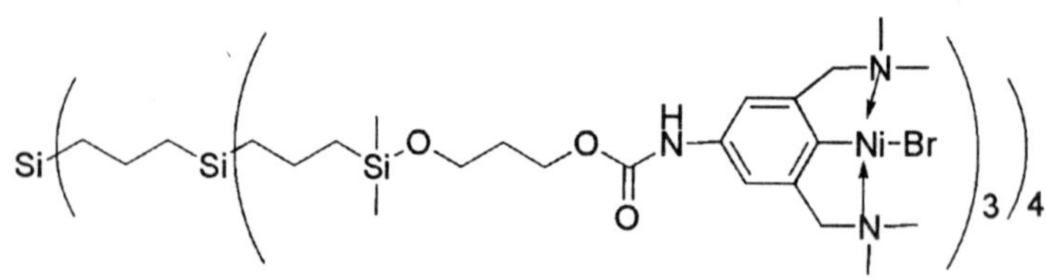

FIG.8.4 Structure of Si-based dendritic Ni-catalyst

Highly charged organometallic carbosilane dendrimers were synthesized upto 2nd generation [74] by hydrosilylation with trichlorosilane, followed by addition of benzyl magnesiumchloride. The products were characterized by NMR spectroscopy, XRD and MALDI-TOF mass spectrometry. The terminal benzyl groups were modified by ruthenium compounds via the reaction of the corresponding benzyl groups with $[CpRu(NCMe)_3]^+OTf$. The structure of the first generation is shown in Fig.8.5. Along with the X-ray, electrospray ionization Fourier transformation cyclo ionization resonance (ESFT-ICR) mass spectrometry confirmed the formation of perfect structure. However, for 2nd generation a mixture of perfect structure with 36-$CpRu^+$ units and dendrimers with only 35 such units had been isolated. A 3rd generation dendrimer bearing 72 $CpRu^+$ units has also been reported [74].

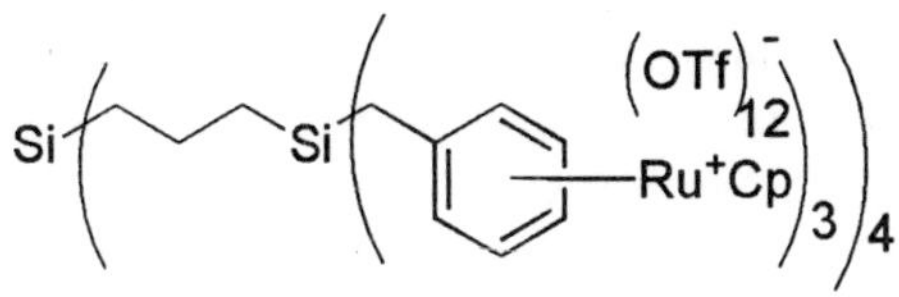

FIG.8.5 Structure of benzyl terminated carbosilane dendrimeric Ru complex

The synthesis of a polyferrocene-branched dendrimer like structure (Fig.8.6), where Si presents in the spacer not as branching point was reported [75]. The dendrimer was synthesized by hydrosilylation of decaallylferrocene with $[(^5\eta\text{-}C_5H_5)Fe(^5\eta\text{-}C_5H_4Si(Me_2)H)]$.

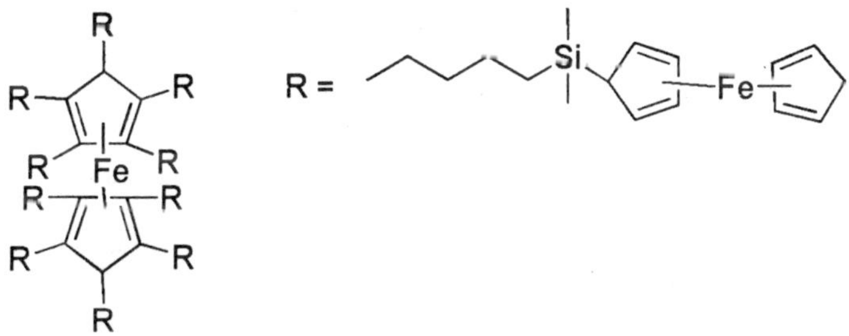

FIG.8.6 Structure of polyferrocene with silicon space dendrimers

Liquid Crystalline (LC) Carbosilanes

The silicon based many dendrimers show the liquid crystalline behavior. The carbosilane dendrimers are the first dendrimers whose solid state properties and mesophase formation have been considered. The coupling of the flexible, dendritic carbosilane scaffolds with rigid mesigenic units as end groups results dendritic liquid crystalline polymers (DLCPs). It is a pecularity of these classes of LC polymers that the attachement of mesogenic units to the flexible carbosilane dendrimer leads to a structural conflict between preferential anisotropic order of the mesogenic units and the spherosymmetric of the dendrimer.

The synthesis of the first liquid crystal carbosilane dendrimer of the 1st generation, which contains terminal photosensitive mesogenic group that undergoes configurational changes in response to light irradiation, was reported [76]. The methoxy cinnamoyl containing LC dendrimer was synthesized by, two step procedure: i) synthesis of fragment with a reactive Si-H terminal group and ii) synthesis of dendrimer with terminal hydroxy and cinamoyl containing groups. The sequence of reactions is shown in scheme 8.10. The dendrimer was characterized by IR, ^{1}HNMR spectroscopies, GPC analysis and DSC studies. The DSC study indicates the liquid crystalline behavior of the dendrimer.

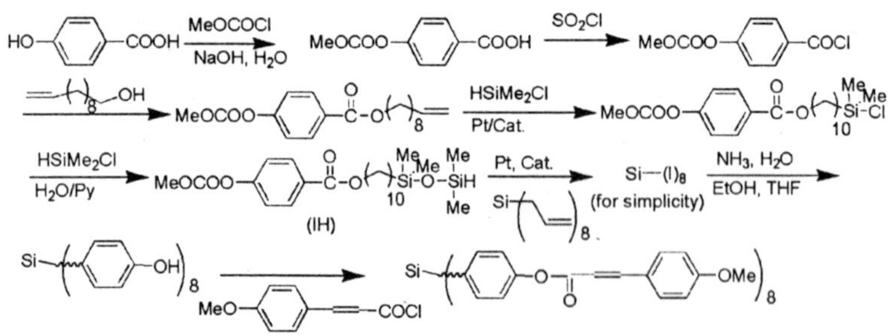

SCHEME 8.10 Synthesis of LC carbosilane dendrimers

The syntheses and characterizations of cyanobiphenyl [11], cholesteryl [12], and perfluoroalkyl [14] end capped carbosilane dendrimers, which showed LC bevavior were also reported. The end capping of the dendrimers was performed by reacting 20-60% excess of tridecafluoro n-octyl mercaptan with the parent dendrimers in the presence of radical initiator, AIBN at 70°C, to obtain fully thioether functionalized dendrimers. The modified dendrimers were characterized by IR, ^{1}HNMR spectroscopic, SEC, WAXS and polarizing microscopic studies. The thermal properties of the dendrimers were studied by DSC. The resulted dendrimers showed interesting generation dependent liquid crystalline behavior. The zeroth generation does not showed any LC behavior but sematic, layer structure and hexagonal packed array of cylindrical shapes were observed for 1st, 2nd and 3rd generations respectively. This may be due to increase in surface density of the dendrimers.

The influences of (i) generation number, (ii) spacer length and (iii) type of mesogen coupled to the dendrimer were investigated [77-78] on the phase behavior of LC carbosilane dendrimers. By the increase of number of generation, which increases the number of end groups or/and reducing the spacer length complicates the formation of well-developed smectic phases. However, in higher generation (>2) the dendrimers with branching multiplicity of 3, no LC behavior was observed. This may be due to dense packing of the mesogens at the dendrimer surface, disabling the formation of smectic layers. Shibaev and his coworkers [79] used cyanobiphenyl, methoxyphenyl benzoate and cholesteryl groups as the mesogenic units for synthesizing LC dendrimers with branching multiplicity 2. The mesogenic units were coupled to carbosilane dendrimers bearing 8 allyl groups by hydrosilylation. The LC properties of the synthesized dendrimers were determined by POM in combination with DSC measurement and XRD studies. In cyanobiphenyl terminated dendrimer, a mesophase transition between two different smectic phases (S_C and S_A) was observed. The LC carbosilane dendrimers upto 5th generation, were also synthesized [80] by functionalization of surface groups with cyanobiphenyl moieties. Dichloromethylsilane was used as the branching reagent and cynobiphenyl moiety containing 11 methylene units were used as the spacer groups. The obtained dendrimers were characterized by POM, DSC and XRD studies [81]. From these studies, it was found that the dendrimers form a lamellar (smectic A) phase at room temperature, which develops in-phase ordering above $40^{o}C$ and above $121^{o}C$ the material transforms into a more disorder mesophase. The cyanobiphenyl functionalized carbosilane dendrimers based on triallylphenyl silane as a core were also reported [82]. The hydroboration and oxidation of the dendrimer gave polyol derivatives, which were then treated with the mesogen, 6-(4-cyanobiphenyl-4-oxy)hexanonyl chloride to get the targeted dendrimer. The synthesized dendrimers were characterized by DSC and POM. The synthesis of carbosilane dendrimers upto 2nd generation with mesogens bearing a chiral tail was reported [83] and they showed smetic phase only. But very interestingly these dendrimers showed formation of smectic C phase, used as the chiral dopant and response times for switching of these phases in electric fields increased with the molecular weight. The perfluoro end capped of these dendrimers were immiscible with lipophilic carbosilane structure. The dielectric relaxation for these perflurinated carbosilane dendrimers was also studied [84] and the results showed a fast relaxation with an Arrhenious type temperature dependence. The photochemical and photooptical properties of the 1st generation LC dendrimer with terminal azobenzene groups, which forms a smectic 'A' mesophase in a wide range of temperature were also reported [85]. Both in solid state and in solution of the dendrimer, the photochemical behavior was studied under UV radiation, which causes E/Z photoisomerization. However, under irradiation with linearly polarized UV light (365 nm), the E/Z isomerization takes place only, no photoorientation was observed. Whereas for linearly polarized green light (488 nm), a well pronounced linear dichroism was observed. Thus, the carbosilane dendrimers permit one to obtain a detail understanding of the behavior of end group induced liquid crystallinity in flexible dendrimers.

The polylithium derivatives of carbosilane dendrimers, which were used as initiators for the anionic polymerization of different monomers, such as styrene, hexamethylcyclo-trisiloxane and ethylene oxide were obtained [86-87] by hydrosilylation of allyl terminated dendrimers with *bis*(decylmethyl)silane. The reaction was derivatized only the half of the end groups and the rest end groups were then treated with secondary butyllithium to get the desired modified dendrimer (Fig.8.7). The important feature of this approach lies due to the location of the lithium atoms in the inner area of the dendrimers, which avoided the main problem of aggregation tendency of general polylithium compounds. A broad applicable one pot reaction in liquid ammonia involving Birch reduction and subsequent SN^2 reaction were utilized to generate a thioether linkage between the monosaccharide moiety and the carbosilane dendrimer [88]. The dendrimers were uniformly functionalized with the monosaccharides such as benzylthioalkyl glycosides of D-glucuronic acid, N-acetyl-D-glycosamine and N-acetylneuraminic acid with good yield. The authors [89] also functionalized carbosilane dendrimers with lactose or silyllactose moities in the saccharide compounds using one pot coupling reaction in MeOH-DMF solvent system. Regiospecific hydrosilylation of 1-

bromo-4-(prop-2-enyl)benzene offers an efficient route to molecular building block precursors that can accommodate sequential divergent and convergent steps for dendritic extension, establishing a modulor methodology for assembly and organization of connectivity used for synthesis of modified carbosilane dendrimers [90].

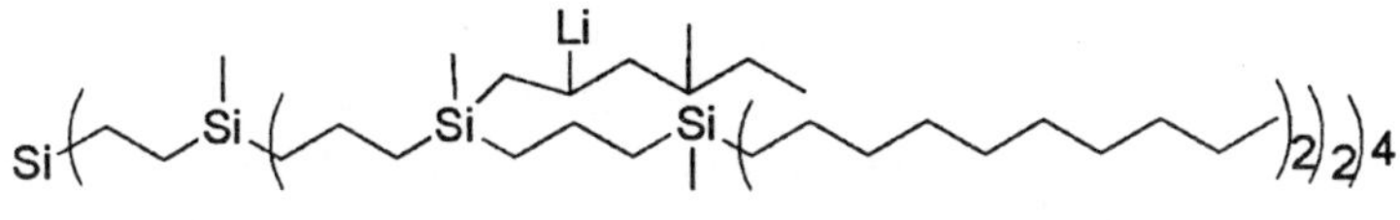

FIG.8.7 Structure of polylithium derivative of carbosilane dendrimer

Organometallic silicon based dendrimers with peripheral Si-cyclopentadienyl Si-O and Si-Fe linkages were synthesized [91]. The cyclopentadienyl functionalized dendrimer $Si[(CH_2)_3SiMe_2(C_5H_5)]_4$ was prepared and used to coordinate, $Co(CO)_2$ moiety. The Si-H and Si-Cl terminated dendrimers reacted with $Co_2(CO)_8$ and $Na^+(^5\eta$-$C_5H_5)Fe(CO)_2]^-$ respectively, to obtain peripheral Si-O and Si-Fe bonds. The first generation unique organosilicon hydride dendrimer was synthesized and characterized [92]. Four folds hydrostannation of tetravinylsilane yield the first generation metallodendrimer $Si(CH_2CH_2SnPPh_3)_4$. Then, the selective bromination results $Si(CH_2CH_2SnBr_3)_4$, followed by reduction by $LiAlH_4$ to give $Si(CH_2CH_2SnH_3)_4$, the desired dendrimer with 12 Sn-H bonds. The products were characterized by elemental analysis, IR, ^{1}HNMR, ^{13}CNMR, ^{119}SnNMR spectroscopic studies and single crystal X-ray structural determination.

A large number of reports on the synthesis of carbosilane dendrimers with allyl end groups have been published [93-96] by using various core units containing allyl or vinyl groups, such as 2,4,6,8-tetramethyl-2,4,6,8-tetravinyl tetrasiloxane, diallylphenylmethylsilane, 1,2-*bis*(triallylsilyl)ethane and triallyl methylsilane. For these syntheses, allyl magnesium bromide as the Grignard reagent and either $HSiCl_3$ or $HMeSiCl_2$ as the hydrosilylation reagent were used. The synthesized dendrimers were characterized by NMR spectroscopy and elemental analysis.

The synthesis of carbosilane dendrimer terminated with phenylethynyl, *p*-bromophenoxy and *p*-phenylphenoxy groups were also reported [97-100]. The products were characterized by NMR, elemental analysis and MALDI-TOF mass spectrometry.

The uncommon carbosilane dendrimers upto 3rd generation [101] were obtained by using tetra*kis*(phenylethynyl)silane as the core, dichloromethylsilane as the hydrosilylation reagent and lithium phenylacetylide as the alkynylating agent. The synthesized dendrimers were characterized by NMR and MALDI-TOF mass spectrometric studies. The glass transition temperatures of these dendrimers were found to be considerably high compared to other common carbosilane dendrimers based on alkenylation. The synthesis of silylacetylene dendrimers was reported [102] by the convergent technique upto 2nd generation with 12 end groups.

8.2.2 Carbosiloxanes

The carbosiloxane dendrimers contain Si-O-Si linkages as well as $Si(CH_2)_nSi$ units in their structures. In general, they are synthesized by sequences of hydrosilylations of terminal double bonds with chlorosilane to form electrophilic silicon species and alcolysis with silanol. The chlorosilanes used are generally $HSiMe_2Cl$, $HSiMeCl_2$ and $HSiCl_3$. The carbosiloxane dendrimers upto 4th generation, were synthesized [103] by using above sequence of reactions (Scheme 8.11). The synthesized dendrimers were characterized by elemental analysis, IR, ^{1}HNMR, ^{13}CNMR, ^{29}SiNMR spectroscopy and GPC studies.

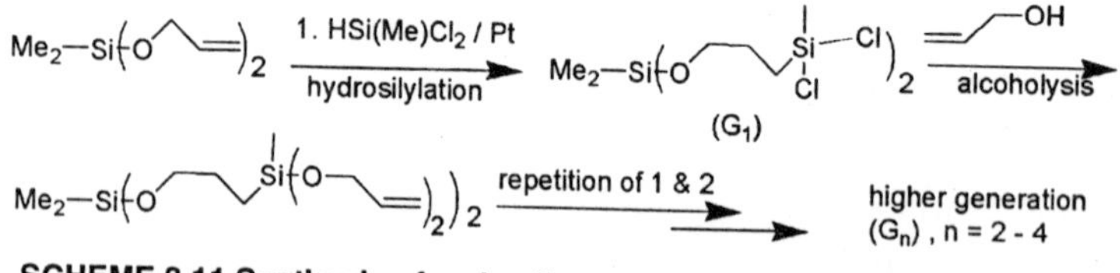

SCHEME 8.11 Synthesis of carbosiloxane dendrimers by hydrosilylation and alcholysis reactions sequence

The carbosiloxane dendrimers upto 5th generation, were synthesized [17] by using siloxane tetramer as the central core unit. However, the building blocks were prepared by the catalytic hydrosilylation and alcoholysis reactions [17, 33]. The core molecule was first reacted with trichlorosilane in the presence of Pt/C catalyst, followed by reaction of the resulted chlorosiloxane with allyl alcohol in the presence of TMEDA, amine as the catalyst. The repetition of these two steps resulted the higher generation dendrimers. The sequence of reactions is shown in scheme 8.12. The synthesized dendrimers were characterized by NMR, IR spectroscopies and SEC measurement. These carbosiloxane dendrimers were used to prepare varieties of end functionalized dendrimers [104-105]. The cholesterol and pyridine end capped dendrimers were synthesized by two step process. The 3rd and 4th generation allyl terminated dendrimers were catalytically hydrosilated with cholorodimethyl silane followed by the reaction of silylated dendrimers with pyridine propanol, pyridine aldoxime and cholesterol in toluene with the presence of excess of TMEDA catalyst, as shown in scheme 8.13. The synthesized dendrimers were characterized by ^{1}HNMR, ^{13}CNMR, IR spectroscopic, elemental analysis and GPC studies.

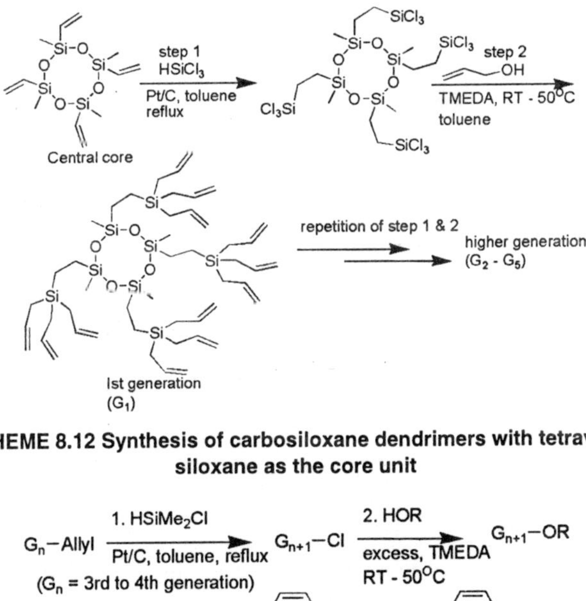

SCHEME 8.12 Synthesis of carbosiloxane dendrimers with tetravinyl siloxane as the core unit

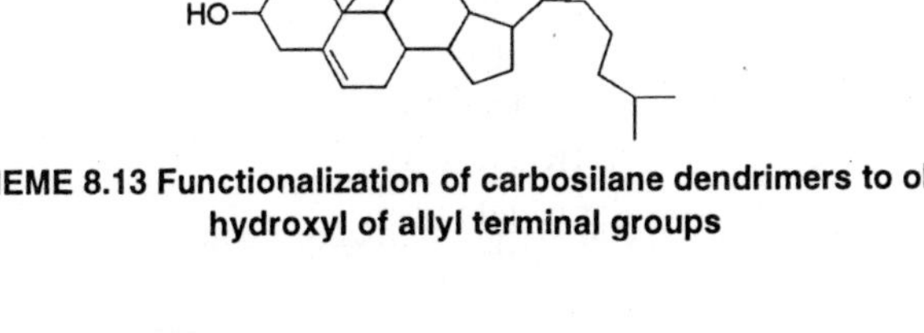

SCHEME 8.13 Functionalization of carbosilane dendrimers to obtain hydroxyl of allyl terminal groups

Similarly, the reactions of the 4th and 5th generations chlorocarbosilane dendrimers with alcohols (ROH), such as 9-anthracene methanol, 8-hydroxyquinoline, 4-hydroxyazobenzene, 2-hydroxymethyl anthraquinone and 5-(2-hydroxy ethyl)-4-methylthiazole in the presence of TMEDA produced Si-O-C bonded end capped dendrimers with very high yield [105], as shown in scheme 8.14. By using the same type of reaction with crown ether (RH) such as 2-hydroxymethyl-12-crown-4 ether and 2-hydroxymethyl-15-crown-5 ether, ether end capped dendrimers were produced [106] (Scheme 8.14). The modified dendrimers in both the cases were characterized by ^{1}HNMR, ^{13}CNMR spectroscopy, elemental and GPC analyses.

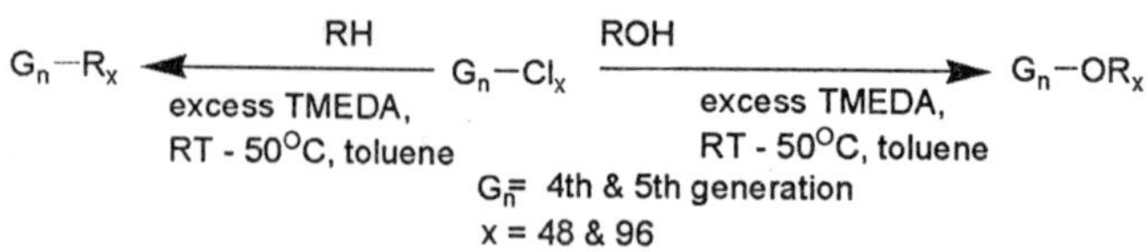

SCHEME 8.14 Functionalization of carbosilane dendrimers with chloro surface groups by various alcoholic and crown ether compounds

A staff type dendritic carbosiloxane was synthesized [107] with the siloxane polymer of type Me_3SiO-$[MeSi(H)O]_n$-$SiMe_3$ as the core unit. The central core grew to higher generation by repetitive reactions of hydrosilation followed by alcolysis, as shown in scheme 8.15. By using the above mention technique, a series of end capped staff type dendritic polymers with ally alcohol, cholesterol, 8-hydroxyquinoline, 5-(2-hydroxyl)-4-methylthiazole, 4-pyridine propanol and 4-pyrindinealdoxime was also synthesized. The synthesized parent dendrimers as well as end capped dendrimers were characterized by ^{1}HNMR, ^{13}CNMR spectroscopy, elemental and GPC analyses. The results are agreed well with the theoretically predicted values.

The novel ferrocenyl silicon containing dendrimers upto 3rd generation, were synthesized [108] by using 1,3,5,7-tetramethylcyclotetrasiloxane as the central core unit. The parent allyl terminated cyclotetrasiloxane based dendrimers were synthesized by repetitive hydrosilylation and alkenylation reactions, as described earlier. The structures of the synthesized dendrimers were characterized by IR, ^{1}HNMR, ^{13}CNMR, ^{29}SiNMR spectroscopy, elemental analysis mass spectrometry. The molecular weight was determined by VPO studies.

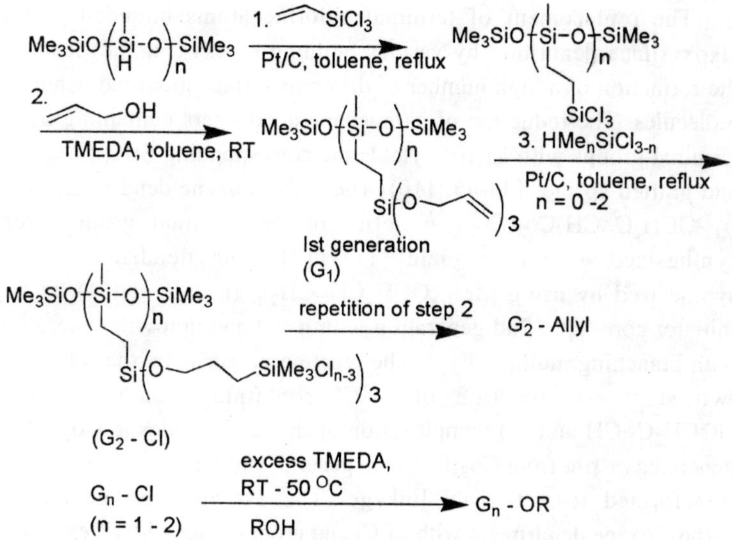

SCHEME 8.15 Synthesis and functionalization of staff type carbosilane dendrimers with polysiloxane as the core unit

The synthesis of first generation carbosiloxane dendrimer from Si-O$(CH_2CH_2CH_2SiMe_2Cl)_4$ with high yield was described [109] by the following steps of reactions as shown in scheme 8.16. The synthesized dendrimer was fully characterized by elemental analysis, IR, ^{1}HNMR, ^{13}CNMR spectroscopy and GPC analysis.

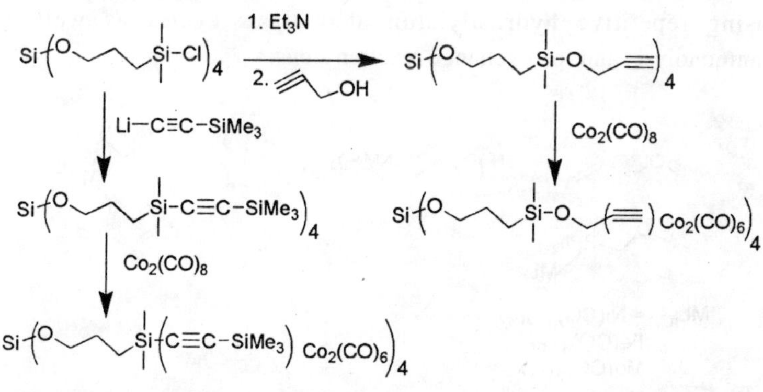

SCHEME 8.16 Synthesis of Ist generation carbosiloxane dendrimer from $Si(OCH_2CH_2CH_2SiMe_2Cl)_4$

The replacement of terminal chlorine atoms bounded to the alkoxysilane dendrimer by various building blocks (Fig.8.8) leads to the formation of a high number of different surface modified dendritic molecules. The reduction of carbosilane dendrimers containing $SiCl_n$ terminal groups with $LiAlH_4$ yields the corresponding SiH_n (n = 1-3) end grafted species [44-45, 110]. The carbosiloxane dendrimers with $(\eta^2\text{-}OCH_2C{\equiv}CH)Co_2(CO)_6$ moieties in the terminal groups were synthesized with high yield [111-112]. The dendrimers were synthesized by using $MeSi(OCH_2CH{=}CH_2)_n$ (n = 2 and 4) as the initiator core, upto 2nd generation with n = 4 and upto 4th generation with branching multiplicity 2. The reaction proceeds by the following two steps: (i) reaction of Si-Cl containing dendrimer with $HOCH_2C{\equiv}CH$ and (ii) complexation of the C≡C entities to $Co_2(CO)_6$, generated *in-situ* from $Co_2(CO)_6$. In general, aminolysis of Si-Cl groups transformed to a C-Si-N linkage. The 1st and 2nd generation carbosiloxane dendrimers with Si-Cl end groups react with an excess of $HN((CH_2)_3NMe_2)_2$ to produce the respective $SiN((CH_2)_3NMe_2)$. The modified dendrimers [113] such as $Si\text{-}[O(CH_2)_3SiMe_2(N(CH_2)_3NMe_2)_2]_4$, $Si\text{-}[O(CH_2)_3Si(Me)O(CH_2)_3Si(Me)(N(CH_2)_3NMe_2)_2)_2]_4$, etc. were obtained. The amino groups in the modified dendrimers can act as triplet chelating ligands towards different transition metal salts such as $AgClO_4$ or $AgBF_4$ producing the corresponding dendritic chelate complexes (Fig.8.9). The synthesis of water-soluble carbosiloxane dendrimers was also reported [114] by the divergent growth approach using repetitive hydrosilylation-alcoholysis cycles as well as ammonolysis and ion exchange reaction steps.

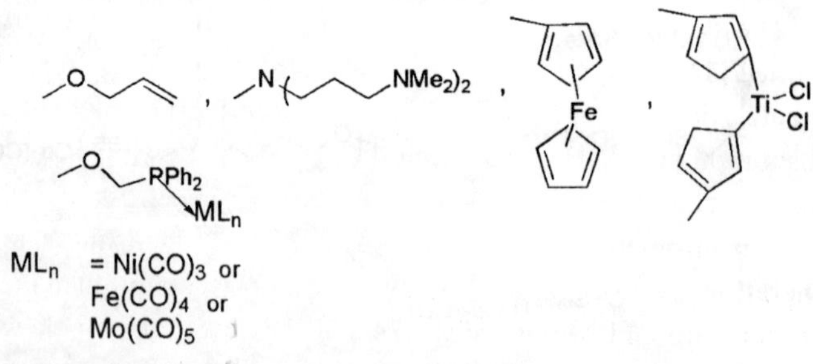

FIG.8.8 Structure of different building blocks for functionalization of alkoxysilane dendrimers

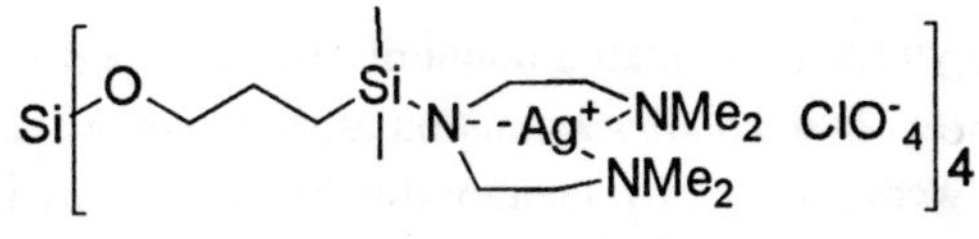

FIG.8.9 Structure of Si-based dendritic-Ag chelate complex

8.2.3 Siloxanes

Although the siloxane polymers are well known for their wide range of applications [115] like resins, oil, elastomers, etc. but the siloxane dendrimers are studied very less. The successful synthesis of the first heteroatom containing siloxane dendrimers upto 4th generation, was reported [116] in 1989. The dendrimers were synthesized by repetition of (i) nucleophilic substitution of the silicon-bound chlorides and (ii) transformation of the silicon-bound ethoxy groups to chlorides. The synthesis of the dendrimers by using methyl trichloro silane ($MeSiCl_3$) as the initiator core unit is shown in scheme 8.17. However, the detail characterization was not reported. Although, these chloro-terminated dendrimers were further surface functionalized by the reactions with sodium or magnesium bromide of different compounds such as Me_3COX, $CF_3(CH_2)SiOX$, (X= Na or MgBr, $C_{10}H_{11}MgBr$) or $Me_3SiOSiOMe_2OH$ to get the desired modified dendrimers [117].

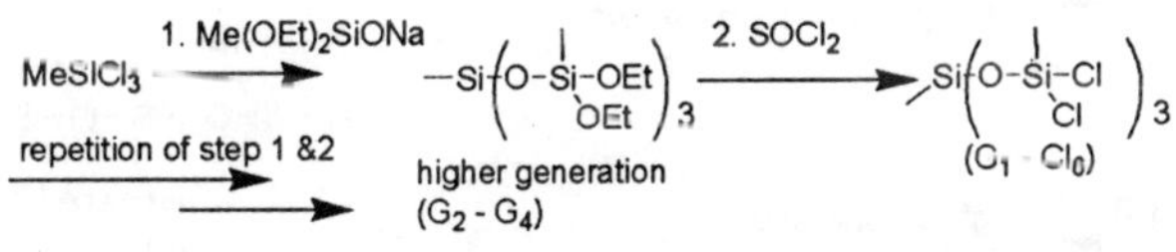

SCHEME 8.17 Synthesis of siloxane dendrimers using trichloro methyl silane as the core unit

The synthesis of a series of dendrimers upto 3rd generation, was reported [118-120] by repetitive two step reactions. These reactions are (i) catalytic oxidation of Si-H units to form Si-O linkages and (ii) substitution of Si-OH group by chlorosilanes, as shown in scheme 8.18. The dendrimers were characterized by NMR spectroscopy SEC and mass spectrometry. The synthesis of siloxane dendrimers upto 3rd generation by using *tris*[(dimethyl phenylsiloxy]methylsilane as the initiator core and $MeSi(OSiMe_2Ph)_3$ as the building block was also reported [19, 118-120]. The whole sequence of reactions is shown in scheme 8.19. The direct reaction of the brominated compound with

$PhMe_2SiONa$ yields very small amount of product whereas significant higher yields are obtained after amination as shown in the scheme 8.19. The products were purified by vacuum distillation followed by characterization by ^{1}HNMR, ^{13}CNMR and ^{29}SiNMR spectroscopic studies. The Mark-Houwink constant 'k' ($4.7X10^{-3}$) and 'a' (0.21) were also determined from the molecular weight and intrinsic viscosity measurements [19].

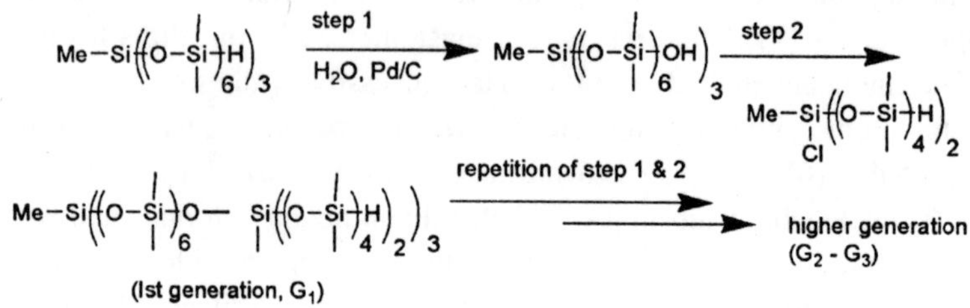

SCHEME 8.18 Synthesis of siloxane dendrimers by oxidation and substitution reactions

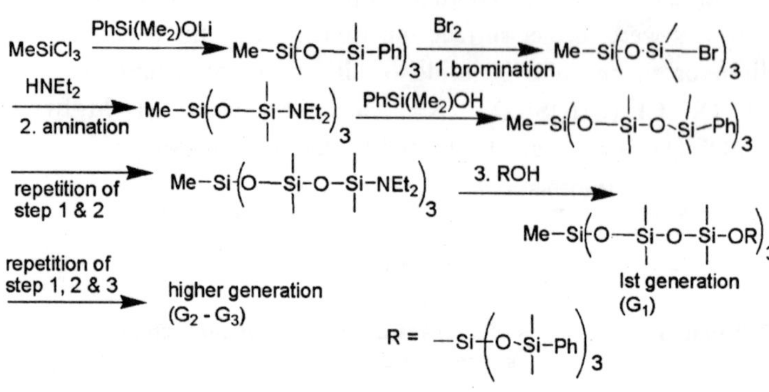

SCHEME 8.19 Synthesis of silaoxane dendrimers with *tris*[(dimethylphenyl siloxy)]methyl silane

Similar to the other types of dendrimers, the siloxane dendrimers are also functionalized at the surface. A new class of liquid crystalline (LC) dendrimer was obtained [121] through the treatment of terminal chlorinated first generation siloxane dendrimer with cholesterol ester $HOSiMe_2(CH_2)_{10}COO$-Chol. (Chol.= Cholesterol and alkane chain is the spacer group) in the presence of pyridine as shown in scheme 8.20. The mesogenic cholesterol derivative was synthesized first by acetylation of cholesterol with the acyl chloride of 10-undecenoic acid followed by hydrosilylation of the terminal double bond with chloromethyl silane and hydrolysis in the presence of ammonia. The large number of methylene groups enhanced the flexibility of the siloxane dendrimer. The bromine terminated siloxane dendrimers were also functionalized with bifunctional silanol type compounds [122], which can be further transformed to polyol terminated dendrimer by catalytic hydrogenation. The sequence of reactions is shown in scheme 8.21.

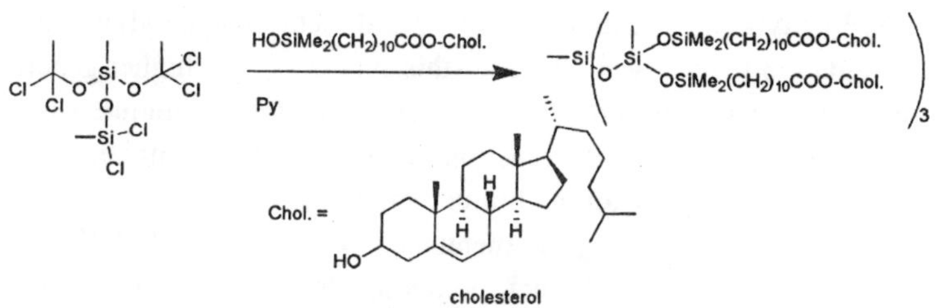

SCHEME 8.20 Functionalization of siloxane dendrimer by end capping with cholesterol units.

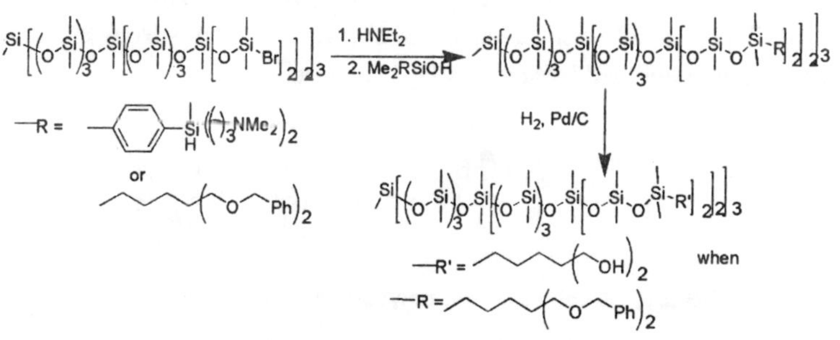

SCHEME 8.21 Functionalization of siloxane dendrimers with aliphatic primary alcoholic end groups

8.2.4 Carbosilazanes

The carbosilazane dendrimers are another important class of silicon based recently developing dendrimers. They contain Si-N linkages in their structures. The planarity of these units as well as Lewis basicities of the nitrogen atom and relative sensitivity of the Si-N bonds, render these dendrimers interesting to investigate. Planarity may induce unique structural geometry. Lewis basicity may enable to bind electron deficient moieties in the dendrimer interior. Also facile cleavage of the Si-N bonds could provide a pathway for deliberate degradation in applications based on dendrimers as templates, e.g., to create nanoporous materials.

The first polycarbosilazane dendrimer was reported [123-124] in 1998. The dendrimers upto 2nd generation, were synthesized by repetition of hydrosilylation and nucleophilic substitution reactions. *Tris*(dimethylvinylsilyl) amine as the core unit, chlorodimethylsilane as the hydrosilylating agent and lithium *bis*(dimethylvinylsilyl)amide as the amidation reagent were used for this synthesis. The higher generation dendrimer was not possible to synthesize by this technique, which may be due to incomplete reactions of chlorosilyl terminated 2nd generation dendrimer with the lithium amide. But the steric hindrance may not be the correct reason as mentioned by the authors [123-124], as the spacers between the branch points are long enough and only a branch multiplicity of two was employed (Fig.8.10). The synthesized dendrimers were characterized by IR, NMR spectroscopy, elemental analysis, VPO and MALDI-TOF mass spectrometry. However, these carbosilazane dendrimers are stable to air, water and anhydrous HCl but they readily degraded by aqueous HCl solution.

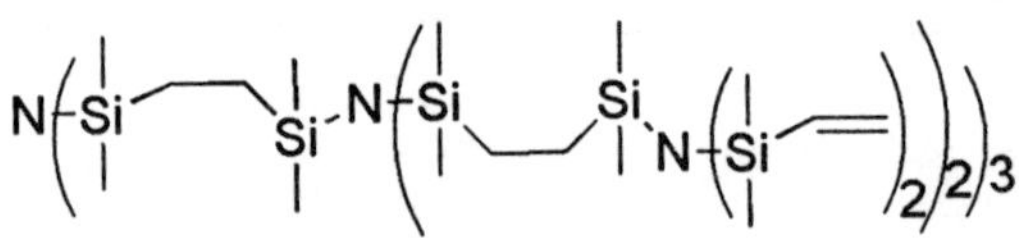

FIG.8.10 Structure of first polycarbosilazane dendrimer

The carbosilazane dendrimers based on carbosilazane cores were also reported [125]. The dendritic carbosilazane cores I to III (Scheme 8.22) were synthesized by reactions of K salt of divinyltetramethyldisilazane with chlorodimethylvinyldisilane, which leads to the central core (I) [126]. Again, hydrosilylation of I with dimethyl chlorosilane and subsequent reaction with K salt of divinyltetramethyldisilazane leads to first generation, III. These parent dendrimers were used for further modification. The reaction of IV (Scheme 8.22) with 1,1,3,3-tetramethyl disiloxane in the presence of Pt catalyst followed by hydrosilylation reactions with the carbosilazane core units, I to III using

Krastedt's catalyst [127] in toluene resulted the desired dendrimers. The whole sequence of reactions is shown in scheme 8.22.

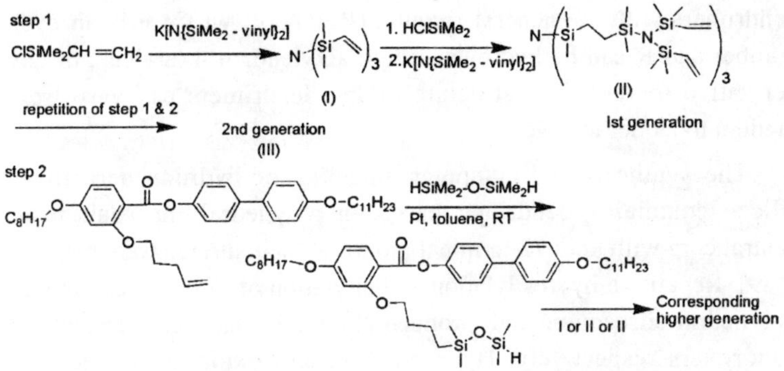

SCHEME 8.22 Synthesis of carbosilazane dendrimers with dimethyl vinyl chloro silane

8.2.5 Siltranes

The silatrane dendrimers are an unusual class of silicon based dendrimers and they are really interesting because of their biological activities. They have immense potential in agricultural chemistry as insecticides and enhancement of crop yield. The synthesis of silatrane dendrimers (Fig.8.11) was first reported [128] by the convergent approach. The wedges were synthesized by the following steps of reactions: (i) reaction of trimethoxy (glycidoxypropyl)silane with triisopropanol amine produces 1-glycidoxypropyl-3,7,10-trimethylsilatrane (I), (ii) two of such compounds reacted with ethanol amine to yield trialkanol amine, which can form a silatrane upon reaction with a trimethoxysilane. The dendrimers were produced branches by reaction of the 'I' with ammonia or diethanol amine instead of ethanol amine. The dendrimers were characterized by NMR spectroscopy and electrospray mass spectrometric studies.

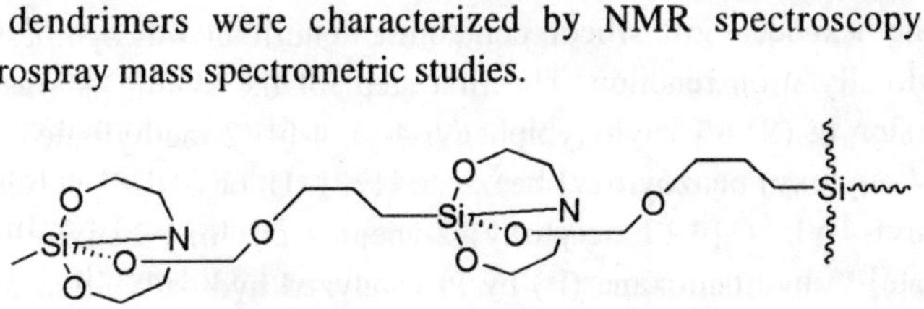

FIG.8.11 Structure of silatrane dendrimer with trioxa-azo silabicyclo[3,3,3] undecane derivative

8.2.6 Silsesquioxanes

Silsesquinoxane·dendrimers are a novel class of silicon containing dendrimers with the general formula $(RSiO_{1.5})_n$, where n is an even number and R can be hydrogen, alkyl, alkylene, aryl, arylene, or any derivative thereof. The structure of the dendrimers can vary from random to ladder to cage.

The synthesis and computer modeling of hydroxy derivatized silicon containing dendrimer based on polyhedral silsesquinoxane central core with upto 72-terminal groups at their surface were reported [129]. Repetitive hydrosilylation and allylation of vinyl functionalized polyhedral silsesquinoxane produce chlorosilyl and allyl-derivatized dendrimers respectively. Hydroboration and oxidation of the allyl functionalized dendrimers produced hydroxy terminated dendrimers. Theoretical studies by using energy minimization and molecular dynamic techniques showed that the dendrimers were relatively spherical, globular molecules with proportion of the terminal hydroxy groups located in a shell at the exterior of the dendrimer. Dendrimer with the increase of generation number i.e. for higher terminal groups back folding also increases.

A novel type of amphiphilic spherosilsesquinoxane [130] derivative, 1-(1,ω-propylene methoxy) oligo(ethylene oxide)-3,5,7,9,11,13,15-heptahydrido-penta cyclo[9.5.1.1.1]octasiloxane was prepared from $(HSiO_{3/2})_8$ and allyl functional oligo(ethylene oxide) by hydrosilylation. The monosubstituted octahydrido silsesquioxane was characterized by IR, ^{1}HNMR, ^{13}CNMR, ^{29}Si NMR spectroscopic and MALDI-TOF mass spectrometric as well as elemental analysis.

The synthesis, characterization and phase behavior of the liquid crystalline silsesquioxanes dendrimers were described [131-132]. At first, the hexadecavinyl silicon containing dendrimer was synthesized by hydrosilylation reaction. The first step for the synthesis was the conversion of (S) 4'-octyloxybiphenyl-4-yl, 4-[4-(2-methylbutoxy)-2-(pent-4-enyloxy) benzoyloxy] benzoate [133] (I), or [(R)-4'-octyloxy-biphenyl-4-yl, 4-[4-(2-octyloxy)-2-(pent-4-enyloxy)benzoyloxy] benzoate] to hydrosiloxane (II) by Pt catalyzed hydrosilylation of 'I' with an excess of tetramethyldisiloxane at room temperature by using Karstedt's catalyst (3% Pt in toluene) with good yield. By using a similar reaction step, the hydrosiloxane (II) was converted to silsequioxane dendrimer after treating with octasilsequioxane dendrimer (III) or octa*kis*(hydridodimethylsiloxyl)octasilsesquioxane in toluene (Scheme 8.23), though, in this case, the yield was low. The term 'octasilsquioxane' has been used in the literature [134] to embody the general class of compounds that contains the 'Si_8O_{12}' cubane unit, which might be derivatized by a substituent as $R_8[Si_8O_{12}]$, where R may be H, halogen, alkyl, alkylene, arylene, etc. or organo-functional derivatives of these as in this case, where R is $SiR_1{}^2R_2R_3$ or $OSiR_1{}^2R_2$. The synthesized silsesquioxane dendrimers were well characterized by ^{1}HNMR, ^{13}CNMR, ^{29}SiNMR spectroscopies, SEC and DSC studies. The optical texture of the compounds was examined by POM. The phase behavior was studied by POM and XRD studies. ^{1}HNMR spectra showed well resolved resonance for all the protons in the dendrimer, indicating rapid relaxation process at room temperature, and point to a high degree of freedom of the internal structure of the silsesquioxane core.

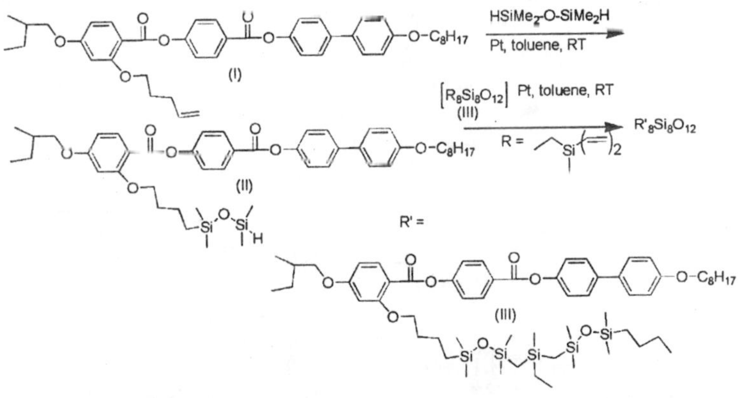

SCHEME 8.23 Synthesis of LC silsesquinoxane dendrimers

The polyhedral oligomeric silsesquioxanes (POSS) dendrimers possessing Cl or vinyl terminal groups were used to synthesize dendrimer-bound phosphines with 24 or 27 arms [135]. These dendrimer-bound phosphines were synthesized either by addition of secondary phosphines to vinyl groups on the periphery of the dendrimer, initiated by AIBN or the reaction of Si-Cl bonds on the periphery with lithiated methyl phosphine (Scheme 8.24). The synthesized dendrimer bound phosphine was characterized by ^{1}HNMR spectroscopy and MALDI-TOF mass spectrometry. The results showed that the dendrimer molecules nominally with 24 phosphine groups have either 23 or 24 arms successfully substituted. In another report, the authors [5] used the alkyl phosphine containing dendrimers based on same POSS core unit for hydrocarbonylation reaction. By applying same reaction on POSS dendrimers of 1st and 2nd generations, alkylphosphine containing dendrimers were prepared [136]. These phosphine containing dendrimers were used as ligands for the hydrocarbonylation of alkene such as hexene, octene, nonene, propene-2-ol, etc. in polar solvents (THF or EtOH) using the complexes [Rh(acac)$(CO)_2$] or [$Rh_2O_2(CMe)_4$] as the metal source to get the corresponding alcohol as the major product. The different functionalized dendrimers showed different properties depending upon the nature of the phosphine and the complexity of the dendrimer itself.

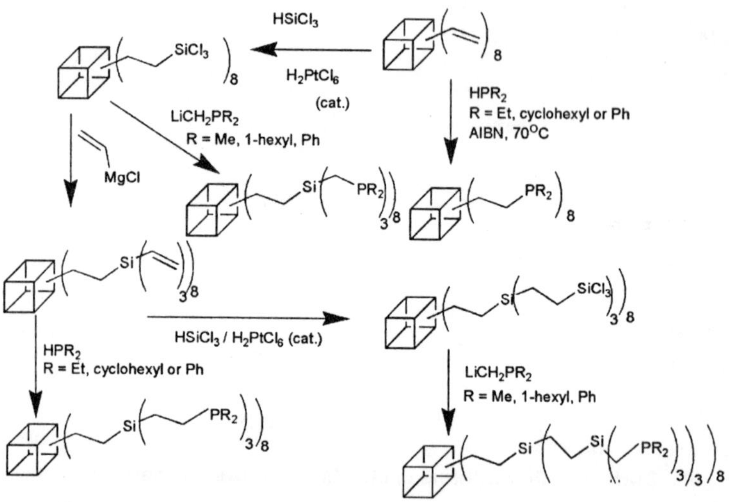

SCHEME 8.24 Synthesis polyhedral oligimeric silsesquinoxane (POSS) dendrimer

The synthesis of various silsesquinoxane dendrimers was reported [137] by using silsesquinoxane octavinyl pentacyclo octasiloxane as the core unit and tricholosilane, dicholomethylsilane and chlorotrimethylsilane as the hydrosilylation reagent. The vinyl and allyl magnesium bromides were used as the Grignard reagents for the synthesis of these dendrimers. The dendrimers were characterized by NMR spectroscopy.

Si-containing dendrimers upto 4th generation were synthesized [138] by the divergent method using octa(vinyldimethyl siloxy) octasilsesquinoxane as the core, $HSiMeCl_2$ as the hydrosilylating agent and CH_2=$CHCH_2MgBr$ as the alkylating agent. Functionalization of the dendrimer with the sandwich complex (η^5-$C_6H_4SiMe_2H$)Fe(η^5-C_6H_5) via hydrosilylation of the allyl terminated dendrimer yields the organometallic end grafted dendrimers [138]. The synthesized dendrimers were thoroughly characterized by IR, NMR spectroscopic and MALDI-TOF mass spectrometric studies. The ferrocenyl-functionalized dendritic polymer was also characterized by cyclic voltammetry, which indicates that the ferrocenyl moieties are independent to each other.

Phosphine containing carbosilane dendrimers upto 2nd generation based on polyhedral silsesquioxane core, were synthesized [139] by the divergent approach. The 1st generation was synthesized by successive hydrosilylation and vinylation or alkylation of vinyl terminated polyhedral silsesquinoxane to form 16 and 24 vinyl arms. Repetitive hydrosilylation/alkylation followed by hydrosilylation/vinylation and addition of phosphine produced the 2nd generation. The synthesized dendrimers were used as ligands for the hydroformylation of octa-1-ene catalyzed by [Rh(acac)$(CO)_2$]. The dendrimer containing alkylphosphine moieties transformed product into alcohol (nonan-1-ol), while the diphenylphosphine counter parts lead to the formation of aldehyde only (nonan-1-al).

Bassindalea and Gentke [140] derivatized octasisesquioxane with alcohol and silanols to form $(ROSiO)_8$ and $(R_3SiOSiO)_8$ by reaction with the corresponding alcohols and silanols in the presence of amines catalyst. Et_2NOH was found to be the most effective catalyst for these derivatization reactions.

8.2.7 Silanes

The dendrimers are based on Si-Si linkages with completely saturated Si atoms with methyl groups in their chemical structures. Further dendritic structure increases the inertness of such molecules by restricting or prohibiting the access of reagents to the inner bonds. Silane dendrimer with a tetrafunctional core and without spacer is not possible to synthesize because of steric reason. The first silane dendrimer was synthesized [141] in 1995. The following steps of reactions were used to synthesize the desired dendrimer. (i) The commercially available *tris* (trimethylsilyl)silane (I) was converted to methyl *tris*(trimethylsilyl)silane (II) by successive reactions with $CHCl_3$ and methyl lithium. (ii) The reaction of (II) with chlorotrimethylsilane and $AlCl_3$ gave methyl[*tris*(chlorodimethyl silyl)]silane. The conversion of this trichlorinated silane with *tris*(trimethyl silyl)silyllithium leads to 2,2,6,6-tetra*kis*(trimethylsilyl)-2',2'-*bis*(trimethylsilyl)-1',1',3',3',3'-pentamethyl trisilyl] undecamethyl-heptasilane, the desired silane dendrimer (Fig.8.12). The dendrimer was characterized by UV, ^{29}SiNMR spectroscopic, mass spectrometric and XRD studies.

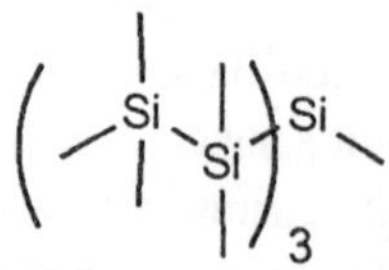

FIG.8.12 Structure of silane dendrimer with methyl substitution

The synthesis of same type of silane dendrimers has also been reported [142-145]. Methyl[*tris*(dimethylsilyl)]silane was synthesized [142] by a Wurtz type of reaction with chlorodimethyl silane and lithium. This silane was chlorinated with CCl_4 to afford methyl[*tris*(chloromethylsilyl)]silane, which was converted into the dendrimer by reaction with *tris*(trimethylsilyl)silyllithium. Sekiguchi *et al.* [143] used 2-lithio-1,3-diphenyl pentamethyl trisilane as a key intermediate for synthesized silane dendrimers upto 2nd generation. The key intermediate was prepared by the reaction of *bis*(1,3-diphenyl pentamethyl trisiloxy)mercury with lithium. The main intermediate was first reacted with chlorodimethylphenylsilane to obtain methyl [*tris*(dimethylphenylsilyl)silane. The phenyl groups of this product were replaced by treatment with trifluoromethane sulfonic acid. The resulted product was then reacted with 2-lithoheptamethyl trisilane to get the first generation (G_1) silane dendrimer. The 2nd generation dendrimer was obtained by treatment of G_1 with trifluoromethane sulfonic acid followed by the reaction with 2-lithioheptamethyltrisilane. The synthesized dendrimers were characterized by UV, NMR spectroscopies and mass spectrometry. By applying the similar synthetic strategy, a variety of first generation silane dendrimers were synthesized [144]. There are variations in core multiplicity, number of spacer atoms between branching points and branching multiplicity in their structures. These silane dendrimers were characterized by UV spectroscopy. Also, some of these dendrimers were characterized by special NMR of 2D ^{29}Si-^{29}Si INADEQUATE NMR experiment [145]. This experiment provided the connectivity for each of the three Si-Si bonds in the molecules and combination with 1D ^{29}Si NMR offer the determination of molecular structures. ^{29}Si NMR was also utilized independently to determine various substitution patterns in these dendrimers [146].

The synthesis and characterization of silane dendrimers [7] with largest silane chain were reported [147] in 1998. The dendrimer is *tris*[2,2,5,5-tetrakis(trimethyl silyll)hexasilyl]methylsilane, with 13 silicon atoms in the longest chain. It was obtained by the reaction of *tris*(chlorodimethyl silyl)methylsilane with 1,1,4,4-tetra*kis*(trimethyl (silyl) 2,2,3,3,5,5,5-heptamethylpentasilyllithium. NMR, UV spectroscopic, mass spectrometry, and XRD studies characterized the dendrimers.

A hybrid dendrimer [148] with alternating silicons and germenium atoms in the structure was synthesized. The reactions involved for this synthesis are as follows: (i) the lithium-mercury exchange reaction of *bis*[*bis*(dimethyl phenyl germyl) silyl]mercury allowed to prepare *bis*(dimethyl phenyl germyl)methyl silyl lithium, as the building reagent, (ii) the reaction of this molecule with chlorodimethylp henyl germane gave *tris*(dimethylp henyl germyl)methylsilane, whose phenyl groups were cleaved with trifluoromethane sulfonic acid, (iii) subsequent reaction with the building reagent yielded the first generation hybrid dendrimer. The permethyl substituted dendrimer was obtained after cleavage of the phenyl groups with trifluoromethane sulfonic acid followed by substitution of the tricholoromethane sulfonate groups by chlorine atoms with NH_4Cl. In the final step, the resulting hexachloride was converted into the permethyl substituted dendrimer by conversion with methyl magnesium iodide (Fig.8.13). The dendrimer was characterized by UV spectroscopy and XRD studies.

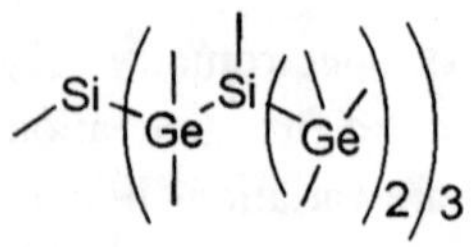

FIG.8.13 Structure of Si-Ge hybrid dendrimer

8.2.8 Others

The Ist generation of an organosilicon dendrimer composed of thiophene ring connected by silicon (Fig.8.14) was synthesized [149] by tetralithiation of tetra-2-thienyl silane followed by reaction with methyl tri-2-thienylsilyl ether. The dendrimer was characterized by NMR spectroscopy and elemental analysis.

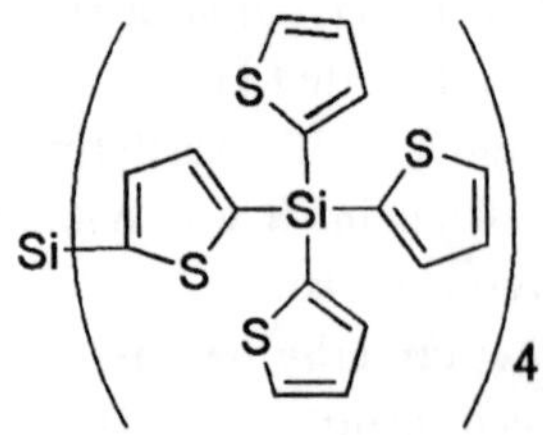

FIG.8.14 Structure of organosilane dendrimer with 4-thiophene units

A number of reports [150-152] on the convergent synthesis of carbosilane-siloxane dendrimers with pyrenyl group at the central core (Fig.8.15) were reported. These fluoresence labeled carbosiloxane dendrimers were synthesized by applying the double exponential growth technique, starting from 1-(triallyl silyl)-3-(dimethyl pyrenyl silyl)propane as the core unit. The fluorescence spectroscopic analysis confirmed the excimer formation in concentrated solution of the dendrimer and the core indicating mutual accessibility of the pyrenyl groups in both the molecules. This was really an exciting result as no excimer was found in a pyrenyl functionalized carbosilane dendrimers [152-153] and this may be due to higher mobility of the appropriate terminal siloxy groups of the carbosilane-siloxane dendrimers.

The double exponential growth approach was also applied to synthesize organosilane dendrimer [154] in 1997. This approach is a

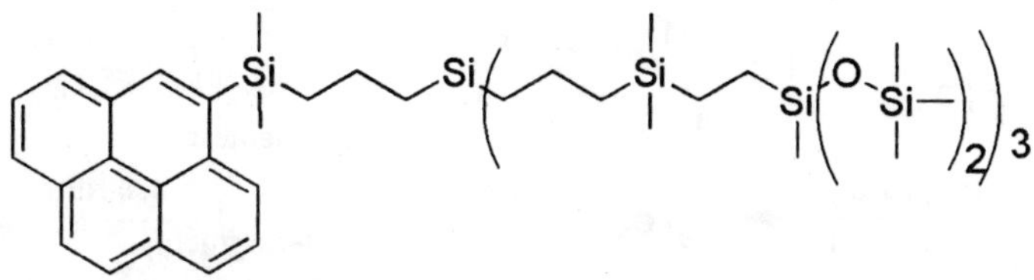

FIG.8.15 Structure of carbosilane-siloxane dendrimer with pyrenyl core unit

combination of divergent and convergent approaches and the authors called this strategy as 'Universal scheme'. They first synthesized the carbosilane dendrimer, $Si[(CH_2)_3\text{-}SiMe(CH_2\text{-}CH{=}CH_2)_2]_4$, which was by the divergent technique, followed by coupling reaction of this dendrimer with carbosiloxane monodendron, $HSiMe_2OSiMe_2\text{-}(CH_2)_2)MeSi\ (OSiMe_2(CH_2)_2SiMe(OSiMe_3)_2)_3$ synthesized via the convergent approach, produced carbosilane-siloxane dendrimer (Fig.8.16). The resulting dendrimer contains 93 silicon atoms and molecular weight of 7513 g/mole as measured by MALDI-TOF mass spectrometric study. The aggregation and film formation behavior of these dendrimers were studied [155] by SFM. The size of the molecules was found to be around 2.5 nm, which is consistent with the size estimated from the structure.

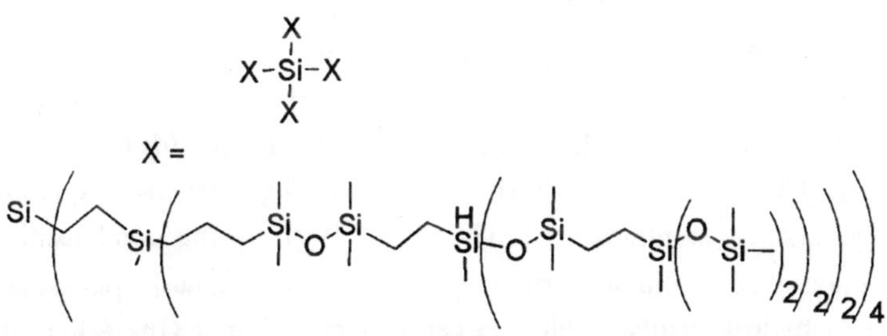

FIG.8.16 Structure of carbosilane-siloxane dendrimer obtained by'Universal scheme'

The synthesis of alkoxysilane dendrimers [156] by the divergent approach, where the reactions involved are Grignard alkylation, hydrosilylation and alcoholysis (Scheme 8.25), were reported. The synthesized dendrimers were characterized by IR, ^{1}HNMR, ^{13}CNMR spectroscopic, elemental analysis, GPC, MALDI-TOF/ESI mass spectrometric studies.

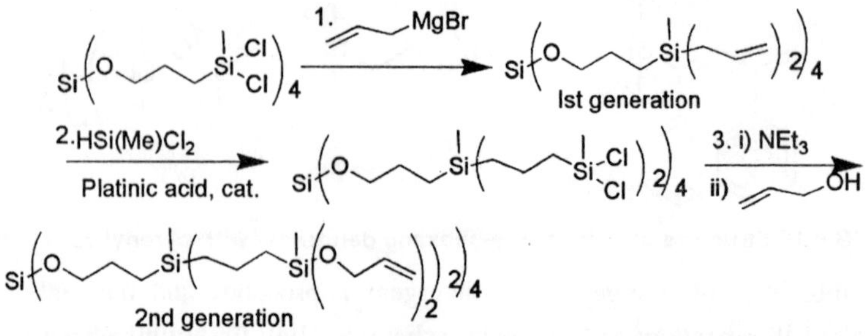

SCHEME 8.25 Synthesis of alkoxy silane dendrimers by divergent approach

8.3 SYNTHESIS, CHARACTERIZATION AND PROPERTIES OF SILICON BASED DIFFERENT TYPES OF HYPERBRANCHED POLYMERS

HB silicon containing polymers have characteristics viz. enhanced solubility, easy end processibility due to their globular structures, single step synthesis and presence of many terminal groups. Thus, they can be further functionalized by varies of other functionalities and these are discussed here.

8.3.1 Polycarbosilanes

For the synthesis of HB polycarbosilanes, a number of AB_n ($n \geq 2$) type monomers such as $ClCH_2SiCl_3$, $HSi(CH_2\text{-}CH{=}CH_2)_n$ (n =2 and 3), $HSi(Me)\text{-}(CH{=}CH_2)_2$, $HSi(Me)\text{-}(CH_2)_9CH{=}CH_2)_2$, $MeHSi\text{-}C_6H_4\text{-}Si(CH{=}CH_2)_3$, etc. have been employed. The first such type of polymers was synthesized [157-158] in 1991 by the condensation reaction of (chloromethyl)trichlorosilane via the Grignard reaction. To obtain hydrolytically stable HB carbosilane polymer (Fig.8.17), the resulting HB polychlorocarbosilane was reduced catalytically by $LiAlH_4$. The molecular weight of the polymer was about 600 g/mole and DP was 13. The polymer was further modified via bromination followed by alkylation using Li or Mg based organometallic compounds [159]. This allyl HB carbosilane (alkylation with allyl magnesium bromide) hydrosilylated was further with dimethyl phenyl silane or chlorodimethyl silane. The synthesized polymer was characterized by NMR, IR spectroscopic and SEC studies.

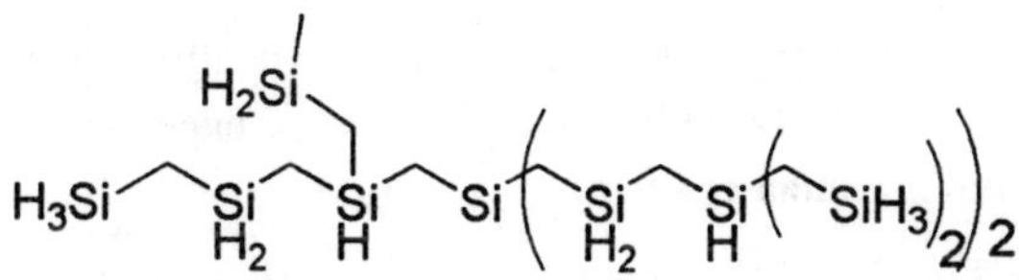

FIG.8.17 Structure of simplest HB polycarbosilane

An oligomeric HB carbosilane [160] was synthesized using (chloropropenyl)dichloro methyl silane as the monomer by the Grignard polymerization process. The product was characterized by NMR spectroscopy and DP was found to be about 8. The synthesis of poly(2,5-silylthiophenes) with alternating silylene and thiophene groups along the branches was reported [161]. 2-Bromo-5-(trimethoxysilyl)thiophene was used as AB_3 type monomer and the HB polymer was prepared by Grignard coupling reactions. The detailed investigation of DB of HB polycarbosilanes [162], which was obtained by well known hydrosilylation of terminal double bonds with trichloro silane and subsequent displacement of the chloride groups by allyl magnesium bromide was reported. The polymers were obtained by Pt-catalyzed polymerization reaction of triallyl silane and functionalization of 2-(10-decen-1-yl)-1,3-oxazoline at the focal point, with triallylsilane prior to hydrosilylation (Fig. 8.18). ^{29}SiNMR spectral study revealed

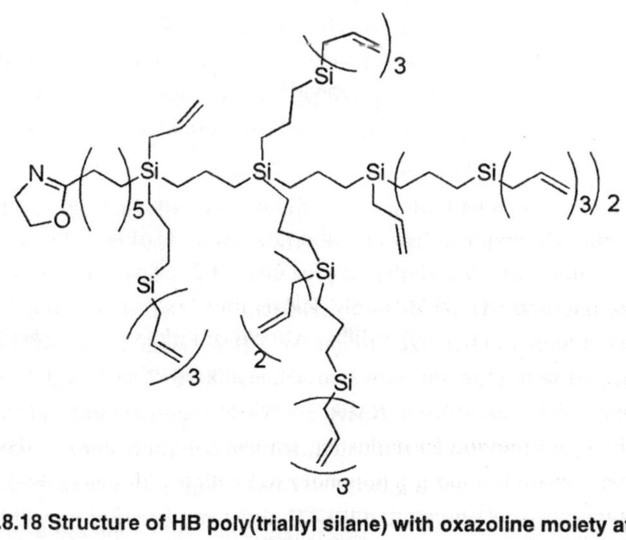

FIG.8.18 Structure of HB poly(triallyl silane) with oxazoline moiety at the focal point

that the polymerization of AB_3 monomer (triallylsilane) occurred randomly with DB 0.42 with well agreement of expected DB [163] of the same type of monomer (0.44).

The polymerization of the oxazoline groups as well as attachment of the oxazoline based HB macromonomers to a trifunctional core affording trimer of the HB fragments [164-165]. The product was a transparent, rubber like solid and strong aggregation in solution was obtained by various techniques. The results support the formation of columnar structures by interaction of the polar core of the trimers with their amide-ester linkages capable of forming H-bonding. These polar centers of the stacks are surrounded by an apolar, disorder exterior.

The hydroxy end capped poly(diallylmethylsilane) derivative [166] was synthesized by coupling of 1-(3-dimethyl silyl)propyloxy-2-trimethyl silyloxyethane with the polymer via hydrosilylation followed by hydrolysis of trimethylsiloxy group. IR study indicates the presence of H-bonded network in the polymer. The first aromatic HB polycarbosilane was synthesized by hydrosilylation was reported [167]. The AB_3 monomer was 1-dimethylsilyl-4-trivinylbenzene in this case. The glass transition temperature of the polymer is very low, (Tg~12°C) inspite of the presence of rigid aromatic groups. The HB poly carbosilanes based on AB_3 type monomers were investigated [168] with respect to the formation of the linear, semidendritic and dendritic unit. The T_g of the polymers depends on the starting monomer and varies from –45°C to 12°C. The DB was 0.44 as determined by ^{29}SiNMR spectroscopy.

The regioselective and steroregular HB poly(methyl silylene) *bis*(1,4-phenyl-trans-vinylene) were synthesized [169] by using a novel AB_2 type monomer, *bis*(4-ethynyl phenyl)methyl silane. The monomer contains one reactive Si-H group and two ethynyl phenyl functional moieties, which were utilized to get the HB polymer by the following sequence of reactions, as shown in scheme 8.26. The products were characterized by IR, UV, ^{1}HNMR, ^{13}CNMR spectroscopic, TGA and GPC analyses. The results indicated that the polymers contain 95% of *trans* vinyl moieties and the polymers were highly thermostable, only 9% weight loss was observed at 900°C under the nitrogen atmosphere. The polymers were also displayed an absorption due to π-π* transition at ~275 nm as a shoulder and a weak absorption at 330 nm due to π to σ charge transfer, which is absent in linear system.

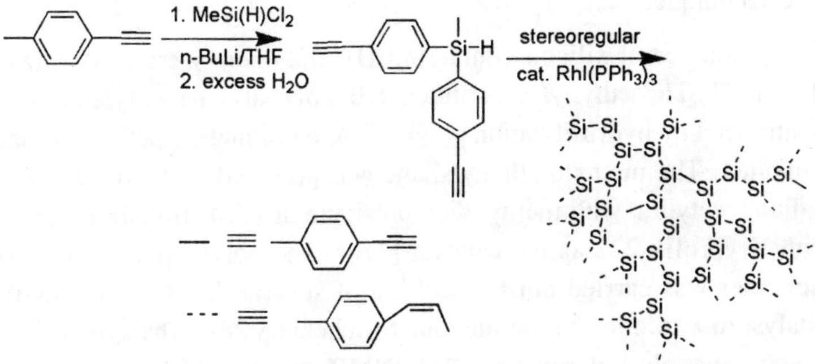

SCHEME 8.26 Synthesis of HB polycarbosilane using bis(4-ethynyl phenyl) methyl silane as AB_2 monomer

The HB carbosilane polymer [24] was synthesized by three step reactions as shown in scheme 8.27. The hydrosilation of triallysilane (AB_3), the central core unit in the presence of Pt/C catalyst to get $(AB_3)_n$, followed by hydrosilation reactions of (AB_3) with hydrosilane ($HSiMe_nCl_{3-n}$, n = 1-2) to obtain AB_3-$SiMe_nCl_{3-n}$ macromolecule. Finally, the chlorine containing silane polymers were treated with allyl magnesium bromide to get the desired HB polymers. The synthesized

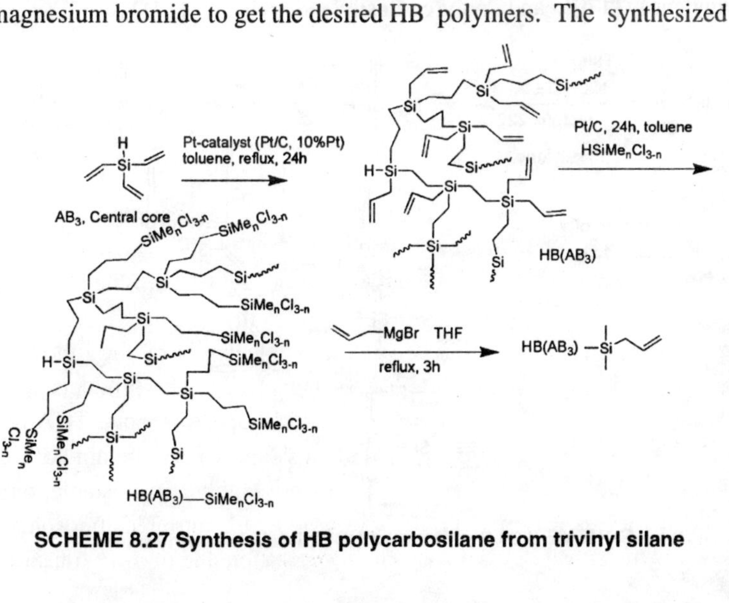

SCHEME 8.27 Synthesis of HB polycarbosilane from trivinyl silane

polymers were purified by the chromatographic technique and characterized by ^{1}HNMR, ^{13}CNMR spectroscopy, elemental analysis and GPC technique.

A number of silicon containing HB polymers were synthesized [170-171]. The ethynyl-substituted HB poly(silylene vinylene) was synthesized by hydrosilylation polymerization of methyldiethynylsilane monomer. The methyldiethynylsilane was prepared by the reaction of sodium acetylide with methyl dichlorosilane in 1,2,4-trimethylbenzene with Kryprofix 222 as the catalyst [170]. The polymerization of this monomer was carried out by addition of several drops of Karstedt's catalyst to a solution of the monomer (Scheme 8.28). The synthesized polymer was characterized by IR, ^{1}HNMR and ^{29}SiNMR spectroscopic studies. The molecular weight increased from 2720 to 3091 Da on keeping in dark for 4 weeks. The HB polymer was soluble in common organic solvents such as diethylether, toluene, chloroform, etc. and remains soluble even on keeping 4 weeks, though molecular weight slight increased. The thermal studies indicated that the T_g was ~14°C and a strong exothermic peak in DSC curve. The TGA data indicated that on heating upto 1300°C both in the nitrogen and in air only 13% and 38% weight losses were found, respectively. The HB poly(silylene vinylene) [170] and poly(carbosilylene arylene) [171] were found

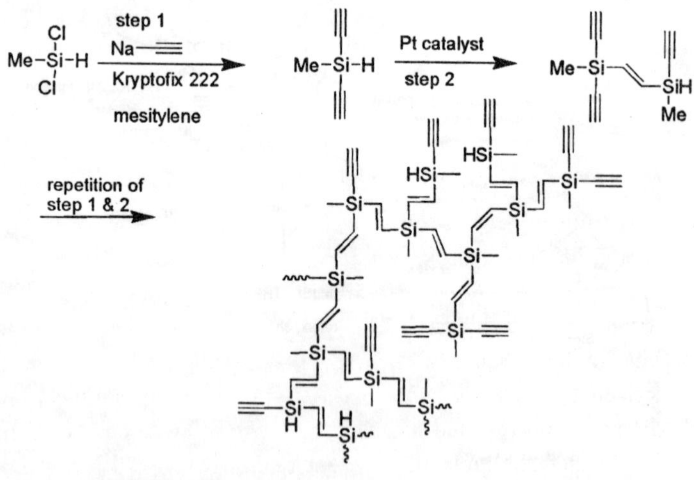

SCHEME 8.28 Synthesis of HB poly(silylene-vinylene)

thermally more stable than the corresponding linear polymers, which may be due to HB structure and crosslinkable terminal groups (e.g. ethynyl and vinyl groups) of the former polymers.

Modification

In this synthetic method, the AB-SiMeCl carbosilane macromolecules [2] were reacted with lithium salts of amides like phenylacetylideamide and *bis*(trimethylsilyl)amide with cholesterol in THF and toluene respectively as shown in scheme 8.29. The yield obtained in each case was more than 80%. The modified polymers were characterized by ^{1}HNMR, ^{13}CNMR spectroscopy and GPC technique.

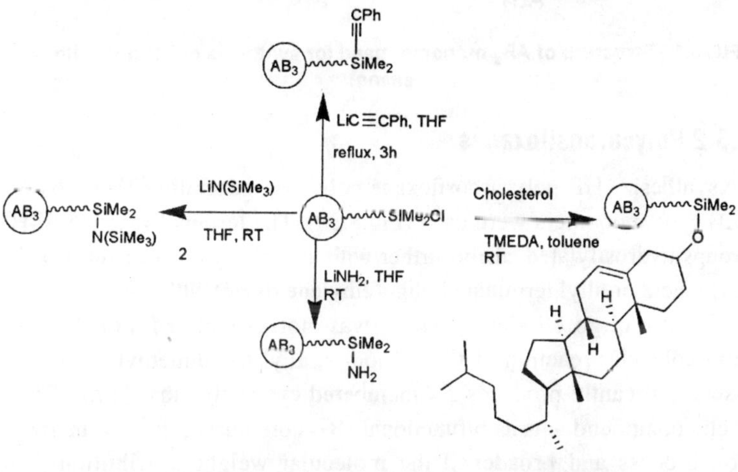

SCHEME 8.29 Functionalization of surface groups of HB polycarbosilane by various compounds

The HB polycarbosilane was synthesized [162, 172] by self-polycondensation through hydrosilylation reaction of AB_3 type monomer, triallylsilane, which was obtained from Grignard reactions of allyl bromide with trichlorosilane. The polymerization reaction was carried out at 95°C in the presence of Pt catalyst. The synthesized polymer was characterized by ^{1}HNMR and ^{29}SiNMR spectroscopic studies. The ^{29}SiNMR spectrum resulted a DB of 0.48 ± 0.05, whereas the GPC analysis indicated the molecular weight ranges from 8-50 X 10^4 Da. The polymer was soluble in common organic solvents.

The HB polycarbosilanes were also synthesized [173] from different AB_n type monomers as shown in Fig.8.19. AB_2-I and II, and AB_3-I monomers were bulk polymerized under controlled reaction conditions such as under low temperature and low catalyst concentration, otherwise crosslinked products were resulted. However, using AB_3-II monomer, where the distance between the double bond and silicon atoms was essentially enlarged and hence, resulted high molecular weight polymer without crosslinking.

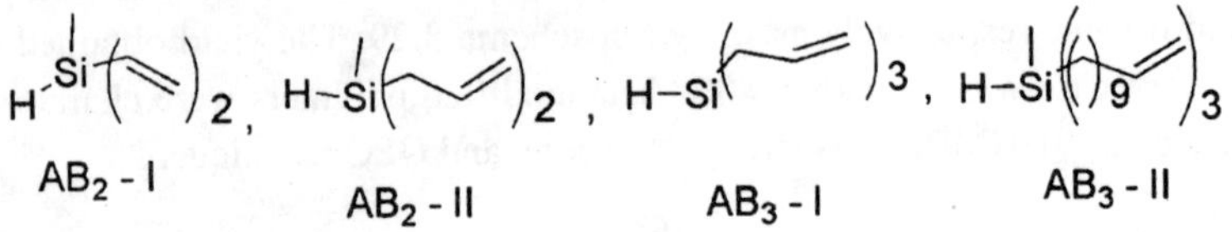

FIG.8.19 Structure of AB_x monomer used for synthesis of HB poly(siloxysilane)

8.3.2 Polycarbosiloxanes

To synthesize HB polycarbosiloxane polymers, generally either AB_3 or AB_4 type monomers were used (Fig.8.20). The terminal reactive Si-H groups hydrosilylated easily further with allylphenylether or react with acylic acid or allyl terminated oligo(ethylene oxide), although the reaction with oligo(ethylene oxide) was not completed [174]. The intramolecular reaction of the monomer, allyl tris(dimethyl siloxane) also significantly produces six membered cyclic structure [175]. This cyclic compound acts as bifunctional (B_2) core during the polymerization process and broadened the molecular weight distribution. To eliminate this undesirable product, the authors also utilized longer alkene containing silane e.g. 6-hex-1-enyl *tris*(dimethyl siloxy)silane but still the products showed broad molecular weight distribution with molecular weight (1.2-2.1 X 10^4) g/mole. The similar type of HB carbosiloxane polymer was also reported [176] by using vinyl *tris*(dimethyl siloxy)silane and *tris*(dimethyl vinylsiloxy)silane as the monomers. In these cases, the product of the intermolecular cyclization was much less, as it was five-member ring and hence the polymer obtained with significantly higher yield. However, the SEC analysis showed a broad PDI. The reaction of vinyl *tris*(dimethyl siloxy)silane with multi Si-H functionalizal core, propyl *tris*(dimethyl siloxy)silane yields HB carbosilane polymer [177]. By this approach, the viscosity of the polymer was controlled but cyclization (10 mole % of monomer) was not avoided.

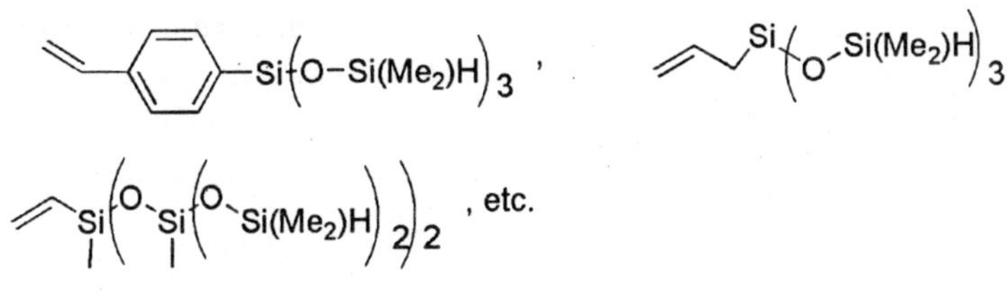

FIG.8.20 Structure of AB_3 and AB_4 type monomers for synthesis of HB carbosiloxane

A series of HB polycarbosiloxane polymers was synthesized [178-179] by using monomers like methylvinyl *bis*(dimethyl siloxy)silane, methylvinyl *bis*(methyl *bis*(dimethyl siloxy)siloxy)silane and vinyl *tris*[methyl bis(dimethyl siloxy) siloxy]silane with branch multiplicities of 2, 4 and 6 respectively. The molecular weight of the products varies widely from oligomers to polymers with maximum molecular weight upto 8900 g/mole. Of course the highest molecular weight is formed from the monomer with branch multiplicity 4. All of these HB polymers have been modified by the reactions of terminal reactive Si-H groups with allyl ether or allyl methyl triethylene glycol. The problem of low molecular weight of the polymers for the unmodified products was overcome by slow addition of the monomers [180-181]. In this case, the resulting polymers have higher molecular weight, 8700-61000 g/mole and high yield (72-79)%. By using macromonomers, poly(dimethyl siloxane)s with DP 10, 50 and 100, a large number of HB polycarbosiloxane polymers was synthesized [182]. The macromonomers contain one vinyl and two Si-H end groups, which on polymerization by hydrosilylation produced long chain HB polymers with molecular weight (1.5-80) X10^4 g/mole. 4-vinylphenyl *tris*(dimethyl siloxy)silane on hydrosilylation yielded HB polycarbosiloxane with molecular weight 9800 g/mole, as measured by SEC [183]. The polymer was modified also by trimethylvinylsilane. The HB polymer shows higher thermal stability than the aliphatic polymer obtained from *tris*(dimethyl siloxy)vinylsilane, may be due to the presence of phenyl group in the structure.

The self-polycondensation of AB_3-II, III and IV monomers in the presence of a heterogeneous Pt-catalyst to form HB poly(siloxy silane)s (Fig.8.21) of high molecular weight with good yields [176, 184]. The addition of Si-H bond to the double bonds was confirmed by ^{13}CNMR spectroscopy. The HB polymer obtained from AB_3-II was soluble in common organic solvents but the Mw was only 9800 Da with PDI 1.6. The T_g of the polymer from AB-IV was substantially higher (T_g = -58°C) than from AB_3-II (T_g= -100°C).

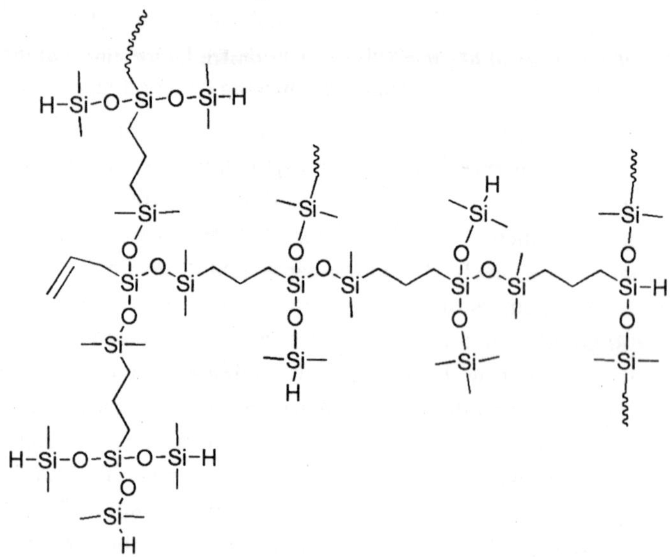

Fig.8.21 Structure of HB poly(siloxysilane)

8.3.3 Polysiloxanes

HB polysiloxanes were synthesized [185] by base catalyzed proton transfer polymerization reaction of a latent $A_2A'B$ type monomer. The latent monomer, 1-(hydroxy dimethyl siloxy)pentamethyl cyclotrisiloxane (I) was treated with a phosphazene, P_4t_{Bu} superbase followed by quenched with vinyl dimethyl chlorosilane/NEt_3 to get the desired HB polymer, as shown in scheme 8.30. The monomer I was first synthesized by a Pd/C catalyzed oxidation of 1-(dimethyl siloxy)pentamethyl cyclotrisiloxane with water. Again, 1-(dimethyl siloxy)pentamethyl cyclotrisiloxane was prepared by the reaction of tetramethyldisiloxane-

1,3-diol with 1,1-dicholoro-1,3,3-trimethyldisiloxane and triethylamine. The whole sequence of reactions is shown in scheme 8.30. FT-IR, ^{1}HNMR ^{13}CNMR and ^{29}SiNMR spectroscopies, elemental analysis, GPC analysis and mass spectrometry characterized all the products. The number average molecular weight of the HB polymers was 11,660 Da as measured by ^{29}SiNMR and 11, 900 Da while analyze by GPC. The T_g of the polymer was ~ -124°C and found to be independent of DB.

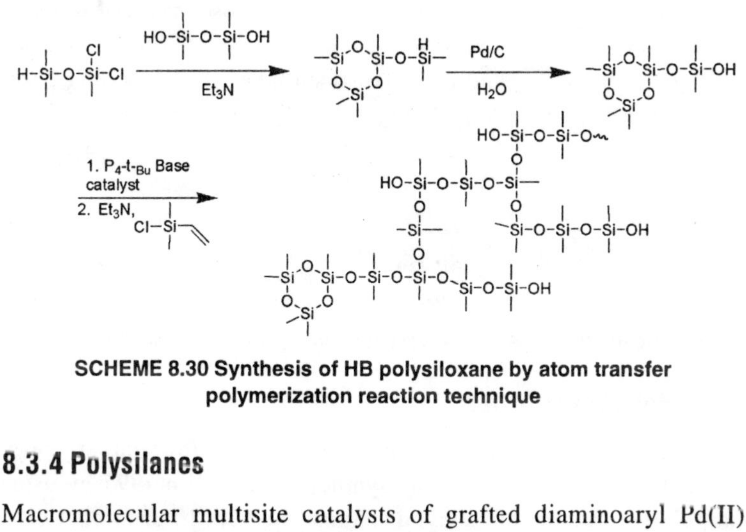

SCHEME 8.30 Synthesis of HB polysiloxane by atom transfer polymerization reaction technique

8.3.4 Polysilanes

Macromolecular multisite catalysts of grafted diaminoaryl Pd(II) complexes onto a HB polytriallysilane support was reported [186]. The HB carbosilane supports neat triallysilane was polymerized with Pt catalyzed polyaddition reaction as shown in scheme 8.31. The polymer was then hydrosilylated with $HSiMe_2Cl$ at ambient temperature followed by treatment with 3,5-*bis*[(dimethyl amino)methyl] phenyl lithium (Li-NCN) at –70°C resulted HB polymer bound $SiMe_2NC(H)N$ product. This product was then lithiated with an excess of $[PdCl_2(SiMe_2)_2]$ in THF offered the desired Pd-complex. Finally it was converted to its cationic derivative by treatment with AgOTf. The whole sequence of reactions is shown in scheme 8.31. The synthesized complex was characterized by ^{1}HNMR spectroscopic studies.

The cationic complex was used as soluble catalyst in aldol condensation of benzaldehyde and methylisocyanoacetate to get oxazoline.

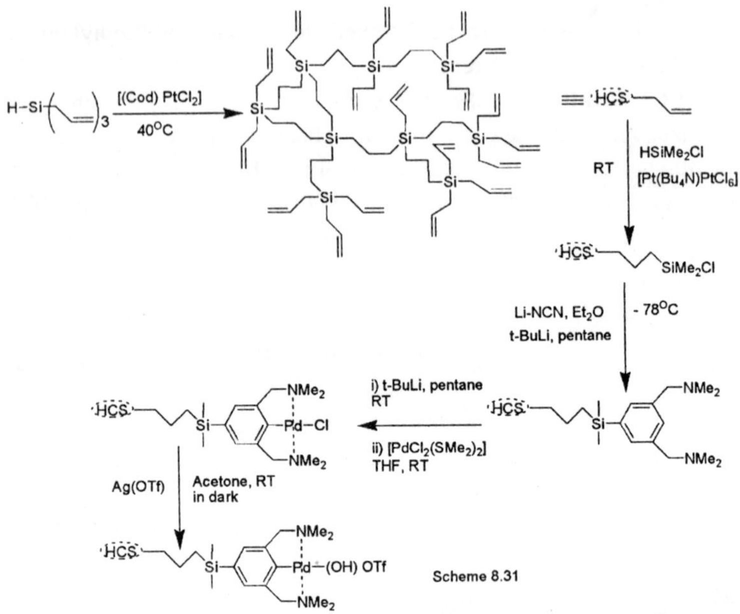

SCHEME 8.31 Synthesis Pd containing HB polycarbosilane

8.3.5 Polyalkoxysilanes

The first such type of polymer was degradable poly *bis*(undecenyloxy)methyl silane synthesized [187] in 1995 by using alkoxyalted reaction of methyl dichlorosilane with 10-undecen-1-ol in pyridine. The monomer, *bis*(undecenyloxy)methylsilane (MeHSi-(O-$(CH_2)_9CH{=}CH_2)_2$) was hydrosilylated under various reaction conditions using H_2PtCl_6 or $Co_2(CO)_8$ as the catalyst to get the desired polymer. The agglomerates of this polymer in a polymethacrylate resin were utilized to create cavities in the matrix by removal of resin after hydrolysis. This result shows that the polymeric structure can be used as a template to create nanometer size cavities in the matrix of other materials. The GPC studies revealed a broad PDI for the polymer. Other type alkoxysilane HB polymers were reported [188-189] by using two different types of monomers. These monomers, triethoxysilane and triethoxytrifluoroacetate were prepared first from tetraethoxysilane. The ethylsilicate HB polymers were obtained by condensation of the monomer with high molecular weight. Finally, the ethoxy end groups of these polymers were converted into trimethylsilyl end groups, which were more stable. The HB polymers were characterized by NMR spectroscopy and SEC.

The HB polysiloxysilane has been synthesized by the hydrosilylation between a double bond and Si-H group of an AB_x type monomer containing Si-O linkages. Several ABx type siloxysilane monomers were used for this purpose (Fig.8.22). HB polysiloxysilane polymers were synthesized [179] from AB_2, AB_4 and AB_6 type monomers at 50°C in the presence of Pt on active charcoal as a catalyst. The resulted polymers were characterized by IR, ^{1}HNMR spectroscopy and GPC analysis. The weight average molecular weight ranges from 5.8–8.9X10^3. However, slow addition of the AB_2 type monomer offered relatively high molecular weight (8.6X10^4) Da polymer. Further end capping reaction was performed by hydrosilylation with a variety of vinyl compounds. The T_g ranges from –126 to –2°C, depending on the nature of the end groups.

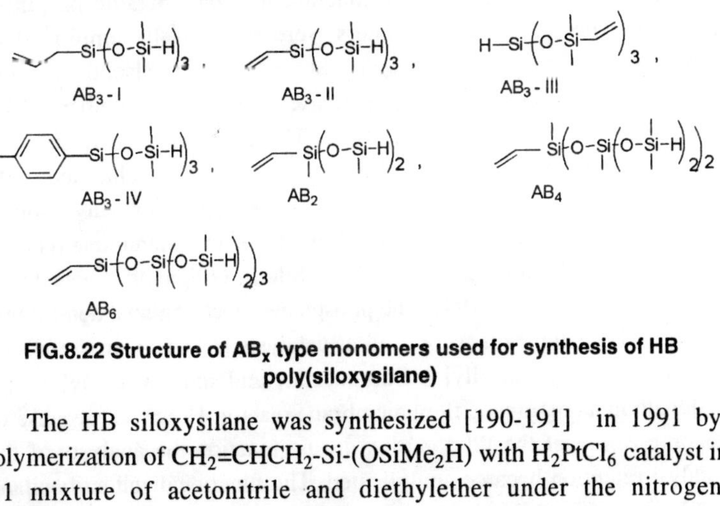

FIG.8.22 Structure of AB_x type monomers used for synthesis of HB poly(siloxysilane)

The HB siloxysilane was synthesized [190-191] in 1991 by polymerization of CH_2=$CHCH_2$-Si-($OSiMe_2H$) with H_2PtCl_6 catalyst in 1:1 mixture of acetonitrile and diethylether under the nitrogen atmosphere at 62°C. The formation of HB polymer was very fast but higher growth was restricted due to steric hindrance of the moieties. The molecular weight of the polymer was 1.9X10^4 g/mole with narrow molecular weight distribution. The IR and ^{1}HNMR spectroscopies were used to characterize the polymer and these studies indicated the total consumption of the double bonds.

8.4 APPLICATIONS OF SILICON BASED DENDRITIC POLYMERS

The silicon based dendritic polymers have an enormous potential in different fields of applications, despite their expensive synthetic routes. Some of these applications are presented here.

8.4.1 Catalysis

One of the most promising applications of carbosilane dendrimers is the use as scaffolds for catalysts or redox active metal complexes. Dendrimer bound catalysts combined the advantages of heterogeneous and homogeneous catalysis. On one hand, they allows the accurate control of the number and structure of active sites, comparable to homogeneous catalysts, on the other hand, they are conveniently removed from a product containing solution using ultra filtration as known from heterogeneous catalysts. The first example of a homogeneous catalyst based on a dendritic carbosilane scaffold [68-69] was used as molecular scaffolds for attachment organometallic Ni pincer complex. The prepared dendrimers were successfully employed as homogeneous catalyst for Kharasch addition reaction. Also, the activity of the catalyst decreases with the increase of generation number, which may be due to steric crowding effect. These dendritic catalysts have turnover frequencies about 30% lower than those obtained with monomeric or polymeric bound analogs [192]. The hydrovinylation of styrene with ethylene to get 3-phenyl-but-1-ene in a membrane reactor, catalyzed by Pd-complexes with hemilabile P, O ligands attached to a carbosilane dendrimer [193]. The phosphane functionalized dendrimers and their Pd-complexes were used as catalysts [194-195] in the allylic alkylation reaction of allyl trifluoroacetate and sodium diethylmethyl malonate in a continuous flow membrane reactor. However, because of decomposition of the Pd-complex during reaction the analysis of the catalyst retention became complicated. The functionalized carbosilane dendrimers [50, 196] were used as soluble supports for the ester-enolate imine condensation reaction, which leads to β–lactum. The reaction turned out to be highly *trans*-selective and led to high conversion. The ester functionalized dendrimers [50] can also be employed as soluble substrate carriers for the production of low molecular weight organic molecules.

The dendrimer bound phosphine based on polyhedral oligomeric silsesquioxane core unit with 16 PPh_2 arms was used [6] as a catalyst for hydroformylation reactions. These types of catalyst resulted much higher linear selectivity (14:1) than that of their small molecular analog (3-4:1) in the hydroformylation reaction of octene. The catalyst was prepared *in-situ* from the dendrimer based phosphine and $[Rh(acac)(CO)_2]$.

A number of ferrocenyl containing carbosilane dendrimers are most promising materials for use in multi electron redox catalyst as these dendrimers containing a controlled number of equivalent redox centers [197]. The diphosphine ligand centered on 1,1'-*bis*(diphenyl phosphine) ferrocene bearing carbosilane dendrimers (Fig.8.23) [69] and their Pd-complexes were used as active catalysts for allylic alkylation reaction. The rate of alkylation reaction (Scheme 8.32) as well as selectivity decreased with the increase of generation number of the dendrimers. This may be due to steric hindrance and change of polarity of the microenvironment. The group IV metal containing carbosilane dendrimers were used as catalyst for the polymerization of olefines and silanes [39]. The methyl aluminoxane activated catalyst was used to prepare polyethylene and found to be very effective.

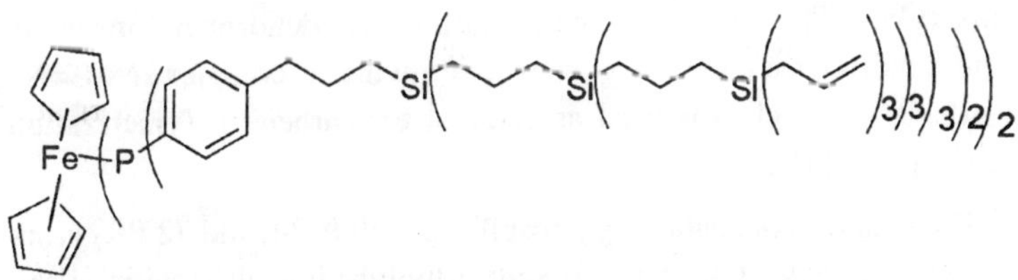

FIG.8.23 Structure of diphosphine ligand based carbosilane dendrimer

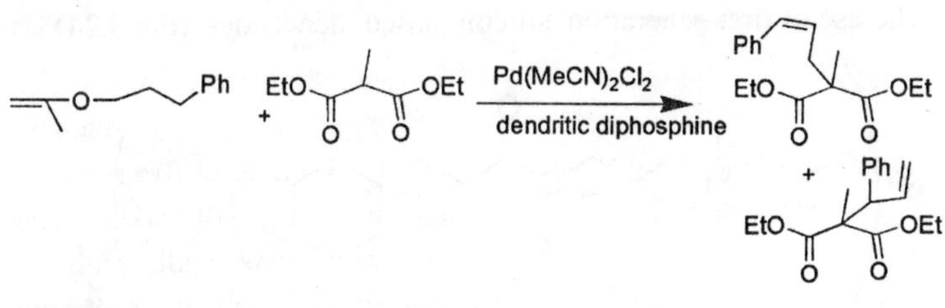

SCHEME 8.32 The reaction of allylic alkylation using diphosphine ligand based carbosilane dendrimer

The carbosilane dendrimers containing organometallics moieties such as metallocene and aryl nickel moieties have potential applications as catalysts [68-69, 198-199]. They have further advantage over homogeneous catalysts as they offer the possibility of accurate introduction of a certain number of catalytic active sites. This may result the successful combination of synergistic and cooperative effects. $^5\eta$–alkyne $Co_2(CO)_6$ containing carbosilane dendrimers, which were used as powerful catalysts in the Pauson-Khand reaction for the preparation of cyclopentanones [200].

The cyclopalladated carbosilane dendrimer as well as that macrocyclic organopalladium cage can be conveniently converted into active Lewis acid catalyst for the aldol condensation reaction [201]. The catalytic activity of the dendrimer based Pd catalyst is much higher than that of mononuclear system.

The perfluoroarylborate end capped carbosilane dendrimers were used [46] to activate stabilized metallocenes to give catalytic systems that polymerize olefins such as ethene, propene, 1-hexene, etc. with high activity. Also, a series of nickel containing carbosilane dendrimers [136, 202] were used successfully as catalyst for ATR reactions. For example, Kharasch's addition reaction [203] of CCl_4 to methyl-methacrylate. The amine terminated carbosilane dendrimer form a Ni bound dendrimeric complex after reaction with Ni-complex, Ni(C=N-^{t}Bu)$_4$(ClO_4)$_2$, which was used as catalyst for carbene polymerization reaction [203-204].

The dendrimers containing phosphines with 8, 24, and 72 PR_2 arms (R= Me, Et, hexyl, Cy or Ph) based on polyhedral oligomeric silsequioxane core [39] were used to catalyze hydroformylation reactions.

The use of first generation silicon based dendrimer (Fig.8.24) as

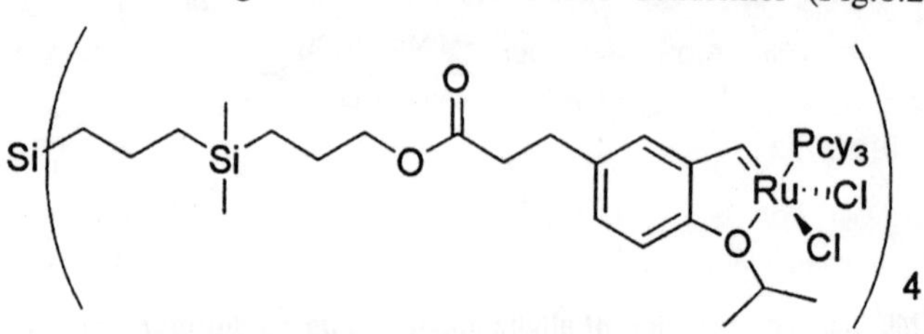

FIG.8.24 Structure of Ru-based silicon dendrimer

highly active, recoverable and recyclable Ru-based catalyst was reported [205]. This catalyst acts as homogeneous metathesis catalyst with high activity.

8.4.2 Xerogels

Carbosilane dendrimers were used as precursors for xerogels. The xerogels were obtained by polycondensation of trimethoxysilyl end capped carbosilane dendrimers upto 2nd generation by the sol-gel procedure [206]. The carbosilane dendrimers upto 4th generation were also used [58, 207] as building blocks for xerogels. By varying hydrolytic solvent and dendritic precursors, the properties of the nanoarchitecture can easily be controlled. Especially, by varying the solvent polarity with acid catalysts, the degree of Si-OH functionality can be controlled. The IR spectroscopy and nitrogen adsorption studies indicated that the total surface area and volume increased with the increase of generation number. These xerogels have potential application as catalysts, chromatographic supports and porous membranes.

The xerogels based on carbosilane dendrimers [58] were used as catalyst supports for epoxidation of cyclohexene. The xerogel X-G_{2HT} and G_{3HT} (X= xerogel, G_2 and G_3 = 2nd and 3rd generation dendrimers, HT = hot toluene i.e. xerogel isolated from hot toluene) were treated with Ti(O^{i}Pr) to yield X-G_{2Ti} and G_{3Ti}. X-G_{2HT} was also treated with the highly active epoxidation catalyst [208] Ti[OSi(^{t}Bu)$_3$]$_4$ to yield X-$G_{2Ti/Si}$. These catalytic materials were used in the epoxidation of cyclohexene, which produced cyclohexene oxide with high yield. The 2nd and 3rd generation alkoxy-terminated carbosilane dendrimers were used [209] as buildings for the synthesis of high surface area xerogels. These xerogels exhibit a high degree of microporosity may be due to the fact that the pores in the wet gel undergo considerable collapse during removal of the polar solvent like THF.

8.4.3 Medicinal

Allyl substituted first generation carbosilane dendrimer was used as hardenable substance for dental application [210]. The silicon based dendrimers are also offering a wide range of applications as new materials in medicine and pharmacy [21, 116]. Because of the presence of large number of functional groups at the periphery, these dendrimers may be the most useful candidates for dendrimer-protein as well as dendrimer-drug conjugates. The dendrimers also offer the possibilities to tailor their surface chemistry and because of their narrow dispersivity and gradation of structural density, they may be potentially used for controlled release drug delivery systems [211]. However, most of the dendrimers are generally toxic, whereas non-toxicity and non-immunogenity are must for such applications, though these could be avoided by using siloxane dendrimers.

8.4.4 Micelles

Dendrimers with amphiphilic structure are regarded as micelle analog material due to their structural characteristics. The flexibility of the carbosilane dendrimers permit to control the size of the inner cavities. The dendrimers of the inner cavities can be controlled within 5-15A^{O} by varying the branch multiplicity and/or spacer length of the dendrimers [212]. It has been found [212] that the higher generation dendrimers possess a dense outer shell with holes in the range of 2-3A^{O}. These results will give an idea about the size of the guest molecules. The surface modified silicon based dendrimers [62] with sulfonate groups were able to solubilize lipophilic alkyl-substituted benzene derivatives in aqueous solution in a micelle-like fashion. The solubilization of the guest molecules partly may be due to aggregate formation similar to ammonium terminated carbosilane dendrimers. The aryl alkylsilane dendrimers with 12 and 36 phenyl groups at the periphery were used [213-214] to form inclusion complexes with THF. The inclusion complexes were also found for carbosilane dendrimers [149] with 16 thiophene moieties.

8.4.5 Others

The polylithium derivatives of carbosilane dendrimers were used [86-87] as initiator for the anionic polymerization reaction of styrene, hexamethyl cyclo*tris*siloxane, ehtylene oxide, etc. The carbosiloxane and siloxane dendrimers are also used as the precursors of ceramic materials for the development of ternary and quaternary ceramic materials [215-217]. The carbosiloxane and siloxane dendrimers could be used [29] as modifier for HB allyl carbosilane polymers. These materials were also very effective reinforcing agents for silicon rubbers [29].

The carbosilane dendrimers without any reactive groups were used [218] for calibration materials in analytical processes and as filler in plastics. A HB polycarbosilane and partially allyl substituted derivative have proven useful as SiC matrix sources for SiC fiber and particulate reinforced composite [219]. The Si-NH groups containing carbosilane dendrimers were also used [197] as the anionic receptors in solution and immobilized into electrode surfaces. The carbosilane dendrimers with silanol end groups were used [218, 220-221] for the preparation of coating with improved scratch resistance and toughness.

The electro-optical, liquid crystalline and rheological properties of silicon based dendrimers are now focal point of interest. The most advantage of using these materials in the above fields is the fact that a globular shape might reduce the bulk viscosity and the switching time of the materials. The ferrocenyl end capped carbosilane dendrimers [222] were utilized for CO gas sensor. The sensor device was used to measure the gas sensing abilities, containing the spin coated dendritic film that responded to the various concentrations of CO gas with the linear correlation.

8.5 SUMMARY

In this chapter the syntheses, characterization and properties of a large variety of silicon based dendritic polymers are discussed. These dendritic polymers were synthesized mainly by using a small set of reactions. The basic synthetic routes towards branched Si-C, Si-O, Si-N, and Si-Si based macromolecules are well established now. These large varieties of structures of the dendritic polymers resulted different important properties for their various applications. These included catalysis, pharmaceutical, surface coating, etc. The unusual properties are obtained because of construction of unusual dendrimer topologies. The variability of the branching multiplicity represents a major advantage for the elucidation of fundamental structure-property relationship of the dendritic polymers.

REFERENCES

[1] C.Kim, S.Kang and E.Park, *J.Korean Chem.Soc.* **43**(1999)393

[2] C.Kim and H.Kim, *J.Polym.Sci. Part A: Polym.Chem.* **39**(2001)3287

[3] L.L.Zhou, P.M.Toporowski, J.Roovers and N.Hadjichristidis, *Rubber Chem. Technol.* **65**(192)303

[4] J.Roovers, P.M.Toporowski and L.L.Zhou, *Polym.Prep.* **33**(1992)182

[5] L.Ropartz, D.F.Foster, R.E.Morris, A.M.Z.Slawin and D.J.Cole-Hamilton, *J.Chem.Soc. Dalton Trans.* (2002)1997

[6] L.Ropartz, R.E.Morris, D.F.Foster and D.J.Cole-Hamilton, *Chem.Commun.* (2001)361

[7] C.Kim and S.Kong, *J.Polym.Sci. Part A: Polym.Chem.* **38**(2000)724

[8] C.Kim and M.Ryu, *J.Polym.Sci. Part A: Polym.Chem.* **38**(2000)764

[9] C.Kim and E.Park, *J.Polym.Sci. Part A: Polym.Chem.* **39**(2001)2308

[10] C.Kim, K.Jeong and I.Jung, *J.Polym.Sci. Part A: Polym.Chem.* **38**(2000)2749

[11] K.Lorenz, D.Holter, B.Stuhn, R.Mulhaupt and H.Frey, *Adv.Mater.* **8**(1996)414

[12] H.Frey, K.Lorenz and R.Mulhaupt, *Macromol. Symp.* **102**(1996)19

[13] H.Frey, R.Mulhaupt, K.Lorenz, U.Rapp and F.J.Mayer-Ponser, *Polym.Mater.Sci.Eng.* **73**(1995)127

[14] K.Lorenz, H.Frey, B.Stuhn and R.Mulhaupt, *Macromolecules* **30**(1997)6860

[15] K.Lorenz, R.Mulhaupt, U.Rapp and F.J.Mayer-Posner, *Macromolecules* **28**(1995)6657

[16] H.Lang and B.Luhmann, *Adv.Mater.* **13**(2001)1523

[17] C.Kim and A.Kwon, *Synthesis* (1998)105

[18] H.Uchida, Y.Kabe, K.Yashino, A.Kanamata, T.Tsumuray and S.Masamune, *J.Am.Chem.Soc.* **112**(1990)7070

[19] A.Morikawa, M.Kakimoto and Y.Imai, *Macromolecules* **24**(1991)3469

[20] M.Vander, W.Alexander, L.Van, W.N.M.Piet, T.C.de Wilde, C.Janine and R.A.C.Bramdes, *Adv.Mater.* **5**(1993)4666

[21] H.Frey and C.Shlenk, *Top.Cur. Chem.* **210**(2000)69

[22] J.P.Majoral and A.M.Caminade, *Chem.Rev.* **99**(1999)854

[23] L.L.Zhou and J.Roovers, *Macromolecules* **26**(1993)963

[24] Y.H.Kim, *J.Polym.Sci. Part A: Polym.Chem.* **36**(1998)1685

[25] N.Hadjichristidis, A.Guyot and L.J.Fetters, *Macromolecules* **11**(1978)668

[26] A.W.van der Made and P.W.N.M.van Leeuwen, *J.Chem.Soc. Chem.Commun.* (1992)1400

[27] A.W.van der Made, P.W.N.M.van Leeuwen, J.C.de Wilde and R.A.C.Brandes, *Adv.Mater.* **5**(1993)466

[28] A.M.Muzafarov, O.B.Gorbatsevich, E.A.Rebrov, G.M.Ignlateva, T.B.Chenskaya, V.D.Myakuschev, A.F.Bulkin and V.S.Papkov, *Polym.Sci.USSR*, **35**(1993)1575

[29] G.M.Ignateva, E.A.Rebrov, V.D.Myakusher, A.M.Muzafarov, M.N.Ilina, I.I.Duborik and V.S.Pypkov, *Polym.Sci. Ser.A* **39**(1997)1302

[30] A.M.Muzafarov, V.D.Myakushev, A.F.Bulkin and V.S.Papkov, *Polym.Sci.* **35**(1993)1575

[31] C.Kim and M.Kim, *J.Org.Chem.* **563**(1998)43

[32] C.Kim and E.Park, *J.Korean Chem.Soc.* **42**(1998)277

[33] C.Kim and J.Park, *Synthesis* (1999)1804

[34] C.Kim and K.An, *Bull.Korean Chem.Soc.* **18**(1997)165

[35] C.Kim and J.Park, *J.Orgnomet.Chem.* **547**(1997)55

[36] N.G.Vasilenko, E.A.Rebrov, A.M.Muzafarov, B.Esswein, B.Striegel and M.Moller, *Macromol. Chem.Phys.* **199**(1998)889

[37] S.Nlate, J.Ruiz, V.Sartor, R.Navarro, J.C.Blals and D.Astruc, *Chem.Eur.J.* **6**(2000)2544

[38] D.Seyferth, D.Y.Son, A.L.Rheingold and R.L.Ostrander, *Organometallics* **13**(1994)2682

[39] D.Seyferth, R.Wyrwa, U.W.Franz and S.Becke, Int. Pat. WO 97/32908(1997), CA **127**:263179

[40] D.Seyferth, T.Kugita, A.L.Rheingold and G.P.A.Yap, *Organometallics* **14**(1995)5362

[41] R.van Heerbeek, J.N.H.Reek, P.C.J.Karmer and P.W.N.M. van Leeuwen, *Tetrahedron Lett.* **40**(1999)7127

[42] R.A.Gossage, E.Munoz-Martinez, H.Frey, A.Burgath, M.Lutz, A.L.Spek and G.van Koten, *Chem.Eur.J.* **5**(1999)2191

[43] R.A.Gossage, E.Munoz-Martinez and G.van Koten, *Tetrahedron Lett.* **39**(1998)2397

[44] M.Chai, Z.Pi, C.A.Tessier and P.L.Rinaldi, *Polym.Prep.* **39**(1998)625

[45] M.Chai, Z.Pi, C.A.Tessier and P.L.Rinaldi, *J.Am.Chem.Soc.* **121**(1999)273

[46] M.Mager, S.Becke, H.Windisch and U.Denniger, *Angew.Chem.Int.Ed.* **40**(2001)1898

[47] M.A.Casado, J.Roovers and S.R.Stobart, *Chem.Commun.* (2001)313

[48] J.J.L.M.Cornelissen, R.van Heerbeek, P.C.J.Kamer, J.N.H.Reek, N.A.J.M.Sommerdijk and R.J.M.Nolte, *Adv.Mater.* **14**(2002)489

[49] C.Kim and I.Jung, *J.Organomet.Chem.* **599**(2000)208

[50] N.J.Hovestad, A.Ford, J.T.B.H..Jastrzebski and G.van Koten, *J.Org.Chem.* **65**(2000)6338

[51] B.A.Omotowa, K.D.Keefer, R.I.Kirchmeier and J.M.Shreeve, *J.Am.Chem.Soc.* **121**(1999)11130

[52] A.Morikawa, M.A.Kakimoto and Y.Imai, *Macromolecules* **25**(1992)3247

[53] B.Alonso, I.Caudrado, M.Moran and J.Losado, *J.Chem.Soc.Chem.Commun.* (1994)2575

[54] B.Alonso, M.Moran, C.M.Casado, F.Lobete, J.Losado and I.Cuodrado, *Chem.Mater.* **7**(1995)1440

[55] I.Caudrado, C.M.Casado, B.Alonso, M.Moran, J.Losado and V.Belsky, *J.Am. Chem.Soc.* **119**(1997)7613

[56] J.Losado, I.Cuodrado, M.Moran, C.M.Casado, B.Alonso, and M.Barranco, *Anal.Chem.Acta* **338**(1997)191

[57] G.E.Oosteron, R.J.van Haaren, J.N.H.Reek, P.C.J.Kamer and P.W.N.M.Leeuwen, *Chem.Commun.* (1999)1119

[58] J.W.Kriesel and T.D.Tilley, *Adv.Mater.* **13**(2001)1645

[59] E.B.Eggeling, N.J.Hovestad, J.T.B.H.Jastrzebski, D.Vogt and G.van Koten, *J.Org.Chem.* **65**(2000)8857

[60] E.V.Getmanova, T.B.Chenskaya, O.B.Gorbatsevich, E.A.Rebrov, N.G.Vasilenko and A.M.Muzafarov, *React.Fun.Polym.* **33**(1997)289

[61] E.V.Getmanova, E.A.Rebrov, N.G.Vasilenko and A.M.Muzafarov, *Polym.Prep.* **39**(1998)581

[62] S.W.Krska and D.Seyferth, *J.Am.Chem.Soc.* **120**(1998)3604

[63] B.Comanita and Roovers, *Des Monomers Polym.* **2**(1999)111

[64] P.I.Coupar, P.A.Jaffers and R.E.Morris, *J.Chem.Soc.Dalton Trans.* (1999)2183

[65] J.Roover, *Macromol.Symp.* **121**(1997)89

[66] D.K.Polyakov, G.M.Ignateva, E.A.Rebrov, N.G.Vasilenko, S.S.Sheiko, M.Moller and A.M.Muzafarov, *Polym.Sci.Ser.A* **40**(1998)876

[67] M.Benito, O.Rossell, M.Seco and G.Segales, *Inorg.Chim.Acta* **291**(1999)247

[68] J.W.J.Knaper, A.W.van der Made, J.C.Wilde, P.W.N.M. van Leeuwen, P.Wijkens, D.M.Grove and G. van Koten, *Nature* **372**(1994)659

[69] G.van Koten and D.M.Grove, *Polym.Mater.Sci.Eng.* **73**(1995)228

[70] R.A.Gossage, L.A.van de Kuil and G.van Koten, *Acc.Chem.Res.* **31**(1998)423

[71] P.Dani, T.Karlen, R.A.Gossage, W.J.J.Smeets, A.L.Spek and G.van Koten, *J.Am.Chem.Soc.* **119**(1997)113117

[72] A.W.Kleij, H.Klein, J.T.B.H.Jastrzebski, W.J.J.Smeets, A.L.Spek and G.van Koten, *Organometallics* **18**(1999)268

[73] J.L.Hoare, K.Lorenz, N.J.Hovestad, W.J.J.Smeets, A.L.Spek, A.J.Canty, H.Frey and G.van Koten, *Organometallics* **16**(1997)4167

[74] J.W.Kriesel, S.Konig, M.A.Freitas, A.G.Marshall, J.A.Leary and T.D.Tilley, *J.Am.Chem.Soc.* **120**(1998)12207

[75] P.Jutzi, C.Batz, B.Neumann and H.G.Stammder, *Angew.Chem.Int.Ed.* **35**(1996)2118

[76] N.Boiko, X.Zhu, A.Bobrosky and V.P.Shibaev, *Chem. Mater.* **13**(2001)1447

[77] H.Frey, K.Lorenz, D.Holter and R.Mulhaupt, *Polym.Prep.* **37**(1996)758

[78] K.Lorenz, D.Holter, H.Frey and B.Stuhn, *Polym.Mater.Sci.Eng.* **77**(1997)168

[79] S.A.Ponomarenko, E.A.Rebrov, N.I.Boiko, A.M.Muzafarov and V.P.Shibaev, *Liq.Cryst.* **21**(1996)1

[80] S.A.Ponomarenko, E.A.Rebrov, N.I.Boiko, A.M.Muzafarov and V.P.Shibaev, *Polym.Sci.Ser.A* **40**(1998)763

[81] R.M.Richardson, S.A.Ponomarenko, N.I.Boiko and V.P.Shibaev, *Liq.Cryst.* **26**(1999)101

[82] D.Terunuma, T.Kato, R.Nishio, K.Matsuoka, H.Kuzuhara, Y.Aoki and H.Nohira, *Chem.Lett.* (1998)59

[83] D.Terunuma, R.Nishio, Y.Aoki, H.Nohira, K.Matsuoka and H.Kuzuhara, *Chem.Lett.* (1999)565

[84] B.Trahasch, B.Stuhn, H.Frey and K.Lorenz, *Macromolecules* **32**(1999)1962

[85] A.Y.Bobrovsky, A.A.Pakhomov, X.M.Zhu, N.I.Boiko, V.P.Shibaev and J.Stumpe, *J.Phys.Chem.B* **106**(2002)540

[86] N.G.Vasilenko, E.V.Getmanova, V.D.Myakushev, E.A.Rebrov, M.Moller and A.M.Muzafarov, *Polym.Sci.Ser.A* **40**(1998)876

[87] N.G.Vasilenko, E.A.Rebrov, A.M.Muzafarov, S.S.Sheiko and M.Moller, *Polym.Prep.* **39**(1998)479

[88] K.Matsuoka, H.Kurosawa, Y.Esumi, D.Terunuma and H.Kuzuhara, *Carbohydr.Res.* **329**(2000)765

[89] K.Matsuoka, H.Oka, T.Koyama, Y.Esumi and D.Terunuma, *Tetrahedron Lett.* **42**(2001)3327

[90] M.A.Casado and R.S.Stobart, *Org.Lett.* **2**(2000)1549

[91] I.Caudrado, M.Moran, A.Moya, C.M.Casado, M.Barranco and B.Alonso, *Inorganica Chimica Acta,* **251**(1996)5

[92] H.Schumann, B.C.Wassermann, M.Frackwiaka, B.Omotowa, S.Schutte, J.Veldera, S.H.Muhlea and W.Krause, *J.Organomet.Chem.* **609**(2000)189

[93] C.Kim, D.D.Bung, D.Chung, E.Park and E.Kang, *J.Korean Chem.Soc.* **39**(1995)789

[94] C.Kim, E.Park and E.Kang, *J.Korean Chem.Soc.* **39**(1995)799

[95] C.Kim, E.Park and E.Kang, *Bull.Korean Chem.Soc.* **17**(1996)419

[96] C.Kim, E.Park and E.Kang, *Bull.Korean Chem.Soc.* **17**(1996)592

[97] C.Kim and K.An, *J.Organomet. Chem.* **547**(1997)55

[98] C.Kim and S.K.Choi, *Main Group Met.Chem.* **20**(1997)143

[99] C.Kim and Y.Jeong, *Bull.Korean Chem.Soc.* **18**(1997)347

[100] C.Kim and A.Kwon, *Main Group Met.Chem.* **21**(1998)9

[101] M.M.K.Boysen and T.K.Lindhorst, *Org.Lett.* **1**(1999)1925

[102] T.Matsno, K.Uchida and A.Sekiguchi, *Chem.Commun* (1999)1799

[103] R.Buschbeck, K.Bruning and H.Lang, *Synthesis* **15**(2001)2289

[104] C.Kim and J.Park, *J.Organomet.Chem.* **629**(2001)194

[105] C.Kim and H.Kim, *J.Polym.Sci. Part A: Polym.Chem.* **40**(2002)326

[106] C.Kim and H.Kim, *Bull.Korean Chem. Soc.* **23**(2002)637

[107] C.Kim and K.Kwark, *J.Polym.Sci. Part A: Polym.Chem.* **40**(2002)976

[108] C.M.Casado, I.Curadrado, M.Moran, B.Alonso, M.Barrranso and J.Losada, *Appl.Organomet.Chem.* **13**(1999)245

[109] K.Bruening and H.Lang, *J.Orgnomet.Chem.* **592**(1999)147

[110] D.Seyferth, D.Y.Son, A.L.Rheingold and R.L.Ostrander, *Organometallics* **13**(1994)2682

[111] K.Bruening, Ph.D.Thesis, Techische Universitat Chemnitz, 2000

[112] K.Bruening, B.Luehmann and H.Lang, *Z.Naturforsch* **54b**(1999)751

[113] B.Luehmann, Diplomarbeit, Techische Universitat Chemnitz, 1998

[114] B.Luehmann, H.Lang and K.Bruening, *Phosphorous Sulfur Silicon Relat.Elem.* **169**(2001)157

[115] R.Bischoff and S.E.Cray, *Prog.Polym.Sci.* **24**(1999)185

[116] E.A.Rebrov, A.M.Muzafarov, V.S.Pabkov and A.A.Zhdanov, *Dokl.Alkad.Nauk.SSSR* **309**(1989)376

[117] A.M.Muafarov, E.A.Rebrov and V.S.Pabkov, *Usp.Khim,* **60**(1991)1596

[118] H.Uchida, Y.Kabe, K.Yoshino, A.Kawamata, T.Tsumuraya and S.Masamune, *J.Am.Chem.Soc.* **112**(1990)7077

[119] H.Uchida, Y.Kabe and K.Yoshino, JP 03263431; CA **116**(1991)236379

[120] H.Uchida, K.Yoshino and Y.Kabe, JP 03263430; CA **116**(1991)130381

[121] S.A.Ponomarenko, E.A.Rebrov, N.I.Boiko, N.G.Vasilenko, A.M.Muzafarov, Y.S.Freidzon and V.P.Shibaev, *Polym.Sci.Ser.A* **36**(1994)896

[122] A.Morikawa, M.A.Kakimoto and Y.Imai, *Polym.J.* **24**(1992)573

[123] J.Hu and D.Y.Son, *Polym.Prep.* **39**(1998)410

[124] J.Hu and D.Y.Son, *Macromolecules* **31**(1998)8644

[125] R.Elsasser, G.H.Mehl, J.W.Goodby and M.Veith, *Angew.Chem.Int.Ed.* **40**(2001)2688

[126] M.Veith, R.Elsasser, R.P.Kruger, *Orgnometallics* **18**(1999)656

[127] B.D.Karstedt, US Pat. 3,814,730 (1994)

[128] T.Kemmitt and W.Henderson, *Chem.Soc.Perkin Trans. 1* (1997)729

[129] X.Zhang, K.J.Haxton, L.Ropartz, D.J.Cole-Hamilton and R.E.Morris, *J.Chem.Soc.Dalton Trans.* (2001)3261

[130] R.Knischka, F.Dietsche, R.Hanselmann, H.Frey, R.Mulhaupt and P.J.Lutz, *Langmuir* **15**(1999)4752

[131] I.M.Saez, J.W.Goodby and R.M.Richardson, *Chem.Eur.J.* **7**(2001)2758

[132] I.M.Saez and J.W.Goodby, *J.Mater.Chem.* **11**(2001)2845

[133] R.Lewthwaite, J.W.Goodby and K.J.Toyne, *J.Mater.Chem.* **3**(1993)241

[134] M.G.Voronkov and V.L.Lavent'yev, *Top.Cur.Chem.* **102**(1982)1199

[135] A.W.Kleij, R.A.Gossage, R.J.M.Gebbink, N.Brinkmann, Ed.J.Reijerse, U.Kragl, M.Lutz, A.L.Spek and G.van Koten, *J.Am.Chem.Soc.* **122**(2000)12112

[136] L.Ropartz, R.E.Morris, G.P.Schwarz, D.F.Foster and D.J.Cole-Hamilton, *Inorg.Chem.Commun.* **3**(2000)714

[137] P.A.Jaffers and R.E.Mooris, *J.Chem.Soc. Dalton Trans.* (1998)2767

[138] A.Mutluay and P.Jutzi, in 'Organosilicon Chemistry', edited by N.Auner, J.Weiss, Vol.4, Wiley-VCH, Weinheim, p.531, 2000

[139] L.Ropartz, R.E.Morris, D.F.Foster and D.J.Cole-Hamilton, *J.Molecular.Cat.A:Chem.* **182-183**(2002)99

[140] A.R.Bassindalea and T.Gentke, *J.Organomet.Chem.* **521**(1996)391

[141] J.B.Lambert, J.L.Pflug and C.L.Stern, *Angew.Chem.Int.Ed.* **34**(1995)98

[142] H.Suzuki, Y.Kimata, S.Satoh and A.Kuriyama, *Chem.Lett.*(1995)293

[143] A.Sekiguchi, M.Nanjo, C.Kabuto and H.Sakurai, *J.Am.Chem.Soc.* **117**(1995)4195

[144] J.B.Lambert, J.L.Pflug and J.M.Denari, *Organometallics* **15**(1996)615

[145] J.B.Lambert, E.Basso, N.Qing, S.H.Lim and J.L.Pflug, *J.Organomet.Chem.* **554**(1998)113

[146] U.Herzog, C.Notheis, E.Brendler, G.Roewer and B.Thomas, *J.Anal.Chem. (Fr.)* **357**(1997)503

[147] J.B.Lambert and H.Wu, *Organometallics* **17**(1998)4904

[148] M.Nanjo and A.Sekiguchi, *Organometallics* **17**(1998)492

[149] J.Nakayama and J.S.Lin, *Tetrahedron Lett.* **38**(1997)6043

[150] V.G.Krasovskii, G.M.Ignateva, V.D.Mykushev, N.A.Sadovskii, T.V.Strelkova and A.M.Muzafarov, *Polym.Sci.Ser.A* **38**(1996)1070

[151] A.G.Vitukhnovsky, M.I.Sluch, V.G.Krasovskii and A.M.Muzafarov, *Synthetic Metals* **91**(1997)375

[152] M.I.Sluch, I.G.Scheblykin, O.P.Varnavsky, A.G.Vitukhnovsky, V.G.Krasovskii, O.B.Gorbatsevich and A.M.Muzafarov, *J.Lumin.* **76**/77(1998)246

[153] V.G.Krasovskii, N.A.Sadovskii, O.B.Gorbatsevich, A.M.Muzafarov, V.D.Myakushev, M.N.Ilina, I.I.Dubovik, T.V.Strelkova and V.S.Pabkov, *Polym.Sci.* **36**(1994)589

[154] G.M.Ignateva, E.A.Rebrov, V.D.Myakushev, T.B.Chenskaya and A.M.Muzafarov, *Polym.Sci.Ser.A* **39**(1997)843

[155] S.S.Sheiko, G.Eckert, G.Ignateva, A.M.Muzafarov, J.Spickermann, H.J.Radar and M.Moller, *Macromol.Rapid Commun.* **17**(1996)283

[156] K.Bruening and H.Lang, *J.Organomet.Chem.* **571**(1998)145

[157] C.W.Whitmarsh and L.V.Interrnte, *Organometallics* **10**(1991)1336

[158] L.V.Interrate, I.L.Rushkin and R.Shen, *Appl.Organomet.Chem.* **12**(1998)695

[159] I.L Rushkin, Q.Shen, S.E.Lehman and L.V.Interrate, *Macromolecules* **30**(1997)3141

[160] J.Yao and D.Y.Son, *J.Polym.Sci. Polym.Ed.* **37**(1999)3778

[161] J.Yao and D.Y.Son, *Organometallics* **18**(1999)1736

[162] C.Lach, P.Muller, H.Frey and R.Mulhaupt, *Macromol.Rapid Commun.* **19**(1997)253

[163] D.Holter, A.Burgater and H.Frey, *Acta Polym.* **48**(1997)30

[164] C.Lach, H.Frey and R.Mulhaupt, *Polym.Mater.Sci.Eng.* **77**(1997)199

[165] C.Lach, R.Hanselmann, H.Frey and R.Mulhaupt, *Macromol.Rapid Commun.* **19**(1997)461

[166] E.V.Getmanova, E.A.Rebrov, N.G.Vasilenko and A.M.Muzafarov, *Polym.Prep.* **39**(1998)581

[167] D.Y.Son and K.Yoon, *Polym.Mater.Sci.Eng.* **80**(1999)200

[168] K.Yoon and D.Y.Son, *Macromolecules* **32**(1999)5210

[169] G.Kwak and T.Masuda, *Macromol. Rapid Commun.* **23**(2002)68

[170] Y.Xio, R.A.Wong and D.Y.Son, *Macromolecules* **33**(2000)7232

[171] Y.Xio and D.Y.Son, *J.Polym.Sci. Part A: Polym.Chem.***39**(2001)3383

[172] M.C.Coen, K.Lorenz, J.Kressler, H.Frey and R.Mulhaupt, *Macromolecules* **29**(1996)8069

[173] C.Drohmann, M.Moller, O.B.Gorbatsevich and A.M.Muzafarov, *J.Polym.Sci. Part A: Polym.Chem.* **38**(2000)741

[174] L.J.Mathias and T.W.Carothers, *Polym.Prep.* **32**(1991)633

[175] T.W.Carothers and L.J.Mathias, *Polym.Prep.* **34**(1993)503

[176] S.Rubinsztajn, *J.Inorg.Organomet.Polym.* **4**(1997)61

[177] C.Herzig and B.Deuber, *Polym.Prep.* **39**(1998)477

[178] J.F.Miravet and J.M.J.Frechet, *Polym.Mater.Sci.Eng.* **77**(1997)141

[179] J.F.Miravet and J.M.J.Frechet, *Macromolecules* **31**(1998)3461

[180] C.Gong, J.F.Miravet and J.M.J.Frechet, *Polym.Mater.Sci.Eng.* **80**(1999)139

[181] C.Gong, J.F.Miravet and J.M.J.Frechet, *J.Polym.Sci.Polym.Chem.* **37**(1999)3193

[182] N.G.Vasilenko, E.A.Rebrov, V.D.Myakushev, A.M.Muzafarov, S.E.Cray, T.Okawa and R.Mikami, *Polym.Prep.***39**(1998)603

[183] S.Rubinsztajn and J.Stein, *J.Inorg.Organomet.Chem.* **5**(1995)43

[184] S.Rubinsztajn, *J.Inorgmetal Polym.* **5**(1995)43

[185] J.K.Paulasaari and W.P.Weber, *Macromolecules* **33**(2000)2005

[186] C.Schlenk, A.W.Kleij, H.Frey and G.van Koten, *Angew Chem.* **112**(2000)3587

[187] A.M.Muzafarov, M.Golly and M.Moller, *Macromolecules* **28**(1995)8444

[188] V.V.Kazakova, V.D.Myakushev, T.V.Strelkova, N.G.Gvazava and A.M.Muzafarov, *Dokl.Chem.* **349**(1996)190

[189] V.V.Kazakova, V.D.Myakushev, T.V.Strelkova and A.M.Muzafarov, *Polym.Prep.* **39**(1998)483

[190] L.J.Mathias and T.W.Carothers, *J.Am.Chem.Soc.* **113**(1991)4043

[191] L.J.Mathias, T.W.Carothers and R.M.Bozan, *Polym.Prep.* **32**(1991)82

[192] G.van Koten and J.T.B.H.Jastrzebski, *J.Molecular Catalysis A: Chiral* **146**(1999)317

[193] N.J.Hovestad, E.B.Eggeling, H.J.Heidbuchel, J.T.B.H.Jastrzebski, U.Kragl, W.Keim, D.Vogt and G.van Koten, *Angew.Chem.Int.Ed.* **38**(1999)1655

[194] D.de Groot, E.B.Eggeling, J.C.Wilde, H.Kooijman, R.J.van Haaren, A.M.van der Made, A.L.Spek, D.Vogt, J.N.H.Reek, P.C.J.Kamer and P.W.N.M.van Leenwen, *Chem.Commun.* (1999)1623

[195] D.de Groot, J.N.H.Reek, P.C.J.Kamer and P.W.N.M.van Leenwen, *Eur.J.Org.Chem.* **2002**(2002)1085

[196] N.J.Hovestad, J.T.B.H.Jastrzebski and G.van Koten, *Polym.Mater.Sci.Eng.* **80**(1999)53

[197] C.M.Casado, I.Cuadrado, B.Alonso, M.Moran and J.Losada, *Electroanal.Chem.* **463**(1999)87

[198] D.Seyferth and R.Wyrwa, US Pat. 96-611495960305 (1999)

[199] D.Seyferth, R.Wyrwa, W.W.Franz and S.Beeke, US Pat. 96-611482960305 (1999)

[200] I.U.Khand, G.R.Knox, P.L.Dauson, W.E.Watts and M.J.Forman, *J.Chem.Soc.Perkin Trans. 1* (1973)977

[201] G.Rodriguez, M.Lutz, A.L.Spek and G.van Koten, *Chem.Eur.J.* **8**(2002)45

[202] A.W.Kleij, R.A.Gossage, J.T.B.H.Jastrzebski, J.Boersma and G.van Koten, *Angew.Chem.Int.Ed.* **39**(2000)176

[203] P.C.J.Kamer, R.J.M.Nolte and W.Drenth, *J.Chem.Soc.Chem.Commun.* (1986)1789

[204] P.C.J.Kamer, R.J.M.Nolte and W.Drenth, *J.Am. Chem.Soc.* **110**(1988)6818

[205] S.B.Garber, J.S.Kingsbury, B.L.Gray and A.H.Hoveyda, *J.Am.Chem.Soc.* **122**(2000)8168

[206] B.Boury, R.J.P.Corriu and R.Nunez, *Chem.Mater.* **10**(1998)1795

[207] J.W.Kriesel and T.D.Tilley, *Chem.Mater.* **11**(1999)1190

[208] M.C. Coles, C.G.Lugmair, K.W.Terry and T.D.Tilley, *Chem.Mater.* **11**(1999)122

[209] J.W.Kriesel and T.D.Tilley, *Chem.Mater.* **12**(2000)147

[210] J.Zech and G.Lechner, Int.Pat. WO 99/09934(1999)

[211] N.Malik, R.Wiwattanapatapee, R.Klopsch, K.Lorenz, H.Frey, J.W.Weener, E.W.Meijer, W.Paulus and R.Duncan, *J.Controlled Release* **65**(2000)133

[212] C.Lach, D.Brizzolara and H.Frey, *Macromol.Theory Simul.* **6**(1997)371

[213] G.Friedmann, Y.Guilbert and J.C.Wittmann, *Eur.Polym.J.* **33**(1997)419

[214] G.Friedmann, Y.Guilbert and J.C.Wittmann, *Eur.Polym.J.* **35**(1999)1097

[215] R.R.Wills, R.A.Markle and S.P.Mukherjee, *Ceran.Bull.* (1983)904

[216] M.Peuckert, T.Vaahs and M.Bruck, *Adv.Mater.* **2**(1990)398

[217] F.Rossignol, Y.Nakata, H.Nagai, T.Okutani, M.Suzuki, N.Kushibiki and M.Murakami, *Chem.Mater.* **11**(1999)358

[218] M.Mager, J.D.Jentsch and C.Schild, Eur.Pat. EP 74,33,13 (1996); CA **126:** 48431

[219] L.V.Interrante, C.K.Whitmarsh and W.Sherwood, *Ceram Trans.* **58**(1995)111

[220] H.Kraus and M.Mechtel, Eur.Pat. EP 866086(1998); CA **129**: 67864

[221] H.M.Meier, M.Mager, K.Dierksen and K.Reuter, Eur.Pat. EP 796882(1997); CA **127**: 264320

[222] C.Kim, E.Park, C.K.Song and B.W.Koob, *Synthetic Metals* **123**(2001)493

CHAPTER 9

PHOSPHOROUS CONTAINING DENDRITIC POLYMERS

9.1 INTRODUCTION

The phosphorous containing dendritic polymers are another important class of heteroatom containing dendritic polymer [1]. They are, in general, highly thermostable, exhibit high solubility in different solvents and also having large functional derivatives with high dipole moments. There are mainly four different types of phosphorous containing dendritic polymers are available depending on the location of the phosphorous atom in their chemical structures. These are: (i) exclusively at the central core, (ii) on the surface, (iii) at the core and at each generation, and (iv) at the core, at each generation and on the surface of the structure. The first report on the synthesis of a novel type dendritic polymer in which the initiator core as well as branching points are quaternary phosphonium ions was published [2] in 1990. Before this report, there was no report on the synthesis of any dendrimer with charged structure. However, in 1994 only, the first report was published on the neutral phosphorous containing dendrimers possessing P-N and P-O linkages [3] or P-C linkages [4]. After that time, the synthesis, characterization and properties of this novel dendritic polymer have been blossoming continuously. There are number of reviews [5-6] and articles [7-10] on the different aspects of phosphorous containing dendritic polymers are available in the literature.

9.2 SYNTHESIS, CHARACTERIZATION AND PROPERTIES OF PHOSPHOROUS CONTAINING DENDRIMERS

In this section the synthesis, characterization and properties of various phosphorous containing dendrimers are discussed.

9.2.1 Synthesis

The synthesis and characterization of phosphorous containing dendrimers of new category, in which the initiator core and subsequent branching points were quaternary phosphonium ion sites were reported for the first time [2]. The initiator core and repeating units were derived from *tris*(*p*-methoxymethyl phenyl) phosphine. The whole sequence of the reactions as route A is shown in scheme 9.1.

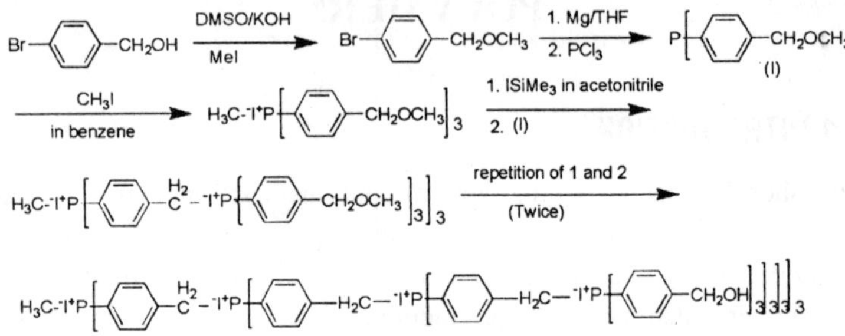

SCHEME 9.1 Synthesis of quaternary phosphonium ion containing dendrimers by route A

All the steps of the reactions were carried out under the dry nitrogen atmosphere. The product in each step was purified by passage through a Dowex 2-X8 column in the iodide form by eluting with 30% aqueous ethanol. The yield of the product in each step was moderately good (22-64%), though, the yield of the 3rd generation dendrimer was very poor (5% only). The first two generations were readily soluble in common solvents like, water, chloroform, acetonitrile, ethyl acetate, etc. but the 3rd generation was only soluble in acetonitrile. All the products were characterized by ^{1}HNMR and ^{31}PNMR spectroscopy and by elemental analysis. The synthesis of phosphorous containing dendrimers upto 40 phosphorous linkages at the core and at branching points with tri- or tetra-directional core units was also reported [11]. The core unit was synthesized by quaternization of *tris*(*p*-methoxymethyl)phenyl phosphane with an alkyl halide or with *p*-methoxymethyl bromobenzene in the presence of nickel bromide to form the core, $[P(PhCH_2OMe)_4]^+$. Then, the synthesis of tri- or tetra-directional dendrimers was carried out by using above core via selective cleavage of the benzylic ether linkages with iodotrimethylsilane, followed by treatment of the resulting benzylic iodide with $[RP(PhCH_2OMe)_3]^+$, offered the first generation with phosphonium salts at the periphery. The repetition of the above two steps resulted the higher generations. The whole reactions sequence for the tetra directional core as route B is shown in scheme 9.2. All the products were purified by passage through an anion exchange column and characterized by ^{1}HNMR, ^{31}PNMR, IR spectroscopic studies and elemental analysis. The solubility behavior of the dendrimer is similar to the earlier case of tridirectional phosphorous dendrimer. Although, all dendrimers are significantly hygroscopic in nature and water solubility decreased with the increase of the generation number.

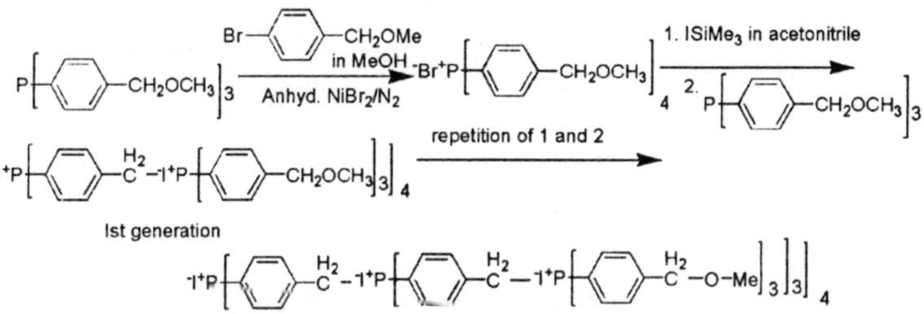

SCHEME 9.2 Synthesis of quaternary phosphonium ion containing dendrimers by route B

As phosphorous exhibits various stable valences, so other type dendrimeric structures were also synthesized [12] with phosphine oxide and phosphine cores with phosphonium ion branch points. The dendrimers with other than phosphonium ion type was synthesized [7-9] from the same fundamental building block, *tris*(*p*-methoxymethyl phenyl)phosphine. It's oxidation leads easily to the corresponding phosphine oxide upon which same reactions sequence as shown in scheme 9.3, offered phosphine/phosphonium ion containing dendrimers. These dendrimers exhibit normal phosphine chemistry at the core site including complexation with metal ions. The whole sequence of reactions as route C is shown in scheme 9.3. A neutral pentaarylphosphorane was also synthesized [7, 9] from $P^+[C_6H_6CH_2OMe]_4$ by treatment with 4-(methoxymethyl)phenyl lithium. This core can also be elaborated to higher generations of phosphorane/phosphonium ion dendrimer by using the same sequence of reactions as route D (Scheme 9.4). This first generation dendrimer is unique in the sense, that it bears a pentadirectional monoatomic core unit, albeit an electrically neutral species.

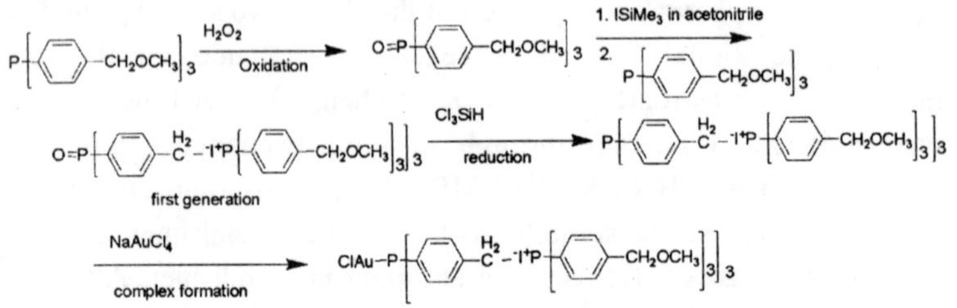

SCHEME 9.3 Synthesis of quaternary phosphonium ion containing dendrimers by route C

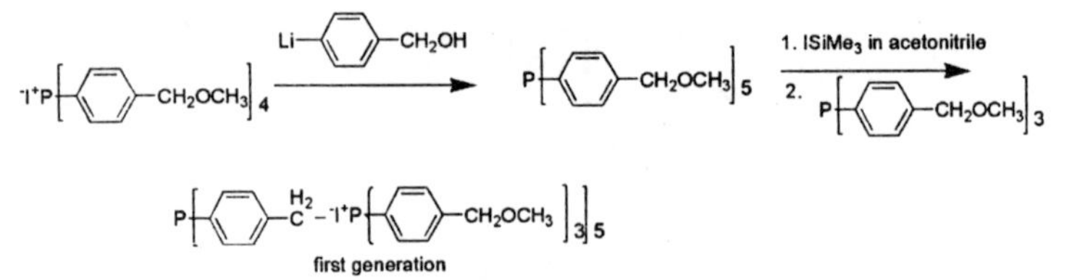

SCHEME 9.4 Synthesis of quaternary phosphonium ion containing dendrimers by route D

The first synthesis of organophosphane dendrimers upto 2nd generation, where phosphine unit was present at each branching point was reported by DuBois and his coworkers [4, 10]. The method involved two sequence of reactions, first was the radical addition of diethylvinyl phosphonate to primary phosphanes followed by reduction of the phosphonates with lithium aluminium hydride to get primary phosphins. The sequence of reactions is shown in scheme 9.5. The metalation of the 2nd generation dendrimer with $[Pd(MeCN)_4](BF_4)_2$ offers complex (A), which exhibits catalytic activity for electrochemical reduction of CO_2.

The synthesis of neutral phosphorous containing dendrimers by using hydrazono derivatives has been intensively studied [1, 6]. Reiteration sequence of three reactions was utilized to get dendrimers upto 3rd generation [13]. Hexapodant N_3P_3-$(OC_6H_4CHO)_6$ was used as the core . The first step was a Schiff's reaction between the core and methylhydrazine followed by treatment of the resulting compound with chlorodiphenyl phosphane or chloroazaphosphozane in the presence of

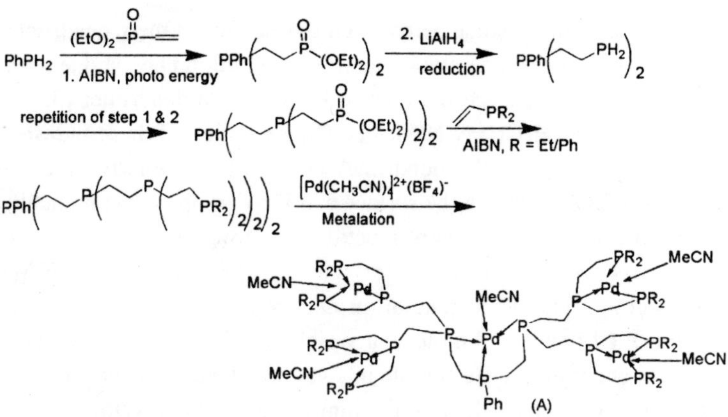

SCHEME 9.5 Synthesis of organo-phosphane dendrimers

a base. The first generation dendrimer was obtained upon the treatment of the resulted compound of the 2nd step with $N_3P(S)\text{-}(OC_6H_4CHO)_2$. The whole sequence of reactions is shown in scheme 9.6. The same reaction can be utilized for other cores such as $X(P)(OC_6H_4CHO)_2$, where X = Cl or S. Although, this synthetic strategy allows introduction of N-P fragment and alternatively three different reactive groups viz. CHO, hydrozone and hydrozonophosphane on the surface, but it suffers from two major draw backs. First, it necessitates the repetition of sequence of three reactions and second, the lack of solubility of the dendrimers prevents construction of higher generation. So, a simple and general strategy for the synthesis of phosphorous containing dendrimers were developed by Majoral's group in 1994 and that was the first method for synthesis of neutral phosphorous dendrimers [3]. It consists of two steps to create one generation. The first step is the reaction of hydroxybenzaldehyde in basic conditions with compounds having P-Cl functions, such as $S{=}PCl_3$. The second step is a condensation reaction of the aldehyde groups with $H_2NN(Me)P(S)Cl_2$. The uniqueness of the strategy are: (i) both the steps are quantitative and (ii) sodium chloride and water are the by-products. Remarkably, no protection or deprotection procedure is also necessary. Thus, it appears that this is the most efficient way to prepare neutral phosphorous containing dendrimers upto 12th generation with 12288 functional groups. This is the highest generation reported till now in dendrimer chemistry, even though it is obvious that there are some defects in the structures of these dendrimers [6, 14]. This highest generation dendrimer formation may be due to the fact that the branches are the longest known for a dendrimer [3, 15], 10°A for P-O-C_6H_6CH=N-N(Me)-P. The repetition of these two steps induce the growth of the dendrimer possessing alternatively either P(S)Cl or CHO groups on the surface and P(S) groups at the branching points. The whole sequence of reactions is shown in scheme 9.7. All dendrimers were well-characterized by ^{1}HNMR and ^{31}PNMR spectroscopy. The 12th generation dendrimer is possessing 12000 terminal P-Cl bonds and molecular weight $3X10^{6}$ Da. The dendrimers upto 11th generation were soluble in a variety of solvents such as THF, $CHCl_3$, CH_2Cl_2, etc. although, the highest generation (12th) has poor solubility and thus, prevented further construction of higher generation.

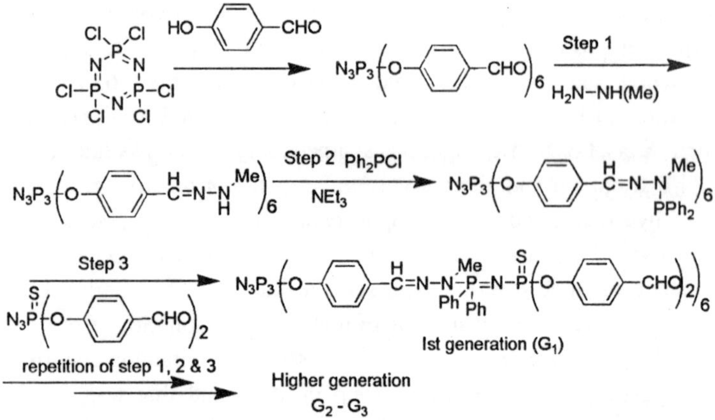

SCHEME 9.6 Synthesis of phosphorous containing dendrimers from hydrazone derivative

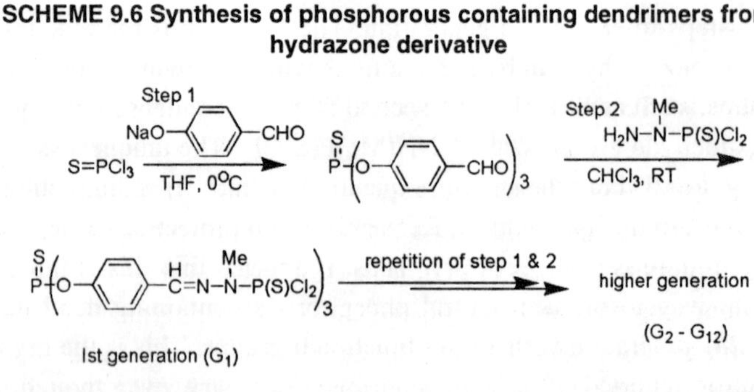

SCHEME 9.7 Synthesis of first neutral phosphorous containing dendrimers

A variety of other cores was also used as the starting materials. The above strategy of synthesis was used to get phosphoryl groups either at each layer or at particular layers by using $H_2NN(Me)P(O)Cl_2$ in place of $H_2NN(Me)P(S)Cl_2$. Some examples are sited here (Fig.9.1) in which the phosphoryl layers are within all the structures (upto 3rd generation) [16], or only on the surface (upto 7th generation) [17] or even two layers (upto 4th generation) [16]. The 1st generation of the first type dendrimers [16] having high reactivity on the surface groups (-Cl and –CHO) and as a result they can be grafted with α, β-unsaturated ester, crown ether, chiral phosphine sulfides or diallyl amine with quantitative yield. The last type dendrimer was the first 'layer block' dendrimer builts with a regular alternation of two types of repeating units upto the 4th generation. The crystal structure of the 1st generation was determined by XRD studies and all dendrimers were characterized by ^{31}PNMR spectroscopy [18]. By using different cores, but similar sequence of reactions, dendrimers upto 8th generation with 1536 aldehyde chain ends were synthesized [19] (Scheme 9.8). All the above dendrimers were characterized by ^{1}HNMR, ^{31}PNMR, IR spectroscopic studies and elemental analysis. Mass spectroscopy (FAB) was also used in few cases.

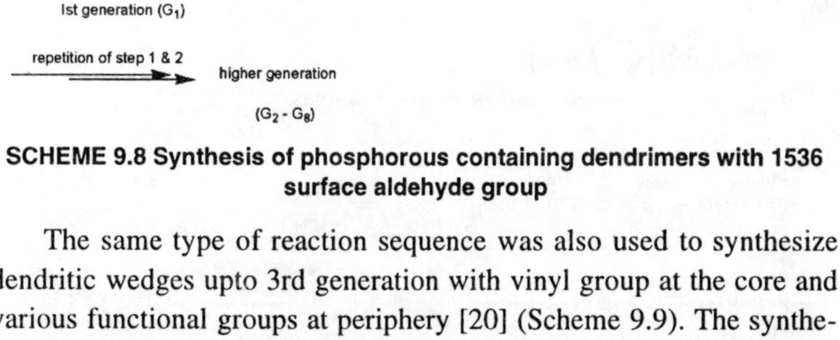

SCHEME 9.8 Synthesis of phosphorous containing dendrimers with 1536 surface aldehyde group

The same type of reaction sequence was also used to synthesize dendritic wedges upto 3rd generation with vinyl group at the core and various functional groups at periphery [20] (Scheme 9.9). The synthesized dendrons were utilized to get desired dendrimers after attaching with suitable core units. Thus, by this methods, P=N-P=S linkages can be created, which possess interesting properties. The vinyl group at the

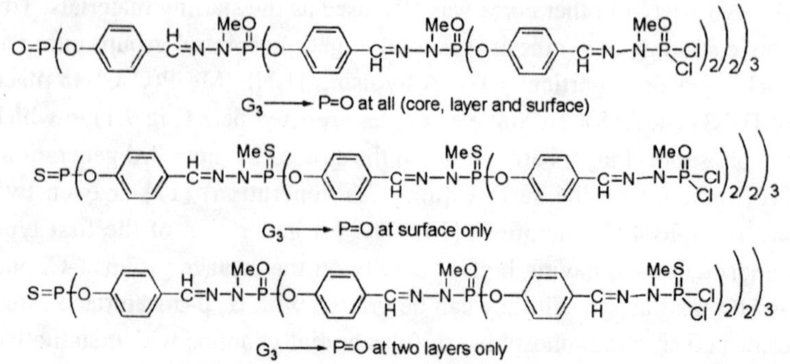

FIG. 9.1 Structures of phosphoryl linkages containing dendrimers

core of the dendrons is activated by the P=N-P=S linkage and thus, shows interesting reactivity towards primary and secondary amines [20-23]. The P=N-P=S linkage is an important linkage for phosphorous containing compounds, as it can be further functionalized easily (discussed in the modification section). The synthesis of phosphorous containing dendrimers possessing P=N-P=S linkage at the specific layer with in the structure of the dendrimer was also reported [13]. The method involved three successive steps, viz. (i) condensation of methylhydrazine on the aldehyde groups, (ii) condensation of the N-H end groups with Ph_2PCH_2OH, leading to phosphine end groups, and finally, (iii) the Staudinger reaction between a phosphine and an azide linked to a P(S) group. The whole sequence of reactions is shown scheme 9.10.

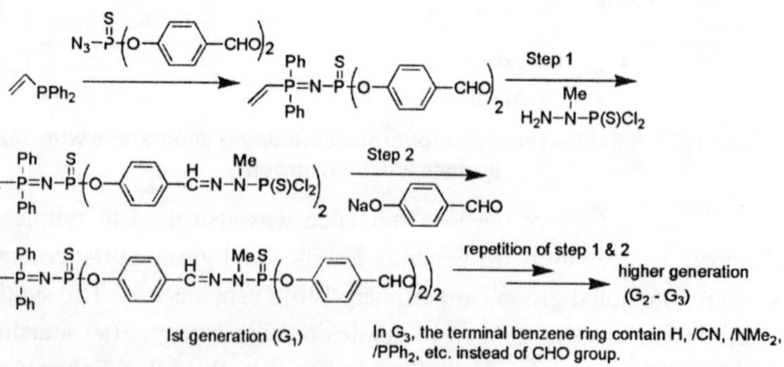

SCHEME 9.9 Synthesis of phosphorous containing dendrimers with vinyl core and various other surface groups

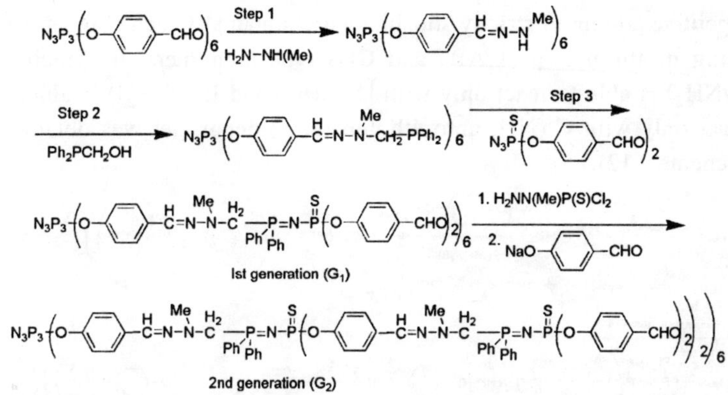

SCHEME 9.10 Synthesis of phosphorous containing dendrimers with P=N-P=S linkages at specific layers

The first generation dendrimer was characterized by XRD studies [15, 19]. The results indicated that the length of each arm is 9A^{O}, distance between terminal P=S or P=O bonds is 17A^{O}. Thus, the molecule resembles as a three bladed propeller when examined in the direction of the central P=S group. This is explained the fact why steric congestion does not disturb for the construction of higher generation and also the terminal groups are readily available for further reactions. Again by using the same sequence of reactions with $Ph_2P(CH_2)Ph_2P$ as the core unit, the phosphorous containing dendrimers upto 4th generation possessing a number of P=N-P=S linkages were also synthesized [24]. In all the dendrimers, $-O-C_6H_4CH=N-N(Me)P(S)$ fragment functions as the linker between each generation. The surface CHO groups of the 4th generation dendrimer were modified by addition of methyl hydrazine to the aldehyde groups and then, treatment of the resulting polyhydrazone with phosphane Ph_2PCH_2OH. Finally, the resulted surface groups reacted with the azide , $N_3P(S)(OC_6H_4CHO)_2$. The whole sequence of reactions is shown in scheme 9.11. All the above synthetic strategies consist of at least two steps to get the next higher generation. Recently, Majoral's groups described [25-26] a new accelerated method, which allows doubling the number of the end groups at each step. In this method, a 4th generation dendrimer was obtained by only 4 steps. Again, the only by-products are water and nitrogen, the reactions are quantitative, hence it is possible to get dendrimers upto 4th generation in a single pot with good purity by repetitive adding of strictly stiochiometric amount of each reagent. By using in alternation of AB_2 and CD_2 type monomers, in which A (NH_2) is able to react only with D (CHO) and B (PRR_2') is able to react only with C (N_3), upto 4th generation dendrimer was obtained (Scheme 9.12).

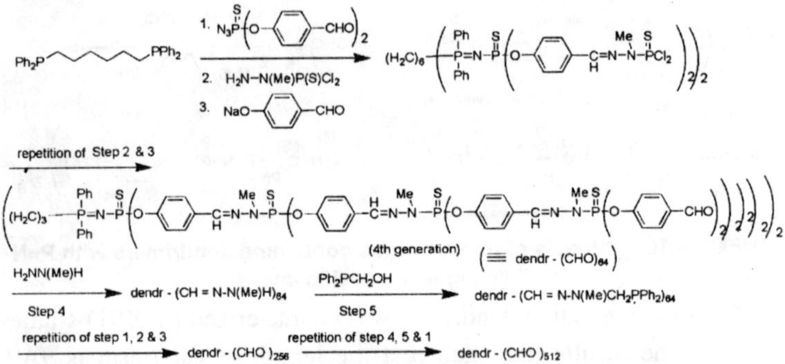

SCHEME 9.11 Synthesis of phosphorous containing dendrimers with $Ph_2P(CH_2)Ph_2P$ as the core unit

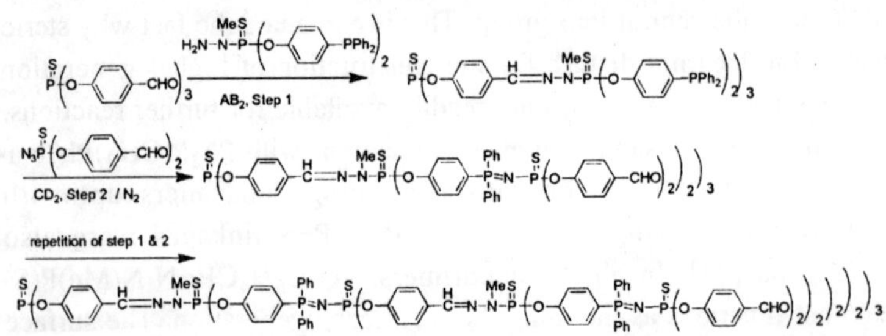

SCHEME 9.12 Synthesis of phosphorous containing dendrimers by accelerated approach, which double the number of end group in each next step

The N-thio phosphorylated and N-phosphorylated imino phosphoranes, (R_3P = N-P(X)R_2; X= O, S) as models for synthesis, reactivity and structural analysis of phosphorous containing dendrimers were reported [27]. By applying Staudinger reaction between phosphorous with azide, $N_3P(X)R_2$ several R_3P = N-P(X)R'_2 and Fe($C_5H_4Ph_2$P=N-P(X)$R'_2)_2$ derivatives were obtained. As in other cases, the P=N-P=X groups were easily alkylated with methyl or isopropyl triflates, which lenthening the P-X bond, as observed by XRD studies. The weaken P-X bond was also cleaved by $P(NMe_2)_3$ to yield P=N-P linkage that to produce tricoordinated phosphorous atoms. This tricoordinated phosphorous atoms can undergo different types of reactions including the reaction with alkyl iodides and functionalized azides. The synthesis of a new phenoxy-diphenyl phosphino derivatives was described [28] for generation of phosphine terminated dendrimer upto 4th generation (32 phosphine groups). The synthetic method involved in the Staudinger reaction with azides and deprotection of borane-phosphine complexes by amines.

Small phosphorous containing dendrimers upto 2nd generation were synthesized [29] by three different methods using the phosphotri-hydrazide, S=P-(N(Me)NH_2)$_3$. These multifunctionalized dendrimers having aldehyde, primary amine, phenoxy, phosphine, or P-Cl function on the periphery. The synthesis of new type of phosphorous containing dendrimers with 1,1'-disubstituted [30] and chiral [31] ferrocenyl building blocks was reported. The structure of one of the synthetic intermediates was determined by XRD analysis. In such case, one single-oxidation was observed for each ferronic layer, corresponding to the independent oxidation of all ferrocenes within one layer to ferrocenium, as observed by electrochemical studies. The oxidation potential for a particular layer depends on their localization within the dendrimers. The synthesis of three series of phosphorous containing dendrimers with azobenzene moieties in the specified internal layers and/or on the surface was reported [32]. The highest generation dendrimer possesses 48 azobenzene groups on the surface. Like other types of azobenzene containing dendrimers, these also undergo E/Z isomerization on irradiation of 350 nm light, what ever are their locations. At room temperature, in dark the azobenzene moieties are remaining as in 'E' form. However, the isomerization rate is independent on the number of azobenzene units and the generation number, when the units are present on the surface. However a different behavior was observed from azobenzene located in the internal layers [32].

9.2.2 Characterization

All phosphorous conatining dendrimers were characterized by help of ^{31}PNMR spectroscopy. It is an extraordinary tool to characterize each step of the synthesis of these dendrimers. The signals, corresponding to each layer are clearly distinguishable for each type of the dendrimers.

The neutral phosphorous containing dendrimers with aldehyde end groups upto 4th generation were characterized [33] by MALDI-TOF mass spectrometry along with SEC, NMR and UV spectroscopies. The mass spectrometric studies although showed the formation of quasi molecular ion but it has been found that the mass spectral pattern, presence of fragments and adducts related to original structure of the dendrimers is largely relevant to the sample preparation. Investigation also showed that the fragmentation occurs mainly due to the relatively strong absorption of the dendrimers at 337 nm. Further more, the fragmentations and formation of adducts involve nitrogen-nitrogen bond cleavage imine metathesis, and reaction of aldehyde groups with internal imino groups.

The Raman vibrational spectroscopy viz. Raman (10-3500 cm^{-1}) and IR (150-3500 cm^{-1}) spectroscopies were utilized [34] to characterize the monomer and first generation phosphorous containing dendrimers. A complete vibrational analysis for the different parts namely, the core, the repeated units, the terminal aldehyde and P-Cl groups of the studied dendrimers is proposed [34]. The assignment was supported by the IR spectral calculation from normal vibrations and band intensities, which was carried out by utilizing force field and electrooptical parameters. The experimental Raman and IR spectra for the dendrimer were interpreted by the help of all of these results. Although, the main features of vibrational spectra of the phosphorous containing dendrimers upto 12th generation were reported [35] for the first time on 2002. The pattern of the spectra is given by the ratio of a number of terminal groups to a number of repeat units, which is nothing but f-1 (f = branching functionality of repeat unit) and become constant for generation > 3 to 5. The analyses are same as described above.

9.2.3 Properties

The phosphorous containing dendrimers are one of the most thermostable dendrimers. Because of their versatility in reactivity of surface, internal layers and core moieties, they can be easily functionalized on their surface, internal layer or even in the core to generate differently modified phosphorous dendrimers. The surface group reactivity, especially the aldehyde group makes them one of the most versatile modified dendrimers. Again, the presence of P=N-P=S linkages in the internal layers allows them to modified uniquely inside the dendrimers. Also, they have some unique properties like very high dipole moment, which is not found in any other categories of dendritic polymers. They also formed complexes with several metals and thereby producing different new catalysts. Their reactivity is so versatile that they can grow another dendritic branch with the parent dendrimer, which is not found in any other kind of dendritic polymers. This is due to uniqueness of phosphorous atom, which can be tri-, tetra-, or penta-coordinated. Majoral's 12th generation dendrimer [25-26] was stable upto 230°C and these dendrimers exhibited high dipole moment. The dipole moment values do not depend on the type of end groups (CHO or $P(S)Cl_2$) but strongly depend on the generation number (Fig.9.2) and the value increases exponentially with the generation. The dipole moment of 11th generation is 328 D, which is extremely high for any known dendritic polymers. The chirooptical properties of the different dendrimer (upto 11th generation) were studied and observed that the molar rotation depends only on the number of stereogenic units not on the number of generation [31].

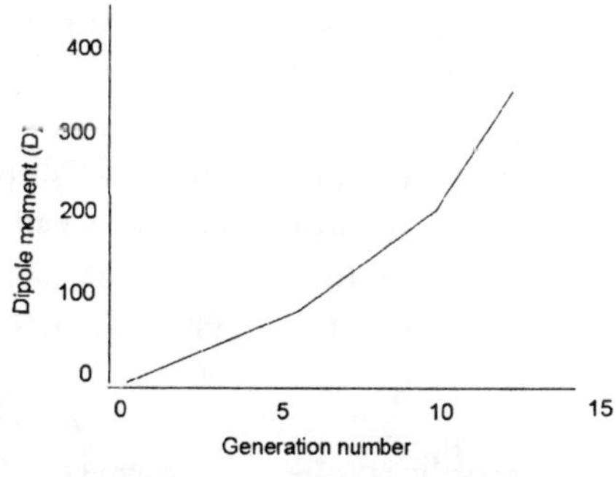

FIG. 9.2 Variation of dipole moment with generation number of phosphorous containing dendrimers

9.3 MODIFICATION OF PHOSPHOROUS CONTAINING DENDRIMERS

The modification of the phosphorous containing dendrimers can be done through surface groups or at internal layers or it can also be possible at the core.

9.3.1 Modification through Surface Groups

A number of modifications through the surface of phosphorous containing dendrimers were reported [36]. $P(X)Cl_2$ terminated phosphorous containing dendrimers upto 7th generation were modified by end capping with allyl amine to it's deivatives [16-17]. The addition of the amines led selectively to monosubstitution, regardless of the number of equivalents of *bis*allylamine used. Even by using 3 equivalents of allyl amine with 3 equivalents Et_3N base, only monosubstituted dendrimers were obtained. Again, even by using propargylamine with the above type $P(X)Cl_2$ terminated dendrimers, the surface group was modified successfully. All the modification reactions are shown in scheme 9.13. These modified dendrimers are very interesting in the sense that these dendrimers have three or even four different types of functionalities simultaneously. This is very special and unique type functionalization and is called 'multiplurifunctionalization'. This is possible only for phosphorous containing dendrimers. Although, the direct di-substitution failed but di-substituted dendrimers were reported through monosubstituted dendrimers [16]. The reaction of proparglyamine with *bis*allylamine, monosubstituted dendrimers resulted di-substituted phosphorous containing dendrimers. The monosubstituted allyl or proparglyamine containing dendrimers undergo disubstitution on treatment with sodium salt of 4-hydroxy benzaldehyde. Even then, surprisingly these disubstituted dendrimers undergo further functionalization [16] and thus, produced four different functionalities at each chain end on treatment with excess $H_2N\text{-}NH_2$ or $Ph_2P\text{=}CH\text{-}CN$. The tetrafunc

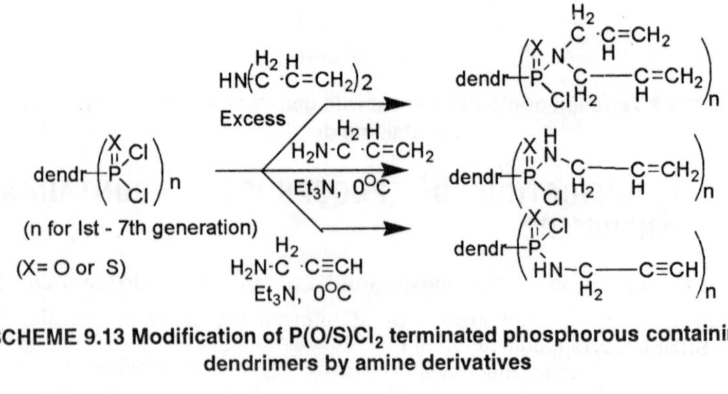

SCHEME 9.13 Modification of $P(O/S)Cl_2$ terminated phosphorous containing dendrimers by amine derivatives

tionalization of the dendrimers is shown in scheme 9.14. The resulted hydrazine modified dendrimers bearing four functionalities viz. NH, NH_2, allyl and hydrazone. All of these groups are exposed on the surface of the dendrimers and hence, they should be highly reactive towards varieties of compounds.

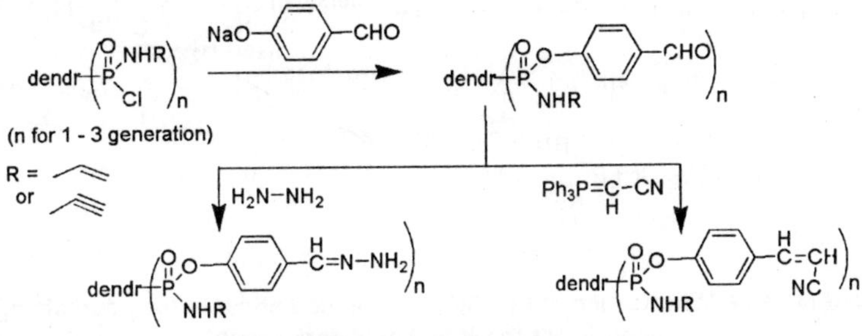

SCHEME 9.14 Modification of CHO terminated phosphorous containing dendrimers by amine derivatives

The direct disubstitution reaction was also reported [37] for $P(X)Cl_2$ (X=O and S) terminated dendrimers by allyl or propargy-lamines. Even cyclic polyamines, such as tetraazacyclo-tetradecane or -pentadecane and tetraazacyclo-tetradecanedione were also used for end grafting of the phosphorous dendrimers upto 3rd generation. It was found that upto 12 macrocyclic amines could be grafted from the X-ray crystallographic studies. The grafting reaction is shown in scheme 9.15.

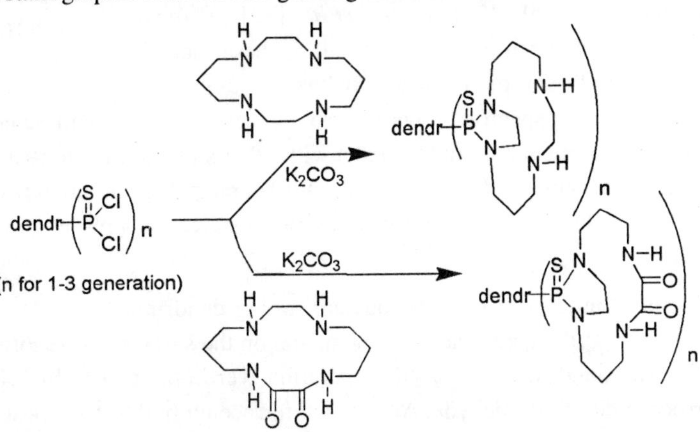

SCHEME 9.15 Modification of P(O/S)Cl_2 terminated phosphorous containing dendrimers by alkyl or propargyl amine derivatives

Another example of multitrifunctionalization was reported [38] starting from difunctionalized dendrimers [37] by substitution of H of one NH group by Ph_2PCH_2OH (Scheme 9.16).

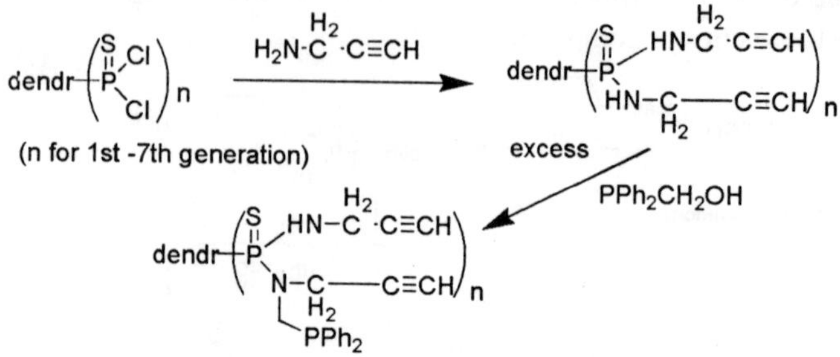

SCHEME 9.16 Modification of $P(O/S)Cl_2$ terminated phosphorous containing dendrimers to trifunctional derivatives

The terminal $S{=}PCl_2$ groups of the phosphorous containing dendrimers were modified with $H_2NCH_2CH_2NEt_2$ to get water soluble dendrimers. Most of the dendrimers are insoluble in water but water-soluble dendrimers are required sometimes specially for biological purpose. The reaction of 2 equivalents of $H_2NCH_2CH_2NEt_2$ with each $P(S)Cl_2$ end group offers directly water-soluble dendrimers (Scheme 9.17). The solubility of $-N^+HEt_2Cl^-$ containing dendrimers is pH dependent, whereas the $N^+(Me)Et_2O^-COMe$ containing dendrimers independent of pH [39]. Loup *et al.* [39] synthesized such type dendrimers upto 5th generation. In all of these modifications through the terminal groups resulted different phosphorous containing dendrimers with specific properties and characteristics. Also, these end capped dendrimers may further react with other suitable functionalities to offer other kind of dendrimers. All of these phenomenons are unique and this is due to uniqueness of phosphorous element present in the dendrimers.

The synthesis of phosphorous containing dendrimers with ferrocenyl units at the core, branching point and on the surface was reported [38]. The dendrimers upto 4th generation were obtained from 1,1'-ferrocene dicarbonaldehyde. At first, the ferrocenyl building block was synthesized from ferrocenyl carbonaldehyde (Scheme 9.18 A and B). The dendrimers with ferrocenyl unit at the core was synthesized from

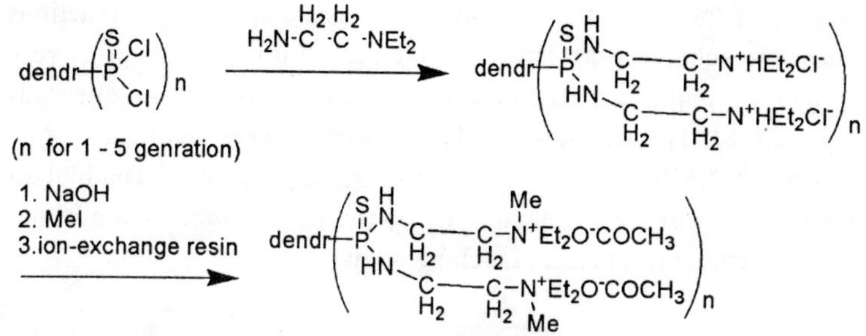

SCHEME 9.17 Modification of $P(O/S)Cl_2$ terminated phosphorous containing dendrimers to get water-soluble derivative

1,1'-ferrocene dicarboxaldehyde by repetition of two step process. First step is the condensation of aldehyde groups with dichloro(methyl hydrazino)phosphine sulfide and the second step is the reaction of the resulted compound with sodium salt of 4-hydroxy benzaldehyde. The sequence of reactions is shown in scheme 9.19. The structural homogenety of each generation was studied by chromatographic studies, ^{1}HNMR, ^{13}CNMR, ^{31}PNMR and IR spectroscopic and MALDI-TOF mass spectrometric studies. The dendrimers upto 2nd generation with ferrocenyl units within the branches were synthesized by using sodium salt of (A) (Scheme 9.18) as the building block and

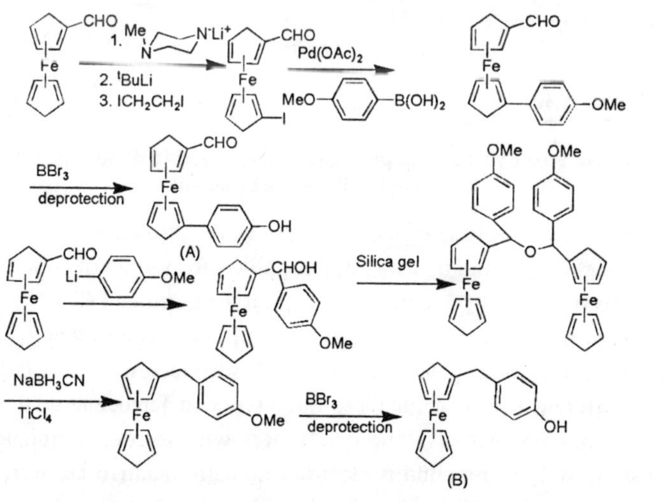

SCHEME 9.18 Synthesis of phosphorous containing dendrimers by using ferrocenyl units as the building block

repetition of the above two reactions. The whole sequence of reactions is shown in scheme 9.20. However, the yield for the each step is very poor and sometimes a partial decomposition of ferrocenyl moieties was observed [38]. The structures of the dendrimers were characterized by ^{1}HNMR, ^{13}CNMR, ^{31}PNMR and IR spectroscopic studies. The highest generation metallodendrimer is one of the largest metallodendrimer with ferrocene units at each branching point.

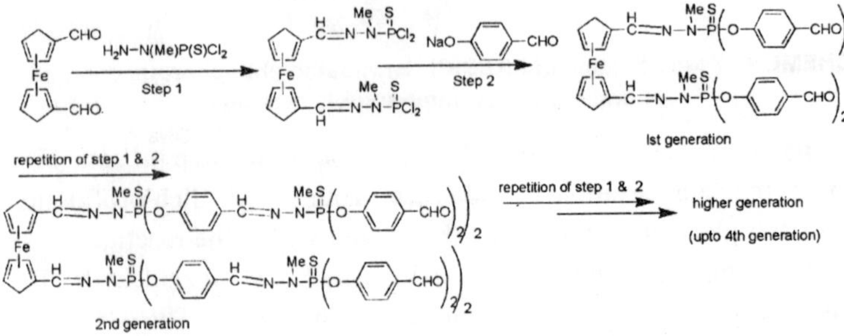

SCHEME 9.19 Synthesis of phosphorous containing dendrimers with ferrocenyl units at the core

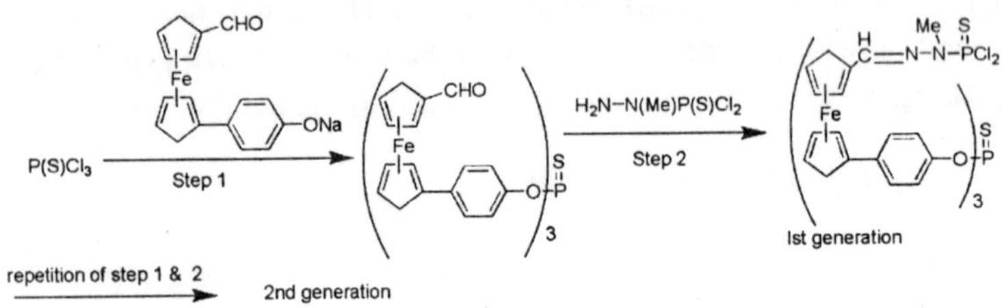

SCHEME 9.20 Synthesis of phosphorous containing dendrimers with ferrocenyl units at the branches

The surface grafting of the phosphorous dendrimers with ferrocene moiety was carried out by using nucleophilic substitution of sodium salt of hydroxybenzaldehyde derivatives (B) on the terminal $P(S)Cl_2$ groups of the dendrimers [3, 6, 15]. One such example is given in scheme 9.21. The electrochemical properties of the dendrimers were studied and found different behavior for these three types of ferrocene containing phosphorous dendrimers. The dendrimers with ferrocene unit at the core showed the compound is electrochemically inactive i.e. ferrocene core is insulated from the outside of the dendrimers. In the dendrimers with ferrocenyl unit at each branching points observed that inner layers are oxidized at the same potential in every case but the outer layer needs a higher potential for oxidation. This may be due to the presence of electron withdrawing aldehyde group in the outer layers. However, the surface modified ferrocene capped dendrimers are the largest redox active phosphorous containing dendrimers. In this case, all the ferrocenyl units are oxidized at the same potential, which indicated that they are all equivalent and electrochemically independent.

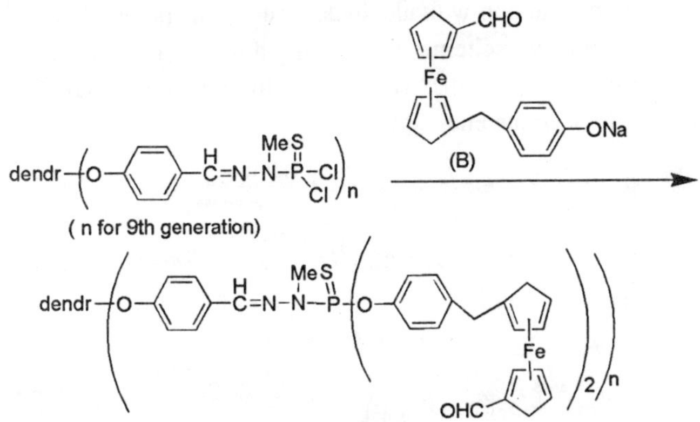

SCHEME 9.21 Synthesis of phosphorous containing dendrimers with ferrocenyl units at the surface

The modifications through the terminal aldehyde groups are also reported. This modification resulted phosphate, phosphite, ylide, phosphonate, etc. types of end capped phosphorous containing dendrimers. The sequence of two reactions namely condensation of terminal aldehyde groups of the dendrimers with hydrazine followed by another condensation of the resulted dendrimers containing hydrazone with aldehyde groups of phosphate containing compounds offers the generation of phosphate end capped dendrimers. Whereas condensation with phosphinite containing compounds offerd phosphite end capped dendrimers. An aldehyde terminated first generation phosphorous containing dendrimer [3] was modified to phosphate and also to phosphinate [40]. The reactions involved in these modifications are shown in scheme 9.22. However, because of poor solubility in the modified dendrimers in both the cases, the modification for higher generation was not possible to carry out. The modification of 5th generation dendrimers by direct condensation of terminal aldehyde groups with $H_2NN(Me)P(O)(OEt)_2$ is shown in scheme 9.23. Most of the phosphonate terminated dendrimers present an unexpected long range 'P-P' coupling constant through seven bonds ($3.8 < {}^7J_{P-P} < 4.5$ Hz) [40]. The methyl hydrazone terminated first generation dendrimers [40] can be further functionalized to olefinic double bonds though well known Wittig reaction. The treatment of triphenyl phosphonylidene ethenone ($Ph_3P{=}C{=}C{=}O$) with the methyl hydrazone dendrimers followed by the reaction with aldehyde resulted the targeted dendrimers [40], as shown in scheme 9.24. The authors [40] also grafted phosphonate units on the surface by the reaction of $HP(O)(OR)_2$ with aldehyde or imine end

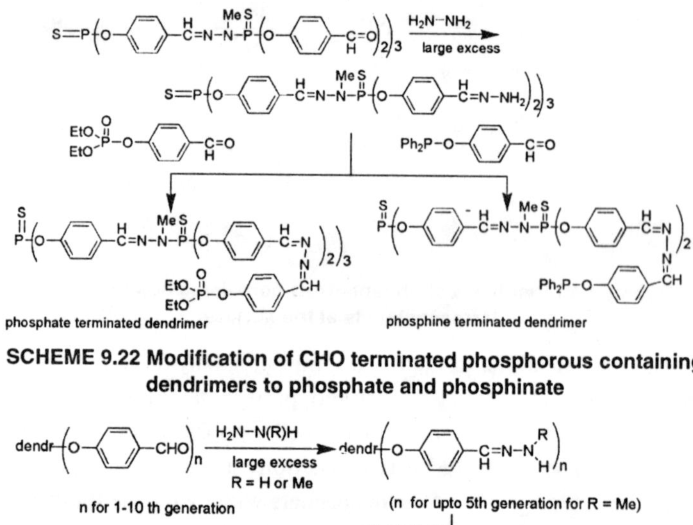

SCHEME 9.22 Modification of CHO terminated phosphorous containing dendrimers to phosphate and phosphinate

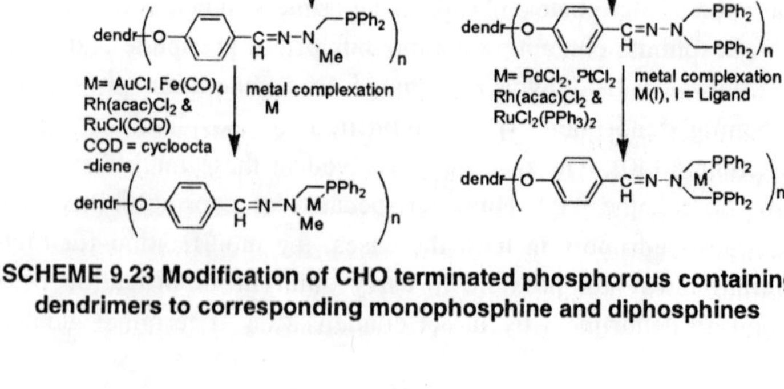

SCHEME 9.23 Modification of CHO terminated phosphorous containing dendrimers to corresponding monophosphine and diphosphines

groups of the phosphorous dendrimers upto 5th generation without any solubility problem. The reactions are shown in scheme 9.25. In all the modified dendrimers [40], it was observed that the solubility of these dendrimers largely depends on the type of substituents on the surface groups. The azine phosphate, azine phosphinite and ylide linkages on the surface significantly reduced the solubility. However, the grafting of long chain hydrocarbons increased the solubility in organic solvents.

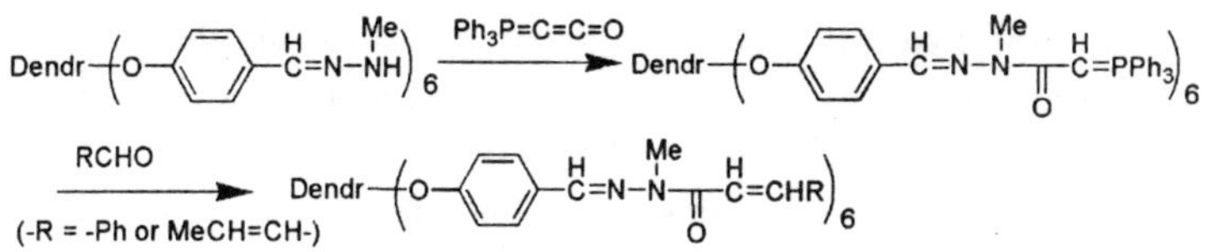

SCHEME 9.24 Modification of CHO terminated phosphorous containing dendrimers by Wittig and condensation reactions

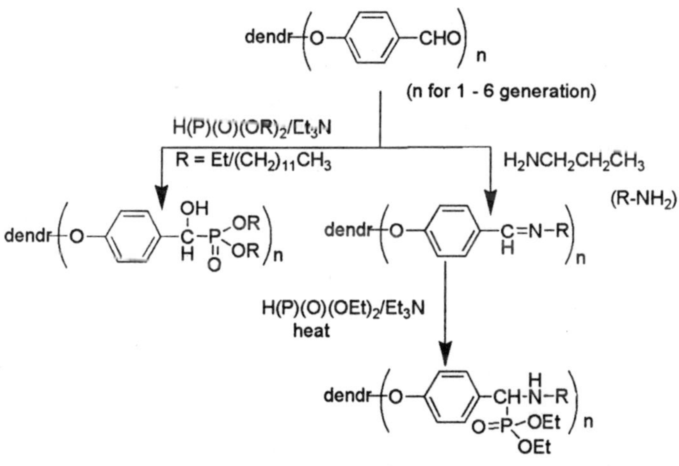

SCHEME 9.25 Modification of CHO terminated phosphorous containing dendrimers to phosphonate derivative

The modification of terminal aldehyde groups of phosphorous containing dendrimers [3] to generate monophosphines or diphosphines end capped dendrimers was also reported [41-45]. The largest surface modified polyphosphine dendrimer was reported [42], and it complexs with various metallic derivatives. The monophosphine end capped dendrimers were complexed with Au-Cl (upto 10th generation) [42], $Fe(CO)_4$, RhCl(Cod) (upto 4th generation) and RhCO(acac) (upto 6th generation) [43]. Whereas the diphosphine end capped dendrimers upto 3rd generation were complexed with $PdCl_2$, $PtCl_2$, Rh(acac) [44], and $RuH_2(PPh_3)_2$ [45]. All these reactions are shown in scheme 9.23. The absence of any trace of free phosphanes in ^{31}PNMR spectra confirmed the complete complexation of all the terminal groups. By visualization using high-resolution electron microscopy, the presence of gold atoms on the surface of these dendrimers was confirmed [42].

The modification and supramolecular autoassembly of that modified phosphorous containing dendrimers are mimicking multisite analogs of galactosyceramide [46]. The aldehyde surface group containing 1st and 2nd generation dendrimers treated with malonic acid for acid end capping. The malonic acid end capped dendrimers were characterized by FTIR, ^{1}HNMR and ^{13}CNMR spectroscopic studies. The self-assembly process between acid terminated dendrimers and the aminolactitol was carried out in water at room temperature with a strict stiochiometric ratio. The completion of assembly formation was confirmed by total solubilization of the insoluble dendrimer and characterized by FT-IR, ^{31}PNMR, ^{1}HNMR and ^{13}CNMR spectroscopy. In FTIR, the disappearance of the conjugated carboxylic acid

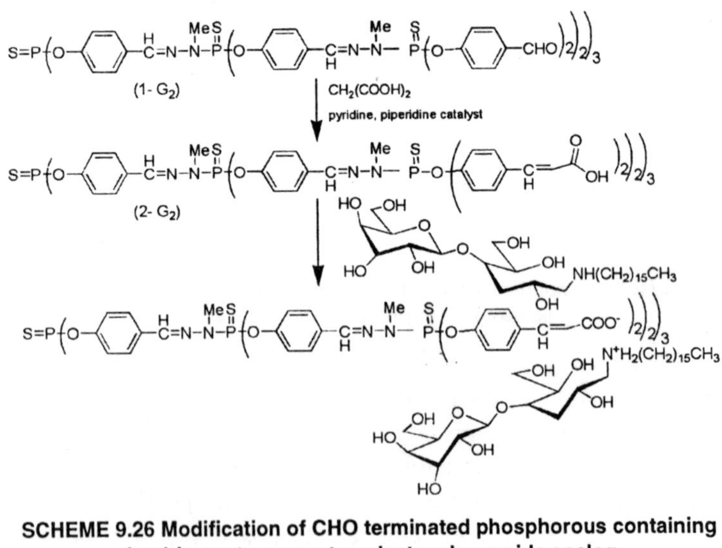

SCHEME 9.26 Modification of CHO terminated phosphorous containing dendrimers to generate galactosylceramide analog

absorption band at 1687 $cm^{-1}$and appearance of the carboxylate absorption bands at 1559 cm^{-1} and 1381 cm^{-1} are very valuable information in this respect. Also, the DLS studies showed aggregate formation and TEM images showed the presence of vesicles. The sequence of above reactions is shown in scheme 9.26.

A series of multifunctionalized oligophosphazenes with upto 4th different types of functional side groups were used as novel cores for synthesis of phosphorous containing dendrimers [47]. The oligophosphazenes were synthesized by the following sequence of reaction as shown in scheme 9.27. All these phosphazenes were characterized by multinuclear NMR spectroscopy and FAB mass spectrometry. However, ^{31}PNMR is the best tool for this purpose. This dendrimers with four different types of functional groups can be easily modified through condensation reaction of the aldehyde functionality (Scheme 9.28). These modified macromolecules are novel dendritic species possessing several types of functional groups at the core in the branches

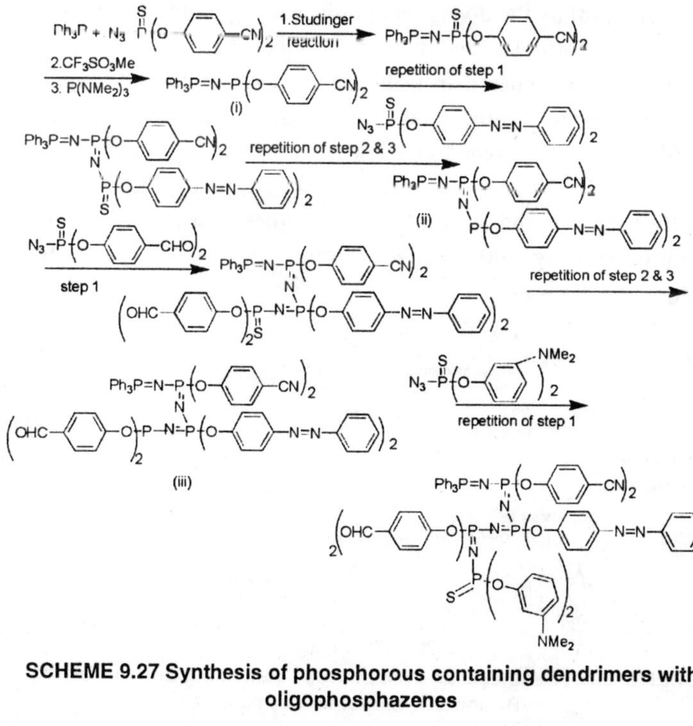

SCHEME 9.27 Synthesis of phosphorous containing dendrimers with oligophosphazenes

and on the surface, and consist of phosphazene linkages not only at the core but also within the branches. This unusual multifunctionalized oligophosphazene core and eight trifluoromethyl groups as end groups are linked to the oligophosphazene chain through phosphazene linkages

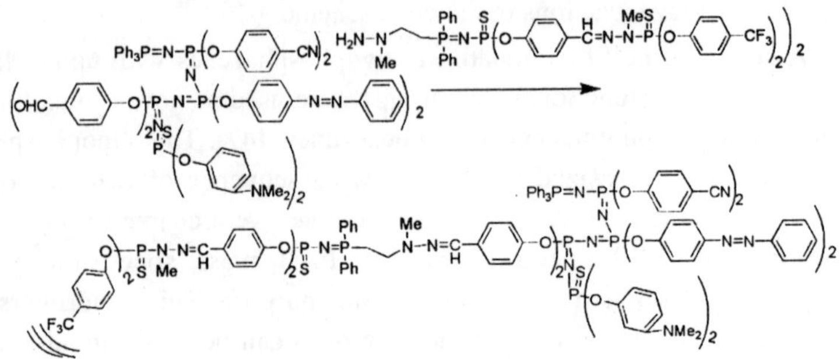

SCHEME 9.28 Modification of CHO groups of phosphorous containing dendrimers with oligophosphazenes

The phosphorous containing dendrimers upto 4th generation, were synthesized [48] by the divergent approach using highly functionalized cores of the form D_6 and new monomers AB_5 and CD_5. These monomers were prepared from hexachlorocyclotriphosphazene. The dendrimers were synthesized by the divergent approach using hexaaldehyde phosphazene as the core unit in three different routes viz. route A, B and C, as shown in schemes 9.29 (A, B and C). This process is really very exciting as it allows to synthesize a dendrimer of 4th generation bearing 600 aldehyde groups by just 4 steps (Scheme 9.29A)!

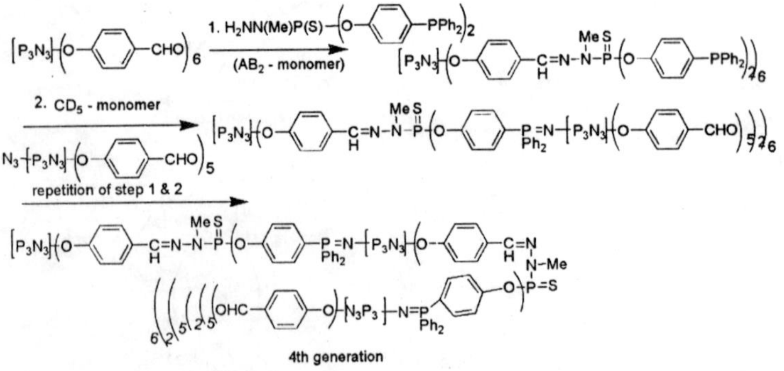

SCHEME 9.29A Synthesis of phosphorous containing dendrimers from AB_2 and CD_5 monomers by route A

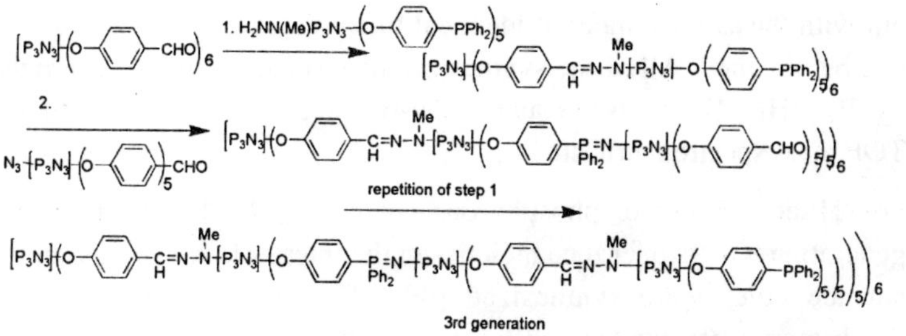

SCHEME 9.29B Synthesis of phosphorous containing dendrimers from AB_2 and CD_5 monomers by route B

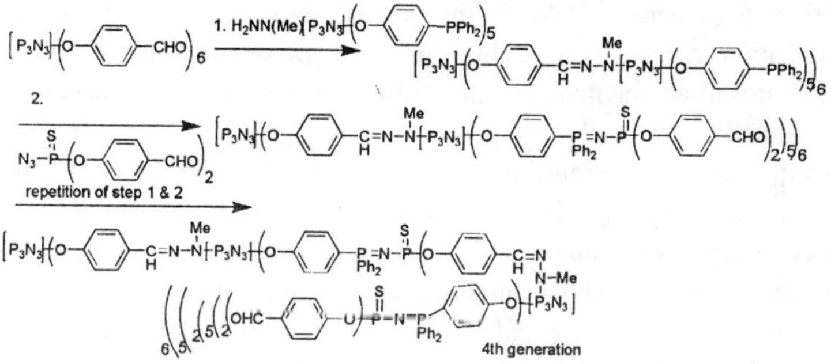

SCHEME 9.29C Synthesis of phosphorous containing dendrimers from AB_2 and CD_5 monomers by route C

This process further increased the number of end groups and indeed just 3rd generation dendrimer bearing 750 phosphonyl groups was prepared by only 3 steps (Scheme 9.29 B). In fact, this methodology may be considered as one of the most convenient method for synthesis of phosphorous containing dendrimer with several advantages. These are: (i) every step offers the next higher generation number, (ii) a maximum number of functional groups can be anchroched to the surface of a dendrimer of a given generation, (iii) purification is easy, just washing is required, (iv) reaction can be easily monitored by ^{1}HNMR and ^{31}PNMR spectroscopy, (v) a minimum amount of solvent is used and the by products are compatible with green chemistry (nitrogen and water), and (vi) the degree of freedom for the construction of dendrimer is very high, i.e. depending on the sequence of monomer additions, dendrimers with different backbones but with the same number of identical functional groups on the surface can be obtained. All of the synthesized dendrimers were characterized by IR, ^{1}HNMR, ^{13}CNMR and ^{31}PNMR spectroscopic and MALDI-TOF mass spectrometric studies.

Three serieses of phosphorous containing dendrimers upto 3rd generation with azobenzene moieties in the interior layers and/or on the surface layer were synthesized [49]. The surface functionalized dendrimers were synthesized by the reaction of chloride or hydrazine derivatives of the different generation phosphorous dendrimers with parahydroxy phenyl azobenzaldehyde in the presence of cesium carbonate (Scheme 9.30A). By this process, the highest generation (4th) dendrimer resulted 48 azobenzene and 48 phenol groups in the structures after functionalization. This functionalized dendrimers were continually grew to higher generations though the surface aldehyde groups using alternating reaction with $H_2N(Me)P(S)Cl_2$ and $NaOC_6H_4CHO$ to get the desired dendrimers with azo group in the interior layer (Scheme 9.30B). Similarly, the dendrimer with azo groups in each layer within the branches and on the surface were synthesized by repetition and alternating reaction of surface functionalized above dendrimers with $H_2N(Me)P(S)Cl_2$ and $HOC_6H_4N{=}NC_6H_4CHO$ in the presence of Cs_2CO_3 (Scheme 9.30C). By using the same reaction sequences but starting from a cyclotriphosphazene core, the dendrimer [49] was also synthesized where azo groups linked directly onto the core unit (Scheme 9.30D). This dendrimer also possesses exactly same 42 number of azo groups and 24 numbers of aldehyde groups in the structure as in the earlier case (Scheme 9.30C). The structures of the above functionalized dendrimers were characterized by ^{1}HNMR and ^{31}PNMR spectroscopy, MALDI-TOF mass spectrometry. The photochemical properties of the dendrimers showed that the position of the azo group within the dendrimers has no influence upon its absorption properties. Again, the value of the molar absorption co-efficient, ε_{max} increases with the number of azobenzene groups in an additive way. On the other hand, the relative amount of E and Z isomers was obtained at the photostationary equilibrium reached by irradiation with a 350 nm light differs greatly depending on the location of the azobenzene moieties.

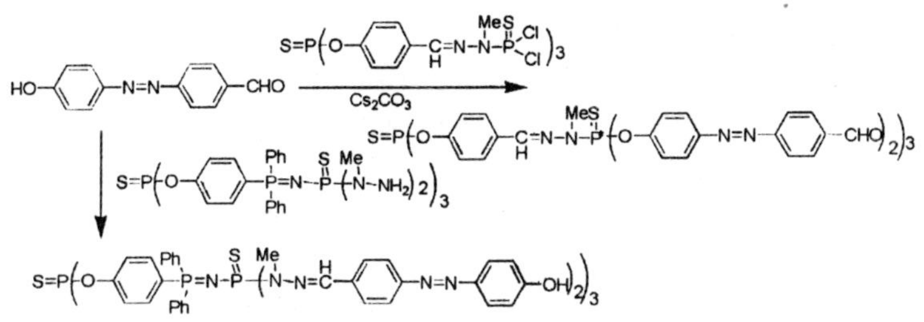

SCHEME 9.30A Modification of phosphorous containing dendrimers with azobenzene derivative at the surface

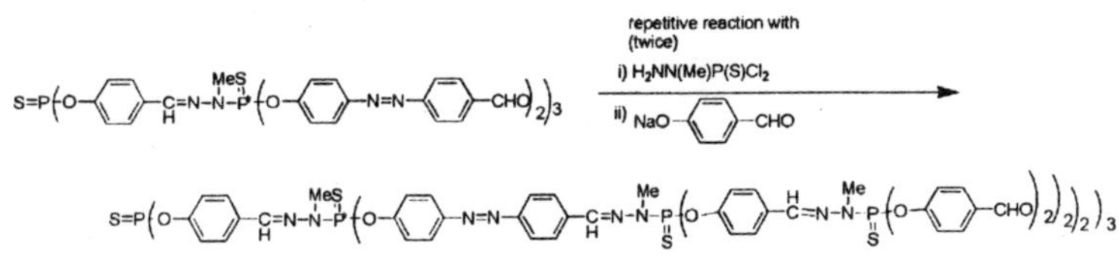

SCHEME 9.30B Modification of phosphorous containing dendrimers with azobenzene derivative at the internal layers

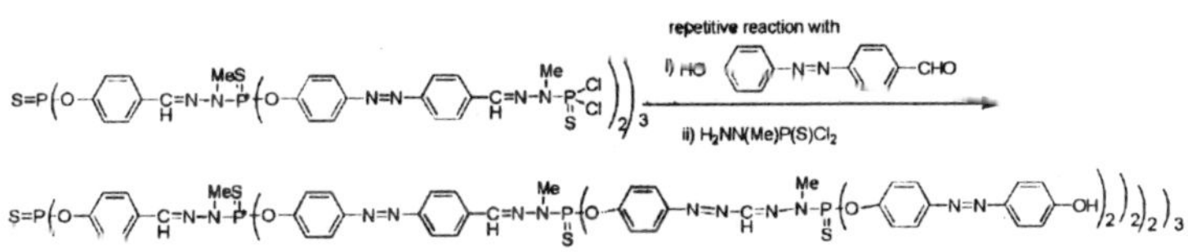

SCHEME 9.30C Modification of phosphorous containing dendrimers with azobenzene derivative at the internal layers and on the surface

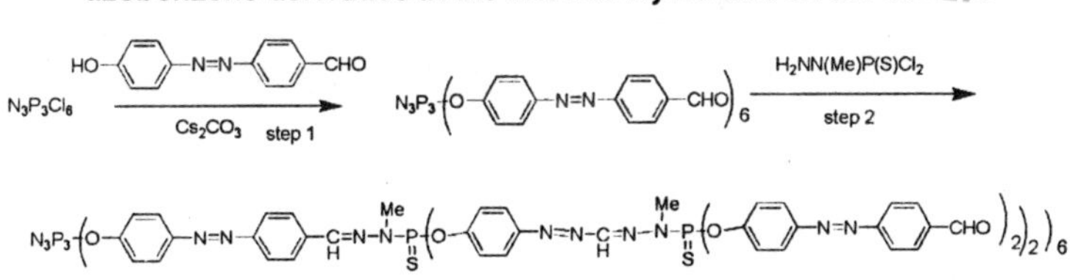

SCHEME 9.30D Modification of phosphorous containing dendrimers with azobenzene derivative at the core

Another series of phosphorous dendrimers with upto 96 redox active TTF (tetrathiafulvalene) moieties on the periphery, which allow the generation of polycataionic species bearing upto 192 positive charges on the surface were synthesized [50]. The dendrimers upto 5th generation, were synthesized by a poly-Wittig olefination reaction between the corresponding polyformyl dendrimers and the phophonium salt (II) through three step sequence of reactions from the corresponding cyanoethyl sulfonyl precursor (Scheme 9.31). All the modified dendrimers exhibited well defined two step reversible oxidation encountered in parent TTF systems (TTF to TTF^{2+}). From the cyclic voltammograms, a clear tendency of the dendrimers to deposit onto the electrode surface was observed.

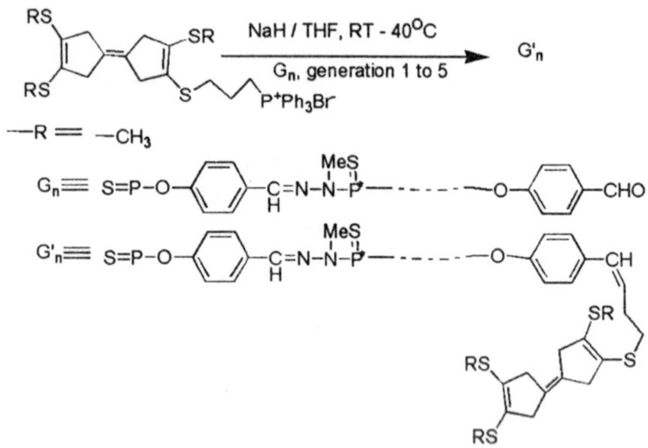

SCHEME 9.31 Synthesis of phosphorous containing dendrimers with 96 redox active moieties

The grafting of 6,12,48 crown ethers at the periphery of phosphorous dendrimers [3] upto 4th generation was reported [51]. The Schiff's reaction between terminal aldehyde groups of the dendrimer and 4-aminobenzene-15-crown-5-ether resulted the desired dendrimers (Scheme 9.32). Also, treatment of these aldehyde groups with fluorenone hydrazone or 1-amino-4-(2-hydroxy ethyl)piperazine affords

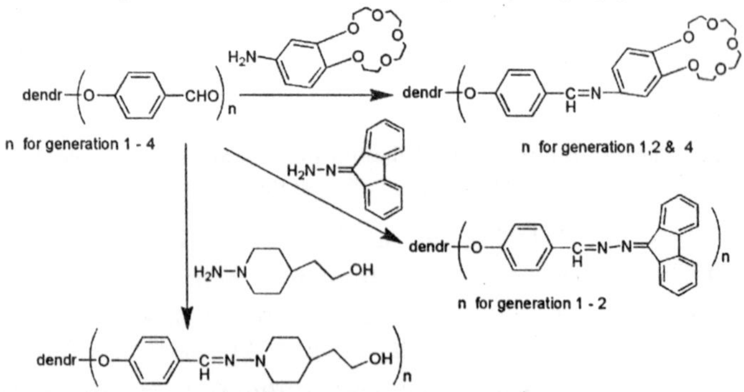

SCHEME 9.32 Modification of CHO terminated phosphorous containing dendrimers by Schiff's reaction

fluorenone and hydroxy terminated phosphorous containing dendrimer (Scheme 9.32). Even by using similar experiment with (cyano-methylene) triphenylphosphorane resulted cinnamonitrile containing end groups for first and 4th generation dendrimers. The reactions for the modification were shown in scheme 9.33.

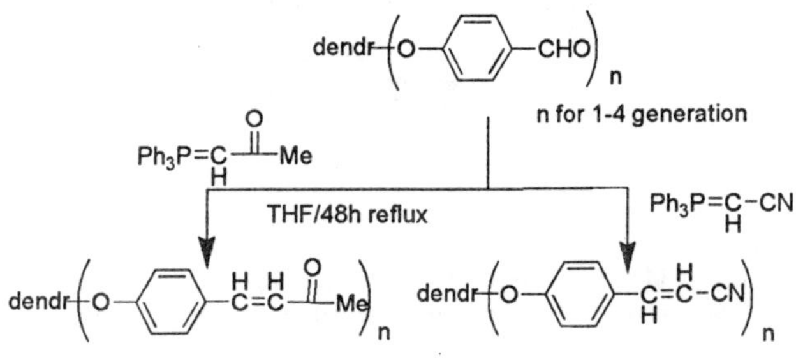

SCHEME 9.33 Modification of CHO terminated phosphorous containing dendrimers by Wittig reaction

Another name-reaction, Horner-Wadsworth-Emmons was applied for the modification [52] of aldehyde terminated phosphorous dendrimers [3]. The terminal aldehyde groups react with sodium salts of the ester, phosphonate, thioamide and carboxylic derivatives of phosphonates to generate dendrimers with moderate yields and possess α,β–unsaturated functional group on the surface. Further, this reaction was also applied to the phosphonate derivative of aminoacids such as glycene, L-alanine, L-phenylalanine and L-methionine. First they produced water soluble phosphorous dendrimers, which on acidification yielded dendrimer with –CH=CH- containing thioamide surface functional groups [52]. The sequence of reactions is shown in scheme 9.34.

The water soluble phosphorous containing dendrimers can also be obtained by surface modification of aldehyde terminated dendrimers using the same synthetic strategy as described earlier [39]. The reaction of hydrazide derivatives of trimethyl ammonium or pyridinium with aldehyde groups of dendrimers affords water soluble phosphorous dendrimers [53] (Scheme 9.35). By formation of carboxylate or alcoholate anions also generates water soluble dendrimers [54]. The reaction of $CH_2(CO_2H)_2$ in the presence of piperidine or pyridine with terminal

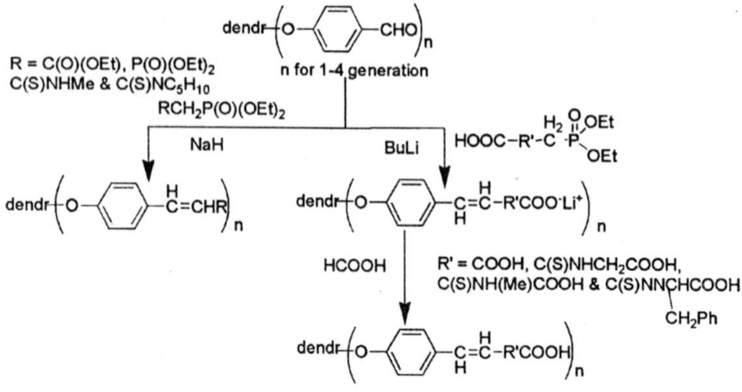

SCHEME 9.34 Modification of CHO terminated phosphorous containing dendrimers by Horner-Wadsworth-Emmons reaction

aldehyde groups of the parent dendrimers followed by treatment of the resulted dendrimers with NaOH affords the water soluble dendrimers (Scheme 9.35). There is also a report [55] on synthesis of water soluble phosphorous containing dendrimers by Schiff's condensation of N,N-disubstituted hydrazide end groups of the parent dendrimers with various benzaldehydes functionalized carboxylic acid, a boronic acid or a carboxylate. The study of the stability of the resulted dendrimers showed that carboxylic acid derivatives were stable for more than 7 months in water, indicating high stability of both hydrazone bonds and the whole structures. However, in all these water soluble phosphorous containing dendrimers possess both hydrophobic interiors with hydrophilic surface structures.

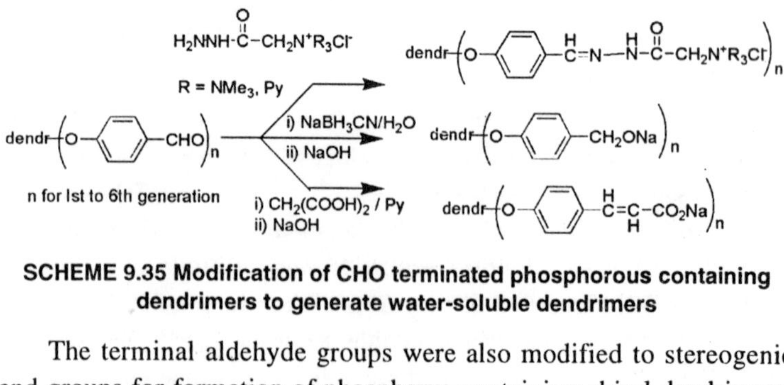

SCHEME 9.35 Modification of CHO terminated phosphorous containing dendrimers to generate water-soluble dendrimers

The terminal aldehyde groups were also modified to stereogenic end groups for formation of phosphorus containing chiral dendrimers.

The modification of 1st - 4th and 7th generation phosphorous dendrimers was reporetd [56] by condensation of surface aldehyde groups with a specific stereoisomer of methylbenzylamine (either S or R) followed by reduction of the resulted imine bonds. The reaction of Ph_2PCH_2OH with the reduced dendrimers resulted the corresponding phosphino-terminated chiral dendrimers. The whole sequence of reactions is shown in scheme 9.36. From the chirooptic study of the above dendrimers revealed that molar specific rotation depends linearly on the number of sterogenic groups present in the dendrimers. However, the value of specific rotation depends on the weight percentage contribution of stereogenic groups to the molecular weight of the chiral dendrimers [56]. This result proved that the terminal stereospecific group behaves like an isolated molecule.

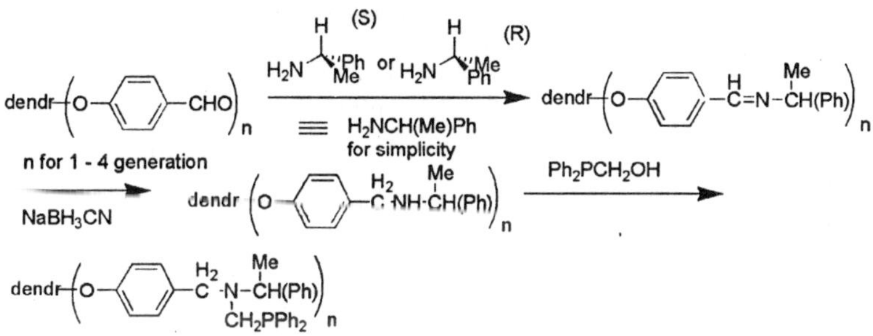

SCHEME 9.36 Modification of CHO terminated phosphorous containing dendrimers by stereospecific compounds

The specific monosubstituted phosphorous containing dendrimers were synthesized upto 7th generation at the surface [57]. This reaction allows to isolate dendrimers, which possess simultaneously upto 192 phosphine, 192 NH and 388 allyl groups. The complexation ability of this dendrimers was also observed [42, 57-58] with $Fe_2(CO)_9$ and $W(CO)_5(THF)$. In all such cases, the complexation occur within a few hours to give the iron and tungsten derivatives. A large number of Pd complexes were prepared by the reaction of $PdCl_2(Cod)_2$, $PdBr_2(Cod)$ and PdMeCl(Cod) with the diphosphines terminated dendrimers. Further more, these metallocene again undergo insertion of CO, norbornene, [44] etc. All of these complexation reactions are shown in scheme 9.37.

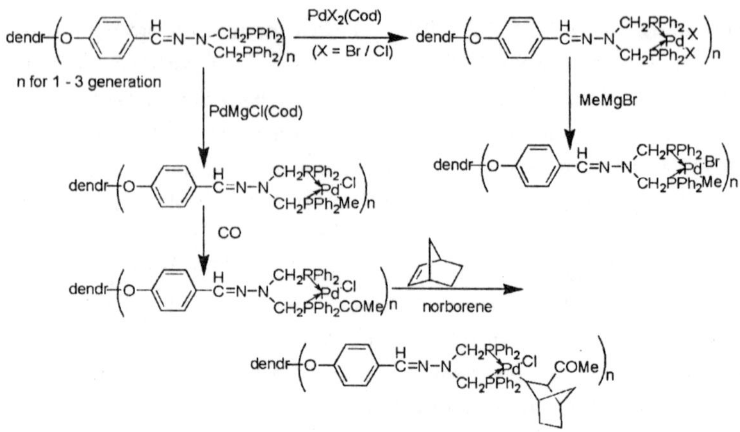

SCHEME 9.37 Modification of CHO terminated phosphorous containing dendrimers to generate metallodendrimers

Three different types of tetraazamacrocycles namely, 1,4,8,11-tetraazacyclotetradecane, 1,4,8,12-tetraazacyclopentadecane and 1,4,8,11-tetraazacyclotetradecane-5,7-dione were grafted [59] on the surface of first and 3rd generation phosphorous containing dendrimers with $P(S)Cl_2$ groups. Both five member rings (diazaphospholanes) and six member rings (diazaphosphorinanes) were formed depending on the nature of macrocycle. These modified dendrimers were characterized by ^{13}CNMR spectroscopy and XRD studies.

9.3.2 Modification through Internal Layers

Modification of most of the dendrimers are generally performed by functionalization on the surface groups. Only a very few examples are there on modification of dendrimers through internal layers. But in case of phosphorus containing dendrimers the post synthetic modification through internal layer is possible. This is due to the fact that in P=N-P=S linkages of the dendrimer are exist as mesomeric form P^+-N=P-S^-, where electronphiles can attract favorably on sulfur. The chemiospecific alkylation of the sulfur in P=N-P=S linkages by methyltrifluoromethane sulfonate [60] or various functionalization triflates [61] resulted the incorporation of charges in the internal layers of the phosphorous containing dendrimers [24]. The other P=S groups remain inactive. If there are several P=N-P=S linkages at different layers of the dendrimer (e.g. 7th generation, where P=N-P=S linkages are there in the core, 5th and 7th layers), the alkylation occurs at all these P=N-P=S linkages containing layers. Since methyl triflate can penetrate upto core of the dendrimers, so this modification reaction confirmed the presence of high porosities or voids in the dendrimer structures. All reactions for the modification are shown in scheme 9.38, for the 1st generation. Similarly, 5th and 7th generations will give 66 and 322 charges after the reaction with CF_3SO_3Me. The modified dendrimers were characterized by ^{31}PNMR spectroscopy, as the alkylation induces shielding of the signal corresponding to thiophophoryl groups from 52 ppm of the parent dendrimers to 28 ppm in the modified dendrimers. A deshielding effect of diphenyl phosphino groups from 21 ppm to 33 ppm was also observed.

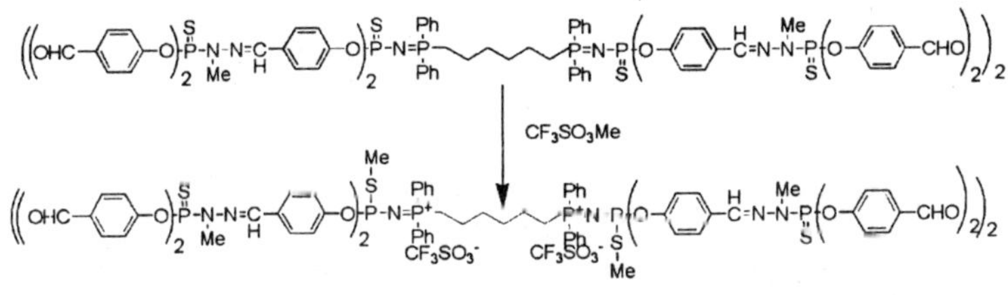

SCHEME 9.38 Modification of phosphorous containing dendrimers at internal P=N-P=S linkages by sulfonate derivatives

The grafting of allyl or propargyl groups at the internal layers containing P=N-P=S linkages of 1st and 4th generation phosphorous containing dendrimers was reporetd [61]. In these cases, allyl or propargyl triflates were used. The synthetic strategy used for this grafting is shown in scheme 9.39, for the Ist generation dendrimer. Similarly, 4th generation dendrimer was also grafted. The treatment of the dendrimer (1st generation) with two equivalents of allyl trifluoromethane sulfonate allows the chemoselective allylation on the sulfur atom of both P=N-P=S internal groups, and the formation of the dication. Similarly, the addition of propargyl triflate to 1st generation leads to the selective functionalization on the sulfur atom of the two P=N-P=S fragments, with quantitative formation of dication. The grafted dendrimers were characterized by ^{1}HNMR, ^{13}CNMR and ^{31}PNMR spectroscopy and X-ray studies. The X-ray crystallographic study confirmed the existence of alkylation on P=N-P=S linkages. As, this type of alkylation reaction induces a weakening of the P-S bond and thus, a nucleophilic phosphine can easily cleave this bond. Further, these reactive tricoordinated phosphorous atoms inside the dendrimers can undergo alkylation, complexation, etc. reactions. By applying these reactions Majoral's group [62-63] modified phosphorous containing dendrimers through their internal layers. They incorporated isothiocyanates, primary amines and that aldehyde inside the dendrimers possessing P=N-P=S linkages by the sequence of alkylation, desulfurization and Staudinger reactions, as shown in scheme 9.40. The tricoordinated phosphorous atoms allow alkylation by methyl or allyl iodide, leading to the formation of second series of functionalization dendrimers including phosphonium salt at some precise internal layers.

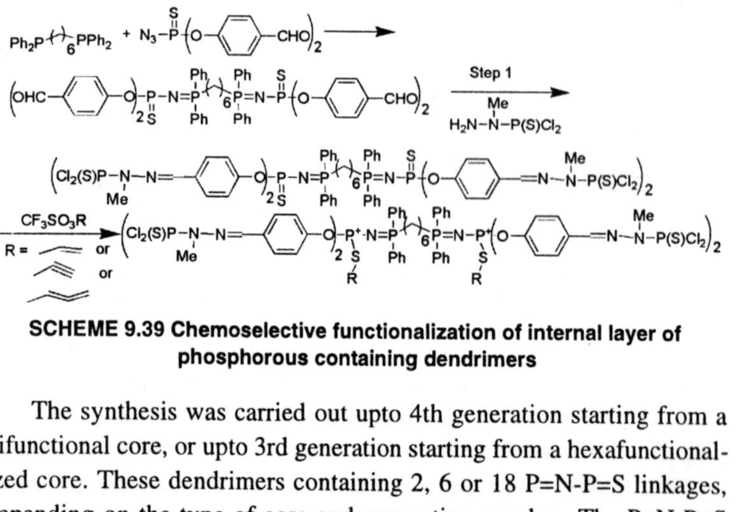

SCHEME 9.39 Chemoselective functionalization of internal layer of phosphorous containing dendrimers

The synthesis was carried out upto 4th generation starting from a difunctional core, or upto 3rd generation starting from a hexafunctionalized core. These dendrimers containing 2, 6 or 18 P=N-P=S linkages, depending on the type of core and generation number. The P=N-P=S linkages were obtained directly by the Staudinger reaction with $N_3P(S)[OC_6H_4CHO]_2$. The functional groups were introduced by various alkyl triflates, alkylation and Staudinger reactions. At first, the strong polarization of the P=N-P=S linkages induced a facile reactivity with various alkyl triflates such as methyl, allyl, and propargyl triflates, leading to the formation of internal amino phosphite groups [P=N-P:]. These highly reactive tricoordinated phosphorous atoms were alkylated by methyl or allyl iodide, leading to a 2nd series of functionalized dendrimers including phosphonium salts at some precise internal

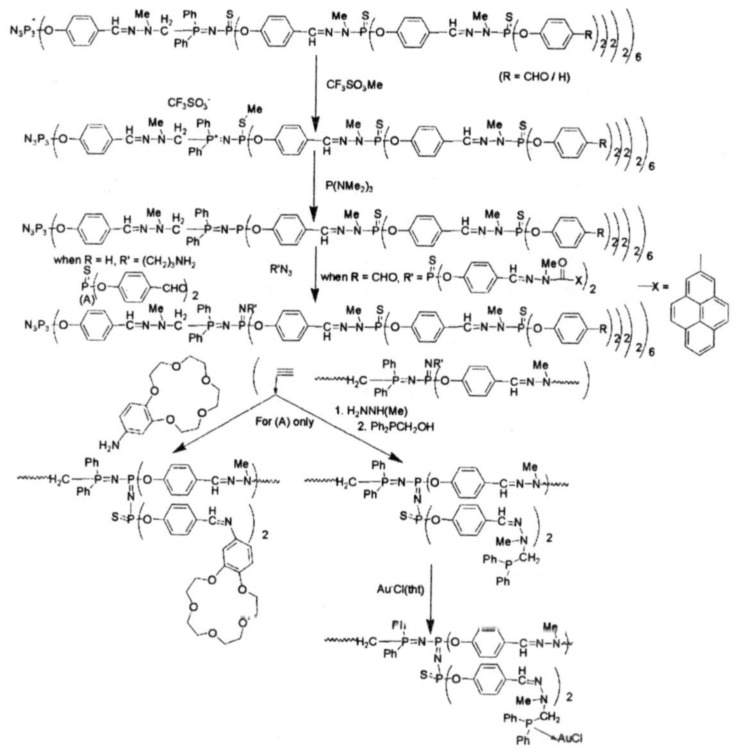

SCHEME 9.40 Modification of phosphorous containing dendrimers at internal P=N-P=S linkages by a series of reactions

layers. A third series of internally functionalized dendrimers was obtained by Staudinger reaction of functionalized azides with the aminophosphite internal groups.

The presence of aldehyde groups inside the dendrimer is really interesting as it can undergo various types of functionalization. In fact, almost all the reactions described above for functionalization of surface aldehyde groups are also applicable here. These inner aldehyde groups were also utilized [62] to functionalize with crown-ether, phosphines and azide. Even this phosphine moieties formed complex with AuCl inside the dendrimers [63] with 90 gold atoms in different layers. The X-ray study confirmed the structure of a 1st generation dinuclear complex. These reactions further support the presence of large amount of space inside the phosphorous dendrimers, as it can accommodate large numbers of groups inside its structure. Similar information was also obtained from the fluorescent label study [64] of the modified dendrimers (Scheme 9.40, where R= CHO and R' = (S)P-(OC_6H_4CH=N-N(Me)-CO(CH_2)$_2$-pyrine)$_2$. The fluoroescence data showed that the fluorescent labels more freely inside the dendrimer. So, it is possible to construct the new branches inside the dendrimer by proper exploitation of interfunctional groups. Thus, by using two steps synthetic process of scheme 9.7 or three steps synthetic process of scheme 9.10, the exploitation of internal aldehyde group of post-synthetic modified dendrimers was reported [65] as shown in scheme 9.40. In the first strategy, the internal aldehyde groups condensate with sodium salts of 4-hydroxy benzaldehyde offered the grafting of new branch dendrimers within the cavities. Reiteration of the two steps permits the growth of higher generations (3rd and 4th) internal dendrimers (Scheme 9.41). In the second strategy [65], first condensation of internal aldehyde group was carried out with methyl hydrazine followed by treatment of the resulted hydrazone with phosphane, Ph_2PCH_2OH and finally then with $N_3P(S)(OC_6H_4CHO)_2$ to get the desired internal dendrimers. The higher generation (3rd and 4th) dendrimers were obtained by repetition of the above three steps] (Scheme 9.42). In both the schemes (9.41 and 9.42), all steps generated

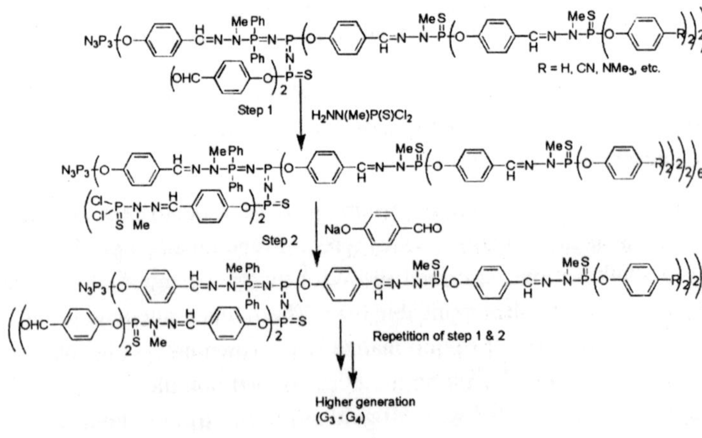

SCHEME 9.41 Modification of phosphorous containing dendrimers at internal P=N-P=S linkages to generate internal dendrimer by route A

some internal functionalities, which can be again utilized for further functionalization or complexation reactions. In the internal layers of such dendrimers (Scheme 9.42) with zirconium derivative, zwitter ions were grafted [66]. The presence of P=N-P=S linkages in the internal branches of the post synthetic modified dendrimers may further open up a new architecture with proper functionalization of the dendrimers. All these reactions along with many other examples concerning the internal reactivity of the dendrimers were reviewed [67].

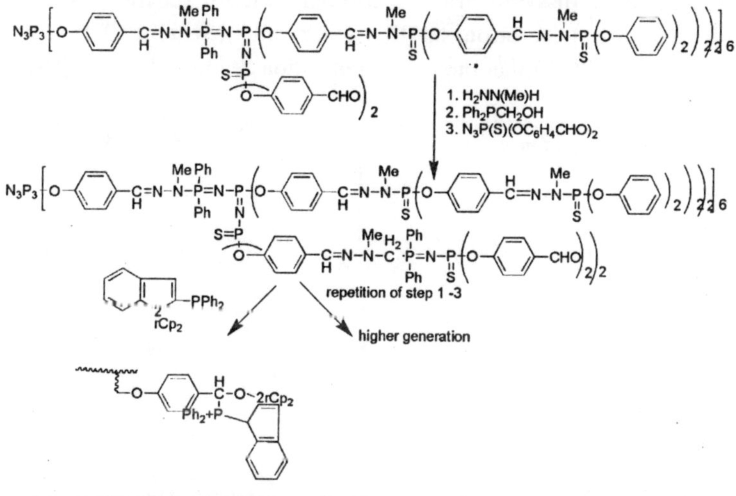

SCHEME 9.42 Modification of phosphorous containing dendrimers at internal P=N-P=S linkages to generate internal dendrimer by route B

9.3.3 Modification through Core

The reactions of the functional groups present in the core are very interesting, as described earlier. The vinyl group located at the core of the dendrons (Scheme 9.9) was activated by the P=N-P=S linkage and undergoes Michael type additions of various functionalized amines and hydrazines [20]. The primary amines of various functionalized dendrons open up the possibility of further reactions, such as condensation reaction to graft diphosphine or azide at the core. These types of reactions offer to generate new dendritic architectures [20] (Scheme 9.43). Further, the azide group can be reacted to other dendrimers possessing diphenylphosphine unit in their structures. For example, the 'layer block' [68] phosphorous containing dendrimer was synthesized [20] by a single process using above 3rd generation azide terminated dendrimer with third generation diphenyl phosphine terminated dendrimer [42]. The 8th generation dendrimer was obtained just by one step (Scheme 9.44). Similarly, a 'surface block' [68] phosphorous containing dendrimer was also synthesized by associating two different type of dendrons through their cores. This dendron possesses two different types of end groups on the two ends of its structure. By using this strategy, a dissymmetrical dendrimer with 3rd generation on one side and 8th generation on the other side (Scheme 9.45) was synthesized [22]. Further more, 3rd generation dendron (Scheme 9.9)

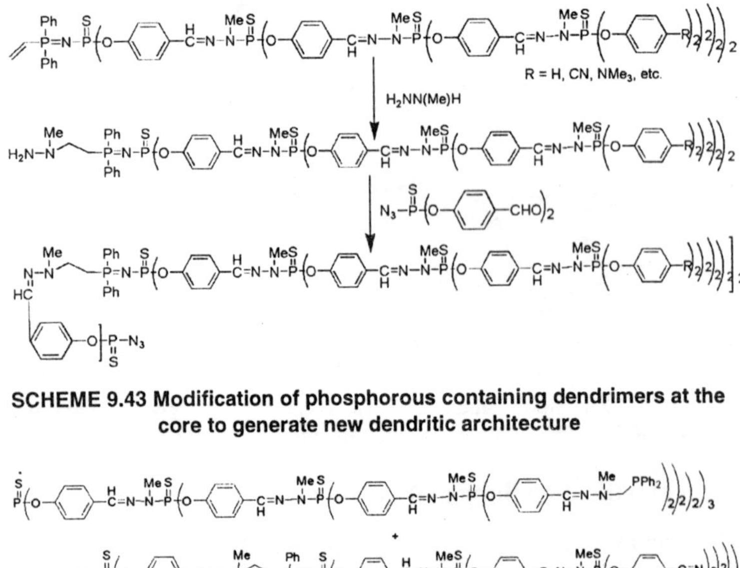

SCHEME 9.43 Modification of phosphorous containing dendrimers at the core to generate new dendritic architecture

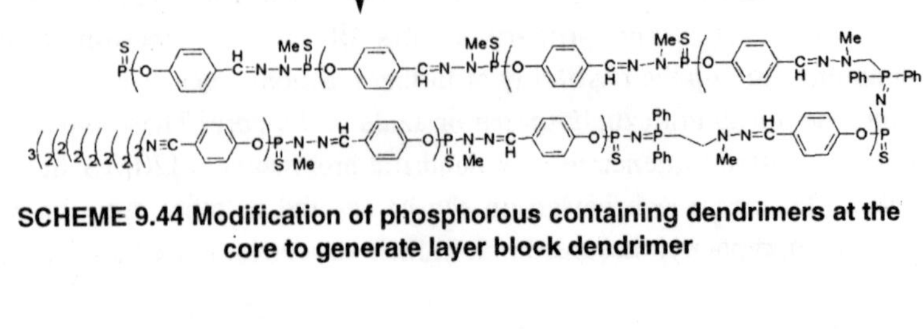

SCHEME 9.44 Modification of phosphorous containing dendrimers at the core to generate layer block dendrimer

was grafted directly in the internal layer of 3rd generation (Scheme 9.41) dendrimer to get G_3-G_3 dendrimer (similar to Scheme 9.42). It is also possible to grow step by step a new dendron starting from the core of a dendron to obtain other type 'surface block' phosphorous containing dendrimers [69].

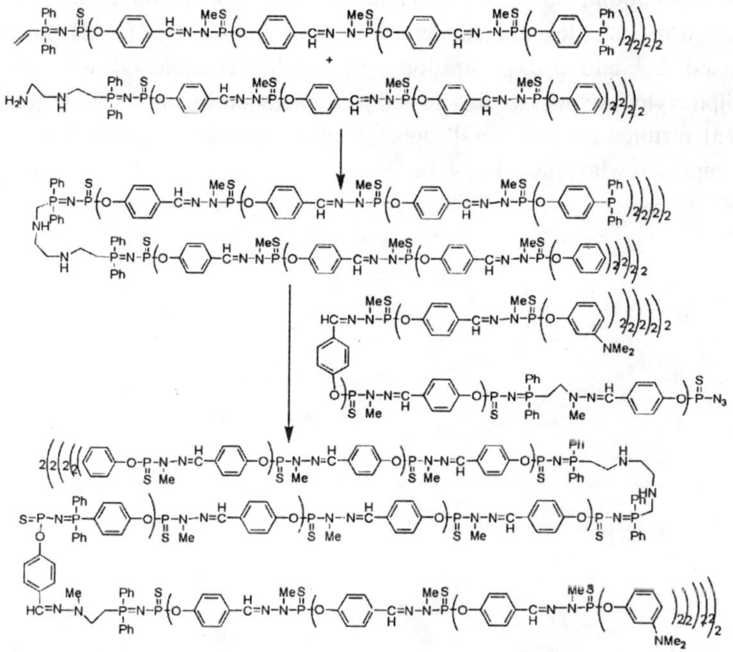

SCHEME 9.45 Modification of phosphorous containing dendrimers at the core to generate surface block dendrimer

There are some other reports where phosphorous atoms are attached at the core or on the surface of other type of dendrimers. These are not truly phosphorous containing dendrimers but the phosphorous atoms have definite role on these dendrimers. Some examples of such dendrimers are presented here. A 2nd generation polyether dendron containing phosphino group at the core, which can coordinate with transition metal complex [*trans*-Ir(CO)Cl] [70]. Again, the complex is reversible with fullerene compound [70] (Fig.9.3). A phosphane substituted polyhedral octasilsesquioxane, $Si_8O_{12}(CH_2CH_2PPh_2)_8$ was used as a core for the building silicon containing dendrimers [71] (discussed in silicon based dendrimer, chapter 8). The coupling of [3,5-*bis*(diphenylphosphinyl) methy]phenyloxy groups to bi-and tri-functional spacer followed by cyclometalation of the ligand precursor compounds with [$NiCl_2$.6HO], Pd[CH_3CN]$_4$(BF_4)$_3$ or *cis*[$PtCl_2(PPh_3)_2$] results Ni, Pd and Pt complexes [72]. The ligand complexes can be used as building blocks for the controlled assembly of both homo- and hetero-multinuclear metallodendrimers. Diaminobutane (DAB) core based 2nd and 3rd generation PPI dendrimers reacted with α–(diphenylphosphino)acetic acid and *p*-(diphenylphosphino)benzoic acid [73] resulted 32 peripheral diphenyl phosphino groups, which were complexed with AuCl (Fig.9.4).

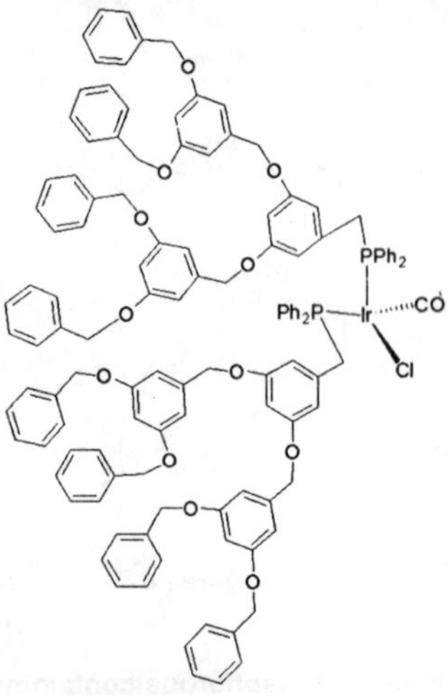

FIG. 9.3 Structure of polyether- phosphorous containing dendrimer-transition metal complex

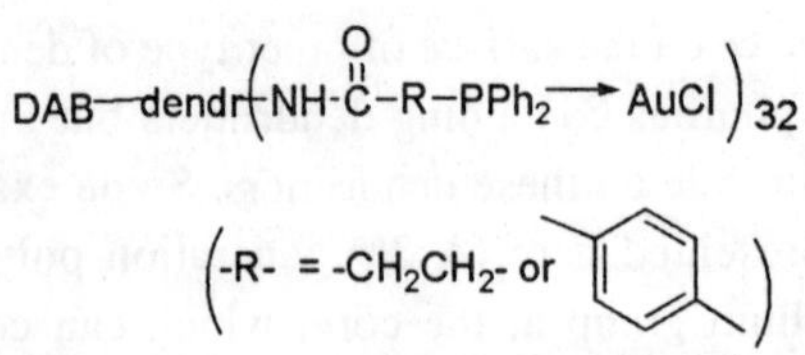

FIG. 9.4 Structure of PPI- phosphorous containing dendrimer-Au complex

Another DAB based 3rd generation PPI dendrimer was phosphinomethylated by using Ph_2PCH_2OH. These terminal $N(CH_2PPh_2)$

groups act as excellent ligands towards $[Pd(CH_3)_2(tmeda)]$, $[NiMe_2(tmeda)]$, $[Ir(Cod)_2BF_4]$ and $Rh(Cod)_2BF_4]$ (Scheme 9.46) [74]. The Pd containing dendrimer complex is an excellent catalyst for the Heck reaction of bromobenzene and styrene with formation of stilbene.

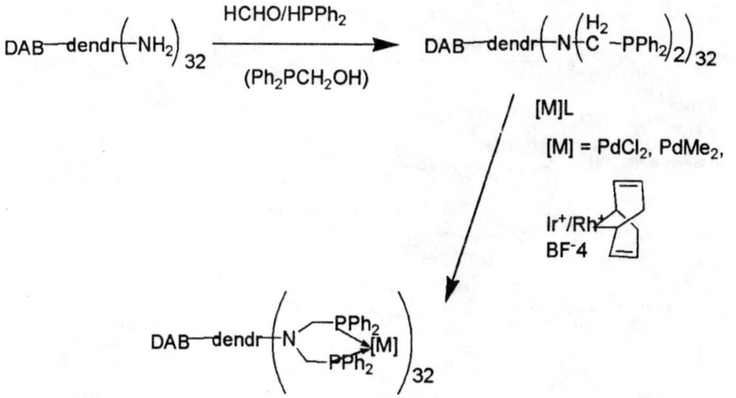

SCHEME 9.46 Modification of PPI dendrimers to generate metal coordinated complex through phosphorous linkages

The synthesis of 2nd generation glycodendrimers based on phosphoramidite building blocks with phosphotriester branching points was reported [75]. The coupling of diisopropylphosphoramidite with alcoholic groups of $O{=}P\text{-}(OCH_2CH_2OCH_2CH_2OH)_3$ in the presence of tetrazole followed by oxidation using tertiarybutylhydroperoxide offers the first generation dendrimer. Again, after deprotection of the end groups by TBAF and repetition of the above sequence of reactions using another phosphoramidite generates 2nd generation, which on grafting with thiolated N-acetylgalactosaminide affords the desired glycodendrimer [75]. The whole sequence of reactions is shown in scheme 9.47. The phosphorus containing dendrimers upto 5th generation based on thiophosphate derivatives at each branching points, were synthesized [76]. The phosphitylation of $S{=}P\text{-}(OCH_2CH_2OCH_2CH_2OH)_3$ with $Et_3N\text{-}P\text{-}(OCH_2CH_2OCH_2CH_2OAc)_2$ followed by oxidation with elemental sulfur resulted the first generation dendrimer with acetate end groups. The transformation of acetate groups to hydroxyl groups by treatment of NH_3 in MeOH afford the first generation dendritic polyol. The repetition of the above sequence of reactions results higher generation dendrimers (Scheme 9.48). By applying the same strategy of reactions, P=Se linkage containing phosphorous dendrimers were also synthesized for the first time [77].

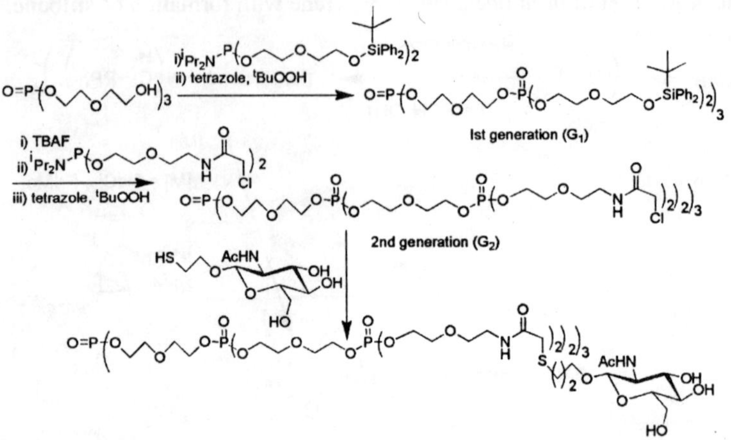

SCHEME 9.47 Synthesis of phosphoramidite containing glycodendrimers

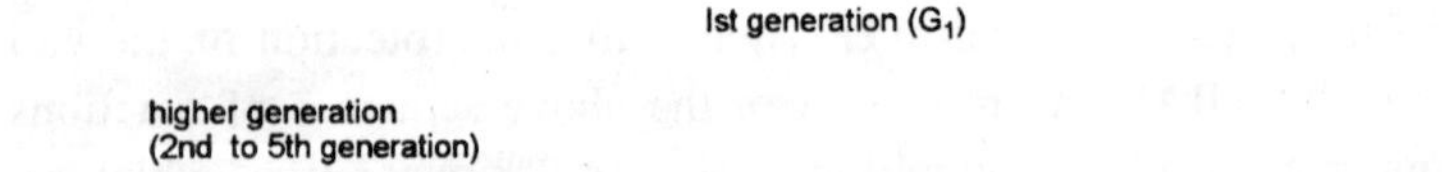

SCHEME 9.48 Synthesis of phosphorous containing dendrimers with thiophosphate derivative

9.4 SYNTHESIS, CHARACTERIZATION AND PROPERTIES OF PHOSPHOROUS CONTAINING HYPERBRANCHED POLYMERS

The synthesis and characterization of linear, HB and dendrimer like structures possessing phosphorous containing polymers were reported [78]. The HB polymers were synthesized by using $OC_6H_4P(Ph_2)$=N-P=S as the repeating unit. Dendrimers upto 5th generation were also synthesized from the same building block. All the polymers were characterized by SEC, solution, intrinsic viscosity and thermal behaviors studies and observed marked differences among the above three types of polymers.

Many functionalized mono-, di-, tri- and tetra-phosphines compounds were synthesized [79]. They utilized these compounds for the synthesis of HB dendrimers with free phosphine groups at each branching point of the structures. The reaction of hexachlorocyclotriphosphazane with hexamethylene diamine resulted a highly branched cyclophosphazene containing polymers upto 8th generation [80-83] (Scheme 9.49). However, it appears [81-84] that the hexachlorocyclotriphosphazene is a suitable core for generation of spherical phosphorous containing dendrimers but with the increase of generation amount of incomplete reactions increased as that was not observed. Even Though efforts were made [80-83] to minimize the incomplete reactions as well as crosslinkings by modifying the reaction conditions and sequence but still one cannot ruled out the complete elimination of the undesirable reactions and hence only HB polymers were obtained.

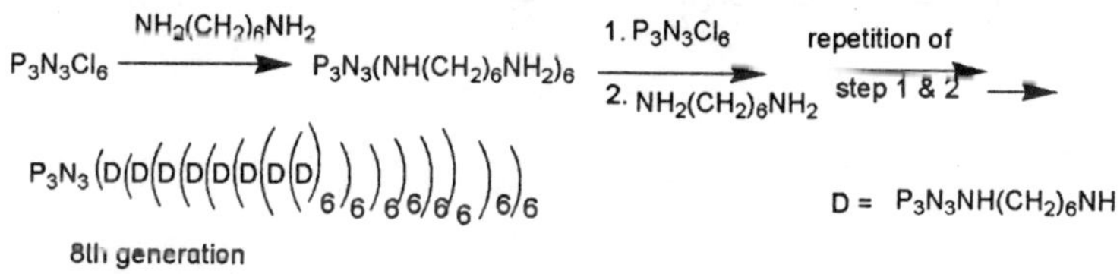

SCHEME 9.49 Synthesis of cyclophosphazene containing HB polymers

The HB phosphorous containing polymer of 4th generation, where phosphine moieties were present in each branching point was synthesized [84]. The synthetic method involved a two steps reaction sequence of which key point was controlled by acid-base hydrolysis of diaminosilanes with phosphine having terminal-OH groups. The reaction sequences are given in scheme 9.50, starting from $P(CH_2CH_2CH_2OH)_3$ core. Although in the scheme, it is found that the polymer having dendritic structure but characterization indicates the HB polymer formation.

Two small phosphorous containing dendrimers, of which one has no reactive functionality on the surface but possesses a vinyl group activated by a P=N-P=S linkage at the core, while the other has

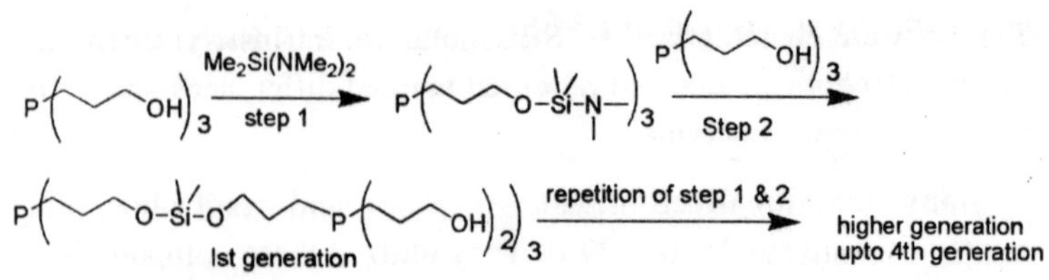

SCHEME 9.50 Synthesis of phosphine containing HB polymers

aldehyde groups on the surface but core has protected phenol group, were synthesized [85]. The first one undergoes Diels-Alder reaction with 2,3-dimethylbutadiene and the protected second one undergoes addition of a diisocyanate on the core. The whole sequence of reactions is shown in scheme 9.51 A and B. All the steps of the reactions were monitored by ^{31}PNMR spectroscopy along with ^{1}HNMR spectroscopy. Mass spectroscopy (FAB) was also used to characterize some of these compounds.

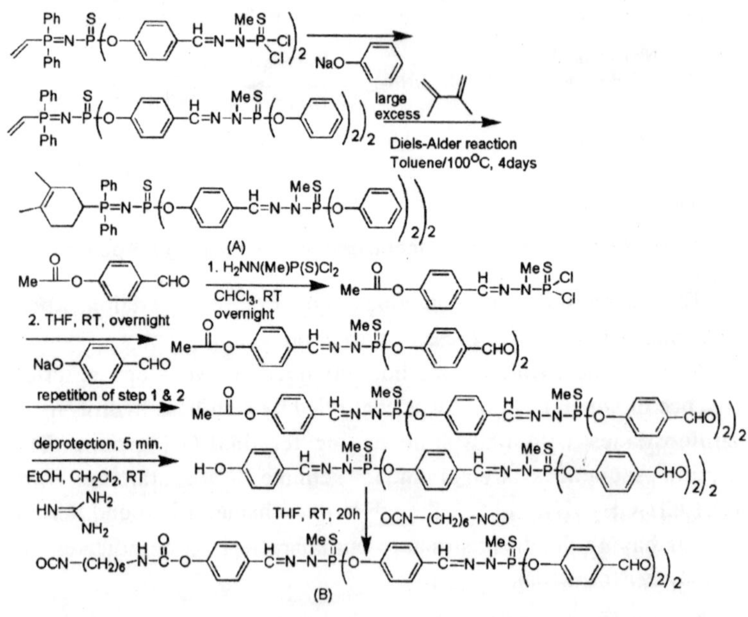

SCHEME 9.51 Synthesis of two phosphorous containing HB polymers with activated vinyl and aldehyde groups at the core and surface separately

9.5 APPLICATIONS OF PHOSPHOROUS CONTAINING DENDRITIC POLYMERS

The most important application of phosphorous containing dendritic polymers is catalysis. The presence of tricoordinated phosphorous atoms, generally on the surface of the dendrimers allows the complexation with various organometallic derivatives having catalytic properties. As in case of other dendritic catalysts, phosphorous containing dendrimeric catalysts are also enjoying both the advantages of homogeneous as well as heterogeneous catalysts. However, the catalytic efficiency of the phosphorous containing dendritic polymers may decrease or increase compared to the analogous monomeric complexes. All types of dendrimers have phosphorous either at the branching points and the surface or only on the surface or only at the core showed above catalytic behavior.

Pd and Ru complexes of phosphorous containing dendrimers (Scheme 9.23) were used [21] for catalysis of Stille coupling reaction and Knoevenagel condensation reactions respectively. The study revealed that a slight increase of catalytic activity from monomer to first generation to the higher generation dendrimers. The third generation diphosphine dendrimers containing 24 Pd or Ru was catalyzed the Stille coupling reaction of methyl-2-iodobenzenoate with 2-[tributylstannyl]thiophene, the Knoevenagel condensation between malononitrile and cyclohexanone leading to the corresponding unsaturated nitrile and the Michael addition between ethyl cyanoacetate and diethylethylid (Scheme 9.12). The catalytic activity of the dendrimer was almost similar or often higher than the corresponding monomeric complex. However, the dendritic catalyst (upto 3rd generation dendrimer) can be easily recovered, just by precipitation with diethyl ether and can be reused at least three times without significant loss of activity [21]. There are other reports [86-90] on the catalytic activity of phosphorous containing dendrimers, where phosphorous present as the complexing agent. Another important use of water soluble phosphorous containing dendrimer is as hydrogel. The water soluble dendrimers (Scheme 9.35) even at very low concentration (1.25% in weight) were readily formed solid hydrogels on heating [53], which allow confinement of acids, metals, etc. It is very surprisingly found that a single dendrimer (4th generation, pyridinum salt, Scheme 9.29) molecule can able to get upto 5 X 10^5 number of water molecules! These gels are translucent, hard and can be crushed into pieces. The reason behind such hydrogel formation may be H-bonding and hydrophillic/hydrophobic interactions. As it allows to confinement of small molecules, so during gelation, several water soluble substance can be incorporated within the gel. A high loading (30%) of nickel acetate was loaded in one case [53].

The phosphorous containing dendrimers were also used for construction of nanocluster compounds. For example, 1st generation alcoholic or carboxylic acid end groups containing neutral phosphorous dendrimer (Scheme 9.35) can react with a titanium cluster, $Ti_{16}O_{16}(OEt)_{32}$ offers meso-structured solid in which the dendrimers act as spacers for the clusters [54]. Triethoxysilyl core based phosphorous containing dendrimers were used [23] to construct mesoporous functionalized silica. A 4th generation thiol terminated phosphorous dendrimer with 96 SH groups and gold cluster compound, $Au_{55}(PPh_3)_{12}Cl_6$ in a 3:1 ratio in dichloromethane results the formation of Au_{55}-clusters. The role of the dendrimers is not only to remove the phosphine and chloride ligands but also to act as an ideal matrix for perfect microcrystals. Again, thiol terminated phosphorous containing dendrimers were used for obtaining the first network of clusters of naked gold, which might have tremendous potential in nano electronic applications [91].

The phosphorous dendrimers containing upto 46 phosphorous atoms with [RhCl($^4\eta$-1,5-C_8H_{12})PR_3] moieties were found to be catalyzed [84] the hydrogenation of 1-decene in a 1:2 metal to substrate ratio, under the reaction conditions of 25°C temperature, 20 bar pressure in THF solvent. The catalytic activity was similar to the corresponding monomeric complex but these dendritic catalysts are recyclable with minor loss in activity and can be easily separated from the reaction mixture.

The electroactive dendrimers viz. ferrocenes [31, 38, 92], TTF-crown-ethers [50] and bithiophenes [93] surface grafted phosphorous dendrimers can be used to modify the surface of electrodes. The modified electrodes may work reversible (e.g. ferrocene or TTF end groups) or irreversible (e.g. bithiophene) depending on the type of the electroactive derivative. TTF-crown-ether surface modified electroactive dendrimers showed sensing properties for detection of barium [50]. Even by grafting of aldehyde terminated phosphorous dendrimers (4th generation, Scheme 9.7) on quartz plates makes that usable for immobilization application [94].

Different biological applications of phosphorous containing dendrimers have tremendous potential. One such field is delivery of active substance in a large sense. An insecticide (fipronil) was delivered [95] by help of water soluble phosphorous containing dendrimers. The pesticide linked to the dendrimer through a linker and depending on the chemical stability of the linker forwards hydrolysis, the release of fipronil can be controlled. The transportation of active substance through these phosphorous containing dendrimers occurs via different types of interactions. These may be (i) strong covalent interaction, (ii) weak interaction through end groups, (iii) interaction in the internal layer or (iv) interaction through a network of the dendrimers. The glycodendrimers (Scheme 9.47) were used as inhibitors of *vicia villosa* binding to asialglycophorin. An enhanced affinity was observed in the study [75] compared to the efficiency of monomeric species. There are some patents [96-97] on different applications of phosphorous containing dendrimers. The phosphorous containing dendrimers upto 5th generation, with protonated or methylated terminal tertiary amine were used as transfection agents after grafting, positively charged functions. The reagents for transfecting are nucleic acid into eukariotic cells, both with and without serum [96]. Such dendrimers upto 5th generation, were examined as transfecting agents of the luciferase gene within 3T3 cells [39]. The capability of transfection depends on the size (generation) of the dendrimer and the chemical nature of the terminal tertiary amines. Unexpectedly, these dendrimers have a better transfection efficiency in the presence of serum than that of without dendrimer one [39]. The phosphorous containing dendrimers were also used as drugs to treat disease link to abnormal aggregated proteins, such as neurodegenrative disease, and particularly Prion's disease [97] in animals.

The phosphorous containing dendrimers are also used as chemical sensor, biosensor, heterogeneous catalysts, nanoelectronic materials, oxidizing agents, aerogels for cosmetic, perfume, aroma, etc., photoactive materials in diagnostic and therapeutic agents. They were also applied for treatment of effluent, in ink formulation, as glues, etc. A series of Cl-Au terminated phosphorous containing dendrimers was used [98] to form Fe_2Au or Fe_3Au clusters by treatment of THF solution of the dendrimer with iron anions, $[Fe_2(CO)_7(PPh_2)]^-$ and $[Fe(CO)_{11}]^{2-}$.

The use of alkyl phosphine containing dendrimers as ligands was also reported [99] for the hydrocarboxylation of alkenes (hexene, octene, nonene, prop-1-ene-2-ol) in polar solvents (EtOH or THF) using the complex $[Rh(acac)(CO)_2]$ or $[Rh_2CO_2\ (CMe_3)_4]$ as the metal source.

9.6 SUMMARY

In this chapter the synthesis, modification, characterization and properties of different types of phosphorous containing dendritic polymers are discussed. The phosphorous containing dendritic polymers are unique, just not as another category of dendritic polymers, but with respect to their synthesis, reactivity, characterization and properties. Due to the versatile behavior of the phosphorous atom, a large number of synthetic methods are possible to use for their syntheses, which resulted a large variety of dendritic polymers. Most interestingly, the highest generation (12th) dendrimer is possible to obtain by one of the most rapid synthetic method with eco-friendly bi-products like water and nitrogen. Even though, they have very complex architecture in their structures like other dendritic polymers but they can be easily characterized at each step by ^{31}PNMR spectroscopy. They are highly thermostable. Because of their versatility in reactivity of surface, internal layers and core moieties, they can be easily functionalized to generate differently modified phosphorous dendrimers. The surface group reactivity, especially the aldehyde group makes them one of the most versatile modified dendrimers. Again, the presence of P=N-P=S linkages in the internal layers allows them to modified uniquely inside the dendrimers. Also, they have some unique properties like very high dipole moment, which is not found any other categories of dendritic polymers. They also formed complexes with several metals and thereby producing different types of new catalysts. Their reactivity is so versatile that they can grow other dendritic branches with the parent dendrimer, which is not found in any other kind of dendritic polymers. This is due to uniqueness of phosphorous atom, which can be tri-, tetra-, or penta-coordinated. The dendrimers have tremendous potential in many useful fields including chemistry, materials, biology, physics and environmental sciences. They can be used for active agents in delivery systems, nanoscale catalysts, immunodiagostics agents, chemical sensors, etc. In a nutshell, phosphorous containing dendrimers open up a new field of research and applications.

REFERENCES

[1] J.P.Majoral and A.M.Caminade, *Chem.Rev.* **99**(1999)845

[2] K.Rengan and R.Engel, *J.Chem.Soc.Chem.Commun.* (1990)1084

[3] N.Launay, A.M.Caminade, R.Lahana and J.P.Majoral, *Angew.Chem.Int.Ed.* **33**(1994)1589

[4] A.Meidaner, C.J.Curtis, R.M Barkley and D.L.Dubois, *Inorg.Chem.* **33**(1994)5482

[5] J.P.Majoral and A.M.Caminade, *Top.Cur.Chem.* **223**(2003)111

[6] J.P.Majoral, A.M.Caminade and V.Maraval, *Chem.Commun.* (2002)2929

[7] R.Engel, K.Rengan and C.S.Chan, *Heteroatom Chem.* 4(1993)181

[8] R.Engel, K.Rengan and C.S.Chan, *Phosphorus, Sulfur and Silicon* **77**(1993)221

[9] R.Engel, Advances in Dendritic Macromolecules, Vol.2, JAI Press Inc. New York, p.73, 1995

[10] A.M.Herring, B.D.Steffey, A.Meidaner, S.A.Wander and D.L.Dubois, *Inorg.Chem.* **34**(1995)1100

[11] K.Rengan and R.Engel, *J.Chem.Soc.Perkin Trans. 1* (1991)987

[12] R.Engel, *Polym.News* **17**(1992)301

[13] C.Galliot, D.Prevote, A.M.Caminade and J.P.Majoral, *J.Am.Chem.Soc.* **117**(1995)5470

[14] A.M.Caminade, M.Slany, N.Launay, M.L.Lartigue and J.P.Majoral, *Phosphorus, Sulfur and Silicon* **109**(1996)517

[15] M.L.Lartigue, B.Donnadieu, C.Galliot, A.M.Caminade, J.P.Majoral and J.P.Fayet, *Macromolecules* **30**(1997)7335

[16] M.L.Lartigue, A.M.Caminade and J.P.Majoral, *Phosphorus, Sulfur and Silicon* **123**(1997)21

[17] M.L.Lartigue, M.Slany, A.M.Caminade and J.P.Majoral, *Chem.Eur.J.* **2**(1996)1417

[18] M.L.Lartigue, N.Launay, B.Donnadieu, A.M.Caminade and J.P.Majoral, *Bull.Soc.Chim.Fr.* **134**(1997)981

[19] N.Launay, A.M.Caminade and J.P.Majoral, *J.Organomet.Chem.* **529**(1997)51

[20] V.Maraval, R.Laurent, B.Donnadieu, M.Mauzac, A.M.Caminade and J.P.Majoral, *J.Am.Chem.Soc.* **122**(2000)2499

[21] V.Maraval, R.Laurent, A.M.Caminade and J.P.Majoral, *Organometallics* **19**(2000)4025

[22] V.Maraval, R.Laurent, S.Merino, A.M.Caminade and J.P.Majoral, *Eur.J.Org.Chem.* (2000)3555

[23] C.O.Turrin, V.Maraval, A.M.Caminade, J.P.Majoral, A.Mehdi and C.Reye, *Chem.Mater.* **12**(2000)3848

[24] C.Larre, A.M.Caminade and J.P.Majoral, *Angew.Chem.Int.Ed.* **36**(1997)596

[25] L.Buruge, G.Margro, A.M.Caminade and J.P.Majoral, *J.Am.Chem.Soc.* **123**(2001)6698

[26] L.Brange, G.Magro, A.M.Caminade and J.P.Majoral, *J.Am.Chem.Soc.* **123**(2001)8446

[27] C.Larre, B.Donnadieu, A.M.Caminade and J.P.Majoral, *Eur.J.Inorgchem.* (1999)601

[28] A.Balueva, S.Merino, A.M.Caminade and J.P.Majoral, *J.Organomet. Chem.* **643**(2002)112

[29] N.Launay, C.Gallios, A.M.Caminade and J.P.Majoral, *Bull.Soc.Chem. Fr.* **132**(1995)1149

[30] C.O.Turrin, J.Chiffre, J.C.Daran, D.de Montauzon, G.Balavoine, E.Manoury, A.M.Caminade and J.P.Majoral, *Comptes Rendus.Chimie* **5**(2002)309

[31] C.O.Turrin, J.Chiffre, J.C.Daran, D.de Montauzon, A.M.Caminade, E.Manoury, G.Balavoine and J.P.Majoral, *Tetrahedron* **57**(2001)2521

[32] R.M.Sebastian, J.C.Blais, A.M.Caminade and J.P.Majoral, *Chem.Eur.J.* **8**(2002)2172

[33] J.C.Blais, C.O.Turrin, A.M.Caminade and J.P.Majoral, *Anal.Chem.* **72**(2000)5097

[34] V.L.Furer, V.I.Kovalenko, A.E.Vandyukovb, J.P.Majoral and A.M.Caminade, *Vibrational Spectroscopy* **31**(2003)71

[35] V.I.Kovalenko, V.L.Furer, A.E.Vandyukovb, R.R.Shagidullin, J.P.Majoral amd A.M.Caminade, *J.Mol.Struct.* **604**(2002)45

[36] J.P.Majoral and A.M.Caminade, *Top.Cur.Chem.* **197**(1998)79

[37] N.Launay, M.Slany, A.M.Caminade and J.P.Majoral, *J.Org.Chem.* **61**(1996)3799

[38] C.O.Turrin, J.Chiffre, D.de Montauzon, J.C.Daran, A.M.Caminade, E.Manoury, G.Balavoine and J.P.Majoral, *Macromolecules* **33**(2000)7328

[39] C.Loup, M.A.Zanta, A.M.Caminade, J.P.Majoral and B.Meunier, *Chem.Eur.J.* **5**(1999)3644

[40] D.Prevote, A.M.Caminade and J.P.Majoral, *J.Org.Chem.* **62**(1997)4834

[41] A.M.Caminade, R.Laurnt, B.Chaudret and J.P.Majoral, *Coord.Chem.Rev.* **178-180**(1998)793

[42] M.Slany, M.Bardaji, M.J.Casanove, A.M.Caminade, J.P.Majoral and B.Chaudret, *J.Am.Chem.Soc.* **117**(1995)9764

[43] M.Slany, M.Bardaji, A.M.Caminade, B.Chaudret and J.P.Majoral, *Inorg.Chem.* **36**(1997)1939

[44] M.Bardaji, M.Kustos, A.M.Caminade, J.P.Majoral and B.Chaudret, *Organometallics* **16**(1997)403

[45] M.Bardaji, A.M.Caminade, J.P.Majoral and B.Chaudret, *Organometallics* **16**(1997)3489

[46] M.Blanzat, C.O.Turrin, E.Pecez, I.Rico-Lattes, A.M.Caminade and J.P.Majoral, *Chem.Commun.* (2002)1864

[47] G.Margo, B.Donnadieu, A.M.Caminade and J.P.Majoral, *Chem.Eur.J.* **9**(2003)2152

[48] V.Maraval, A.M.Caminade, J.P.Majoral andJ.C.Blais, *Angew.Chem.Int. Ed.* **42**(2002)1822

[49] R.M.Sebastain, J.C.Blais, A.M.Caminade and J.P.Majoral, *Chem.Eur.J.* **8**(2002)2172

[50] F.LeDerf, E.Levilain, G.Trippe, A.Gorgues, M.Salle, R.M.Sebastain, A.M.Caminade and J.P.Majoral, A.M.Caminade and J.P.Majoral, *Angew.Chem.Int.Ed.* **40**(2001)224

[51] N.Launay, A.M.Caminade and J.P.Majoral, *J.Am.Chem.Soc.* **117**(1995) 3283

[52] D.Prevote, S.LeRoy-Gourvennec, A.M.Caminade, S.Masson and J.P.Majoral, *Synthesis* (1997)1199

[53] C.Mamillon, F.Gauffre, T.Gulik-Krzywicki, C.Loup, A.M.Caminade, J.P.Majoral, J.P.Vors and E.Rump, *Angew.Chem.Int.Ed.* **40**(2001)2626

[54] M.K.Boggiano, G.J.A.Asoler-Illia, L.Rozes, C.Sanchez, C.O.Turrin, A.M.Caminade and J.P.Majoral, *Angew.Chem.Int.Ed.* **39**(2000)4249

[55] R.M.Sebastian, G.Magro, A.M.Caminade and J.P.Majoral, *Tetrahedron* **56**(2000)6269

[56] M.L.Lartigue, A.M.Caminade and J.P.Majoral, *Tetrahedron Asym.* **8**(1997)2697

[57] M.Slany, A.M.Caminade and J.P.Majoral, *Tetrahedron Lett.* **37**(1996)9053

[58] M.Bardaji, M.Slany, M.L.Lartigue, A.M.Caminade, B.Chaudret and J.P.Majoral, *Main Group Chem.* **2**(1997)133

[59] D.Prevote, B.Donnadieu, M.Moreno-Manas, A.M.Caminade and J.P.Majoral, *Eur.J.Org.Chem.* (1999)1701

[60] C.Larre, A.M.Caminade and J.P.Majoral, *Angew.Chem.Int.Ed.* **36**(1997)596

[61] C.Larre, B.Donnadieu, A.M.Caminade and J.P.Majoral, *J.Am.Chem.Soc.* **120**(1998)4029

[62] C.Larre, D.Bressolles, C.Turrin, B.Donnadieu, A.M.Caminade and J.P.Majoral, *J.Am.Chem.Soc.* **120**(1998)13070

[63] C.Larre, A.M.Caminade and J.P.Majoral, *Chem.Eur.J.* **4**(1998)2031

[64] L.Brauge, A.M.Caminade, J.P.Majoral, S.Slomkowski and M.Wolszczak, *Macromolecules* **34**(2001)5599

[65] C.Galliot, C.Larre, B.Donnadieu, A.M.Caminade and J.P.Majoral, *Science* **277**(1997)1981

[66] V.Cadierno, A.Igau, B.Donnadieu, A.M.Caminade and J.P.Majoral, *Organometallics* **18**(1999)1580

[67] J.P.Majoral, C.Larre, R.Laurent and A.M.Caminade, *Coord.Chem.Rev.* **190-192**(1999)3

[68] C.J.Hawker and J.M.J.Frechet, *J.Am.Chem.Soc.* **114**(1992)9405

[69] V.Maraval, R.M.Sebastian, F.Ben, R.Laurent, A.M.Caminade and J.P.Majoral, *Eur.J.Inorg.Chem.* (2001)1681

[70] V.J.Catalano and N.Parodi, *Inorg.Chem.* **36**(1997)537

[71] B.Hong, T.P.S.Thoms, H.J.Murfer and M.J.Lebrun, *Inorg.Chem.* **36**(1997)6146

[72] W.T.S.Huck, B.SnellinkRuel, F.C.J.M.van Veggel and D.N.Reinhoudt, *Organometallics* **16**(1997)4287

[73] P.Lange, A.Schier and H.Schmidbaur, *Inorg.Chem.* **35**(1996)637

[74] M.T.Reetz, G.Lohmer and R.Schwickardi, *Angew.Chem.Int.Ed.* **36**(1997)1526

[75] R.Roy, *Top.Cur.Chem.* **187**(1997)241

[76] G.Salamonczyk, M.Kuznikowski and A. Skowronska, *Tetrahedron Lett.* **41**(2000)1643

[77] G.Salamonczyk, M.Kuznikowski and E.Poniatowska, *Chem.Commun.* (2001)2202

[78] S.Merino, L.Brauge, A.M.Caminade, J.P.Majoral, D.Taton and Y.Gnanon, *Chem.Eur.J.* **7**(2001)3095

[79] V.Huc, A.Balueva, R.M.Sebastian, A.M.Caminade and J.P.Majoral, *Synthesis* (2000)726

[80] F.Sournies, F.Crasnier, M.Graffeuil, J.P.Faucher, R.Lahana, M.C.Labarre and J.F.Labarre, *Angew.Chem.Int.Ed.***36**(1997)1526

[81] F.Sournies, L.Labrousse, M.Graffeuil, F.Crasnier, J.P.Faucher, M.C.Labarre and J.F.Labarre, *Phosphorus, Sulfur and Silicon* **89**(1994)47

[82] F.Sournies, C.Fonga, M.Graffeuil, F.Crasnier, J.P.Faucher, M.C.Labarre and J.F.Labarre, *Phosphorus, Sulfur and Silicon* **90**(1994)159

[83] F.Sournies, F.Crasnier, C.Vidal, M.C.Labarre, and J.F.Labarre, *Main Group Chem.* **1**(1996)207

[84] M.Pefrucci-samiya, V.Guillemette, M.Dasgupta and A.K.Kakhar, *J.Am.Chem.Soc.* **121**(1999)1968

[85] L.Brauge, V.Maraval, R.Laurent, A.M.Caminade and J.P.Majoral, *ARKIVOC* (2002)151

[86] R.Kreiter, A.W.Kleij, R.J.M.K.Gebbink and G.van Koten, *Top.Cur.Chem.* **217**(2001)163

[87] D.Astruc and F.Chardac, *Chem.Rev.* **101**(2001)2991

[88] C.E.Oosteron, J.N.K:Reek, P.C.J.Kamer and P.W.N.M.van Leeuwen, *Angew.Chem.Int.Ed.* **40**(2001)1828

[89] L.J.Twyman, A.S.H.King and I.K.Martin, *Chem.Soc.Rev.* **31**(2002)69

[90] A.M.Caminade, V.Maraval, R.Laurent and J.P.Majoral, *Cur.Org.Chem.* **6**(2002)739

[91] G.Schmid, W.Meyer-Zaika, R.Pugin, T.Sawitowski, J.P.Majoral, A.M.Caminade and C.O.Turrin, *Chem.Eur.J.* **6**(2000)1693

[92] C.O.Turrin, J.Chiffre, D.de Montauzon, G.Balavoine, E.Manoury, A.M.Caminade and J.P.Majoral, *Organometallics* **21**(2002)1891

[93] R.M.Sebastian, A.M.Caminade, J.P.Majoral, E.Levillain, L.Huchet and J.Ronacali, *Chem.Commun.* (2000)507

[94] B.Misksa, S.Slomkowski, M.M.Chehimi, M.Delamar, J.P.Majoral and A.M.Caminade, *Colloid Polym.Sci.* **277**(1999)58

[95] R.Goeller, J.P.Vors, A.M.Caminade and J.P.Majoral, *Tetrahedron Lett.* **42**(2001)3587

[96] J.P.Majoral, B.Meunier, A.M.Caminade, C.Loupand, M.A.Zanta-Boussif, Fr.Pat. WO 0138335 (2001)

[97] J.P.Majoral, J.Solassol, C.Loupand, J.Leclair, S.Lehmann, A.M.Caminade and B.Meunier, Fr. Pat. Applied (2002)

[98] M.Benito, O.Rossell, M.Seco, G.Segales, V.Maraval, R.Laurent, A.M.Caminade and J.P.Majoral, *J.Organomet.Chem.* **622**(2001)33

[99] L.Ropartz, D.F.Foster, R.E.Morris, A.M.Z.Slawin and D.J.Cole-Hamiltton, *J.Chem.Soc.Dalton Trans.* (2002)1997

CHAPTER 10

OTHERS DENDRITIC POLYMERS

10.1 INTRODUCTION

The dendritic polymers, which are not possible to include in the earlier chapters, are discussed in this chapter. As these dendritic polymers are also very important for different applications such as catalysts, electronic materials, bioactive materials, etc. so their syntheses, characterizations and properties evaluation are also equally important like other categories of dendritic polymers and hence are discussed here.

10.2 SYNTHESIS, CHARACTERIZATION AND PROPERTIES OF OTHER DENDRIMERS

In this section the synthesis, characterization and properties of the following dendrimers are discussed.

10.2.1 Polycarbonates

Although the HB polycarbonates were reported first [1] but the the synthesis of first ideal polycarbonate dendrimers was reported [2] in 2000 by an one-pot multiaddition convergent method. 2,6-dimethyl-4-heptanol or *tert*-butyl alcohol was used as the end capping compound, although any secondary or tertiary alcohol may be used. Initially, glycerol was used as a building block but it formed cyclic carbonate through the selective reaction of the imidazole carboxylic ester with the primary hydroxyl group and subsequent intermolecular cyclization with the secondary hydroxyl group and finally elimination of the parent alcohol. So, 1-[N,N-*bis*(2-hydroxyethyl)amino]-2-propanol (HEAP) was used as the building block, and no cyclized product was formed as hydroxyl groups, are five atoms apart. The first step of this synthetic approach was the imidazole carboxylic ester formation by the reaction of 2,6-dimethyl-4-heptanol (DMH) with CDI in the presence of KOH as the catalyst. This resulted ester was then reacted with HEAP by using the same reaction conditions to get 1st generation dendron. Finally, the activated imidazole carboxylic ester focal group containing different generation dendrons coupled with a 10 folds excess of 1,5-pentanediol to get the desired polycarbonate dendrimers. The whole sequence of reactions is shown in scheme 10.1.

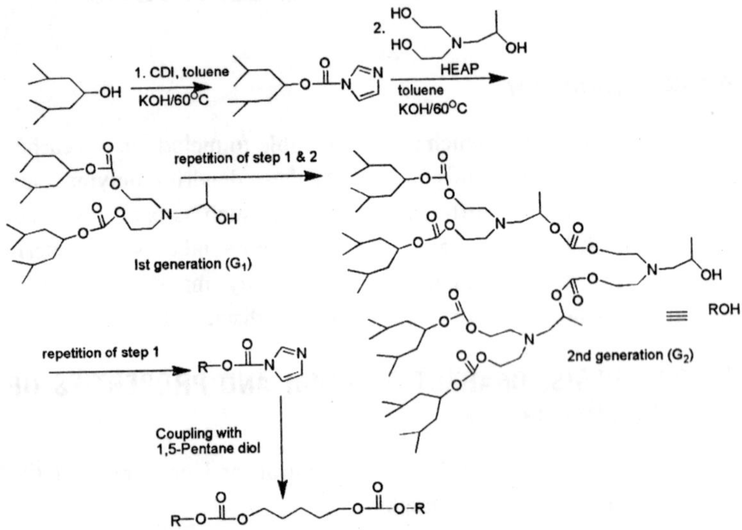

SCHEME 10.1 Synthesis of polycarbonate dendrimers by selective esterification reactions

There are so many advantages of these approaches while CDI was used [2]. The most important among these are: (i) no purification is required and (ii) high selectivity of the reactions. The synthesized product in each step was obtained with very high yield (85-77.6)% and characterized by ^{1}HNMR, ^{13}CNMR spectroscopy and electrospray mass spectrometry. The molecular weight obtained from mass spectra for the 2nd generation dendrimer was 2319.56 Da.

10.2.2 Quaternary Ammonium Ions

The 'balloon' and 'star' shaped ammonium dendrimers were synthesized in 1992 [3] by repetitive sequence of reactions from triethanolamine. The dendrimers were synthesized by alkylation of triethanolamine with an excess of appropriate alkyl halide in acetonitrile. Methyl iodide and benzyl bromide or 2-chloroethanol were used as the quarterizing agents for the core. The salts were obtained by concentration of the reaction mixture under reduced pressure. To get higher generations dendrimers, the salts were then tosylated using an excess of tosylchloride with pyridine in acetonitrile. The resulted tosylates were then immediately treated with an excess of triethanolamine in acetonitrile by using the same procedure as of initial quarternization. The addition of further generations was accomplished by repetition of tosylation and alkylation reactions. The dendrimers upto 3rd generation (Fig.10.1) were obtained by this technique with very high yield in each stage. The dendrimers were characterized by ^{1}HNMR and IR spectroscopic studies.

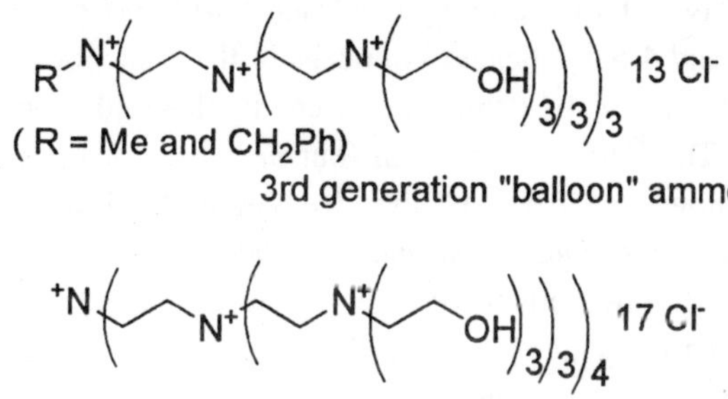

3rd generation "balloon" ammonium dendrimer

3rd generation "star" ammonium dendrimer

FIG.10.1 Structure of ammonium dendrimers based on triethanol amine core

The dendrimeric ion exchange resin based on ammonium dendrimer was also synthesized [4]. For this synthesis, the Merrified resin (2% crosslinked chloromethylated styrene/divinylbenzene copolymer, 200-400 mesh size and ~1 mequiv. Clg^{-1}) was treated with a tertiary amine, triethanol amine or 1,4-diazabicyclo[2,2,2]octane (*Dabco*) in 1:1 CH_2Cl_2/CH_3CN at reflux condition for 20h, followed by reaction of the resulted resin with *p*-toluene sulfonyl chloride in pyridine. The tosylated product was then treated with an excess of triethanolamine in 1:1 CH_2Cl_2/CH_3CN to get the 1st generation dendrimer (Fig.10.2). The repetition of the tosylation and alkylation results 2nd generation dendrimer with 13 quaternary ammonium sites. The dendrimers were characterized by combustion analysis [4], which indicated that higher than 2nd generation without defect was not possible to synthesize. This may be due to steric reason. The dendrimers were tested as ion exchange resin and found that chloride ion could be exchanged with a variety of other monovalent anions like F^-, Br^-, I^- chloroacetate, toluenesulfonate but not acetate or divalent sulfate.

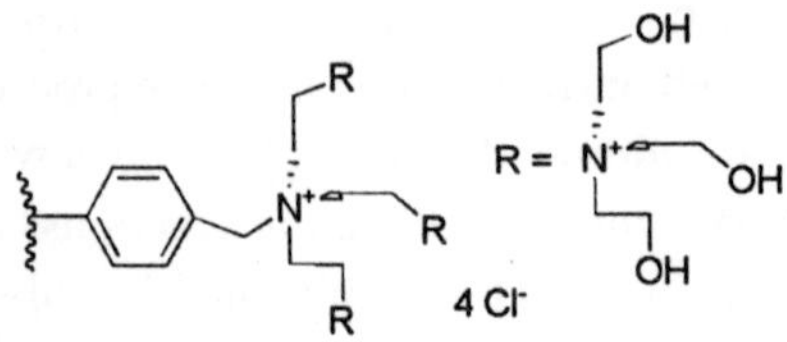

FIG.10.2 Structure of ion exchange quaternary ammonium dendrimers

The synthesis of two charge trapping dendrimers with one phenyl viologen in the core and 4 or 12-benzyl viologens in the branches and periphery was reported [5]. The dendrimers were synthesized by the divergent approach. The initiator core with four peripheral hydroxyl groups was first synthesized from 3,5-dihydroxymethylaniline and *bis*(2,4-dinitro-phenyl)bipyridinium hexafluoro phosphate. The initiator core was treated with HBr in acetic acid to get the corresponding tetra-bromide. This compound was then reacted with the nitrogen of pyridine of the branching unit to yield after substitution of hydroxyl for bromide and anion exchange, the four folds substituted octabromide PF^-_6. Finally, the ethyl terminated bipyridinium hexafluorophosphate was four fold attached on intermediate tetrabromide and eight folds on intermediate octabromide to yield again after ion exchange zero and first generations electron trapping dendrimers. The whole sequence of reactions is shown in scheme10.2. ^{1}HNMR spectroscopy, elemental analysis and MALDI-TOF mass spectrometry characterized the synthesized dendrimers. The electrochemical behavior of the dendrimers revealed a linear growth of size with the generation number from the diffusion coefficients measurement by rotating disk experiment and the corresponding hydrodynamic radii measurement. Results for spectro-electrochemical study indicate fast reduction and slow oxidation of the phenyl viologen core in the dendrimers. This slow oxidation was explained by the activation barrier of ≈ 0.25 eV related to the tailored endergonic electron transfer from the core phenyl viologen to peripheral benzyl viologens.

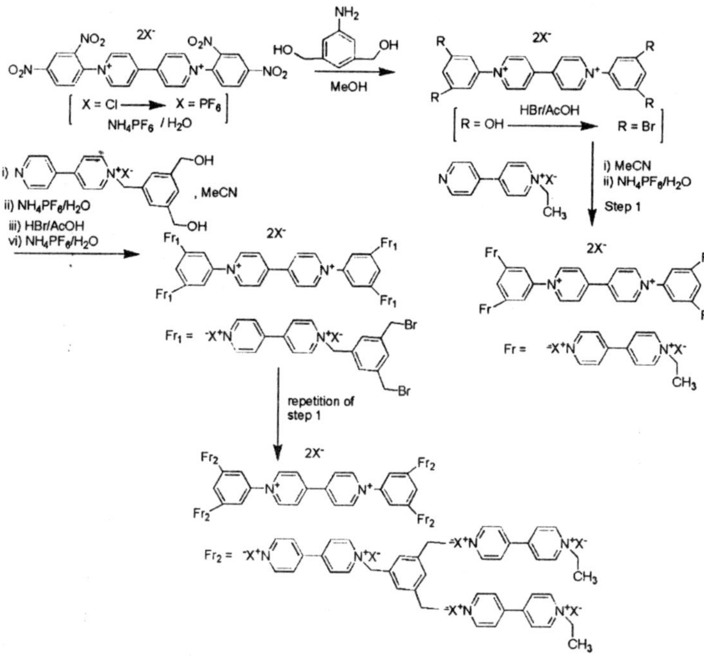

SCHEME 10.2 Synthesis of quaternary pyridinium ion containing charge trapping dendrimers

Similar electroactive dendrimers upto 3rd generation were also reported [6]. These dendrimers consist of a viologen skeleton that contains upto 45 electrochemically accessible 4,4'-bipyridinium subunits, which cause electron sponge properties. The synthesis of the dendrimers was performed by the divergent method using almost similar synthetic strategy. Here, the initiator core consists of a mesityl derivative linked to three 4,4' -bipyridinium units. The three peripheral nitrogens react quantitatively with primary benzyl bromides of 1,3-di(hydroxymethyl)-benzyl bromide to form the branching unit. The resulted hexol was then transformed to the corresponding active hexabromide, which was reacted with pyridine-nitrogen of 4,4'-bipyridinium derivative to get the different generations of the dendrimer after substitution of hydroxyl for bromide and amino exchange to PF_6^- ion, and activation by bromide formation. The whole sequence of reactions for the synthesis of the dendrimers, zeroth to 3rd generation is shown in scheme 10.3.

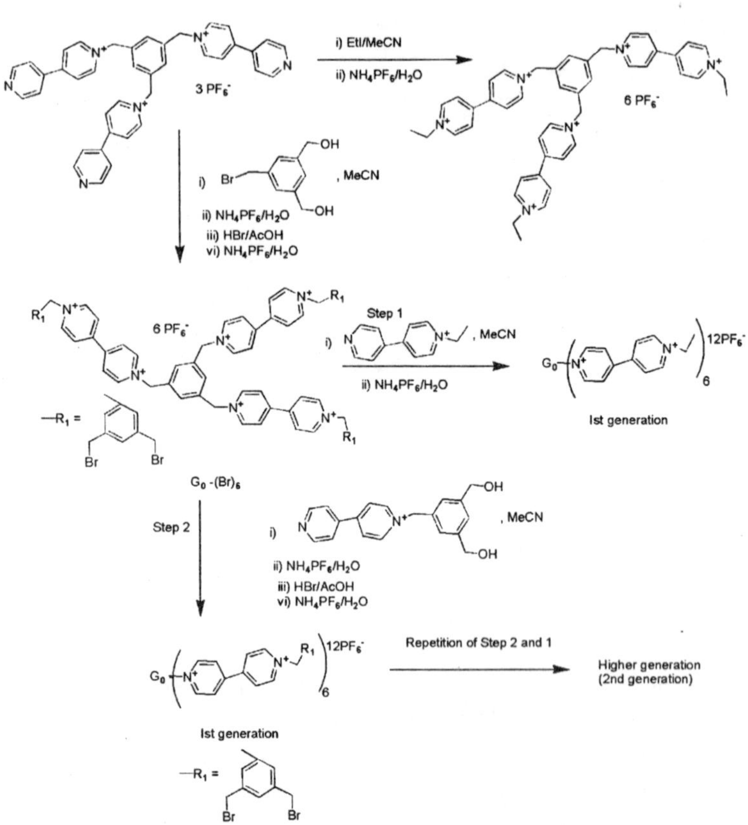

SCHEME 10.3 Synthesis of pyridinium dendrimer with mestyl derivative core unit

The synthesized dendrimers were characterized by ^{1}HNMR, ^{13}CNMR spectroscopy and elemental analysis. However, MALDI-TOF mass spectrometry was not useful tool for these dendrimers as they have high charges. The electrochemical studies revealed an excellent correlation between the hydrodynamics radius and the generation number of the dendrimers. Also, an electrochemically switchable and generation dependent charge transfer complexity were found in the cation radical oxidation state. The growth of dendrimers completes decoloration at 0 V and blue to violet coloration if reduced at -0.48 V against Ag/AgCl system.

10.2.3 Carbohydrates

As the dendrimers containing large number of functionalities, so saccharides could be utilized easily as the building blocks for construction of carbohydrate containing dendrimers. The saccharide can also be considered as a suitable building block that controlled various characteristics of dendrimer, such as their sizes, shapes, topologies, flexibilities and surface properties, and are likely to confer biocompatibility upon the molecules.

Two β-D-glucopyranoside based dendrimers, one containing three tetra- and the other three hepta-saccharide dendrons attached to a central trifunctionalized monosaccharide core. The hepta saccharide containing a 6-aminohexyl aglycone and appropriate protecting groups are necessary to synthesize in this approach. At first, the disaccharide derivative 3, was synthesized by regioselective coupling of the diol 1, with the trichloroacetimidate 2, in the presence of TMSOTf catalyst at -20°C in CH_2Cl_2. The benzoylation of 3 offered the fully protected derivative, which was treated with 90% TFA in CH_2Cl_2 to remove the benzylidene group and to get new glucosyl acceptor, the diol 5. Diol 5 was then treated with a slight excess of glucosyl bromide 6, in the presence of AgOTf as a promoter and molecular sieves 4A° as an acid scavenger to get regioselectively only product, trisaccharide derivative 7. Further, glucosyl bromide 6 was coupled with N-benzyloxy carbonyl aminohexanol 8 in the presence of $HgBr_2/Hg(CN)_2$ in CH_2Cl_2 to offer β-glycoside 9. Debenzoylation of 9 by NaOMe in MeOH offered qualitatively tetranol 10, which was converted to 4,6-O-benzylidene acetal, 11 by the treatment of 10 with $PhCH(OMe)_2$ in the presence of (+)camphorsulfonic acid catalyst. Then, the selective protection of the 2-OH group in 11 was carried out by partial benzoylation of 11 under phase-transfer conditions ($BzCl/Bu_4NI/K_2CO_3$ in CH_2Cl_2) to get the corresponding monobenzoate 12. The treatment of 12 with the thioglucotrioside 7 by NIS/TfOH in CH_2Cl_2 at room temperature resulted the tetra saccharide derivative 13. Debenzylidenation of 13 by 90% TFA in CH_2Cl_2 offered the corresponding diol 14, which was once again treated with 7 under the same conditions as those used for the glycosylation of 11, but at 0°C to get 6-O-regioselectively heptasaccharide 15. The removal of amine protecting group by hydrogenolysis resulted the desired dendron, 16 with free amine group at the focal point. The other types of dendrons were also synthesized [7]. The iodonium-ion-protected monoglycosylasion of the diol 11 with two molar equivalents of perbenzoylated thioglycoside 17 in the presence of NIS/TfOH in

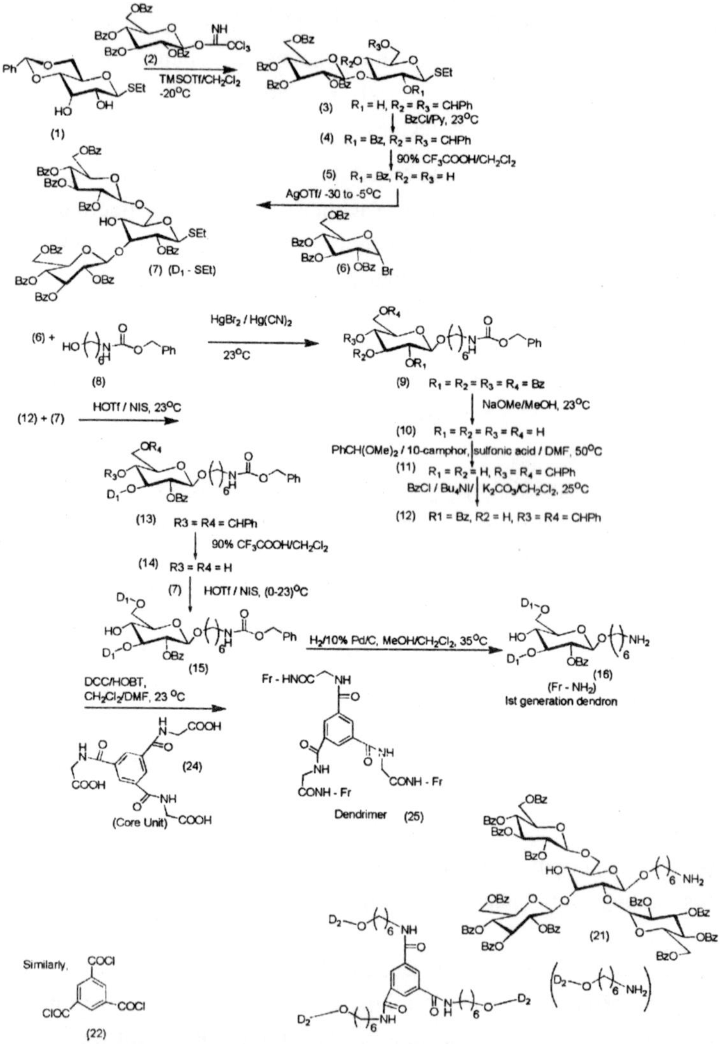

core unit is used for 2nd generation dendron to get the corresponding dendrimer (23). The reaction scheme is described in the text.

SCHEME 10.4 Synthesis of oligosccharide dendrimers

CH_2Cl_2 resulted trisaccharide 18. Debenzylation of 18 by 90% TFA in CH_2Cl_2 afforded the diol 19, which was glycosylated by NIS/TfOH to get 6-O-regioslective tetrasacchanride derivative 20. The desired dendron 21 was than obtained by deprotection of amine group by hydrogenolysis reaction. This dendron was then coupled with benzenetricarboxyl trichloride 22, which resulted the glycol oligosaccharide dendrimer 23. However, the same coupling reaction was found to be unsuccessful for the dendron 16. So, the other trifunctional core 24 was used. The coupling of 16 with 24 in the presence of DCC/HOBT resulted the desired dendrimer 25. The whole long sequence of reactions showed in scheme 10.4. However, the overall yield for this synthesis was very poor. The synthesized dendrimers were characterized by ^{1}HNMR spectroscopy. The yields for the final coupling reactions were 20% and 8% only respectively for the dendrimers 23 and 25.

The α-D-mannopyranoside containing dendrimers were synthesized [8] by the convergent approach (Scheme 10.5). The *tris*-branched mannopyronoside derivative was prepared first and after group manipulation was protected, the '3-mer' was further reacted with branched carboxylic acid derivatives to form higher generation dendrons. Following the same synthetic strategy, the α-D-mannopyranosides containing '9 mer', '18-mer' and '36-mer' were synthesized in 90%, 88% and 30% yields respectively. The dendrons and dendrimers were characterized by ^{1}HNMR, ^{13}CNMR spectroscopy and MALDI-TOF mass spectrometry. The synthesis and characterization of four dendritic ferrocene derivatives incorporating fully protected and deprotected β-D-glucopyranosyl residues on one or both cyclopentadienyl rings, the complex formation between two of these carbohydrate coated ferrocene derivatives and β-cyclodextrin and the electrochemical and energy transfer properties of the dendrimers were reported [9]. The dendrimers were synthesized by the reaction of mono- and di-substituted ferrocene acid chloride with benzyl or acetyl protected amino derivative of β-D-glucopyranosyl in the presence of Et_3N as shown in scheme 10.6. The deprotection reaction was quantitatively carried out by the treatment of NaOMe in MeOH. All the protected and deprotected dendrimers were characterized by ^{1}HNMR, ^{13}CNMR spectroscopy and low and high resolution liquid secondary ion mass spectrometry (LSIMS).

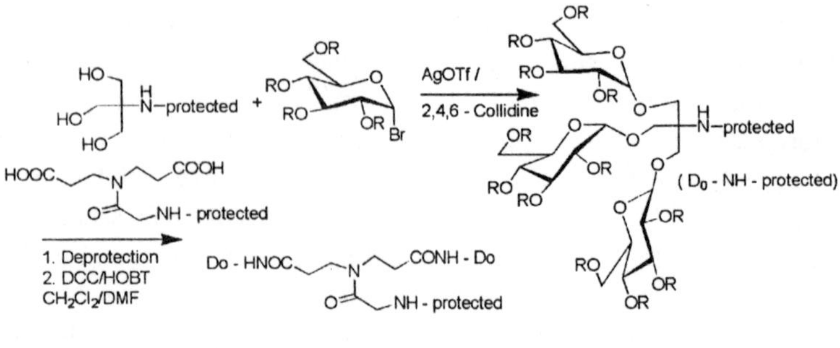

3 mer dendron

SCHEME 10.5 Synthesis of mannopyranoside containing dendron

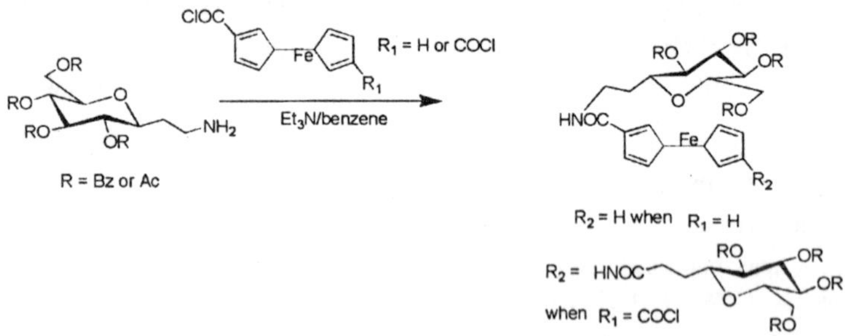

SCHEME 10.6 Synthesis of ferrocene containing carbohydrate dendrimers based on β-D- glucopyranosyl

The study showed that only mono substituted ferrocene containing deprotected dendrimer formed strong complexes with β-cyclodextrins in water. The electrochemical behavior indicated that the diffusion co-efficient decreases with the increase of molecular weight of the dendrimer. The oxidation potential of the ferrocene unit, the rate constant of heterogeneous electron transfer, and the rate constant for the energy transfer reaction with the luminescent excited state of the $[Ru(bpy)_3]^{2+}$ complex were strongly affected by the numbers of substituent and carbohydrate branches present in the substituents of the dendrimers.

The synthesis of lactose dendrimers based on 3,5-di(2-aminoethoxy)benzoic acid repeating unit upto 3rd generation with 8 end groups was reported [10]. The lactose coupling was done through lactose isothiocyanate, resulting thiourea-linked carbohydrate dendrimers. The synthesized dendrimers were characterized by

^{13}CNMR and mass spectrometry. The binding ability to the cholera toxin B subunit by the lactose-functionalized dendrimers was examined by the fluorescence assay. The binding affinities ranging from 18 mM for monovalent lactose to 33 mM for an octavalent lactose dendrimer.

10.2.4. Organometallics

An overview of various organometallic dendrimer is discussed in this subsection. The ruthenium containing dendrimers are very interesting among these not only because of the facts that the ruthenium metal acts as a connecting unit between the interior building blocks and the outermost branching layers but also on the application point of view. The ruthenium containing dendrimer with tripyridine ligands on the surface, where 12 ruthenium atoms were present in the inner layer surrounded by 36 terminal hydroxyl groups (Fig.10.3) with pentaerythritol core, was synthesized [11].

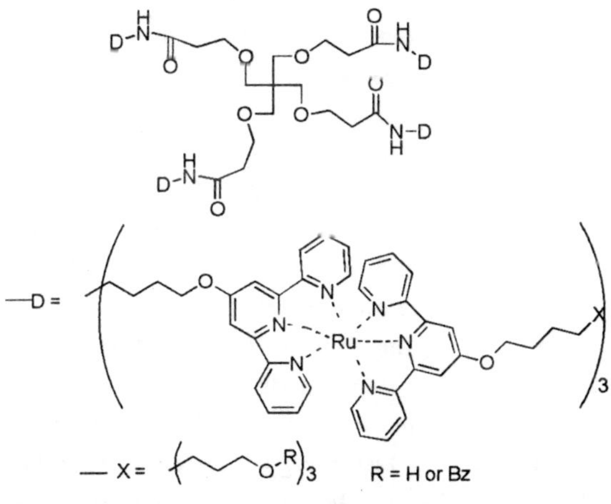

FIG.10.3 Structure of Ru containing dendrimers with pentaerythritol core

A new type of hexa coordinated ruthenium metal centered luminescent dendrimers was synthesized [12-17] using same core, branching units and terminal groups by divergent and convergent methods (Fig.10.4). These trilayer dendritic macromolecules containing upto 22 metal centers and all branching points rely on non-covalent bonding. The molecular weight of a dendrimer was 10,890 Da with estimated size 5 mm and 24 terminal and 21 bridging ligands. The dendrimers can form different isomers as the two coordinating N-atom of each chelating site bridging ligand are not equivalent. They can also form several diastereoisomers as metal centers are also stereogenic centers. The studies of photophysical and electrochemical properties of the dendrimers indicated that these properties were dominated by the contributions of the peripheral metal centers only with little or no effects from the interior metal centers. This may be due to spherical shape of the dendrimer molecules, which shielded the interior layer by the surface layer. However, the photophysical measurements showed 100% efficiency for energy transfer between the interior metal and peripheral metal centers. This is also quite obvious, as light transfered between the different layers should be very efficient owing to the layered covalently bonded structure. In this polypyridine ligands containing metallodendrimers, the metal units are linked together by bridging ligands. The choice of suitable bridging ligands is very crucial and the final properties of dendrimers depend on that choice. This is due to fact that (i) their coordinating sites influence the spectroscopic and redox properties of the active metal based units, (ii) their structures and the orientation of their coordinating sites determine the architecture of the dendrimers and (iii) their chemical nature controls the electronic communication between the metal based units. In general, each dendrimer of this type exhibits (i) intense ligand centered absorption bonds in the UV region and moderately intense MLCT bonds in the visible region, and (ii) a relatively long-lived luminescence in the red spectral region, due to the lowest ^{3}MLCT level. Because of the chemical nature of the bridging ligands, there is only a small electronic interaction was observed in this dendrimer but this small interaction is sufficient to cause a very fast energy transfer between nearby units. This leads to the quenching of the potentially luminescent units having higher energy ^{3}MLCT levels and the sensitization of the luminescence of the units having lower energy ^{3}MLCT levels.

A docosanuclear dendritic metal complex of nanometer size, containing an osmium(II) based core and 21-ruthenium(II) based branching units was synthesized by divergent and convergent approaches. The key building blocks were $[Os(2,3\text{-dpp})_3]^{2+}$, 'complex ligand' and $[Ru(2,3\text{-Medpp})Cl_2]^{2+}$ and $[\{Ru(bpy)_2(m\text{-}2,3\text{-dpp})\}_2 RuCl_2]^{4+}$ 'complex metals'.

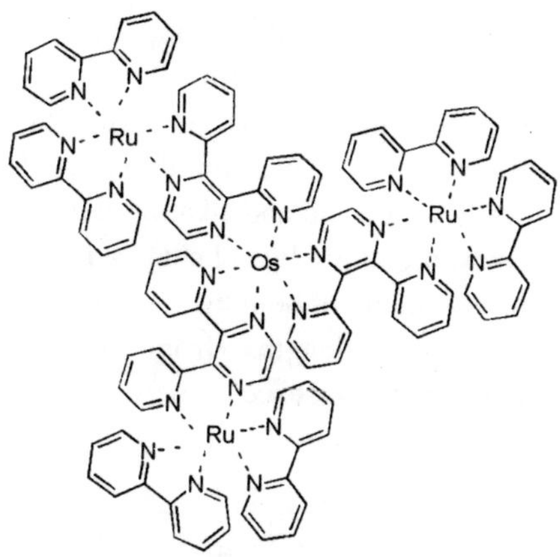

FIG.10.4 Structure of Os centered Ru containing dendrimer

Almost, similar type of ruthenium containing dendrimers, where *tris*(bipyridine)ruthenium(II) complexes were bound within the inner layer of the dendrimer was also synthesized [18]. The dendrimers containing same core unit but *tert*-butyl ester terminated peripheral groups (Fig.10.5). The metallodendrimers were charactorized by using UV-vis. spectroscopy, cyclic voltammetry and MALDI-TOF mass spectrometry.

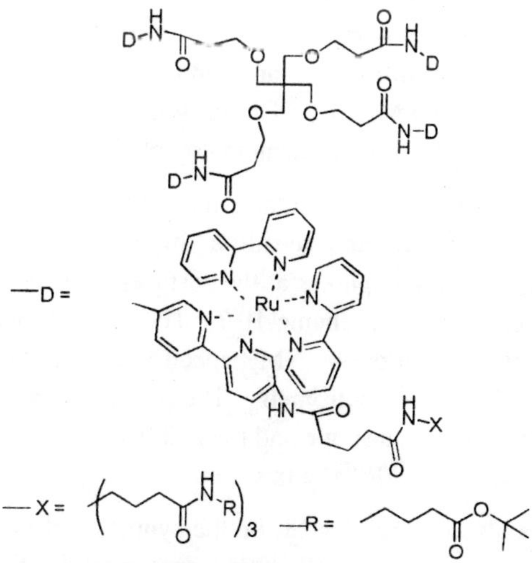

FIG.10.5 Structure of *tris*(bipyridine) Ru containing dendrimer with pentaerythritol core

A star like 1st generation dendrimer with a *tris*(bipyridine) ruthenium(II) core and six benzo- or aza-crown ether units on the surface (Fig. 10.6) was reported [19]. The alkynyl or vinyl linked systems in this dendrimer were electropolymerizable and produced films capable of recognizing alkali and alkaline earth metal cations. The absorption study showed that the ligand centered absorption band and the MLCT emission maximum of both the monomer and polymer films were shifted to lower wave lengths (from 686 to 670 nm for the monomeric dendrimer) in the presence of these ions.

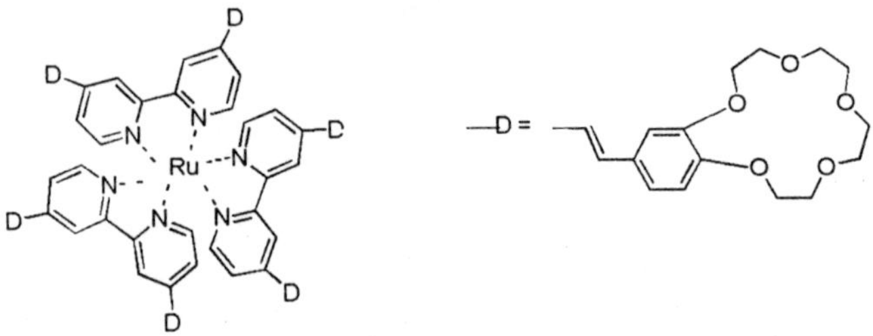

FIG.10.6 Structure of Ru containing dendrimer with azacrown ether surface groups

The photoactive and redox active bipyridinium ruthenium complex containing dendrimers upto 3rd generation were also synthesized [20-22]. The larger dendritic complex exhibited more intense emission and a longer excited state life time than common $[Ru(bpy)_3]$ complexes. NMR spectroscopy and MALDI-TOF mass spectrometry characterized the structures of the dendritic ruthenium complexes.

The star like ruthenium containing dendrimers with $[(\eta\text{-}C_5H_5)Fe^{11}\text{-}(\eta\text{-}C_6(CH_3)_6)]$ core unit and pyridine, bipyridine and terpyridine containing ruthenium complexes at the surface, were synthesized [23] by the divergent method (Scheme 10.7). The characterizations of the dendrimers were carried out by NMR spectroscopy, MALDI-TOF and electrospray (ES) mass spectrometry. The electromechanical study of the dendrimers indicates a single and reversible anodic wave at +0.82 V and +0.79 V for the Ru^{II}/R^{III} redox system.

There are also reports [24-28] on the synthesis of organoplatinum and mixed organoplatinum-palladium dendrimers by the convergent approach. The dendrimers were built up by converting a four coordi-

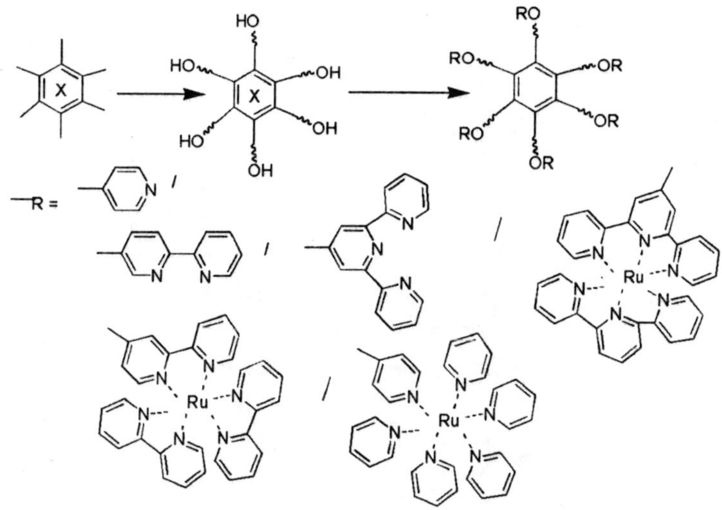

SCHEME 10.7 Outline of Synthesis of Ru-pyridine terminated dendrimers with metallocene core unit

nate, square planar platinum(II) center into a six co-ordinate, octahedral platinum(IV) center by oxidative addition of an aryl bromide to the repeat unit. For an example, Fig. 10.7 shows an oraganoplatinum dendrimer with four arms in which transition metal ions are present in every layer of the dendrimers excepts in the core [25-26]. The synthesis of the dendritic complexes upto 4th generation with 28 coordination centers was carried out by two steps iterative procedure, where oxidative addition of [$PtMe_2$(bpy)] into the monomer, 5,5"-*bis*(bromomethyl)-2,2'-bipyridine) was followed by the treatment with [$Pt_2Me_4(SMe)_2$] to transform the new bipyridine moieties into dimethyl platinum complexes (Scheme 10.8). UV-visible spectroscopy, NMR spectroscopy and GPC analysis characterized the synthesized dendrimers.

The synthesis of palladium containing dendrimers was reported [29-31] by applying coordination chemistry and hydrogen bonding interactions. The method was based on the self-assembly of the branching units in which an unstable coordinative CN group and two tetradentate diorganylsulfide palladium complexes were covalently linked. A Cl^- ion that occupies the fourth co-ordination sites of the palladium atom prevents the CN groups from being bonded. After the building

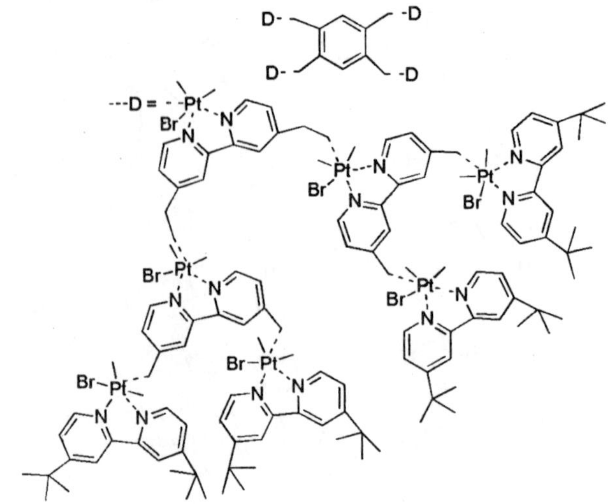

FIG.10.7 Structure of organoplatinum containing dendrimer

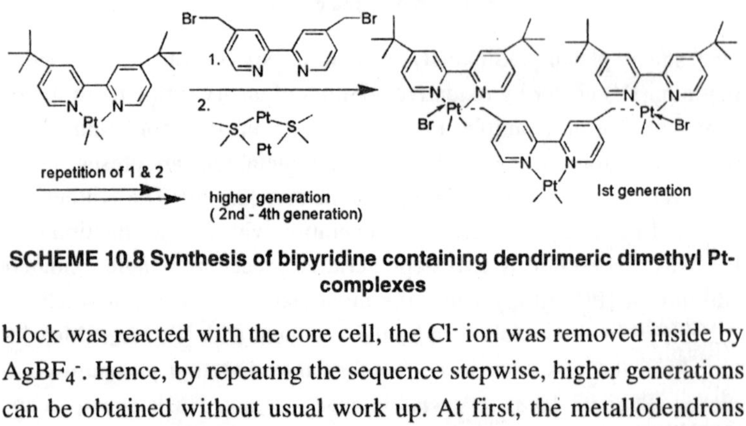

SCHEME 10.8 Synthesis of bipyridine containing dendrimeric dimethyl Pt-complexes

block was reacted with the core cell, the Cl⁻ ion was removed inside by $AgBF_4^-$. Hence, by repeating the sequence stepwise, higher generations can be obtained without usual work up. At first, the metallodendrons upto 3rd generation, were synthesized [29-31] by the one pot divergent method (Scheme 10. 9). The dendrons were carrying a barbuturic acid moiety at the focal point, which was connected to three melamine derivatives to get the desired dendrimers (Fig.10.8). The synthesis of another palladium containing metallodendron was also reported [32] by the convergent method using alternative cyano and pyridine building blocks. The dendrons contain two palladium chloro complexes, which were linked to the periphery of the dendrons.

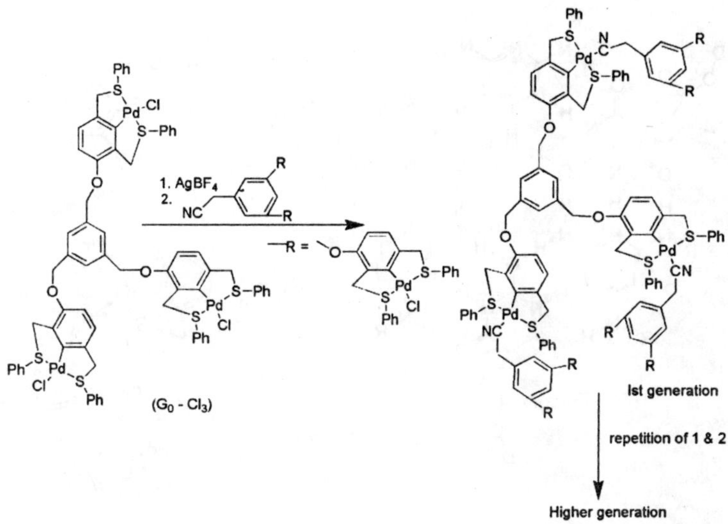

SCHEME 10.9 Synthesis of Pd containing dendrimers by the divergent method

The synthesis of hexanuclear and decanuclear dendritic assemblies of ruthenium-phenanthroline complexes, which form robust, structurally rigid and well defined nanoscopic complexes and exhibit intriguing luminescent properties, was reported [33]. The binuclear, tetranuclear, hexanuclear and decanuclear ruthenium dendrimers were obtained from the reaction of the oxidized Ru-orthoquinone complex (A) with Ru-orthodiamine complex (B) and subsequent reactions of the oxidized dimer and tetramer with a slight stoichiometric excess of $(Ru(phen)_2$ (phendiamine)$]^{2+}$ complex. The dimer (1) and the corresponding dendrimer were synthesized by the following sequence of reactions from 1,10-phenanthroline-5,6-dione (phendione) and 1,10-phenanthroline containing Ru complexes (A and B) 5,6-diamine (phendiamine). Phendiamine was also obtained from phendione by treatment of NH_2OH/HCl, $BaCO_3$, EtOH, reflux to form oxime followed by reduction with 10% Pd/C, $H_2N\text{-}NH_2$, EtOH reflux. Similarly, the Ru tetramer was obtained by the following sequence of reactions as shown in scheme 10.10 from $[Ru(phen)_3]^{2+}$. SEC purified the products. The dendrimers were characterized by ^{1}HNMR spectroscopy and MALDI-TOF or ESI mass spectrometry. The dendrimeric

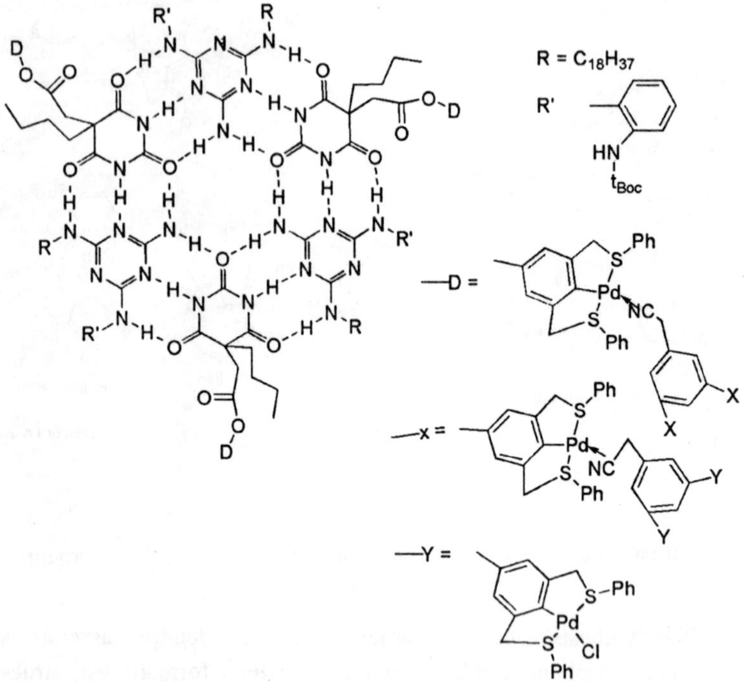

FIG.10.8 Structure of Pd containing dendrimer formed through H-bonding and coordination bonding

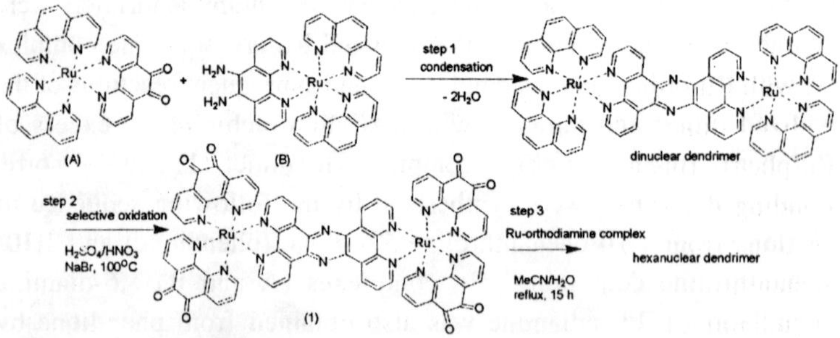

SCHEME 10.10 Synthesis of oxidized Ru containing dendrimers

complexes were soluble in CH_3CN and DMSO (for PF_6^- salts) and in H_2O and MeOH (for Cl^- salts).

The synthesis and characterization of two isomeric metallodendrimer based on *bis*(2,2',6',2''-terpyridine)ruthenium(II) were reported

[34]. The dendrimers are spectrally alike and showed almost similar physical and chemical properties. However, cyclic voltammetric study revealed that the internal density and voids were different. The neutral metallodendrimers upto 2nd generation, were also synthesized [34] by using the similar synthetic strategy. These dendrimers possess no external counter ion.

The ultra fast cyclic voltammetry at scan rates of megavolt per second was utilized [35] to adjust precisely the width of diffusion layer within the outer shell of a 4th generation dendrimer of containing 64 $[Ru^{II}(tpy)_2]$ redox centers. The analysis showed that the apparent diffusion co-efficient featuring election hopping between the $[Ru^{II}/Ru^{III}(tpy)_2]$ redox centers distributed on the dendrimer surface. The study also showed that the $[Ru^{II}/Ru^{III}(tpy)_2]$ redox centers were extremely labile in their potential wells so that they may cross talk considerably more easily than they would do in solution at an equivalent concentration.

The metallodenerimers with ten chiral Ru centers were synthesized in a stereospecific fashion [36]. The homochiral decamer dendritic complex was synthesized (Scheme 10.11) by the same way as shown in scheme 10.10. The dendrimeric complexes show good stability even under harsh conditions like heat, concentration of acids and bases and were stable to racemization in the absence of intense irradiation. The decamer complex containing 448 non-hydrogen atoms with 20^+ net charge and have diameters >5 nm. The diameters of these dendrimeric complexes are ranging from 1.6 nm to over 5 nm for dimers to decamers complexes. These dendrimers were conformationally rigid and exhibit well defined global topologies. Some of the diasteroisomers exhibit macroscopically chiral structures, other show a disk like topology. This difference in global or tertiary structure was exemplified by difference in their colloidal behavior, as observed in electric birefringence measurements. A series of enantiomerically and diastereomerically pure ruthenium(II) dimers, tetramers, hexamers, decamers dendritic complexes was also synthesized [37] by using same synthetic strategy as described earlier.

The synthesis and luminescence properties of a new tetranuclear mixed metals (Ru and Os) dendrimeric complex with six pyrenyl

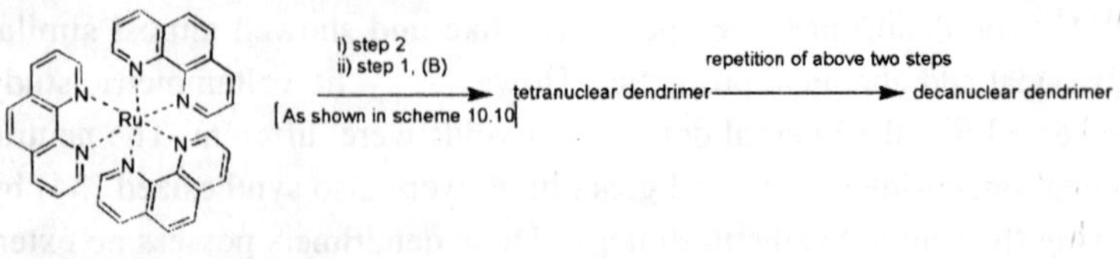

SCHEME 10.11 Synthesis of homochiral decamer dendritic complex

chromophores at the periphery (Fig.10.8) were reported [38]. The synthesis of the 1st generation dendrimer with pyrenyl subunit was carried out by 'complexes as ligands/complexes as metals' synthetic approach. The absorption study showed that at ~ 550 nm absorption is due to spin allowed MLCT transition and a long tail, which extended towards the IR region assigned to spin forbidden MLCT transition involving the Os(II) chromophore. The absorption peaks at 310-350 nm region and 239 nm were assigned to pyrene-centered transitions. The band peak appeared at 279 nm may be the contributions from both the peripheral pyrenyl chromophores and the four metal subunits of the tetranuclear structure. The luminescence properties of the dendrimer showed a λmax at 860 nm with relaxation time 16 ns. Thus, the pyrene containing dendrimer showed better antennaee properties compared to the dendrimer with pyrene units due to the absorption of the additional organic chromophores and also to new absorption features originating from the interaction of the pyrene peripheral subunits with the tetranuclear metal based core in the former.

The 'complexes as metals/complexes as ligands' synthetic strategy was utilized both for divergent and convergent approaches [17]. This strategy is based on reactions of the metal complex building blocks possessing free chelating sites with metal complexes containing labile ligands. The complexes with free chelating sites play the role of ligands ('complex ligands') and those with labile ligands play the role of metals ('complex metals'). In the divergent approach, the first step was the formation of tetranulcear $[Os\{(m\text{-}2,3\text{-dpp})Ru(2,3\text{-Medpp})_2\}_3]^{14+}$ species in which peripheral ligands were 2,3-dpp with the second chelating site inactivated (protected) by methylation. These species were obtained from the reaction of the $[Os(2,3\text{-dpp})_3]^{2+}$ 'complex ligand' core with three equivalents of the $[Ru(2,3\text{-}Medpp)_2Cl_2]^{2+}$ 'complex metal'. The repetition of deprotection and reaction with 'complex metal' offered higher generation dendritic complexes (Scheme 10.12). Whereas in the convergent approach, the deprotected units with $[\{Ru(bpy)_2(m\text{-}2,3\text{-}dpp)\}_2RuCl_2]^{4+}$ yields different generation dendritic complex. For example, the reaction of the deprotected tetranuclear species with six equivalents of the trinuclear $[\{Ru(bpy)_2(m\text{-}2,3\text{-}dpp)]_2RuCl_2]^{4+}$ leads to the docosanuclear dendritic complex (Scheme 10.12). All the dendritic complexes were characterized by elemental analysis and studies of their absorption spectral, luminescence and electrical properties.

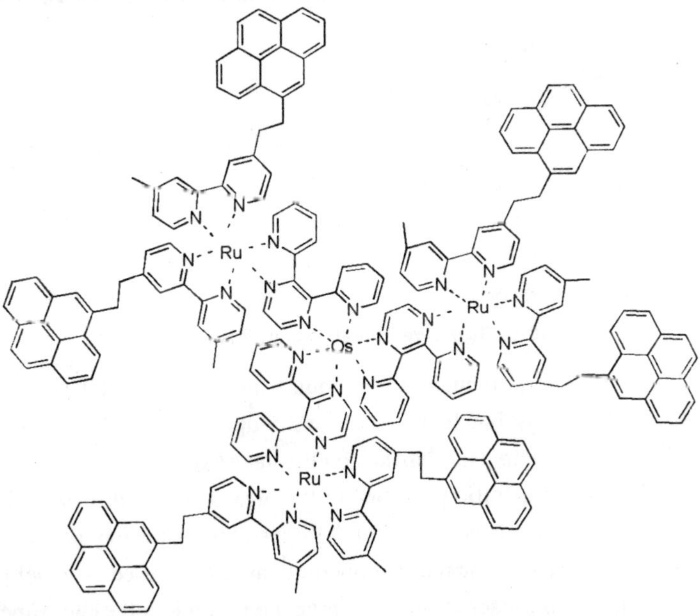

FIG.10.9 Structures of Os and Ru containing dendrimers with pyrrine units

The electrochemical behavior of dendritic polynuclear Ru(II) and Os(II) polypyridines complexes at varying positive potential was studied [39] in liquid SO_2. All the dendrimers contain 2,3-or 2,5-*bis*(2-pyridyl) pyrazine as the bridging ligand and 2,2'-bipyridine as the terminal ligand (Fig.10.10). The study showed the following order of

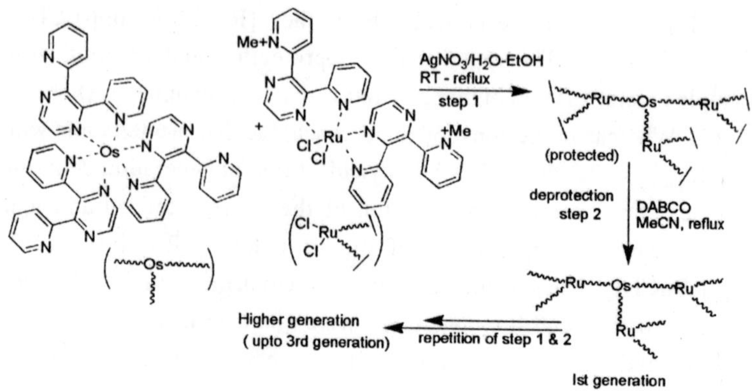

SCHEME 10.12 Synthesis of Os centered Ru-dendritic complexes

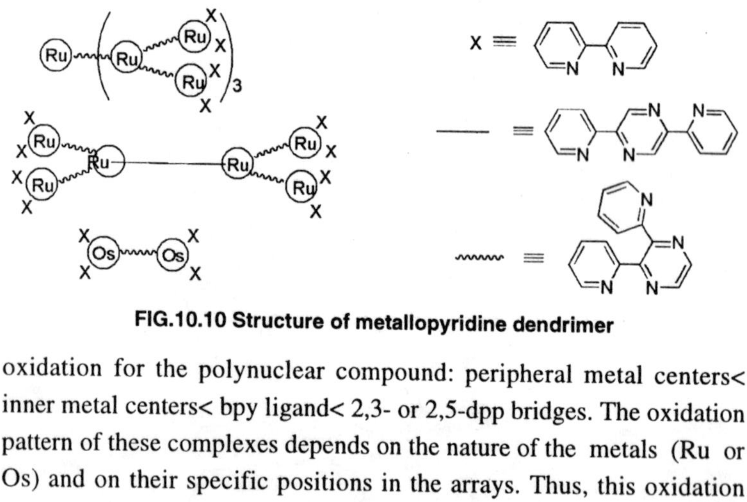

FIG.10.10 Structure of metallopyridine dendrimer

oxidation for the polynuclear compound: peripheral metal centers< inner metal centers< bpy ligand< 2,3- or 2,5-dpp bridges. The oxidation pattern of these complexes depends on the nature of the metals (Ru or Os) and on their specific positions in the arrays. Thus, this oxidation pattern can be used as a sort of ' finger-print' of each particular dendrimer. Therefore, the redox properties can also be used as a tool for the chemical characterization of large and highly charged super-molecules.

The synthesis of germanium containing dendrimers was first reported both by divergent and convergent approaches [40]. In the divergent technique, germanium tetrachloride was used as the core, and successive alkenylation and hydrogermylation reactions were used to get the desired dendrimers (Scheme 10.13). However, by using

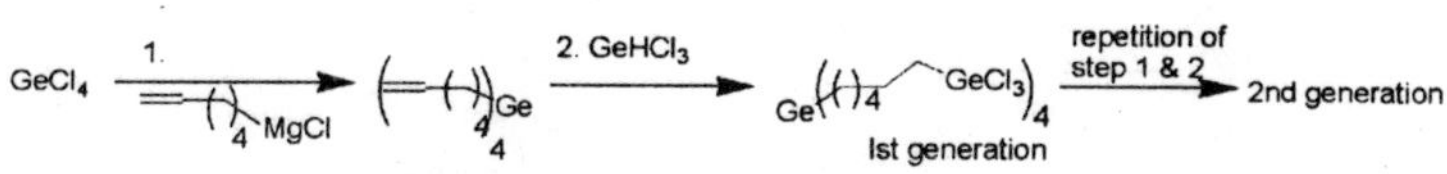

SCHEME 10.13 Synthesis of Ge containing dendrimers from $GeCl_4$

different route, the first generation germanium dendrimer was also synthesized. In this technique, hydrozirconation of organogermaniun polyalkenes followed by *trans*-alylation and alkenylation offered the desired dendrimer (Scheme 10.14).

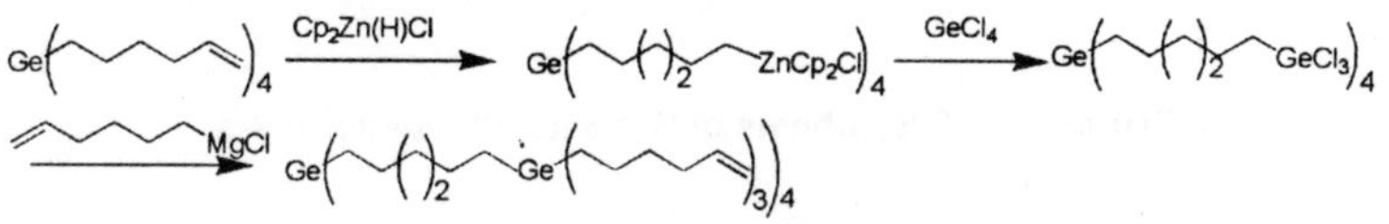

SCHEME 10.14 Synthesis of Ge containing dendrimers from polyalkene germanium compound

Though the purity of the product is less compared to the earlier method. Whereas, in the convergent method, the dendron possesses a chlorohexyl chain linked to the core and reacts with germanium tetrachloride to get the desired dendrimer of 1st generation (Scheme 10.15). However, 2nd generation is difficult to synthesize because of steric reason.

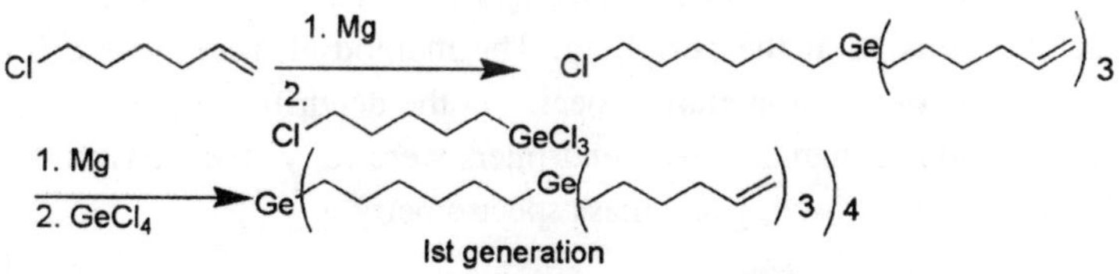

SCHEME 10.15 Synthesis of Ge containing dendrimers from chlorohexene

The first bismuth containing dendrimers up to 2nd generation were synthesized [41] by utilizing direct ortholithiation of aryl bismuthanes bearing a sulfonamide function at ortho position of the aromatic ring, followed by reaction with *bis*[2-(diethylaminosulfonyl)phenyl]bismuth iodide (Scheme 10.16). However, the overall yield of the reactions is only 20%.

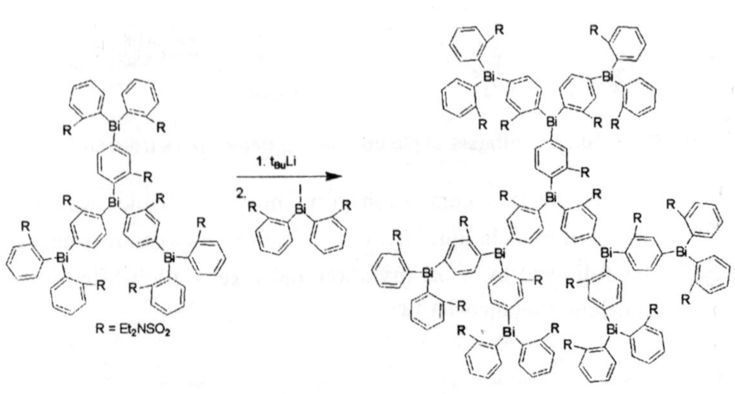

SCHEME 10.16 Synthesis of Bi containing dendrimers

10.2.5 Polyureas and Polyurethanes

A novel, fast and simple synthetic strategy were reported [42] for synthesis of polycarbamate/urea dendrimers, based on an AB-CD_2 coupling reaction with a diisocyanate building block. The reactivity difference of the two-isocyanate functionalities of the AB building block allows the construction of these dendrimers without the necessity of activation or deprotection step. This technique required only stirring, heating and accurate dosing. Further, no work up step is required for the purification of the compounds. Even, it allows the use of a wide variety of polyols, amines and amino-alcohols in the core, branching units or at the periphery. The method also applicable for incorporation of organometallic species in the dendritic structure. The synthesized polycabamate/urea dendrimers were fully characterized by ^{13}CNMR, IR spectroscopy and mass spectrometry.

The synthesis of dendritic polyurethenes (PUs) by reacting diisocyanates and/or polyisocyanates with compounds containing at least two groups, which are reactive toward isocyanates was reported [42] in a patent. The di- and poly-isocyanates used were 2,4-/2,6-toluene diisocyanate, MDI, IPDI, 2-butyl-2-ethylpentamethylene diisocyanate, 2-isocyanatopropyl cyclohexyl isocyanate, dicyclohexyl methane 2,4-diisocyante and 4-methyl cyclohexane 1,3 -diisocyante along with other types of isocyanates. The compounds having at least two groups, which are reactive with isocyanates used were propylene glycol, glycerol, mercaptoethanol, ethanolamine, diethanolamine, 2-amino-2-methylpropane-1,3-diol, etc. The dendritic polyurethanes were synthesized by iterative reaction of a diisocyanate with an amino diol at (-10 to 30)°C followed by heating in the range of (30-80)°C in suitable solvents in the presence of a suitable catalyst. The details for the different dendrimers are described in the patent [43].

The synthesis of structurally uniform polyurethanes in which dendritic branches were constructed from a diisocyanatobenzyl chloride and a protected 3,5-dihydroxybenzyl alcohol and were coupled to a polyfunctional alcohol as the central core was also reported [44].

The characterization and monitoring of reaction of PU dendritic wedges and dendrimers by using MALDI-TOF mass spectrometry were reported [45]. The 1st and 3rd generation PU-porphyrin dendrimers and tri-core dendrimers were obtained by attachment of the corresponding dendritic wedges to a porphyrin unit and to 1,3,5-benzene tricarboxylic acid. The reactions were monitored by MALDI-TOF, proved to be powerful tool for designing and optimizing such type of reactions. This tool was also utilized for the post decay analysis of PU dendrons upto 3rd generation. The cleavage took place at both inner and outer layers.

The polyurethane dendrimers were synthesized [46] via the Curtius reaction. The urethane branches were synthesized from aromatic phenol dicarboxylic acids by treatment with alcohol and diphenylphosphoryl azide using protective group technique and then these were coupled convergently to a triurethane derivative, derived from benzene tricarboxylic acid

10.2.6 Dendrimers of Mixed Chemical Linkages

The dendrimers having more than one type of chemical linkages and difficult to categorize into a particular class, are included here. For example, a dendrimer having, ester amide, Si, P along with metal ion, was reported [47-49]. The dendrimer is based on chiral ferrocenyl diphosphine ligands to be used as the dendritic catalyst. By using different cores dendrimers with 3,6,8 (Fig.10.11) 12-16 and 24, 'Josiphos' units were obtained [47-49]. The rhodium complexes of these systems were obtained by stirring one equivalent of ligand with

$[Rh(Cod)_2]BF_4$ in CH_2Cl_2 and used as the precursors in the hydrogenation of dimethylitaconate. The enantioselectivity of the catalysts was very high ($\geq$ 98%) and comparable to monomeric Rh-Josiphos complexes. The separation of these dendrimeric catalysts is much more easier compared to monomeric one.

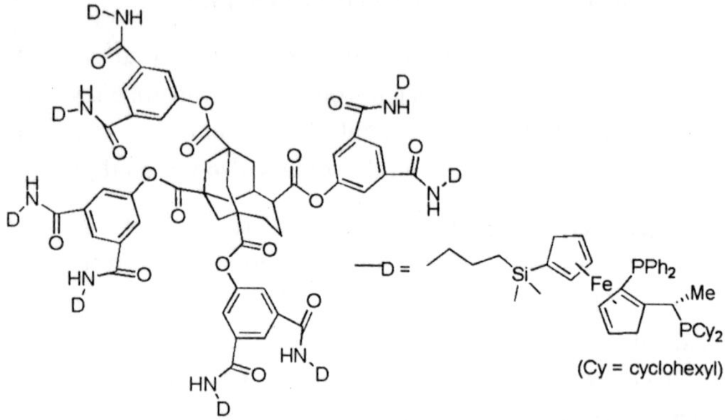

FIG.10.11 Structure of mixed linkages chiral dendrimer with ferrocenyl diphosphine ligand as the surface groups

10.2.6.1 Poly(ester amide)s

The synthesis of enantiomerically pure dendrimers upto 2nd generation based on a chiral *trans*-3,4-dihydroxypyrrolidine was reported [50]. In this synthetic strategy, benzene polycarboxylic acids were used as the central nucleus and terephthalic acid was used as the spacer unit between the pyrrolidine nuclei. The AB_2 monomer, pyrrolidine derivative reacted with the corresponding acyl chloride in Schotten-Baumann conditions, as shown in scheme 10.17, to get the first generation dendrimer. The synthesis of 2nd generation was carried out by pyrrolidine nuclei (Scheme 10.17). The structures were characterized by IR, ^{1}HNMR, ^{13}CNMR, spectroscopy and elemental analysis. The chiroptical property measurement showed that the dendrimers are optically active.

The preparation and characterization of asymmetric redox active poly(ester amide) dendrimers containing ferrocene unit were reported. The dendrimers were prepared by the reaction of chlorocarbonyl ferrocene with Behera's amine [51]. The hydrolysis of the peripheral

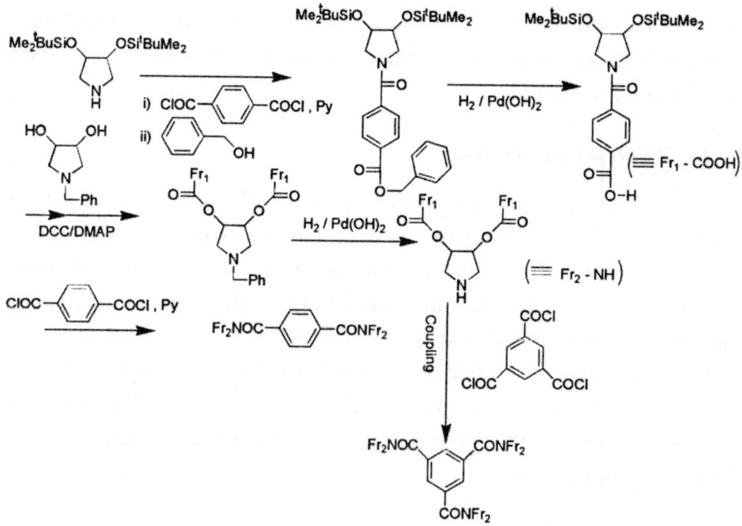

SCHEME 10.17 Synthesis of pyrrolidine derivative containing dendrons and dendrimers

ester groups and the reaction of the resulting triacid with Behera's amine produced the 2nd generation dendrimer. The repetition of the above two steps resulted 3rd generation. However, by applying a combination of convergent and divergent synthetic approaches, 2nd generation Behera's amine allows to prepare the (n+1)th generation dendrimers from (n-1)th generation polyacid. The structure of the prepared dendrimers (Fig. 10.12) were confirmed by ^{1}HNMR, ^{13}CNMR, UV-visible and IR spectroscopies and MALDI-TOF mass spectrometry. The molecular weight varies from 627 to 4725 Da as determined by mass spectrometry. The electrochemical behavior of the dendrimers showed that they act as redox active compounds. The anodic electrochemistry is characterized by the one electron oxidation

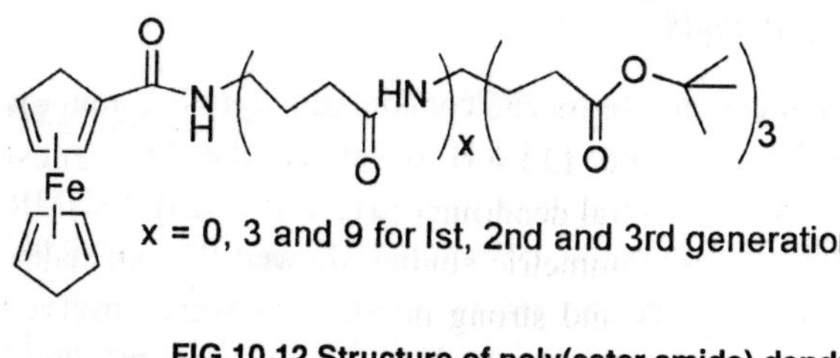

x = 0, 3 and 9 for Ist, 2nd and 3rd generations respectively

FIG.10.12 Structure of poly(ester amide) dendrimer

of the ferrocene nucleus. The heterogeneous electron transfer rate constant decreases with the increase of generation number as observed in voltammetric studies.

10.2.6.2 Poly(ester amine)s

A novel class of aliphatic poly(ester amine) dendrimers upto 3rd generation was synthesized [52] by the divergent method using piperazine, diethanolamine and acryloyl chloride as monomeric building blocks. The dendrimers were synthesized by the following sequence of reactions as shown in scheme 10.18. The synthesized dendrimers were characterized by IR, ^{1}HNMR, ^{13}CNMR spectroscopy and ESI-mass spectrometry. The molecular weights of the highest generation dendrimer were 2315 Da for hydroxyl and 3183 Da for acrylate peripheral dendrimers.

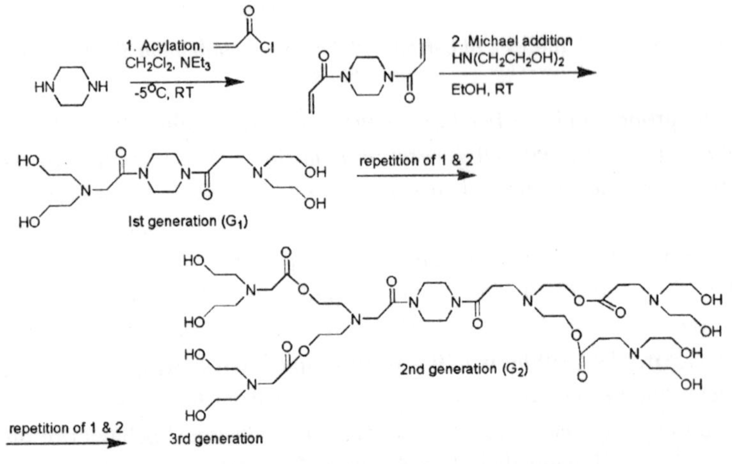

SCHEME 10.18 Synthesis of poly(ester amine) dendrimers

10.2.6.3 Poly(ether amine)s

Both neutral and cationic ferrocene containing poly(ether amine)s dendrimers were synthesized [53-54] to study the host-guest interactions. In the case of neutral dendrimer [31] (Fig.10.13), ^{1}HNMR spectroscopy and cyclic voltrammetric studies showed that all redox centers behave independently and strong interactions were observed between dendrimers, especially the higher generations and small inorganic anions like $H_2PO_4^-$, HSO_4^-. Cl^- and NO_3^-. These may be due to electrostatic interactions involving ferricinium cation and anion and H-bonding interaction of the amide hydrogen with the anion. However, only H-bonding interaction was not sufficient. Again, for the neutral ferrocene containing dendrimers, recognition was very weaker than the cationic state, as the neutral diamagnetic 18-electron form does not usually interact efficiently with anions. Whereas for the polycationic ferrocene functionalized dendrimers [54] better association with the increase of generation for the recognition of Cl^- and Br^- was observed. However, the chelating anions like $H_2PO_4^-$ were not recognized in contrast to neutral dendrimers. This may be due to the fact that the neutral dendrimers contain amide linkages, whereas cationic one contains secondary amines that preferentially interact with single atom halogen anions, like Cl^- and Br^-. Also, the synergistic effects between the single X^- ...HN hydrogen bond, the electrostatic interaction and the shape selectivity of the dendritic structure account for the recognition of the Cl^- and Br^- ions.

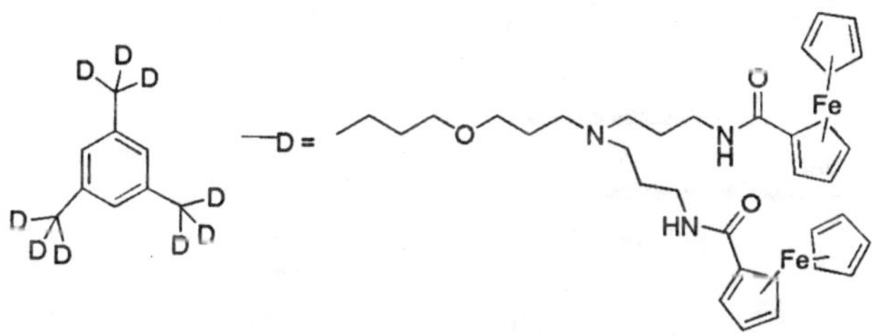

FIG.10.13 Structure of ferrocene containing dendrimer with ether, amine and amide linkages

10.2.6.4 Poly(ether imide)s

The synthesis of dendritic poly(ether imide) were reported [55] by the step wise convergent approach using 1-(4-amino phenyl)-1,1-*bis*(4-hydroxyphenyl)ethane as the building block. The aromatic nucleophilic substitution of the building blocks with 3-nitro-N-phenyl phthalimide results the first generation dendron with amino phenyl group at the focal point. This focal group can be transformed to activate nitro group by treatment with 3-nitrophthalic anhydride. The subsequent generations were obtained by the sequence of reactions: (i) the aromatic nucleophilic displacement of the activated nitro group with phenoxide nucleophile to form the ether linkage and (ii) the condensation of aminophenyl group with nitro–substituted phthalic anhydride, followed by ring closure to yield the imide ring containing the reactive nitro group. The synthetic strategy upto 2nd generation poly(ether imide) from *p*-nitro aectophenone is shown in scheme 10.19. These dendrons were then reacted with 1,1,1-*tris*(4-hydroxy phenyl)ethane to get the desired dendrimers (Scheme 10.19).

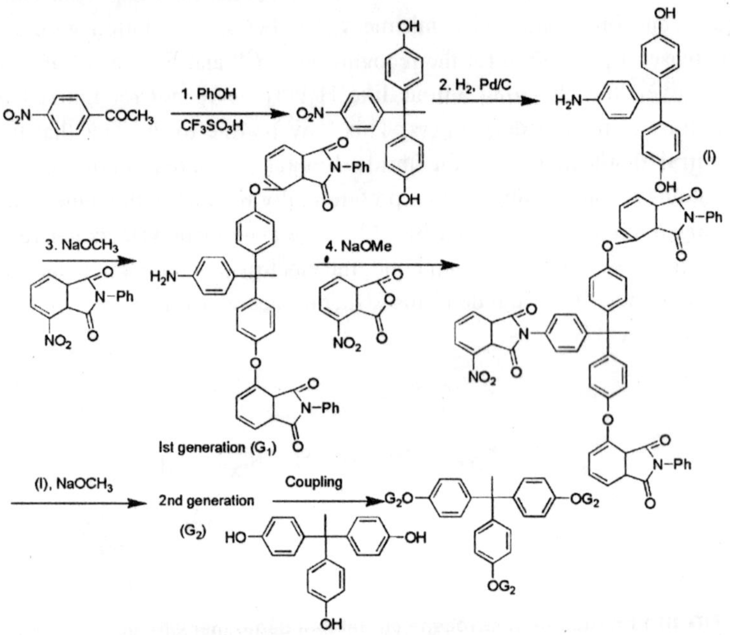

SCHEME 10.19 Synthesis of poly(ether imide) dendrimers

The synthesized dendrimers were characterized by IR, ^{1}HNMR, ^{13}CNMR spectroscopies, SEC and mass spectrometry. The thermal behavior of the dendrimers was studied by DSC and TGA. The molecular weight of the dendrimers were 2973 Da and 6901 Da respectively for the 1st and 2nd generation and were monodisperse as measured by MALDI–TOF mass spectroscopy. The dendrimers were soluble in common organic solvents like DMSO THF, $CHCl_3$, CH_2Cl_2, etc. The T_g, (183–208)°C increases with the increase of generation. The dendrimers were thermally stable upto 430°C, with a 5% wt loss at 453°C for G_1 and of 462°C for G_2.

10.2.6.5 Poly(alkylnyl ester-amide)s

The dendrimers of 1st generation were synthesized [56] from benzene-bridged tri- and hexa-alkylnyl phenylalanines. The palladium-mediated coupling of *p*-ethynyl-phenylalanine with hexabromobenzene resulted benzene-bridged hexa alkynyl α-amino acid in high yield. The tripodal phenylalanine (1) was synthesized first by the reaction of 1,3,5-triethynylbenzene with *p*-iodo-L-phenylalanine, whereas hexapodal compound (2) was synthesized from hexabromobenzene as shown in scheme 10.20. The deprotection of tripodal compound (1) followed by treatment with the amine, $H_2NC(CH_2OCH_2\ CH_2CO_2Me)_3$ yielded the desired 1st generation dendrimer (Scheme 10.20). Similarly, the 1st generation of the hexapodal containing dendrimer was also obtained as shown in scheme 10.20. The synthesized dendrimers were characterized by FTIR, ^{1}HNMR and ^{13}CNMR spectroscopies and electrospray mass spectrometry. The coordination of these type 1st generation dendrimers was also reported [57] with ferrocene aldehyde and $[(Ph_3P)_2Pt\ (\mu\text{-}Cl)_2Pt(PPh_3)_2]^{2+}$ as shown in Fig. 10.14.

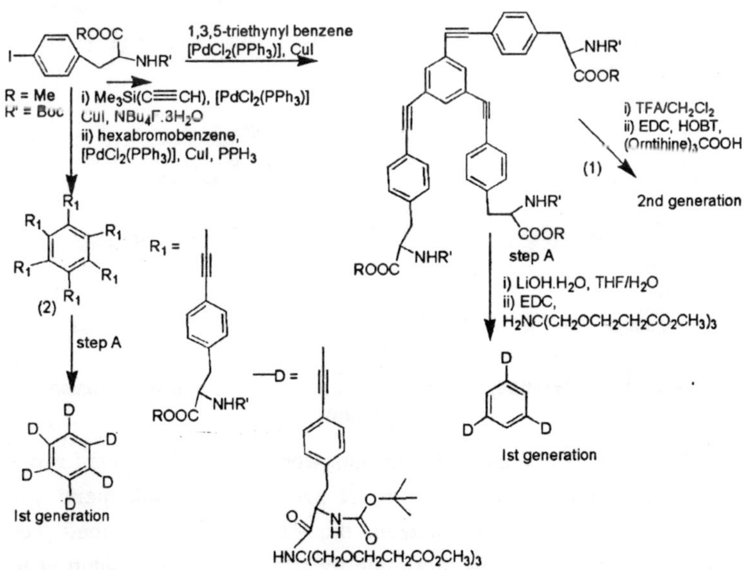

SCHEME 10.20 Synthesis of alkynyl phenylalanine containing dendrimers

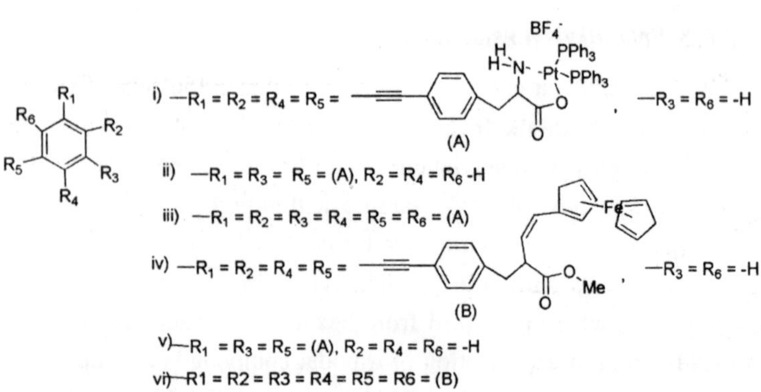

FIG.10.14 Structures of coordination complexes of alkynyl phenylalanine based dendrimers

10.2.6.6 Others

An improved synthesis of pure ethereal tetraamine core with high yield and subsequent reaction with acrylonitrile, followed by catalytic reduction offered a polyamine dendrimer (Scheme 10.21) were reported [58]. The products in each step were well characterized by IR, ^{1}HNMR, ^{13}CNMR and ESI-MS spectrometry. The ethereal tetraamine core can form complex with Zn-tetraphenyl porphyrin shift reagent (1:4), which confirmed it's purity.

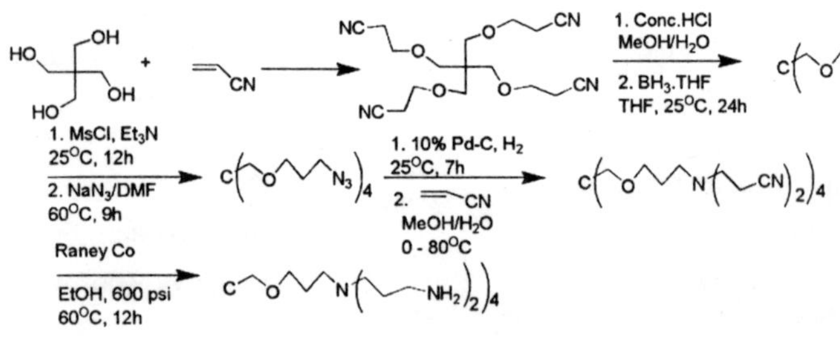

SCHEME 10.21 Synthesis of poly(ether amine) dendrimer with tetraamine core unit

The novel azobenzene dendrimers were synthesized [59] for non-linear optical materials. The novel azobenzene dendrimers were modified by adding an azobenzene branching unit as a second order non-linear optical chromophore and by placing aliphatic chain at the chain end. The molecular dynamic calculations for the dendrimers showed that azobenzene dendrons tend to form a molecularly assembled structure rather than a spherical one. The first order molecular hyperpolarizability was measured to be 3101x 10^{-30} esu for the dendrimer with 15 azobenzene chromophores. These experimental results indicate that the chromophoric units were oriented non-centrosymmtrically along the molecular axis and each of the units coherently contributed to the second harmonic generation. The molecular structures of the novel azobenzene dendrimers upto 4th generation, were also theoretically studied [60] using molecular dynamics and molecular orbital theory. The study revealed that the second order molecular hyperpolarizability value of the dendrimers in $CHCl_3$ solution is large and they have favorable rod shape structures. The theoretical study for nitro azobenzene dendrimers upto 4th generation in nonpolar solvent like CCl_4 and polar solvents like $CHCl_3$ and acetonitrile by using the molecular dynamic method was also done [61]. The second order hyperpolarizibility value is large in polar solvent with large dielectric constant. However, the experimentally observed enhancement of the second harmonic generation activity is not sufficiently reproduced by theoretical calculation. The hyperpolarizabilities of the 2nd, 3rd and 4th generations nitro azobenzene dendrimer are 4.8, 8.0 and 21.7 times that of the first one (98x10^{-30}esu) in acetonitrile solution.

10.2.7 Polyamines

Two types of novel dendrimers with a 1,3,5-triazine ring at the branching point were synthesized [62] by both divergent and convergent methods. In the divergent approach, 2,4,6-*tris*(*p*-niotroanilino)-1,3,5-triazine was used as a trifunctional central core (Scheme 10.22) and each cycle involved the reduction of NO_2 groups followed by reaction with 2-chloro–4,6-*bis*(*p*-nitroaniline)–1,3,5-triazine. Whereas in the convergent approach, the wedges were formed by reaction of 2,4-dianilino–6-chloro-1,3,5-triazine with *p*-nitrophenol, followed by reduction of nitro group and coupling with 2,4-dichloro–6-(*p*-nitrophenoxy)–1,3,5-triazine (Scheme 10.23). The synthesized polymers were characterized by ^{1}HNMR, IR, UV spectroscopy, elemental (CHN), GPC analyses and mass (MALDI-TOF) spectrometry. The thermal behavior of the dendrimers was studied by TGA/DTA and DSC. GPC indicates monodispersity of the dendrimers (1.05-1.16) with molecular weight ranging from 350 to 2850 Da. The thermal analysis confirmed the high heat resistance and the first generation had the best stability (temperature for 50% decomposition > 600°C), may be due to presence of alternate imine like nitrogen linked 1,3,5-triazine and benzene rings.

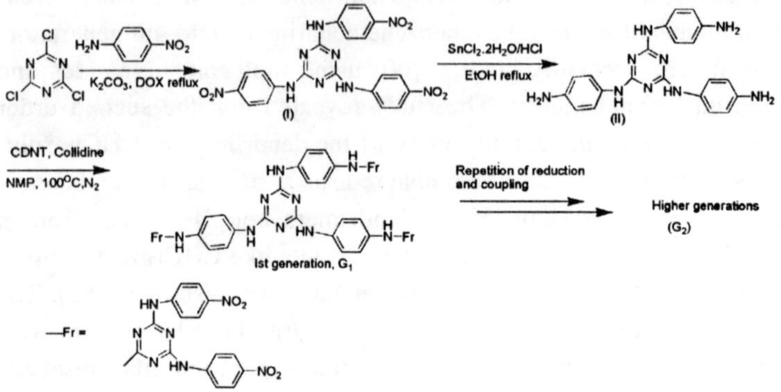

SCHEME 10.22 Synthesis of polyamine dendrimers based on 1,3,5-triazine moieties by the divergent method

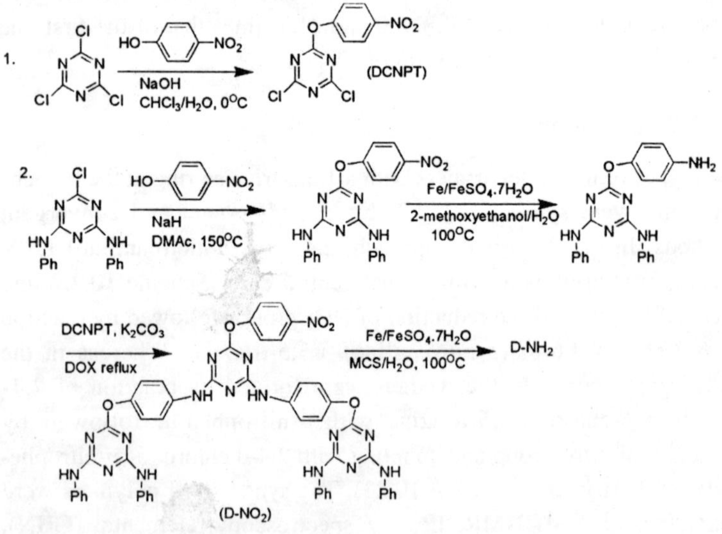

SCHEME 10.23 Synthesis of polyamine dendrimers based on 1,3,5-triazine moieties by the convergent method

The analytically pure triallyamine dendrimers upto 3rd generation, which are highly electron rich and that can produce highly delocalized radical cations were synthesized [63]. The synthetic method has two special features. First, biphenyl methyl protective groups were used on the amines to assist in deprotective hydrogenolysis of larger structure. Second, highly active catalysts were used for formation of both di- and triethyl amines that are based on a 1:1 ratio of $P(t\text{-}Bu)_3$ and $Pd(dba)_2$ to improve reaction yield of the C-N bond formation and decreased reaction times. Further, the low catalyst loading and high molecular weight of ligand allow for simpler purification of the materials. The whole reaction sequence is shown in Scheme 10.24.

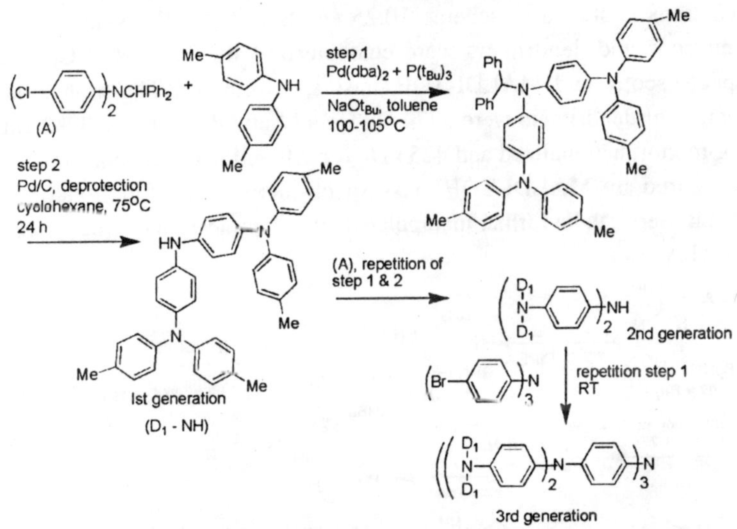

SCHEME 10.24 Synthesis of triallyl amine (TAA) dendrimers

The products in each step were formed with very high yield (≈80%), specially the 3rd generation, where the yield is 89%. The synthesized dendrimers were characterized by UV, ^{1}HNMR ^{13}CNMR spectroscopies, MALDI-TOF mass spectrometry and GPC analysis. The GPC indicates the monodispersity (1.02) of the dendrimers. The ultra fast fluorescence anisometry measurements were used to investigate the energy migration processes in the dendrimer. The result showed that all three dendrimers decayed to a residual value within 100 fs and higher the generation the lower is the residual anisotropy value.

This fast anisotropy decay to smaller residual value with the increase of generation is an indication of the exciton delocalization in multi chromophore branched system.

10.2.8 Polyhydrazones

Melamine based polyamine dendrimers with one or two 'Boc' protected amine sites on the surface were synthesized and characterized [64]. The synthesis of the dendrimers was carried out by the orthogonal convergent approach. In this synthetic approach, *p*-aminobenzyl amine was used to interconnect triazines and to avoid functional interconversions and protecting group manipulations. The whole sequence of reactions is shown in scheme 10.25 (A, B and C). The synthesized dendrons and dendrimers were characterized by ^{1}HNMR, ^{13}CNMR spectroscopy and MALDI-TOF mass spectrometry. The molecular weight of dendrimers were 3774 Da for all butyl dendrimers, 3992 Da for monofunctionalized and 4253 Da for difunctionalized dendrimer as measured by MALDI-TOF mass spectrometry. The functionalized dendrimers can be further manipulated after deprotection of Boc groups by TFA.

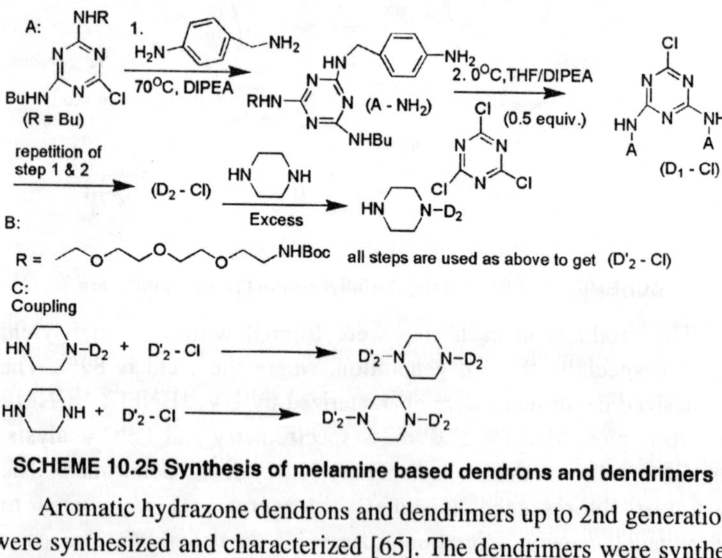

SCHEME 10.25 Synthesis of melamine based dendrons and dendrimers

Aromatic hydrazone dendrons and dendrimers up to 2nd generation were synthesized and characterized [65]. The dendrimers were synthesized convergently using repetitive protection of hydrazine group by phthalic anhydride and the Vilsmeyer formulation of aromatic ring

(Scheme 10.26) from biphenyl hydrazine hydrochloride [65]. The dendrimers were soluble in common organic solvents such as PhCl, $CHCl_3$, DMF, etc. These hydrazone dendrimers were amorphous materials, which can form excellent optical films by spin coating, and the films were clear, transparent, homogeneous and mechanically tough. The luminescence properties of the dendrimers indicated high quantum efficiency. The electroluminescence (El) devices fabricated from the dendrimers showed a strong green emission.

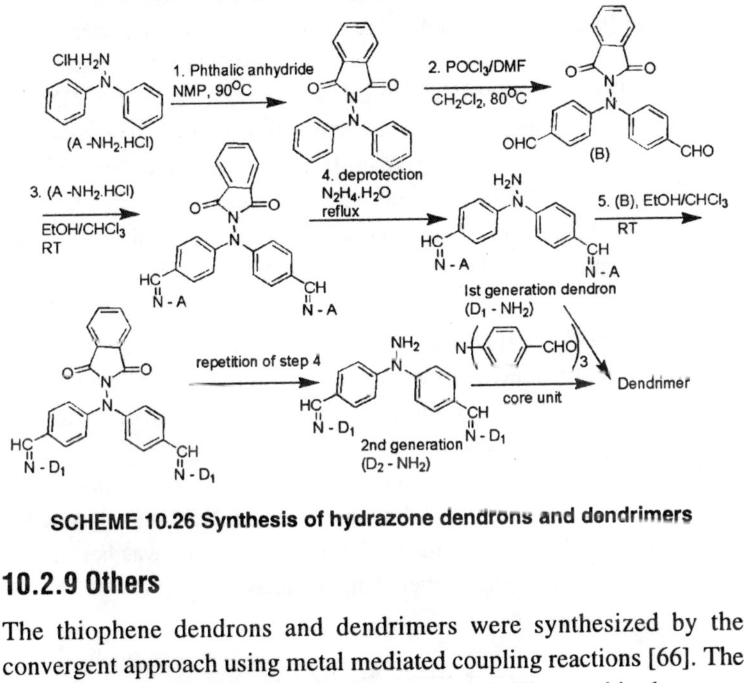

SCHEME 10.26 Synthesis of hydrazone dendrons and dendrimers

10.2.9 Others

The thiophene dendrons and dendrimers were synthesized by the convergent approach using metal mediated coupling reactions [66]. The starting compound for this synthesis was 2,3-dibromothiophene, an AB_2 type monomer. The Stille coupling [67] of this monomer with 2-bromo–5-hexylthiophene resulted first generation, and repetition of activation and coupling reactions with the AB_2 offered higher generation dendrimers. The whole sequence of reactions is shown in scheme 10.27. The direct coupling of 1st and 2nd generation dendrons afforded the corresponding dendrimers (Scheme10.27). However, 3rd generation dendron can not be coupled by this route, because of poor reactivity and higher steric hindrance.

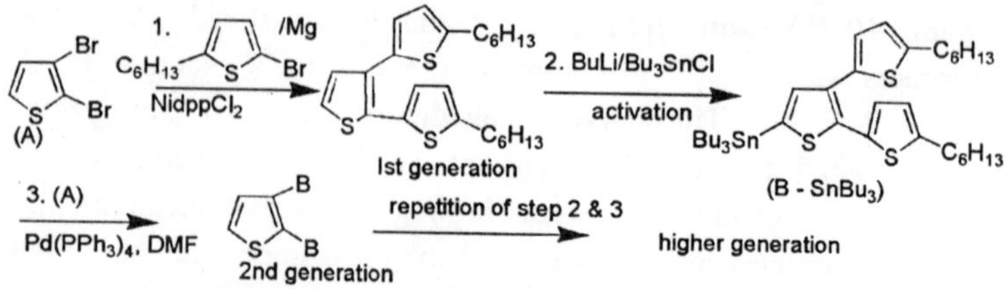

SCHEME: 10.27 Synthesis of polythiophene dendrimers

The same dendrimers, both 2nd and 3rd generations were also synthesized [67] by using 4,5,4′,5′-tetrabromo-[3,3']-bithiophene in the third step of the above scheme of reactions, in stead of dibromide AB_2 monomer. The synthesized dendrimers were characterized by ^{1}HNMR spectroscopy, elemental analysis and MALDI-TOF mass spectrometry. The molecular weights (Da) obtained from mass spectra were 1823.6, 1705.6 and 3809.2 for 1st, 2nd and 3rd generation, dendrimers respectively. All the dendrimers and dendrons were highly soluble in common organic less polar solvents such as $CHCl_3$, THF and even in hexane.

The dendritic polyphenyl azomethines containing π-conjugated backbone upto 4th generation were synthesized [68] by the convergent approach via dehydration of aromatic ketones with aromatic amine in the presence of $TiCl_4$. The dendrimers were synthesized by iterative sequence of reactions of ketone with 1,4-diamino benzophenone in the presence of $TiCl_4$ and 1,4-diazabicyclo[2,2,2]octane (DABCO) (Scheme10.28). The starting material for this synthesis was benzophenone and the core unit was *p*-phenylene diamine.

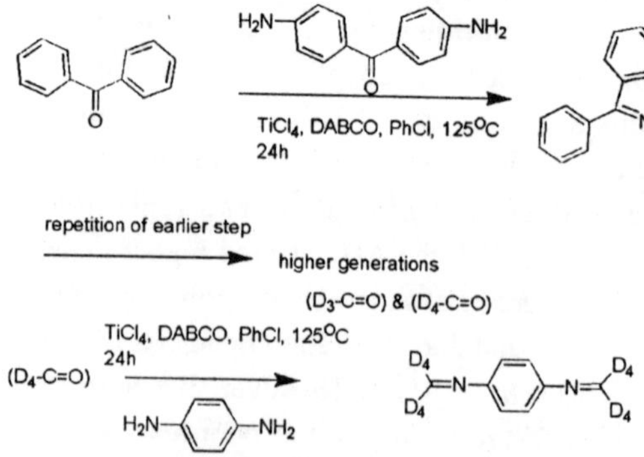

SCHEME 10.28 Synthesis of azomethine dendrimers

The synthesized dendrimers were characterized by IR and ^{1}HNMR spectroscopies, elemental analysis, GPC analysis and mass spectrometry. The highest molecular weight obtained was 3800 Da with polydispersity 1.02 by GPC analysis, while mass spectrum indicates the highest molecular weight 5451 Da. The azomethine dendrimers were highly soluble in common organic solvents such as $CHCl_3$, THF, DMSO, etc. in contrast to their linear analogs, which have very poor solubility characteristic. The dendrimers exhibited also high thermostability and only 10% weight loss was found above 500°C (511-521)°C. This is due to the presence of highly thermostable azomethine linkage (bond energy 1615 kJ/mole). The TEM and AFM picture showed that the highest generation dendrimer molecules have a sphere like structure with diameter 2.3 nm and regularly assemble on a plate.

The synthesis of dendrimer with 24-nitrile groups and octaferrocene complex of the dendrimer were reported [69]. The regiospecific chlorocarbonylation of the polybenzyl cores, in the para position of benzyl groups offers the chlorocarbonyl derivatives, which were reacted with Newkome's aminotripod $NH_2C(CH_2OCH_2CH_2CN)_3$ to get dendrimers. These cores were further reacted with (5-aminopentyl)-1-ferrocene to get the octaferrocene complex. At first, the hexabenzyl and octabenzyl cores were synthesized by $CpFe^+$-induced benzylation reaction. Then, the cores were chlorocarbonylated followed by reaction with Newkome's aminotripod resulted the desired dendrimers. Similarly, the cores were reacted with ferrocene unit to get the complexes. The whole sequence of reactions is shown in scheme 10.29. The nitrile groups terminated dendrimer having molecular weight 3328 Da, as determined by MALDI-TOF mass spectrometry. The octaferrocene dendritic complex has outstanding properties as sensor for various anions with dendritic effect just like earlier reproted 24-amidoferrocen and 24-[Cpfeη^6-(N-alkylaniline)]$^+$ dendrimers [54].

The synthesis and properties of a highly fluorinated dendritic nonlinear optical chromophore were reported [70]. The core unit was based on phenyl tetra cyanobutadienyl (Ph-TCBD) thiophone-stilbene containing chromophore. The dendritic chromophore was synthesized by the esterification between the carboxylic acid groups of the fluorinated dendrons and the three hydroxylic groups of the chromophore

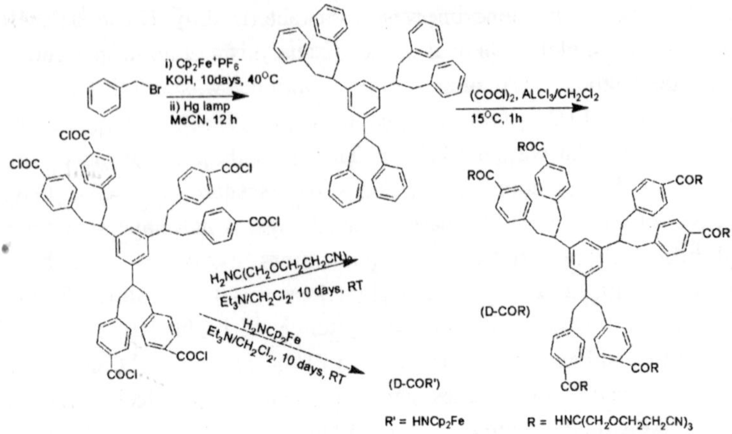

SCHEME 10.29 Synthesis of ferrocenyl and nitrile end groups containing dendrimers

precursor. The product was then reacted with tetracyanoethylene (TCNE) to get the desired dendrimer (Scheme 10.4). The structure of dendrimer was fully characterized by ^{1}HNMR, ^{19}FNMR and UV-visible absorption spectroscopies and MALDI-TOF mass spectrometry. The dendrimer was highly soluble in common organic solvents such as $CHCl_3$, THF, and cyclopentanone. It was relatively thermo stable and started to decompose only at 265°C under the N_2 atmosphere, as measured by seat-pan DSC. The dendritic chromophore displays absorption maximum for π-π* charge transfer of 574 nm (λmax) in 1,4-dioxane solution with high electro-optic co-efficient (30 pm/V at

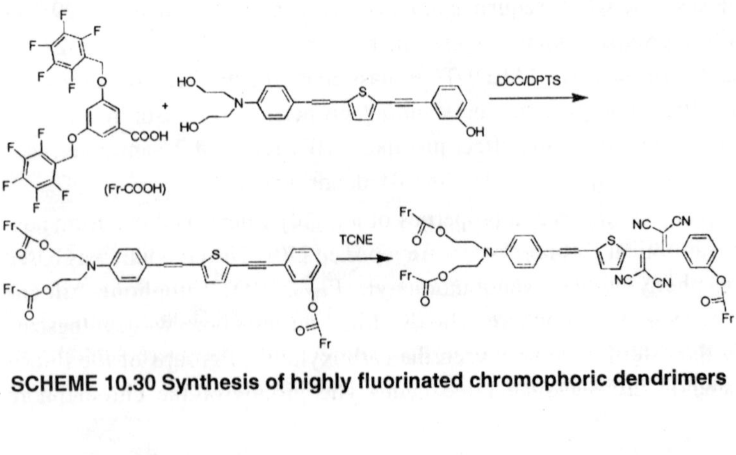

SCHEME 10.30 Synthesis of highly fluorinated chromophoric dendrimers

1.3 μm). This combination of properties of the dendritic chromophores makes it as a novel molecular design for the next generation electro-optic material.

A new class of dendritic polyuracils (Fig. 10.15) was reported [71]. These were crosslinked by a [2+2] photodimerization of the uracil units and hence resulted 'locked' dendrimers upon irradiation at 240 nm in CH_3CN, whereas on irradiation at 240 nm they were unlocked. The structures of these dendrimers were confirmed by ^{1}HNMR spectroscopy and MALDI-TOF-MS spectrometry. The mass spectrum and SEC profile showed only a single peak for the dendrimers. The polyuracil dendrimers were intramolecularly locked via photoinduced dimerization of the uracil units, as confirmed by SEC and ^{1}HNMR spectroscopy. The ^{1}HNMR shows signals for cyclobutane rings and SEC profile showed no substantial change in the chromatogram before and after irradiation. A continuous irradiation for 80h resulted a 38% decrease in the absorbance at 275.6 nm, which corresponds to the dimerization of six uracil units (approx.) on average per dendritic molecule. However, on irradiation of shorter wavelength light (240 nm) on the crosslinked dendrimer resulted an increase in absorbance at 275.6 nm, indicating a photo cleavage of the cyclobutane ring.

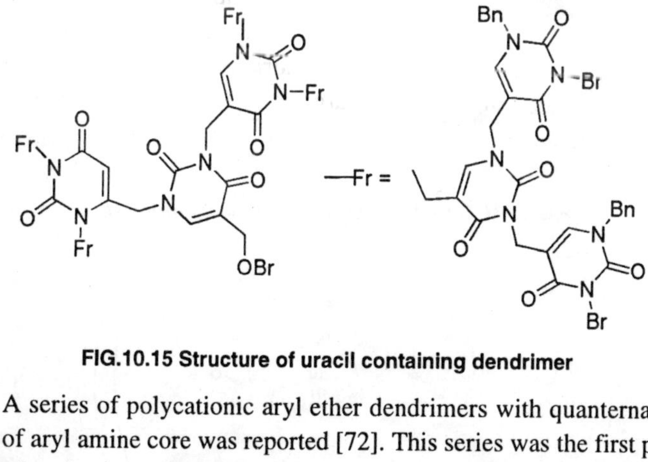

FIG.10.15 Structure of uracil containing dendrimer

A series of polycationic aryl ether dendrimers with quanternarization of aryl amine core was reported [72]. This series was the first polycationic mixed core-shell dendrimers. The structural characteristics of these dendrimers consist of an apolar core with a peripheral ionic layer, which is surrounded by a less polar layer of dendrons. Two types of polycationic species were designed [72]. The first one was constructed from polyaryl diamine core molecules and contains an ionic core shell and less polar outer shell. The second type is derived from a relatively large carbosilane dendritic core producing polycationic mixed core shell dendrimer species with a potentially variable but with a distinct distance between the core of the molecule and the ionic layer. The polycationic four aryl ether dendrimers (Fig.10.16) were constructed from tetra arylsilane building blocks and Frechet type dendrons under mild reaction conditions. The synthesized dendrimers were characterized by ^{1}HNMR, ^{13}CNMR spectroscopy, ESI mass spectrometry and combustion analysis. The dendrimers are soluble in solvents like CH_2Cl_2 and toluene but insoluble in water. The mixed core shell dendrimer (Fig. 10.17) was constructed by the treatment of the known carbosilane dendrimer with slight excess of a first generation Frechet dendron. The structure of the dendrimer was characterized by ^{1}HNMR, ^{13}CNMR spectroscopy and ESI mass spectrometry.

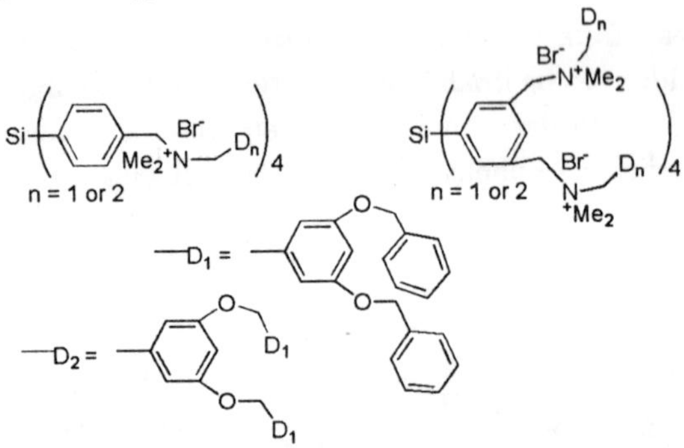

FIG.10.16 Structures of polycationic dendrimers

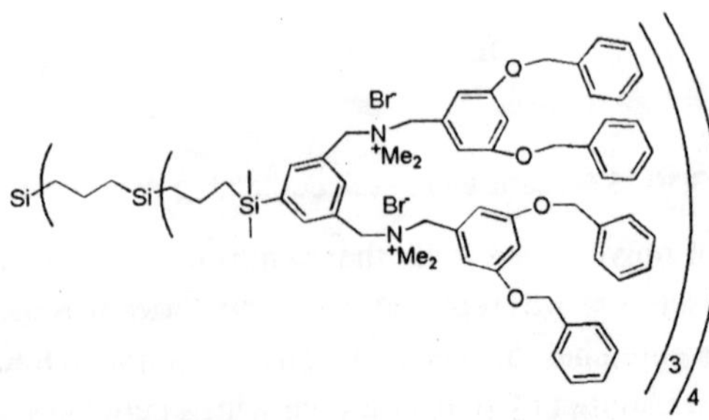

FIG.10.17 Structure of mixed core shell dendrimer

These dendrimers were utilized to form assemblies with anionic guest molecules such as methyl orange (MO). The results obtained from UV-visible and NMR spectroscopies and MALDI-TOF mass spectrometry demonstrated that all cationic sites throughout the core of dendrimers were involved in ion pair formation with anionic guest for guest/host ratios upto 24. Further, the assembly of anionic, organic molecules with these cationic dendrimers was fully reversible and could be regulated by addition of other anions or lowering the pH of the medium.

These dendrimers (Fig.10.16) were also used [72] as phase transfer catalysts for the SN^2 reaction between KCN and $PhCH_2Br$ to get $PhCH_2CN$ under basic conditions. It was found that, in the presence of these dendrimers (0.01 mole %), a much higher conversion of 40-59% compared to 6% in the absence of this catalyst. This dendrimer can also be used as substrate (MO dye) delivery over a nanofiltration membrane.

10.3 SYNTHESIS, CHARACTERIZATION AND PROPERTIES OF OTHER HYPERBRANCHED POLYMERS

In this section the synthesis, characterization and properties of other HB polymers, which are not reported earlier in this book, are discussed here.

10.3.1 Polycarbonates

The HB aryl polycarbonates were synthesized [1, 73] as potential engineering materials. The HB polycarbonate with phenol end capping structure was synthesized [1] by polymerization of an A_2B type monomer, in which 'A' group was carbonylimidazolide and 'B' was a *tert*-butyl dimethyl silyl ether protected phenol. The high molecular weight polycarbonate was synthesized by the reaction of this A_2B monomer with silver fluoride, but some difficulties were observed in this method. An incomplete reaction due to the loss of the reactive species by hydrolysis of the carbonylimidazolide group was unavoidable. Thus, the overall yield was low and also the unreacted monomer and oligomeric products have to be removed to get the pure high molecular weight HB polycarbonate. Again, the purified polymer showed low solubility might due to crosslinked product formation through phenolate imidazolide exchange reactions. However, by the in corporation of new A_2B type monomer, in which the carbonylimidazolide, A_2 groups were replaced with chloroformates, the synthetic stage improved to a greater extent [73]. Another AB_2 type monomer, containing a single chloroformate and two *tert*-butyl dimethyl silyl ether protected phenols, was also used to determine the efforts of the ratio of reactive groups on the DB and to modify the chain end functionalities of the final HB polycarbobates [73]. At first, the monomers were synthesized from 1,1,1-*tris*(4'-hydroxyphenyl) ethane by the following sequence of reactions as shown in scheme 10.31. The polymerization of the monomers were carried out at 23-70°C in 20% CH_3CN/THF in the presence of a stoichiometric excess of AgF for 20-30 h to get the HB polycarbonate (Scheme10.31). SEC monitored the reactions. All the products were characterized by ^{1}HNMR, ^{13}CNMR and FTIR spectroscopies and elemental analysis. The thermal characterization of the polymers was studied by TGA and DSC analyses. The weight average molecular weight for the HB polymers was obtained upto 10^6 Da with polydispersity 2-3, as measured by SEC. The DB was determined both by degradation and ^{1}HNMR analysis techniques and found to be 0.53-0.55. The DSC study demonstrated a large difference between the T_gs of the HB polycarbonates. It varies from 205°C for hydroxy terminated polymers to 180°C for *tert*-butyl dimethyl silyl ether end group polymers. The fluoroformate terminated polymer, however, exhibited no T_g before decomposition, which was started at 290°C. The TGA study in air indicated the high thermal

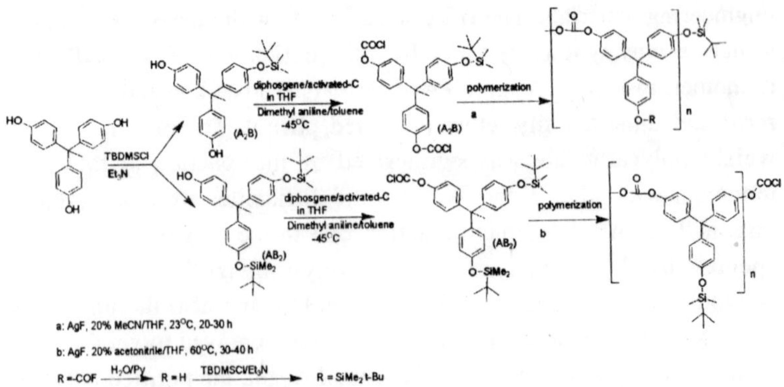

SCHEME 10.31 Synthesis of HB polycarbonate

stability of the polymers, as the OH terminated HB polymers started to decompose at 350°C, whereas *tert*-butyldimethyl silyl terminated polymers stable upto 400°C.

10.3.2 Carbohydrates

The synthesis of HB amino polysaccharides was reported [74-75] from AB_2 type monomer. The HB aminosaccharide was synthesized by the 10-comphorsulfonic acid (CSA) catalyzed polymerization of 2-methyl-(6-O-tosyl-1,2-dideoxy-α-D-glycopyrano)-[2,1-d]-4,5-dihydrooxazole (1), AB_2 type monomer with two hydroxyl groups at 3,4-position (Scheme 10.32). The ^{1}HNMR, ^{13}CNMR spectroscopy and GPC analysis were used to characterize the HB polymers. The number average molecular weight for the polymer was found to be 6300 g/mole as measured by GPC, whereas the weight average molecular weight was 4.5 X 10^5 Da determined by light scattering technique. The DB varies from 0.45-0.51 as determined by ^{1}HNMR spectroscopy. The polymers were soluble in DMF, DMSO, pyridine, etc. but insoluble in water and low polarity solvents. The HB polysaccharides showed unusual thermal properties, the decomposition of tosyl group started at 166°C and only a 45% weight loss was found upto 337°C. The decomposition completed at 600°C by two stage decomposition process. The tosyl free HB aminopolysaccharide was also produced by refluxing tosylated polymer under alkaline (NaOH) condition in aqueous ethanol. The IR spectrum confirmed the complete removal of tosyl groups in the polymer. Interestingly, the thermal behavior of tosyl

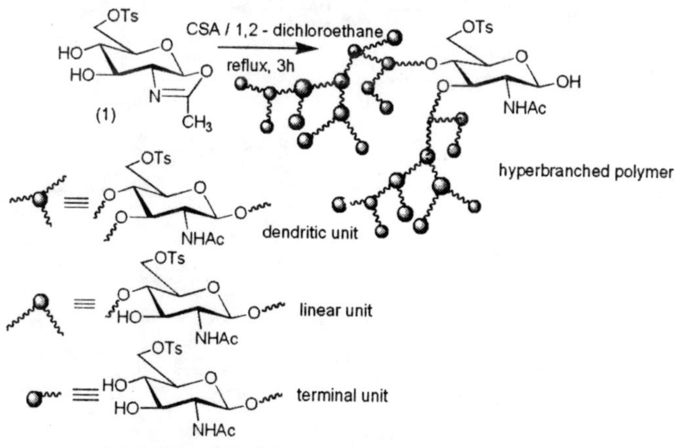

SCHEME 10.32 Synthesis of HB polyaminosaccharide

free polymer was different from the tosylated polymer [75]. It was decomposed gradually, instead of two-stage decomposition and only 35% wt. loss was observed at 600°C.

10.3.3 Polyureas and Polyurethanes

HB polyurethanes were prepared [76] from the AB_2 type monomer with one alkyl primary alcoholic group and two-aryl carbamate protected isocyanate functional groups. The aryl-aryl carbamate groups are very useful for the protection of aromatic isocyanates, since the carbamate bonds are easily cleaved by heating. The polymerization was proceeded by refluxing THF in the presence of dibutyl tin dilaurate as the catalyst (Scheme 10.33). The monomer was prepared from 3,5-dinitrobenzyl alcohol in a four stage reaction by means of various protective group techniques. The molecular weight of the polymer as well as the crosslinked structure formation depends on the concentration of the monomer. For example, the soluble polymer with a weight average molecular weight 34000 Da was obtained for a molecular concentration of 1 mole/liter, whereas a crosslinked gel type product, which only swelled in THF, DMF and DMSO with the increase of concentration, was found. However, by using an aliphatic alcohol, which offered end-capping reaction during polymerization of the AB_2 monomer, this

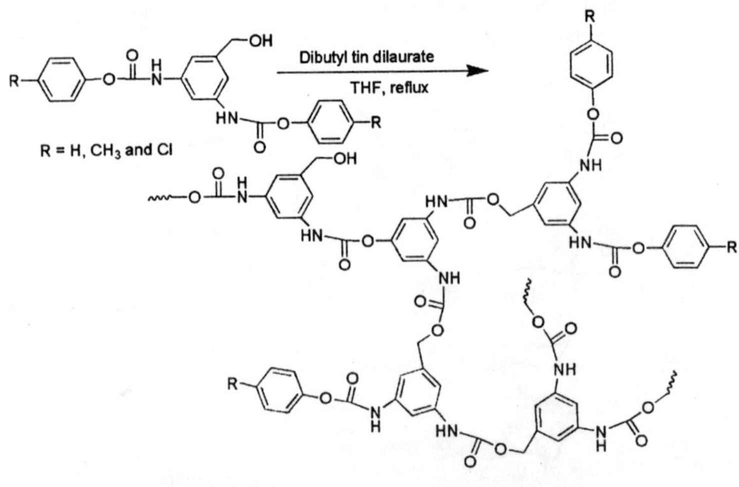

SCHEME 10.33 Synthesis of HB polyurethane

crosslinked network formation was avoided. Also, by this end-capping thermostable polymer was obtained. The end-capped polymer was thermally stable upto 200°C, whereas the un-endcapped polymer degraded above 120°C only. The T_gs vary from 119 to 57°C depending on the functional groups. All the endcapped polymers were soluble in THF, DMF and DMSO. However, the process is expensive as the monomer is not available commercially require more steps for synthesis and also causes toxilogical and hygienic problem due to elimination of phenolic compounds.

Fully aromatic HB polyurethane (PU) was synthesized [77-78] by the self-polycondensation of AB_2 and A_2B type monomers (Fig.10.18). A dispersion polymerization of the insoluble monomer (AB_2) in refluxed toluene resulted toluene insoluble HB-PU with high yield. The weight average molecular weight of the polymer was 9100 Da as measured by GPC and ^{1}HNMR indicated that the DB was 0.59. The polymerization of 5-hydroxyisophthaloyl azide as A_2B type monomer resulted a carbamate terminated PU after an end capping reaction with ethanol before isolation. However, the synthesized HB-PU was insoluble in most of the organic solvents may be due to crosslinked product formation by the side reactions, of isocyanate groups. Again a soluble polymer was obtained by polymerization of the A_2B monomer in the presence of an equivalent amount of an alcohol. The IR and ^{1}HNMR studies indicated that azide groups were completely removed and the carbamate groups were formed. When oligoethyleneoxy segment containing monomer was used, the intrinsic viscosities of the HB-PUs increased with the increase of length of the oxyethylene segments and ranged from 0.2 to 0.3 dL/g in DMSO. Whereas the

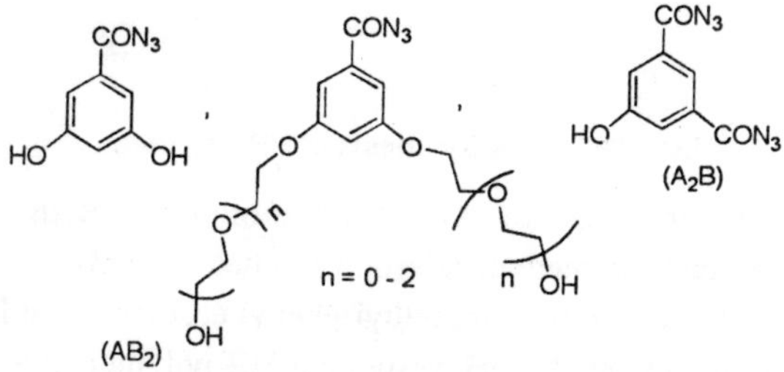

FIG.10.18 Structure of AB_2 and A_2B monomers with carbonyl azide groups for synthesis of polyurethane dendrimers

T_g decreases remarkably with the introduction of oxyethylene units in the range of 106–13°C. However, the oxyethylene units improved thermal stability by about 100°C in comparison with the corresponding fully aromatic polyurethane. However, here also the monomers are not commercially available and phenylurethanes are thermally unstable, in addition azides are difficult to prepare and handle.

The synthesis of HB polyureas was reported [79] by the polycondensation of AB_2 type monomer containing carbonyl azide and amino groups (Scheme 10.34). The self-polycondensation of 3,5-diaminobenzoyl azide in NMP at 110°C generated the corresponding isocyanate *in-situ* and ultimately resulted the HB polyurea with high yield. During polymerization, initially a rapid evolution of N_2 gas was observed. The synthesized polymer was characterized by IR and NMR spectroscopies and GPC analysis. IR spectra indicated the absence of any residual azide or isocyanate and showed the formation of urea linkages. The ^{1}HNMR measured the DB, which was 0.55. The weight average molecular weight and molecular weight distribution were found to be 19,500 Da and 1.56 respectively. The polymer was soluble in DMSO, NMP and DMF solvents.

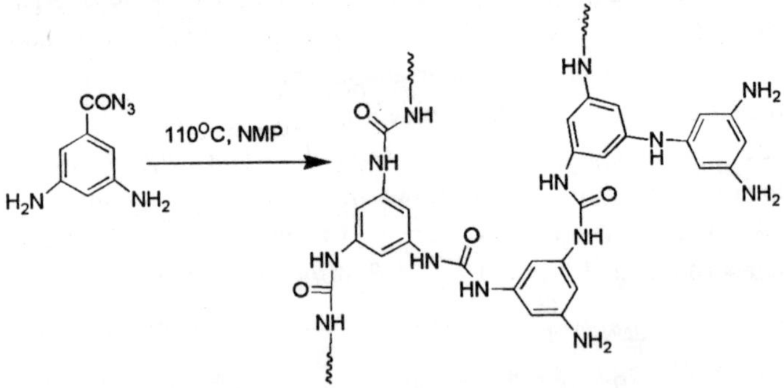

SCHEME 10.34 Synthesis of HB polyurea

The polymerization of AB_2 type monomers with activated methylene and dichloromethylene group resulted HB polymers [80-81]. The polymerization of *p*-(chloromethyl)phenyl acetonitrite in DMSO in the presence of aqueous NaOH, resulted a HB polymer. Interestingly, the activated methylene units attached with electron withdrawing CN groups were converted to branching point during the polymerization process. Even though, the molecular weight obtained by this process was not high enough but the polymerization of the AB_2 monomer in the presence of tetra-n-butylammonium chloride (TBAC) as the catalyst resulted gel product. However, the polymerization of 4-(4'-chloromethylbenzloxy)phenyl acetonitrile monomer offered relatively high molecular weight product without gelation in DMSO-aqueous NaOH solution in the presence of TBAC catalyst (Scheme 10.35) with high yield. The number average molecular weight of this polymer was 9100 Da as determined by ^{1}HNMR spectroscopy and a broad molecular weight distribution was observed in GPC analysis. However, by the addition of monofunctional monomers, such as methoxybenzylchloride, the molecular weight and its distribution was controlled. The synthesized HB polymers were soluble in DMF, DMSO and THF.

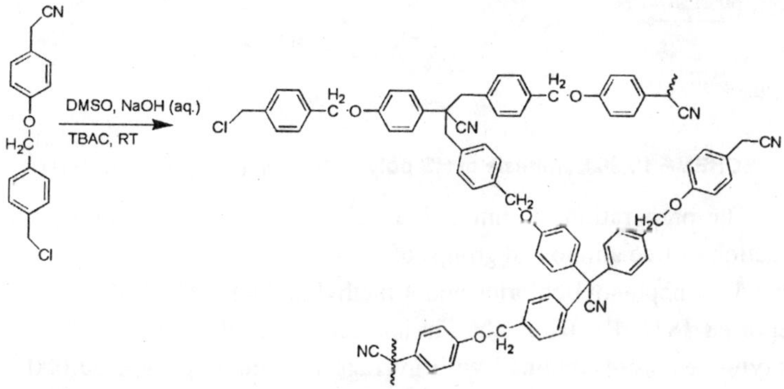

SCHEME 10.35 Synthesis of HB polymer through activated methylene groups

10.3.4 Polyimides

The HB aromatic polyimides were synthesized [82] from polyamic acid methyl ester by using AB_2 type monomer. The monomer of an isomeric mixture of 3,5-*bis*(4-amino phenoxy) diphenyl ether 3,4-dicarboxylic acid mono methyl ester was synthesized first by using a series of sequence of reactions starting from 3,5- dimethoxy phenol. The direct self-condensation of the monomer in presence of (2,3-dihydro–2-thioxo–3-benzoxazolyl) phosphoric acid diphenyl ester (DBOP) as a condensing agent in NMP at room temperature for 3h followed by imidization at 300°C for 1h under vacuum or at 100°C for 24h in the presence of acetic anhydride and pyridine in DMSO, resulted the desired product (Scheme 10.36). The resulted polymers were soluble in NMP, DMF, DMSO, etc. indicated these were not crosslinked products. The synthesized polymers were characterized by IR and ^{1}HNMR spectroscopies and GPC analysis. The thermal characteristics of the polymers were studied by TGA and DSC. The polymers are highly thermostable with only 10% weight loss at 470°C under the nitrogen atmosphere, as indicated by TGA study. The DSC revealed that T_g of the HB polymer is ≈ 193°C under the nitrogen atmosphere.

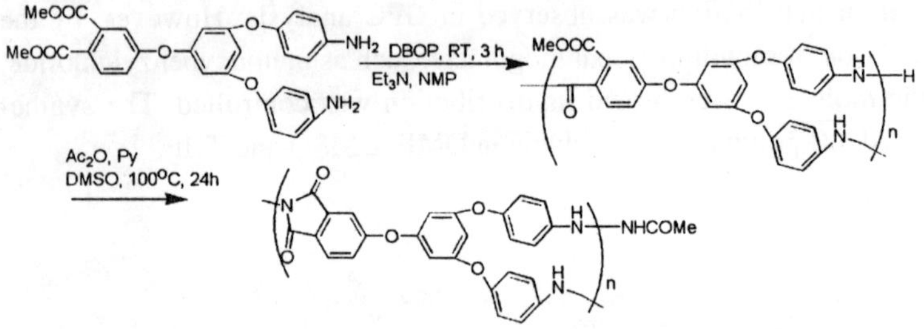

SCHEME 10.36 Synthesis of HB polyimide from polyamic acid ester

The preparation of similar HB polyimides by using end capping reaction of free amino end groups of the same precursor (as above) with acetyl-, n-heptanoyl-chloride and 4-methyl phthalic anhydride was also reported [83]. By using the similar sequence of reactions, the HB polyimides were obtained with average molecular weight 37,000 to 1,88,000 Da and degree of breaching 0.5. The thermal study indicates only 5% weight loss at about 400°C with T_gs 189°C, 138°C and 186°C for the end groups of acetoamide, n-heptanoamide and 4-methyl phthalimide respectively. The interesting result was obtained in the density of the transparent film (<1.35), which is lower than their linear analogs. They also have lower dielectric constant. These results confirmed the looser packing structures and thereby higher solubility of the HB polyimides.

A series of aromatic HB polyimides was synthesized [84] by polycondensation of *tris*(4-aninophenyl) amine (TAPA) with dianhydride such as 2,2-*bis*(3,4-dicarboxy phenyl)hexafluoropropane dianhydride (6FDA), 3,3',4,4'-biphenyl sulfonates tetracarboxylic dianhydride (DSDA) and pyromellitic dianhydride (PMDA) (Scheme 10.37). When a dianhydride solution was added to the TAPA solution, an amine terminated HB polyimides was obtained, whereas the reverse addition yielded anhydride terminated HB polyamide. The resulted HB polymers were characterized by IR, ^{1}HNMR spectroscopy and GPC technique. From ^{1}HNMR spectral analysis, the DB of the amine terminated HB polyimides were found to be 0.64, 0.72 and 0.68 for 6FDA-TAPA, DSDA-TAPA and PMDA-TAPA respectively. The GPC analysis indicates the broad molecular weight distribution (5.8-1.8) with molecular weight 37,000-1,60,000 Da. The thermogravimetric analysis of the HB polyimides indicated good thermostability with only 5% wt. loss at about 450°C under the N_2 atmosphere.

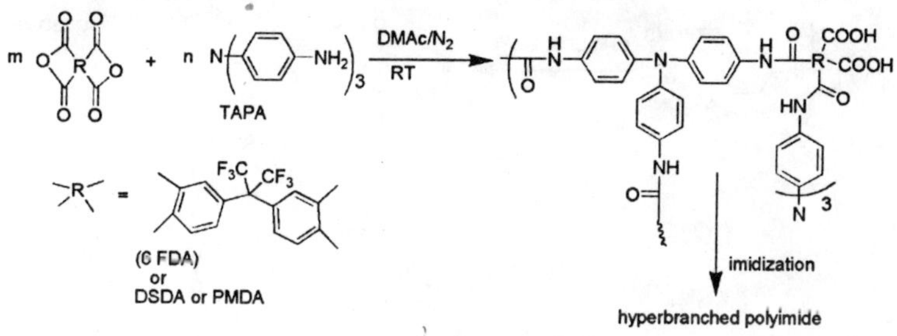

SCHEME 10.37 Synthesis of HB polyimide by polycondensation of triamine with dianhydrides

10.3.5 Poly(ether imide)s

A series of HB poly(ether imide)s was synthesized [85-86] from AB_2 monomer, N-[3,5-di-(*tert*-butyl dimethyl silyloxy) phenyl]–4-fluorophthalimide by a rapid fluoride-catalyzed acylation reaction. The polymerization of the AB_2 monomer at 240°C proceeded very rapidly (2.5 min) [85], yielding high molecular weight HB poly(ether imide) (Scheme 10.38) with DB 0.66. It was also observed [86] that DB decreases with the increase of reaction time. The rheological properties of the HB polymers were varying with DB. The monomers were polymerized with different reaction times viz. 2.5, 5.0, 7.5, 10.0 and 20.0 min.

The synthesized HB polymers were characterized by spectroscopic and analytical techniques. The molecular weight varies from 82 to 295 kDa with polydispersity 1.6 to 4.1 and DB 0.66 to 0.42 with different reaction times (2.5–20 min.). The rheological behavior of 40% solution of HB polymers in 1,2-dichlorobenzene showed that the magnitude of the viscosity and first normal-stress co-efficient increases with the decrease of DB. The onset of shear thining (accompanied by a rise of normal stress effects) appears at lower shear rates with the decrease of DB. However, the substantial increase in the weight average molecular weight with reaction time suggests that the observed difference in rheological properties may be a function of the change in molecular weight.

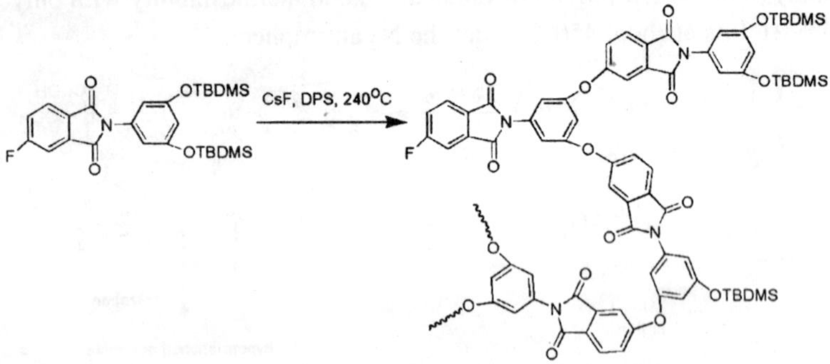

SCHEME 10.38 Synthesis HB poly(ether imide)

10.3.6 Poly(arylene ether phosphineoxide)s

A series of HB poly(arylene ether phosphine oxide)s was synthesized [87] from (4-fluorophenyl)-(4-hydroxyphenyl)phosphine oxide (A) in the presence of different core molecules with symmetrically altered reactivity of the aryl fluoride groups. The molecular weight of the polymers was controlled by the concentration of the core molecules. The core molecule (C) used for this synthesis was *tris*(4-fluorophenyl)phosphine oxide, *tris*(3,4-difluorophenyl)phosphine oxide and *tris*(3,4,5-trifluoro phenyl)phosphine oxide. The HB polymers were obtained by nucleophilic aromatic substitution reactions (Scheme 10.39) of (A) with 3, 5 and 10 mole percentage of (C). The synthesized polymers were characterized by ^{1}HNMR, ^{13}CNMR, ^{31}PNMR spectroscopy and GPC analysis. The number average molecular weight was ranging from 3270 to 8100 Da with PDIs 1.25 to 2.28 and DB 0.54 to 0.61.

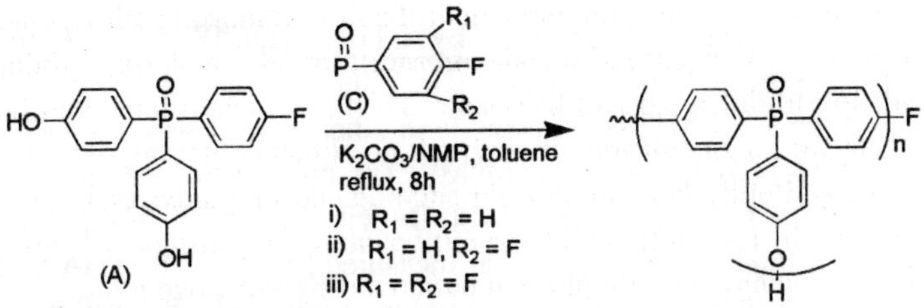

SCHEME 10.39 Synthesis of HB poly(arylene ether phosphine oxide)s

10.3.7 Poly(ether sulfone)

The HB poly(ether sulfone) with 1,3,5- s- triazine moiety was synthesized [88] by the direct polymerization of AB_2 type monomer, 2,4-*bis*(4-hydroxyphenyl)–6-(4-(4-(4–fluoro benzene)sulfonyl phenoxy)phenyl)–1,3,5-s-triazine. The synthesis of the monomer and the HB polymer is shown in scheme 10.40. The structure of the synthesized

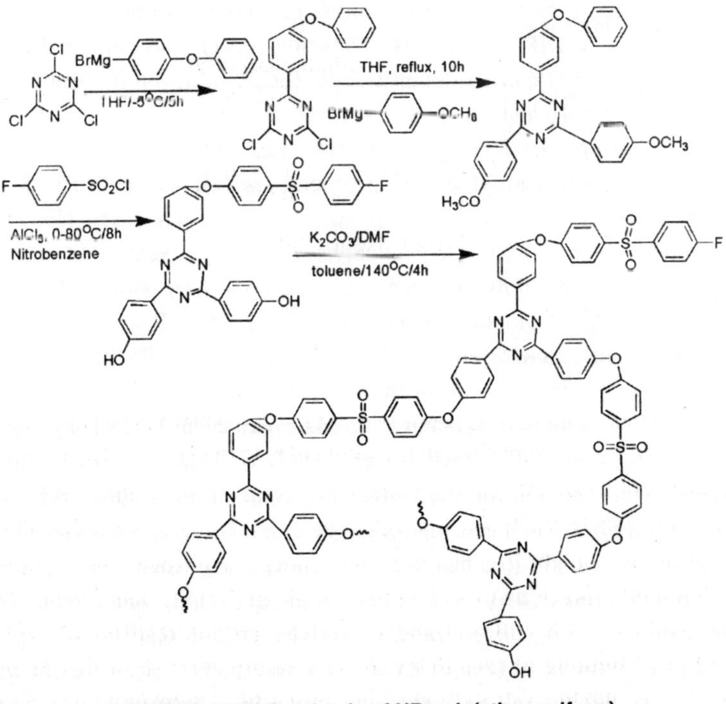

SCHEME 10.40 Synthesis of HB poly(ether sulfone)

HB polymer was characterized by FTIR, ^{1}HNMR and ^{13}CNMR spectroscopies and GPC analysis. The polymer was soluble in THF, 1,4-dioxane and DMSO but insoluble in other organic solvents such as CH_3COCH_3, CH_2Cl_2, DMF and NMP. The number average molecular weight of the HB polymer was 5200 Da with PDI = 2.0. The polymer was amorphous with Tg ≈ 295°C as measured by DSC. The TGA study showed high thermostability of the polymer with only 5% weight loss at 480°C.

10.3.8 Polyamines

A few aromatic HB polyamines with s-triazine rings have been synthesized [89-90] from A_2 [aromatic diamine e.g. 4,4'-sulfonyl dianiline, 4,4'-methylene dianiline, 4,4'-oxy dianiline, 4,4'-(1,4-phenylene diispropylidene) *bis*-aniline, etc.] and B_3 [2,4,6-trichloro-s-triazine (CYC)] types of monomers by using a single step nucleophilic displacement polymerization technique in a polar solvent (DMAc) in the presence of DIPEA as the base (Scheme 10.41). This polycondensation reaction was carried out at (0-5)°C for 2h followed by heating at (40-45)°C for 1h and the final step was carried out at (80-85)°C for 6-8h. No gel product formation was observed under the used reaction conditions of this polymerization process, which was confirmed by solubility test in different polar solvents for each case. The HB polymers obtained by this technique, have very high yield (~ 82-90 %). The synthesized polymers were characterized by solubility test, inherent viscosity measurement, FTIR, UV, ^{1}HNMR and ^{13}C-NMR spectroscopic, XRD, etc. studies. The inherent viscosity of the polymers was found to be in the range of 0.20-0.45 dLg^{-1} in 0.5 % (w/v) N,N'-dimethyl acetamide (DMAc) at 27±0.1°C. They were soluble only in highly polar solvents such as DMAc, DMF, DMSO, NMP, etc. The DB of the polymers varies from 0.56 to 0.60 as measured by ^{1}HNMR spectroscopy. The TGA data of these polyamines indicate no weight loss below 250°C under the nitrogen atmosphere, while DSC shows that the T_g values at about 240 °C and 230 °C, for polyamines with sulfonyl and methylene groups respectively. The modified limiting oxygen index (LOI) measurements show that all the polymers having self-extinguishing property. These polymers were used as multipurpose polymeric additives viz. as antidegradants, process aids, flame retardants, semireinforcing agents, etc. in industrial linear polymers like LDPE and plasticized PVC [90a]. One of these hyperbranched polymers was used as a crosslinker and performance improver for a vegetable oil based poly(ester-amide) resin [90b] as well as bisphenol A based epoxy resin [90c]. These polyamines also exhibit fluorescence behavior in the range of 400-650 nm [90d]. They are also used as matrices for preparation of nanoparticles such as Ag and Cu by in-situ reductive technique. These nano-composites colloidal solutions have potential antibacterial activity [90e].

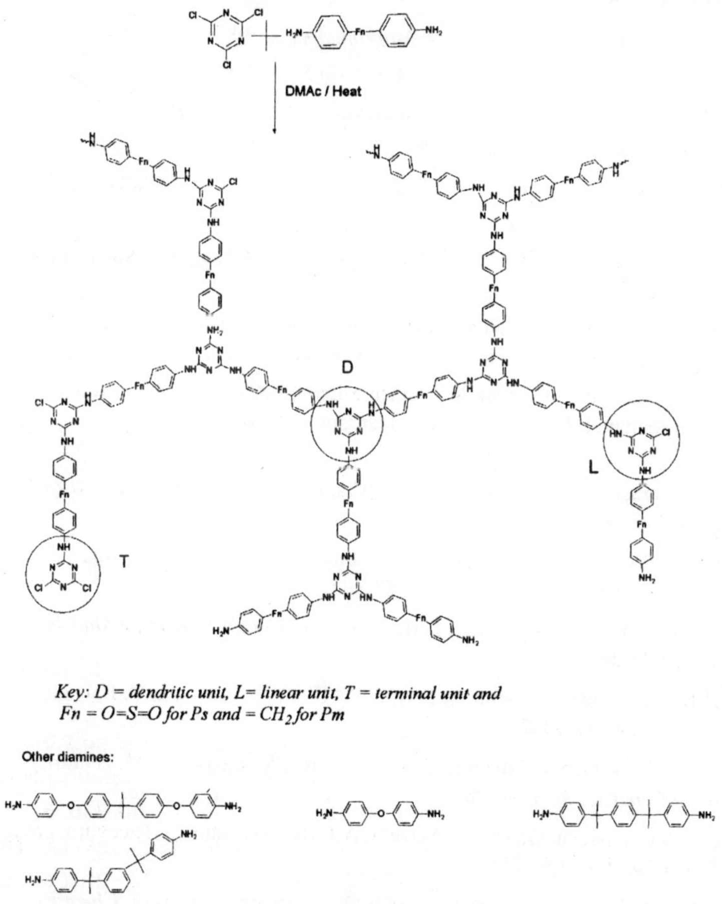

SCHEME 10.41 Synthesis of HB polyamines with 1,3,5-triazine moieties

10.4 SUMMARY

In this chapter dendritic polymers such as polycarbonates, polyurethanes, polyurea, metallodendrimer, mixed chemical linkages dendritic polymers, carbohydrates, etc. are discussed. They are also important in different fields of applications such as catalysts, electronic, multipurpose polymeric additives, bioactive materials, etc. like other classes of dendritic polymers.

REFERENCES

[1] D.H.Bolton and K.L.Wooley, *Macromolecules* **30**(1997)1890

[2] S.P.Rannard and N.J.Davis, *J. Am.Chem.Soc.* **122**(2000)11729

[3] K.Rengan and R.Engel, *J.Chem.Soc.Chem.Commun.* (1992)757

[4] A.Cherestes, T.October and R.Engel, *Heteroatom Chem.* **9**(1998)485

[5] S.Heinen, W.Meyer and L.Walder, *J.Electroanal.Chem.* **498**(2001)34

[6] S.Heinen and L.Walder, *Ange.Chem Int.Ed.* **39**(2000)806

[7] B.Colonna, V.D.Harding, S.A.Nepogoviev, F.M.Raymo, N.Spencer and J.F.Stoddart, *Chem.Eur.J.* **4**(1998)1244

[8] P.R.Ashton, E.F.Hownsell, N.Jayaraman, T.M. Nilsen, N.Spencer, J.F.Stoddart and M.Young, *J.Org.Chem.* **63** (1998)3429

[9] P.R.Ashton, V.Balzani, M.Clemente-Leon, B.Clolonna, A.Credi, N.Jayaraman, F.M.Raymo, J.F.Stoddart and M.Venturi, *Chem.Eur.J.* **8**(2002)673

[10] I.Vrasidas, N.J.de Mol, R.M.J. Liskamp and R.J.Pieters, *Eur.J.Org.Chem.* **2001**(2001)4685

[11] G.R.Newkome, F.Gardullo, E.C.Constable, C.N.Moorefield and A.M.W.C.Thompson, *J.Chem.Soc.Chem.Commun.* (1993)925

[12] G.Denti, S.Campagna, S.Serroni, M.Ciano and V.Balzani, *J.Am.Chem.Soc.* **114**(1992)2944

[13] S.Campagna, G.Denti, S. Serroni, M.Ciano, A. Juris and V. Balzani, *Inorg.Chem.* **31**(1992) 2982

[14] V.Balzani, S.Campagna, G.Denti, A.Juris, S.Serroni and M.Venturi, *Acc.Chem.Res.* **31**(1998)26

[15] S.Campagna, G.Denti, S.Serroni, A.Juris, M.Venturi, V.Ricevuto, *et al.* *Chem.Eur.J.* **1**(1995)211

[16] V.Balzani, A.Juris, M.Venturi, S.Campagna and S.Serroni, *Chem.Rev.* **96**(1996)759

[17] S.Serroni, A.Juris, M.Venturi, S.Campagna, I.R.Resino, G.Denti, A.Vredigl and V.Balzani, *J.Mater.Chem.* **7**(1997)1227

[18] G.R.Newkome, A.K.Datri and L.A.Godinez, *Chem.Eur.J.* **5**(1999)1445

[19] P.D.Beer, O.Kocian, R.J.Mortimer and C.Ridgway, *J.Chem.Soc.Chem. Commun* (1991)1460

[20] J.Issberner, F.Vogtle, L.De Cola and V.Balzani, *Chem.Eur.J.* **3**(1997)706

[21] M.Plevoets, F.Vogtle, L.DeCola and V.Balzani, *New.J.Chem.* (1999)63

[22] F.Vogtle, M.Plevoets, M.Nieger, G.C.Azzellini, A.Credi, L.De Cola, V.De Marchis, M.Venturi and V.Balzani, *J.Am.Chem.Soc.* **121**(1996)1821

[23] V.Marvand and D.Astruc, *Chem. Commun.* (1997)773

[24] S.Achar and R.J.Puddephatt, *Angew. Chem.Int. Ed.* **33**(1994)847

[25] S.Achar and R.J.Puddephatt, *J.Chem.Soc.Chem Commun.* (1994)1895

[26] S.Achar, J.J.Vittal and R.J.Puddephatt, *Organometallics* **15**(1996)43

[27] G.X.Liu and R.J.Puddephatt, *Organometallics* **15**(1996)5257

[28] S.Achar, C.E.Immoos, M.G.Hill and V.J.Catalano, *Inorg.Chem.* **36**(1997)2314

[29] W.T.S.Huck, F.C.J.M. van Veggel, B.L.Kropman, D.H.A.Blank, E.G.Keim, M.M.A.Smithers and D.N.Reinhoudt, *J.Am.Chem.Soc.* **117**(1995)8293

[30] W.T.S.Huck, F.C.J.M. van Veggel and D.N.Reinhoudt, *Angew. Chem.Int. Ed.* **35**(1996)1213

[31] W.T.S.Huck, R.Hulst, P.Timmerman, F.C.J.M. van Veggel and D.N.Reinhoudt, *Angew.Chem Int.Ed.* **36**(1997)1006

[32] W.T.S.Huck, L.J.Prins, R.H.Fokkens, N.M.M.Nibbaing, F.C.J.M. van Veggel and D.N.Reinhoudt, *J.Am.Chem.Soc.* **120**(1998)6240

[33] F.M. MacDonnell, M.J.Kim and S.Bodige, *Coord.Chem.Rev.* **185-186**(1999)535

[34] G.R.Newkome, E.He, L.A.Godinez and G.R.Baker, *J.Am.Chem.Soc.* **122**(2000)9993

[35] C.Amatore, Y.Bouret, E.Maisonhaute, J.I.Goldsmith and H.D.Abruna, *Chemistry* **7**(2001)2206

[36] M.J.Kim, F.M.MacDonnell, M.E.Gimon-Kinsel, T.D.Bois, N.Asgharian and J.C.Griener, *Angew.Chem.Int.Ed.* **39**(2000)615

[37] F.M.MacDonnell, M.D.Meser Ali and M.J.Kim, *Comments Inorg.Chem.* **22**(2000)203

[38] N.D.McClenaghan, F.Loiseau, F.Puntoriero, S.Serroni and S.Campagna, *Chem.Commun.* (2001)2634

[39] P.Ceroni, F.Paolucci, C.Paradis, A.Juris, S.Riffia, S.Serroni, S.Campagna and A.J.Bard, *J.Am.Chem.Soc.* **110**(1998)5480

[40] T.D.James, H.Shinmori, M.Takenchi, S.Shinkai, *Chem.Commun.* (1996)705

[41] H. Suzuki, H. Kureata and Y.Chen, *Chem.Commun.* (1997)23

[42] H.W.I.Peerlings, R.A.T.M. van Benthem and E.W.Meijer, *J.Polym.Sci. Part A Polym.Chem.* **39**(2001)3112

[43] B.Bruchmann, U.Ehe, F.Winggerter, K.Stieferhofer and U.Treuling, US pat. 6,376,637 B1(2002)

[44] R.Spindler and J.M.J.Frechet, *J.Chem.Soc.Perkin Trans.1* (1993)913

[45] R.T.Taylor and U.P.Puapaiboon, *Tetrahedron Lett.* **39**(1998)8005

[46] U.Puapaiboon and R.T.Taylor, *Rapid Commun.Mass Spectrometry* **13**(1999)508

[47] C.Kollner, B.Pugin and A.Togni, *J.Am.Chem.Soc.* **120**(1998)10274

[48] R.Schneider, C.Kollner, I.Weber and A.Togni, *Chem.Commun.* (1999)2415

[49] A.Togni, N.Bieler, U.Burckhardt, C.Kollner, G.Pioda, R.Schneider and A.Schnyder, *Pure Appl.Chem.* **71**(1999)1531

[50] S.Cicchi, A.Goti and A.Brandi, *Eur.J.Org.Chem.* (1998)2591

[51] C.M.Cardona and A.E.Kaifer, *J. Am.Chem.Soc.* **120**(1998)4023

[52] Y.Sha, L.Shen and X.Hong, *Tetrahedron Lett.* **43**(2002)9417

[53] C.Valenio, J.L.Fillaut, J.Ruiz, J.Guittard, J.C.Blais and D.Astruc, *J.Am.Chem.Soc.* **119**(1997)2588

[54] C.Valerio, E.Alonso, J.Ruiz, J.C.Blais and D.Astruc, *Angew.Chem.Int.Ed.* **38**(1999)1747

[55] C.M.Leu, Y.T.Chang, C.F.Shu, C.F.Teng and J.Shiea, *Macromolecules* **33**(2000)2855

[56] B.Kayser, J.Altman and W.Beck, *Chem.Eur.J.* **5**(1999)754

[57] B.Kayser, J.Altman, H.Noth, J.Knizek and W.Beek, *Eur.J.Inorg.Chem.* (1998)1791

[58] G.R. Newkome, A.Mishra and C.N.Moorefield, *J.Org.Chem.* **67**(2000)3957

[59] S.Yokoyama, S.Mashiko, Y.Yamaguchi and Y.Yokomichi, *Polym.Adv.Technol.* **11**(2000)692

[60] Y.Yamaguchi, Y.Yokomichi, S.Yokoyama, and S.Mashiko, *Polym.Adv.Technol.* **11**(2000)674

[61] Y.Yamaguchi, Y.Yokomichi, S.Yokoyama and S.Mashiko, *J.Molecular Structure (Theochem.)* **57**(2000)35

[62] K.Takagi, T.Hatton, H.Kunisada and Y.Yuki, *J.Polym.Sci. Part A Polym.Chem.* **38**(2000)4385

[63] M.I.Ranasinghne, O.P.Vamavski, J.Paulas, S.I.Hauck, J.Louie, J.F.Hartuiq and T.Goddson III, *J.Am.Chem.Soc.* **124**(2000)6520

[64] W.Zhang, D.T.Nowlan III, L.M.Thomson, W.M.Lackowski and E.E.Simanck, *J.Am.Chem.Soc.* **123**(2001)8914

[65] H.Nam, D.Ho Kong, J.K.Kim and S.Y.Park, *Chem.Lett.* (2000)1298

[66] C.Xia, X.Fan, J.Locklin and R.C.Advincula, *Org. Lett.* **4**(2002)2067

[67] J.K.Stille, *Angew.Chem. Int. Ed.* **25**(1986)508

[68] M.Higuchi, S.Shiki, K.Ariga and K.Yamamoto, *J.Am. Chem.Soc.* **123**(2001)4414

[69] C.Valerio, F.Moulines, J.Ruiz, J.C.Blais and D.Astruc, *J.Org.Chem.* **65**(2000)1996

[70] J.Luo, H.Ma, M.Haller, A.K.Y Jen and R.R.Barto, *Chem.Commun.* (2002)888

[71] M.Tominaga, K.Konishi and T.Aida, *Chem.Lett.* (2000)374

[72] A.W.Kleij, R.van de Coevering, R.J.M.K.Gebbink, A.M.Noordman, A.L.Spek and G.van Koten, *Chem.Eur.J.* **7**(2000)181

[73] D.H.Bolton and K.L.Wooley, *J.Polym. Sci. Part A Polym.Chem.* **40**(2002)823

[74] J.Kadokawa, M.Sato, M.Karasu, H.Tagaya and K.Chiba, *Angew.Chem.Int.Ed.* **37**(1998)2373

[75] J.Kadokawa and H.Tagaya, *Polym. Adv. Technol.* **11**(2000)122

[76] R.Spindler and J.M.J.Frcchet, *Macromolcules* **26**(1993)4809

[77] A.Kumar and S.Ramakrishnan, *J.Chem.Soc.Chem.Commun.* (1993)1453

[78] A.Kumar and S.Ramakrishnan, *J.Polym.Sci. Part A: Polym. Chem.* **34**(1996)839

[79] A.Kumar and E.W.Meijer, *Chem Commun.* (1998)1629

[80] R.H. Jin and Y.Andou, *Macromolecules* **29**(1996)8010

[81] R.H.Jin, S. Motokucho, Y.Andou and T.Nishikubo, *Macromol. Rapid Commun.* **19**(1998)41

[82] K.Yamanaka, M.Jikei and M.Kakimoto, *Macromolecules* **33**(2000)1111

[83] K.Yamanaka, M.Jikei and M.Kakimoto, *Macromolecules* **33**(2000)6937

[84] J. Fang, H.Kita and K.Okamoto, *Macromolecules* **33**(2000)4639

[85] D.S.Thamposon, L.J.Markoski and J.S.Moore, *Macromolecules* **32**(1999)4764

[86] D.S.Thamposon, L.J.Markoski, J.S.Moore, I.Sendijarevic, A.Lee and A.J.McHugh, *Macromolecules* **33**(2000) 6412

[87] D.P.Bernal, L.Bedrossian, K.Collins and E.Fossum, *Macromolecules* **26**(2003)333

[88] Y.Chang, Y.C.Kwon, K.Park and C.Kim, *Korean Polym.J.* **8**(2000)142

[89] S.S.Mahapatra and N.Karak, *J.Appl.Polym.Sci.* **106**(2007)95

[90] S.S.Mahapatra and N.Karak, *Polym.Degrd.Stab.* **92**(2007)947

[90a] S.S.Mahapatra and N.Karak, *Euras.Chem.Technol.J.* **9**(2006)29

[90b] S.S.Mahapatra and N.Karak, *Prog.Org.Coat.* **60**(2007)328

[90c] S.S.Mahapatra and N.Karak, (submitted, 2007)

[90d] S.S.Mahapatra and N.Karak, (submitted, 2007)

[90e] S.S.Mahapatra and N.Karak, (submitted, 2007)